1st International Conference on Computational Engineering and Material Science (ICCEMS 2020)

IOP Conference Series: Materials Science and Engineering
Volume 925

Online
17 - 18 July 2020

ISBN: 978-1-7138-4764-9
ISSN: 1757-8981

Printed from e-media with permission by:

Curran Associates, Inc.
57 Morehouse Lane
Red Hook, NY 12571

proceedings.com

Some format issues inherent in the e-media version may also appear in this print version.

Institute of Physics
Dirac House, Temple Back
Bristol BS1 6BE UK

Phone: 44 1 17 929 7481
Fax: 44 1 17 920 0979

techtracking@iop.org

Additional copies of this publication are available from:

Curran Associates, Inc.
57 Morehouse Lane
Red Hook, NY 12571 USA
Phone: 845-758-0400
Fax: 845-758-2633
Email: curran@proceedings.com
Web: www.proceedings.com

TABLE OF CONTENTS

IOP Conf. Series: Materials Science and Engineering **925** (2020) 011001 doi:10.1088/1757-899X/925/1/011001

1st International Conference on Computational Engineering and Material Science (ICCEMS-2020), 17-18 July 2020, GM Institute of Technology, Davangere, India

Preface

This volume of the IOP Conference Series: Materials Science and Engineering contains papers from the 1st International Conference on Computational Engineering and Material Science, which took place virtually from the 17th to the 18th of July 2020 at the GM Institute of Technology, Davangere, India.

Due to the pandemic of COVID-19 situation, in India there was complete Lockdown announced by the government. In this pandemic gathering was not allowed, not only India, but throughout the world. So, the conference was conducted using Microsoft Teams to connect all the researchers and speakers from the different parts of the world in a single platform and executed successfully the conference without postponing this.

The Proceedings presents novel and original work in various engineering domains and its applications by the experts and budding researchers. This conference provided opportunities for researchers to exchange new ideas and application experiences to establish research relations and to find global partners for future collaboration. It covers a variety of topics, including Communication, Concrete, Datamining, Electrical, FEM (Finite Element Method), Image Processing, IoT, Machine Learning, Manufacturing Technology, Material Science, Renewable Energy, Signal Processing, VLSI, WSN. It also provides a platform for researchers, engineers and academicians from all over the world to present their research results and development activities based on these domains. The conference has received paper from We have received papers from Australia, Chile, Philippines. Also from various parts of India: Karnataka, Tamilnadu, Telangana, Punjab, Andhra Pradesh, Uttar Pradesh, Bihar, Assam, Noida, New Delhi. This proceedings volume contains 77 papers presented at the conference. All the papers have been subjected to peer review by at least two referees.

Acknowledgements

On behalf of the GM Institute of Technology, Davangere, India I would like to thank all the invited speakers, session chairs and members of the discussion panels for making the conference such a great success.

IOP Conf. Series: Materials Science and Engineering **925** (2020) 011002 doi:10.1088/1757-899X/925/1/011002

Peer review declaration

All papers published in this volume of *IOP Conference Series: Materials Science and Engineering* have been peer reviewed through processes administered by the Editors. Reviews were conducted by expert referees to the professional and scientific standards expected of a proceedings journal published by IOP Publishing.

- **Type of peer review:** Single Blind
- **Conference submission management system:** Easy Chair
- **Number of submissions received:** 139
- **Number of submissions sent for review:** 139
- **Number of submissions accepted:** 77
- **Acceptance Rate (Number of Submissions Accepted / Number of Submissions Received X 100):** 55%
- **Average number of reviews per paper:** 2
- **Total number of reviewers involved:** 26
- **Any additional info on review process:** Initial scrutiny of the paper was to check the plagiarism percentage of the paper. So, after receiving the paper, it was checked in iThenticate. We have maintained this plagiarism rate at 20% excluding reference portion. Once the paper passed from this limit then the paper was sent for technical review.
- **Contact person for queries:** Dr. Anand Paul (anandpaul.editor@outlook.com)

GOF-SLFN- An Intelligent Attack Detection System against Denial of Service (DoS) attacks based on Glow Worm Swarm optimized Single Layer Feed Forward Networks for vehicular Cyber Physical Systems (VCPS)

Rama Mercy Sam Sigamani[1] and Padmavathi Ganapathi[2]

[1,2] Department of Computer Science, Avinashilingam Institute for Home Science and Higher Education for Women, Coimbatore, India
[1] mercyrama@gmail.com, [2] padmavathi.avinashilingam@gmail.com

Abstract. Vehicular cyber-physical network use either wireless communications or internet to communicate with each other for an efficient data transfer. Data includes transportation details, safety, mobility and sustainability. However, these communication networks for vehicular transportation are susceptible to different cyber-attacks, which may lead to severe damages or huge casualties in vehicular infrastructures. Hence the intelligent attack detection system needs to be obtaining the lime light of research based on recent machine and deep learning algorithms. In this paper, the GOF-SLFN (Glowworm optimized features for Single-layer feed-forward network) is proposed to detect the attacks among the vehicular cyber physical systems. Since DoS attacks can happen at any layer of OSI Model such as application, network and transport the foremost purpose of the paper is to detect the DoS attacks using the proposed intelligent algorithm. The glowworm algorithm extracted the optimum features based on the probabilistic mechanism and SLFN has been proposed using glowworm to detect the DoS attacks. The proposed system outperformed in detection when compared with artificial neural network, SLFN, Multi-layer Perceptron and random forest. The proposed GOF-SLFN has been again compared with firefly, Bat and Latent Dirichlet Allocation (LDA) and found the proposed intelligent system provided promising results for detection of DoS attacks in VCPS. The proposed system has been tested on the real time scenarios created with SUMO and OMNET++ environment integrated with CIC IDS 2018 datasets. The performance metrics considered were based on accuracy, sensitivity and specificity.

Keywords: Cyber Physical Systems, DoS Attacks, GOF-ULFN, OMNET+, SUMO, transportation.

1. Introduction

A vehicular Information system is a kind of intelligent automobile equipment which provides the drivers and customers with the security, infotainments, environmental conditions and other information services [1]. These kinds of systems consist of body train, power transmission systems, information systems and security systems. Nowadays, with an advent of Internet of things (IoT), vehicular systems have got its more insight in providing the more applications to the customers. Even though vehicular information system has reached its new dimension, effect of various security vulnerabilities has also reaching its new peak. Many experts, scientists and researchers from the various countries [2] have shown that the security vulnerabilities are common on the vehicles which have adverse effect on authentication, communication and even accident damages. Hence the research on vehicular vulnerabilities is considered to be more important for maintaining the social stability.

Machine learning and Deep learning are playing a very important role in the classification and prediction of the data. To design, an intelligent intrusion system we need the most powerful algorithm to detect and predict the attacks in accordance to take its countermeasures. Several techniques such as "Artificial Neural Networks (ANN), Support Vector Machine (SVM), and Ada boost Classifier" were used for designing an intelligent intrusion system, but detection and measurement need improvisation. Nowadays better machine learning algorithms are replaced by the deep learning algorithms for better detection [1], Since a lot of samples are required for the deep learning framework, machine learning attracts more researchers in dealing with small samples for designing an intelligent intrusion detection system. However, the machine learning based intrusion systems needs more lime light in terms of accuracy and time complexity.

The paper focuses on the implementation of most powerful single feed forward learning algorithm integrated with the Glow worm optimization to detect the different attacks in vehicular networks. The proposed system consists of three different stages such as Data collection, Feature Extraction, Feature Selection and Prediction of attacks. The real time vehicular cyber physical networks are simulated by integrating the SUMO in OMNET Environment and tested with the proposed algorithms on selection of appropriate features thus enabling the better detection of attacks based on the measure of detection accuracy. Moreover, the paper also elaborates the attack detection of the proposed and the existing machine learning algorithms and highlighted the analysed outcome of detection of attacks on comparison.

The organization of the paper is as follows: Section 2 presented the literature study undergone for VANET and its security techniques. Section-3 deals with overview of Uni-layer feed-forward learning machines and Glow worm algorithms. The working methodology such as data collection unit, feature selector and extractor, classifier is detailed in Section-4. Section -5 deals with the experimental design, performance evaluation with comparative analysis between the different learning algorithms. Finally, conclusion is presented in Section-6.

2. Literature Survey

Paul Darby described Decentralized Security Exchange (DSE) to proficiently bolster digital physical security for vehicles, including any electric, self-governing or traditional vehicle having a CAN transport. The exhibition of DSE for different degrees of theoretical registering replication forcefulness is dissected. The significance of the DSE answer for vehicles, particularly during debacle situations is clarified, similar to its pertinence to open trust inefficient power vitality advances.

Huangmiao CHEN dissected security vulnerabilities, dangers and uncommon security prerequisites of the present vehicle data framework. This framework includes data resources in-vehicle framework, divides vehicle data framework into two classes/levels: family and business, and characterizes target and necessity of security assurance for two sorts of vehicle data frameworks, separately. At last, a progression of doable assessment strategies and instruments are exhibited for assessment practice. A major commitment of this paper is to investigate characterized security assurance assessment for the vehicle data framework and tops off the hole of assessing security condition of car data framework everywhere throughout the world [3].

Simona Gifei underlined the significance of managing in an integrative way with security and wellbeing guidelines, meaning to blend a rational procedure for the advancement of self-sufficient vehicles. Reciprocal, it is emphasized the significant coordinating procedure to guarantee intentional ranges for cyber security dangers. The creators propose such a unitary idea for creating self-governing vehicles by coordinating quality, wellbeing and (cyber) security into one framework, called Quality Safety Cyber Security Integrated Management System (QSCSIMS). It displayed the framework and a a numerical model of the related procedures for the standard-based design. The performance metrics such as improvement of procedures, diminishing of waste, lower costs, protected and secure frameworks lastly, fulfilled clients [4].

Madhusudan Singh examined the accommodate security prerequisites designing to give digital security in IV. To satisfy the proposed goal, objective demonstrating manner of RE, the Knowledge Acquisition in Automated Specification (KAOS) Model methodology utilized for the design of utilizing Objective instrument locating the accommodating prerequisites. Self-compromise is gained by five structure squares, for example, Monitor, Analyze, Plan, Execute and Knowledge for example [MAPE-K] circle engineering, which provided the accommodating abilities to propose accommodating prerequisites of digital security in [5].

Pero Škorput described the malicious attacks and its manifestation on vehicles connected in Cooperative ITS environment. "Public Key Certification Authority Model in Cooperative ITS" model was proposed that lowered the odds of malicious malware attacks with predefined basic set of decisions. The divided five level decisions are for the processing of the security solution in the environment. The Cooperative ITS has a varied challenge in ensuring safety and security due to dissimilar transport modes. The literature study on securing connected vehicles revealed that this kind of environment leads to increase in the harmful rate and fatality rate in transportation when attacked than in other traffic domains. The various intentions both harmless and malicious of attackers in the cyber domain are differentiated as they reflect in the injury and fatalness in large for the road transportation [6].

Muhammad Shafique presented the security threats and their respective threat models in the layers of CPS and identified the relevant research challenges for the development of secure CPS. The challenges identified are addressed with a comparison of the state-of-the-art static and adaptive techniques and analyzed for detection and prevention with associated limitations. This paper further discussed the security techniques against characterized attacks on CPS layers with open research problems in developing the intelligent security measures using machine learning algorithms for CPS. In addition, the paper provided an outlined view of on-going project for securing CPS and the research problems and its motivational analysis aspects were discussed [7].

3. Glow-Worm Swarm Optimization Algorithm

Krishna et.al proposed this swarm based intelligence algorithm for optimizing the multi-model problems. Glow-worm (GLW) swarm optimization technique is adopted in many recent applications for minimizing the search time of optimum to improve the objective function. Likewise, in this article, we adopted this swarm algorithm for attack detection system in vehicular cyber physical systems. Basically glow-worm optimization technique is adopted from natural species of glowworms. GLW is developed using its attraction factor called "luciferin". Interaction between glow-worms are highly depends on the luciferin factor.

The proposed model initiated with development of fitness function and developed the GLW algorithm. The objection or fitness function is to detect the various attacks in the vehicular systems. The targeted attack in this paper is "DDOS- Distributed Denial of Service" includes diverse modules such as volume attacks, application layer attacks etc. GLW is highly utilized for these attacks detection in the specified location to enhance the vehicular system efficiency.

3.1. GLW Stages

GLW initially starts with random value in the search space to detect the optimum value based on the defined fitness function. The glow-worm algorithm involves three stages in each iteration to detect the local optimum value in the current space. Three stages are given below.

- Glow-worm attraction factor update stage
- Glow-worm movement stage
- Glow-worm neighbourhood stage

Glow-worm Attraction Factor Update Stage:

In the GLW algorithm, attraction factor is "luciferin" which produces the light enzyme on its own. Based on this attraction factor, each glow-worm interacts with neighbor glow-worms. The capability of luciferase enzyme is directly proportional to the present location of glowworm in the given search space. In this stage, it detects the local optimum value in the given search area rather than global value. Likewise, each glowworm attracted each other based on the highest luciferin values and updated with the current value. Similarly, the glowworm updates its attraction factor values nearer the optimum. The key rule of the attraction factor is given in the below equation 1.

$$F_j(S+1) = (1-\delta)F_j(S) + \omega O_j(S+1) \tag{1}$$

where δ is the luciferin decay constant [0, 1] and ω denotes the luciferin enhancement constant which is set as 0.6. The objective function is Oj (S+1) at 'j' location at sample time 'S'.

Glow-worm Movement Stage

In this second stage, "probabilistic mechanisms" plays a major role in glowworm technique. Naturally, glowworm attracts its neighbor based on the above mechanism. It checks the brightness of its own with its neighbor and attracted each other. Likewise, it moves entirely in the search space and detected the nearer optimum value. Fig.2 describes the probabilistic mechanism in detail with six glowworms. The probability movement of each glowworm 'i to j'is given below.

$$\varphi_j(S) = l_f(S) - l_i(S) / \sum_{k=Mi(S)} l_k(S) - l_i(S) \tag{2}$$

where $j \in M_i(S)$

$$M_i(S) = \begin{Bmatrix} j: d_{ij} < r_d^i(S); \\ l_i(S) < l_i(S) \end{Bmatrix} \tag{3}$$

'S' denotes the step index and dij represents the Euclidean distance. Mi (S) represents that movement towards the neighbour j from i by computing the Euclidean distance between i and j based on the luminescence level associated with glowworm j at time t.
Discrete-time movements are given below.

$$G_i(S) = G_i(S) + \rho(G_{ij}(S) - G_i(S)) / \|G_i(S) - G_i(S)\| \tag{4}$$

Glowworm Neighborhood Stage

Normally, local optimum values are detected in the previous stages. In this phase, the local optimum values are ignored and the global optimum values are selected from the given workspace. It detects the multiple sensor locations and continues the attraction process in the neighborhood range. Multiple peaks are identified by varying the sensor range. For this purpose, glowworm agent 'i' is within the local-decision domain with dynamic radial range $0 < r_d^i \leq r_s^i$. The enhancement movement is obtained using equation 5.

$$r_d^i(S+1) = min\{r_2, max\{0, r_d^i(S) + \beta(S_t - |M_i)(S)|)\}\} \tag{5}$$

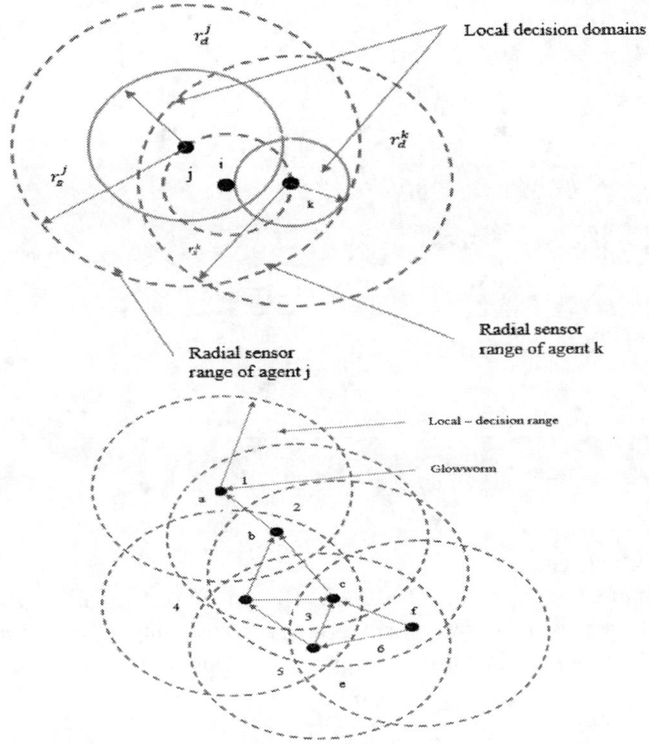

Figure 1. Glowworm Neighborhood Stage

Hence the optimization technique is used for the best feature selection in order to get the best features that enable to detect attacks based on their change in their values.

3.2. UNI-Layer ELM Network

The proposed uni-layer neural network follows the "ELM-extreme learning mac hine model". As per [12] [13], ELM provides high speed, exactness, preparing velocity, speculation and approximation functions. In this paper, we developed this uni-layer elm network for detecting DDoS attacks in the vehicular CPS. Each node in the hidden layer process the various vehicles input features and detects the output layer. The novel of the proposed network is automatic tuning is not required for this network. Hidden nodes need not tuning. The activation function utilized in the proposed model is sigmoid function. Let 'L' denotes the hidden neurons and output layer is denoted by 'z'.

The algorithm of Uni-layer ELM is as follows:

Initial Phase

Input: The arrived data, a kernel function $k(x_i, x_j)$and a regularization parameter C.

Output: φ_i

1. $\Omega_i = (C^{-1} + \Delta ELM - i)^{-1}$
2. $\varphi_i = \Omega_i . T_i$

Sequential Phase

Input: The (i+1)[th] sampling data

Output: φ_{i+1}

1. $K_{i+1} = [K(x_1, x_{i+1}), \ldots\ldots, K(x_i, x_{i+1})]^T$

2. $z_{i+1} = \Omega_i . k_{i+1}$

3. $r_{i+1} = C^{-1} + K_{(x_{i+1}, x_{j+1})} - z_{i+1}^T k_{i+1}$

4. $\Omega_{i+1} = r_{i+1}^{-1} \begin{bmatrix} \Omega_i r_{i+1} + z_{i+1} z_{i+1}^T & -z_{i+1} \\ -z_{i+1}^T & 1 \end{bmatrix}$

5. $e_{i+1} = t(x_{i+1}) - k_{i+1}\varphi_i$

6. $\varphi_{i+1} = \begin{bmatrix} \varphi_i - z_{i+1} r_{i+1}^T e_{i+1} \\ r_{i+1}^T e_{i+1} \end{bmatrix}$

4. Proposed Architecture

The proposed architecture consists of three different phases such as Data collection unit, feature Extractor and Selector using Glow worm optimizer and Single layer feed forward Networks. The working methodology of the three different phases is detailed in this section.

4.1. Data Collection Unit

The real time scenario has been created with the SUMO software. The position of the vehicles at a particular time t on the road is computed using a microscopic traffic based on vehicles and road characteristics. These positions are used to show vehicles on a graphical user interface (GUI) or written in a file (dump file). The different real time scenarios for the vehicular systems are created as shown in Fig. 2 and Fig. 3.

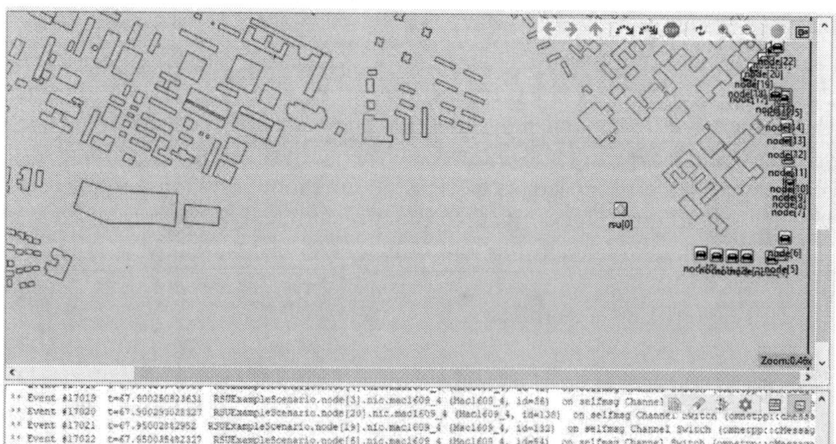

Figure 2. Real Time Scenario for the Vehicles cyber systems created by the SUMO-OMNET Environment

IOP Conf. Series: Materials Science and Engineering **925** (2020) 012001 doi:10.1088/1757-899X/925/1/012001

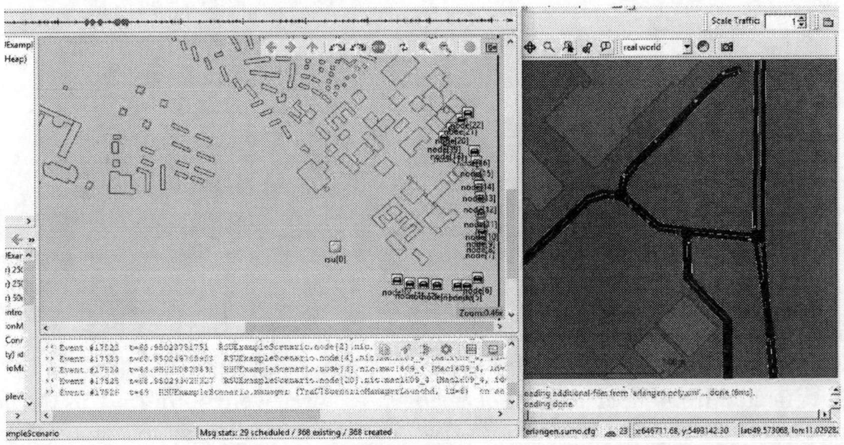

Figure 3. Real time Road Scenario with the Four Lane /Two Lane Road System using SUMO-OMNET Environment

The input data are the initials positions, vehicle characteristics (acceleration, maximum speed), road network, directions, position of destination, speed limits, etc. SUMO [15] a simulator with a feature to support car-following model (issue of [16-19]) used in this paper creates a road side scenario. These are simulated as the nodes in an OMNET ++ environment. The list of parameters which are created by SUMO are listed in Table I.

Table 1. Parameters of SUMO Interface

S.NO	List of parameters	Specifications
01	No of vehicles	100
02	Speed	35-40 km/hr
03	Direction	Bi directional
04	Type of roads	Two-lane /Four Lane Roads
05	Network used	IPv6
06	No of attacks used	02

Table I shows the different parameters collected in the SUMO Interfaces. After converting as the nodes using OMNET++ Environment, different features such as no of TCP packets transmitted, No of packets received, Signal Strength, latency, Speed of the vehicles, Direction, throughput, time stamp, Input Message ID and layer ID were calculated manually. The calculated features are listed in Table II.

4.2. Feature Selector and Extractor

As discussed above, various features were calculated from the network scenario to detect the various attacks. Since the number of features is more, it may cause the instability in the classifier which results in inaccurate predictions. Hence the optimization technique is used for the best feature selection in order to get the performance. The fitness function for calculating the best features is given as in equation 6.

Table 2. List of the Features calculated for the Vehicular Systems

S.NO	List of parameters	Specifications
01	No of TCP packets Transmitted	8 bytes
02	No of TCP Packets Received	8 bytes
03	Speed of the vehicles	35-40km/hr
04	Signal Strength	-10 to 25 dbm

IOP Conf. Series: Materials Science and Engineering **925** (2020) 012001 doi:10.1088/1757-899X/925/1/012001

05	Latency	2 ms
06	Direction	Omni direction
07	Throughput	40-80
08	Time Stamp	4 ms
09	Input Message ID	ID1
10	Layer ID	ID2

$$Fitness\_Function \; = \; W_B \; + \{\, i = 1 \sum n \; G_i \, X \, F_i\}^{-1} \qquad (6)$$

where WB is the actual no of features used, Gi= Cost function Fi= No of feature to be selected. From the 10 features collected from the real time scenario, 05 features were optimized as the best features for the high accurate classification. The obtained five features are taken as the inputs for the SLFN for classification and prediction of attacks. The labeled features used for the prediction of attacks are listed in table III.

4.3. SLFN Classifier

The labeled features are then feed to the SLFN classifier which consists of seven input layers and three output layers which depend on number of attacks. The proposed classifier classifies the different attacks based on the above equation 6. The pseudo code of the proposed classifier is as follows

Pseudo Code

%Initialization
Start the process
Let 'N' be the Glowworm input features for training with an Activation Function and n Hidden neurons
Assigning the Input weights and biases with random values
Calculate the hidden Matrix H
Calculate the Output weight Matrix β
If ($\beta \leq = \beta_1$)
Then /Attack_1 is Classified (DoS Attacks)
Else if ($\beta \leq = \beta_2$)
Then /Attack_2 is Classified (Botnet Attacks)
Else if ($\beta \leq = \beta_n$)
Then Attack_n is classified
End

5. Results and Discussions

5.1. Experimental Setup

As discussed in above section-II, SUMO integrated with the OMNET has been used for the real time data collection under normal conditions. To identify the different attacks, CIC-IDC2018 datasets were used as the reference in the real time scenario created. Nearly 1500 data were collected and used for the experimenting the proposed algorithm.

5.2. Performance Evaluation

The Glowworm features which are extracted from the datasets are used for training and testing process. For evaluation, the data collected are segregated as training and test data with proportions of 70% and 30% respectively. The evaluation is carried out for the different glow worm datasets with the performance parameters as in equations 7,8 and 9.

$$Accuracy = \frac{DR}{TNI} \times 100 \qquad (7)$$

$$\text{Sensitivity} = \frac{\text{TP}}{\text{TP+TN}} \times 100 \quad (8)$$

$$\text{Specificity} = \frac{\text{TN}}{(\text{TP+TN})} \times 100 \quad (9)$$

Where TP and TN represents True Positive and True Negative values and DR & TNI represents the Number of Detected Results and Total number of Iterations. The proposed GOF-SLFN algorithm has been evaluated by different cases based on the performance.

Accuracy Evaluation

For accuracy evaluation, the proposed algorithm has been executed for the 100 trials and mean of the results has been considered as final accuracy for the classification of different attacks in the transportation vehicular networks. The accuracy has been calculated for classification of different attacks. The accuracy of the proposed algorithm has been evaluated using the different learning kernels are shown in the Table 4.

Table 3. List of Labelled Features used by the Proposed Architecture for Classification of different attacks

No of TCP packets Transmitted	No of TCP packets Received	Throughput	Latency (ms)	Time of arrival to OBU (ms)	LayerID	Msg ID	Label	Label Details
5	5	48	0.7	0.45	186	457	0	Normal
5	5	62	0.8	0.458	174	427	0	Normal
5	5	25	0.3	0.459	60	138	0	Normal
5	5	60	0.5	0.45	66	153	0	Normal
5	5	46	0	0.45	30	63	0	Normal
5	5	49	0.1	0.45	42	93	0	Normal
5	5	8	3.2	0.78	36	78	2	Botnet
5	5	29	4	0.45	36	525	0	Normal
5	5	46	0	0.457	84	547	0	Normal
5	0	7	5	0.88	101	496	1	DOS
5	0	9	5.6	0.90	107	498	1	DOS
5	0	5	10	0.92	113	500	1	DOS
5	0	7	12	0.94	119	502	1	DOS
5	0	7	17	0.96	125	504	1	DOS
5	0	9	25	0.98	131	506	1	DOS
5	0	5	6.7	0.95	137	514	1	DOS
5	0	64	0.2	0.50	66	153	0	Normal

Table 4. Accuracy Comparison for different learning function used in Proposed Algorithm

S.No	Dataset details	Learning Kernel	Training Accuracy	Testing Accuracy
01	Five Feature Dataset	Sigmoid	95.5%	95.4%
		Sine	95.0%	94.5%
		Tanh	95.0%	94.5%

From TABLE 4, it is evident that the accuracy of 94.5 % obtained when the sigmoidal function is adopted for the proposed algorithm for CIC IDC datasets is maximum. Again the accuracy of the proposed GOF-SLFN along with the Sigmoidal learning function is evaluated with randomly chosen hidden neurons are shown in Table 5.
TABLE 5 clearly shows that the sigmoidal learning along with 75 neurons has a maximum accuracy of 95.5 % in the classification of different attacks of vehicular cyber physical systems. The highest accuracy obtained from GOF-SLFN has been compared with other learning algorithms for the evaluation of the performance based on the classification of attacks. The different scenarios of attacks have been considered for comparing the proposed algorithm with other existing learning algorithms.

Table 5. Accuracy Comparison for different neurons used in gof-slfn algorithm

Sl.No	Dataset details	No of neurons	Training Accuracy	Testing Accuracy
01	Five Feature Dataset	10	93.5%	93.0%
		20	93.5%	91.5%
		50	92.5%	92.5%
		75	95.5%	94.5%
		100	91.5%	92.4%
		150	92.4%	91.5%
		200	91.5%	92.5%

5.3. Scenario-I Analysis
In the first scenario, DoS attacks were considered for the classification and different comparative analysis are shown in following Fig. 4.

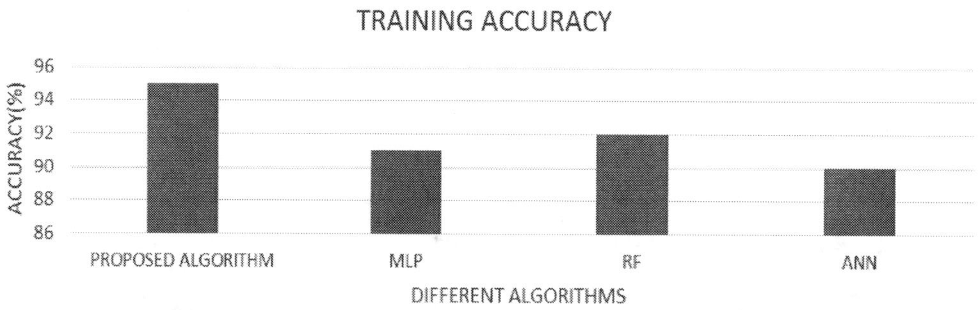

Figure 4. Training Accuracy Analysis for the different Algorithms in classifying/predicting the DoS Attacks

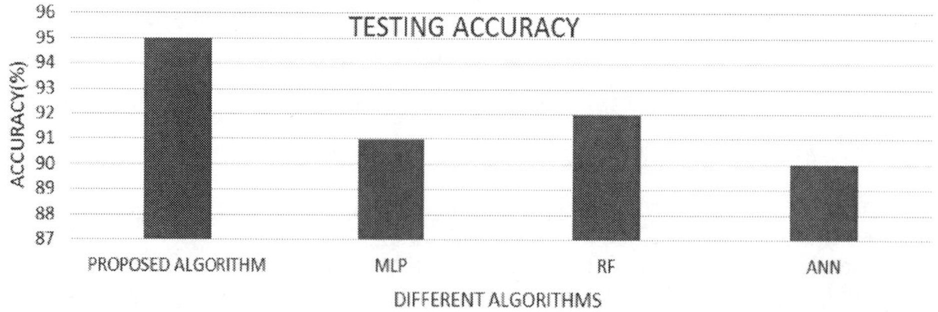

Figure 5. Testing Accuracy Analysis for the different Algorithms in classifying/predicting the DoS Attacks

Fig. 4 and Fig. 5 clearly shows training and testing accuracy of the proposed algorithm as 95.5% when compared with the other algorithms such as Multilayer perceptrons (MLP), Artificial neural networks (ANN), and Random Forest (RF).

Sensitivity and Specificity Analysis

Again the sensitivity and specificity has been calculated and compared with the learning algorithms in which the proposed algorithm outperforms the other existing algorithms. Fig.6 and Fig. 7 shows this analysis made.

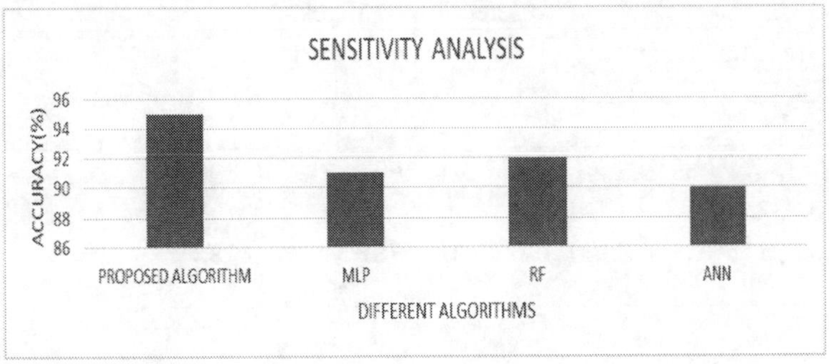

Figure 6. Sensitivity Analysis for the different Algorithms in classifying/predicting the DoS Attacks

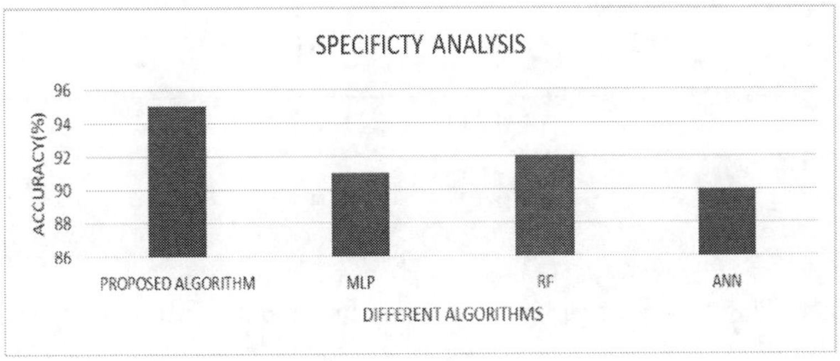

Figure 7. Selectivity Analysis for the different Algorithms in classifying/predicting the DoS Attacks

5.4. Scenario-II Analysis

In the second scenario, Bomb net attacks were considered for the classification and evaluation. The comparative analysis between the proposed algorithm are depicted in following TABLE 6.

Table 6. Comparative Analysis between the different learning algorithms in classifying the Bomb Net attacks

Attack	Algorithm	Accuracy		Sensitivity	Specificity
		Training	Testing		
Botnet Attacks	ANN	88.5%	88.0%	89.0%	89.0%
	RF	92%	91.5%	92%	91.5%
	MLP	91%	90%	91%	90.5%
	Proposed algorithm GOF-SLFN	95.5%	95.0%	94.0%	95.0%

From the table 7 it is clear that performance of proposed algorithm has outperformed the existing learning algorithms in terms of accuracy, sensitivity and specificity.

5.5. Time Analysis

The training and testing time has been calculated for the proposed GOF-SLFN and compared with the other existing algorithms. The TABLE 7 shows the comparative analysis of training time and testing time for different networks with respective to different attacks.

Table 7. Comparing Training Time of Proposed GOF-SLFN Model with the existing algorithms in detection of attacks

Types of Attacks	Training Time(secs)				
	GOF-SLFN	SLFN	MLP	RF	ANN
DoS attacks	0.233	0.345	0.1269	0.298	4.56
BotNet attacks	0.231	0.3300	0.1278	0.288	4.67

Table 8. Comparing Testing Time of Proposed GOF-SLFN Model with the existing algorithms in detection of attacks

Types of Attacks	Testing Time (secs)				
	GOF-SLFN	SLFN	MLP	RF	ANN
DoS attacks	0.109	0.108	0.1117	0.118	2.39
BotNet attacks	0.112	0.111	0.121	0.121	2.78

From the above tables 7 & 8, time complexity is also reduced in the proposed algorithm on comparison with the machine learning algorithms in detection of different attacks in real time vehicular scenario. Table 9 depicts the comparative analysis of the proposed algorithm with different optimization algorithms namely, firefly, Bat and LDA. From the table, the proposed algorithm provides promising detection accuracy than other optimization algorithms.

Table 9. Comparing Testing Time of Proposed GOF-SLFN Model with the existing algorithms in detection of attacks

S.No	Algorithms	Performance Metrics	DoS Attacks	Botnet Attacks
1	Swarmfly-SLFN	Training Accuracy	87.5%	86.45%
		Testing Accuracy	87.0%	86.0%
		Sensitivity	86.0%	84.0%
		Specificity	85.0%	83.0%
2	BAT-SLFN	Training Accuracy	93%	92%
		Testing Accuracy	92.5%	91.0%
		Sensitivity	92%	91%
		Specificity	91.5%	90.5%
3	LDA-SLFN	Training Accuracy	94%	94%
		Testing Accuracy	93%	93%

		Sensitivity	90%	90%
		Specificity	90.5%	90.5%
4	Proposed Algorithm GOF-SLFN	Training Accuracy	95.5%	95.5%
		Testing Accuracy	95.0%	95.0%
		Sensitivity	94.0%	94.0%
		Specificity	95.0%	95.0%

6. Conclusion

The proposed GOF-SLFN algorithm has been designed based on the glowworm optimizer integrated with Single layer feed forward networks with 75 neurons for the detection of different DoS attacks. The performance of the proposed algorithm has better accuracy in detection, better sensitivity and selectivity than the existing machine learning algorithm. The proposed classifier seems to be more intelligent in detection of the attacks and finds its application in vehicular cyber physical systems. The algorithm can be improvised by testing with the more number of attacks since the paper deals with the only two different types of attacks. The implementation of bio-inspired algorithms and deep learning algorithms will improvise the accuracy of detecting the different attacks. The future scope of this work can be considered based on the implementation of deep learning algorithms which will improvise the accuracy of detecting the different attacks.

References

[1] Syed J, Gardezi S, Awais M, Faye I and Meriaudeau F 2017 Mammogram Classification using Deep learning features *IEEE Int. Conf. on Signal and Image Processing Applications(ICSIPA).*

[2] Darby P and Gottumukkala R 2019 Decentralized Computing Techniques in Support of Cyber-Physical Security for Electric and Autonomous Vehicles *IEEE Green Technologies Conf. (GreenTech)* 1-5.

[3] Chen H, Chen X, Fan L and Chen C 2015 Classified security protection evaluation for vehicle information system *Int. Conf. on Cyber Security of Smart Cities, Industrial Control System and Communications (SSIC)* 1-6.

[4] Gifei S and Salceanu A 2017 Integrated Management System for quality, safety and security in developing autonomous vehicles *10th Int. Symp. on Advanced Topics in Electrical Engineering (ATEE)* 673-676.

[5] Singh M and Kim S 2017 Reconcile security requirements for intelligent vehicles *17th Int. Conf. on Control, Automation and Systems (ICCAS)* 1646-1651.

[6] Kim H, Ju B and Na J 2018 An Access Control Method for Vehicle Management System *Int. Conf. on Information and Communication Technology Convergence (ICTC)* 949-952.

[7] Mansfield K, Eveleigh T, Holzer TH and Sarkani S 2013 Unmanned aerial vehicle smart device ground control station cyber security threat model *IEEE Int. Conf. on Technologies for Homeland Security (HST)* 722-728.

[8] Jin PJ, Zhang G, Walton CM, Jiang X and Singh A 2013 Analyzing the impact of false-accident cyber-attacks on traffic flow stability in connected vehicle environment *Int. Conf. on Connected Vehicles and Expo (ICCVE)* 616-621.

[9] Javaid A, Sun W, Devabhaktuni VK and Alam M 2012 Cyber security threat analysis and modeling of an unmanned aerial vehicle system *IEEE Conf. on Technologies for Homeland Security (HST)* 585-590.

[10] Škorput P, Vojvodić H and Mandžuka S 2017 Cyber security in cooperative intelligent transportation systems *Int. Symp. ELMAR* 35-38.

[11] Shafique M, Khalid F and Rehman S 2018 Intelligent Security Measures for Smart Cyber Physical Systems *21st Euromicro Conf. on Digital System Design (DSD)* 280-287.

[12] Huang GB, Zhu QY and Siew CK 2006 Extreme learning machine: theory and applications *Neurocomputing* **70** 489-501

[13] Lian C, Zeng Z, Yao W and Tang H 2013 Displacement prediction of landslide based on psogsa-elm with mixed kernel *Sixth Int. Conf.* on *In Advanced Computational Intelligence (ICACI)* 52-57.
[14] Wang B, Huang S, Qiu J Liu Y and Wang G 2015 Parallel online sequential extreme learning machine based on MapReduce *Neurocomputing* **149** 224-232.
[15] Lu S and Lu Z 2016 A pathological brain detection system based on kernel-based ELM *Multimed Tool Appl.* 1-14.
[16] Krauss S 1998 Microscopic modeling of traffic flow: Investigation of collision free vehicle dynamics.
[17] Wagner P and Lashevsky I 2003 Empirical basis for car-following theory development.
[18] Kerner BS 2009 Introduction to modern traffic flow theory and control: the long road to three-phase traffic theory *Springer Science & Business Media.*
[19] Treiber M, Ansgar H and Dirk H 2000 Congested traffic states in empirical observations and microscopic simulations *Physical review E.* **62** 1805.

Power Allocation Based on Channel State Information in Massive MIMO System

SNCA Boovarahan[1] and K Umapathy[2]

[1,2] Sri Chandrasekharendra Sarawathi Viswa Mahavidyalaya Kancheepuram, Tamilnadu, India.
[1] itsmeboovar@gmail.com

Abstract. A system with large number of carriers doing the transmission in many sub-channels (OFDM) or multiple antennas (MIMO) sequentially the power was allocated. However, the information of the channel parameters is predictable, then successive power allocation for notable advancement in system productions can be attained. Broadcasting power allocation might upgrade system performance by means of comprehensive capacity increase or bit error rate reduced. Power Allocation (PA) for Heuristic Beamforming with Maximum Ratio Transmission and Zero-Forcing Beamforming method was used. The relay to earth transmission over1000K users with Signal to Noise ratio of 5 dB, and not correlated Rayleigh fading channels. Maximum ratio and Zero forcing these two linear methods of pre-coding can be taken. This MATLAB simulation differentiated the Spectral Efficiency (SE) acquired when having precise CSI with pilot sequences of length τp. The conclusions are based on movability of the user signals. Simulation results produces the balancing effect between PA and Spectral Efficiency, and neatly displayed the variation with existing work, and precisely displayed that comparatively greater transmit power will be a huge energy efficient level is reached.

Keywords: Channel state information, Spectral Efficiency, Power Allocation, Maximum Ratio Transmission, zero forcing.

1. Introduction

Power and Channel allocation is specifically complex when large number of antennas are established at all the base stations. Power & channel allocation expressed as distributing transmit power encompassed by clients, involving bearings, whereas satisfying a set of control the limitations of power that have Synchronized with inferences economically. A major difficulty in the channel and power allocation is the inter-user intrusion that stand up and restrict the performance when large number of users were distributed in parallel. Power and channel allocation is specifically complex when Large number of antennas were installed at each earth base station.

The Power and channel allocation with unique antenna transmitters, present time large number of antenna techniques shows the power and channel allocation with exact spatial division of users. Mathematically, channel and power allocation correlates to the choosing of a signal correlation matrix for every user. A base transceiver station that are positioned close together having coordination was propose for a huge cellular MIMO system, which takes entire adjacent position systematization–to increase adding rate, restricted outer position closed together cell coordination–to limits interference for the cluster edge users. Many cell block non zero elements along diagonal sides is recycled to correspond the broadcasted signals alone several Basin identical same closed together cells. A Large feasible precoding method for transmit MIMO channels is Block Diagonalization (BD), That gives each consumer a disturbance-free channel with precise linear precoding matrix architecture. The resource distribution with power restriction per-BTS (PBPC) [1], there are several reasons for doing optimal power allocation scheme. Primarily, with some weight factors it reduces the difficulty for evaluating the most effective problem differentiate with the ideal plan. Secondly for each user equal power distribution concluded only with very small capacity loss distinguish to optimum water-supply, specifically at elevated Signal to interferences noise ratio, and with tracking the power allocation covering users have a critical part to play than over streams of every user. Later device scaling did it simple to tune the transmit power in the middle of various users, such as connecting to fixed rate limitations.

Depend on zero- forcing precoding [2], the power & channel allocation issue for energy-effective communication in large MIMO systems question of optimization aimed at improving energy efficiency (EE) and at the same time providing the necessary spectral efficiency (SE). The optimization of power distribution issue [3], Primarily develop the clear limits of average per user Signal interference noise ratio and use these basic limitations to made the SINR restrictions for the proposed optimization problem. The power distribution matter which reduces the final transmit power at a period of time when meeting the Signal interference noise ratio limitations, as mentioned by the evaluated average SINR basic threshold value both for receiver MRC and precoder MRT [23].

The rate of power saving reduces the essential enlarging of SINR per-user. That should have been noted that is gain of installing a huge amount of antennas BS leads to develop into at the edge, as the eventual SINR production is bounded by intrusion, channel prediction fault and transmit power limitations. A taleEE power allocation makes most effective use of the problem has been found it depends on the evaluated SINR basic limiting according to SINR pre-user demand and per-user power utilized limitations.

2. Existing System

Hao Xu et al. [4] proposed the basic limit of average SINR of D2D links, and systematic way of a power influence issue to reduce D2D links' the power of data transmission extending SINR target limitations. A computational process intersects to the particular ideal result is proposed. The mid-SINR systemic under bound almost equates the measured average SINR. f a D2D receiver conducts co-channel nosiness is massive, the SINR limitations might not be proceed towards applying GCPA algorithm. The bottommost leap of D2D links' average SINR, and evaluates a power influence issue to reduces D2D links' data transmit power extending the SINR target limitations. The proposed technique uses the Power Allocation for Heuristic Beamforming and also the Lagrange multiplier and Lagrange dual function gives a leap on the best value.

Ying Zou et al. [5] proposed the codebook-situate beamforming is utilized extensively to recompense the critical channel attenuation. But the overall codebook size makes it complicated to explore the optimal precoder and combiner in service directly. The plan proposed uses the rapid start tactics to initialize. Through treating any spectrum output inspection as a random variable, the proposed scheme uses the other approach of discrete stochastic optimization that uses probabilistic forecasting to assess the ease of reducing the computational complexity.

P. Sreesudha et al. [6] displays the presentation of CDMA framework relies upon the qualities of scattering arrangement. The scattering arrangements should be remarkable and have improved auto connection and cross relationship esteems. Generally Multiple Access Interference (MAI) emerges that restricts the framework limit. The proposed work improves the presentation of multi-cell frameworks relies upon the asset allotment; that is, the manner by which the time, power, recurrence, and spatial assets are isolated among clients.

Do Dung Nguyen et al. [7] examined based on the signal to noise interference results and the proposed work inspiration is, for instance, not to upset neighbouring frameworks and the comparing limitations are called delicate moulding on the grounds that the condition of the signals that sent is feasibly determined if power in absence of the necessity will have excelled the edge. The insistent for maximum data rates have been one over longer distance in every primary consideration beyond the improvement of MIMO orthogonal- frequency-division-multiplexing (OFDM) communications systems. For extensive bandwidth channels, primary incomplete result a combination of narrowband overlapping subcarriers is used to address the multipath problem. Use of overlapping OFDM subcarriers will not only increase spectral efficiency, power allocation however lower symbol rates employed by narrowband subcarriers lowers the influence of multipath signal processes.

N. Kirubananda Sarathy et al. [8] describes the mixed radix FFT in MIMO system design including space time block code techniques, channel estimation and signal processing algorithm which is used for performing only frequency and time synchronization. These are all the main drawbacks of the existing system.

3. Proposed System

The multi-cell relay to earth signals has tempted a huge attraction; the system-efficiency will be upgraded furthermore in the event that the recurrence & reuse designs are re-established from added activity with transmitters. In a possible way this could make the entire network act as one large virtual cell that uses every type of resources.

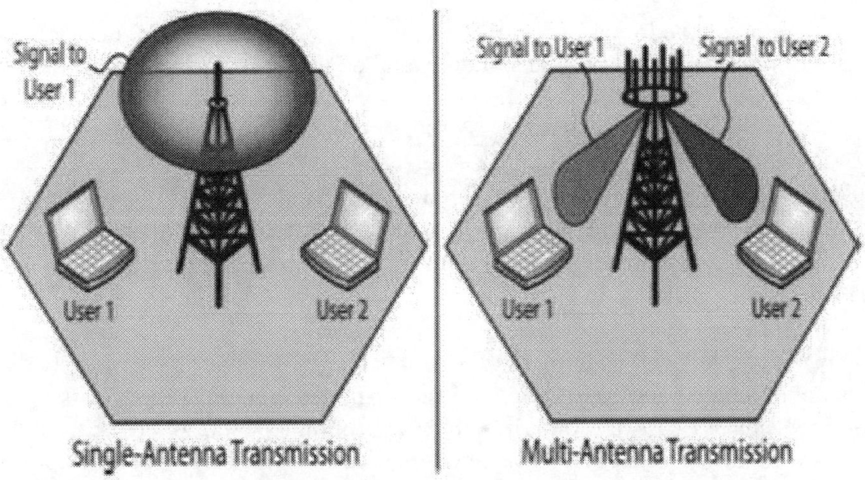

Figure 1. Difference in single and multi-antenna transmission.

From as it were one receiving wire, the flag creates concurring to a settled radio wire plan of the antenna (e.g., equally strong in all directions) and creating all disturbances in all the ways where the deliberate user is not detected.

4. System Model

Guillen Many-cell structure, the channel from all BS to MSk is indicated
Where is the channel from BSj.
Dynamic cooperation clusters (DCC) specified that:
channel samples of base stations to clients in Cj⊆ {1,...,Kr}, whereas impedances made to clients i_∈Cj exceptionally low values and it will worked in the portion of the Gaussian noise;
• BSj set out the users in Dj⊆Cj with data. Specific channel components of hk will bring data and/or bigger interference. These will be chosen by the diagonal matrices

These can be chosen by the diagonal matrices

$$D_k \epsilon \mathbb{C}_{i'}^{N\times N}$$

$$C_k \epsilon \mathbb{C}_{i'}^{N\times N}$$

which are stated as

$$Dk = \begin{bmatrix} D_{1k} & \cdots & 0 \\ \vdots & \ddots & \vdots \\ 0 & \cdots & D_{1k} \end{bmatrix} where\ D_{jk} = \begin{cases} I_{Nj},\ if\ k\epsilon\ D_j, \\ O_{Nj,}\ otherwise, \end{cases} \quad (1)$$

$$Ck = \begin{bmatrix} C_{1k} & \cdots & 0 \\ \vdots & \ddots & \vdots \\ 0 & \cdots & C_{1k} \end{bmatrix} where\ C_{jk} = \begin{cases} I_{Nj},\ if\ k\epsilon\ C_J, \\ O_{Nj,}\ otherwise, \end{cases} \quad (2)$$

Thus, $h_k^H D_k$channel shifts information to MSk and $h_k^H C_k$ channel that takes obstructions. It's vital to have both **D**k and **C**k, to affirm that as it were the proper base stations transmit to MSk at the time of evaluating controlled allocation of power (Equation 1 and 2).

The image inspected troublesome baseband-maintained signal at MSk is $y_k \in C_\epsilon$ and is designated by the linear i/o model

$$y_k = h_k^H x + n_k \quad (3)$$

Where $n_k \in C$ is the merged vector of additive noise and disturbances from nearby systems. Spreading out the one-cell i/o model in Eq. (3), the symbol sampled difficult baseband sustained signal at MSk is

$$y_k = h_k^H C_k \sum_{i=1}^{K_r} D_i s_i + n_k \quad (4)$$

Term nk~CN(0,σ2 k) is anticipated to demonstrate with clamor and frail obstructions from all BSj with k _∈Cj. Forecast for confined level of CSI required in the assessment of the transmission and is selected the sensible signals only by the users and stored signals that are capable than the additional signals that are not needed but they included in Cj.

The BS correlates to every cell edge users of adjoin cells.The variance σ_k^2 is usually dissimilar with the users and is approximated and followed using the sustained signals. The σ_k^2 k is precisely fixed with control confinements; in case the broad control utilization systems is created, at that point signals disturbances too created. This association without particular effect on power limitations is steady, prime significance in any asymptotic assessment since bigger cell systems essentially obstructions bounded within the SNR method. BSj *assumed* for the necessity of channels h_{jk} and σ_k^2 variances exactly to every users k ∈Cj.

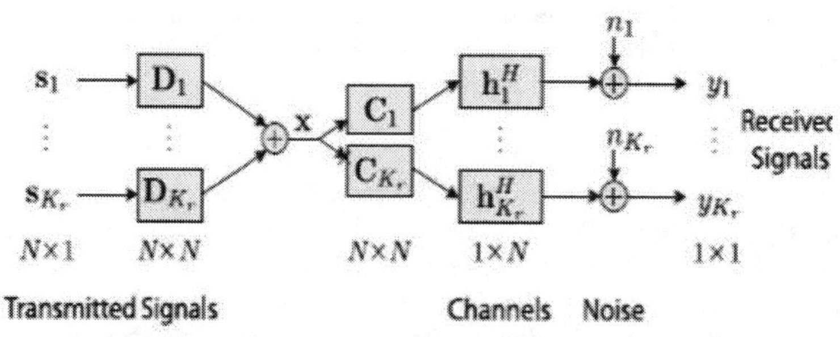

Figure 2. Schematic diagram of the system model for down to earth signals multi-cell communications with Kr unique antenna users are distributed by N antennas.

With the unique cell outline, the transmission is bounded by the L power limitations in

$$\sum_{K=1}^{K_r} tr\ (Q_{lk} S_k) \leq ql \quad l = 1, \ldots., L,$$

Where $Q_{lk} \epsilon \mathbb{C}^{N \times N}$ are Hermitian positive fractional particular matrices and level $ql \geq 0$ for all l,k. If **Q***lk* is proportionate then *ql* was considered in mW, disperses as overlying limit on allowed transmit power. Subspace navigated by **Q***lk* (Equation 5). To make the control is constrained in every heading, these networks *L* for K users.

$$\sum_{l=1}^{L} Q_{lk} > 0_N \tag{5}$$

An influential contrast is that the exact transmitted signals are D_{ksk}. Thus, every weighting matrix **Q***lk* should please the extra condition that $Q_{lk} - D_k^H Q_{lk} D_k$ is diagonal for all being zero with l and k. This scientific prediction confirms the control will not be doled out for subspaces that is not allowed for rationale of diminishing of scaled power with subspaces used in transmission which is conceivable only when **Q***lk* is non two opposite corners of a square of a matrix. The major cell plot of power limitations that affect signals from chosen base station.

Every transmitter power obstruction is analyzed and it having $L = Kt$ and the level of threshold that affecting the BS*l* is Eq. (6)

$$Q_{lk}^{per} - BS = D_k^H diag\ (0_N + \cdots . +. N_t - 1, 1_{Nt} + 1 + \cdots . N_{kt}) D_k\ \forall l \tag{6}$$

5. Heuristic Beamforming Power Allocation

We vectors of beamforming are destroyed as $V_k = \sqrt{p_k v_k}$for all *k*, where $V_k$are normalized beamforming directions and $p_k \geq 0$are the factors of power allocation (Equation 7,8).The common objective power allocation

$$subject\ to\ SINR_k = \frac{|h_k^H C_k D_k v_k|^2}{\sigma_k^2 + \sum_{i \neq k} |h_k^H C_k D_k v_k|^2} \quad \forall k \tag{7}$$

$$\sum_{K=1}^{K_r} v_k^H Q_{lk} v_k \leq ql\ \forall l \tag{8}$$

Authoritative outline in signal processing is to identification of a magnitude of data symbol *sk* which is noted under channel deformation, white noise and interferences. If many channels observed values are available for a particular data symbol, this plot can be stated as (Equation 9)

$$y = \sum_{K=1}^{K_r} h_k s_k \quad + n \tag{9}$$

Where h_k is the channel for character *sk(Equation 10)*,

$$\mathrm{E}\{s_k\} = 0, \mathrm{E}\{|s_k|^2\} = 1 \tag{10}$$

The character *sk* can be calculated from the vector valued calculation y as
$s_k = V^H\ y$ Utilizing a linear sustained combining filter
The n number objective power allocation problem in Eq. (11) then becomes

$$subject\ to\ SINR_k = \frac{|\sqrt{p_{jk}k}h_{jk}^H C_{jk}k\ v_{jk}|^2}{\sigma_k^2 + \sum_{i \neq k} \left|\sqrt{p_{ji}i}h_{jk}^H C_{ji}k\ v_{ji}i\right|^2}\ \forall k \quad (11)$$

$$\sum_{k \in D_j} p_{jk}k \leq q_j\ \forall j \quad (12)$$

Where jk is the indicator of the base station that distributes MS k and the beamforming vectors are decayed as, $V_k = \left[0 \dots 0\ \sqrt{p_{jk}k\ V_{jk}^T k}\ \ 0..0\ \right]^T$ is the uniform beamforming direction and $V_{jk}k \in C_{jk}^N \times 1$ and $p_{jk}k \geq 0$ is the power allocated by $BS_{jk}$for transmission to MS_k

6. Zero-Forcing Beam Forming(ZFBF)

Zero forcing mention to signal processing that fully removes the interference. A conceptual stimulation is that zero-forcing at the same time reduces the MSE in middle of the received signal and the transmitted symbol sk (13).

$$MSE_k = E\left\{\left|\sum_{i=1}^{K_r} \sqrt{p_{ji}i\ h_{ji}^H k\ C_{ji}k\ V_{ji}i\ s_i} + n_k - s_k\right|^2\right\} \geq E\{|n_k|^2\} \quad (13)$$

In the perfect case without transmit power limitations. Zero-forcing is only applicable to the users who are all active, which are stated as follows.

The beam-forming directions are called *zero-forcing* beam forming *(ZFBF)* in the direction of the users in the organized set *S*.

7. SLNR & Maximizing Beam Forming

MRT and ZFBF within the combining factors from coordinate expansion of the particular criteria for: huge SNR and low obstructions and power, respectively. This norm separates the picking of beamforming directions by ignoring or removing interference. The method to static between SINR is enhance the SLNR (14) that will be denoted as

$$\mathrm{SLNR}_k = \frac{\frac{1}{\sigma_k^2}|h_{jk}^H\ k\ C_{jk}k\ v_{jk}k\ |^2}{\frac{1}{\eta_{jk}} + \sum_{i \neq k}\ \frac{1}{\sigma_k^2}\left|h_{ji}^H i\ C_{jk}i\ V_{jk}k\right|^2}\ \forall k \quad (14)$$

8. SLNR Maximizing Beam Forming

The beamforming direction (15)

$$V_{jk^k}^{SLNR} = \frac{\left(\frac{1}{\eta_{ik}} I_{N_{jk}} + \sum_{i=1}^{K_r} \frac{1}{\sigma_i^2} C_{jk^i}^H h_{jk}^H\ k\ C_{jk^i}\right)^{-1} h_{jk^k}}{\left\|\left(\frac{1}{\eta_{ik}} I_{N_{jk}} + \sum_{i=1}^{K_r} \frac{1}{\sigma_i^2} C_{jk^i}^H h_{jk}^H\ k\ C_{jk^i}\right)^{-1} h_{jk^k}\right\|_2}\ \forall k \quad (15)$$

9. Power Allocation

The subsections that came before prediction characterized by MRT, ZFBF, and SLNR-MAX has heuristic methods to selecting the beam forming headings and$\{p_{jk^k}\}_{k=1}^{K_r}$ essentially finding the analyzing part reached by the way of transmitting way. The SINRs becomes Eq. (16)

$$\mathrm{SINR_k} = \frac{\mathrm{p_{jk}k}\ \rho_{\mathrm{kk}}}{\sigma_{\mathrm{k}}^2 + \Sigma_{\mathrm{i \neq k}}\ \mathrm{p_{jk}k}\ \rho_{\mathrm{kk}}} \tag{16}$$

with fixed $\rho ik = \left|h_{jik}^{H} C_{jik} v_{jii}\right|^2$ for all *k, i.*

Power allocation will be explained in a detailed manner under ZFBF, because all the essential interfering channels are zero: $\rho_{ik} = 0$ for $i \neq 0$

10. Result and Discussion

The Spectral Efficiency is stated as a role of number of the Base Stations antennas depicted as M, and we measure TDD mode having the value of τp= 1000K with FDD mode utilizing either τp = 10/τp= M, or τp= min(M;50), where the concluding forms a randomly Average pilot period selected by having 50.

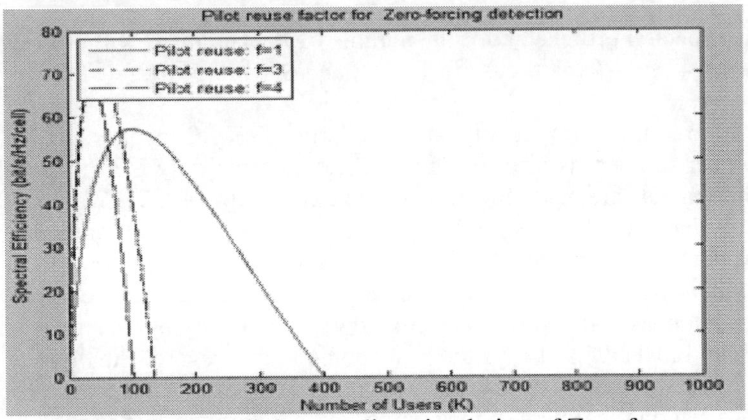

Figure 3. Downlink precoding simulation of Zero force.

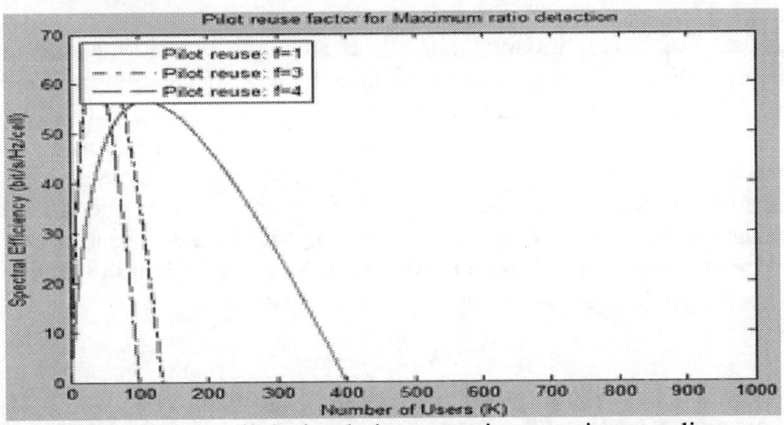

Figure 4. Downlink simulation – maximum ratio precoding.

Fig.4, 5 Mid Spectral efficiency downlink as a justification for few BS antennas, With different processing systems and with various CSI types accessible at the Base stations. The red line indicates the Perfect CSI, here simulation is tested under the Number of user (K) in the range of 1000 and path loss component 2.5.

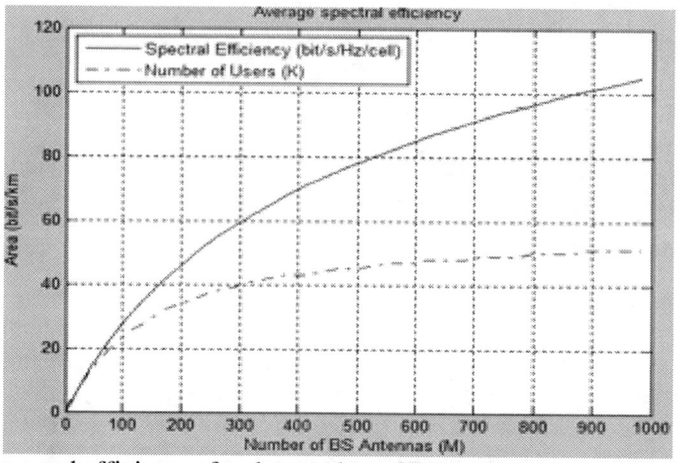

Figure 5. Average spectral efficiency, for the number of BS antennas, with Zero Forcing processing, pilot reuse factor f=3 and SNR of 5 dB.

Particular users are rearranged to capitulate the larger SE for every antenna number, and the corresponding number of users is also displayed. FDD systems only gain from the addition of more antennas if the pilot signals are also longer, as in the case of $\tau p = M$. There is no sake with $\tau p = 10$ to have larger than 10 antennas, meanwhile the execution saturates with 50 antennas with min(M;50) in the process. TDD operation is entirely observable, equivalent to the number of BS antennas, while FDD operation can only hold largeantennas by also increasing the overhead of the pilot. Installation of FDD systems with large antennas is practically possible, for slowly changing channels where we can take on a huge overhead pilot but TDD is always the best choice in this regard. Finally, the channel evaluation is uncomplicated when working in time division duplexing mode, so the concatenation of pilot signals only desired to be K length regardless of the number of base stations antennas M.

11. Conclusion

The analysis was based on a scientific model of the device that estimated the hardware failure of each antenna by an additive distortion noise relative to the antenna's signal strength. This model has many appealing features: it's also mathematically workable, has been validated experimentally in previous works, and can be influenced scientifically by structures utilizing correction algorithms to mitigate physical impairments. The ZF issued double the amount of spectral efficiency as MR. The cause for the more tolerable performance gap is that also ZF experienced from interference in larger cell case, From the pilot contamination and large number of inter-cell interferers made it not possible to remove all interference. In summary, the factors of pilot reuse signals area powerful design limit in Massive MIMO networks and the better choice determined on the load of the users.

Reference

[1] Hsu YC and Lin YD 2002 Multihop cellular: A novel architecture for wireless data communications *Journal of Communications and Networks* **4** 30-39.

[2] Zhang J and Li Y 2011 A New Load-Balancing Algorithm for Multihop Wireless Networks *Int. Conf. of Information Technology, Computer Engineering and Management Sciences* **2** 162-166.

[3] Srinivasan R 2008 IEEE 802.16 m evaluation methodology document (EMD) IEEE 802.16 m-08/004r3.

[4] Sheu JS, Lyu SH and Lyu SJ. 2014 Cell sectorization and power management for inter-cell interference coordination in OFDMA-based network MIMO systems. *In IEEE Wireless Communications and Networking Conference (WCNC)* 1638-1642.

[5] Proakis JG and Salehi M 2001 Digital communications. *New York: McGraw-hill* **4**.

[6] Femenias G, Riera-Palou F and Thompson JS 2016 Robust scheduling and resource allocation in the downlink of spatially correlated MIMO-OFDMA wireless systems with imperfect CSIT *IEEE Transactions on Vehicular Technology* **65**.

[7] Lei L, Yuan D, Ho CK and Sun S 2014 Resource scheduling to jointly minimize receiving and transmitting energy in OFDMA systems *In 2014 11th Int. Symp. on Wireless Communications Systems (ISWCS)* 187-191.

[8] Zhang L, Lu P, Yu Z, Cao H, Sun C and Wu C 2013 A utility-based adaptive resource scheduling scheme for multiple services in downlink multiuser mimo-ofdma systems *In 2013 IEEE 77th Vehicular Technology Conf. (VTC Spring)* 1-5.

[9] Wei L, Hu RQ, Qian Y and Wu G 2015 Energy efficiency and spectrum efficiency of multihop device-to-device communications underlaying cellular networks *IEEE Transactions on Vehicular Technology* **65** 367-80.

[10] Kaneko M, Nakano T, Hayashi K, Kamenosono T and Sakai H 2016 Distributed resource allocation with local csi overhearing and scheduling prediction for OFDMA heterogeneous networks *IEEE Transactions on Vehicular Technology* **66** 1186-99.

[11] Einhaus M, Klein O and Walke B 2008 Comparison of OFDMA resource scheduling strategies with fair allocation of capacity *In 2008 5th IEEE Consumer Communications and Networking Conf.* 407-411.

[12] Li M, Li P, Huang X, Fang Y and Glisic S 2014 Energy consumption optimization for multihop cognitive cellular networks *IEEE Transactions on Mobile Computing* **14** 358-72.

[13] Tuncel NO and Koca M Joint mobility load balancing and inter-cell interference coordination for self-organizing OFDMA networks *In 2015 IEEE 81st vehicular technology conference (VTC Spring)* 1-5.

[14] Wu SL and Wang WW 2015 Energy-efficient multimedia multicast scheduling and resource allocation algorithms for OFDMA-based systems *In 2015 IEEE Int. Conf. on Computer and Information Technology; Ubiquitous Computing and Communications; Dependable, Autonomic and Secure Computing; Pervasive Intelligence and Computing* 1377-1382.

[15] Jangsher S and Li VO 2014 Power control in multihop cellular networks with multiple radio access technologies *In 2014 IEEE 25th Annual International Symposium on Personal, Indoor, and Mobile Radio Communication (PIMRC)* 1674-1678.

[16] Ahmad SA, da Silva CR and DaSilva LA 2914 Marginal Power Efficiency Considerations in Power Control Adaptations for Multihop Cellular Networks *IEEE Transactions on Vehicular Technology* **63** 4718-23.

[17] Ramji T 2015 Adaptive resource allocation and its scheduling for good tradeoff between power consumption and latency in OFDMA based wireless distributed computing system. *In2015 Int. Conf. on Computation of Power, Energy, Information and Communication (ICCPEIC)* 0496-0501.

[18] Hoang TD, Le LB and Le-Ngoc T 2016 Joint prioritized link scheduling and resource allocation for OFDMA-based wireless networks *In 2016 IEEE Int. Conf. on Communications (ICC)* 1-6.

[19] Farazmand Y and Alfa AS 2013 A game theoretic power allocation and relay load balancing in OFDMA-based DF cellular relay networks *In 2013 IEEE 78th Vehicular Technology Conf. (VTC Fall)* 1-6.

[20] Zhao Y, Fang X, Huang R and Fang Y 2012 Joint interference coordination and load balancing for OFDMA multihop cellular networks *IEEE Transactions on mobile computing* **13** 89-101.

[21] Zhang W, Wang Y, Peng F and Yuan Y 2013 Interference coordination with vertical beamforming in 3D MIMO-OFDMA networks *IEEE Communications Letters* **18** 34-7.

[22] Prabhu Chandran K and Ramesh GP 2017 Power and resource allocation for based two- tier heterogeneous network using small cell clustering algorithm *International Journal of Engineering & Technology* **7** 135-140.

[23] Subramani P, Rajendran GB, Sengupta J, Pérez de Prado R and Divakarachari PB 2020 A Block Bi-Diagonalization-Based Pre-Coding for Indoor Multiple-Input-Multiple-Output-Visible Light Communication System *Energies* **13** 3466.

A Novel Approach for Improved Load Balancing and Interference Reduction in Wireless Networks using OFDMA

SS Saravana Kumar[1] and K Umapathy[2]

[1,2] Sri Chandrasekharendra Saraswathi Viswa Maha Vidyalaya, Kanchipuram
[1] sssk.saravana@gmail.com, [2] umapathykannan@gmail.com

Abstract. The quality of service offered to the user equipment in Multihop cellular networks with proper connection is affected severely due to the fading effects. These fading effects can lead to link loss between the user equipment (UE) and base station. This problem is severe when the UE traveling inside of tunnel, forests and big buildings. The duration of the link loss may vary from seconds to hours. Link loss management can control the link loss, increase the coverage area and signal quality in cell boundaries. It requires predictive power and link scheduling in cell sectors and signal quality enhancement at cell boundaries using filters. Interference Coordination and Load Balancing at base station and UE algorithm is used. This paper introduces a link loss management algorithm with load balancing to provide seamless service to users and this increases the Quality of Service (QoS) of safety critical communications.

Keywords: Base Station, Cellular networks, Inter channel interference, Power distribution, Base station, Signal to interference plus noise ratio, MIMO channel model.

1. Introduction

Technologies for load balance between surrounding cells of a mobile wireless communication network can reduce traffic congestion and improve network system capacity. The load balancing can be achieved by changing different parameters of the radio access network. These modifications may be made at cell or adjacent cell level. The modifications can be made iteratively in response to different operational statistics gathered. The changes will change cell size and cell shape, as well as adapt the transfer to optimize system resources and the use of hardware. An iterative optimization method will obtain performance statistics and network configuration from a mobile wireless network on a regular basis. The collected information can be periodically analyzed to determine parameter adjustments. Additional capacity configuration from the communication network can prevent or significantly delay the acquisition of additional hardware resources to mitigate device capacity problems. A widely used approach for evaluating the efficiency of a load balancing system is whether it provides a min-max balanced load solution [1]. Self-organized cellular networks include a set of functions for automated cellular network setup, optimization, and maintenance. Because mobile end-users continue to use network resources when traveling from one cell boundary to another, traffic within a cell is not constant.

Load balancing has thus become one of the most involved and evolving research fields in the Cellular Network. It includes moving the load from overloaded cells to neighboring cells with free resources for

IOP Conf. Series: Materials Science and Engineering **925** (2020) 012003 doi:10.1088/1757-899X/925/1/012003

a more balanced distribution of the load to preserve acceptable end-user experience and network efficiency. In Multihop Cellular Network (MCN), MCN's architecture parallels that of Single-hop Cellular Networks (SCN), except that bases and mobile stations are limited in transmission range. The base and mobile stations are therefore not always available in one hop [2, 22]. In this paper, A load balancing system that only applies to a small number of algorithms is not beneficial because many new schemes will have to be built. Section 2 explains the architecture of the network, section 3 discusses interference coordination (IC) and adaptive MCN resource scheduling.

2. Related Work

Tuong Due Hoang et al. [18] discusses On mutual prioritized connection scheduling and allocation of resources for OFDMA-based wireless networks. They are classifying the user equipments into priority based user equipments and non-priority based user equipments. The link scheduling and modified power allocation algorithms will be served to the classified user equipments including link removal procedures. The link scheduling algorithm maximizes the scheduled links. The relation removal procedure is then only applied if the solution obtained from the changed power allocation could not accommodate the minimum required levels of all connections. The authors are not discussing about how to compensate for the removed links.

Guillen Femenias et al. [6] discusses about Scheduling and allocation of resources, assigning data transmission speeds, bandwidth and power optimally to the transmitter and the receiver under inaccurate channel state information. Megumi Kaneko et al. [10] gives information about resource allocation for the down link of an OFDMA based macrocell/femtocell overlaid Heterogeneous network (HetNet). In their work Femto cell base station predicts the sub-channels likely to be used in the corresponding frame by neighboring MUs based on the channel state information, Signal for interference plus ratio of noise and signal to noise. Lidong Zhang et al. [8] giving much information on resource scheduling and sharing scheme based on minimization of utility function. They are selecting a particular antenna sequence as per the channel state information at the transmitter (CSIT) for maximum spectral efficiency. They are also discussing about carrier frequency allocation to user for maximum spectral efficiency.

3. Proposed System

The proposed system addresses the problem of link loss. It executes the link loss management algorithm, when the channel becomes noisy. It manages the power in allocated channels for interference coordination and mitigation. The developed OSI model architecture gives high performance for link loss management, interference cancelation, resource allocation and power scheduling. Simulation results verify the same.

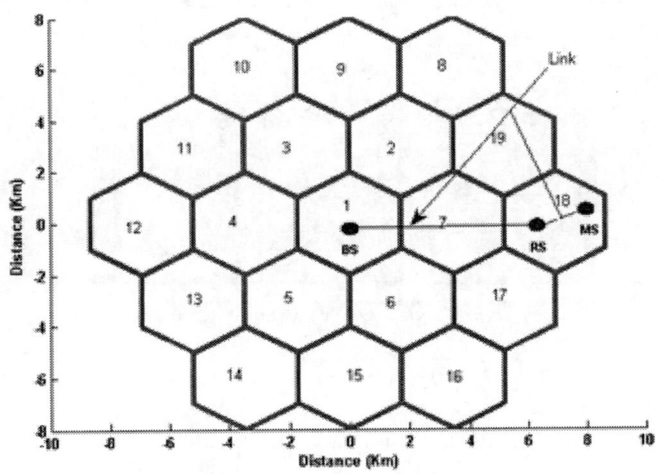

Figure 1. Cell distribution.

We are considering a single cell of a 19 cell MCN that has n number of user equipments and a base station at the center. The shape of the cell is hexagonal and contains three sectors. Figure 1 shows the distribution of cells in Multihop cellular networks. As per the open system reference model architecture, the components of the down link of MIMO OFDMA networks shown in Figure 2 [3, 4, 5]. Cross layer scheduler and resource allocator is a module that control time frequency resource block allocation in resource blocks. In this system the allocated bandwidth is divided into N bands according to the number of user terminals. One message will be transmitted in each band. The data link layer identifies the received messages from the network layer; then it transmits the messages to physical layer. The physical layer does multiplexing, power allocation and resource allocation. T_N is the number of transmit antennas to transmit the message and R_N is the number of receiving units. According to the IEEE 802.16j / m specification, the downlink subframe and uplink subframe - time division duplex frame is constructed. Sub frames are further divided into time frequency zones and each time frequency zone will be allocated to $MS - RS$ communication (access zone) or $RS - MS$ communications (relay zone) etc.

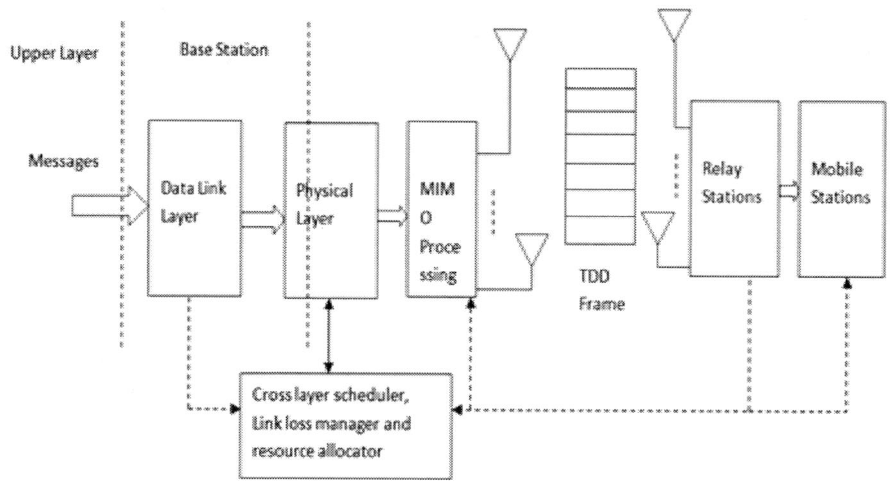

Figure 2. Multihp Cellular Network Architecture.

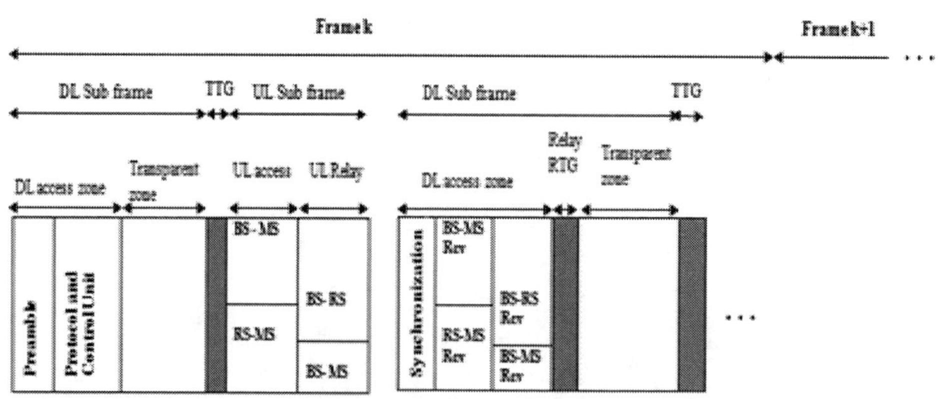

Figure 3. TDD Frame structure.

The standard defines two relay modes of operation. First one is transparent mode of operation in which each frame does not transmit header information. Second one is non transparent mode of operation in which each frame transmits header information instead of synchronous information. TDD frame is also called resource blocks. Cross layer scheduler allocates resource blocks to channels of MCN. This resource allocation should be compatible with frequency reuse techniques for interference coordination.

Efficient Self-ReconfigurationandRouteSelectionforWirelessSensorNetworkstomitigate link failures are described in [22].

3.1. Channel Model

The channel modeling needs the identification of channel coefficients which are identified from the radio signal propagation characteristics and long term and short term fading effects. Channels can be modeled based on deterministic approach and stochastic approach. Deterministic approach requires site characteristics, transmitter and receiver location, channel frequency, signal frequency and channel bandwidth. Because of the difficulties involved in the accurate estimation of channel coefficients from radio signal propagation characteristics and environment, we preferred stochastic approach to model the channels. Infinite and finite impulse response models [3, 5, 6, 20] are the examples of stochastic models. Consider a signal vector S = [S_1, S_2, S_3…S_n] at the base station and Y =[Y_1,Y_2,Y_3…Y_n] is the signal vector at the mobile station. These two vectors are related by the relation

$$y = h * s + n \quad (1)$$

Where h is a matrix which describes the characteristics of the channel and n is the channel noise. The value of h can be obtained as

$$H=\begin{bmatrix} h_{1,1} & \cdots & h_{1,n} \\ \vdots & \vdots & \vdots \\ h_{m,1} & \cdots & h_{m,n} \end{bmatrix} \quad (2)$$

The spatial correlation effects $R_{(m,n)}$are incorporated in to the channel model using the below equation

$$y = R_{(m,n)} * h * S \quad (3)$$

The signal vector for channel r S_r is modeled by using the below equation

$$S_r = \sqrt{\frac{P_n}{N}} v_n(t)\, s_n(t)$$

Where P_n , v_n (t) and s_n (t) are the transmitted power, filter and signal component on nth sub channel respectively. The order of the impulse response models are selected using suitable criterions, since deviated order amplifies noises. The channel impulse response over the whole frame period t can be written as

h_m^(nR,nT) (t)= ∑_(l=0)^(L_p-1) 〖h_m^(nR,nT) (t)δ(t-t_l)〗 (4.b)

In equation (4), m is the sub channel number, nR is the number of receivers, nT is the number of transmitters, L_p is the frame period and t_l is the latency. The corresponding frequency response when evaluated over a sub band m can be safely approximated by

H_m^(nR,nT) (t)= ∑_(l=0)^(L_p-1) 〖h_m^(nR,nT) (t) e^(j2πf_b τ_L) 〗 (5)

Moreover, the impulse response model may not be accurate at all times due channel estimation noise. Thus noisy channel model can be modeled as

h_m^(nR,nT) (t)=ζ*h_m^(nR,nT) (t)+(1-ζ)*h_m^(nR,nT) (t-1) (6)

Where ζ is a variable which controls the channel noise. Interference The desired signal is communing from the BS and the interference is coming from the co channels. The Interference received by the ith base station is affected by the distance between the base station and neighboring cells. We consider that fractional frequency reuse and sectoring has been employed to control the interference effects.

Thus the final interference calculated by equation (7) gives the interference level after interference control. As mentioned in the system model, the base station is located in the center and the interference effects are calculated by using the below equation

$$I_i = I_{ia} + I_{ico} \qquad (7.a)$$

Where I_{ia} is the adjacent channel interference of ith cell, and I_{ico} is the co-channel inference of ith cell. The adjacent channel interference can be estimated by using

$$I_{BR} = \sum_{(i,j)} H(B(i,j,k), R(i,j)) \qquad (7.b)$$

In equation (8), B(i,j,k) is the base station k at j^{th} cell of ith sector and R is the relay station. Cellular networks are analyzed in two levels: system level and link level. In link level, the simulation is designed to simulate the connection between MS and BS/RS. In system level, in addition to link conditions, the occurrence probability, large scale path loss, shadowing effects and relationship among multiple point to point links are modeled. Now SINR at the link between BS and RS can be calculated by using the below equation [20, 21,11]

$$\Gamma_{az-br} = \frac{H[N_B(i,j,k), N_R] \cdot P_B}{I_i + N_0} \qquad (8.a)$$

Where H(x,y) denotes the channel gain between x and y, P_B the power transmitted by the base station, and N_0 is the channel noise. The channel gain H(x,y) can be obtained from the following equation

$$H(x,y) = [PL(x,y)]^{-1} \cdot 10^{\xi(x,y)/10} \cdot G_t(x,y) \cdot G_r(x,y) \qquad (8.b)$$

Where PL(x,y) is the path loss between x and y, $\xi(x,y)$ is the shadow fading, G_t is the transmitter gain and G_r is the receiver gain.

Link loss management Algorithm

The algorithm is designed and given in Table 1 based on the knowledge of channel state information at the transmitter and receiver and the channel information are assumed to lie in a band. The data rates in the channel can be obtained from [8,10,2,17]

$$R_k = \sum_{n \in N} \log_2(1+\Gamma_k) \qquad (8.c)$$

Where k is the channel in $n \in N$ channels of the system. Γ_k is the signal to interference plus noise ratio in k^{th} channel. Now l is the total number of available links in a system and $l(Th_1)$ is the total number of links having Γ greater than Th_1, $l(Th_2)$ is the number of links having Γ greater than Th_2.

Table 1. Link loss management system.

Algorithm

Step 1: Initialization
$,\Gamma, N_0, I$

Step 2: Iteration
1) Calculate the channel noise N_0 on each channel
2) Calculate the interference I on each channel
3) Calculate the path loss between MS and BS
4) Calculate the SINR on each channel
(1) When SINR is less than Th_1, transmit the same information simultaneously on other channel C_2
(2) When SINR is less than Th_2, switch to channel C_2 from C_1
(3) When SINR is greater than Th_1/Th_2, continue on C_1
5) go to step 2.

Step 3: Finalization

Now consider an user u communicating with BS by link l_k^u, after time t= t_1 ,Γ_k less than⟦Th⟧_1 , Now the user equipment starts the signal transmission in two links l_k^u with data rate R_k and l_(k+1)^u with R_(k+1). After time t_1, assume that Γ_k less than t_2, the disconnection threshold, now the user equipment removes link l_k^u and switch to l_(k+1)^u. Like the above, After time t_2, Assume that Γ_k greater than ⟦Th⟧_1 |⟦Th⟧_2, now the user equipment continues on link l_k^u. scheduled to maximize the SINR on each channels.

The power in each channel is scheduled as follows [1, 7, 8, 9, 13, 15, 16, 18]. Let P_n^F denotes the power for sub channel n and define p^F=⟦[p_1^F,p_1^F...p_1^F]⟧^T ,Power allocation is a process of assigning power to each channel as per the channel state information. SINR is parameter of CSIT and optimum power is now the power in each channel is scheduled by.

$$p_{OPT}^{F} = \underset{p_F}{argmax} \sum_{n=1}^{N} \mathrm{B} * \log(1 + P_n^F \Gamma_n)$$

$$\sum_{n=1}^{N} P_n^F \leq P_{max}^F$$

$$P_n^F \geq 0 \quad \forall n$$

The solution is obtained after executing an iteration algorithm for equation [9]. After determining the optimum value of p_F, coordinated beam will be generated in the down-tilt (or) up-tilt from the base station/mobile station. Power allocation is a function the physical layer, cross layer link scheduling and resource allocation module controls the power transmitted by the transmitter. Power modified the directionality of the beam and sharpens the beam in the down-tilt and up-tilt direction.

4. Interference coordination and load balancing

4.1 Initialization
L, r, S_{in}, S_{out}

4.2 Handover execution
Wherever Step1: (measurement and report):
The BS is responsible for occasionally estimating and processing R ̄_l^n (Tw) of the users associated with the BS. When any new user accesses the BS, the BS will compare R ̄_l^n (Tw) with Rmin and determine whether the BS itself is overloaded or not.
Step2: (decision and execution):
If BS is overburden discovered if new users appear, To prevent substantial traffic loads, the transfer system will be implemented.
3:(final) To avoid large traffic loads, the transfer system will be executed.

5. Results and Discussion

System level and link level simulation was conducted to verify the performance of link loss management algorithm. The system level and link level simulations accurately capture the air interface, path loss and shadow fading effects. The SINR value is calculated as given in equation (7) on each channel in the simulated link. Then the data rate of the channel was calculated from the power assigned to the channel and the calculated SINR value. To assess the handover dormancy and service interruption time, Tmeasure, Tsignal, TBS_proc, and TRS_proc are thought to be 5 msec in light of the fact that a normal edge size of an IEEE 802.16e framework is 5 msec. Trng and Tassoc are thought to be 20 msec and 50 msec, separately. The estimation of Tassoc is moderately extensive in light of the fact that it requires extra spine organize correspondence.

Table 2. Simulation parameters.

Sl.No.	Simulation parameter	Values
1.	Cell	19 cells (6 Directional Antennas sectors), Carrier frequency 1.9GHz.
2.	Carrier frequency	1.9GHz
3.	BS Power constraint	30 dB
4.	RS Power constraint	20dB
5.	Cell radius	1km
6.	In MCN 1 difference between a BS and an RS	800 m
7.	T measure	5ms
8.	Tmg_Bs and Tmg_RS	22ms
9.	Tassoc	50ms
10.	TBS_proc and TRS_proc	5ms
11.	T signaling	5ms
12.	Neighbours_BS in SCN, MCN1 and MCN2	4
13.	NeighbourBS in SCN	1
14.	NeighbourBS in SCN MCN 1	6
15.	Neighbour RS in MCN 2	5

Figure 4 demonstrates the channel information rate and the comparing complete number of client supplies. The information indistinguishable for each channel since each channel is having diverse estimation of bearer recurrence, proliferation steady and way misfortune type. Impedance and clamors happening in cell systems are the fundamental purpose behind doing this examination. The obstruction happening in each channel was figured as given in condition (7) and (8). In the 19 cell arrangement of multihop cell systems, given in figure 1, radio signs proliferated from every phone is going about as a wellspring of impedance for the radio signs engendered from different cells.

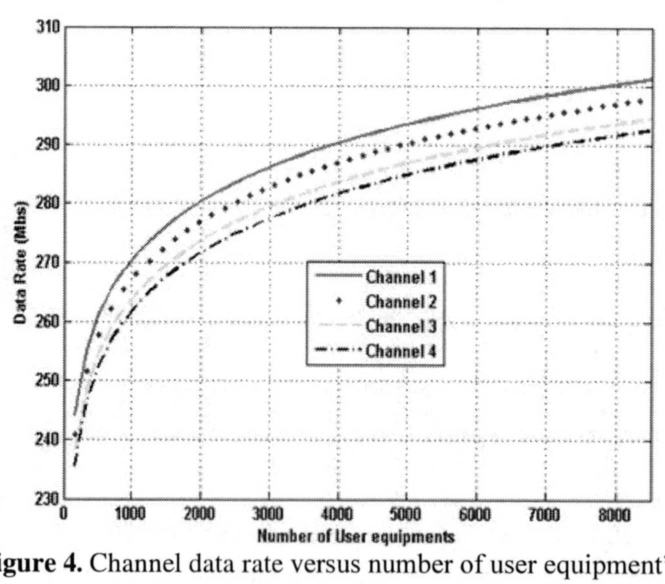

Figure 4. Channel data rate versus number of user equipment's.

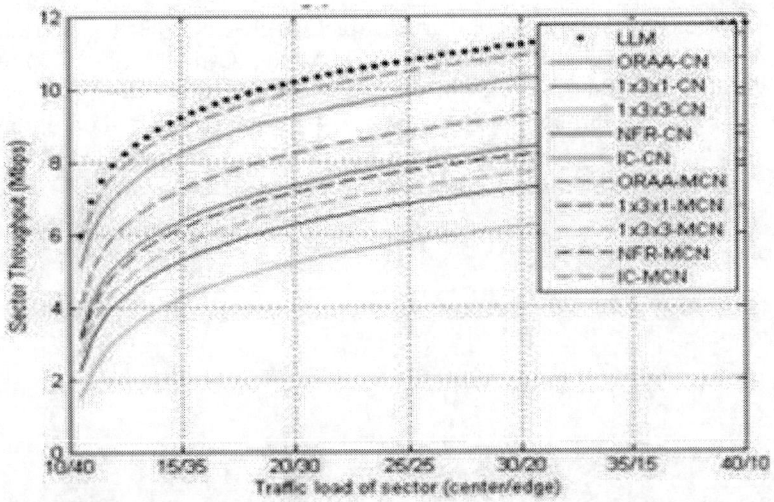

Figure 5. Sector throughput with different user distribution.

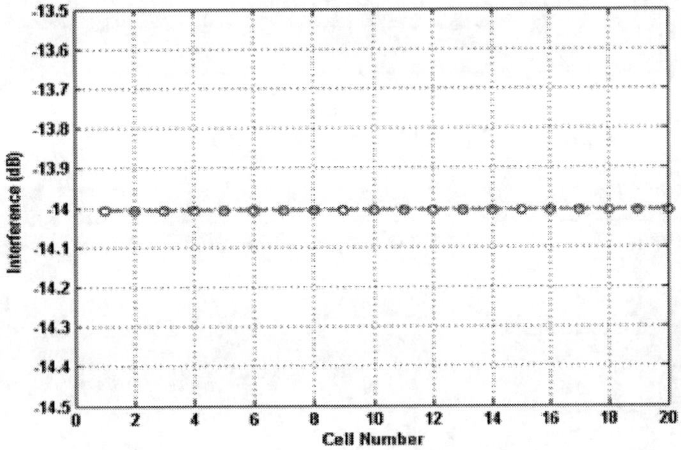

Figure 6. Interference in a cell due to neighbour cells.

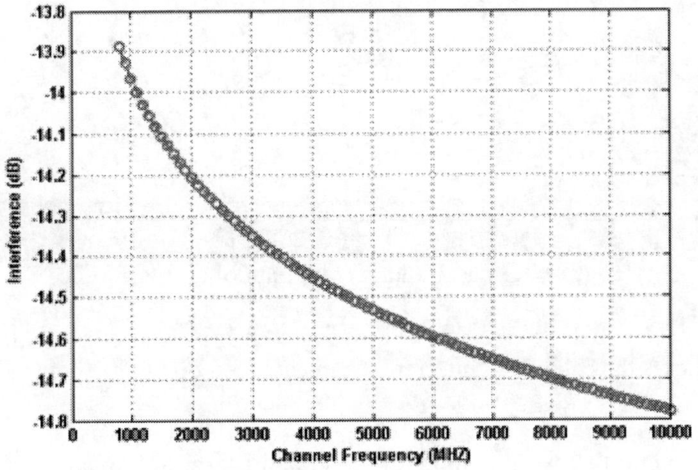

Figure 7. Channel frequency versus Interference.

IOP Conf. Series: Materials Science and Engineering **925** (2020) 012003 doi:10.1088/1757-899X/925/1/012003

Figure 5. shows the sector through put comparison for ORAA-CN, 1x3x1-CN, 1x3x3-CN, NFR-CN, IC-CN, ORAA-MCN,1x3x1-MCN, 1x3x3- MCN, NFR-MCN, IC-MCN. The LLM technique is having the sector through put in the range of 12Mbps compared with other. The LLM uses the overloaded situation (ρ=60), of the sector throughput can increases with the handover conditions can be satisfied. The higher ratio greater 70 increases the handover conditions can be satisfied. The sector throughput will continue to rise until the conditions for the handover are not met.

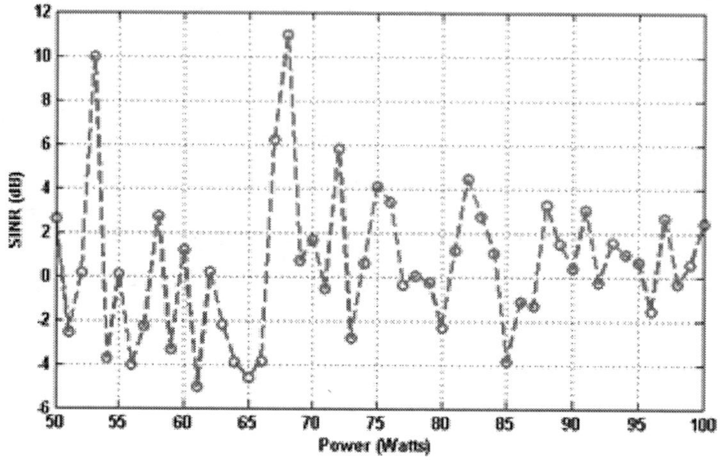

Figure 8. Channel SNIR versus Channel power.

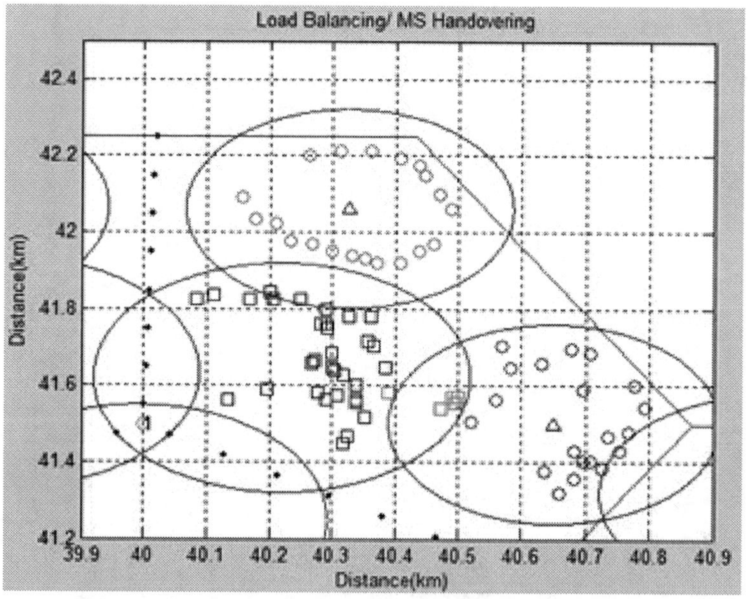

Figure 9. Load Balance with handover interference.

The interference in each cell due to the signals propagated from other cells calculated using equation (7) and (8). Figure 6 shows the calculated interference in each cell due to the neighboring cell.

6. Conclusion

The main concept behind this work is long term and short term fading and varies with respect to the channel frequency. We modelled the large scale path loss and shadowing factor, and identified the effect of channel frequency. Figure 7 shows the effect of channel frequency on interference. The allocated power as per equation (9), (10) and (11) decides the interference on each channel. The power allocated to the channels did not affect the channel SINR value in high level as displayed in Figure 8. The Figure

9 shows the how load balancing and mobile station Handovering. It is used by both foundations and EU algorithms to interact with and balance load. This article gives users a load balance algorithm to offer a smooth service and thus improve the quality of service critical communication (QoS).

Reference

[1] Hsu YC and Lin YD 2002 Multihop cellular: A novel architecture for wireless data communications *Journal of Communications and Networks* **4** 30-9.

[2] Zhang J and Li Y 2011 A New Load-Balancing Algorithm for Multihop Wireless Networks. *In 2011 Int. Conf. of Information Technology, Computer Engineering and Management Sciences* **2** 162-166.

[3] Srinivasan R 2008 IEEE 802.16 m evaluation methodology document (EMD) *IEEE 802.16 m-08/004r3*.

[4] Sheu JS, Lyu SH and Lyu SJ 2014 Cell sectorization and power management for inter-cell interference coordination in OFDMA-based network MIMO systems *In 2014 IEEE Wireless Communications and Networking Conf. (WCNC)* 1638-1642.

[5] Proakis JG and Salehi M 2001 Digital communications *New York: McGraw-hill* **4**.

[6] Femenias G, Riera-Palou F and Thompson JS 2016 Robust scheduling and resource allocation in the downlink of spatially correlated MIMO-OFDMA wireless systems with imperfect CSIT *IEEE Transactions on Vehicular Technology* **65** 614-629.

[7] Lei L, Yuan D, Ho CK and Sun S 2014 Resource scheduling to jointly minimize receiving and transmitting energy in OFDMA systems *In 2014 11th International Symposium on Wireless Communications Systems (ISWCS)* 187-191.

[8] Zhang L, Lu P, Yu Z, Cao H, Sun C and Wu C 2013 A utility-based adaptive resource scheduling scheme for multiple services in downlink multiuser mimo-ofdma systems *In 2013 IEEE 77th Vehicular Technology Conf (VTC Spring)* 1-5.

[9] Wei L, Hu RQ, Qian Y and Wu G 2015 Energy efficiency and spectrum efficiency of multihop device-to-device communications underlaying cellular networks *IEEE Transactions on Vehicular Technology* **65** 367-80.

[10] Kaneko M, Nakano T, Hayashi K, Kamenosono T and Sakai H 2016 Distributed resource allocation with local csi overhearing and scheduling prediction for OFDMA heterogeneous networks *IEEE Transactions on Vehicular Technology* **66** 1186-99.

[11] Einhaus M, Klein O and Walke B 2008 Comparison of OFDMA resource scheduling strategies with fair allocation of capacity *In 2008 5th IEEE Consumer Communications and Networking Conf.* 407-411.

[12] Li M, Li P, Huang X, Fang Y and Glisic S 2014 Energy consumption optimization for multihop cognitive cellular networks *IEEE Transactions on Mobile Computing* **14** 358-72.

[13] Tuncel NO and Koca M 2015 Joint mobility load balancing and inter-cell interference coordination for self-organizing OFDMA networks *In 2015 IEEE 81st vehicular technology conference (VTC Spring)* 1-5.

[14] Wu SL and Wang WW 2015 Energy-efficient multimedia multicast scheduling and resource allocation algorithms for OFDMA-based systems *In 2015 IEEE Int. Conf. on Computer and Information Technology; Ubiquitous Computing and Communications; Dependable, Autonomic and Secure Computing; Pervasive Intelligence and Computing* 1377-1382.

[15] Jangsher S and Li VO 2014 Power control in multihop cellular networks with multiple radio access technologies *In 2014 IEEE 25th Annual Int. Symp. Personal, Indoor, and Mobile Radio Communication (PIMRC)* 1674-1678.

[16] Ahmad SA, da Silva CR and DaSilva LA 2014 Marginal Power Efficiency Considerations in Power Control Adaptations for Multihop Cellular Networks *IEEE Transactions on Vehicular Technology* **63** 4718-23.

[17] Ramji T 2015 Adaptive resource allocation and its scheduling for good tradeoff between power consumption and latency in OFDMA based wireless distributed computing system *Int. Conf. on Computation of Power, Energy, Information and Communication (ICCPEIC)* 0496-0501.

[18] Hoang TD, Le LB and Le-Ngoc T 2016 Joint prioritized link scheduling and resource allocation for OFDMA-based wireless networks *In 2016 IEEE Int. Conf. on Communications (ICC)* 1-6.

[19] Farazmand Y and Alfa AS 2013 A game theoretic power allocation and relay load balancing in OFDMA-based DF cellular relay networks. *In 2013 IEEE 78th Vehicular Technology Conf. (VTC Fall)* 1-6.

[20] Zhao Y, Fang X, Huang R and Fang Y 2012 Joint interference coordination and load balancing for OFDMA multihop cellular networks *IEEE Transactions on mobile computing* **13** 89-101.

[21] Prabhu Chandran K and Ramesh GP 2017 Power and resource allocation for MIMO based two-tier heterogeneous network using small cell clustering algorithm *International Journal of Engineering & Technology* **7** 135-140.

[22] Subramani P, Rajendran GB, Sengupta J, Pérez de Prado R and Divakarachari PB 2020 A Block Bi-Diagonalization-Based Pre-Coding for Indoor Multiple-Input-Multiple-Output-Visible Light Communication System *Energies* **13** 3466.

IOP Conf. Series: Materials Science and Engineering **925** (2020) 012004 doi:10.1088/1757-899X/925/1/012004

Channel Estimation of MIMO-OFDM through Wavelet Transform to Improve the Spectral Efficiency and Data Transmission Rate

S Yogeeswaran[1], S Parasuraman[2] and A Manimaran[3]

[1] Department of ECE, P.T. Lee. CNCET, Kanchipuram, India.
[2,3] Department of ECE, Karpaga Vinayaga College of Engineering and Technology, Chengalpattu, India.
[1] yogash28s@gmail.com, [2] parasuengineers@gmail.com, [3] lims0181@gmail.com

Abstract. We define a new wavelet transform that is based on a recently defined family of scaling functions: the fractional B-splines. The interest of this family is that they interpolate between the integer degrees of polynomial B- splines and that they allow a fractional order of approximation. The approximation can determine noise in communication signals and help in signal reconstruction and channel estimation in MIMO receiver system. In free space optical communication links, the turbulent atmospheric channel causes fluctuations in both intensity and phase of the transmitted optical signal which degrades their performance. In order to make free space optical communications commercially viable, there is a need for characterizing the atmospheric effects on the received signal statistics, and possibly correcting for the signal degradation using physical optics principles and advanced real-time signal processing techniques. As part of this work, we have constructed an experimental free space optical link over 100 meters so as to characterize the effects of inner-scale and outer-scale turbulence effects, and used it as the platform for carrying out the above objectives. Through these techniques, we have demonstrated bit error rate (BER) reduction and propose to estimate channel accurately.

Key words: OFDM, MIMO, BPSK, BER.

1. Introduction

Free space optical communication has drawn considerable attention recently for a variety of high data rate applications due to its large bandwidth potential (> 10 Gbps), availability of fast switching optical sources and detectors, unregulated spectrum, ease of re-deployment and the potential for low bit error rate ($BER<10^{-9}$) through coding techniques [1]. However, such optical links are susceptible to atmospheric wave propagation effects such as scattering (dust, rain, fog, and snow), scintillations and beam wandering. Out of these potential impairments, scintillations and beam wandering are due to refractive index fluctuations caused by thermal gradients in the atmosphere along the beam propagation path as shown in Figure 1. Such refractive index fluctuations result in random variation in the received signal. These signal variations at the receiver leads to increase the bit error probability, thereby limiting the system performance in free space optical links. In order to optimize the system performance, one has to characterize the atmospheric turbulence effects and build a model that relates the scintillations and beam wandering induced signal variance to atmospheric parameters. A key challenge is to separate out the relative contributions of scintillations and beam wandering from the overall received signal. Subsequently, specific techniques have to be developed to mitigate the atmospheric turbulence effects and thereby demonstrate improved link performance. In this paper, we have characterized the link with respect to the atmospheric conditions and also provided a solution for mitigating the atmospheric turbulence effects on received signal, alternatively on signal statistics [2-9].

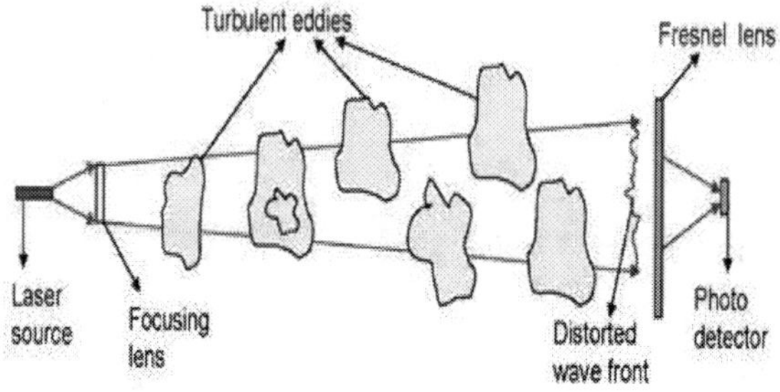

Figure 1. Schematic diagram of a free space optical link.

The specific issues addressed in this paper are as follows:
1. Characterization of atmospheric turbulence effects. Setting up an outdoor free space optical link and automation data collection. Characterization of signal variance due to scintillations and beam wandering. Correlating the above variances with atmospheric parameters.
2. Mitigation of signal degradation effects due to atmospheric turbulence. Aperture averaging to reduce signal variance due to scintillations. Wavelet-based signal processing to counter beam wandering-related signal variance.

2. Multiple Antenna techniques

SISO
Single-Input Single-Output is the classical method in wireless communication and the most common antenna configuration, using one antenna at transmitter and one at the receiver. It is used in radio, TV broadcast and in technology as WiFi, Bluetooth.

SIMO
Single-Input Multiple-Output is the system using one antenna at transmitter and multiple antennas at the receiver. It provides receiver diversity which receive the strongest signal from several transmit antennas. Generally, it is used in Uplink environment. In Multiple-Input Single-Output two or more number of antennas are used in the transmitter and one antenna at the receiver. It provides transmit diversity because of multiple antenna at a transmitter side. MISO technology has applications in WLAN, MAN and digital television (DTV). Commonly, it is used in downlink scenarios.

MIMO
Multiple-Input Multiple-Output uses multiple antennas at both sides which provides transmit diversity and receiver diversity. It's applicable in every kind of networks like PAN, LAN, WLAN, WAN, MAN. MIMO system can be applied in different ways to receive either a diversity gains, capacity gain or to overcome signal fading. The typical aspirations of a system designer are high data rate, low bit error rate, low power consumption, low cost and easy implement ability [2]. The MIMO system ensures us very high data rates even more than 1Gbps while minimizing the bit error rate. By Shanon's theorem the rate of transmission is always less than or equal to the capacity. Practically it is less than the capacity. The capacity depends on the bandwidth of the channel and SNR of the channel. Both the bandwidth and signal to noise ratio are characteristics of the channel.

2.1 MIMO-OFDM

Orthogonal Frequency Division Multiplexing (OFDM) is one of the most promising physical layer technologies for high data rate wireless communications due to its robustness to frequency selective fading, high spectral efficiency, and low computational complexity. OFDM can be used in conjunction with a Multiple-Input Multiple-Output (MIMO) transceiver to increase the diversity gain and/or the system capacity by exploiting spatial domain. Because the OFDM system effectively provides numerous parallel narrowband channels, MIMO-OFDM is considered a key technology in emerging high-data rate systems such as 4G, IEEE 802.16, and IEEE 802.11n. MIMO communication uses multiple antennas at both the transmitter and receiver to exploit the spatial domain for spatial multiplexing and/or spatial diversity. Spatial multiplexing has been generally used to increase the capacity of a MIMO link by transmitting independent data streams in the same time slot and frequency band simultaneously from each transmit antenna, and differentiating multiple data streams at the receiver using channel information about each propagation path, future standards need to specify both bandwidth requirements and type of signaling that achieves the data rate required for minimal predefined qualities of services for future applications. That is the average bit error rate (BER) can be made arbitrarily close to zero by use of channel coding, for transmissions up to the maximum achievable rate. For mobile channels, that are time-varying and dispersive in time and frequency however the channel capacity derivation is still an open research area.

3. Literature survey

3.1 Transmission scheme for MIMO-OFDM system

This contribution introduces a new transmission scheme for multiple-input multiple-output (MIMO) orthogonal frequency division multiplexing (OFDM) systems. The new scheme is efficient and suitable especially for symmetric channels such as the link between two base stations or between two antennas on radio beam transmission. The principle is based on the estimation of channel parameters of a pilot data send by the receiver to the transmitter. Then, the transmitter codes the transmitted signal using the estimated channel parameters to adapt the signal to the channel variations. Conducted Monte-Carlo simulation results show that the proposed scheme has better performance, in terms of bandwidth efficiency and complexity, compared to the conventional MIMO-OFDM scheme methods in the case of a symmetric channel.

3.2MIMO-OFDM wireless systems: basics, perspectives, and challenges

Multiple-input multiple-output (MIMO) wireless technology in combination with orthogonal frequency division multiplexing (MIMOOFDM) is an attractive air-interface solution for next-generation wireless local area networks (WLANs), wireless metropolitan area networks (WMANs), and fourth-generation mobile cellular wireless systems. This article provides an overview of the basics of MIMO-OFDM technology and focuses on space-frequency signaling, receiver design, multiuser systems, and hardware implementation aspects. We conclude with a discussion of relevant open areas for further research.

3.3 Evaluation of BER for AWGN, Rayleigh and Rician fading channels under various modulation schemes

Several transmission modes are defined in IEEE 802.11 a/b/g WLAN standards. A very few transmission modes are considering for IEEE 802.11 a/b/g in physical layer parameters and wireless channel characteristics. In this paper, we evaluated the performance of available transmission modes in IEEE 802.11b. However, the performance analysis can be done Straight forward using the evaluation of IEEE 802.11b. The performance of transmission modes is evaluated by calculating the probability of Bit Error Rate (BER) versus the Signal Noise Ratio (SNR) under the frequently used three wireless channel models (AWGN, Rayleigh and Rician). We consider the data modulation and data rate to analyze the performance that is BER vs. SNR. We also consider multipath received signals. The simulation results had shown the performance of transmission modes under different channel models and the number of antennas. Based on simulation results, we observed that some transmission modes are not efficient in IEEE 802.11b. The evaluation of performance confirms the increase in the coverage area of the physical layer in the 802.11b WLAN devices.

3.4 BER comparison of Rayleigh fading channel with Almouti space time block coding and the method of MRC

This paper proposes a technique which uses chaotic communication system combined with Rayleigh fading channel, for secure communications and to improve the system performance by mitigating interference. For secure communications, chaotic sequences are used. Many chaotic communication systems have been proposed, but most of them show a poor performance of MIMO-OFDM system, Rayleigh fading channel is used. Evaluation and comparison of the performances of MIMO- OFDM system in the AWGN (Additive White Gaussian Noise) channel, Rayleigh fading channel and the MRC diversity methods are provided. Results are verified and analyzed for two cases, one when we used 1*2 and 2*1 diversity scheme for Rayleigh fading channel with Almouti scheme is used in the proposed system and second when diversity method under realistic channel conditions (i.e. noise and multipath fading). This paper proposes a wireless communication structure based on two coupled chaotic systems. In order to enhance the error rate of MRC is used in the proposed system. Computer simulations are done to verify the performance of the proposed approach. A simulation tool with a Graphical User Interface (GUI) which implements these algorithms is also developed to provide ease in the execution.

4. Methodology

The broad objectives of this work were to characterize the behavior of the turbulent atmospheric channel and to mitigate the turbulence-induced degradation of the received signal. To carry out the work experimentally, we have analyzed signal form free space optical communication link of 100 meters' length with automated data collection. A key aspect of our work related to the atmospheric channel characterization is the separation of the variance due to scintillations and beam wandering [3]. We observed such signal variance at different times of the day and correlated them to atmospheric parameters such as temperature, wind velocity; pressure and humidity with fractional spline wavelet transform technique. We have also investigated schemes for mitigating the turbulence-induced degradation of the received signal by analyzing signal losses in low frequency and high frequency with fractional spline wavelet transform.

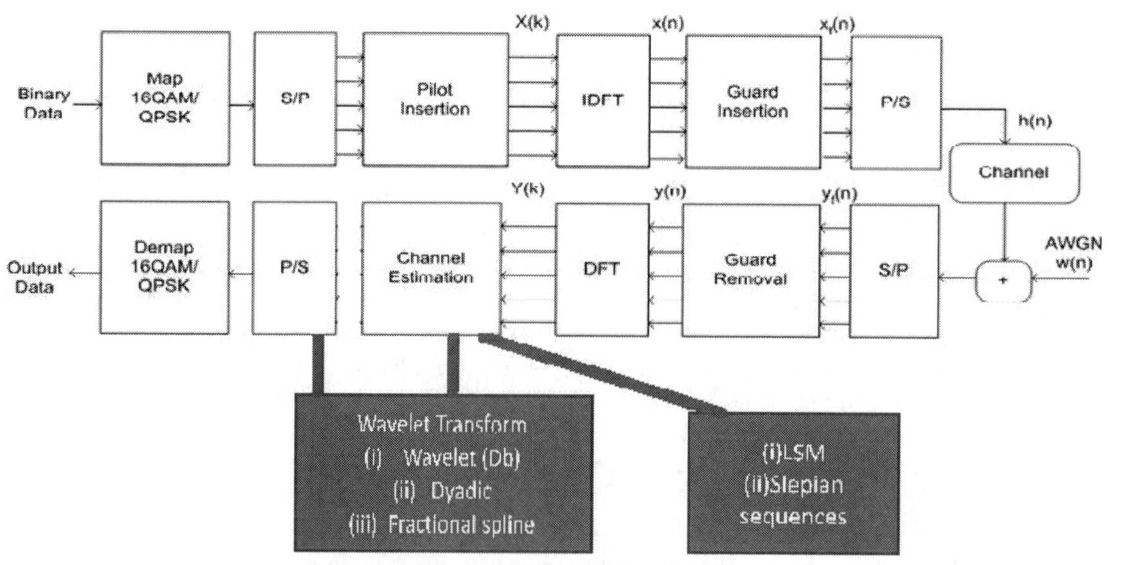

Figure 2. Block Diagram of OFDM Transceiver

5. Design Circuit

5.1 Binary Format

Binary format means that the sign (positive or negative) is in the leftmost bit of the field and the integer value is in the remaining bits of the field. Positive numbers have a zero in the sign bit; negative length of the field is from one to four digits; the compiler assumes a binary field length of 2 bytes. If the length of the field is from five to nine digits, the compiler assumes a binary field length of 4 bytes.

5.2Input and Output Sizes

If the convolutional code uses an alphabet of $2n$ possible symbols, this block's input vector length is L*n for some positive integer L. Similarly, if the decoded data uses an alphabet of $2k$ possible output symbols, this block's output vector length is L*k. This block accepts a column vector input signal with any positive integer value for L. For variable-sized inputs, the L can vary during simulation. The operation of the block is governed by the operation mode parameter."

5.3 Input Values and Decision Type

The entries of the input vector are either bipolar, binary, or integer data, depending on the Decision type parameter.

5.4 Fixed-Point Viterbi Decoding Examples

The following two example models showcase the fixed-point Viterbi decoder block used for both hard- and soft-decision convolutional decoding. If you are reading this reference page in the MATLAB® Help Browser, click Fixed-point Hard-Decision Viterbi Decoding and Fixed- point Soft-Decision Viterbi Decoding to open the models numbers have a one in the sign bit and are in twos complement form. In binary format, each field must be either 2 or 4 bytes long [4]. A binary field can be from one to nine digits in length and can be defined with decimal positions.

6. MATLAB

MATLAB is a high-performance language for technical computing. It integrates computation, visualization, and programming in an easy-to-use environment where problems and solutions are expressed in familiar mathematical notation. Typical uses include:

- Math and computation

Experimental specifications

- Algorithm development
- Modeling, simulation, and prototyping
- Data analysis, exploration, and visualization
- Scientific and engineering graphics
- Application development, including Graphical User Interface building
- MATLAB is an interactive system whose basic data element is an array that does not require dimensioning. This allows you to solve many technical computing problems, especially those with matrix and vector formulations, in a fraction of the time it would take to write a program in a scalar non-interactive language such as C or Fortran. The name MATLAB stands for matrix laboratory. MATLAB was originally written to provide easy access to matrix software developed by the LINPACK and EISPACK projects, which together represent the state-of-the-art in software for matrix computation. MATLAB has evolved over a period of years with input from many users. In university environments, it is the standard instructional tool for introductory and advanced courses in mathematics, engineering, and science [5]. In industry, MATLAB is the tool of choice for high productivity research, development, and analysis MATLAB features a family of application-specific solutions called toolboxes. Very important to most users of MATLAB, toolboxes allow you to learn and apply specialized technology.

6.1 The MATLAB mathematical function library

This is a vast collection of computational algorithms ranging from elementary functions like sum, sine, cosine, and complex arithmetic, to more sophisticated functions like matrix inverse, matrix eigenvalues, Bessel functions, and fast Fourier transforms.

6.2 The MATLAB Application Program Interface (API):

This is a library that allows you to write C and Fortran programs that interact with MATLAB. It includes facilities for calling routines from MATLAB (dynamic linking), calling MATLAB as a computational engine, and for reading and writing MAT-files.

7. BER v/s SNR

In digital transmission, the number of bit errors is the number of received bits of a data stream over a communication channel that has been altered due to noise, interference, distortion or bit synchronization errors. The bit error rate or bit error ratio (BER) is the number of bit error divided by the total number of transferred bits during a studied time interval. BER is a unit less performance measure, often expressed as a percentage. The bit error probability is the expectation value of the BER. The BER can be considered as an approximate estimate of the bit error probability. This estimate is accurate for a long time interval and a high number of bit errors. The BER may be improved by choosing strong signal strength by choosing a slow and robust modulation scheme or line coding scheme, and by applying channel coding schemes such as redundant forward error correction code. Binary symmetric channel which is used in analysis of decoding error probability in case of non bursty bit errors on Additive white Gaussian noise (AWGN) channel without fading [6]. A worst case scenario is a completely random channel, where noise totally dominates over the useful signal. In a noisy channel, the BER is often expressed as a function of the normalized carrier-to-noise ratio measured denoted Eb/N0 that is energy permit to noise power spectral density ratio, or Es/N0 that is energy per modulation symbol to noise spectral density. As the name implies, a bit error rate is defined as the rate at which errors occur in a transmission system. This can be directly translated into the number of errors that occur in a string of a stated number of bits. The definition of bit error rate can be translated into a simple formula:

BER = number of errors / total number of bits sent

Table 1. Comparisons of parameters

Comparative analysis:

MODULATION SCHEME	SNR	BER
BPSK	10dB	10^-5
QPSK	10dB	10^-2.9
QAM	10dB	10^-3
64-QAM	10dB	10^-1.8

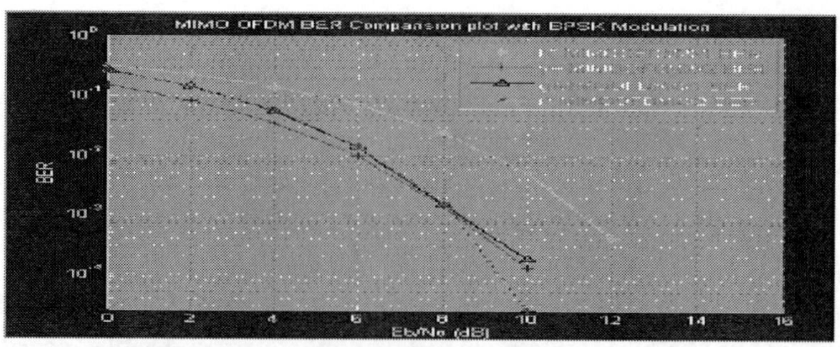

Figure 3. Compare BER vs BPSK modulation

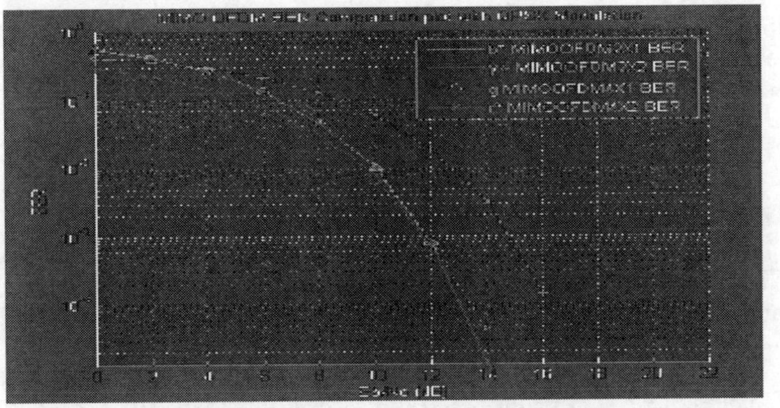

Figure 4. Characteristics of Bit Error Rate

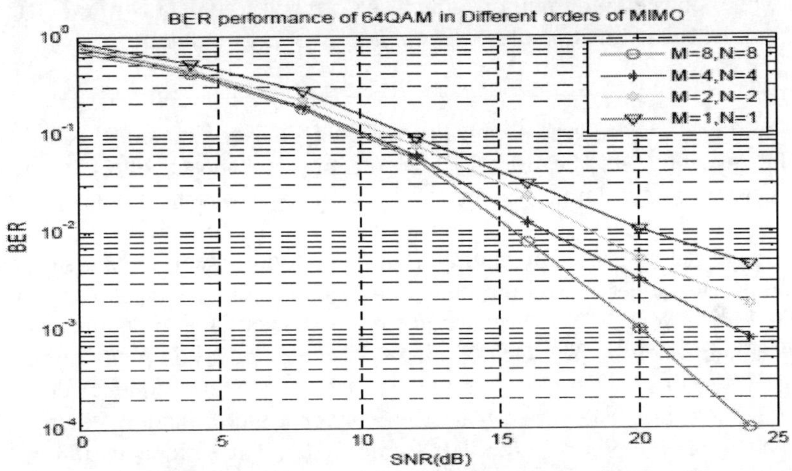

Figure 5. Characteristics of BER vs SNR 64 QAM

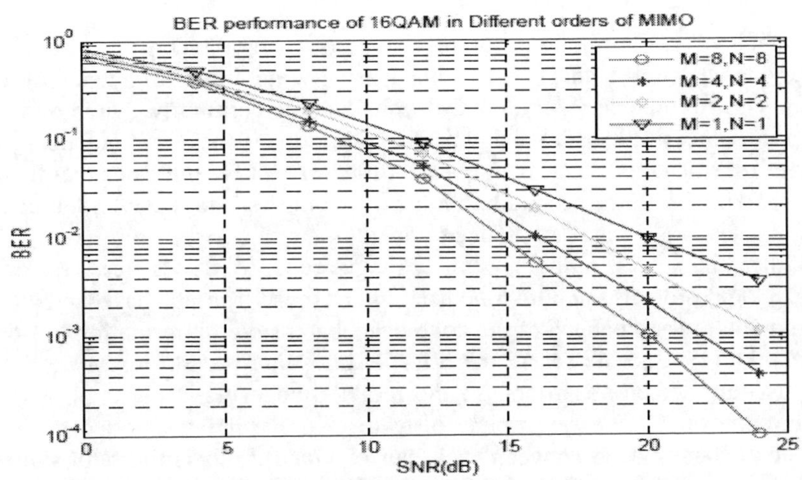

Figure 6. Performance of 64 QAM with MIM

8. Comparisons between Hard and Soft-Decision Decoding

The two models are set up to run from within BER Tool to generate a simulation curve that compares the BER performance for hard-decision versus soft-decision decoding.

8.1 Serial to parallel converter

This is a simple method to convert a serial data that have been entered to the circuit into parallel one. The reverse parallel to serial can be done in a similar manner. In fact, the first one is useful in computer systems architecture. For example, in ALUs, data can be entered serially and the input, for ALU operations, can be applied in parallel.

8.2 QPSK Modulation and Demodulation

The QPSK (also sometimes called PRK, phase reversal keying, or 2PSK) is the simplest form of phase shift keying (PSK). It uses two phases which are separated by 180° and so can also be termed 2-PSK. It does not particularly matter exactly where the constellation points are positioned, and in this figure they are shown on the real axis, at 0° and180°. This modulation is the most robust of all the PSKs since it takes the highest level of noise or distortion to make the demodulator reach an incorrect decision. It is, however, only able to modulate at 1 bit/symbol and so is unsuitable for high data-rate applications [8]. In the presence of an arbitrary phase-shift introduced by the communications channel, the demodulator is unable to tell which constellation point is which. As a result, the data is often differentially encoded prior to modulation. BPSK is functionally equivalent to 2-QAM modulation.

8.3 FFT and IFFT

The OFDM transmitter and receiver contain Inverse Fast Fourier Transform (IFFT) and Fast Fourier Transform (FFT), respectively. The IFFT/FFT algorithms are chosen due to their execution speed, flexibility and precision. For real time systems the execution speed is the main concern. The IFFT block provides orthogonality between adjacent subcarriers. The orthogonality makes the signal frame relatively secure to the fading caused by natural multipath environment. As a result, OFDM system has become very popular in modern telecommunication systems. The main objective of this paper is to design IFFT/FFT blocks for OFDM, because are main blocks for modulation and demodulation in OFDM transmitter and receiver. The OFDM signal is generated by implementing the Inverse Fast Fourier Transform (IFFT) at the transmitter which is used to convert frequency domain to time domain and Fast Fourier Transform (FFT) which is used to convert time. Equation Where, N is the transform size or the number of sample points in the data frame. X(k) is the frequency output of the FFT at kth point where k=0, 1, …, N-1 and x(n) is the time sample at nth point with n=0,1,…, N-1.

8.4 Pilot signal

"Pilot tone" redirects here. For pilot tones in motion picture sound recording systems, see Pilottone. In telecommunications, a pilot signal is a signal, usually a single frequency, transmitted over a communications system for supervisory, control, equalization, continuity, synchronization, or reference purposes. Spectrum of an FM broadcast signal. The pilot tone is the orange vertical line on the right of the spectrogram. In FM stereo broadcasting, a pilot tone of 19 kHz indicates that there is stereophonic information at 38 kHz (19×2, the second harmonic of the pilot). The receiver doubles the frequency of the pilot tone and uses it as a phase reference to demodulate the stereo information. If no 19 kHz pilot tone is present, then any signals in the 23-53 kHz range are ignored by a stereo receiver. A guard band of ±4 kHz (15-23 kHz) protects the pilot tone from interference from the baseband audio signal (50 Hz-15 kHz) and from the lower sideband of the double sideband stereo information (23-53 kHz). The third harmonic of the pilot (19×3, or 57 kHz) is used for Radio Data System. In AM stereo, the bandwidth is too narrow to accommodate subcarriers, so the modulation itself is changed, and the pilot tone is infrasonic (below the normal hearing range, instead of above it) at a frequency of 25 Hz. In color television, the color burst placed between each pair of video fields is the pilot signal to indicate that there are color subcarriers present. In the NTSC television system, a pilot tone of 15.7342657 kHz issued to indicate the presence of MTS stereo.

8.5 Parallel to serial

When data is to be sent over a large distance then the parallel data are send serially over single wire (plus ground). To send parallel data serially from microprocessor or microcontroller or DSP chips, parallel to serial converter is required. Not only is parallel to serial converter required at the microprocessor chip level but also in other parts of communication system. For telephone system for example, parallel to serial and serial to parallel converters are required for implementing modulation/demodulation.

8.6 Add cyclic prefix

The term ***cyclic prefix*** refers to the prefixing of a symbol with a repetition of the end. Although the receiver is typically configured to discard the cyclic prefix samples, the cyclic prefix serves two purposes. As a guard interval, it eliminates the inter symbol interference from the previous symbol. As a repetition of the end of the symbol, it allows the linear convolution of a frequency selective multipath channel to be modeled as circular convolution, which in turn may be transformed to the frequency domain using a discrete Fourier transform. This approach allows for simple frequency-domain processing, such as channel estimation and equalization. In order for the cyclic prefix to be effective (i.e. to serve its aforementioned objectives), the length of the cyclic prefix must be at least equal to the length of the multipath channel. Although the concept of cyclic prefix has been traditionally associated with OFDM systems, the cyclic prefix is now also used in single carrier systems to improve the robustness to multipath propagation.

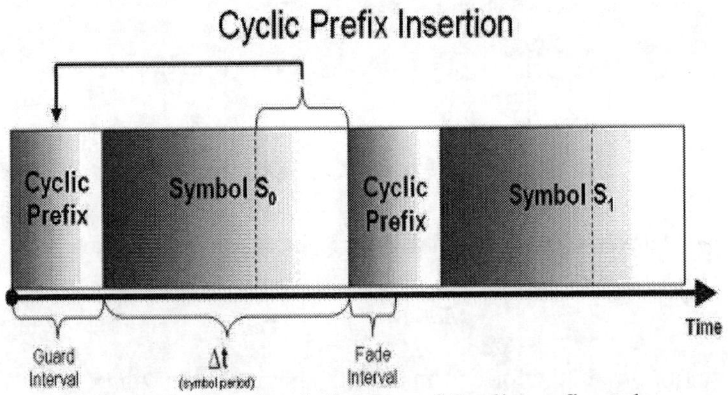

Figure 7. Functional diagram of Cyclic prefix code

8.7 OFDM Transmitter IFFT and Receiver FFT

The IFFT transform a spectrum (amplitude and phase of each component) into a time domain signal. An IFFT converts a number of complex data points, of length that is power of 2, into the same number of points in time domain. Each data point in frequency spectrum used for an FFT or IFFT operation is called a bin. The Inverse Fast Fourier Transform(IFFT) performs N-Point IFFT operation for the received constellation points from the QPSKM appear. The output is of N time domain samples. After N-point computation these values are passed through parallel to serial convertor FFT Converters time domain to frequency domain. The parallel symbols which are received from serial to parallel converter perform N- Point FFT operation and sends to de mapped.

8.8 Symbol Detection

Symbol detection in multi-input multi-output (MIMO) communication systems using different particles warm optimization (PSO) algorithms is presented. This approach is particularly attractive as particles warm intelligence is well suited for real-time applications, where low complexity and fast convergence is of absolute importance. While an optimal maximum likelihood (ML) detection using an exhaustive search method is prohibitively complex, PSO assisted MIMO detection algorithms give near-optimal bit error rate (BER) performance with a significant reduction in ML complexity. The input serial binary data will be processed by a data scrambler and then channel coding is applied to the input data to improve the BER (bit error rate) performance of the system. The encoded data stream is further interleaved to reduce the burst symbol error rate. Dependent on the channel condition like fading, base modulation modes such as BPSK (binary phase shift keying), QPSK (quadrature phase shift keying) and QAM are adaptively used to boost the data rate. The modulation mode can be changed even during the transmission of data frames. The resulting complex numbers are grouped into column vectors which have the same number of elements as the FFT size, N. To combat the multipath delay spread in wireless channels, the time-domain samples s(m) is cyclically extended by copying the last Ng samples and pasting them to the front, as shown in figure 8.

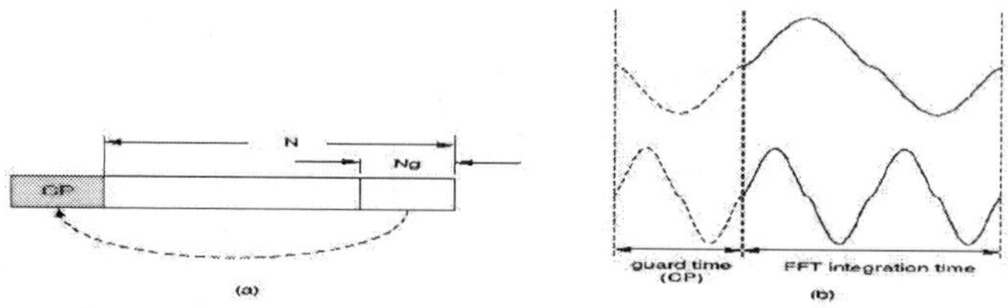

Figure (a) Concept of CP.
(b) OFDM symbol with cyclic extension.

Let $u(m)$ denote the cyclically extended OFDM symbol as

$$\vec{u}(m) = \begin{bmatrix} u(mN_{tot}) \\ \vdots \\ u(mN_{tot}+N_{tot}-1) \end{bmatrix} = \begin{bmatrix} CP \\ \vec{s}(m) \end{bmatrix}_{N_{tot}\times 1}$$

Figure.8. OFDM symbol with cyclic extension with its wave form

$$Ntot=N+Ng \tag{1}$$

Where; is the length of $u(m)$. In the form of matrix, the CP insertion can be readily expressed as a matrix product of $s(m)$ and an N×N matrix ACP. By straight computation, it holds that: One of the challenges from the harsh wireless channels is the multipath delay spread. If the delay spread is relatively large compared to the symbol duration, then a delayed copy of a previous symbol will overlap the current one which implies severe ISI. To eliminate the ISI almost completely, a CP is introduced for each OFDM symbol and the length of CP, Ng must be chosen longer than the experienced delay spread, L, i.e., $Ng > L$. In addition, CP is capable of maintaining the orthogonality among subcarriers which implies zero ICI. It is because the OFDM symbol is cyclically extended and this ensures that the delayed replicas of the OFDM symbol always have an integer number of cycles within the FFT interval, as long as the delay is smaller than the CP. No matter where the FFT window starts, provided that it is within the CP, there will be always one or two complete cycles within FFT integration time for the symbol on top and at below respectively. For the wireless channel, it is assumed in this thesis as a quasi-static frequency-selective Rayleigh fading channel. It indicates:

$$r(mN_{tot}+n) = \sum_{l=0}^{L-1} h_{l,m}u(mN_{tot}+n-l)+v(mN_{tot}+n) \tag{2}$$

that the channel remains constant during the transmission of one OFDM symbol. Suppose that the multipath channel can be modeled by a discrete-time baseband equivalent $(L-1)^{th}$-order FIR (Finite impulse response) filter with filter taps.

$$\{h_0,h_1,.....,h_l,...h_{L-1}\} \tag{3}$$

It is further assumed that the channel impulse response, i.e., the equivalent FIR filter taps, are independent zero mean complex Gaussian

$$\{p_0, p_1, \ldots p_l, \ldots, p_{L-1}\} \tag{4}$$

random variables with variance of per dimension. The ensemble of is the PDP (power delay profile) of the channel and usually the total power of the PDP is normalized to be 1 as the unit average channel attenuation. Denote the CIR (channel impulse response) vector *hm* as

$$\overrightarrow{h_m} = \begin{bmatrix} h_{0,m} \\ . \\ . \\ . \\ h_{L-1,m} \end{bmatrix}_{L\times 1} \tag{5}$$

Where, the subscript *m* is kept to imply that the channel may vary from one OFDM symbol to the next one. Then the complex baseband equivalent received signal can be represented by a discrete-time convolution.

9. Results and Discussion

For Eb/No=10dB, Bit Error Rate=3.87x10^{-06}

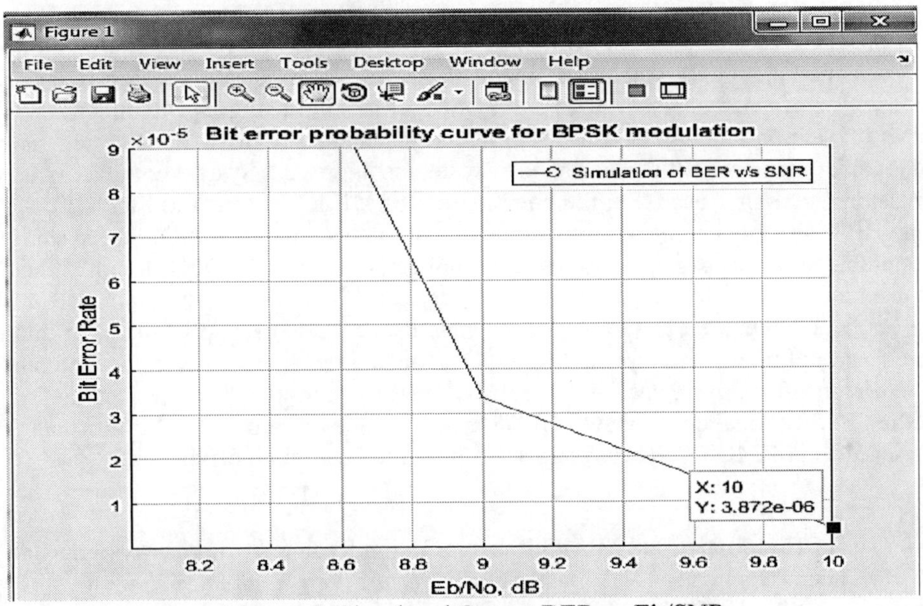

Figure 9. Simulated Output BER vs Eb/SNR

IOP Conf. Series: Materials Science and Engineering **925** (2020) 012004 doi:10.1088/1757-899X/925/1/012004

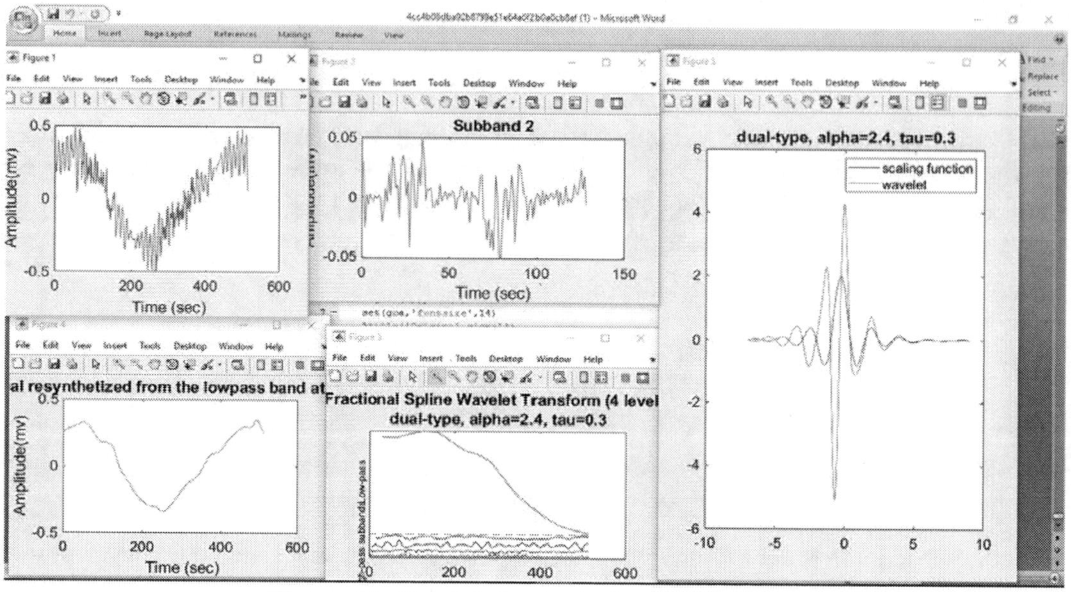

Figure 10. Simulated output of various errors

10. Conclusion

MIMO-OFDM is a powerful modulation technique used for high data rate, and is able to eliminate ISI. It is computationally efficient due to the use of FFT techniques to implement modulation and demodulation functions. The performance of using MATLAB/SIMULINK tool box. Computer simulations are done to verify the performance of the proposed approach. A simulation tool with a Graphical User Interface (GUI) which implements these algorithms is also developed to provide ease in the execution. Performance analysis of FFT-OFDM and is done using BPSK under AWGN channel. STBC-MIMO system is implemented for 2*2 diversity scheme. STBC-MIMO system as we increase number of receive antennas BER performance of system is improved. System complexity is reduced and the transmission rate is increased. So, the combination of MIMO system and FFT-OFDM system will improve the BER performance of wireless communication system. MIMO-OFDM system with BPSK scheme is suitable for low capacity, short distance application. While the OFDM with higher modulation scheme is used for large capacity, long distance application at the cost of slight increase in Eb/No. BER is large in BPSK as compared to QAM and it generally depends on applications. In this project, we implemented and optimized the 2x2 MIMO-OFDM transmitter and receiver for different embedded platform. We reduced the bit error rate of MIMO-OFDM from 10^-2 to 10-5 using 2x2 array. We concluded that BPSK modulated MIMO - OFDM system achieves better BER results than QPSK/QAM and other modulated MIMO - OFDM systems for the same bandwidth efficiency.

References

[1] Kho YH and Taylor DP 2008 MIMO Receiver Using Reduced Complexity Sequence Estimation with Channel Estimation and Tracking *IEEE Trans. Veh. Technol.* **58** 682–691.

[2] Kim JG and Lim JT 2008 MAP-Based Channel Estimation for MIMO amp OFDM Over Fast Rayleigh Fading Channels *IEEE Trans. Veh. Technol.* **57,** 1963-1968.

[3] Ramesh GP 2020 Microstrip Patch Antenna for Peripheral Arterary Disease Diagnosis *In Recent Trends and Advances in Artificial Intelligence and Internet of Things* 241-251.

[4] Adeogun RO, Teal PD and Dmochowski PA 2014 Extrapolation of MIMO mobile-to-mobile wireless channels using parametric-model-based prediction *IEEE Transactions on Vehicular Technology* **64** 4487-4498.

[5] Katselis D, Kofidis E and Theodoridis S 2008 On training optimization for estimation of correlated MIMO channels in the presence of multiuser interference *IEEE transactions on signal processing* **56** 4892-4904.

[6] Pendem S and Ramesh GP 2020 *75 GHz 5G Frequency Spectrum Analysis.* In *Recent Trends and Advances in Artificial Intelligence and Internet of Things* 165-176.

[7] Ganesh RS, Jayakumari J and Akhila IP 2011 Channel estimation analysis in MIMO-OFDM wireless systems. In *2011 International Conference on Signal Processing, Communication, Computing and Networking Technologies* 399-403.

[8] Flashy AM and Ramesh GP 2020 Multi Band Antenna System for Quality Evaluation Application of Apple Fruit *In Recent Trends and Advances in Artificial Intelligence and Internet of Things* 199-206.

[9] Subramani P, Rajendran GB, Sengupta J, Pérez de Prado R and Divakarachari PB 2020 A Block Bi-Diagonalization-Based Pre-Coding for Indoor Multiple-Input-Multiple-Output-Visible Light Communication System *Energies* **13** 3466.

Design of Microstrip Patch Antenna with improved characteristics and its performance at 5.1GHz for Wireless Applications

S Parasuraman[1], **S Yogeeswaran**[2] **and G P Ramesh**[3]

[1] Department of ECE, Karpaga Vinayaga College of Engineering and Technology, Chengalpattu, Chennai, India.
[2] Department of ECE, P.T. Lee Changalvaraya Naicker College of Engineering and Technology, Kanchipuram, India.
[3] Department of Electronics and Communication Engineering, St. Peter's Institute of Higher Education and Research, Chennai, India.
[1]parasuengineers@gmail.com, [2]yogash28s@gmail.com, [3]rameshgp@yahoo.com

Abstract. Despite the many advantages of microstrip patch antennas, they do have some considerable drawbacks. One of the main limitations with patch antennas is their inherently narrowband performance due to its resonant nature. With bandwidth as low as a few percent; broadband applications using conventional patch designs are limited. The paper presents a broadband microstrip patch antenna for wireless communication. In its most basic form, a microstrip patch antenna consists of a radiating patch on one side of a dielectric substrate which has a ground plane on the other side. The patch is generally made of conducting material such as copper or gold and can take any possible shape. A rectangular patch is used as the main radiator. There are several advantages of this type of broadband antenna, such as planar, small in size, simple in structure, low in cost, and easy to be fabricated, thus attractive for practical applications. This rectangular microstrip patch antenna is designed for wireless communication application that works at 2.4 GHz with gain 11 d B for outdoor place. It also has a wide angle of beam in its radiation pattern. The results obtain that microstrip patch antenna can be used as client antenna in computer and workable antenna for wireless fidelity.

Keywords: Microstrip Antenna, Wireless Fidelity, Frequency

1. Introduction

One of types of wireless communication at 2.4 GHz is Wireless Fidelity (WiFi). A WiFi enabled device such as a personal computer, video game console, smartphone or digital audio player can connect to the Internet when within range of a wireless network connected to the Internet [1]. The coverage of one or more (interconnected) access points (hotspot) can round up an area as small as a few rooms or as large as many square miles. With the development of MIC and high frequency semiconductor devices, microstrip has drawn the maximum attention of the antenna community in recent years. In spite of its various attractive features like, light weight, low cost, easy fabrication, conformability on curved surface and so on, the microstrip element suffers from an inherent limitation of narrow impedance bandwidth [2, 13].

2. Methodology Adapted

2.1 Antenna Shape

In its most basic form, a Microstrip patch antenna consists of a radiating patch on one side of a dielectric substrate which has a ground plane on the other side as shown in Figure 1. The patch is generally made of conducting material such as copper or gold and can take any possible shape. The radiating patch and the feed lines are usually photo etched on the dielectric substrate. Microstrip patch antennas radiate primarily because of the fringing fields between the patch edge and the ground plane. For good antenna performance, a thick dielectric substrate having a low dielectric constant is desirable since this provides better efficiency, larger bandwidth and better radiation [3].

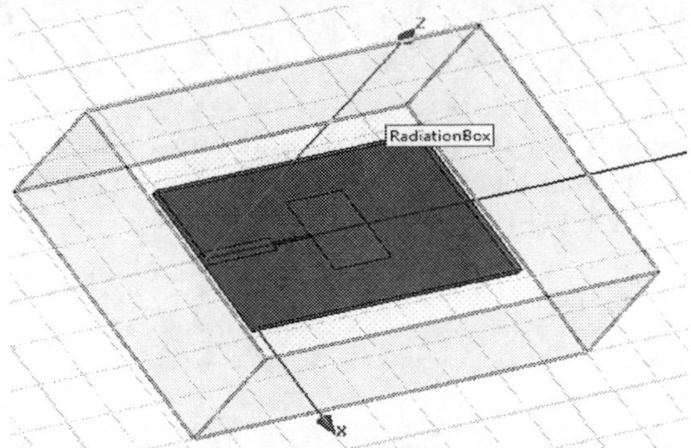

Figure 1. Structure of Microstrip Patch Antenna

Consider Figure 2, below, which shows a rectangular microstrip patch antenna of length L, width W resting on a substrate of height h. The co-ordinate axis is selected such that the length is along the x direction, width is along the y direction and the height is along the z direction.

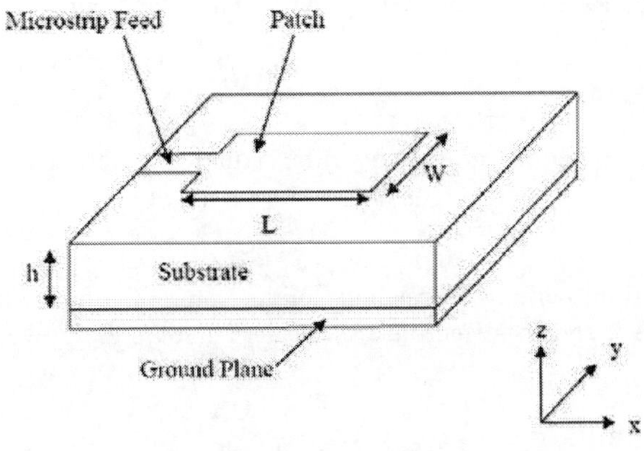

Figure 2. Microstrip Patch Antenna

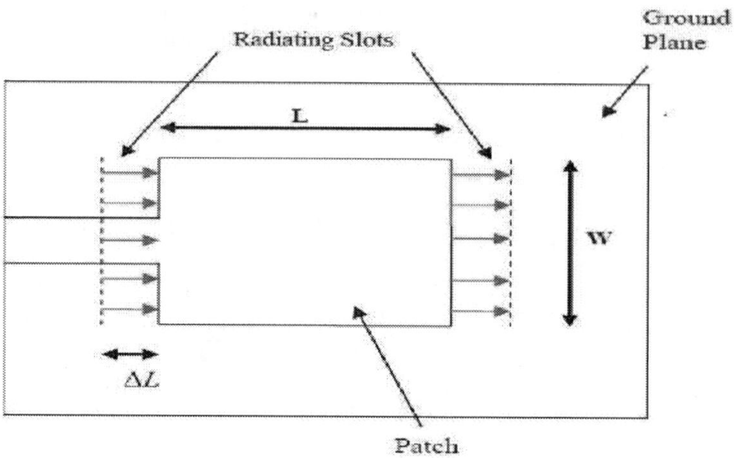

Figure 3. Top View of Antenna

An antenna is an electrical conductor or system of conductors Transmitter Radiates electromagnetic energy into space Receiver-Collects electromagnetic energy from space The IEEE definition of an antenna as given by Stutzman and Thiele is, "That part of a transmitting or receiving system that is designed to radiate or receive electromagnetic waves". The major parameters associated with an antenna are defined in the following sections.

2.2 Antenna Gain

Gain is a measure of the ability of the antenna to direct the input power into radiation in a particular direction and is measured at the peak radiation intensity. Consider the power well as gain at various points in space. It serves as the signature of an antenna and one look at it is often enough to realize the antenna that produced it. Density radiated by an isotropic antenna with input power P at a distance R which is given by $S = P/4\pi R^2$. An isotropic antenna radiates equally in all directions, and its radiated power density S is found by dividing the radiated power by the area of the sphere $4\pi R$. An isotropic radiator is considered to be 100% efficient. The gain of an actual antenna increases the power density in the direction of the peak radiation:

2.3 Antenna Efficiency

The surface integral of the radiation intensity over the radiation sphere divided by the input power P of the elative power radiated by the antenna, or the antenna efficiency.

2.4 Effective Area

Antennas capture power from passing waves and deliver some of it to the terminals. Given the power density of the incident wave and the effective area of the antenna, the power delivered to the terminals is the product.

2.5 Directivity

Directivity is a measure of the concentration of radiation in the direction of the maximum.

2.6 Path Loss

We combine the gain of the transmitting antenna with the effective area of the receiving antenna to determine delivered power and path loss.

IOP Conf. Series: Materials Science and Engineering **925** (2020) 012005 doi:10.1088/1757-899X/925/1/012005

2.7 Return Loss

It is a parameter which indicates the amount of power that is "lost" to the load and does not return as a reflection. Hence the RL is a parameter to indicate how well the matching between the transmitter and antenna has taken place. Simply put it is the S11 of an antenna. A graph of s11 of an antenna vs frequency is called its return loss curve [4]. For optimum working such a graph must show a dip at the operating frequency and have a minimum d B value at this frequency. This parameter was found to be of crucial importance to our project as we sought to adjust the antenna dimensions for a fixed operating frequency (say 5.1GHz). A simple RL curve is shown in figure 4.

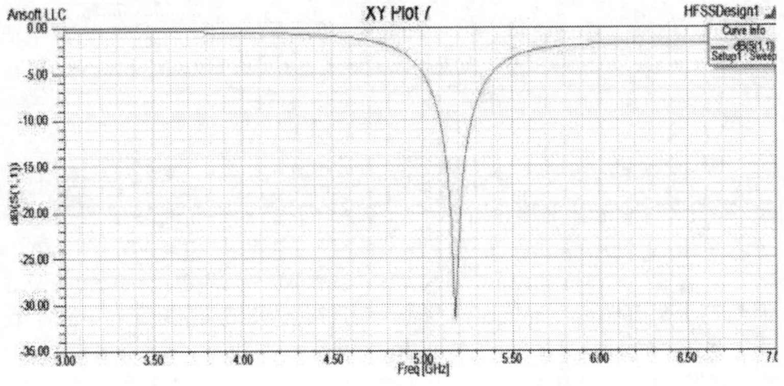

Figure 4. RL curve of an antenna

2.8 Radiation Pattern

The radiation pattern of an antenna is a plot of the far-field radiation properties of an antenna as a function of the spatial co-ordinates which are specified by the elevation angle (θ) and the azimuth angle (φ). More specifically it is a plot of the power radiated from an antenna per unit solid angle which is nothing but the radiation intensity. It can be plotted as a 3D graph or as a 2D polar or Cartesian slice of this 3D graph. It is an extremely parameter as it shows the antenna's directivity as

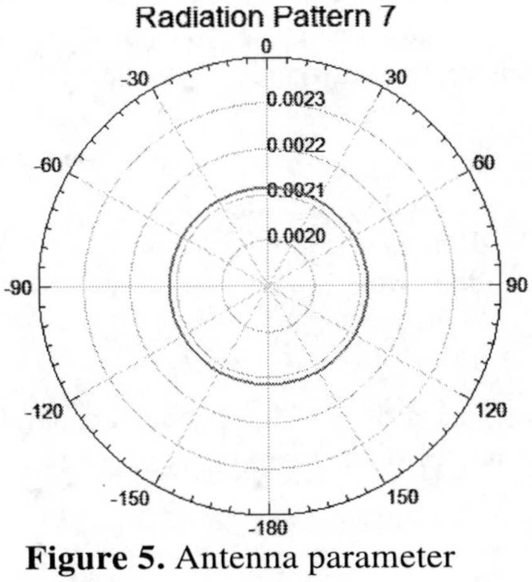

Figure 5. Antenna parameter

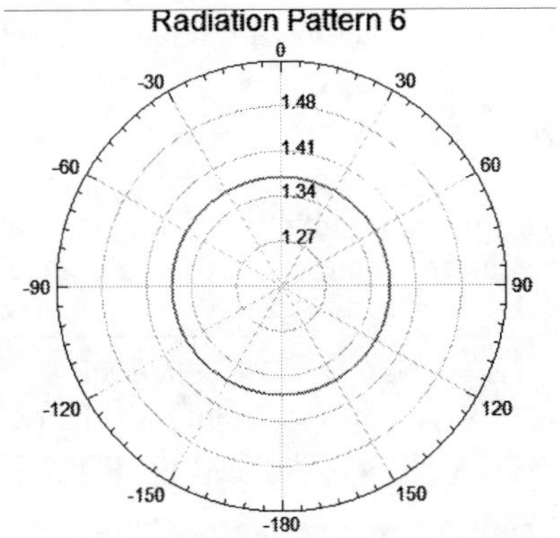

Figure 6. Design parameter

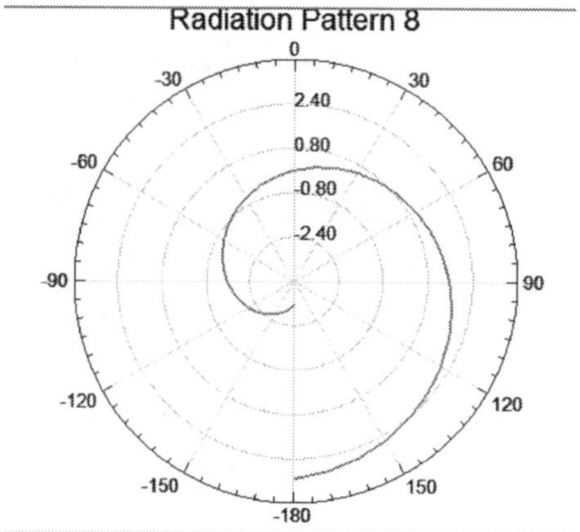

Figure 7. Variable parameter

2.9 Applications of Microstrip Patch Antennas

Microstrip patch antennas are increasing in popularity for use in wireless applications due to their low-profile structure. Therefore, they are extremely compatible for embedded antennas in handheld wireless devices such as cellular phones, pagers etc. The telemetry and communication antennas on missiles need to be thin and conformal and are often microstrip patch antennas. Another area where they have been used successfully is in satellite communication.

2.10 Feed Techniques

Microstrip patch antennas can be fed by a variety of methods. These methods can be classified into two categories - contacting and non-contacting. In the contacting method, the RF power is fed directly to the radiating patch using a connecting element such as a microstrip line. In the non- contacting scheme, electromagnetic field coupling is done to transfer power between the microstrip line and the radiating patch. The four most popular feed techniques used are the microstrip line, coaxial probe (both contacting schemes), aperture coupling and proximity coupling (both non-contacting schemes).

2.10 Microstrip Line Feed

In this type of feed technique, a conducting strip is connected directly to the edge of the microstrip patch. The conducting strip is smaller in width as compared to the patch and this kind of feed arrangement has the advantage that the feed can be etched on the same substrate to provide a planar structure.

2.11 Coaxial Feed

The Coaxial feed or probe feed is a very common technique used for feeding Microstrip patch antennas. The inner conductor of the coaxial connector extends through the dielectric and is soldered to the radiating patch, while the outer conductor is connected to the ground plane.

2.12 Aperture Coupled Feed

In this type of feed technique, the radiating patch and the microstrip feed line are separated by the ground plane coupling between the patch and the feed line is made through a slot or an aperture in the ground plane.

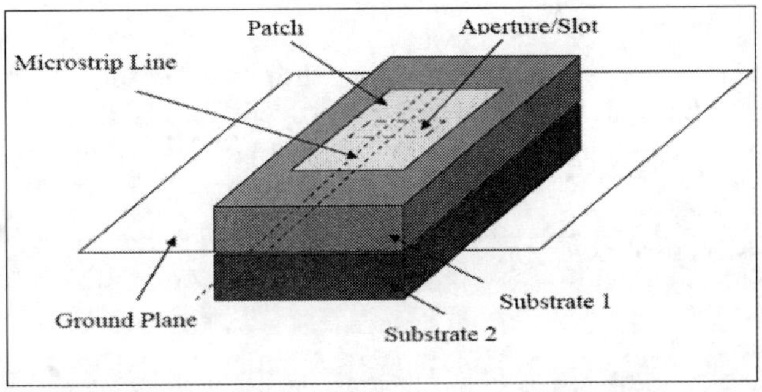

Figure 8. Aperture Feed Patch Antenna

The most common type of microstrip antenna is the patch antenna [5-6]. Antennas using patches as constitutive elements in an array are also possible. A patch antenna is a narrowband, wide-beam antenna fabricated by etching the antenna element pattern in metal trace bonded to an insulating dielectric substrate, such as a printed circuit board, with a continuous metal layer bonded to the opposite side of the substrate which forms a ground plane. Common microstrip antenna shapes are square, rectangular, circular and elliptical, but any continuous shape is possible. Some patch antennas do not use a dielectric substrate and instead are made of a metal patch mounted above a ground plane using dielectric spacers; the resulting structure is less rugged but has a wider bandwidth. Because such antennas have a very low profile, are mechanically rugged and can be shaped to conform to the curving skin of a vehicle, they are often mounted on the exterior of aircraft and spacecraft, or are incorporated into mobile radio communications devices. Another type of patch antenna is the Planar Inverted-F antenna (PIFA) [7] The PIFA is common in cellular phones (mobile phones) with built-in antennas. The antenna is resonant at a quarter-wavelength (thus reducing the required space needed on the phone), and also typically has good SAR properties. This antenna resembles an inverted F, which explains the PIFA name. The PIFA [8] is popular because it has a low profile and a unidirectional pattern. These antennas are derived from a quarter-wave half-patch antenna. The shorting plane of the half-patch is reduced in length which decreases the resonance frequency. Often PIFA antennas have multiple branches to resonate at the various cellular bands. On some phones, grounded parasitic elements are used to enhance the radiation bandwidth characteristic. The folded inverted conformal antenna (FICA) [9-11] has some advantages with respect to the PIFA, because it allows a better volume reuse

3. Testing and Results

In testing, microstrip patch antenna is used as a substitute for external omnidirectional antenna from wireless LAN card in Ad-hoc mode. In a wireless computer network, ad-hoc mode is a method for wireless devices to directly communicate with each other. Operating in ad-hoc mode allows all wireless devices within range of each other to discover and communicate in peer-to-peer fashion without in to broadband wireless routers).

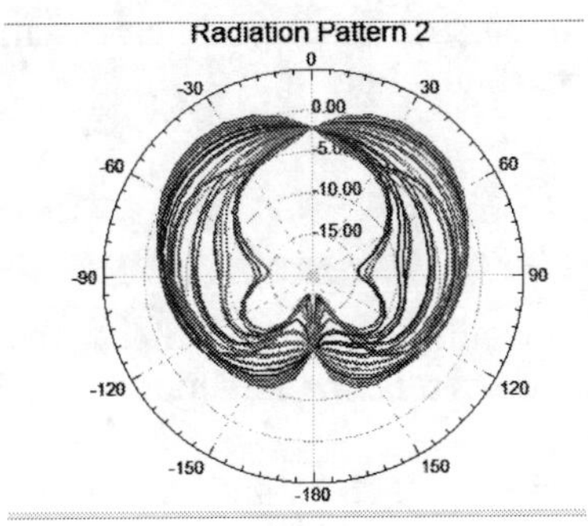

Figure 9. Gain

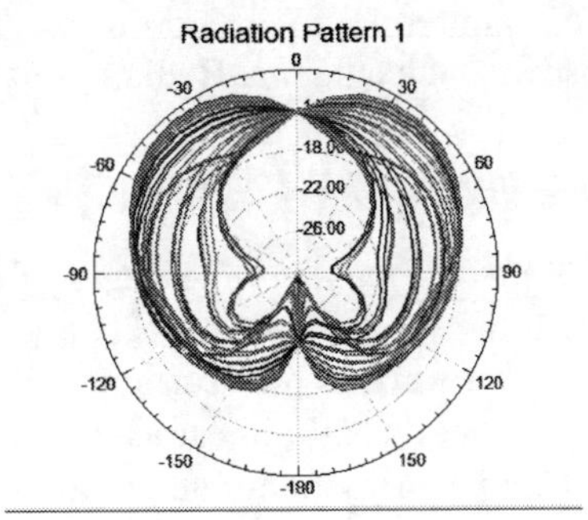

Figure 10. Directivity

A computer is used as a producer of wireless signals, while other computers installed microstrip patch antenna as of omnidirectional antenna. In wireless card will be used as testing of microstrip patch antenna radiation pattern by detecting the signal strength of wireless signal sources.

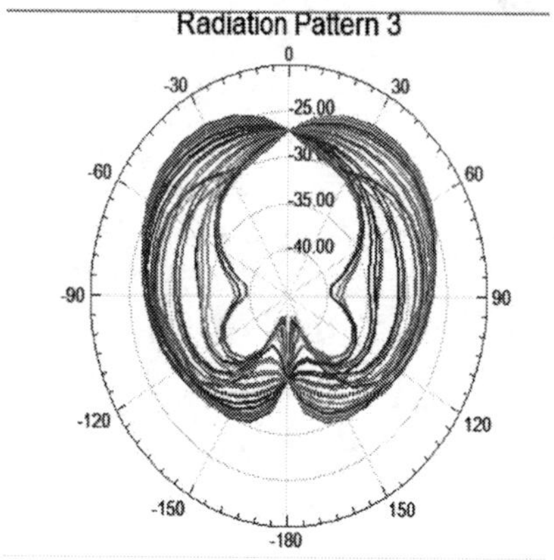

Figure 11. Realized Gain.

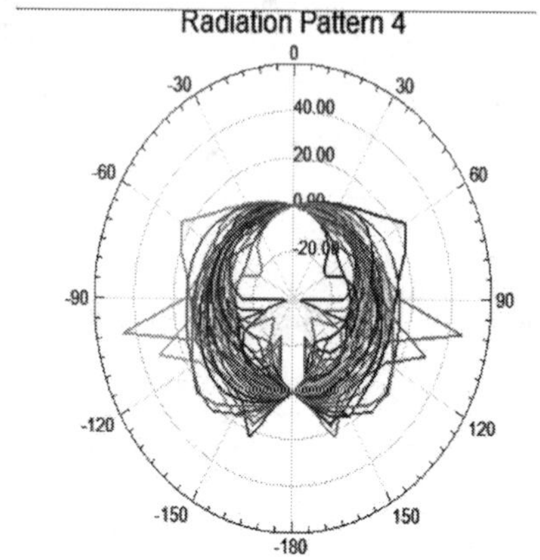

Figure 12. Polarization Ratio

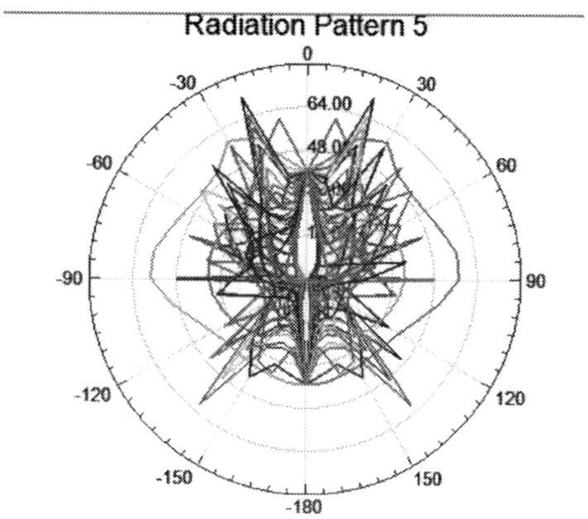

Figure 13. Axial Ratio

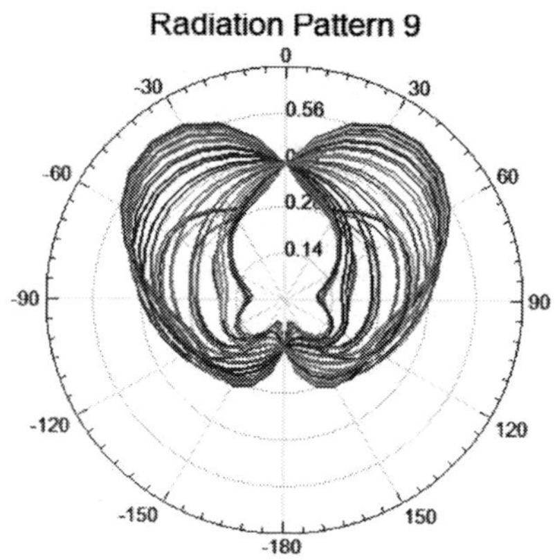

Figure 14. rE

Signal strength information is recorded in the table. After that, micro strip patch antenna is shifted by 5, and then the signal strength is displayed on the computer recorded back in the table. This is done repeatedly for every different angle to 360^0. When omnidirectional antennas are connected to the wireless LAN card, we obtained by the same signal strength to all direction of -50 dBm. If the data in the table plotted in polar coordinates, it will be obtained microstrip patch antenna radiation pattern of microstrip patch antenna. Radiation pattern of microstrip patch antenna.

3.1 Advantages and Disadvantages of Patch Antennas:

Some of their principal advantages of microstrip patch antennas are given below:

- Light weight and low volume, Low profile planar configuration which can be easily made conformal to host surface, Low fabrication cost, hence can be manufactured in large quantities, Supports both, linear as well as circular polarization.
- Can be easily integrated with microwave integrated circuits (MICs) and Capable of dual and trip le frequency operations and Mechanically robust when mounted on rigid surfaces.

Microstrip patch antennas suffer from a number of disadvantages as compared to conventional antennas. Some of their major disadvantages are given below:

- Narrow bandwidth, Low efficiency and Low Ga in and Extraneous radiation from feeds and junctions
- Poor end fire radiator except tapered slot antennas, Low power handling capacity and Surface wave excitation.

3.2 Apparature coupled feed for microstrip antennas

Microstrip patch antenna is connected to the wireless LAN card in the computer via cable RG-58 which has a characteristic impedance of 50Ω. Signal strength information is recorded in the table. After that, the microstrip patch antenna is shifted by 5, and then the signal strength is displayed on the computer recorded back in the table [12]. This is done repeatedly for every different angle to 360^0. When omnidirectional antennas are connected to the wireless LAN card, we obtained by the same signal strength to all direction of -50 dBm. If the data in the table plotted in polar coordinates, it will be obtained microstrip patch antenna radiation pattern of microstrip patch antenna.

4. Conclusion

A theoretical survey on microstrip patch antenna is presented in this paper. After study of various research papers, it concluded that wide bandwidth and low power handling capacity can be overcome through an array configuration and slotted patch. Some characteristics of feeding technique and various antenna parameters are discussed. Particular microstrip patch antenna can be designed for each application and different merits are compared with conventional microwave antenna and wireless communication.

References

[1] Constantine J and Balanis A 1997 Antenna Theory, Analysis and Design *John Wiley & Sons Inc. 2nd edition.*
[2] Garg R, Bhartia P, Bahl IJ and Ittipiboon A 2001 Microstrip antenna design handbook *Artech house*.
[3] Ramesh GP 2020 Microstrip Patch Antenna for Peripheral Arterary Disease Diagnosis *In Recent Trends and Advances in Artificial Intelligence and Internet of Things* 241-251.
[4] Nakar PS 2004 Design of a compact microstrip patch Theantenna for use in wireless/cellular devices *Master's thesis, Florida State University*.
[5] Pendem S and Ramesh GP 2020 75 GHz 5G Frequency Spectrum Analysis *InRecent Trends and Advances in Artificial Intelligence and Internet of Things* 165-176.
[6] Kraus JD 1988 Antennas and analysis *2nd Edition. Mc Graw Hill International.*
[7] Flashy AM and Ramesh GP 2020 Multi Band Antenna System for Quality Evaluation Application of Apple Fruit *In Recent Trends and Advances in Artificial Intelligence and Internet of Things* 199-206.
[8] Lee KF and Man LK 2011 Microstrip Patch Antennas *World Scientific* 8-12.
[9] Yang F, Zhang XX, Ye X and Rahmat-Samii Y 2001 Wide-band E-shaped patch antennas for wireless communications *IEEE transactions on antennas and propagation* **49** 1094-100.
[10] Rosu I 2011 PIFA-Planar Inverted F Antenna *YO3DAC/VA3IUL, RF Technical Articles.*
[11] Di Nallo C and Faraone A 2005 Multiband internal antenna or mobile phones *Electronics Letters* **41** 514-515.
[12] Meade RD, Winn JN and Joannopoulos JD 1995 Photonic crystals: Molding the flow of light *Pinceton Univ. Press.*
[13] Subramani P, Rajendran GB, Sengupta J, Pérez de Prado R and Divakarachari PB 2020 A Block Bi-Diagonalization-Based Pre-Coding for Indoor Multiple-Input-Multiple-Output-Visible Light Communication System **Energies 13** 3466.

IOP Conf. Series: Materials Science and Engineering **925** (2020) 012006 doi:10.1088/1757-899X/925/1/012006

Design and Fabrication of Contour Shape Microstrip Patch Antenna for Biomedical Application

S Parasuraman[1], S Yogeeswaran[2] and G P Ramesh[3]

[1] Department of ECE, Karpaga Vinayaga College of Engineering and Technology, Chengalpattu, Chennai, India.
[2] Department of ECE, P.T. Lee Changalvaraya Naicker College of Engineering and Technology, Kanchipuram, India.
[3] Department of Electronics and Communication Engineering, St. Peter's Institute of Higher Education and Research, Chennai, India.
[1]parasuengineers@gmail.com, [2]yogash28s@gmail.com, [3]rameshgp@yahoo.com

Abstract. This paper presents the design of a contour shape reconfigurable microstrip patch antenna. The reconfigurable microstrip patch antenna is one of the excellent candidates for portable wireless device very simple because of its low profile, light weight and low cost. However, the requirement of frequency bandwidth is becoming greater in present wireless communication system. It is presenting for biomedical application band which covering from 8GHZ to 12GHZ frequency range. This microstrip patch antenna is capable of operating on both linear and circular polarization. This type of patch antenna fabricated on FR4 substrate and also fed by feed line is same substrate. By controlling the bias voltage of four PIN diodes, the polarization of antenna can be switched between four states two state of linear polarization for (vertical and horizontal) and two states for circular polarization for (RHCP & LHCP). There are several steps to design this antenna including determine the antenna specific characteristics antenna simulation construction and analysis. The explanation detail for design procedure simulation results, methods and techniques are discussed in this paper

Keywords: Reconfigurable patch antenna, p-i-n diodes, contour shape slot antenna.

1. Introduction

Polarization reconfigurable antennas are attractive for many researches in wireless communication and also rapid development in mobile communication. It can be expands the capability of communication system. In this reconfigurable multifunctional antennas despite being single and compact. A polarization reconfigurable dielectric resonator antenna (DRA) fed by a compact coplanar waveguide (CPW)-to-slot line feeding structure of the vertical and horizontal polarizations of the DRA are excited by CPW feed and slot line feed, which are switched by two p-i-n diodes [1]. This proposed antenna has the merits of concise Structure and low cost and supports wide applications in wireless communication systems [2]. Good directivity and about 6.5dB gain are obtained in past literature [3].

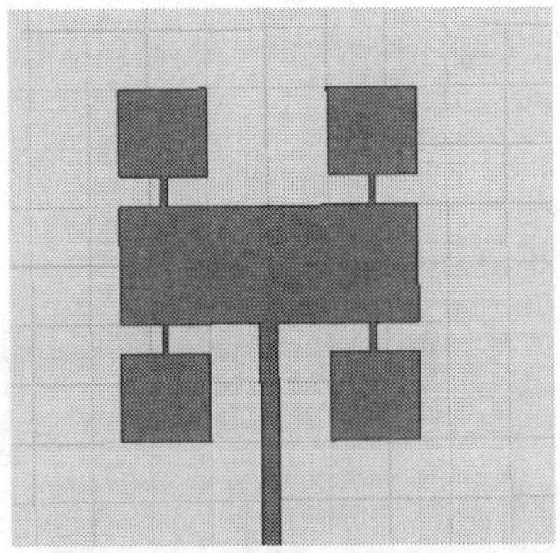

Figure 1. The geometric proposed antenna

Figure 2. Substrate model

A novel reconfigurable microstrip antenna with radiation pattern selectivity and polarization diversity is presented by controlling the two states of switches [4]. The authors introduce the utilization of ferromagnetic compounds into an antenna design and investigate the capability of controlling the antenna properties by means of an external magnetic field and also patch antenna design with a part of the dielectric substrates replace by YIG compound in past literature [5]. In this paper a single feed reconfigurable polarization printed monopole antenna is presented. The proposed antenna consists of a monopole patch and a narrow L-shaped slot created on a ground plane. Good omnidirectional pattern is achieved through this design [6].

Some system a linear-polarization (LP)/circular-polarization (CP) switchable reconfigurable antenna using a Radio Frequency Micro- Electro-Mechanical-System (RF-MEMS) switch is proposed and a novel package platform is introduced [7]. The AR bandwidth of the antenna can be further improved by increasing the substrate thickness, which may help reduce the sensitivity of its performance to fabrication tolerances [8]. In this two polarization are orthogonal to each other at the same frequency is used [9]. The computed return loss for each case is better than 10dB over the 2.4-2.484GHz frequency range [10].

In this communication, to present the design of a single feed rectangular microstrip antenna with reconfigurable polarization capability and to analyze its simulation results. The basic structure utilizes an X shaped slot at the center of a rectangular cross-shaped patch. The X shape is chosen to induce symmetric current distributions for TM10 and TM01 modes as well as to obtain greater area reduction [11, 19]. The polarization can be switched between CP (LHCP&RHCP) and LP (Horizontal & Vertical) by changing the shape of the slot according to the ON or OFF state of the PIN diode without changing the geometrical parameters and is devoid of any impedance matching. Furthermore, the antenna has an added advantage of reduced size with low levels of cross-polarized radiation in LP state and CP state. In addition, the antenna is simple because it requires only four PIN diodes as well as less area to occupy the patch and dc-bias circuit compared to conventional polarization diversity antennas available in literature.

2. Antenna Geometry

The geometry of the proposed antenna is shown in Fig. 1.1. The antenna is fabricated on a substrate of thickness h (1.6 mm) and relative permittivity ε_r (4.4). A contour shape of arm length l_x mm and width w_x is then carved at the center of the cross patch. The antenna is coupled using a 50 Ω microstrip line fabricated using the same substrate material. The dimension of the ground plane is 100×100 mm^2. Four PIN diodes are inserted into the center of the slot is oriented normal to the feed line. There is a printed crossed section in the center of the contour shape that connects both the diodes to the patch. For the proper biasing of the diodes, three narrow slot lines are carved in the patch. Three small SMD capacitors C1, C2 and C3 are soldered at these slot lines which block the dc bias current as well as provide good

IOP Conf. Series: Materials Science and Engineering **925** (2020) 012006 doi:10.1088/1757-899X/925/1/012006

RF continuity [12]. The PIN diode requires a bias voltage of 1.1 V which is supplied from a battery through chip inductors. The dc bias circuit, used to control the ON/OFF state of diodes, is located on the right side of the patch.

3. Description of contour shape antenna

Table 1. Switching condition

Antenna	DI	D2	D3	D4	Type of Polarization
1	ON	OFF	ON	OFF	Horizontal
2	OFF	ON	OFF	ON	Vertical
3	OFF	ON	ON	OFF	RHCP
4	OFF	ON	ON	OFF	LHCP

The fundamental resonant modes (TM_{10} and TM_{01}) of the un-slotted cross shaped patch are between (2-3) GHz with orthogonal polarizations. The appropriate selection of the slot size modifies the horizontal and vertical electrical lengths of the patch equally so that the two resonant frequencies are lowered to 2.32 GHz and 2.39 GHz. The simulated surface current distribution of the cross patch antenna with contour shape is depicted in Fig. 2. It is well evident from the figure that the insertion of the patch increases the current path thereby lowering the resonant frequency. The contour shape length is optimized to achieve maximum area reduction using Ansoft HFSS. The antenna exhibits good radiation characteristics for both resonant frequencies with an area reduction of 77% and 64% for the first and second frequency respectively when compared to a standard rectangular patch operating at the same frequencies. Four PIN diodes are inserted at the center of the X-slot to achieve the reconfigurable polarization capability. The orthogonally polarized tri frequency cross patch antenna can be reconfigured for different polarization with respect to the bias voltage applied to the diodes. The bias circuit consists of three dc block capacitors, RF chokes, two switches and input voltage. The dc bias lines are connected to the patch through RF chokes. The ON state of the diode is represented by a series resistor, R=1.35 Ω, while the OFF --state is represented by a capacitor of C=0.35 pF. Three dc block capacitors of C=33 pF are chosen to isolate the RF components from the dc signal and RF choke inductor isolate the RF signal from flowing into the dc signal [13]. The ON/OFF states of the diodes are controlled with respect to the potential applied to the terminals, which is described in Table I.
Antenna feeding is a very important design factor and it directly affects antenna properties, system level performance, and realistic prototyping. Among many feeding methods, three are more suitable for slot antennas. The three feeding methods are

- Simple probe
- Coplanar Waveguide (CPW)
- Microstrip Line (ML) feed

Finally, the microstrip line feed is used because it is simple and ease in matching. The slot antenna geometry is as follows: the ground plane is backed by a cavity filled with a substrate with require relative permittivity. Ansoft's High Frequency Structure Simulator (HFSS) was used to design and study antenna properties.

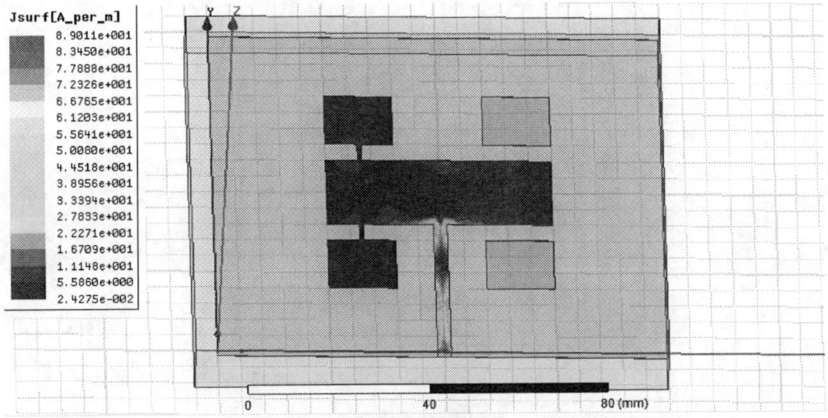

Figure 2(a)

Figure 2(b)

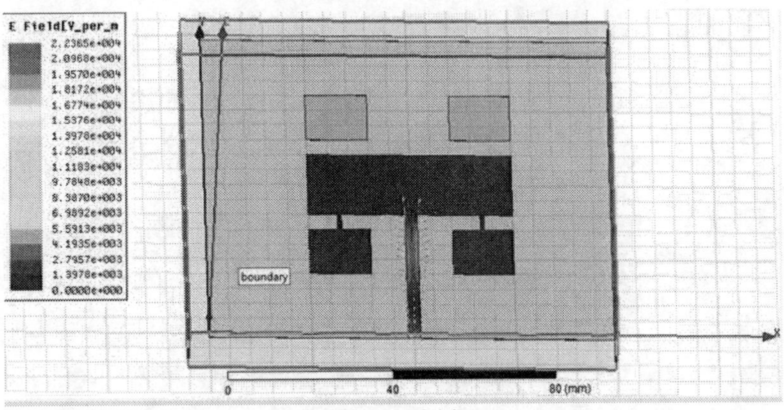

Figure 2(c)

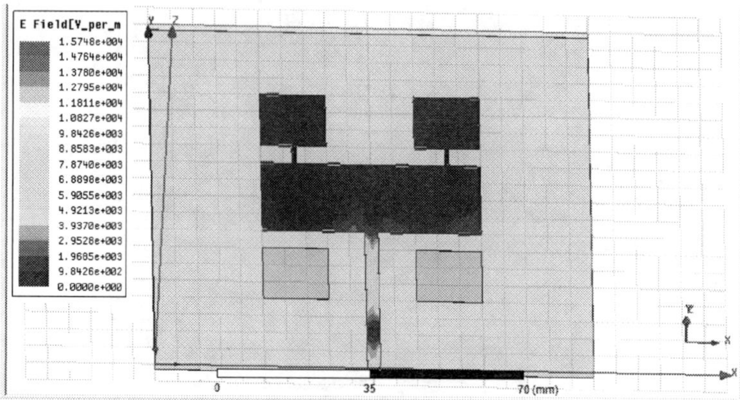

Figure 2(d)

Figure 2. Simulated vector field current distribution for both vertical and horizontal polarization of an antenna at 2-3 GHZ

A vector field is an assignment of a vector to each point in a subset of Euclidean space. A vector field in the plane, for instance, can be visualized as a collection of arrows with a given magnitude and direction each attached to a point in the plane.

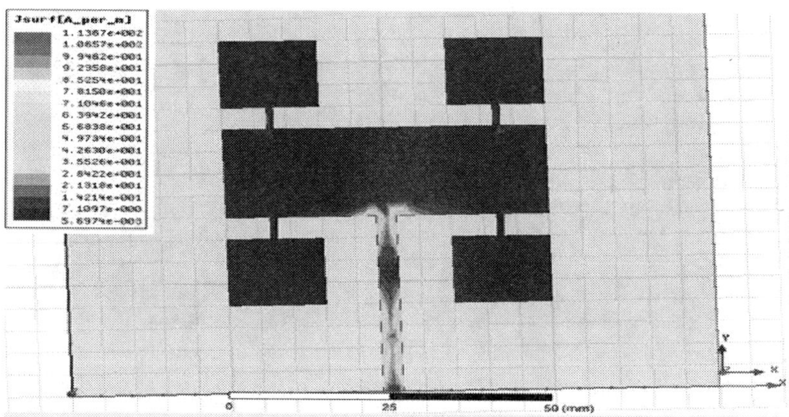

Figure 3(a)

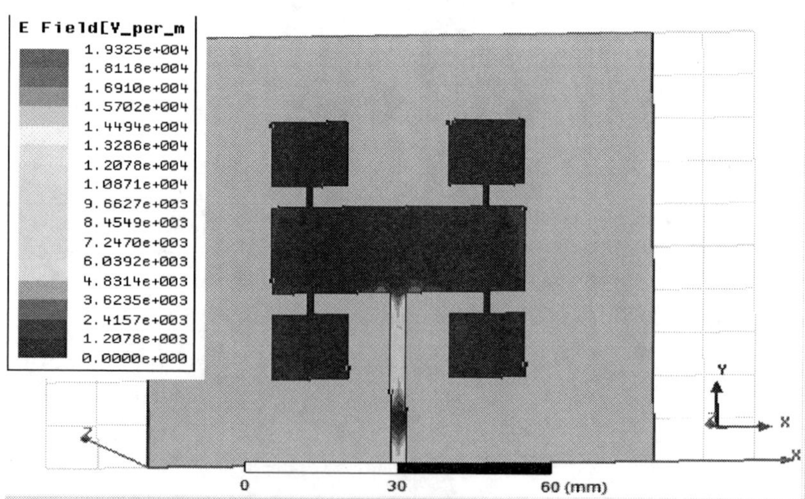

Figure 3(b)

Figure 3(c)

Figure 3. Simulated magnetic field current distribution for both vertical and horizontal polarization of an antenna at 2-3 GHZ

4. Results and Discussion

When a current field represents force, the line integral of a current field represents the work done by a force moving along a path, and under this interpretation conservation of energy is exhibited as a special case of the fundamental theorem of calculus [14-16]. The return loss is another way of expressing mismatch. It is a logarithmic ratio measured in dB which compares the power reflected by the antenna to the power that is fed into the antenna from the transmission line. The relationship between SWR and return loss is given as follows:

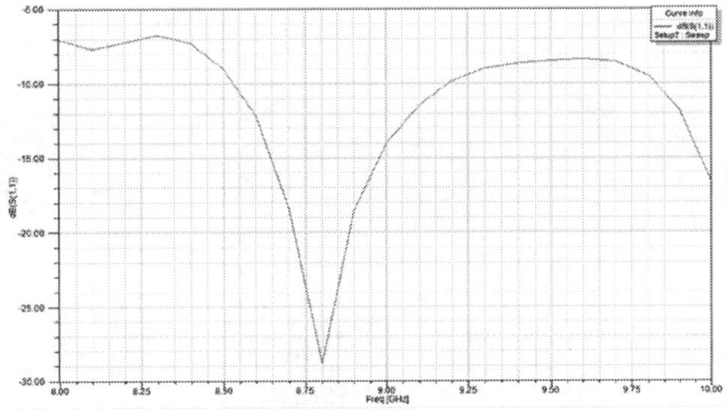

Figure 4. Return Loss

Voltage Standing Wave Ratio (VSWR) is a measure of how well an antenna is matched to the transmission line it is connected to.

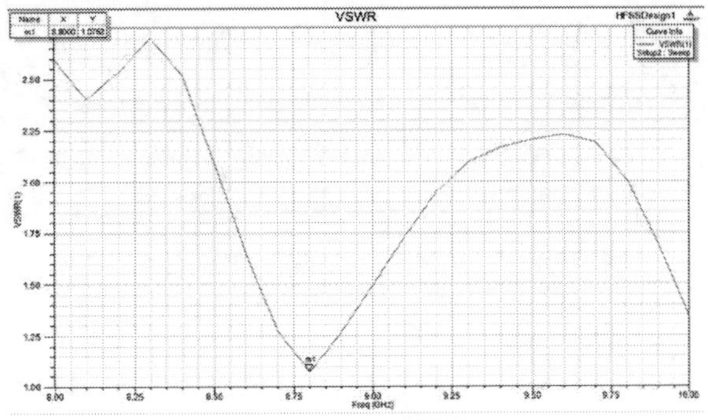

Figure 5. VSWR

The energy radiated by any antenna is contained in a transverse electromagnetic wave that is comprised of an electric and a magnetic field. These fields are always orthogonal to one another and orthogonal to the direction of propagation. The electric field of the electromagnetic wave is used to describe its polarization and hence, the polarization of the antenna. In general, all electromagnetic waves are elliptically polarized. In this general case, the total electric field of the wave is comprised of two linear components, which are orthogonal to one another. Each of these components has a different magnitude and phase. At any fixed point along the direction of propagation, the total electric field would trace out an ellipse as a function of time [17-18]. This concept is illustrated in Figure 6, where, at any instant in time, Ex is the component of the electric field in the x- direction and Ey is the component of the electric field in the y-direction. Generally, the polarization is categorized into two types such as

1. Linear polarization
Horizontal and Vertical polarization
2. Circular polarization
Left hand side and right hand side polarization

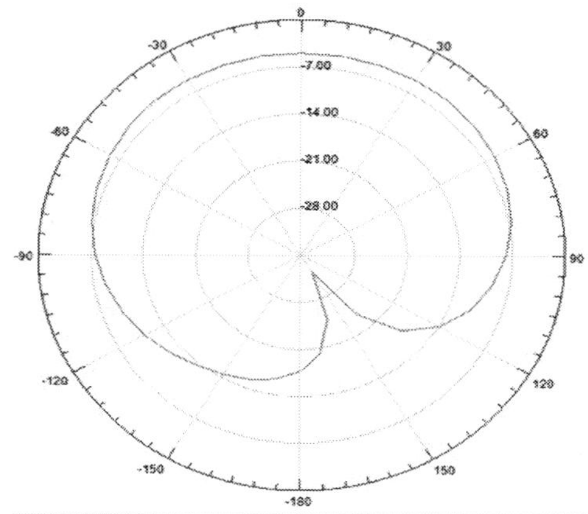

Figure 6. Radiation pattern for RHCP

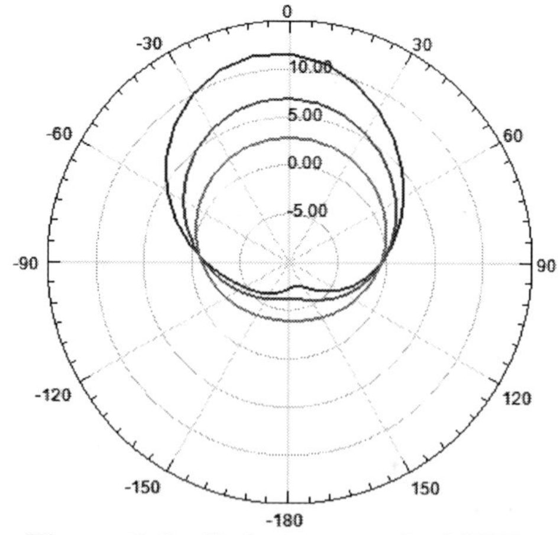

Figure 7. Radiation pattern for LHCP

5. Conclusion and Future Work

A reconfigurable microstrip patch antenna with switchable slots for polarization diversities has been presented in this communication. The antenna can produce both linear and circular polarization by controlling the bias conditions of four PIN diodes. The proposed design achieves a cross polar level better than -10 dB in both linear and circular polarization broadside radiation characteristics and moderate gain. The first phase is study and understands the parameter that showed the behavior and how it can be used and interfaces with other circuit. The microstrip patch antenna consist four major parts: the conductive patch, the dielectric substrate, the ground plane and the feed line. The most widely used patch conductive for low cost, low-profile patch antenna is coppers. To design and simulate patch antenna by using high frequency structure simulator. There are three methods that are widely used in the analysis, design, and modeling of patch antenna is transmission, cavity and full-wave models. In future, we are going to fabricating patch antenna by using FR4 substrate and analyze ±45° orientation of polarization. To test and analyze patch antenna by using signal generator, network analyzers and etc.

References

[1] Liu J and Feng Z 2012 Polarization reconfigurable dielectric resonator antenna with a CPW-to-slotline feed *In 2012 Int. Conf. on Microwave and Millimeter Wave Technology (ICMMT)* **1** 1-4.

[2] Cao SX, Yang XX, Gong B and Shao BC 2011 A reconfigurable microstrip antenna with agile polarization using diode switches. *In 2011 IEEE Int. Symp. on Antennas and Propagation (APSURSI)* 1566-1569.

[3] Ramesh GP 2020 Microstrip Patch Antenna for Peripheral Arterary Disease Diagnosis. *In Recent Trends and Advances in Artificial Intelligence and Internet of Things* 241-251.

[4] Cao W, Zhang B, Liu A, Yu T, Guo D and Pan K 2012 A reconfigurable microstrip antenna with radiation pattern selectivity and polarization diversity *IEEE Antennas and Wireless Propagation Letters* **11** 453-6.

[5] Zervos T, Alexandridis AA, Lazarakis F, Pissas M, Stamopoulos D, Angelopoulos ES and Dangakis K 2012 Design of a polarisation reconfigurable patch antenna using ferrimagnetic materials *IET microwaves, antennas & propagation* **6** 158-64.

[6] Amini MH, Hassani HR and Nezhad SM 2012 A single feed reconfigurable polarization printed monopole antenna *In 2012 6th European Conf. on Antennas and Propagation (EUCAP)* 1-4.

[7] Pendem S and Ramesh GP 2020 75 GHz 5G Frequency Spectrum Analysis *In Recent Trends and Advances in Artificial Intelligence and Internet of Things* 165-176.

[8] Wu B and Okoniewski M 2012 A novel scheme for realizing a microstrip antenna with switchable circular polarization *In 2012 6th European Conference on Antennas and Propagation (EUCAP)* 3278-3282.

[9] Sung YJ 2012 Frequency and polarisation reconfigurability from an open-loop square ring antenna *IET microwaves, antennas & propagation* **6** 505-9.

[10] Ali Ramadan, Mohammed Al-Hussein, Karim Y. Kabalan and Ali El-HajjGeorgios, Atmatzakis and Christos G. Christodoulou (2011)," A Directional Polarization Reconfigurable Microstrip Antenna," IEEE transactions on microwave.

[11] Flashy AM and Ramesh GP 2020 Multi Band Antenna System for Quality Evaluation Application of Apple Fruit *In Recent Trends and Advances in Artificial Intelligence and Internet of Things* 199-206.

[12] Kim B, Pan B, Nikolaou S, Kim YS, Papapolymerou J and Tentzeris MM 2008 A novel single-feed circular microstrip antenna with reconfigurable polarization capability *IEEE Transactions on Antennas and Propagation* **56** 630-8.

[13] Long S, McAllister M and Shen L 1983 The resonant cylindrical dielectric cavity antenna *IEEE Transactions on Antennas and Propagation* **31** 406-12.

[14] Hsu SH and Chang K 2007 A novel reconfigurable microstrip antenna with switchable circular polarization *IEEE antennas and wireless propagation letters* **6** 160-2.

[15] Chen YB, Jiao YC and Zhang FS 2007 Polarization reconfigurable CPW-fed square slot antenna using pin diodes *Microwave and Optical Technology Letters* **49** 1233-6.

[16] Ma KP, Qian Y and Itoh T 1999 Analysis and applications of a new CPW-slotline transition *IEEE Transactions on Microwave theory and Techniques* **47** 426-32.

[17] Wu SJ and Ma TG 2008 A wideband slotted bow-tie antenna with reconfigurable CPW-to-slotline transition for pattern diversity *IEEE transactions on antennas and propagation* **56** 327-34.

[18] Li Y, Zhang Z, Chen W and Feng Z 2010 Polarization reconfigurable slot antenna with a novel compact CPW-to-slotline transition for WLAN application *IEEE Antennas and Wireless Propagation Letters* **9** 252-5.

[19] Subramani P, Rajendran GB, Sengupta J, Pérez de Prado R and Divakarachari PB 2020 A Block Bi-Diagonalization-Based Pre-Coding for Indoor Multiple-Input-Multiple-Output-Visible Light Communication System *Energies* **13** 3466.

Compressive Strength of PPC Based Quaternary Blended Concrete

G Swamy Yadav[1,2], Ayyappa R A[1], M Guruprasad[1], G Hari Prasad[1], S Vyshnavi[1] and Ch.Pragnya[1]

[1] Department of civil engineering, SR Engineering College, Warangal, Telangana, India
[2] Centre for Construction Methods and Materials, SR Engineering College, Warangal, Telangana, India
swamy_yadav-g@srecwarangal.ac.in, gunda.hari.9@gmail.com, ayyappa_reddy_a@srecwarangal.ac.n, guru_prasad_m@srecwarangal.ac.in, vyshnavisamala.21@gmail.com, pragnyachennoju02@mail.com

Abstract. Production of sewage sludge is increasing day by day all over the world. By disposing of the sewage sludge the environmental issues take place in the society. Hence the research for replacement of cement using sewage sludge ash is essential to reduce the use of natural resources. Certainly as pond ash also coal ash as waste which compresses of bottom ash and fly ash. They contain the major chemical components which cement contain. So the major aim in this research study is to the best use of Biochar and pond ash in concrete. Concrete M20 specimens are prepared according to IS10262-2019 and by replacing cement with 5% Biochar and sand with 2-30% of pond ash by varying percentages in steps of 2 and the compressive strength was determined finally observed that 20% replacement of pond ash and 5% Biochar is giving the optimum value of compressive strength.

Keywords: Biochar, Pond ash, Flyash, Sewage sludge, Compressive Strength.

1. Introduction

Over the last few years' replacement of cement and sand is being experimented by researchers from all over the world by silica fume, fly ash, GGBS, rice husk ash on cement, and pond ash with sand in certain percentages. Septage ash is a by-product of the sewage sludge and wood pallets which absorbs the water in sewage sludge during incineration at 500oc-700oc to become ash and the pond ash in the bottom ash in the thermal power plants which are obtained after the fly ash is obtained from the bottom of plant by spraying water which is not finer than the fly ash so it is used in the replacement of fine aggregate. All these researches are done to maintain the sustainability of natural resources that can be maintained or kept in existence by replacing natural resources with renewable resources or manmade resources like fly ash, M sand, Risk husk, ash recycled aggregates, and with some waste materials. The alternative solution for sewage sludge disposal is incineration which makes sewage sludge to SSA (sewage sludge ash). In this process, main components are sewage sludge in presence of high temperature (500 to 700 degree Celsius) such as SiO_2, CaO, Al_2O_3 (Wojciech piasta 2016) [1], (Ciarán J. Lynn 2015) [2] are the components of OPC (Ordinary Portland cement). So, it can be used in place of cement content in concrete without any objection with the current population and urbanization, it is clear that the demand for cement and fine aggregate will increase in the future. Sustainability of natural resources can be maintained or kept in existence by replacing natural resources with renewable resources or manmade resources like fly ash, M sand, Risk husk, ash recycled aggregates, and with some waste materials (Sankh 2018) [3]. It is clear that the demand for cement will increase in the future. (Ing 2016) [4]. The use of industrial by-products (waste material) as an alternate to cement and sand in concrete production (Barbora Krejcirikova 2018) [5]. When sewage sludge is used in agriculture within a certain timeframe (Nakic.D 2018) [6]. Sewage sludge contains SiO2, CaO, Al2O3, Fe3O4, and other components of cement it can be also used as a raw material for cement production.

2. Literature review

Researchers like (Prabhanjan 2020)[10] were studied by taking waste ballast by crushing to 10mm, 12.5mm, and 20mm sizes which are used for road pavements and added 5%EDTA solution and found that increased strength than the normal aggregate. (Laxmi 2020) [9] Made an attempt using rubber tyre chips (20mm) of 5%, 10%, 15%, and 20% as replacement of aggregate in concrete and found that about 5% to 8% can be used without loss of strength. (Ayyappa 2020) [7] Stu died by using coconut shells by 2,4, and 6 percentages in place of coarse aggregate and eggshell power were used in place of cement by percentages of 5, 10 and 15 and observed optimum at 6% both eggshell powder and coconut shells and also observed that weight of the concrete was reduced. (Yadav 2020) [12] Used waste steel fibers by varying from 1 to 6% in steps of 1 and 1% polypropylene fibers constantly and observed that using waste steel fibers workability reduced and all the mechanical properties were increased. It also resulted in reduced cracks due to the presence of polypropylene fibers. (Guruprasad 2019) [8] Attempted an experiment using SIFRON fibers in concrete as slurry infiltrated hybrid fiber reinforced concrete and kept for curing in an Acid media for 60 days and compared with slurry infiltrated mono fiber reinforced concrete and found that in first case concrete is having better resistant to acid attack. The experimental work was done to observe the effects taking place during the replacement of the GGBS, Limestone powder, and pond ash on fresh and solidified properties of SCC. Replacement of pond ash by weight of fine aggregates in various percentages such as 20%, 40%, 60%, GGBS& limestone powder of same percentages such as 20%, 40%, and 60% by weight of cement. By observing the experimental results 40% replacement is optimum. As a part of using pond ash, we noticed settlement undergoes long time to settle. After that 60% replacement of any byproduct was not good for usage. Pond ash-poor to use as fine aggregate due to lack of initial setting time and strength obtained. By providing the confinement at different percentages use to increase strength at 28 days.
Sewage sludge is a byproduct of domestic sewage sludge treatment plant. Due to urbanization and population growth, the amount of SSA (sewage sludge ash) is also increased and also it will increase in the future. Before the installation of domestic sewage sludge treatment plants in urban areas, the public use to dispose of sewage sludge in open areas, lakes, and directly in damping areas. In this present study on replacement of cement with SSA (sewage sludge ash) and fine aggregate with pond ash in percentages such as SSA is kept at constant 5% as the cement is PPC already consists of 30-35% of fly ash so additional % of ash may result in a decrease in early strength and also 5% SSA gives the concrete higher early strength than any other % of SSA being replaced (Swamy 2019) [11] so keeping the SSA as constantly changing the fine aggregate % with pond ash such as 2%,4%,...30% was found.

3. Methods and Materials

3.1. Materials Used

3.1.1. Cement

The most commonly available Portland pozzolana cement(PPC) was used for the investigation. All the tests on cement are conducted as per IS 1489-1(1991). Cement was bought from the same source throughout the research work. While storing cement, all possible contact with moisture was avoided.

The specific gravity of cement	=	3.16
The fineness of cement	=	7.16

3.1.2. Fine Aggregates

Locally available river sand is used in the present study

The specific gravity of F.A	=	2.71
Fineness Modulus = 246/100	=	2.46

The sand used is Fine Sand (Zone-II)

3.1.3. Coarse Aggregate

Crushed granite used as Coarse aggregate

The specific gravity of Coarse Aggregate	=	2.65
Water absorption	=	1%

3.1.4. Water

The important ingredient is water for concrete which participates in chemical reactions with cement. Portable water was used in the present work throughout.

3.1.5. Pond ash

In India, thermal power plants adopt wet techniques to remove the debris in large ponds. They have washed away with water in the form of slurry and forms Pond ash. If it is coarser and lees pozzolanic cannot be used. Hence researchers working worldwide are focussing to use it alternatively in the concrete industry is one of the effective methods to utilize. Thermal Power plant Ramagundam is producing plenty of pond ash which is made used in the present study.

3.1.6. Biochar

Septage or "septic tank matter" is mostly rewarded slime that is amassed and put away in a septic tank or comparative on location sewage facility. Septage debris is a side-effect from the pretreatment of family wastewater (sewage) in a septic tank. Following a couple of long stretches of collection, septage is normally siphoned out of a septic tank by a vacuum truck. Septic tanks receive black water from flush toilets, as well as greywater. To overcome this environmental impact by dumping of untreated sludge, sewage sludge ash is used as a replacement of cement in concrete in this project. The sewage treatment plant in Warangal producing Septage ash(Biochar) by the process of incineration of Septage. It is made use in the present work.

3.2. Methodology

Previous researchers used binders and fillers are replaced with flyash, GGBS, Rice husk ash, Biochar, pond ash, etc, separately but in the present work, the combination of binder and filler replaced with Biochar and pond ash together. Concrete cubes of 150mm of 6 numbers were cast per each Mix out of which 3 were tested for 7-day strength and remaining for 28 – day strength. In each Mix of Concrete cement is replaced with 5% Biochar constantly throughout the project and replacement of fine aggregate by varying percentages of the pond ash from 2% to 30% in steps of 2% each incrementally.

Figure 1. below shows the flow of the present work involved in various stages. The methodology follows initially material testing the basic properties of cement, fine aggregate, and coarse aggregate, and based on the results mix design for M20 concrete was made using IS10262-2019 and the quantities were estimated.

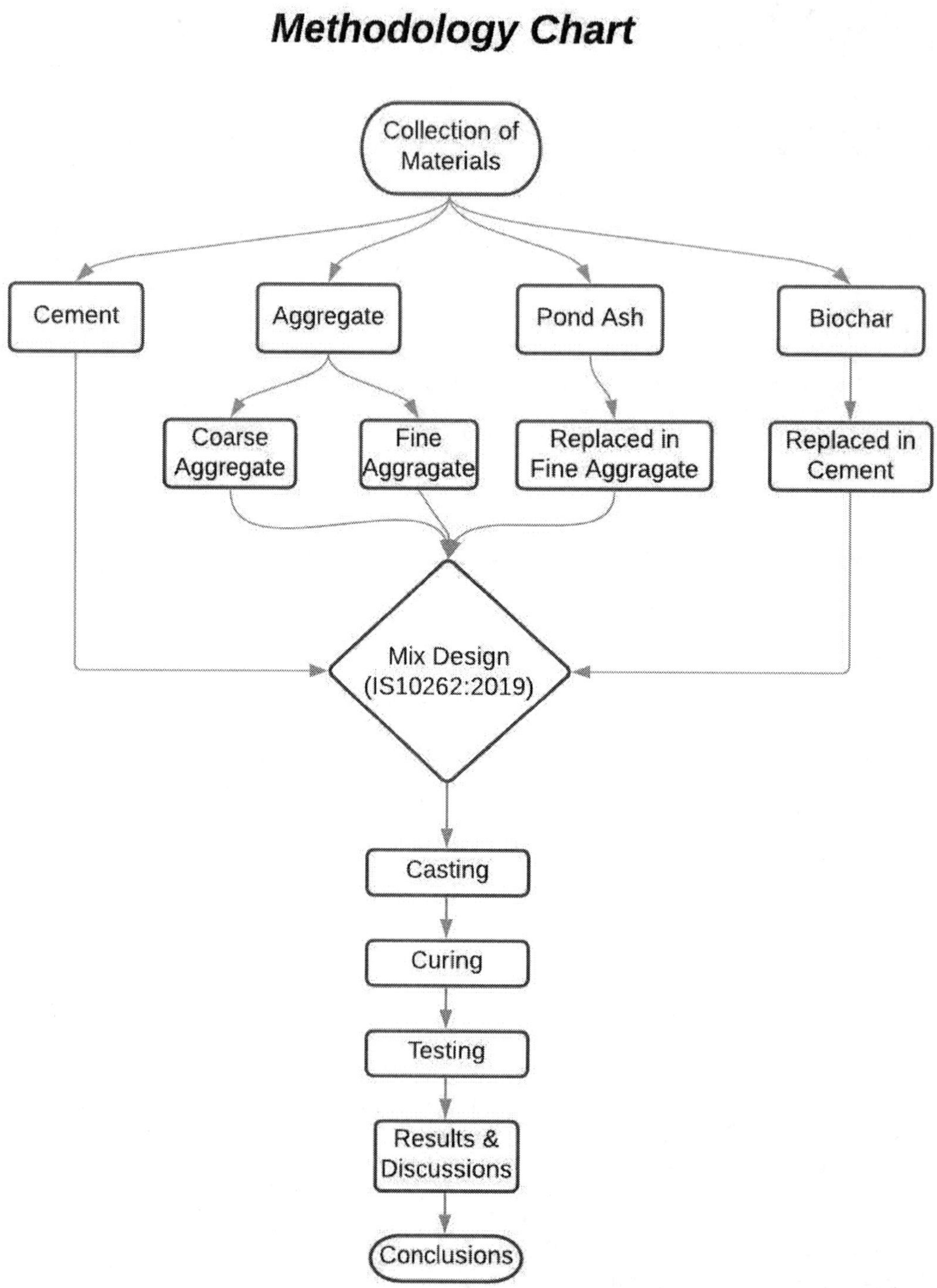

Figure 1. The flow of the present work involved in various stages.

The estimated quantities are represented in **Table 1.** given below calculated as per the IS10262-2019 mix design code.

IOP Conf. Series: Materials Science and Engineering **925** (2020) 012007 doi:10.1088/1757-899X/925/1/012007

Table 1. Quantities of Materials Used.

Mix	Cement (kg/m^3)	Biochar (kg/m^3)	Fine Aggregate (kg/m^3)	Pond ash (kg/m^3)	Coarse Aggregate (kg/m^3)	Water (kg/m^3)
M0	450.0	0.0	640.75	0.00	1204.053	202.74
M2	427.5	22.5	627.93	12.81	1204.053	202.74
M4	427.5	22.5	615.12	25.63	1204.053	202.74
M6	427.5	22.5	602.30	38.44	1204.053	202.74
M8	427.5	22.5	589.50	51.26	1204.053	202.74
M10	427.5	22.5	576.67	64.07	1204.053	202.74
M12	427.5	22.5	563.86	76.89	1204.053	202.74
M14	427.5	22.5	551.04	89.70	1204.053	202.74
M16	427.5	22.5	538.23	102.52	1204.053	202.74
M18	427.5	22.5	525.41	115.33	1204.053	202.74
M20	427.5	22.5	512.60	128.15	1204.053	202.74
M22	427.5	22.5	499.78	140.96	1204.053	202.74
M24	427.5	22.5	486.97	153.78	1204.053	202.74
M26	427.5	22.5	474.15	166.59	1204.053	202.74
M28	427.5	22.5	461.34	179.41	1204.053	202.74
M30	427.5	22.5	448.52	192.22	1204.053	202.74

The estimated quantities are mixed thoroughly as shown in **Figures 2** and **3** in dry condition and then by the addition of water the ingredients were mixed to get a good consistency.

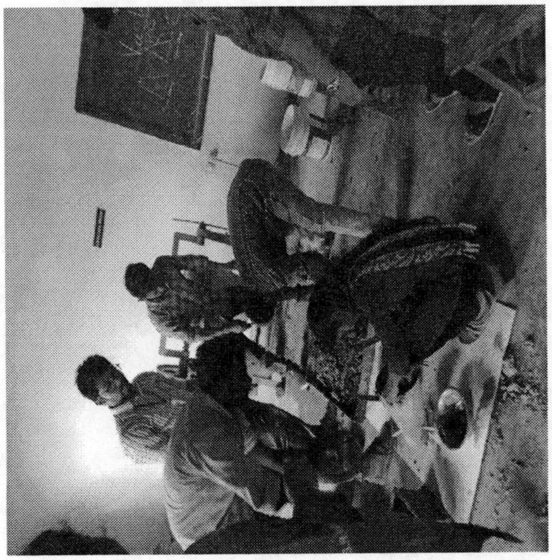

Figure 2. Mixing of concrete ingredients by hand.

Figure 3. Mixing of concrete ingredients by Concrete mixer.

IOP Conf. Series: Materials Science and Engineering **925** (2020) 012007 doi:10.1088/1757-899X/925/1/012007

After mixing required consistency the concrete was placed in the prepared molds of size 150x150x150 mm cubes shown in F**igure 4.** and are placed in a curing tank for 28-days, Then they were tested in standard compression testing machine shown in **Figure 5** to get compressive strength.

Figure 4. The casting of cubes.

Figure 5. Testing of cubes.

4. Results and Discussions

Based on the results obtained from the experimental data the compressive strength of M20 concrete with 5% of Biochar being replaced with Cement and alternate proportion replacing fine aggregate with pond ash at 2%, 4%,...30%. Plots were drawn and conclusions were made. **Figure 6** shows the variation of compressive strength with varying percentages of pond ash after 7-days.

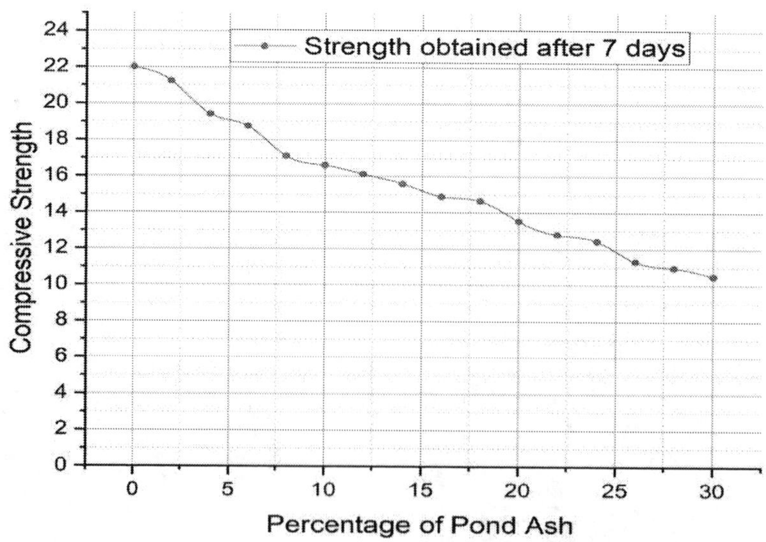

Figure 6. Compressive Strength at 7-Day.

The variation of 7-Day compressive strength shows decreasing throughout as the percentage of pond ash increases. This indicates low early strength. After 28-day the compressive strength was observed to be increasing by increasing the percentage of pond ash indicating the pozzolanic nature of pond ash.

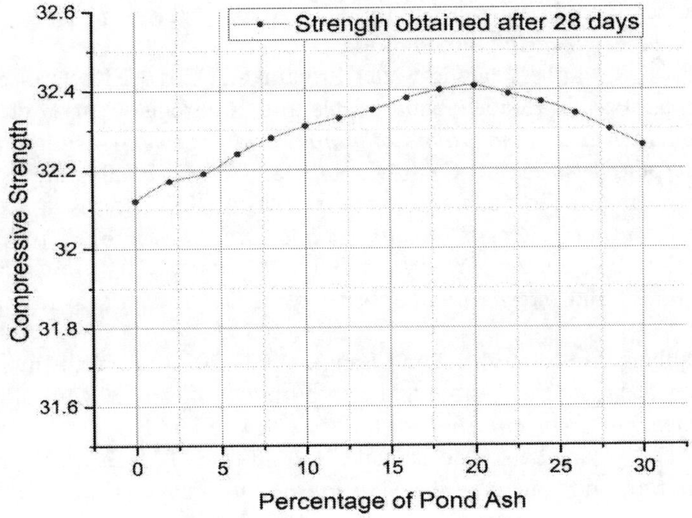

Figure 7. Compressive Strength at 28-Day.

The variation of compressive strength shown in **Figure 7** after 28 – day increasing initially and found maximum at 20% replacement of pond ash and compressive strength is reducing by the addition of pond ash further.

5. Conclusion

The following conclusions were made based on the results using Biochar and Pond ash together in concrete, it is giving low early strength and later showing better results. The optimum percentage of pond ash in place of fine aggregate was found to be 20%. As giving later strength this combination may be useful for durable structures. It may be observed that the cost of concrete was reduced, as the cement content is reducing about 5% makes the concrete eco-friendly. Use of pond ash in place of fine aggregate we conclude that sustainability of non-renewable resources can be maintained. Further research can be done by applying the same concrete to structural members. Durability Properties can be checked for the same concrete. It can be used for different mixes using the same combinations.

References

[1] Piasta W and Lukawska M 2016 The effect of sewage sludge ash on properties of cement composites *Procedia engineering* **161** 1018-24.

[2] Lynn CJ, Dhir RK and Ghataora GS 2018 Environmental impacts of sewage sludge ash in construction: leaching assessment *Resources, Conservation and Recycling* **136** 306-14.

[3] Sankh AC, Biradar PM, Naghathan SJ and Ishwargol MB 2014 Recent trends in replacement of natural sand with different alternatives *In Proc. of the Int. Conf. on Advances in Engineering and Technology* 59-66.

[4] Ing DS, Chin SC, Guan TK and Suil A 2016 The use of sewage sludge ash (SSA) as partial replacement of cement in concrete *ARPN Journal of Engineering and Applied Sciences* **11** 3771-5.

[5] Krejcirikova B, Kolarik J and Wargocki P 2018 The effects of cement-based and cement-ash-based mortar slabs on indoor air quality *Building and Environment* **135** 213-23.

[6] Nakic D, Vouk D, Serdar M and Cheeseman CR 2018 Use of MID-MIX® treated sewage sludge in cement mortars and concrete *European Journal of Environmental and Civil Engineering* **16** 1-6.

[7] Ayyappa RA, Sandeep Reddy B, Swamy Yadav G, Sudarshan DS 2020 Partial Replacement of Cement and Coarse Aggregate by Egg Shell Powder and Coconut Shells *International Journal of Innovative Technology and Exploring Engineering* **4** 1242-1246.

[8] Guruprasad M, Priyanka A, Prabhanan N and Swamy G 2019 Behaviour of Slurry Infiltrated Hybrid Fiberreinforced Concrete Subjected to Acidic Attack *International Journal of Recent Technology and Engineering* **4** 6598-6602.

[9] Laxmi E, Yadav GS, Patil K, Jalade S and Priyamka A 2020 Effects of Scrap Rubber TYRE Aggregates as Partial Replacement of Natural Aggregates in High Strength Concrete. *International Journal of Innovative Technology and Exploring Engineering* **4** 2039-2043.

[10] Prabhanjan N, Yadav GS, Sahithi G, Sravanthi B and Tipraj B 2020 Assesment to Increase the Mechanical Properties of used Ballast by EDTA Solution as a Construction Material *International Journal of Recent Technology and Engineering* **5** 157-160.

[11] Swamy G, Krishna EV, Kumar SY, Prasad GH and Pavan R 2019 Potential Use of Biochar as Construction Material *International Journal of Recent Technology and Engineering* **1** 1570-1572.

[12] Yadav GS, Prabhanjan N, Sahithi G and Sangeetha G 2020 Strength Investigation of Fly Ash Based Concrete Waste Steel Fibre and Polypropylene Fibre as Reinforcing Materials *In Int. Conf. on Emerging Trends in Engineering (ICETE)* 155-161.

[13] Nakic D, Vouk D, Serdar M and Cheeseman CR 2018 Use of MID-MIX® treated sewage sludge in cement mortars and concrete *European Journal of Environmental and Civil Engineering* **16** 1-6.

[14] Arumugam K, Ilangovan R and Manohar DJ 2015 A study on characterization and use of pond ash as fine aggregate in concrete *International Journal of Civil & Structural Engineering* **2** 466-74.

[15] Sumeet, Munnur S 2015 Study on Strength and Durability of Concrete with Partial Replacement of Fine Aggregate by Pond Ash *IJSRD - International Journal for Scientific Research & Development* **2** 296-299.

[16] Bagwan KM and Kulkarni DS 2014 A Study of Characteristic and Use of Pond Ash for Construction. *International Journal of Emerging Technology and Advanced Engineering* **4** 835-840.

[17] Haldive SA and Kambekar AR 2013 Experimental study on combined effect of fly ash and pond ash on strength and durability of concrete *International Journal of Science and Engineering Research* **4** 81-6.

[18] Patel J, Patel K and Patel G 2013 Utilization of pond fly ash as a partial replacement in fine aggregate with using fine fly ash and alccofine in HSC *IJRET: International Journal of Research in Engineering and Technology* **2** 2319-1163.

[19] Arumugam K, Ilangovan R and Manohar DJ 2011 A study on characterization and use of pond ash as fine aggregate in concrete *International Journal of Civil & Structural Engineering* **2** 466-74.

[20] Ganesh B, Bai HS and Nagendra R 2011 Effective utilization of pond ash for sustainable construction–need of the hour *International Journal of Earth Sciences and Engineering* **4** 151-4.

[21] Stolaroff JK, Lowry GV and Keith DW 2005 Using CaO-and MgO-rich industrial waste streams for carbon sequestration *Energy Conversion and Management* **46** 687-99.

[22] Torres A, Bartlett L and Pilgrim C 2017 Effect of foundry waste on the mechanical properties of Portland Cement Concrete *Construction and building materials* **135** 674-81.

[23] Toutanji H, Delatte N, Aggoun S, Duval R and Danson A 2004 Effect of supplementary cementitious materials on the compressive strength and durability of short-term cured concrete *Cement and Concrete Research* **34** 311-9.

[24] Van der Lugt P, Van den Dobbelsteen AA and Janssen JJ 2006 An environmental, economic and practical assessment of bamboo as a building material for supporting structures *Construction and building materials* **20** 648-56.

[25] Vaughn SF, Kenar JA, Tisserat B, Jackson MA, Joshee N, Vaidya BN and Peterson SC 2017 Chemical and physical properties of Paulownia elongata biochar modified with oxidants for horticultural applications *Industrial Crops and Products* **97** 260-7.

[26] Vegas I, Frías M, Urreta J and San José JT 2006 Obtaining a pozzolanic addition from the controlled calcination of paper mill sludge. Performance in cement matrices *Materiales de Construcción* **56** 49-60.

Comparative review on mechanical properties of concrete with replacing curing water by self-curing compounds

BL Niranjan Reddy[1] **and M Vinod Kumar**[2]

[1,2] Vel Tech Rangarajan Dr. Sagunthala R&D Institute of Science, Chennai, India.

Abstract. The curing of concrete requires large amount of water. Self-curing concrete is one kind of concrete, which fix itself by holding water (dampness content) in it. A body of literature on the different self-curing compounds is uses for curing purpose without externally curing the concrete is currently available, but a systematic review is lacking. Therefore, this paper reviews the published literature on the use of different self-curing compounds in concrete and past work analysis on self-curing concrete. Various chemicals were uses to attain this curing. It was found out that various chemical admixtures like Poly ethylene glycol, Sodium lignosulphonate, Polyvinyl alcohol, Super absorbent polymer and normally accessible and regularly utilizes materials like Light weight aggregates, Wood powder and Light expanded clay aggregate were uses to fabricate self-curing concrete. This paper summarizes the collected literatures on replacement of curing water with self-curing compounds and how these affects the fresh properties and hardened (solidified) properties of concrete, furthermore, we expected that this work may advise further examination regarding the utilization of curing compounds to upgrade the properties of cement and concrete items.

Keywords: Light weight aggregates, Poly Ethylene Glycol, Polyvinyl alcohol, Sodium Lignosulphonate, Self-curing compounds.

1. Introduction

Self-curing concrete (SCC) is generally called internal curing concrete. " self-curing might be characterized as the procedure by which the hydration of concrete happens in view of the accessibility of extra inner water that isn't a piece of the blending Water [1]. The Characteristics of concrete is majorly considered in strength and durability point of view. Optimum strength depends on major parameter Curing. Therefore, it is important to cure for good hydration for at least 14-28 days and to meet the target intensity. Curing of concrete is the process for the preservation of moisture and temperature conditions for the normal hydration reaction is to be defined as the process for the concrete to develop durable properties over time [2]. Improper curing will quickly have an effect on concrete efficiency and longevity. Coming to High-performance concrete falling into the so called modern concrete category. One of the key issues with such high intensity concrete formulations is their propensity to crack at an early age [3]. Concrete with a high cement content and low water to cement ratio can result, not only in delayed strength production and reduced longevity but also in autogenous shrinkage and cracking, in an internal drying of concrete by consuming water through hydration, called self- desiccation [4]. The continued availability of inner humidity to maintain hydration reactions at moderate temperatures and when external forces are not available at an early age are essential to a proper curing [5], this is achieved with external curing after mixing, placing and finishing in conventional curing. Water is the most widely used construction material and because of regular use of water supplies in modern world, water shortage zones are gradually increased. Furthermore continuous cure is also very difficult with work done at heights, vertical components, sloped towers and floors [6].

Internal cure technique (IC) provides extra concrete moisture for better cement hydration and decreased self-desiccation; this is an extremely promising technique [7]. Conventional application eliminates water leakage from the concrete's surface even when internal treatment produces inner storage tanks to fulfill the hydration needs of the whole concrete structure and thus assume a stronger result [8]. The self-curing process is based on a chemical admixture / curing agent which can reduce water evaporation through a retaining function [9]. Many admixtures are used to achieve efficient curing outcomes for self-curing concrete such as poly-ethylene glycol (PEG), lightweight aggregates, poly vinyl acryl (PVA), sodium lignosulphonate (SL) and poly acryl amide (PAM), light expanded clay aggregate, wood powder and coco pith [1] . The writers have collected much of the available literature and are quickly explored in the context of descriptions of various properties. A review had been embraced to know the impact of using different self-curing agents in producing concrete equivalent to the conventionally cured concrete and presented. Besides, this paper at long last gives a viewpoint that may be useful for the further examination.

2. Types of Curing

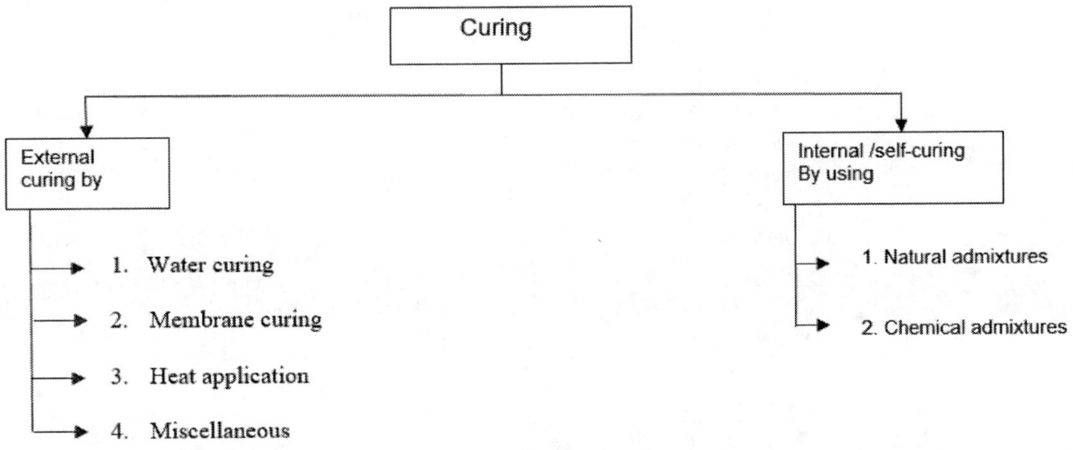

Potential materials used in self curing concrete:

Self-curing can be done by two types of admixtures 1) Chemical admixtures, 2) Natural admixtures.

Natural admixtures:
a) Wood powder
b) Light weight aggregates
c) Spinacea pleracea
d) Calatropis gigantean

Chemical admixtures:
a) Poly ethylene glycol (PEG)
b) Poly vinyl alcohol (PVA)
c) Poly acryl amide (PAM)
d) Sodium Lignosulphonate (SL)
e) Poly acrylic acid (PAA)

Table 1. Specifications of PAA [6]

properties	value
Average Molecular Weight	5000000 g/mole
Stability	Stable
Water Solubility	Soluble
Density at 25°C	1.189 g/ml
Shape and appearance	Powder
Colour	White

Table 2. Specifications of LECA [2]

Properties	Values obtained
Type	Crushed
Specific gravity	1.29
Fineness modulus	2.62
Bulk density (kg/m3)	669
Water absorption	16%

Table 3. Specifications of PEG-400 [21]

Properties	Values
Specific gravity	1.12 at 27°C
pH	>6
Molecular weight (g/mole)	400
Appearance	Clear liquid
Colour	White
Hydroxyl value (mg KOH/g)	300
Nature	Water soluble
Molecular formula	H(OCH2CH2)n OH
Density g/cm^3	1.125

Table 4. Specifications of spinacia oleracea and Calotropis gigantea [9]

Name of the admixture	pH value
Spinacia oleracea	6.59
Calotropis gigantea	5.17

Table 5. Specifications of SL [36]

Properties	Standard Values
Specific gravity	1.25
Appearance	Yellow Brown
Lignosulfonate	55%(min)
Dry matter	95%(min)
Water-insoluble	1.5%(max)
Water Reducing capacity	8% min
Sulphate	2%-5%
Calcium and magnesium	0.5% max
Moisture	7% max

3. Literature Survey

R.Malathy., (2017) Conducted experimental work on the new and hardened properties of organic self-curing concrete with 30% fly ash as part of a cement substitute. In this work mostly focuses on the action of internally cured concrete of M20, M30 and M40 by utilizing naturally available vegetative materials like Spinacea pleracea(SP), Calatropis gigantean(CG) and chemical admixture like polyethylene glycol(PEG) at 0.6%, 0.24% and 0.3%(by weight of cement) respectively. Self-curing concrete with these ideal values had a strength activity index more than 1 at 28 days and got 1.15 at 58 this is due to pozzolonic action of fly powder at later ages. This study concentrates on the study of concrete performance in aggressive environments, and all such tests have been positive compared to standard concrete. From the point of view of durability, the weight loss is 15%, 10% and 5% less than regular concrete for ideal values of SP, PEG and CG respectively. The reason behind this is that the cube's pore structure is tight and does not enable additional acid intake. Finally, it was concluded that the vegetative materials added along with poly ethylene glycol as self-curing agents execute good strength, workability and durability characteristics in fly ash mixed concrete, and also these synthetic materials are used in pavements, RCC related works, overhead tanks and pre-stressed concrete structures to accomplish long term strength and high performance.

NageshT.Suryawanshi and Sunil B.Thakare., (2018) is used as a combination of metakaolin as mineral admixture, poly-vinyl alchahol(PVA) and poly acrylic acid(PAA) as chemical admixtures, for self-curing concrete. The experiment was conducted on high strength concrete and a target mean strength of 61.51 N/mm^2 fixed by various trail mixes. In the study cement is replaced with metakoline at differing proportions between 5% and 15%, and the optimum value was found at 5% and up to 7.5% strength equals to control mix. This regulation mixture includes a variety of different amounts of poly-vinyl alchahol(PVA) and poly-acrylic acid(PAA) applied individually by weight of cement at different percentages of 0.02, 0.04, 0.06, 0.08 and 0.1. The tests have shown that Concrete unified with poly vinyl alcohol of dose 0.04% was found to produce optimum strength of 65.40 N/mm^2 and Concrete unified with poly acrylic acid of dose 0.02% was found to produce optimum strength of 64.20 N/mm^2.

Ahmad Mustafa Sabaoon and Navinderdeep Singh., (2019) The primary purpose of this study was to examine the mechanical properties of self-curing concrete by incorporating natural and chemical compounds as self-curing agents, such as wood powder and poly ethylene glycol 400(PEG-400). The concrete they used is M30 with a particular combination of wood proportion (2%, 4%, 6%, 7%, and 8%) and PEG proportions (0.5%, 1%, 1.5%, 2%, and 2.5%). Lastly, 6% wood powder was found to be an excellent test and again the constant percentage of wood powder (6%) with different PEG combinations was tested in room and outdoor temperature. It has been reported that at room temperature the blend of 6% wood powder with 1.5% PEG obtained excellent performance. In addition to the 6% wood powder and 2.5% PEG blend, the outdoor climate achieved optimum performance.

The work carried out by G. Thrinath, P. Sundara Kuma., (2017) on internal curing concrete by Polyethylene Glycol (PEG) as cement replacement would investigate the compressive and bending qualities of the M30 grade concrete. Cement is partially superseded by PEG-400 as 0.5 %, 1 %, 1.5 % and 2 % by weight of cement. The experimental program was accomplished for assess the compressive, split tensile and the flexural strength of concrete for given mix. 500 * 100 * 100 mm prisms are designed for 3, 7 and 28 days bending studies. As the percentage of PEG-400 increases, the workability of concrete is also rising. The study showed that the ideal PEG-400 dosage is 1% for compressive and split tensile strength, in the bending test 0.5%.

M. V. Jagannadha Kumar, K.Jagannadha Rao , B. Dean Kumar , V. Srinivasa Reddy.,(2018) The aim of this study is to discover an ideal mix proportion of self-curing concrete for different grades M20, M40 and M60 by supplementing poly ethylene glycol (PEG-400)as internal curing agent. The ideal measurements of PEG (communicated in rate by weight of concrete) embraced for M20, M40 and M60 grades internal cured concrete are 1%, 0.5% and 0.5% separately. In this study, various fresh and solidified properties at different ages such as 28, 60, 90, 180, and 365 days are evolved. The tests of

NDT, such as Rebound Hammer and Ultrasonic Pulse Velocity, are also used for the quality and properties, such as structure integrity and compression strength. It is observed that there is a reduction in the weight loss and better pore structure of concrete, as the grade and the ages of self-curing concrete with PEG are increased. It was concluded that rebound numbers indicated that the surface hardness of self cured concrete was superior when compared to traditionally cured concrete.

Daud Mohamad, Wani Mohd Sapuan., (2017) this paper exhibits the impact of using baby diapers polymer as an internal curing admixture in self curing concrete. In this process, baby diaper powder was used in different proportions (1%, 2%, 3 %, 5 % and 10 %) of cement weight. Various tests are carried out using 100mm*100mm*100 mm cube moulds on the new and solidifying characteristics of concrete. Hardened concrete characteristics are performed at room temperature at 3, 7, 28 and 90 days of air curing. With expanding the level of infant diapers flowability of concrete is expanded got maximum slump at 10%. Eventually it is found that at 1% diaper polymer concrete got optimum values in terms of compressive strength. Due to the polymer inside the concrete, the concrete that has a higher percentage of polymers had absorbed large amounts of water compared with the fewer polymers.

Cvk chaitanya, Priya Prasad, D. Neeraja, A. Ravitheja (proceedings) A study is being carried out in order to report on the use of pre-saturated lightweight aggregates(LECA) has been well established to counteract self-desiccation and autogenous shrinkage. In this research properties of concrete with M30 grade are carried out with different dosages of LECA (0%, 10%, 15%, and 20% by volume) as self-curing agent for water retention. They are concentrated primarily on the compressive, split and flexure properties and not involved on fresh and durability properties. Ultimately, the study concluded that the solidified properties of concrete are raised up to 15% of LECA. At 15% of LECA the compressive, split tensile and flexure strengths are raised by 1.67%, 1.29%, and 2.47% respectively compared to control mix.

Sanjay Kumar R, Suganya Devi K, (2016) deals with self-curing reinforced concrete beams enclosed with steel I section by utilizing Poly ethylene glycol (PEG-400) as internal curing agent and comparisons have been made between basic curing concrete beams. In this study, an attempt was also made to replace sand with barite powder ($BaSO_4$) by up to 30%. RC beams with size 150 mm x 200 mm x 1200 mm are enclosed with the ISMB 100 hot rolled steel section of 1 m long. Flexure strength tests are carried out on these beams after 28 days and have seen that the self-curing beam and self-curing beam enclosed with I section withstand 13 percent and 50 percent progressively extreme burden contrasted with the checking beam. What's more, going to the self-curing beam with barite powder was 17 per cent lower strength than checking beam. Since, barite powder was able to absorb water more easily.

Mohammad Balapour A, Weijin Zhao (2020) the essential objective of this assessment was to assesses the likely utilization of a novel lightweight total (LWA), spherical permeable responsive aggregate (SPoRA), delivered from squander coal bottom debris, for self-curing of concrete. SPoRA's engineering properties needed for concrete internal curing were evaluated including specific gravity, porosity, sphericity, water absorption, and desorption of water. In comparison with commercial LWAs, SPoRA has three main features which make it better: a round shape to expand the concrete workability, better absorption properties to assimilate and desorb water for inside curing, and a shiny surface with crystalline interior structure to give surface reactivity the cementitious grid while having attractive inner crystalline quality. Through X-ray processed tomography (XCT), the high porosity of SPoRA (39.6%–57.8%, by volume) was estimated and verified, XCT assessment additionally demonstrated that the pore structure of SPoRA is very much associated, permitting put away water to travel through the pore structure to the external surface during the self-desiccation of concrete. It can be inferred that SPoRA made from low- and high-calcium bottom ashes is successful LWAs which can be used in concrete for internal curing purposes.

Vaisakh G, Dr. M S Ravi Kumar, P Siva Bala, (2018) the primary goal of this research was examine the various properties of M50 grade self- curing concrete by using poly ethylene glycol(PEG-400) as internal curing agent. The impact of PEG content on quality pointers like compressive, split tensile and elasticity modulus and durability markers like electrical resistivity, water sorptivity, chloride permeability, and water sorptivity were concentrated by differing percentage of PEG 400 from 0 to 2% with 0.5% of span. As the PEG content increased, there is an increasing trend in the mechanical properties due to hydration reaching its completeness up to a PEG content of 1.5 %. In terms of compressive, tensile and elasticity modulus, the specimens exhibited less strength beyond 1.5 %. Moreover, with increasing PEG content to 1.5 %, water absorption decreases and then increases. Water sorptivity and permeability to the chloride also show similar trends. Concrete resistivity increases with PEG content increase to 1.5 % and decreases with further PEG increase.

S. El-Dieband, T. A. El-Maaddawy, (2018) conducted experiment on assessment of reinforcement corrosion protection of self-curing concrete by utilizing water-soluble polymers like poly ethylene glycol (PEG) and poly acryl amide (PAM). They performed numerous experiments such as electrical resistivity, permeability checking, chloride ion penetrability, and durability indices on concrete with separate dosages of PEG+PAM together were held at 0.025 % by cement weight. The PEG and PAM weight ratios were varied; the first ratio was 80% PEG and 20% PAM (SC-80PEG-20PAM), and the next ratio was 60% PEG and 40% PAM (SC-60PEG-40PAM). Rapid chloride ion penetrability (RCPT) and water permeability tests were conducted on disc specimens cut from concrete cylinders, with a diameter of 100 mm and thickness of 50 mm. The electrical resistivity was conducted on concrete cylinders with a diameter of 100 mm and a length of 200 mm. it is noticed that all the tests results are got better results at SC mixtures SC-100PEG and SC-80PEG-20PAM compared to SC-60PEG-40PAM.

Zaheer Ahmed, Harshith H J, Dr. Mushtaq Ahmed Bhavikatti and Akshatha B A., (2019) made analysis on the mechanical properties of self-curing concrete with the application of light expanded clay aggregate (LECA) as a natural mixture. In this study, coarse aggregates are replaced by light expanded clay aggregates up to 20% for M30 grade concrete. These light expanded clay aggregates are immersed in water around 24 hours while using LECA as a curing agent. This pre wetted aggregates are used for preparing moulds. They perform compressive cube strength testing by replacing LECA aggregates by 5 %, 10%, 15 %, and 20%. However, it has been found that low compressive strength is attained beyond 5 % replacement of LECA compare to control concrete.

T. A SajanaKhader and T.SShabana., (2018) investigate the Mechanical Properties of internal curing Concrete by adding Sodium Lignosulphonate(SL) and Poly vinyl alcohol(PVA) as self-curing agents.

M35 grade concrete is used at a rate of PVA 0.24%, 0.48%, 0.96% and SL 0.5%, 1%, 1.5%, 2% by weight of cement. PVA and SL can retain moisture for a long period of time. The investigation revealed that the mix of 0.48% PVA and 1% SL gives optimum values of mechanical properties when compared to all other mixes. And also in the view of durability aspect it gives better resistance against acid and chloride attack. Finally, it was concluded that Compressive strength and Split Tensile strength expanded by 23.94% and 50.7% respectively for 0.48% PVA and 1% SL at 28 days.

Table 6. A summary of the various research works published on internal curing concrete.

Author	Admixture	Quantity of admixture in percentages	Type of concrete and grade	Mechanical properties	Durability properties
Cvk chaitanya et al., [2]	LECA	0%, 10%, 15%, and 20% by volume of aggregate.	M30	Optimum results obtained at 15% of LECA.	
Vaisakh G et al., [4]	PEG-400	0.5%, 1%, 1.5%, and 2%.	M50	In terms of compressive, tensile and elasticity	Water sorptivity chloride permeability and

				modulus, the specimens exhibited optimum results at1.5 %.	Concrete resistivity also exhibits maximum results at 1.5%.
Sanjay Kumar.R et al., [7]	PEG-400 and barite powder	Barite powder-30%.		self-curing beam and self-curing beam enclosed with I section withstand 13 % and 50% more strength compared to control beam.	
R.Malathy et al., [9]	Spinacea pleracea, Calatropis gigantean and PEG	Spinacia oleracea and Calotropis gigantea at 0.6% and 0.24% respectively by weight of cement, and PEG-0.3%.	Flyash based concrete, M20, M30, M40.	Got better results compared to conventional concrete at optimum percentages of curing agents.	The percentage loss in strength is less when compared to conventional concrete is about 15% for Spinacia oleracea , 10% for PEG and 5% for Calotropis gigantean.
Ahmad Mustafa Sabaoon et al., [10]	Wood powder and PEG-400	Wood powder-2%, 4%, 6%, 7%, and 8% and PEG-0.5%, 1%, 1.5%, 2%, and 2.5%.	M30	The combination of wood powder and PEG at 6% and 2% respectively have beneficial outcome.	
Vaseem Akram.N et al., [11]	PVA	0%, 0.5%, 1%, 1.5%, and 2%.	M30	The properties of concrete got optimum results at 15% of PVA.	
B.Ajitha et al., [12]	PVA	0%, 0.03%, 0.06%, 0.12% and 0.24%.	M60	0.24% of PVA gives the optimum values of compressive strength and workability.	
NageshT.Suryawanshi et al., [13]	PVA and PAA	0.02%, 0.04%, 0.06%, 0.08% and 0.1%.	High strength concrete	Metakaolin by 7.5% and combine with PVA and PAA solely improve mechanical properties of concrete.	
BalaSubramanian et al., [14]	PEG-400	0.2%, 0.3% and 0.4%	Silica fume based concrete, M60, M70 and M80.	Got optimum results at 10% of silica fume and 0.4% of PEG.	Rapid chloride permeability and water sorpitivity values got better values at 5% of silica fume and 0.4% of PEG.
T.A SajanaKhader et al., [15]	SL and PVA	SL- 0.5%, 1%, 1.5%, 2% and PVA- 0.24%, 0.48%, 0.96%.	M35	Got optimum values of Compressive strength and Split Tensile strength at 0.48% PVA and 1% SL at 28 days.	Better resistance against acid and chloride attack by adding self-curing agents.

IOP Conf. Series: Materials Science and Engineering **925** (2020) 012008 doi:10.1088/1757-899X/925/1/012008

RiyazAhamed.K et al., [16]	SL	0.5%, 1%, 1.5%, 2%, 2.5% and 3%.	M-20	Optimum results obtained at 0.5% of SL.	
S. El-Dieband et al., [17]	PEG and PAM	PEG+PAM - 0.025% by weight of cement.	M25		Better indices of durability compared to traditional air-cured concrete.
G. Thrinath et al., [20]	PEG-400	PEG- 0.5%, 1%, 1.5%, 2%.	M30	Got optimum values for compressive and flexural strength at 1% and 0.5% respectively.	
M. V. Jagannadha Kumar et al., [21]	PEG-400	0.5%, 1%	M20, M40, M60.	M-20 grade concrete got optimum values at 1%, remaining both grade concretes got optimum values at 0.5%.	
Zaheer Ahmed et al., [22]	LECA	5%,10%, 15% and 20%.	Normal concrete, M30	By increasing LECA percentage the strength of concrete is gradually decreased.	
Vishnu T et al., [23]	Light weight fine aggregate and PEG-600	Light weight fine aggregate- 10%,20%,30% and PEG- 1%,2%,3%,4%	Normal concrete, M40, 0.39	At the mix of 20% light weight fine aggregate and 2% PEG got optimum results.	
Daud Mohamad et al., [27]	baby diapers	1%, 2%, 3%, 5%, and 10%.		At 10% and 1% of diaper polymer concrete got optimum values in terms of slump and compressive strength respectively.	

4. Research Gap

An exhaustive report was experienced by gathering a wide scope of research articles. By directing a comprehensive literature survey, it was identified that huge experimentation has been done on basic properties of concrete by adding self-curing natural and chemical admixtures. In spite of the fact that there has been broad work done on the mechanical properties of self-curing concrete with the assistance of moulds however there was no relevant research regards to the durability of concrete and reinforced moulds however there was no relevant research regards to the durability of concrete and reinforced structures like beams, columns and slabs. To utilize self-curing concrete, it is necessary to survey the behavior of the concrete in terms of durability and load carrying capacity of RC structures. This territory has not yet been investigated.

5. Scope for Future Research

Despite a broad assessment has been done in the basic properties of concrete. Future research is important to comprehend the advancement in durability of concrete and also to study the different properties of reinforced concrete structures.

IOP Conf. Series: Materials Science and Engineering **925** (2020) 012008 doi:10.1088/1757-899X/925/1/012008

6. Conclusion

Based on the literature review, following ends can be acquired:

- The solution to many problems posed by lack of curing water is self-curing / inner curing concrete.
- The self-curing material has a greater hydrating effect than traditional concrete under drying conditions.
- It is observed that when the quantity of poly ethylene glycol (PEG) increases the slump value rises.
- It has been observed that using 1- 2% of poly ethylene glycol (PEG) as chemical admixture in different grades of concrete gives strength equal to normal concrete and also rises the strength up to 8-15%.
- Self-curing concrete is more effectual of reducing autogenous shrinkage.
- The mix of 0.48% Poly vinyl alcohol (PVA) and 1% Sodium Lignosulphonate(SL) gives optimum values in point of strength and durability as compared to all other mixes.
- Some vegetative materials like Spinacea pleracea and Calatropis gigantean added along with poly ethylene glycol as self-curing agents perform better mechanical properties.
- From the study it is noticed that by adding LECA the compressive and flexural strength of concrete is decreased, but the workability of concrete is increased.
- To order to improve the properties of concrete, other chemical admixtures were also applied to concrete along with LECA.

Acknowledgement

It is with massive appreciation that I acknowledge the technical support and specialized assistance of Dr. M Vinod Kumar, associate professor at Vel Tech Rangarajan Dr.Sagunthala R&D Institute of Science.

References

[1] Khan RA, Gupta C 2020 A Review on the study of self-curing concrete *Journal of engineering science* **11**1-7.

[2] Prasad P, Neeraja D and Ravitheja A 2019 Effect of LECA on mechanical properties of self-curing concrete *Materials Today: Proc.* **19** 484-8.

[3] Bentur A, Igarashi SI and Kovler K 2001 Prevention of autogenous shrinkage in high-strength concrete by internal curing using wet lightweight aggregates *Cement and concrete research* **31** 1587-91.

[4] Vaisakh G, Kumar MR and Bala PS 2018 An experimental study on properties of m50 concrete cured using peg 400 *International Journal of Civil Engineering and Technology (IJCIET)* **9** 725–32.

[5] Dhir RK, Hewlett PC, Lota JS and Dyer TD 1994 An investigation into the feasibility of formulating 'self-cure'concrete *Materials and structures* **27** 606.

[6] Hameed AM, Rawdhan RR, Al-Mishhadani SA 2017 Effect of self-curing agents on the different properties of cement and mortar *International Journal of Current Research* **9** 54586-94.

[7] Sanjay Kumar R and Suganya Devi K 2016 Flexural behavior of self-curing concrete beam encased with steel section *International Journal for Scientific Research & Development(IJSRD)* **4** 1351-54.

[8] Han B, Zhang L and Ou J 2017 Smart and multifunctional concrete toward sustainable infrastructures *Singapore:: Springer.*

[9] Malathy R 2017 Experimental Study on Strength and Durability Properties of Bio-Self-Cured Fly Ash Based Concrete under Aggressive Environments *International Journal of Structural and Construction Engineering* **11** 328-33.

[10] Sabaoon AM and Singh N 2019 Experimental investigation of self-curing concrete by using natural and chemical admixtures *Indian Journal of Science and Technology* **12** 1-6.

[11] Vaseem Akram N and Balachandiran P 2018 Experimental study of self curing concrete by using poly vinyl alchahol *International Journal of Scientific & Engineering Research* **9** 102-05.

[12] Ajitha B and Nirupama G 2017 Estimation of desired properties of self curing concrete using admixture *International Journal of Engineering Technology, Management and Applied Sciences* **5** 581-89.

[13] Suryawanshi NT and Thakare SB 2018 Self-curing assessment of metakaolin based high strength concrete Using super absorbent polymer *International Journal of Civil Engineering and Technology (IJCIET)* **9** 1082-87.

[14] K. Bala Subramanian, A. Siva, S. Swaminathan and Arul. M. G. Ajin, "Development of high strength self curing concrete using super absorbing polymer", *International Journal of Civil and Environmental Engineering,* **Vol. 9**, pp. 1536-41, 2015.

[15] Sajana Khader TA and Shabana TS 2018 Experimental investigation on the mechanical properties of self curing concrete using sodium lignosulphonate and polyvinyl alcohol *International jouranl of innovative research in science, engineering and technology* **7** 2580-2593.

[16] Riyaz Ahamed K, Pradeep Kumar A, Priyadarshini D, Kalaivani K and Kingsta Beautlin M 2015 Experimental study on self curing concrete using sodium lignosulphonate *International Journal of Emerging Technologies and Engineering (IJETE)* **2** 74-78.

[17] El-Dieb AS and El-Maaddawy TA 2018 Assessment of reinforcement corrosion protection of self-curing concrete *Journal of Building Engineering* **20** 72-80.

[18] Maruthachalam D 2018 Experimental study on mechanical characteristics of self-curing concrete using natural fibre *International Journal of Pure and Applied Mathematics* **119** 1075-81.

[19] Dayalan J 2016 Compressive strength and durability of self curing concrete *International Research Journal of Engineering and Technology (IRJET)* **3** 1013-17.

[20] Thrinath G and Sundara Kuma P 2017 Eco-friendly self-curing concrete incorporated with polyethylene glycol as self-curing agent *International journal of engineering* **30** 473-78.

[21] Jagannadha kumar MV, Jagannadha Rao K, Dean Kumar B and Srinivasa Reddy V 2018 Effect of polyethylene glycol on the properties of self-curing concrete *Internationaljournal of engineering & technology* **7** 529-32.

[22] Ahmed Z, Akshatha BA, Bhavikatti MA and Harshith HJ 2019 An experimental study on internal curing of concrete using light expanded clay aggregate *International Journal of Research and Scientific Innovation (IJRSI)* **6** 1-5.

[23] Vishnu T and Beena BR 2016 An Experimental Investigation of Self-Curing Concrete Incorporated with Light Weight Fine Aggregate and Polyethylene Glycol *International Journal for Innovative Research in Science & Technology* **3** 116-122.

[24] Balapour M, Zhao W, Garboczi EJ, Oo NY, Spatari S, Hsuan YG, Billen P and Farnam Y 2020 Potential use of lightweight aggregate (LWA) produced from bottom coal ash for internal curing of concrete systems *Cement and Concrete Composites* **105** 103428.

[25] Karthik A, Sudalaimani K and Kumar CV 2017 Investigation on mechanical properties of fly ash-ground granulated blast furnace slag based self curing bio-geopolymer concrete *Construction and Building Materials* **149** 338-49.

[26] Memon RP, Mohd. Sam ARB, Awang AZ, Huseien GF and Memon U 2018 A review: mechanism, materials and properties of self-curing concrete *ARPN Journal of Engineering and Applied Sciences* **13** 9397-09.

[27] Mohama D, Beddu S, Sadon SN 2017 Performance of self-curing concrete using baby diapers polymer *International Journal of Applied Engineering Research* **12** 14687-92.

[28] Ghourchian S, Wyrzykowski M, Lura P, Shekarchi M and Ahmadi B 2013 An investigation on the use of zeolite aggregates for internal curing of concrete *Construction and Building Materials* **40** 135-44.

[29] Mahendra K and Bala Krishna P 2018 To study the strength characteristics of concrete by replacing curing water with self-curing compounds *International Research Journal of Engineering and Technology (IRJET)* **05** 2620-24.

[30] Hasholt MT, Jensen OM, Kovler K and Zhutovsky S 2012 Can superabsorent polymers mitigate autogenous shrinkage of internally cured concrete without compromising the strength? *Construction and Building Materials* **31** 226-30.
[31] Sandeep MV, Ravi Kiran K and Anukesh CH 2018 A study on effect of curing compounds on recycled aggregate concrete on strength and other properties *Indian J.Sci.Res.* **17** 528-35.
[32] Gopi R, Revathi V and Kanagaraj D 2015 Light expanded clay aggregate and fly ash aggregate as self curing agents in self compacting concrete *Asian journal of civil engineering (BHRC)* **16**1025-35.
[33] Ketter MK and Ong'ayo EO 2017 Compressive Strength Study of Self-Curing Concrete Using Red Clay and Black Cotton Soil Lightweight Expanded Clay Aggregates (LECA) *International Journal of Science and Research (IJSR)* **6** 1719-22.
[34] Kumar MM and Maruthachalam D 2013 Experimental Investigation on Self-curing concrete *International journal of advanced scientific and technical research* **2** 300-6.
[35] Dayalan J 2016 Compressive strength and durability of self-curing concrete *International Research Journal of Engineering and Technology (IRJET)* **03** 1013-17.
[36] Arundhava Priya E, Gopalan A and Mohanraj N 2019 Comparative study on self-curing concrete using sodium lignosulfonate and light weight aggregate *International Journal of Innovative Technology and Exploring Engineering (IJITEE)* **8** 309-11.

Erosion studies of Plasma-Sprayed NiCrBSi, Mo and Flyash Cenosphere coating

Nagabhushana N[1], Rajanna S[2] and MR Ramesh[3]

[1] Department of Mechanical Engineering, New Horizon College of Engineering, Bengaluru, Affiliated to VTU, Belgaum, Karnataka, India
[2] Department of Mechanical Engineering, Government Engineering College, Kushalnagar, Karnataka, India
[3] Department of Mechanical Engineering, National Institute of Technology Karnataka, Surathkal, India

Abstract. The erosion studies of plasma sprayed NiCrBSi / Mo / Flyash Cenosphere and NiCrBSi / Flyash Cenosphere on Superni 76 at room temperature for 30° and 90° impact angle were studied. The microstructure, adhesion properties, microhardness and porosity were investigated. Using SEM, EDS and EDAX techniques, the effect of phase change in the coating during the erosion was analyzed. The NiCrBSi/Mo/Cenosphere coating exhibits greater erosion resistance, and its loss of volume accounting nearly to 50% of the NiCrBSi/ Cenosphere coating. Compared to a 90° impact angle, the smallest erosion loss was observed at 30°. The erosion process assessed using SEM micrographs showed that the coating suffered ductile fracture, exhibited acute deformation and had abnormal oxide cracks. The enhanced metal oxide has a shielding effect, can resist erosion, and thus has better erosion resistance.

Keywords: Impingement, Molybdenum, Hard facing, Cenosphere, Harness ratio.

1. Introduction

The phenomenon used to remove particles from the surface by the impingement is called erosion. The NiCrBSi coating has a high wear resistance and low coefficient of friction, used for hard bearing surfaces, abrasive wear resistant coatings and hardfacings. Among spray materials, nickel-based alloys are widely used because they show oxidation, high-temperature corrosion, good wear resistance, and low cost. NiCrBSi alloy is one of the better performing alloys and is usually used in mechanical components, such as rollers in hot plate cooling stations, rolling mills and pump bushings [1]. However, for example, there is still a need to improve the wear resistance of these alloys. In fact, the tribological properties of this coating seem to be strongly influenced by the nature, presence, morphology and quantity of the second phase [2]. The composition of silicon and boron reduces the melting point of these alloys, and gives them "self-melting" characteristics, especially suitable for spray coating [3]. Thermal spraying method is widely used processes for coatings, and finds applications in industries to a great extent.

Among the deposition processes, of which the most commonly used are arc wire spray, high speed oxygen fuel (HVOF) and flame spray plasma spray processes [4]. The advantage of plasma spraying technology is that it can process a variety of low-grade ores to get high value-added products, and can deposit metals, ceramics, and respective mixtures to produce a near-uniform composite coating with desired microstructures on substrate. In fact, thermal spraying is a process in which the coating material is heated, melted and then pushed on to the substrate, forming coating. For this, metals, cermets, certain polymeric materials and ceramics can be used in the type of wires, rods or powders [5]. In spite of high temperature produced by the spray process, a portion of powder particles will melt incompletely because they spend short span during the stay in plasma stream. Due to drop in kinetic energy, temperature and oxidation cause uneven coatings with porosity more than 3% and reasonably poor adhesion between the substrate and the coating, this plays major role in tribological properties [6]. Porosity and micro cracks are undesirable micro structural characteristics as they enhance permeability in applications where corrosion resistance is required [7]. The coating quality based on several criterions, such as deposition rate, spray distance, angle (continuous or intermittent), combustion gas, deposition temperature, deposition technique, feed rate, process, sprayed particle size, the pressure applied and substrate temperature. All these criterions should be selected cautiously in order to achieve the most excellent coating performance for each purpose [8]. The best values of the above said criterions can provide the mechanical properties and best structure of the coating, which is necessary to meet the requirements.

In addition, few studies have added industrial waste, such as fly ash Cenosphere (hereinafter referred to as Cenosphere), to nickel-based composite coatings. Fly ash available in powder form from thermal power plants is available easily and treated as industrial waste. Fly ash is mainly composed of Fe2O3, SiO2, Al2O3 and traces of TiO2, Cao, etc. Among them Al2O3, SiO2, improves the resistance of wear of the coating. Using such industrial waste as a raw material for coatings can reduce the cost of coatings and minimize environmental pollution. This paper analyzes the structure, mechanical properties and corrosion properties of Cenosphere-Mo-NiCrBSi and Cenosphere-NiCrBSi coatings deposited by atmospheric plasma spraying (APS).

2. Materials and Methods

2.1. Substrate

In the current work, Superni 76 alloy (ASTM Grade- Hastelloy X, Composition- Ni-18.5Fe-21Cr-2.5Co-9Mo-1Mn-0.1C-0.5W) is used as substrate. Superni 76, cold rolled plates are used in manufacturing turbine blades is purchased from Mishra Dhatu, Hyderabad. Sample of 25x25x3 mm^3 and 25x20x3 mm^3 dimensions were obtained from the plates with the help of milling machine and sharp edges removed by grinding. The powders NiCrBSi, Mo and Cenosphere with their weight percentages are blended using planetary ball milling spherical balls of tungsten carbide with a speed of 300 rpm, 10mm diameter for 10-minute interval to have correct mixture. The density of the coating is measured using pycnometer. Figure 1 presents the mechanically blended powder mixture of uniform distribution of Cenosphere and Molybdenum in NiCrBSi network clearly shows the Suitability and possibility of blending adopted mechanically.

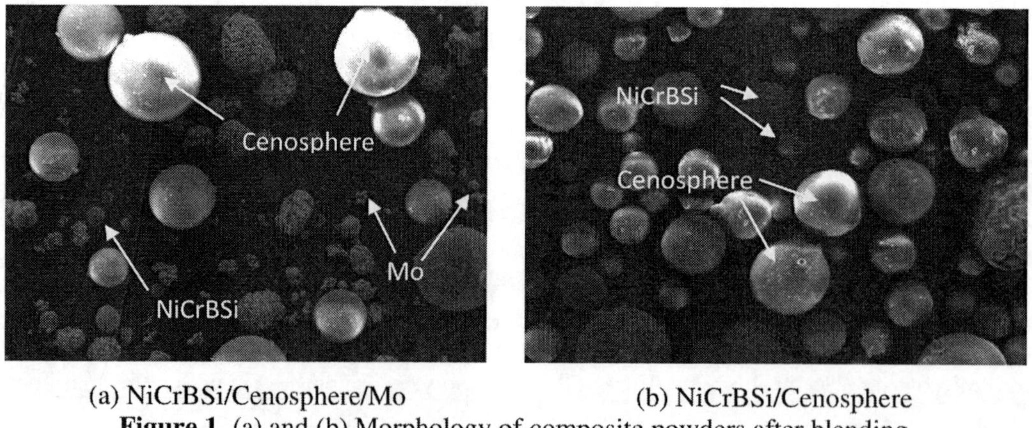

(a) NiCrBSi/Cenosphere/Mo (b) NiCrBSi/Cenosphere

Figure 1. (a) and (b) Morphology of composite powders after blending

2.2. *Coating deposition*

To ensure good bonding between substrate and coating, the surface has been grit blasted with alumina particles of 150μm grit size. The two powders, such as NiCrBSi/20%Mo/20%Cenosphere and NiCrBSi/40%Cenosphere, are deposited over the substrate. Plasma spray is carried out using METCO USA 3 MB equipment. In plasma spray, the feeder supplies the powder mixed with argon gas from compressor at desired pressure. In turn the powder melts in the plasma zone and gets deposited over the substrate. Coating deposited 20-25μm in each pass with deposition efficiency of around 40%. The thickness of the coating is measured using SEM. The variables selected throughout spray deposition are recorded in Table 1. JOEL-JSM-6380LA scanning electron microscope (SEM) used to examine the morphology of the coatings. ARTRAY, AT 130, JAPAN microscope is used to find out the porosity percentage.

In case of NiCrBSi/20%Mo/20%Cenosphere coating, due to the presence of Mo (20%), unmelted particles and as well as pores are observed as shown in Figure 2(a). Molybdenum forms MoO2 in the plasma zone which is well is known anti friction material. MoO2 has lower coefficient of friction than that of pure molybdenum. It also has the ability to prevent expansion and softening in extreme temperature environment. White region indicating NiCrBSi splats and grey region showing Mo phases. In the same way, Cenosphere particles are dispersed along the Ni alloy splat boundaries and become visible as dark grey region.

(a) NiCrBSi/Cenosphere/Mo (b) NiCrBSi/Cenosphere

Figure 2. (a) and (b) SEM micrographs of top surface of composite coatings

Table 1. Plasma spray parameters

Argon	Pressure	100-120 psi
Hydrogen	Pressure	50 psi
Current		1350 amps
Voltage		60-70V
Powder feed		120-150 gm/min
Stand of Distance		100-125 mm

2.3. *Erosion Tests*

ASTM G76-13 standard is used to carry out erosion tests using solid particle air jet erosion instrument (Ducom-TR-471-800) at room temperature. Diagrammatic presentation of erosion testing machine is shown in Figure 3, and conditions of erosion testing are listed in Table 2. The samples were acetone cleaned, weighed using electronic balance with least count of 0.001g. The erosion testing is carried out for five cycles of five minutes each and for respective angle of impact. Once after each cycle, the eroded surfaces are cleaned using acetone, dried and found weight loss. The erosion rate is assessed by the ratio of mass loss of coating to mass loss of erodent. This procedure is continued until the gradual erosion rate accomplished a sustained value independent of testing time or erodent particles. The sustained value of incremental erosion rate is defined as steady state erosion rate. Volume loss due to erosion is determined by the difference of volumes before and after erosion.

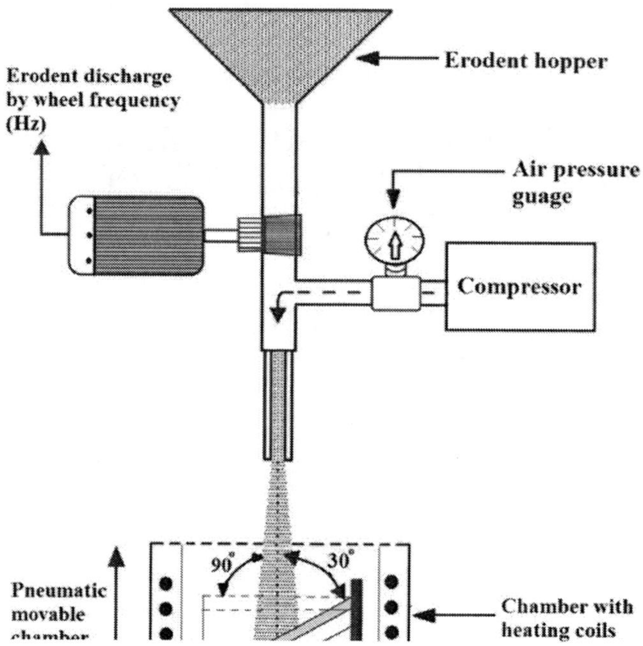

Figure 3. Schematic of erosion test rig

Table 2. Erosion test parameters.

Test Parameters	Variables	
Impact Angle in Degrees	30	90
Temperature in ^{0}C	Room Temperature	
Erodent Velocity in m/s	30	
Test Time (mins)	5	

3. Results and Discussion

3.1 Coatings Characterization

The microstructure of coating across the cross section of NiCrBSi/20%Mo/20%Cenosphere and NiCrBSi/40%Cenosphere is presented in Figure 3(a) and Figure 3(b). The individual splats produced by the melted, partially melted coating powders and their morphology are shown in Figure 4(a) and Figure (b) and are well bonded to the substrate. The coating thickness across the section is in the range of 250-320 µm. The percentage porosity of NiCrBSi/20%Mo/20%Cenosphere and NiCrBSi/40%Cenosphere coating is measured and found to be 2.90±0.20 and 3.70±0.21%, respectively. Figure 5(a) and 5(b) depicts EDS analysis and also element and percentage of weight, which shows splats rich in oxygen co-exists with silicon, iron and aluminium showing respective oxides. Chromium and nickel are distributed uniformly surrounding molybdenum and Cenosphere.

IOP Conf. Series: Materials Science and Engineering **925** (2020) 012009 doi:10.1088/1757-899X/925/1/012009

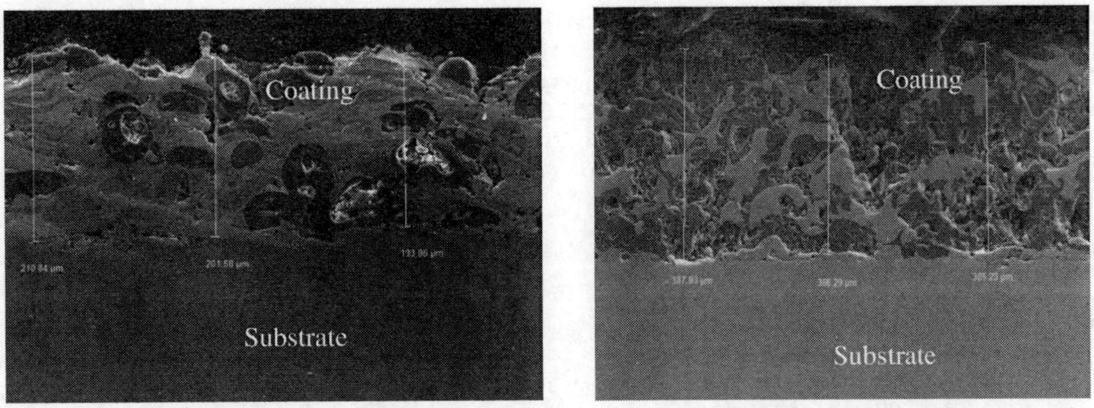

(a) NiCrBSi/Cenosphere/Mo (b) NiCrBSi/Cenosphere

Figure 4. Cross-sectional SEM micrographs of composite coatings

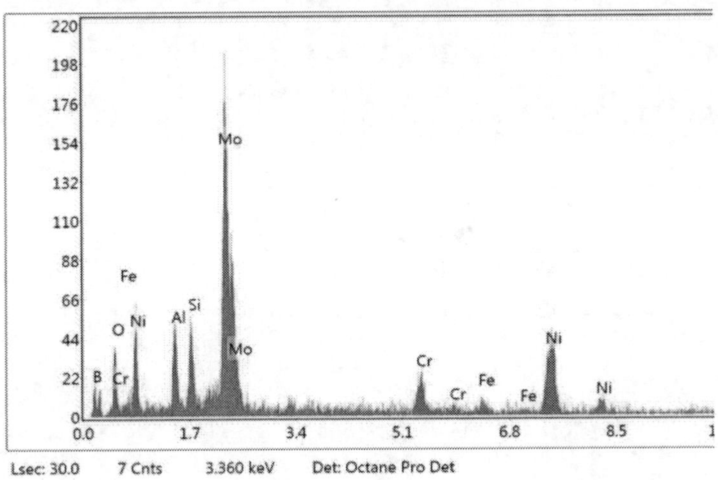

Element	Weight %
B	0.71
O	21.00
Al	6.46
Si	5.21
Mo	39.47
Cr	5.11
Fe	1.87
Ni	20.17

(a)

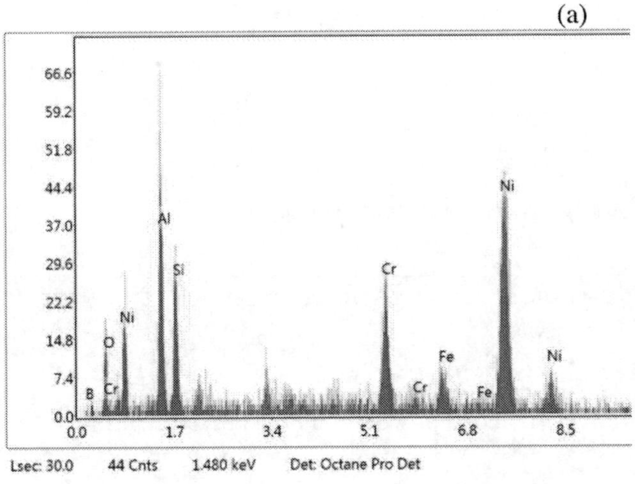

Element	Weight %
B	38.55
O	8.01
Al	9.01
Si	4.44
Cr	7.43
Fe	3.71
Ni	28.85

(b)

Figure 5. (a) and (b) EDS analysis of NiCrBSi/Mo/Cenosphere with respective oxide percentages

3.2. XRD analysis

XRD model of the as-coated and powder samples of NiCrBSi/20%Mo/20%Cenosphere and NiCrBSi/40%Cenosphere composites is exhibited in Figure 6(a) and Figure 4(b), respectively. The phases in powder and as-sprayed coating were analyzed using x-ray diffractometer (DX GE-2P, JEOL,

JAPAN). For NiCrBSi/20%Mo/20%Cenosphere, the major peaks found in as-coated and powder patterns corresponds to Ni3B, MoNi4, Al8Mo3 and MoS2 and similarly minor peaks corresponds to NiCr2O4, Cr7C3, Sio2, Al2O3 and BMO2. In the case of NiCrBSi/40%Cenosphere the major peaks in powder and as-coated are SiO2, Al2O3, Ni2Si, Cr7C2, Ni3B and NiCrO4, the minor peaks correspond to Al2O3, Ni2Si, Cr7C2, Ni3B, NiCrO4. In case of NiCrBSi/20%Mo/20%Cenosphere, during oxidation Mo reacts with oxygen to form MoO2, which acts as a good self-lubricating agent.

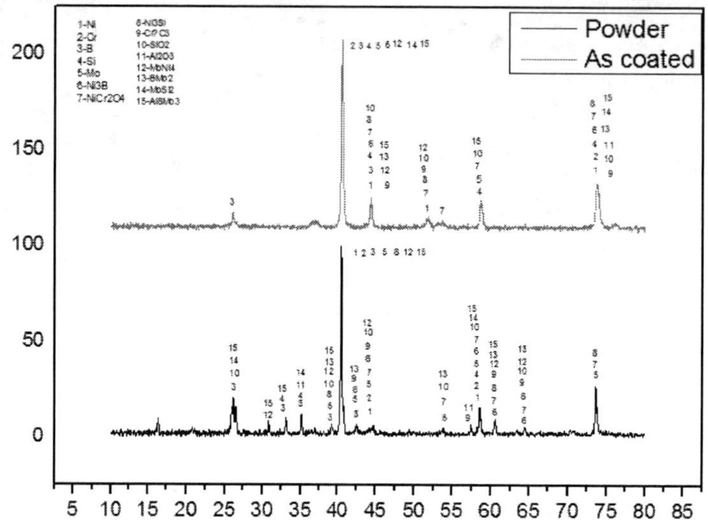

(a) NiCrBSi/Cenosphere/Mo

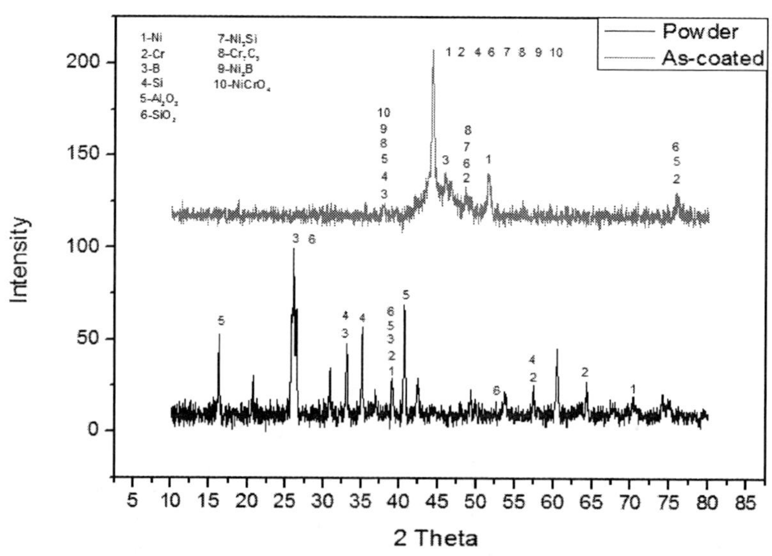

(b) NiCrBSi/Cenosphere

Figure 6. (a) and (b) X-ray diffraction patterns of blended powders and as sprayed coating

3.3 Micro hardness measurement

It is reported that in flight time is more in plasma spray than other spray processes due to which metal oxides are formed during oxidation. The hard reinforcement Al2o3 and SiO2 phases stay as it is even after Plasma spray. Micro hardness of coating and substrate is shown in Figure 7 and factors affecting microhardness are, non-uniformity in the cross section due to porosity, partially melted, melted and micro structural in homogeneity of the coating powders. The microhardness calculated with Clemex Vickers tester RS-232 (9-25 pins) under 200 g loads and a dwell time of 15 s. Average hardness value of NiCrBSi/20%Mo/20%Cenosphere and NiCrBSi/40%Cenosphere coatings was measured as 617 and

825 HV, respectively. It is noticed that the substrate hardness shows steady increment towards the interface due to strain hardening of grit blasting before coating and also compressive stresses induced at the interface during the impact of molten or partially melted particles in the spray process.

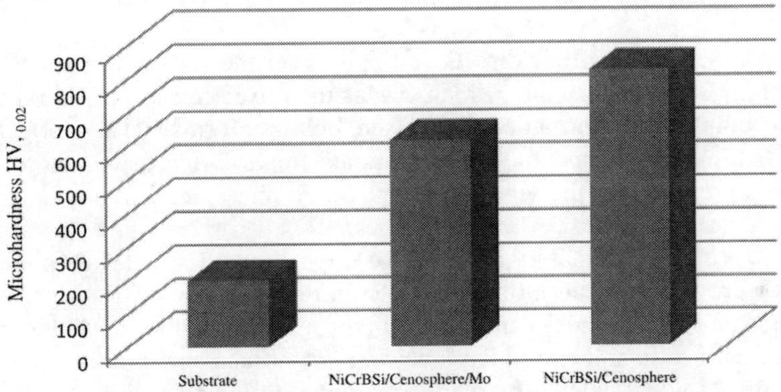

Figure 7. Microhardness of substrate and coatings

3.4 Erosion Behaviour

Erosion primarily depends on mass and number of particles impacting the surface with impact velocity. Brittle fracture and plastic deformation are the mechanisms by which the material is removed from the surface. Also depend on impact angle, velocity, environment and particle size. Therefore, if dominates plastic deformation, at low angles seen maximum wear. Similarly, at high impact angles brittle fracture occurs in which wear occurs at maximum. At low angles, Impact with round particle yield surface ploughing, but in case of angular particles yields are cutting. At 900 the damage usually occurs by chipping and cracking of surface material due to low strength of inter splats.

The erosion rate is obtained by the ratio of weight loss in the coating material to mass of eroded particles. During the initial cycle, the erosion rate was higher, and it was found that in subsequent cycles, the wear rate decreased. Sunderarajan and Roy [9] revealed that the erosion rate on the impact angle depends largely on the properties of the target.

Effect of volume loss and erosion rate with respect to two coatings is presented in Figure 8. The erosion volume loss at 90° impact angle is higher than that at 30° implying brittle mode of erosion. At lower impact of 30°, an erodent particle slide on the surface and ploughs the material.

Figure 8. Erosion wear rate and volume loss of coating at 30° and 90° impact angles.

3.5 Erosion Mechanism

At an impact angle of 30°, when the eroded particles hit with a high tangential force, severe plough marks appeared on the eroded surfaces of the coatings as shown in Figure 9(a) and 9(b). Plough, micro-cuts, and pits were found on the surface of eroded sample; substantiate the ductility and the removal of

IOP Conf. Series: Materials Science and Engineering **925** (2020) 012009 doi:10.1088/1757-899X/925/1/012009

the material in platelet form. Large groves are formed by the impact of angular erodent with large particle size by ploughing mechanism. Ploughing mainly occurs in the softer nickel-based matrix area, and micro-cutting is caused by the influence of tiny corner erosion particles.

At 90° angle of impact, erodent particle directs vertically over the surface as shown in Figure 9(c) and 9(d). The constant impact creates indents, which causes the raised lips around the indentation. At high strain rate because of plastic deformation the raised lips are formed. The protective effect of the reinforcing agent (Mo) is weakened by the continuous impact of erosion, thereby weakening the boundary of the crack, resulting in the separation of the hard phase and leaving the crater.

Maximum volume loss experienced at higher angle of impact resulting in brittle mode of material loss. Cracks and craters were found because of localised strain resulting in brittle fracture. Since one-third area is bounded in plasma coating cracks are formed in the interface of hard particles and matrix. Further progress of erosion leads to lateral and radial cracks at subsurface and surface of coating. The cracks formed interlock with the non bonded area, leads to removal of lamellae.

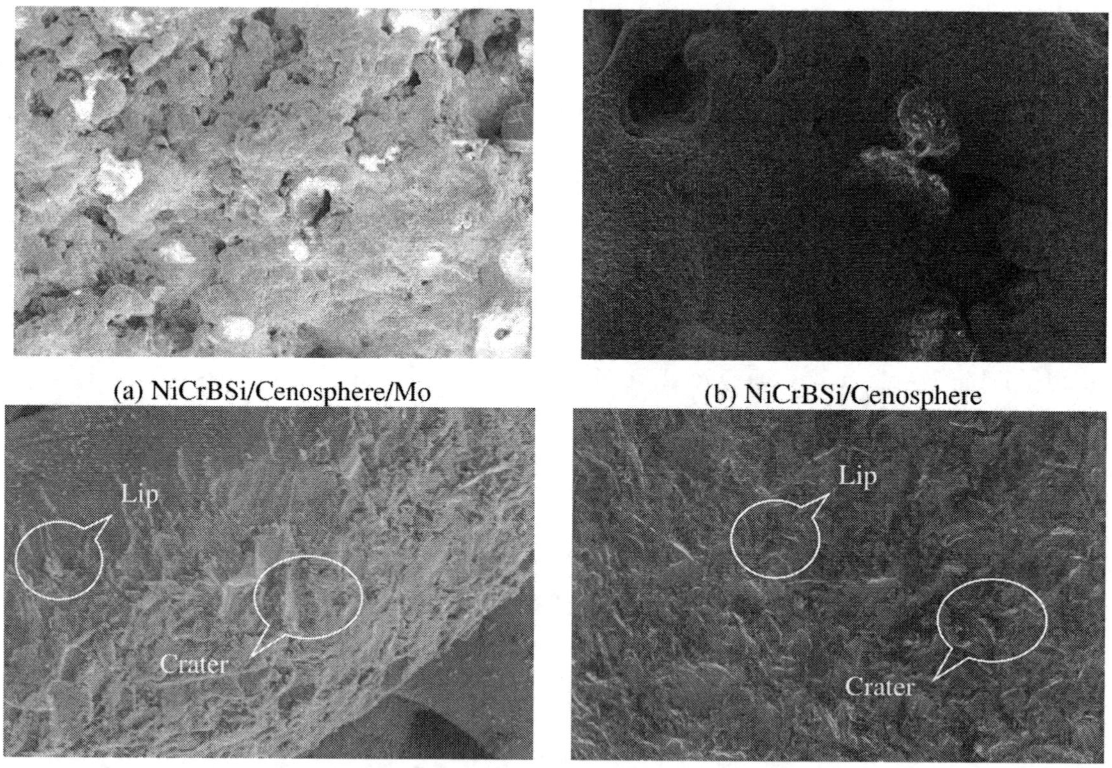

(a) NiCrBSi/Cenosphere/Mo (b) NiCrBSi/Cenosphere

(c) NiCrBSi/Cenosphere/Mo (d) NiCrBSi/Cenosphere

Figure 9. SEM images of eroded surface at impact angles of 30° (a, b) and 90° (c, d).

4. Discussion

From the Figure 6d it is noticed more amorphous nature is accomplished in NiCrBSi / 40% Cenosphere coating. The high strain generated during the erosion is conducive to the formation of shear bands. In case of NiCrBSi / 20% Mo / 20% Cenosphere coating, MoO2 can be used as a surface active element together with other molten particles, thereby greatly reducing the surface tension by dispersing along the splash boundary. Due to surface tension good wettability of MoO2 formed which enhances bonding between splats yield in superior erosion performance of NiCrBSi / 20% Mo/ 20% Cenosphere coating. Found less erosion volume in NiCrBSi / 40% Cenosphere coating as compared to that of NiCrBSi / 20% Mo / 20%Cenosphere.

5. Conclusions

1. Plasma spraying is used to obtain thick coating with thickness of 210 and 375 μm respectively for NiCrBSi/20%Mo/20%Cenosphere and NiCrBSi/40%Cenosphere coatings.
2. Found the erosion rate is minimum at 30° angle of impact and is maximum at 90° when alumina is used as erodent.
3. Mo surface activity is the most important factor, because it minimizes the uniformity of similar molten particles in the spraying process and make the particles dispersed along the splat boundary of the coating.
4. Thus, NiCrBSi/20%Mo/20%Cenosphere coating exhibited good resistance to volumetric erosion at both 90° and 30° impact angle than NiCrBSi/ /40%Cenosphere coating.
5. 5. At lower impact angles, plough, micro-cut, and hard-phase pullout features were observed on the eroded surface of the coating, indicating that ductile type of fracture.

References

[1] González R, Cadenas M, Fernández R, Cortizo JL and Rodríguez E 2007 Wear behaviour of flame sprayed NiCrBSi coating remelted by flame or by laser *Wear* **262** 301-307.

[2] González R, Cadenas M, Fernández R, Cortizo JL and Rodríguez E 2007 Wear behaviour of flame sprayed NiCrBSi coating remelted by flame or by laser *Wear* **262** 301-307.

[3] Planche MP, Liao H, Normand B and Coddet C 2005 Relationships between NiCrBSi particle characteristics and corresponding coating properties using different thermal spraying processes *Surface and Coatings Technology* **200** 2465-2473.

[4] Vencl A, Mrdak M and Cvijović I 2006 Microstructures and tribological properties of ferrous coatings deposited by APS (atmospheric plasma spraying) on Al-alloy substrate *FME Transactions* **34** 151-157.

[5] Samur R and Çetiner BG 2005 Quantitative determination of microcracks on ceramic coatings by computer image analysis *Metalurgija* **44** 41-44.

[6] González R, García MA, Peñuelas I, Cadenas M, del Rocío Fernández Ma, Hernández Battez A and Felgueroso D 2007 Microstructural study of NiCrBSi coatings obtained by different processes *Wear* **263** 619-624.

[7] Felgueroso D, Vijande R, Cuetos JM, Tucho R and Hernández A 2008 Parallel laser melted tracks: Effects on the wear behaviour of plasma-sprayed Ni-based coatings *Wear* **264** 257-263.

[8] Gómez-del Río T, Garrido MA, Fernández J.E. Cadenas M and Rodrígueza J 2008 Influence of the deposition techniques on the mechanical properties and microstructure of NiCrBSi coatings *Journal of Materials Processing Technology* **204** 304-312.

[9] Sundararajan T, Kuroda S and Abe F 2004 Steam Oxidation Studies on 50Ni-50Cr HVOF Coatings on 9Cr-1Mo Steel: Change in Structure and Morphology Across the coating/ Substrate Interface *Mater. Trans.* **45** 1299-1305.

[10] Nagabhushana N, Rajanna S, Mathapati M, Ramesh MR, Koppad PG and Reddy NC 2019 Microstructure and tribological characteristics of APS sprayed NiCrBSi/flyash cenosphere/Cr2O3 and NiCrBSi/flyash cenosphere/Mo composite coatings at elevated temperatures *Materials Research Express*. **6** 086451.

[11] Nagabhushana N, Rajanna S and Ramesh MR 2020 Microstructure and mechanical properties of the Mo/Flyash/NiCrBSi coating deposited by atmospheric plasma spraying *International Journal of Mechanical and Production Engineering Research and Development* **10** 16-22.

IOP Conf. Series: Materials Science and Engineering **925** (2020) 012010 doi:10.1088/1757-899X/925/1/012010

Mode-I fracture toughness evaluation of bio based short areca fibers reinforced polymer composites

GB Manjunatha[1], KN Bharath[2], H M Harsha[3] and P Sivaprakash[4]

[1,3,4] Department of Mechanical Engineering, GM Institute of Technology, Davanagere, Karnataka, India.
[2] Department of Mechanical Engineering, GM Institute of Technology, Davanagere, Karnataka, India.
*Corresponding email: manjugb4444@gmail.com

Abstract. Naturally accessible materials have recently become appealing to researchers and scientists due to their suitability as an ideal reinforcement for polymer composites reinforced with fiber. Low cost, relatively strong mechanical properties, qualities of non-abrasive and bio-degradability, exploited as a switch to the standard fiber. This research investigates the effect of areca fiber reinforcement of fracture load and fracture toughness. Epoxy resin used as a matrix material. Type I fracture hardness checks are conducted in a universal servo hydraulic control system. The findings revealed a significant improvement in fracture load values and durability of fracture with improved fibre content.

1. Introduction

Natural fibre-reinforced polymer composites become more relevant as environmental consciousness and compel the industries to look for biodegradable and recyclable materials [1]. Natural fiber comprises numerous ingredients such as cellulose, lignin, pectin, and other products. We have distinctive signature for the inclusion of such elements. The use of natural fibers to reinforce the material in epoxy is therefore a simple and positive approach to boost the mechanical properties of composite materials. Many research on the application of natural fibers coir, hemp, rice husk, jute, bamboo, cotton, wood, etc. as the reinforcements in composites have been mentioned in literature [2]. Composites strengthened by areca / betel nut fiber consider greater benefits in the new production of composite products such as electrical insulation applications [3]. Different materials are manufactured, and flaws are not completely preventable on a regular basis. The presence of faults contributes to the proliferation of cracks and structural fractures. Brittle fracturing produces sudden and devastating failures and may trigger major accidents. It is therefore important to research the actions of a defective substance. Predicting the failure of materials having flaws fracture mechanics has been introduced [4, 5]. Due to their complexity during performance prediction, fracture is one of the most challenging distresses. Heterogeneous structure, nonlinear viscoelastic material characteristics, specific loading conditions, and attachment of the fiber matrix are very problems for using basic research and techniques. In fracture mechanics, essential to consider the basic rules such as energy accumulated below the crack tip reaches the material's crack resistance, then cracking initiates in the vicinity of the crack tip. Using Linear Elastic Fracture Mechanics (LEFM) principles, a crack initiation begin with the Fracture Plastic Zone (FPZ) near the crack tip is studied [6]. Mode I and Mode II deformation can occur for a cracked part that is subjected to in-plane loading. The broken systems often undergo mixed-mode loading under realistic circumstances, i.e. a Combination of mode I and Mode II. Brittle fracture in mixed mode is one of the most frequent forms of mechanical failure in broken framework built from brittle materials. The structural integrity of the fractured component under mixed loading mode is indeed important [7]. Researchers also recommended various theoretical and experimental approaches for exploring brittle fracture in mixed fashion. Although experimental fracture studies on individual components are always costly and complicated, researchers tend to perform their experiments on specimens from laboratory. However, under their complex service loading conditions, acceptable fracture parameters are often required to equate the experimental results obtained from the basic laboratory experiments to the fracture case in cracked structures. Suitable research conditions may verify the fracture criteria by conducting a sequence of tests on appropriate test materials [8,9]. For this research, notched semi-circular bending (SCB) test was designed to illustrate the fracture properties of the areca-fibre-reinforced polymer composite. SCB test was initially used to identify the characteristics of fracture resistance. Nowadays more and more studies have used the SCB method to study the susceptibility to cracking [10].

The objective of this study is to fracture characterize that account for the variation in mode-dependent fracture resistance of areca fiber reinforced polymer composites through notched SCB and SCAB test. These tests produced maximum load. Furthermore, the relationship between fracture toughness and different composition were found.

2. Materials and Experimentation methods

To fulfil the objective of this research work, the fracture behaviour areca fiber reinforced polymer composites (AFRPC) manufactured with three different compositions were analysed.

2.1. Materials

Areca is empty fruit bunch Fibers (husk) were obtained from local Farm House near davanagere, Karnataka, India. The Epoxy resin L-12 and the hardener (K6) are supplied by Yuje Enterprises Malleswaram, Bangalore.

2.2. Methods

2.2.1 Fiber extraction

Areca fiber was used to fabricate the composites. About five days, dry areca husk was drenched in deionized water. The process of soaking releases the fibers and these can be easily taken out. Lastly, the fibers were splashed with deionized water and again dried for almost 15 days at room temperature.

2.2.2 Alkali Treatment

First, the areca fibers were processed in a 10% NaOH alkali solution, where the overall solution amount was used. The fibers are kept at room temperature in an alkaline solution for 36 hours. Instead, at room temperature, these fibers were dried for 48 hours to get alkali-treated fibers [11].

2.2.3 Preparation of composites

Volume fraction of fiber is one of the most significant factors that affect the properties of the composite. The randomly dispersed areca husk fibers with epoxy resin for different compositions are used to prepare the composite as shown in table 1. Initially; the mould is refined with wax as releasing agent on the surface to easy abstraction of composite after curing. Epoxy resin is mixed with hardener in a ratio of 10:1 by weight. Methodically this can be mixed with reinforcement areca fiber before filled into mould and compressed at the room temperature in a hydraulic press about 30 minutes [12]. Specimens were removed from the mould after 24 hours of curing in the laboratory at standard atmosphere. The process of fabrication of composite material is as shown in 'Figure.1'.

Table 1. Composition of Testing Specimens

Sl.no	Composition	
	Areca Fiber	Epoxy Resin
1	40	60
2	50	50
3	60	40

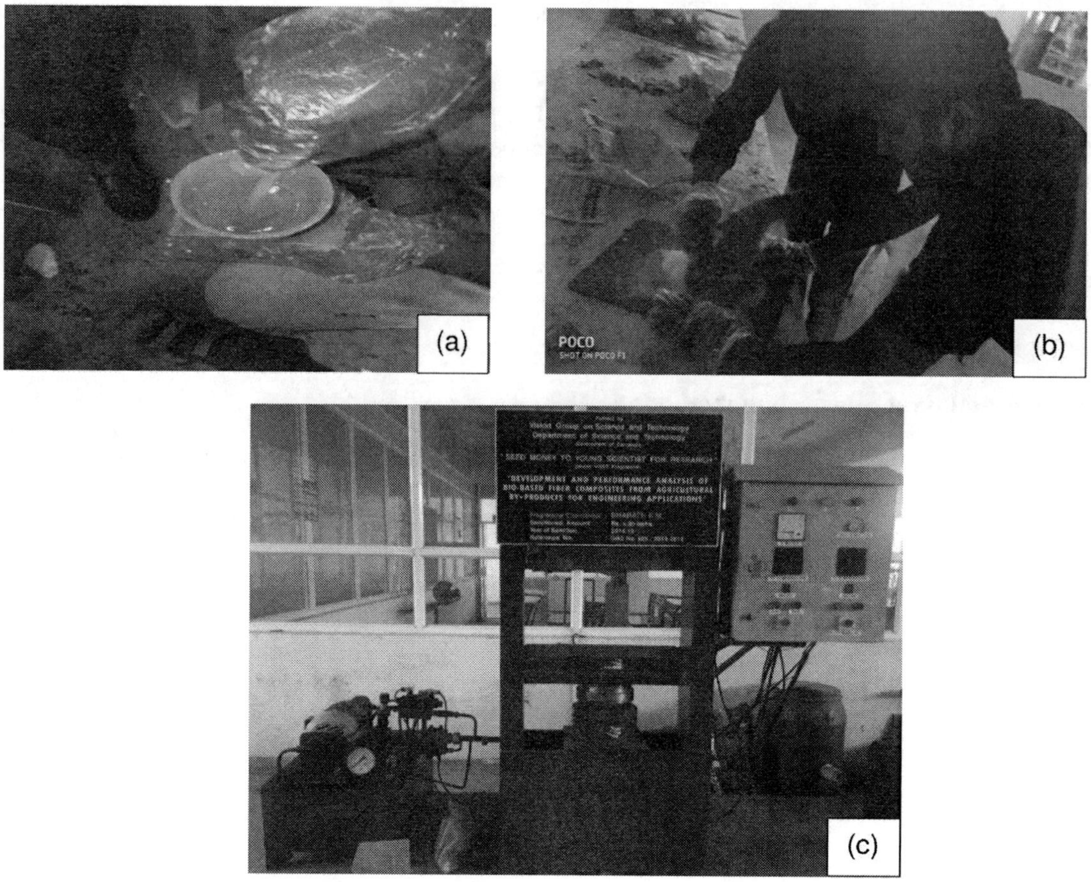

Figure 1. Fabrication of laminated composite: (a) Mixing of resin with hardener (b) Mixing of areca fibers and resin (c) Hydraulic hot press machine

2.3. Fracture toughness test

2.3.1 Semi-circular Bend (SCB) and Semi-circular Arc Bend (SCAB) Test

Semi-circular bend (SCB) specimen configurations are advised by various test standards for the toughness test in mode I, mode II and mixed mode I / II fracture. For the classical SCB specimen, one major problem is related to the practical complications of creating a crack in the specimen. SCAB is also appropriate for mode I, II and mixed-mode experiments and this form of specimen configuration will save more than 50 per cent of the material relative to SCB specimens [12]. The SCB and SCAB test design are seen in 'Figure.2'.

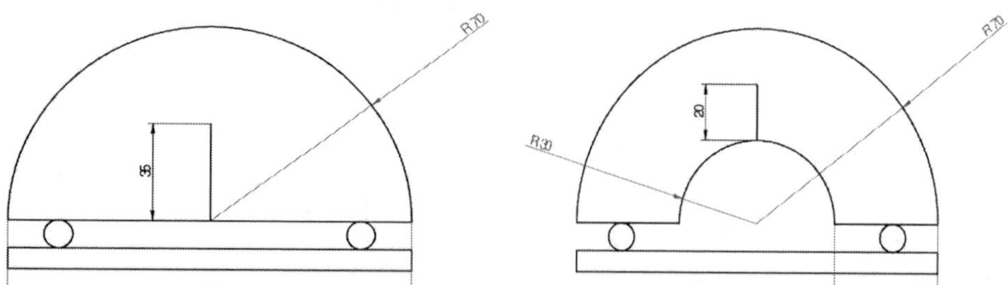

(a) (b)

Figure 2. Standard Specimen Configuration (a) SCB (b) SCAB

(a) (b)

Figure 3. Experimental setup for (a) SCB (b) SCAB

Fracture toughness testing is conducted in computerized universal testing machine (UTM) at room temperature. Single notched SCB and SCAB specimens are introduced for fracture experimentation to this end. 'Figure.3' shows the geometry of the SCB and SCAB specimen experiment setup. During the tests the failure load data is recorded. All of the test samples quickly fractured with negligible non-linear deformation from the crack tip. It indicates the brittle behaviour of the samples tested on fracture [13]. The toughness of the fracture is measured using the fracture load derived from each test, and the findings are reported.

3. Results and Discussion

Throughout the present study, the toughness of the fracture is calculated for specific composition from the SCB specimen and SCAB specimen with (α=0 °), and the experimental findings were described in this portion.

3.1. Semi-circular Bend (SCB) Fracture Test

The SCB fracture method was used to obtain the fracture toughness of composite materials reinforced with areca fiber. The SCB fracture test is easy to execute and allows the preparation of test specimens simply through core moulds [14]. A lot of studies that have examined post-failure behaviours related to brittle material, where the failure rate tends to decrease before the separation takes place [15]. Experimental tests are used to determine the fractures toughness for different compositions of composites reinforced with areca fiber.

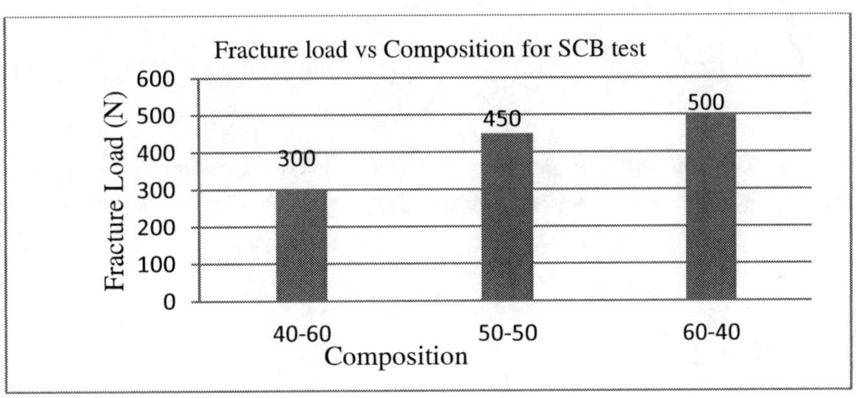

Figure 4. Fracture load vs. Composition for SCB specimens

IOP Conf. Series: Materials Science and Engineering **925** (2020) 012010 doi:10.1088/1757-899X/925/1/012010

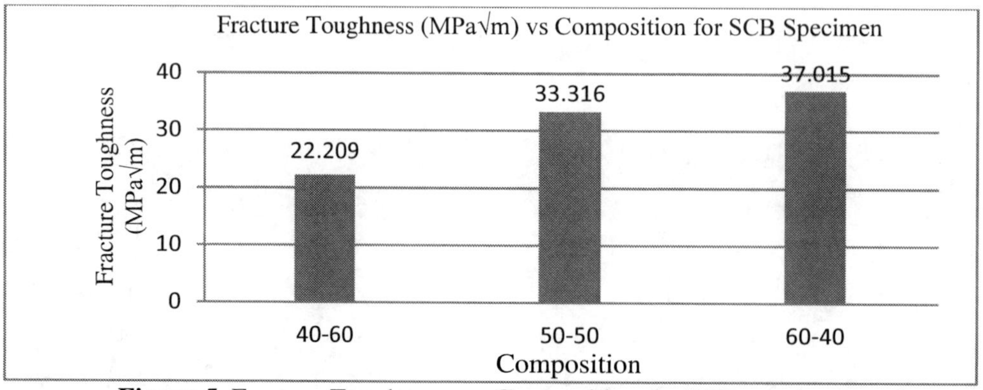

Figure 5. Fracture Toughness vs. Composition for SCB specimens

'Figure 4' presents the test results by plotting fracture load for different composition. Observations can be clearly noted from that; Fracture load was highest for 60-40 composition as compare to other compositions. SCB specimens were high fracture load to the increased reinforcements. This was estimated with the greater fracture load, more area to resist the failure of the material [16]. It also shown to have greater fracture toughness values at 60-40 composition, which illustrate the larger reinforcement, is more resistance to the cracking process [14]. It is seen that, as the percentage of reinforcement increases from 40% to 60%, the fracture toughness also increases.

3.2. Semi-circular Arc Bend (SCAB) Fracture Test

The fracture load results found in this research for different compositions of areca fiber reinforced composites are shown in "Figure.6". The 60-40 composition shows higher value of fracture load of 300N as compare to other two compositions. Specimen with larger reinforcement withstands more loads before fracture and more area to resist the failure of the material. The specimens failing by fracture indicated an increase in strength with increasing fiber composition. Fracture load is used to determine the fracture toughness of different composite material and it is shown in "Figure 7". It is clear that areca fiber reinforced Composite with 60-40 composition shows highest fracture toughness than the other two composite. It is seen that as the fracture toughness increases with increasing the percentage of reinforcement. Which shows that the larger reinforcement is more resistance to the cracking process [14]. This was also estimated with the greater fracture load, more area to resist the failure of the material [16].

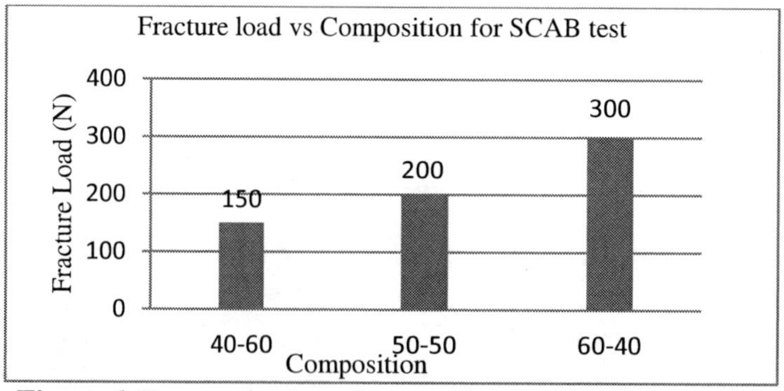

Figure 6. Fracture load vs. Composition for SCAB specimens

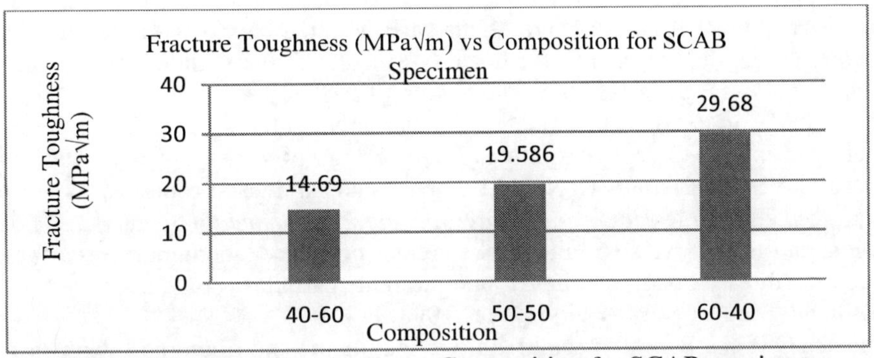

Figure 7. Fracture Toughness vs. Composition for SCAB specimens

4. Conclusion

In this study, the effects of different composition on the fracture properties of areca fiber reinforced polymer composites were investigated. A laboratory experiment was conducted to characterize the fracture resistance of AFRP composites with fibre volume percentages that ranged from 40 to 60%. The specimen had higher reinforcement had more significant effect on fracture toughness. The main findings are summarized as follows:

For Semi-circular Bend (SCB) Fracture Test:
Fracture load and fracture toughness increases with increasing reinforcement percentage. The (60–40) composition composite specimens give a failure load of 500 N and fracture toughness of 37.015 MPa√m, which is better than (40–60) and (50-50) composition composites.
For Semi-circular Arc Bend (SCAB) Fracture Test:
Fracture load and fracture toughness increases with increasing reinforcement percentage. The (60–40) composition composite specimens gives a failure load of 300 N and fracture toughness of 29.68 MPa√m, which is better than (40–60) and (50-50) composition composites.

References

[1] Hegde R, Kirthan L J, Girish V A and Girishkumar R 2018 Development and mechanical properties investigation of areca husk fiber reinforced polymer composite *International Journal of Mechanical Engineering and Technology* **9** 913–7
[2] Gope P C, Singh V K and Rao D K 2015 Mode i fracture toughness of bio-fiber and bio-shell particle reinforced epoxy bio-composites *Journal of Reinforced Plastics and Composites* **34** 1075–89
[3] Ashok R B, Srinivasa C V. and Basavaraju B 2018 A review on the mechanical properties of areca fiber reinforced composites *Science and Technology of Materials*
[4] Analysis F 2016 Optimization of Notch Parameter on Fracture Toughness of Natural Fiber Reinforced Composites Using Taguchi Method
[5] Leonard L W H, Wong K J, Low K O and Yousif B F 2009 Fracture behaviour of glass fibre-reinforced polyester composite *Proceedings of the Institution of Mechanical Engineers, Part L: Journal of Materials: Design and Applications* **223** 83–9
[6] Saha G and Biligiri K P 2015 Fracture damage evaluation of asphalt mixtures using Semi-Circular Bending test based on fracture energy approach *Engineering Fracture Mechanics* **142** 154–69
[7] Ayatollahi M R, Aliha M R M and Saghafi H 2011 An improved semi-circular bend specimen for investigating mixed mode brittle fracture *Engineering Fracture Mechanics* **78** 110–23
[8] Aliha M R M and Fattahi Amirdehi H R 2017 Fracture toughness prediction using Weibull statistical method for asphalt mixtures containing different air void contents *Fatigue and Fracture of Engineering Materials and Structures* **40** 55–68

[9] Fakhri M, Haghighat Kharrazi E and Aliha M R M 2018 Mixed mode tensile – In plane shear fracture energy determination for hot mix asphalt mixtures under intermediate temperature conditions *Engineering Fracture Mechanics* **192** 98–113

[10] Huang B, Shu X and Zuo G 2013 Using notched semi circular bending fatigue test to characterize fracture resistance of asphalt mixtures *Engineering Fracture Mechanics* **109** 78–88

[11] Srinivasa C V. and Bharath K N 2011 Impact and hardness properties of areca fiber-epoxy reinforced composites *Journal of Materials and Environmental Science* **2** 351–6

[12] Gope P C and Rao D K 2016 Fracture behaviour of epoxy biocomposite reinforced with short coconut fibres (Cocos nucifera) and walnut particles (Juglans regia L.) *Journal of Thermoplastic Composite Materials* **29** 1098–117

[13] Manjunatha G B and Bharath K N 2018 Investigating the contribution of geometrical parameters and immersion time on fracture toughness of jute fabric composites using statistical techniques *Frattura ed Integrita Strutturale* **12** 14–24

[14] Limón-Covarrubias P, Cueva D A, Vidal G V, Ortiz O J R, Hernández R O A and González J R G 2019 Analysis of the behavior of SMA mixtures with different fillers through the semicircular bend (SCB) fracture test *Materials* **12**

[15] Kuruppu M D and Chong K P 2012 Fracture toughness testing of brittle materials using semi-circular bend (SCB) specimen *Engineering Fracture Mechanics* **91** 133–50

[16] Ban H, Im S and Kim Y R 2015 Mixed-mode fracture characterization of fine aggregate mixtures using semicircular bend fracture test and extended finite element modeling *Construction and Building Materials* **101** 721–9

IOP Conf. Series: Materials Science and Engineering **925** (2020) 012011 doi:10.1088/1757-899X/925/1/012011

A Review on Tribological Behaviour of Natural Fiber Reinforced Polymer Composites

C B Manjunath[1], C V Srinivasa[2], Basavaraju B[3], G B Manjunatha[4] and Ashok R B[5]

[1,2,4] Department of Mechanical Engineering, GM Institute of Technology, Visvesvaraya Technological University, Davanagere, Karnataka, India.
[3] Department of Chemistry, Alva's Institute of Engineering and Technology, Visvesvaraya Technological University, Moodbidri, Mangalore, Karnataka, India.
[5] Department of Mechanical Engineering, PES Institute of Technology & Management, Visvesvaraya Technological University, Shivamogga, Karnataka, India.
Corresponding email: srinivasacv@gmit.ac.in

Abstract. In the past, asbestos and copper were preferred as friction materials because they have good ability to dissipate heat, but have proven to be harmful to the environment. Recently, more researchers are focused on non-asbestos friction composite materials due to its non-toxicity and biodegradability. Despite synthetic fiber composites having eco-friendly nature, because of its cost and pollution most of the researchers show interest on natural fiber composites. Hence, there is a need to explore the analysis on the tribological behavior of composite materials. The availability of natural fibers and the ease of manufacturing have tempted researchers to study their feasibility of their application as reinforcement and the extent to which they satisfy the required specifications in tribological applications. Hence, the aim of this review is to demonstrate the tribological behavior of natural fiber reinforced composites and find knowledge about their usability for various applications that tribology plays a dominant role. This review presents the reported work on natural fiber reinforced composites with special reference to the type of fibers, matrix, polymers, treatment of fibers and test parameters to understand their usability for various automotive applications.

Keywords: Natural fibers; Friction materials; Tribological behaviour; Natural fiber reinforced polymer composites

1. Introduction

The modern world today is preoccupied with the speed and efficiency of machines and structures. It needs quality fabrics in the designer's hands to be used for the applications. Scientists have produced and are improving newest and older technologies for many specific applications such as space, aircraft, marine, and industrial, and structures, and it is important to study existing products in order to further understand them in service condition. One third of our global energy consumption was wastefully devoured in friction. Beyond the main energy saving, very important additional economics will be rendered by rising the costs involved with producing and removing unnecessarily burned out parts. Wear dissipation of energy strongly impairs the national economy and most people's lifestyle. Therefore, the effective diminution and control of metal wear is always desired.

1.1. Animal fibers
Yunhai Ma [1] researched the impact of wool fiber on wear and friction properties as the friction material reinforced. The fabrics strengthened with 3%wt wool fiber is stable with coefficient of friction and lower wear intensity. While temperature rises, reliability of the coefficient of friction increased. Rockwell stability reduced, and the impact force rose as the contents of wool fibers grew. Concluded the wool-fibre-enhanced fabrics were effective in elastic use and abrasive use. I OOladele [2], investigated the effects of fiber fraction on the abrasion properties of reinforced polyester composites in treated cows hair fiber. Abrasion resistance is optimum at 4 weight% CHF content. T Narendiranath Babu[3], tribological and brinell hardness of Angora, Kenaf and Ramie by combining the standard fibers in a particular proportion with Epoxy LY556 pitch and Hardener HY951. It has been discovered from the observation that Kenaf natural composite exhibits superior tribological properties compared to Angora and Ramie.

1.2 Mineral Fibers
U K Navin Chand [4] Studied tribological behaviour of ultra-high molecular weight polyethylene (UHMWPE) filled with epoxy gradient composites. Samples were prepared for different centrifugation time periods. The rising wear rate of composites at an increasing centrifugal period. Sliding wear rate reduction in composites was attributed due to the lubricating effect and smooth surface of UHMWPE. It is strongly entangled chain structure.

B Naga Raju [5] investigate the tribological properties of ZnO filled polymer nano composites. Materials were prepared by mixing 1wt%, 2wt%, 4wt%and 6wt% of ZnO nano particles with polyester resin. polyester composite of 1% ZnO has outstanding wear properties relative to pure polyester. The loading of nano ZnO particles in polyester improved the polyester microstructure and avoided the degradation of banded polyester structure during the friction cycle which may be one of ZnO is anti-wear mechanisms. The composite 's enhanced and better Tribo-performance can be attributed due to unique mechanical properties of nano particles. Antaryami Mishra [6] Studied wear behavior rubber dust- epoxy composite. Under dry sliding condition, materials exhibited very low friction coefficient and low wear levels. The lowest wear levels by weight are 10% rubber mud. Nevertheless, a strange finding was obtained, i.e. declining tendency while rubbing at different speeds against steel with 20% rubber reinforcement in epoxy. This may have arisen from the high surface finish of the disks and the composites, creating a thin film of transition. A Ayman[7] Reviewed extensively the friction and wear of polymer composites filled with nano-particles and concluded that composites formed by the addition of nano-scale particles to a polymer matrix result in improved electrical , mechanical and thermal properties of the composites..

Punyapriya Mishra [8] developed a mathematical model for predicting an abrasive wear behaviour of bagasse fiber-reinforced polymer composite. A second order polynomial model for predicting wear loss has been achieved. The model was developed using Surface Response Method (RSM). Analysis of the variance technique (ANOVA) at the confidence level of 95% was used to check the model 's validity. results are strong correlation, which means that analytical models derived from the response surface method can be used to describe the composite's tribological behavior. S Supreeth [9] Studied the effect of fiber length on the tribological behavior of short pulp reinforced bisphenol-a composite and the result shows that wear levels increase with load increases for composite specimens with lower interfacial bond strength. For this experimental research, it was observed that the wear properties of strengthened composites are greatly affected by the fiber length.

N.Mohan [10] Studied SiC tribological properties of glass-epoxy composites filled at high temperatures. The introduction of these ceramics (Al2O3, SiC, TiC, etc.) as in the matrix significantly improves the friction coefficient and eliminates wear loss. In both composites the wear rate increases with higher temperature / applied load and under the same conditions. Silicon carbide particulate composite packed with G-E experiences a lower wear intensity relative to virgin G-E composite with a higher friction factor. Mahdi Alajmi [11] researched the interaction between Fillers / Epoxy composite mechanical properties and tribological behaviour. Performance has affected tribological resources favorably.

F Chegdani [12], Investigated effect of fiber type on tribological actions of natural fibre reinforced plastics during the profile milling process. Three different reinforced polypropylene (PP) composites of short natural fibers (bamboo, sisal and miscanthus) are estimated. The consistency of the NFRP machined surface is quantified using a multi-scale, wavelet decomposition-based process. Bamboo fibers reinforced plastics shows in height contact stiffness after machining shows smoother surface finish.

T Narendiranath [13], Reviewed Tribological Properties, SiO2, TiO2, BaSO4, Al2O3, CaO, MgO, K2O, Na2O, Fe2O3 Strengthened with Basalt Fibers. The mineral fiber reinforced with polymer such as thermoplastic and thermoset, metal and concrete, for an overview of the application of this fiber in biodegradable matrix was presented. the studies have been reviewed on the industrial applications of composites reinforced with basalt fibre. B S. Kanthraju [14] systematically investigated the wear resistance using bidirectional fibers (E-glass and basalt) and polytetrafluoroethylene fillers (PTFE) and graphite. It indicates that the introduction of PTFE and graphite in G-B / E hybrid composites has a major impact on wear behavior under varying sliding distance / loads. PTFE filled G-B / E composites revealed lower similar wear levels than unfilled and graphite filled G-B / E hybrid composites. The most important opinion of this work is to highlight the significance of fibers and practical fillers in the design of wear-resistant polymer matrix composites.

Jyothish J Nair [15] Studied vibration analysis and wear characteristics of shell powder filled glass fiber reinforced epoxy composite. Sea shell powder percentage was varied by 5%, 10 %, and 15% by volume. With more load, sliding velocity, and distance the weight loss increased Modal analysis is conducted using the accelerometer of the Fast Fourier Transform analyser and an impact hammer to assess the vibrational characteristics of the Glass / Epoxy composite. NX Nastran gets the analytical results. The experimental and analytical results are compared, and they have been in good agreement. Glass / Epoxy showed good vibration characteristics with 5 per cent sea shell powder filler.

A lakshumu [16], Studied Chemical treatment of natural fibers that can clean the surface of the fibre, chemically modify the surface, stop the absorption of moisture and increase surface roughness. For chemical modifications of these composites a variety of different chemicals are used. Thus, to ensure their long-term stable behaviour in practical applications, it becomes necessary to take care of the chemical properties of composites. G Vikas [17] studied an overview of tribological properties of polymer composites enhanced by basalt fibre. Yusuf Sahin [18], Investigated the tribological behaviour of carbon-reinforced epoxy composite (CFRC) and its nano-based composites, including Al_2O_3 and PTFE powders developed using molding techniques. Response surface methodology was used to model the impact on weight loss under dry sliding conditions of different variables such as material types,applied load and rotational speed.

C H Shreenidhi [19], Conducted a tribological study of polymer composite content with different filler materials and examined the positive impact of commercially usable nano-additives on the tribological properties of the polymer composite. This describes the tribological actions of specific forms of composites of glass, jute, metal powder as filler material, epoxy resin, and numerous filler styles varying from particulates, chips, and chopped branches. P Gopal [20], developed hybrid natural fiber epoxy materials that are used as body shells for the automotive. Composite material used as epoxy composite of bio-material characterisation with the lowest density is used. It finds that tribological properties are improving significantly.

Robin Zuluaga [21], Studies Tribological work has an impact on the global production of natural composites. The composites filled with Musaceae fiber bundles and cured with two hardeners; rises of about 15% in global styrene double bond conversion observed using both hardeners for tidy matrix and composites, indicating that any changes in resin structure occur during the tribological study. Determining the usable alternatives for controlling the therapeutic environment to monitor tribological behaviour. P A Udaya Kumar [22], Studied the effect of hybrid reinforcement on dry sliding wear behaviour of vinyl ester (VE) composites. With the coconut shell powder (CSP) content the specific wear rate fell more rapidly. Percentage CSP is more protective against wear from sliding Reinforcement form, geometry and dispersion lead to improved wear resistance. These outcomes are very important for the manufacture a new composite integrated with surface-treated reinforcing materials such as Carbon fiber and CSP that will have foremost application in automotive sliding or bearing components.

AnetaKrzyżak [23], Analyzed the abrasive wear in different abrasion conditions of selected aircraft materials. In contemporary aircraft construction, the mass percentage of epoxy resin composites is generally higher than 50% and these materials must meet increasingly demanding requirements. Investigated the influence of external factors on the abrasion process, each group of samples was subjected to abrasion in an insulated environment under different external conditions, in the presence of water and loose abrasives, brown fused alumina (BFA) and white fused alumina (WFA). The results allow us to conclude on which type of filler and in which concentration it contributes most to improving the abrasion resistance of the composite material. Additionally, it has been found that the conditions in which abrasion occurs have a very important effect on the course of this process.

1.3 Plant Fibers

1.3.1 Bast Fiber

A S Singha [24], Studied on new series of green composites involving Hibiscus sabdariffa fiber as a reinforcing material in urea formaldehyde (UF) resin based polymer matrix. Initially urea formaldehyde resin prepared was subjected to evaluation of its optimumal properties. Then reinforcement of the resin with Hibiscus sabdariffa fiber was attained using engineered resin in the particle sizes of three different types, short fiber and long fiber. Reveals traction resistanc. N Vijaya Kumar [25], Investigated Properties of Wear on Jute Reinforced Polypropylene Composites at different loads and at different sliding velocities. The composites with varying percentage of fiber weight (5%, 7.5%, 10%, 12.5% and 15%) using injection moulding process. The friction and wear loss coefficient is studied for different loads and sliding speeds. A Thimmana Gouda [26], Studied wear characteristics on Natural Fiber Hybrid Polymer Composite for Orthopedic Implants. Identified foremost changes in tribological properties of 36% Hybrid Natural Fiber Polymer Composite compared to 12% and 24%. Finally, 36% Material can be used for orthopaedic implants or replacement of human bone.
Temesgen [27], Studied Sliding Wear Properties of Jute Fabric Reinforced Polypropylene Composites. Adding woven Jute fabric to the PP matrix increases the wear resistance properties and the coefficient of friction was reduced by 45-65%. R Karthik [28], studied tribological behavior of rice husk and eggshell with particulate coir-polyester. The effects of egg shell and rice husk polyester composites on surface roughness properties were investigated.

Priyadarshi [29], Studied Abrasive Wear Behavior of Surface Modified Jute Fiber Reinforced Epoxy Composites. Three-body abrasive wear research was conducted in conjunction with ASTM-G65 using a rubber wheel abrasion system (dry sand wears tester) at room temperature through varying factors such as fiber quality, abrasive particle scale, standard load and sliding speed. The result indicates that the wear-resistance ability is increased by chemical treatment of jute fiber. B Sudhabindu [30], Studied the effect of tri-biological properties of polymer composites reinforced with jute / e-glass fibre. Wear loss is very minimal and is observed for 180 minutes. F Z Alshammaria, [31], Studied the tribological efficiency of Jute Fiber Reinforced Epoxy Composites with Fiber Orientation. Fiber orientation has a very significant effect on the wear and frictional efficiency. The composite is testing in both parallel and antiparallel orientations. better wear resistance found in the antiparallel orientation

Sandeep Kumara [32], Investigated sliding wear properties of Bauhinia-vahlii-weight (BVW) and Bauhinia-vahlii-weight / sisal (BVWS) hybrid composites. The rice husk was hybridized by with BVW and BVW-Sisal fiber reinforced epoxy composites. The evaluated results show that the addition of optimally varied rice husk fillers reveals the increase in wear efficiency of natural fiber based epoxy composites.

1.3.2 Fruit Fiber

L Boopathi[33], Studied the wear properties of Borassus fruit fiber reinforced epoxy composites. wear properties improved with the alkali treatment of the fibres. Alkali treated fiber-reinforced composites exhibited greater wear properties than others. P Chandu [34], Experimentally Evaluated Wear Behavior of Banana/ Pineapple Hybrid Composites. It shows friction coefficient remains almost constant. Find that there is a direct relationship between the sliding distance and the wear rate. Hemant Patel [35], Studied on Mechanical Behaviors of Banana/ Sisal Hybrid Composite Reinforced with Epoxy Resin. The properties, such as tensile and flexural properties of hybrid bananas and sisal-reinforced epoxy composites used in combination with plastics.

Abhishek Kumar [36], Studied Natural Fiber Composites Tribological Characterization. There have been attempts to explore the potential use of natural fiber in composites as a filler material. To evaluate their (banana fiber) potential for strengthening wear performance and mechanical properties; There are numerous human, mechanical and wear efficiency studies of the composites. From the experiments conducted, it can be seen that as far as wear performance is concerned a banana fiber is concerned, as the study shows that erosion wears performance of composites based on banana fibre. Banana fiber also has a very low specific gravity, and the composites are very low in density. From other mechanical characteristics it can be said that as in potential application, banana fiber-epoxy composites are also used in terms of strength.

1.3.3 Grass and Reeds Fiber

Mukesh Kumar Mann [37], Studied Kans grass fiber (KGF) polyester composites friction and wear behaviour. It was found that at virtually all sliding conditions, the tidy polyester specimen had less wear resistance than the KGF composites. Friction activity was significantly affected by the fiber quality and variability in sliding velocity. The rise in the fiber Vol percentage from 10 to 20.08 slightly increased the KGF composite specimens average friction coefficient at 0.5 m / s and 1.5 m / s, while fluctuating behavior was observed at moderate sliding level.

1.3.4 Leaf Fiber

Kamal Kumar Basumatary [38], Studied the impact of fiber loading on Ipomoea carnea's abrasive wear actions in reinforced epoxy composite. The incorporation of Ipomoea carnea into the epoxy significantly reduces the loss of abrasive wear. Addition of fiber (40 weight percent) removed particulates from the matrix resin due to poor interfacial adhesion. Abrasive wear is very sensitive to normal load relative to sliding speed and increases slightly with increasing sliding speed. The average wear intensity increases with a rise in the sliding distance, when the debris occupies the area between the abrasive that limits the extent of abrasive surface penetration through the composite sample.

With through sliding space, wear rate slowly decreases and in multi-pass condition achieves an Almost steady state. The composite 's average wear rate decreases with an increase in the sliding gap as the debris fills the space between the abrasive, which increases the degree of penetration of abrasive particles into the composite sample. Further, surface modification is found to have a significant impact on wear performance. Manu Prasad [40], investigated the effect of PALF Reinforced Bisphenol-A Composite Fiber Orientations on tribological behavior. Upon PALF reinforcement, the wear resistance of pure Bisphenol-A resin is found to be increased. Among three types of composite fiber orientation, bidirectional composite shows the least specific wear rate and friction coefficient The Coefficient of friction for all three types of composite fiber orientations decreases with an increase in normal load, As the usual load rises, more material is removed due to self-lubrication of the test samples and the removed material is placed on the mating surfaces of the composite and abrasive disk resulting in self-lubricating.

The tribological behavior of PALF reinforced Bisphenol-A composite is thus greatly influenced by the orientation of fibre.

TMadhusudhan [41], Reviewed study of tribological behavior on polymer hybrid composites reinforced by natural glass fibres. In this study the tungsten carbide (WC) filler enhances erosion resistance. T Narendiranath Babu [42], Studied mechanical and tribological behavior of epoxy composites strengthened by sisal, abaca, and sun-hemp hybrids. It is observed from the tests that the Sun-Hemp exhibited superior Tensile and Flexural properties, while Sisal exhibited high resistance to impact and hardness. Sun-hemp has better wear properties when compared with two other composites with the lowest wear rate of 85 micrometers. With these findings design engineers will be able to pick the correct composite for industrial purposes. Nitin Mukesh Mathur (2017) [40], Reviewed on Sugar Cane Bagasse Composite Material and scope. Discussed on various recent work or review report on natural fibers and fillers used in the manufacture and application of cost-effective composite materials.

This research focuses on the degree to which sugarcane bagasse particles are used as reinforced material for producing composite material. Researchers have done much research on natural fiber-reinforced polymer composites, but SCB dust particles have never been used at low cost in the manufacture and application of epoxy-based composite. There is definite scope of study of mechanical behavior under various test conditions for polymer composite reinforced by SCB dust. Cost and quality control of natural composite reinforced filler is the major stone to be used by product designers and manufacturers as an alternative material. In addition to all these, the main motive is the manufacture for commercial use of an economic natural fiber based composite material.

S Vigneshkumar [43], studied the tribological actions of composites reinforced by fibre. Polyester was used as the binding medium for the preparation of a composite model reinforced with compositions from sisal, banana, kenaf, carbon fiber and rice husk. The Tribological operating parameters illustrate the effect on the friction and wear efficiency of fibre-reinforced composites by incorporating load, sliding velocity and sliding width. The specimens' wear reaction was found to be determined by the load applied, the sliding distance and the velocity.

Nitin Doifode [44], Studied Epoxy Composite Friction and Wear Behavior Reinforced under Dry Condition with Ipomoea Carnea Fiber. This paper describes the effect of normal wear load (micron) and friction coefficient of reinforced / reinforced Ipomoea Carnea epoxy composites with different weight percentage of the fiber fraction (10 percent, 20 percent, 30 percent, 40 percent wt). To measure the micron wear and friction coefficient of Ipomoea Carnea reinforced epoxy composites, the 12 mm diameter fiber composite pin was manufactured and tested for each fiber weight concentration under three separate loading conditions (10N,20N,30N). Measurement of friction and wear was done by load cell and LVDT, respectively. The Pin-on - Disk Tribometer acquired wear data in real time, and time (Sec) friction coefficient. Each test was performed under dry condition for 10 min (600 Sec), and data for wear (micron) and friction coefficient with time variation (100 Sec interval) were recorded from the monitor.

1.3.5 Seed Fiber

H Aireddy [45], Examined tribological actions of polymer matrix composites reinforced by Bio-waste. The composite is observed to show the maximum corrosion intensity at an impingement angle of 90 °. Weared rate of erosion is minimised by increasing coir dust volume. Abrasive wear resistance declines with decreased usual load, and improves with increased amount of fibre. JT Siva [46], Studies Effect of Naturally Woven Coconut Sheath Polyester Composite on Tribological Properties and treated with Alkali. Hybrid composites (polyester resin with naturally woven coconut sheath (N) and glass fibers (G)) were tested in a pin on the disc configuration against the hardened steel counter-face. From the results all hybrid combinations excluding (NNN) with the alkaline treatment show deteriorated wear properties. For all combinations, except for NGN and GGG hybrids, the friction properties are changed by having low friction coefficient. In untreated specimen for which the fibers are highly visible, partial removal of individual phase (resin) is prevail. Such phenomenon, however, is not dominant in the alkali treated material, which shows better reinforcing behavior that compliments low friction properties. The alkali treated specimen the the size of the fiber relative to the untreated specimen resulting in low resistance to wear. The untreated fibers are best suited for tribological applications by a compromise between friction and wear properties.

Umar Nirmala [47], Studies the impact of betel nut fiber treatment and contact conditions on the adhesive wear and friction performance of polyester composites. Composite wear and traction quality were seen to be increased by about 54% and 95% relative to dry under wet contact conditions.

R Abraham [48], Examined the impact of date palm seeds on tribological polyester composite behavior under Various Research Conditions. It revealed that the friction coefficient for polyester composites packed with date seed powder falls by 15% with a rise in filler under heavy contact strain. In addition to rising the size, the friction coefficient rises by up to 10%. Unfortunately, the date seed filler continuously decreases from 15% to 25% lower the wear resistance of the planned composites under heavy contact pressure and fast slipping pace. The load applied (contact pressure) and the sliding velocity have a direct impact on the polyester composite wear rate which is filled with date seed.

Prafull Sharma [49], Evaluated the Impact Resistance of Cotton Fiber Reinforced Polymer Composites Cotton fabric is used as an epoxy-based polymer composite reinforcement material. it shows that for engineering applications the manufactured cotton fiber-reinforced composite has promising characterstics. G Rajasekhar [50], Studied on Coconut Fiber Reinforced Polyamide Matrix Composite to its Wear Behavior. The coconut fiber reinforced composites have substituted the most commonly produced plastic fiber (Glass, Kevlar) strengthened polymer composites in many applications. Concluding with an increase in normal force results in an increase in wear rate due to bonding between the coconut fiber and matrix is inadequate to withstand normal force resulting in damage to the matrix thereby increasing wear rate. The coefficient of friction decreased with regular force, but it declines less with extreme force.

1.3.6 Stalk Fiber

Yunhai Ma [51], Studied tribological characterization of Stalk-reinforced friction composites from waste grain. The composites for traction were formulated with specific amounts of maize stalk fibre. The addition of corn stalk fibers was found to have a positive impact on the coefficients of friction and the wear levels of friction composites. The results revealed that the satisfactory performance of wear resistance was related to the secondary plateaus formed on the worn surfaces. The wear output of coconut shell powder (CSP) and coir fiber reinforced polyester resin composites was experimentally investigated by K Vignesh [52]. The effect of the concentration of coconut shell powder and coir fiber and the sliding distance in the composite weight loss was analyzed. The result shows that the friction coefficient increases with load increases in composites containing polymer matrix in the CSP and coir fiber. The wear rate is observed to increase with the increases in the load applied. With the addition of CSP and coir fibre, the wear rate decreases. When the load applied increases, friction also increases at the material's contact surface and spinning disk. For the polymer matrix composites, the addition of the CSP and coir fiber serves as reinforcement. The wear resistance improves thanks to this addition.

1.3.7 Wood Fiber

S S Mahapatra [53], Analyzed abrasive composite wear performance using Taguchi approach. Fiber length has been found to play a significant role in the phenomenon of wear. Using a popular evolutionary technique known as particle swarm optimization (PSO), and neural network, the fiber length was optimized. The study suggests that for minimal composite wear, the duration of the fiber should be 7-8 mm. Abrasive wear (weight loss) increases with increased load as high temperature is produced at higher load, which causes degradation of the composite surface. The factor, sliding speed, doesn't contribute significantly to weight loss. Nevertheless, abrasive wear plays a significant role in surface degradation due to the size of the chopped sugar cane fibre. At the lowest fire scale, i.e. 1 mm, limited wear resistance is observed. Again, the wear (weight loss) increases with the maximum length of fiber, i.e. 10 mm. Minimum wear is observed 5 mm long fiber. The explanations for these actions can be clarified as the fibers don't get support from the matrix at lower length due to weak interface adhesion. As the fiber length becomes too large, the composite's anisotropy may increase due to the fiber's random orientation, which causes weight loss to increase. Vijay Kumar Thakur[54], studied on Characterization of Lignocellulosic Fibre-Made Phenolic Composites. properties of composites were observed to increase on reinforcement with lignocellulosic Saccaharum cilliare fibres.

P SnehaLatha [55], investigated the impact of stacking tribological properties series of reinforced polymer hybrid composites woven cloth of bamboo-glass. The results revealed that the erosion rate is lower when hybridized with glass fiber compared with pure glass fiber, the composite of bamboo is more susceptible to erosion. Narendra Kumar [56], Studied on Short Fibers Reinforced Polymer Composites for Tribological Characterization and Damage Analysis. Centered on the erosion wear actions of short bamboo fiber reinforced composites filled with alumina (Al_2O_3) particulate matter.

2. Conclusions

Natural fibre-reinforced polymer composites are desirable and challenging materials to replace traditional materials and address important environmental issues. As demands for the use of biodegradable materials are growing due to environmental issues and government regulations and policies, many industries are attempting to substitute traditional automotive materials with biodegradable materials where friction and wear are important. However, there is much less information available in the literature on the tribological performance of bio-polymer composite material reinforced by natural fibre.

This review discusses the ability of natural reinforced polymer composites, which tests structure which tribological properties at various operational conditions. Many experiments are linked to the dry friction state, but lubrication may be implemented in practical action.

- The main element affecting a composite's porosity is humidity, which influences tribological properties although experiments are performed without knowledge of humidity.
- The tribological properties studied with the benefit of lubrication will contribute to improved use of natural fibers as friction material.
- The tribological properties studied with a lubrication effect can enhance the application of natural fibers as friction materials.
- The Tribological behavior of Natural fiber reinforced polymer composites produced with three different compositions was analysed to fulfil the objective of this research work.

References

[1] Ma Y, Liu Y, Gao Z, Lin F, Yang Y, Ye W and Tong J 2014 Effects of wool fibers on tribological behavior of friction materials *Journal of Thermoplastic Composite Materials* **27** 867–80
[2] Oladele IO, Agbeboh NI, Omokafe SM and Ibrahim OI 2018 Effects of fiber fraction on the mechanical and abrasion properties of treated cow hair fiber reinforced polyester composites *Tribology in Industry* **40** 254–62
[3] Narendiranath Babu T, Aravind SS, Naveen Kumar KS and Sumanth Rao MS 2018 Study on mechanical and tribological behaviour of Angora, kenaf and ramie hybrid reinforced epoxy composites *International Journal of Mechanical Engineering and Technology* **9** 11-20
[4] Chand N, Dwivedi U K and Sharma MK 2007 Development and tribological behaviour of UHMWPE filled epoxy gradient composites *Wear* **262** 184–90
[5] Naga Raju B, Ramji K and Prasad VSRK 2011 Studies on tribological properties of ZnO filled polymer nanocomposites *ARPN Journal of Engineering and Applied Sciences* **6** 75–82
[6] Mishra A 2012 Dry Sliding Wear Behavior of Epoxy-Rubber Dust Composites *International Journal of Mechanical and Mechatronics Engineering* **6** 1218–23

[7] Aly A A, Zeidan E-S B, Alshennawy A A, El-Masry A A and Wasel W A 2012 Friction and Wear of Polymer Composites Filled by Nano-Particles: A Review *World Journal of Nano Science and Engineering* **2** 32–9
[8] Mishra P 2012 Statistical analysis for the abrasive wear behavior of bagasse fiber reinforced polymer composite *International Journal of Applied Research in Mechanical Engineering* **2** 7-11
[9] Supreeth S, Vinod B and Sudev L J 2014 Influence of Fiber Length on the Tribological Behaviour of Short PALF Reinforced Bisphenol-A Composite **2** 825–30
[10] Mohan N, Mahesha C R and Raja R 2016 Tribo-mechanical behaviour of SiC filled glass-epoxy composites at elevated temperatures *International Journal of Engineering, Science and Technology* **6** 44
[11] Alajmi M and Shalwan A 2015 Correlation between mechanical properties with specific wear rate and the coefficient of friction of graphite/epoxy composites *Materials* **8** 4162-75
[12] Chegdani F, Mezghani S, El Mansori M and Mkaddem A 2015 Fiber type effect on tribological behavior when cutting natural fiber reinforced plastics *Wear* **332** 772–9
[13] Babu TN 2016 A Review on Mechanical and Tribological Properties of Epoxy Reinforced with Basalt Fibres *International Journal of ChemTech Research* **9** 131–9
[14] Kanthraju BS and Suresha B 2016 Indian Journal of Advances in Chemical Science Enhancement of Mechanical Properties and Wear Resistance of Epoxy : Glass Fiber, Basalt Fiber, Polytetrafluoroethylene and Graphite *Indian Journal of Advances in Chemical Science* **107** 113
[15] Nair JJ, Augustine A and Davanageri MB 2016 Wear Characteristics and Vibration Analysis of Shell Powder Filled Glass Fiber Reinforced Epoxy Composite Laminates *International Journal of Research in Engineering and Technology* **5** 215–21
[16] Naidu A and Rao P 2016 A Review on Chemical Behaviour of Natural Fiber Composites *International Journal of Chemical Sciences* **14** 2223–38
[17] Vikas G and Sudheer M 2017 A review on properties of basalt fiber reinforced polymer composites *American Journal of Materials Science* **7** 156–65
[18] Şahin Y, Erdil E and Şahin H 2017 The tribological wear behavior of carbon fabric-reinforced epoxy composites *Proc. of the Int. Symp. of Mechanism and Machine Science* 174-181
[19] Shreenidhi CHBN 2017 Tribological test of polymer composite material with different filler materials : A Review *International Research Journal of Engineering and Technology(IRJET)* **4** 2859–62
[20] Gopal P, Bupesh Raja VK, Chandrasekaran M and Dhanasekaran C 2017 Wear study on hybrid natural fiber epoxy composite materials used as automotive body shell *ARPN Journal of Engineering and Applied Sciences* **12** 2485–90
[21] Correa C E, Zuluaga R, Castro C, Betancourt S, Vázquez A and Gañán P 2017 Influence of tribological test on the global conversion of natural composites *Polimeros* **27** 339–45
[22] Udaya Kumar P A, Ramalingaiah, Suresha B and Hemanth R 2018 Mechanical and tribological behavior of vinyl ester hybrid composites *Tribology in Industry* **40** 283–99
[23] Mucha M, Racinowski D, Pindych D and Krzyżak A 2018 Analysis of abrasive wear of selected aircraft materials in various abrasion conditions *Journal of KONES* **25** 217–22
[24] Singha A S and Thakur V K 2008 Mechanical properties of natural fibre reinforced polymer composites *Bulletin of Materials Science* **31** 791–9
[25] Gouda AT, Jagadish SP, Dinesh DK, Virupaksha Gouda H and Prashanth DN 2014 Wear study on hybrid natural fibre polymer composite materials used as orthopaedic implants *International Journal of Recent Development in Engineering and Technology* **3** 25-33
[26] Gouda AT, Jagadish SP, Dinesh DK, Virupaksha Gouda H and Prashanth DN 2014 Wear study on hybrid natural fibre polymer composite materials used as orthopaedic implants *International Journal of Recent Development in Engineering and Technology* **3** 25-33
[27] Yallew T B, Kumar P and Singh I 2014 Sliding wear properties of jute fabric reinforced polypropylene composites *Procedia Engineering* **97** 402–11

[28] Sathiyamurthy S and Jayabal S 2014 Tribological Behaviour of Rice Husk and Egg Shell Hybrid Particulated Coir-Polyester Composites *IOSR Journal of Mechanical and Civil Engineering* 75–80
[29] Swain PTR and Biswas S 2017 Abrasive wear behaviour of surface modified jute fiber reinforced epoxy composites M*aterials Research* **20** 661–71
[30] Sudhabindu B, Ramarao PV and Udaya Kiran C 2018 Effect of tri-bological properties on jute/E-glass fibre reinforced polymer composites *ARPN Journal of Engineering and Applied Sciences* **13** 2619–22
[31] Alshammari F Z, Saleh K H, Yousif B F, Alajmi A, Shalwan A and Alotaibi J G 2018 The influence of fibre orientation on tribological performance of jute fibre reinforced epoxy composites considering different mat orientations *Tribology in Industry* **40** 335–48
[32] Kumar S, Mer KKS, Gangil B and Patel VK 2019 Synergy of rice-husk filler on physico-mechanical and tribological properties of hybrid Bauhinia-vahlii/sisal fiber reinforced epoxy composites *Journal of Materials Research and Technology* **8** 2070–82
[33] Boopathi L 2012 Influence of Fiber Length in the Wear Behaviour of Borassus Fruit Fiber Reinforced Epoxy Composites *International Journal of Engineering Science and Technology* **4** 4119–29
[34] Chandu KVPP, Krishna SVG, Subrahmanyam BV and Rao KV 2018 Experimental Analysis on Wear Behaviour of Banana - Pineapple Hybrid Natural Fiber Composites *International Journal of Engineering Science Invention (IJESI)* 31-37
[35] Mann MK, Ahlawat V, Nain S, Gir M and Beniwal A 2019 Wear and friction behaviour of kans grass fiber/polyester composites *International Journal of Innovative Technology and Exploring Engineering* **8** 918-21
[36] Basumatary KK, Mohanta N and Acharya SK 2014 Effect of fiber loading on abrasive wear behaviour of Ipomoea carnea reinforced epoxy composite *International Journal of Plastics Technology* **18** 64-74
[37] Mohanty JR, Das SN and Das HC 2014 Effect of Fiber Content on Abrasive Wear Behavior of Date Palm Leaf Reinforced Polyvinyl Pyrrolidone Composite *International Scholarly Research Notices* **2014** 1–10
[38] S. Supreeth VB and DLJS 2014 Effect of Fiber Orientations on Thermal Properties of PALF Reinforced Bisphenol: A Composite *The International Journal of Science And Technology* **2** 275–80
[39] Madhusudhan T, Senthilkumar M and Athith D 2016 Mechanical Characteristics and Tribological Behaviour Study on Natural - Glass Fiber Reinforced Polymer Hybrid Composites : a Review *International Research Journal of Engineering and Technology (IRJET)* **2016** 3
[40] Mathur NM and Bairwa K 2017 A literature review on Composite material and scope of Sugar cane Bagasse *International Journal of Engineering Development and Research* **5** 125–33
[41] Narendiranath Babu T, Talla HK, Kokkirala S and Batta K 2018 Study on mechanical and tribological behaviour of sisal, abaca and sun-hemp hybrid reinforced epoxy composites *International Journal of Mechanical Engineering and Technology* **9** 865–75
[42] Vigneshkumar S and Rajasekaran T 2018 Experimental analysis on tribological behavior of fiber reinforced composites *Mater. Today Proc.* **3** 2155-62
[43] Doifode N and Nalawade P M 2018 Friction and Wear Behavior of Epoxy Composites Reinforced With Ipomoea Carnea Fiber under Dry Condition *Journal of Engineering Research and Application* **8** 22-7
[44] Aireddy H and Mishra SC 2011 Tribological behavior and mechanical properties of Bio-waste reinforced polymer matrix composites *Journal of Metallurgy and Materials Science* **53** 139–52
[45] Ra I 2015 Effect of Date Palm Seeds on the Tribological Behaviour of Polyester Composites under Different Testing Conditions *Journal of Material Science & Engineering* 4
[46] P. Sharma, Pandey A and Srivastava A 2017 Tensile and Impact Strength Analysis of Cotton Fiber Reinforced Polymer Composites *ELK Asia Pacific Journals* **6** 1-5

[47] Rajasekhar P, Ganesan G and Senthilkumar C 2017 Wear Behavior of Coconut Fiber Reinforced Polyamide Matrix Composites *International Journal of Recent Technology and Engineering* **6** 2277–3878

[48] Ma Y, Wu S, Zhuang J, Tong J, Xiao Y and Qi H 2018 The evaluation of physio-mechanical and tribological characterization of friction composites reinforced by waste corn stalk *Materials* **11** 901

[49] Vignesh K, Natarajan U and Vijayasekar A 2014 Wear Behaviour of Coconut Shell Powder and Coir Fibre Reinforced Polyester Composites *Journal of Mechanical and Civil Engineering (IOSR-JMCE)* **14** 53-7

[50] Mahapatra S and Chaturvedi V 2010 Modelling and analysis of abrasive wear performance of composites using Taguchi approach *International Journal of Engineering, Science and Technology* **1** 123–35

[51] Thakur VK, Singha AS, Kaur I, Pudukotai RN and Liping Y 2011 Studies on analysis and characterization of phenolic composites fabricated from lignocellulosic fibres Polymers and Polymer Composites **19** 505-12

[52] Latha PS and Rao MV 2017 Erosion Wear Behaviour of Bamboo-Glass Hybrid Fiber Reinforced Polymer 2438–46

[53] Kumar N, Pandey A, Singh VP and Hussain SM 2017 Mechanical Behavior, Tribological Characterization and damage analysis of Short Fibers Reinforced Polymer Composites *IOSR Journal of Mechanical and Civil Engineering (IOSR-JMCE)* **14** 16-20

[54] Patel H, Parkhe A and Shrama PK 2016 Mechanical Behaviors Of Banana And Sisal Hybrid Composites Reinforced With Epoxy Resin *International Journal of Research–Granthaalayah* **4** 206-16.

[55] Mukti PS and Choubey AK 2017 Tribological Characterization of Natural Fiber Composites *Ijesc* **7** 1–12

[56] Siva I, Winowlin Jappes J T, Szakal Z and Sukumaran J 2014 Alkali treatment and its effect on tribological properties of natrually woven coconut sheath polyester composite International *Journal Sustainable Construction & Design* **4** 1–7

[57] Nirmal U, Yousif BF, Rilling D and Brevern PV 2010 Effect of betelnut fibres treatment and contact conditions on adhesive wear and frictional performance of polyester composites *Wear* **268** 1354–70

Principal Component Analysis and its Application to Nominal Gross Domestic Product

Suchitra Hiregowda[1,4], **GH Pujar**[2,4*] **and VS Betageri**[3,4]

[1] Department of Computer Science & Engineering, GM Institute of Technology, Davangere, Karnataka, India.
[2] Research Centre, Department of Physics, GM Institute of Technology, Davangere, Karnataka, India.
[3] Research Centre, Department of Chemistry, GM Institute of Technology, Davangere, Karnataka, India.
[4]Visvesvaraya Technological University, Belagavi, Karnataka, India
[1]suchitra_18cs065@gmit.ac.in, [2]puttpuja@gmial.com, [3]virupaxb@gmail.com

Abstract. Gross Domestic Product refers to financial or market value of all goods and services which are produced within the countries in a specific time duration. As expansive measure of universal domestic production, it functions as a comprehensive record of the country's economic health. Nominal Gross Domestic Product is the computation of raw data and is more utilitarian comparing national economies on the international market. Some countries have very differing Gross Domestic Product per capita between its regions. A countries region tend to intersect overtime, the discrepancy between the poorer and richer regions is kept over decades in other cases. Here, we considered progress in the nominal Gross Domestic Product per capita across 19 regions in Spain to provide the analysis of progress. This could answer few questions with respect to Nominal Gross Domestic Product viz., Have the regions converged? Which is the spread between regions? The central theme of this article is to answer whether can we make a cluster analysis of the regions after applying principal component analysis?

Keywords. Cluster Analysis, GDP, Nominal GDP, PCA, Spanish regions.

1. Introduction

GDP calculation gives an overall record of the state of the economy to that of a satellite in the space which helps in surveying the weather across an entire continent. This record also provides the policy makers and the central banks to judge if the economy is contracting or expanding and also if the needs required to be boosted or restrained. It helps to understand if there are any threats like recession or rampant inflation loom on the horizon.

Generally, the dataset can be used to calculate GDP which has been compiled by the INE (Spanish institute of statistics) and published on Kaggle - Nominal GDP per capita of Spam. This dataset contains information on the Nominal GDP per unit of population covering the 19 Spanish zones (autonomous communities and cities) from 2000 to 2016. To answer the questions that have been highlighted in the Abstract and to figure out whether there has been a convergence, we need to calculate whether the nominal GDP has been steady by looking at the spread between the maximum and the minimum values of the nominal GDP per unit of population covering the Spanish zones. But our true goal is to perform a cluster analysis pertaining the 19 nominal GDP per capita across the 19 regions/zones of Spain. So as to perform cluster analysis we first need to select the appropriate features to visualize the different cluster classes and their effects on the Nominal GDP Since our data set consists of 18 columns (here dimensions), we will have to perform PCA for feature extraction before the visualization [1-6].

PCA is a powerful statistical technique in Linear Algebra that will reduce the proportions of the data and assist us to understand, organize the data with a minor proportion as compared to the original data.The chief constituents are primarily the vectors that are linearly unrelated and have variation within the data. From the fundamental constituents, the top p picked out which has the most difference. In PCA our goal is to separate the data by illustrating a line or a plane between data by puzzling out the dimension which has the highest variance [7-15].

2. Background of PCA Module

PCA uses the Rotation tools to analyze the multiband datasets. There is a high correlation in the Data bands due to their analogous spectral domains. It could remove superfluous spectral information using multiband datasets. Remote sensing application uses this PCA to create a smaller dataset from multiple bands by retaining original spectral information as much as possible [14, 15].

2.1. Linear Algebra Background Theorem

Theorem 1. If the matrix A is symmetric (meaning $A^T = A$), then A is orthogonally diagonalizable and has only real eigenvalues. In other words, there exist real numbers λ1, λ2..., λn (the eigenvalues) and orthogonal, non-zero real vectors v#»1,v#»2,v#»3,...,v#»n(the eigenvectors) such that for each i#» = 1,2,...,n

$$Av_i = \lambda_i v_i \tag{1}$$

2.2. Implementation

We have implemented PCA on the basis the minimization of the reconstruction error. The Frobemus standard of dissimilarity b/w original &reconstructed matrix is called as the reconstruction error.

Particularly, among all the feasible linear projections from dimensions' *n* to *m*, taking the first k-components of PCA transformation of **X** reduces the reconstruction error.

2.3. Proof

The input to the problem is m data points of the form $x_i = (x_1^i, x_2^i, \ldots x_n^i)$, i = (1 , 2, .. m). The aim of the PCA algorithm is to transform this n-dimensional feature vector to a k-dimensional space with an orthonormal basis $u_1, u_2, \ldots u_k$. The new set of transformed co-ordinates in the k-dimensional space IS represented as z_1 , z_2, .., z_k where $z_i = (x - \bar{x})$. The mean of **x** over all the data points is u_i and $\bar{x}$. From the new set of co-ordinate axes, we attempt to reconstruct the data **X**. The reconstructed data example $\hat{x}_i$ is given as

$$\hat{x}^i = \bar{x} + \sum_{j=1}^{n} z_j^i u_i \tag{2}$$

The objective of PCA IS to minimize the error created by this reconstruction, given by

$$\text{error}_m = \sum_{j=1}^{n}(x^j - x^{-j}) \tag{3}$$

Substituting the above values in the error equation, we get the following result.

$$\text{error}_k = \sum_{j=k+1}^{n} u_j^{\gamma} \Sigma u_j \tag{4}$$

$$\text{where, } \Sigma = \frac{1}{m}\sum_{j=1}^{n}(x^j - \bar{x})(x^j - \bar{x})^{\gamma} \tag{5}$$

IOP Conf. Series: Materials Science and Engineering **925** (2020) 012012 doi:10.1088/1757-899X/925/1/012012

3. Algorithm to Perform PCA

To implement PCA, it is required to get Singular Value Decomposition of matrix A that results the eigenvalues &eigenvectors of $\boldsymbol{A^T A}$ immediately.

Part 1: *Function to return the orthogonal vector*

```
def householder_ reflection (a, e) ;
    assert a.ndim == 1
    assert np.allclose(1, np.sum(e**2))
    u = a - np.sign(a[0]) * np.linalg.norm(a) * e
    v = u / np.linalg.norm(u)
    H = np.eye(len(a)) - 2 * np.outer(v, v)
return H
```

The Householder reflection is an alternate method of orthogonal transformation that transforms a vector x into a unit vector y parallel to x. The Householder reflection matrix **H** with a normal vector v takes the form:

$$H = 1 - 2vv^T \tag{6}$$

A matrix **H** must be constructed such that $\boldsymbol{Hx} = \propto e_i$ for a constant $\propto$ and $e_i = [1\ 0\ 0]^T$. Since the matrix **H** is orthogonal, $|Hx| = |x|$ and $|\propto e_i| = \propto$ and therefore $\propto = \pm|x|$

The symbol is so chosen that it has the contrasting symbol of $x1$, hence the vector u is

$$u = \begin{bmatrix} x1 + \text{sign}(x1)|x1| \\ x2 \ldots \ldots. \\ xn \end{bmatrix} \tag{7}$$

With the unit vector v interpreted as $= \frac{\gamma}{|\gamma|}$, the corresponding Householder reflection matrix is

$$\mathbf{H} = 1 - 2vv^T = 1 - 2\frac{uu^T}{u^Tu} \tag{8}$$

Part 2: *Function to return the QR Decomposition*

```
def qr_decomposition(A):
   n, m = A.shape
   assert n >= m
   Q = np.eye(n)
   R = A.copy()
   for i in range(m - int(n==m)):
      r = R[i:, i]
      if np.allclose(r[1:], 0):
      continue
      # e is the i-th basis vector of the minor matrix.
      e = np.zeros(n-i)
      e[0] = 1
      H = np.eye(n)
      H[i:, i:] = householder_reflection(r, e)
      Q = Q @ H.T
      R = H @ R
      return Q, R
```

The Householder reflection technique of QR decomposition is done by finding appropriate H matrices and multiplying them from the left by the actual matrix A to construct the upper triangular matrix **R**. As we have seen earlier, unalike the GramSchmidt policy, the Householder reflection approach doesn't accurately form the matrix **Q**. Although, the matrix **Q** can be found by taking the dot product of each consecutively formed a Householder matrix.

$$\mathbf{Q} = H1H2\ldots Hm2Hm - 1 \tag{9}$$

IOP Conf. Series: Materials Science and Engineering **925** (2020) 012012 doi:10.1088/1757-899X/925/1/012012

Part 3: *Function to return Eigen decomposition of matrix A*

```
def eigen_decomposition(A, max_iter=100):
    A_k = A
    Q_k = np.eye( A.shape[1] )
    for k in range(max_iter):
        Q, R = qr_decomposition(A_k)
        Q_k = Q_k @ Q
        A_k = R @ Q
    eigenvalues = np.diag(A_k)
    eigenvectors = Q_k
    return eigenvalues, eigenvectors
```

This algorithm is Iterative: In every step, we have to calculate A_{k+1} by considering the QR decomposition of A_k, reversing the order of Q &R, and multiplication of matrices with each other. The off-diagonals become meagre and at the end, we return the respective eigenvalues and eigenvectors.

Part 4: *Implementation of PCA Algorithm*

```
class PCA:
    def __init__(self, n_components=None):
        self.n_components = n_components
    def fit(self, X):
        n, m = X.shape
        # subtract off the mean to center the data.
        self.mu = X.mean(axis=0)
        X = X - self.mu
        # Eigen Decomposition of the covariance matrix
        C = X.T @ X / (n-1)
        self.eigenvalues, self.eigenvectors = eigen_decomposition(C)
        # truncate the number of components
        if self.n_components is not None:
            self.eigenvalues = self.eigenvalues[0:self.n_components]
            self.eigenvectors = self.eigenvectors[:, 0:self.n_components]
    descending_order = np.flip(np.argsort(self.eigenvalues))
    self.eigenvalues = self.eigenvalues[descending_order]
    self.eigenvectors = self.eigenvectors[:, descending_order]
    return self

    def transform(self, X):
        X = X - self.mu
        return X @ self.eigenvectors

    def proportion_variance_explained(self):
        return self.eigenvalues / np.sum(self.eigenvalues)
```

The following steps are followed before we implement PCA In the idealized form in the fit() function.

1. Making sure that the data is centered
2. Optionally we "whiten" the data such that each attribute has unit variance.
3. Keeping Eigenvalues in declining order

This is done by subtracting off the mean to center the data and figuring out the Eigen covariance matrix. Since we are performing dimensionality reduction we also truncate the number of components and finally sort the eigenvalues in the descending order to return them.

4. Results and Discussions

First performing the PCA for feature extraction is needed to reduce the dimensionality for cluster analysis. This part is implemented in the PCA module. Using the PCA module, the dataset was transformed to generate the first 3 components of PCA. Using this transformed data, the cluster analysis was performed with the K-means clustermg.

Algorithm:

```
pca = PCA(n_components=3)
pca.fit(x)
x_prime = pca.transform(x)
x_prime = pd.DataFrame(x_prime)
```

Front-end implementation of PCA

K indicates that algorithm is a iterative one which separates dataset into K pre-defined definite diverging clusters in which everydata is owned by only onecluster. Here, k value is 3. It makes the intracluster data points which are as analogous as feasible and paralllely considering the groups also as different or as far as possible. It allocates data points to a group so that total sum of $(\text{distance})^2$ between data points and groups centroid (AM of all data points belonging to group) is at minima. The minimum variation is observed within clusters (see Figure 1).

```
from sklearn.cluster import KMeans
kmeans = KMeans(n_clusters=3)
clusters = kmeans.fit(x_prime)
```

K-Means implementation

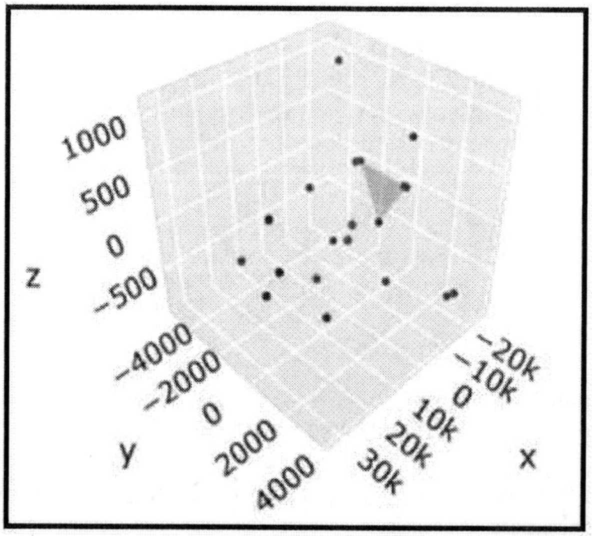

Figure 1. Similarity proportions between the regions of spain.

Visualizations:

Our dataset contains the compilation of the nominal GDP per unit of population for each zone/autonomous entity of Spain and the national average (in € / from 2000 to 2016). We first analyze the territorialized evolution ofthe nominal GDP per unit of population in Spain, Identify the economic circles, the spread between regions, and examine if there has been a convergence (in GDP per capita) between regions from 2000 to 2016 (Figure 2).

So we create a BOXPLOT to visualize the distribution of the nominal GDP per capita between the spanish regions (time period: 2000-2016)

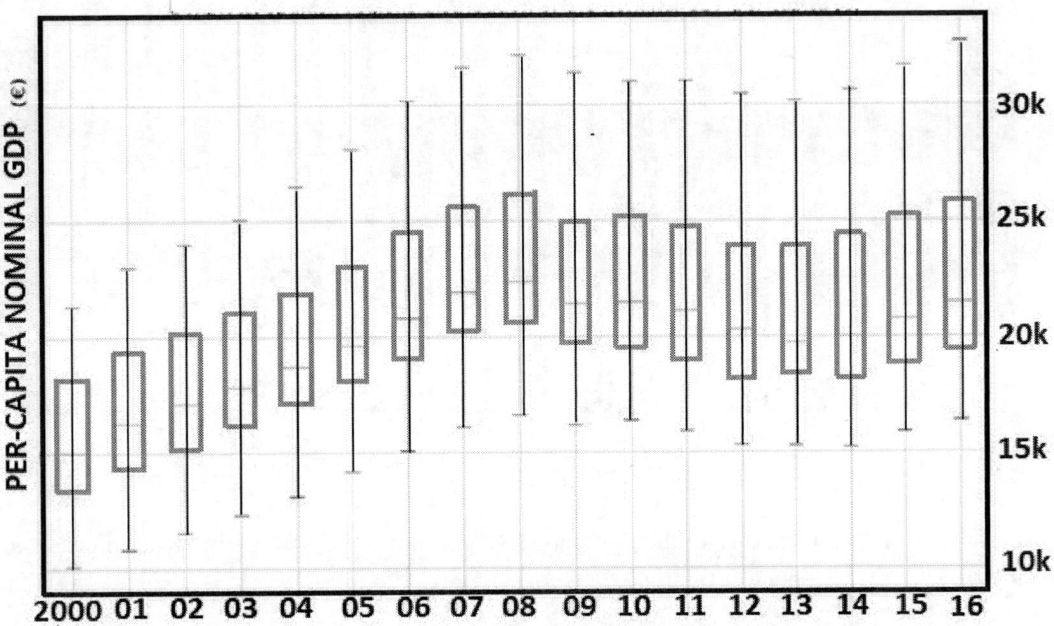

Figure 2. Boxplots, Spanish regions distribution of the nominal GDP per capita.

And we created a cloud PLOT to visualize the trend and spread of the regions as shown in Figure 3.

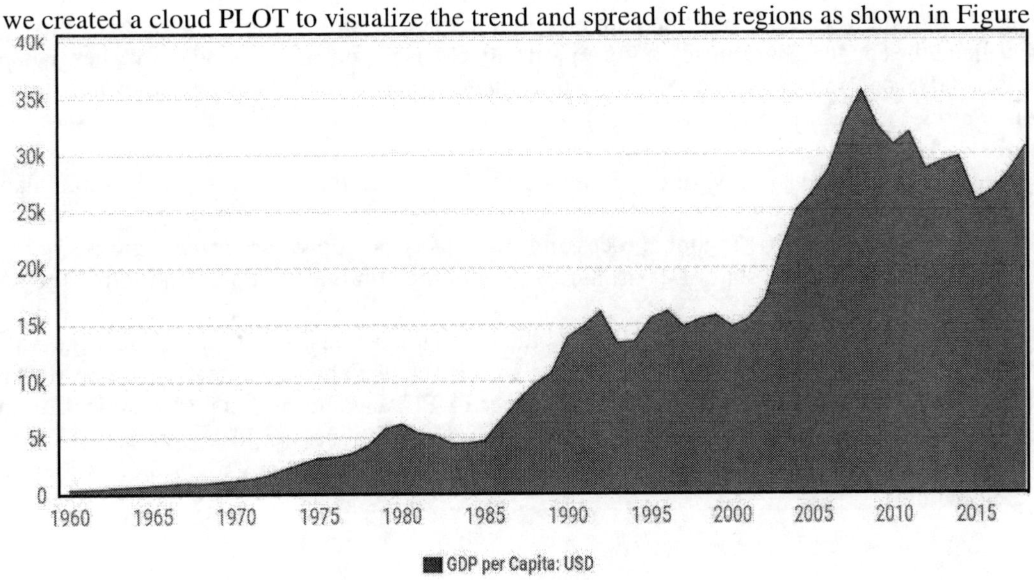

Figure 3. Nominal GDP per capita trend and differences across Spanish regions.

Both plots emanate a relatively high spread in the nominal GDP per unit of population between the sectors. To generate a visualization/examination of the divergence between regions we can concatenate the series the percentage difference of both the maximum and the minimum values by creating a new data frame. We can infer from the plots that the series is regular. The plot that describes the MIN and MAX nominal GDP per Capita across Spanish regions is shown in Figure 4.

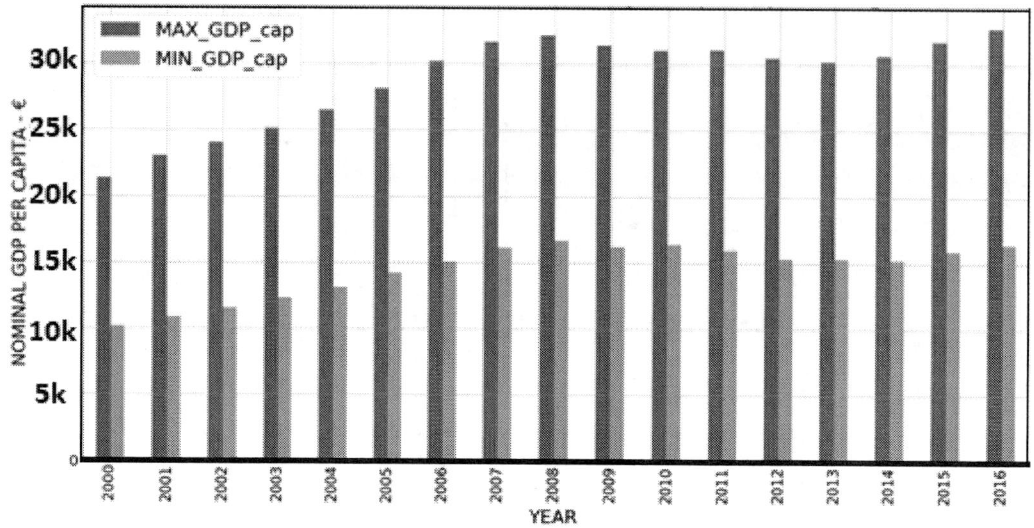

Figure 4. MIN and MAX values of the nominal GDP per unit of population in the Spanish sectors.

5. Conclusions

During the last 16 years, the spread between the maximum and the minimum values in the nominal GDP per capita over all the Spanish regions has remained very steady. The differences in 2016 are almost the same as those of the year 2000. We can conclude that in spite of the efforts by the Spamsh government to redistribute the country's wealth across its regions; there has not been a regional convergence in phrase of nominal GDP per unit of population. From the Cluster Analysis and the Plots, we can distinguish 3 economic circles (3 classes from the clusters)

1. From 2000 to 2008: a period of continuous growth of the nominal GDP per unit of population in all the regions.
2. beginning of 2008 to 2013: global recession started a strong downturn in all regions.
3. From 2013 to 2016: the start of a partial recovery of the nominal GDP per capita In all regions

PCA refers to a statistical approach which uses the orthogonal transformation to transfigure a set of observations of possibly corresponding variables into a set of values of linearly non-corresponding variables. Implementing neural networks with a layer of PCA would help us to optimize the model. There will be fewer chances of over fitting and the visualizations would be Improved. The mam advantage of PCA is to optimize the data and that would be helpful in any algorithm as it eradicates correlated features.

IOP Conf. Series: Materials Science and Engineering **925** (2020) 012012 doi:10.1088/1757-899X/925/1/012012

References

[1] Jolliffe IT and Cadima J. 2016 Principal component analysis: a review and recent developments *Philosophical Transactions of the Royal Society A: Mathematical, Physical and Engineering Sciences* **374** 20150202.

[2] L Li Y, Wang Nand Carroll RJ 2013 Selecting the number of principal components in functional data *Journal of the American Statistical Association* **108** 1284-94

[3] Sara Noori Mohammad Ali 2016 Risk Factor of Heart Attack in Sulaimany City *IOSR Journal of Nursing and Health Science* **5** 10-13

[4] Vinod HD 2011 Hands-on Matrix Algebra Using R: Active and Motivated Learning with Applications *World scientific publishing Co. Pte. Ltd, Singapore*

[5] Lolliffe IT 2002 Principal Component analysis, Second Edition *Springer Verlag, New York, USA, (Chapter 3 - Properties of Sample Principal Components* 29 - 59

[6] Vander Plas J 2016 Python Data Science Handbook First Edition *O'Reilly, Sebastopol, USA, (Chapter - Machine Learning - In Depth: Principal Component Analysis* 433 - 445

[7] Papi M and Caracciolo G 2018 Principal component analysis of personalized biomolecular corona data for early disease detection *Nano Today* **21** 14-7.

[8] Polo E, Collado M, Pelaz B and del Pino P 2017 Advances toward more efficient targeted delivery of nanoparticles in vivo: understanding interactions between nanoparticles and cells *ACS nano* **11** 2397-402.

[9] Boselli L, Polo E, Castagnola V and Dawson KA 2017 Regimes of biomolecular ultrasmall nanoparticle interactions *Angewandte Chemie* **129** 4279-82

[10] Caracciolo G, Farokhzad OC and Mahmoudi M 2017 Biological identity of nanoparticles in vivo: clinical implications of the protein corona *Trends in biotechnology* **35** 257-64

[11] Westmeier D, Chen C, Stauber RH and Docter D 2015 The bio-corona and its impact on nanomaterial toxicity *European Journal of Nanomedicine* **7** 153-68

[12] Mahmoudi M, Bertrand N, Zope H and Farokhzad OC 2016 Emerging understanding of the protein corona at the nano-bio interfaces *Nano Today* **11** 817-32

[13] Mirshafiee V, Kim R, Park S, Mahmoudi M and Kraft ML 2016 Impact of protein pre-coating on the protein corona composition and nanoparticle cellular uptake *Biomaterials* **75** 295-304

[14] Palchetti S, Digiacomo L, Pozzi D, Peruzzi G, Micarelli E, Mahmoudi M and Caracciolo G 2016 Nanoparticles-cell association predicted by protein corona fingerprints *Nanoscale* **8** 12755-63

[15] Schöttler S, Becker G, Winzen S, Steinbach T, Mohr K, Landfester K, Mailänder V, Wurm FR, 2016 Protein adsorption is required for stealth effect of poly(ethylene glycol)- and poly(phosphoester)-coated nanocarriers *Nat. Nanotechnol.* **11** 372-377.

An Efficient Association Rule Mining Algorithm Based on Animal Migration Optimization Processing of Unknown Incidents in Crime Analysis Brance

Jagadeesh K R[1] and Kumanan T[2]

[1] Department of Computer Science, Meenakshi Academy of Higher Education and Research, Chennai, India.
[2] Faculty of Engineering and Technology, Meenakshi Academy of Higher Education and Research, Chennai, India
[1]jaganbeema@gmail.com, [2]kumananvetri@gmail.com

Abstract. A new protection scheme was proposed to avoid this problem. It is intended that each node uses a salted and a not salted HLL. If their estimates differ considerably, an attacker attempts to manipulate the estimates of HLL. In addition to avoiding manipulation, the proposed salted and unsalted (SNS) regime can also detect attempts at manipulation. A practical configuration showing how manipulation attempts can be detected in a low false positive probability has been shown to be applicable to this SNS scheme. Therefore, if merge ability is to be preserved it can be an interesting approach to protect HLLs from avoidance. In this paper the proposal for a new mining algorithm based on Animal Migration Optimization is made to decrease the number of Association Rules called ARM-AMO. The idea is to remove from the data rules which are not highly supportive and unnecessary. First of all, common item sets and association rules are generated with an Apriori algorithm. AMO also reduces the number of association rules incorporated in a new fitness function. In here, we provide a well-organized mechanism for incident derivation under the unwanted incident. This mechanism very useful for measure the heavy load of an incoming incident and exact calculation of the probability. In additional method is a Select-ability mechanism, which performs an important responsibility in incident derivation under the unwanted incident in both the settled and the unknown incident. A model for signifying derivative incident introduced jointly with an Advanced Sampling Technique that come close to the derived incident probabilities. This augmentation executed the prioritization techniques. In this prioritization techniques, recognize such cases in which the order of incident finding is strong-minded and mechanism for the definition of a settled detection execution.

Keywords: Unknown incident prediction, crime analysis, prioritization, computation overhead, rule reduction.

1. Introduction

Although crimes are a thought with different definitions, the definition of crime, generally accepted, consists of actions beyond the moral values of society which are fated by society as well as which require lawful penalty from administration [1]. In studying various scientific fields such as sociology, psychology and criminology, the concept of crime is considered [2]. In order to address crime and detect offenders, furthermore, the advantages of technology as well as the analysis of assistance of informatics is important. Due to its increasing complicated nature, new approaches to case analysis and the detailed identification of similarities among crime records are a major problem for criminology [3] as criminal records have become more complicated than before. Big data stacks which include before verified crimes can be investigated using big data approach and data mining techniques to disclose the linkages and the relationships between the characteristics. This offer significant advantages for offenders, resolving criminal cases as well as taking appropriate defences before illegal cases are taken.

Although an illegal case is generally instantaneous as well as unpredictable, a possible prevention exists in certain areas if corruptions of the same type are dedicated by offenders of the similar profile in the same manner [4]. Analysis of crimes may be used to identify a wide range of links and deductions on records of crime, including the characteristics of offenders, victims, characteristics of cases, weapons and type of behavior of crime, geographical distribution and causal relations Various studies on the analysis of crime and profiling of offenders [5] have been carried out. Neuronal reproduction network as well as social media analytics are used to analyze crime-based research on data mining techniques [6].

The main contribution of this research work is to detect the unwanted incidents happening on the scenario through association rule mining algorithm. In this research work crime analysis is considered where unwanted incidents are happening through database analysis. This is attained by generating the association rules using apriori algorithm and reducing the number of generated rules using animal migration algorithm. The performance assessment is carried out in the matlab to ensure the performance improvement in the proposed techniques.

The complete organization of the research effort is given as follows: In this section, thorough overview about the crime analysis, unwanted incidents happening in crime department and the need of association rules are given. In section 2, discussion of different related research works is given. In section 3, detailed discussion about the planned technique along with suitable examples as well as diagrams. In section 4, performance assessment of the research work is mentioned. Lastly in section 5, complete conclusion of the research work is given.

2. Related Works

Sudhakar in the process of unifying crime data, Chau et al. [7] have industrialized a new method based on police reports as well as dictionary search, rules, statistics as well as machine learning. Smearing the inputs from the extraction model built on the artificial neural network, the report attempted to recognize the usable objects.

Lin and Brown's research [8] seeks identify criminals by using rules on the crime of theft. This study uses the Apriori algorithm as well as association rules are found to produce fine results on data from crime. Their study shows that the true rate of prediction increases according to the increase in data in the dataset.

Kaza et al. [9] developed the predicting illegal relationship algorithms that used to predict automatically the buses that are a co-offender to prevent the upcoming attacks. They used the lively social network analysis (SNA) methods as well as multivariate survival analysis.

Sharma [10] proposed an enhanced ID3 algorithm, an improved feature selection technique as well as an attribute-importance feature to classify e-mails as either maybe-suspicious or non-suspicious e-mails. Also they used a tool that is named as zero crime to help the system identify e-mails in relation to illegal actions.

Li et al. [11] has presented comparing the performance of the event ontology method as the priori information and the method based on Support Vector Machine to analyze the attributes as well as relations in web pages. Also these methods are used to reconstruct the scenario for crime mining.

Fogel et al. [12] proposed the use of changed neural networks and evolved rule-based classifiers. Both methods are useful to distinguish between toxic via narcotic and reactive mechanisms of action (MOAs) of small molecules.

Sathyadevan [13] anticipated the use of naive bayes algorithm through the concept of named entity recognition (NER), also recognized as entity or element removal, to classify the news trainings into the crime type and to create a crime model. In addition to that, Apriori algorithm is used to find and create frequent patterns in crime by training crime data from the different web sites. For prediction in crime, they used the decision tree concept.

Thongsatapornwatana and Chuenmanus [14] proposed the suspect vehicle recognition system using association rule to analyze the vehicles with forged license plate crossing the barrier that potentially involved the illegal activity.

3. Unknown crime incident prediction using optimization procedure

This work proposes to decrease the number of association instructions by proposing a new mining algorithm founded on the Animal Migration Optix (AM O) known as ARM-AMO. The idea is that rules that are not highly supportive as well as needless are removed from the data. The first thing to be done is to produce common itemsets and association rules using Apriori algorithm.

3.1. Apriori algorithm to generate frequent item sets plus association rules

An algorithm to learn relational databases as well as rule. In 1994, Agrawal as well as Srikant planned the Apriori algorithm. Apriori is intended for use in transaction databases. Other procedures are planned to discovery association rules for data that do not have transactions or time marks (DNA sequencing). (Winepi and Minepi). A set of items (a set of items) are perceived each transaction. The Apriori algorithm uses a threshold C to identify the articles that constitute a subset of C transactions at least.

The applicant length k of the item k-1 sets is produced. The candidates with a pattern that is unfamiliar are then prunes. According to the lower end lemma, the candidates set comprise all the frequent k-length sets. The database is then scanned for common item sets between the applicants.

For an operation database T, in addition to for a provision threshold of μ, the pseudo code for the procedure is assumed below. The common set notation is used, but note that T is a multi-set. Ck is the level k applicant. The procedure is supposed to produce the applicant sets from the huge item sets of the previous level, which are responsible for the lower closure of the lemma. Count [c] accesses the field of the candidate's data structure c assumed to be zero in the beginning. Many details are omitted below, the data structure used for storing candidate sets and for counting their frequencies is typically the main part of implementation.

Apriori (T, ε)
    $L_1 \leftarrow$ {large 1 - itemsets}
    $K \leftarrow 2$
    While $L_{k-1} \neq \emptyset$
$C_k \leftarrow \{c = 1 \cup \{b\} | a \in L_{k-1} \wedge b \notin a, \{s \subseteq c | |s| = k-1 \subseteq L_{k-1}\}\}$ For transaction $t \in T$
$.D_t \leftarrow \{c \in C_k | c \subseteq t\}$
For candidates $c \in D_t$
Count [c] $\leftarrow$ count [c] +1

$L_k \leftarrow \{c \in C_k \mid count[c] \geq \varepsilon\}$
$K \leftarrow k+1$
Return $\cup_k L_k$

This procedure follows three steps:
1. For l after 1 to I do
2. For each set JI such that for both h ϵ JI happens in at least k bags do
3. Inspect the data to determine whether the set JI occurs in at least k bags

Greatest of its period was spent in accessing the database for this algorithm pending it results in a common association match. On this basis, two quantities of the probability model were calculated:
1. Success rate: the possibility that the set is a success
2. Failure rate: the possibility that the set is a failure

With this possibility model, it brings out the main performance features of the procedure.

3.2. Animal migration optimization to reduce number of association rules

The above section will be resultant with the set of association rules for the crime incident patterns. These generated rules will be huge in numbers which will increase the computation overhead of crime pattern detection. It is resolved in this research work by reducing the number of rules generated considerably. This is achieved through the introduction of the method of optimizing animal migration. Relocation is a common wonder in ecology of animal behavior, which has been intensely studied in the animal kingdom. The migration is persistent and streamlined and influenced by the locomotive efforts of the animal to new habitations. It be contingent on an impermanent inhibition of the response of the station, but encourages its eventual disinhibition and recurrence. The relative prolonged movement of individuals is animal migration, usually on a cyclical basis. This is an omnipresent wonder found in all main groups of animals, such as birds, fish, mammals, reptiles, insects, amphibians plus crustaceans. Local weather, local supply of food as well as the season of the year can be the trigger for the relocation. The typical picture of relocation is that of northern landslides, like swallows plus birds of prey, which take long tropical journeys.

There are two processes for optimizing animal migration: migration and updating the population. At the beginning, the procedure simulates how the animal collections move. The person is referred to as a true coded vector of the D dimensions. Started that uses the vector of the NP D dimension parameter to the minimum as well as maximum limits prescribed, where NP indicates the population size:

$$\vec{X}_{min} = \{x_{1,min}, x_{2,min}, \cdots, x_{D,min}\}$$
$$\vec{X}_{max} = \{x_{1,max}, x_{2,max}, \ldots, x_{D,max}\}$$

Therefore, we can initialize the jth component of the ith vector as

$$x_{j,i,0} = x_{j,min} + rand_{i,j}[0,1].(x_{j,max} - x_{j,min})$$

Where, rand i,j[0, 1] is a uniformly distribution random number between 0 as well as 1, i = 1, …, NP and j = 1, …, D.

The animal should follow three rules during the migration phase: (1) to escape crashes with its neighbours; (2) to move in the similar direction as its neighboring animals and (3) to stay near to the neighbours. For the main rule, we need to change each person's position in the group. We require the latter two rules that the person be placed in new positions according to his neighbours' current positions. Particular animals leave the collection as well as then new animals joint the new population during the population update process. We assume that the number of animals available is determined and that a new person with probability Pa will replace the animals. Depending on the fitness quality the probability is used. The Pa probability is 1. For best fitness. The possibility is 1/NP for the nastiest fitness.

Algorithm: Animal migration optimization algorithm

IOP Conf. Series: Materials Science and Engineering **925** (2020) 012013 doi:10.1088/1757-899X/925/1/012013

Begin
Set the generation counter G=1; and randomly initialize a population of NP animal X_i
    Evaluate the fitness for each individual in P
    While stopping criteria is not satisfied do
        For i=1 to NP do
            For j=1 to D do
                $.X_{i,G+1} = X_{i,G} + \delta.(X_{neighbourhood,G} - X_{i,G})$
            End for
        End for
        for i=1 to NP do
            Evaluate the offspring $X_{i,G+1}$
            If $X_{i,G+1}$ is better than X_i then
                $X_i = X_{i,G+1}$
            End if
        End for
        for i = 1 to NP
            For j=1 to D
                Select randomly r1 ≠ r2 ≠ i
                If rand > P_a then
                    $X_{i,G+1} = X_{r1,G} + rand\,(X_{best,G} - X_{i,G}) + rand(X_{r2,G} - X_{i,G})$
                End if
            End for
        End for
        For i=1 to NP do
            Evaluate the offspring $X_{i,G+1}$
            If $X_{i,G+1}$ is better than X_i then
                $X_i = X_{i,G+1}$
            End if
        End for
        Memorize the best solution achieved so far
    End while
End

By using the above algorithm number of rules generated will be reduced, thus the unwanted incidents can be predicted accurately with lesser computation overhead.

4. Results and discussion

In this segment, the proposed model of associative classifier is evaluated in MATLAB. The performance metrics that are considered in this research method for the estimation of the planned and existing research method are "Accuracy, Precision, Recall, F-measure, Number of rules and Error rate".

Accuracy: Precisions are the most intuitive measurements of performance and are simply a ratio of accuracy to the total observations.

Accuracy = (TP + TN) / (TP + FP + FN + TN)

The graphical comparison representation for the above simulation values are illustrated in the following figure 2.

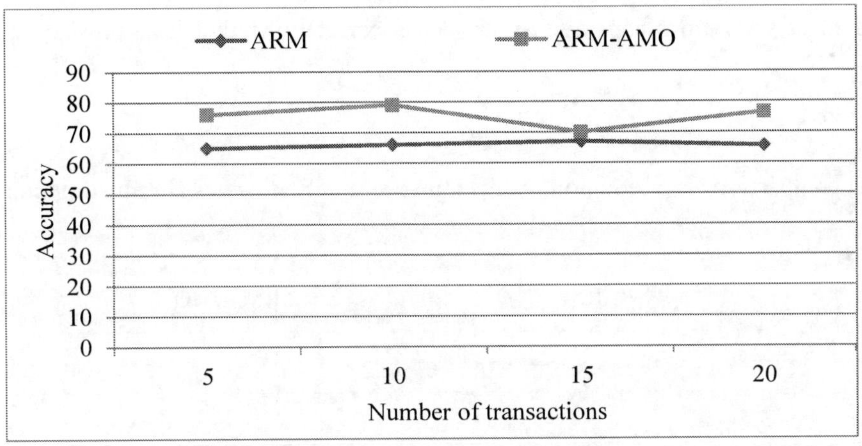

Figure 1. Accuracy comparison

In figure 1, graphical comparison illustrated for the performance metrics accuracy between the proposed ARM-AMO method and the existing ARM method. From this comparison it can be proved that the planned method ARM - AMO is increased in its accuracy value which is 14.42% better than the existing ARM method.

Precision: Precision is the ratio of positive observations correctly anticipated to the total positive observations forecast. The low fake positive rate refers to high precision.

Precision = TP / TP + FP

The graphical comparison representation for the above simulation values are illustrated in the following figure 2.

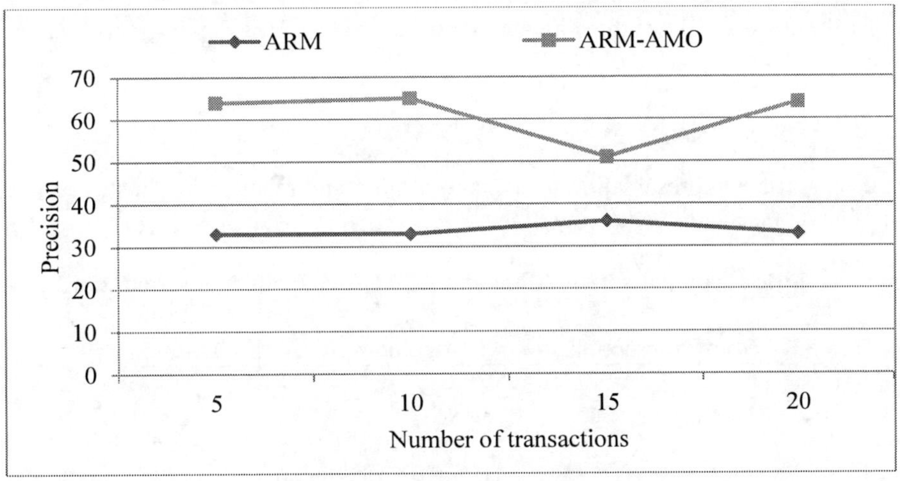

Figure 2. Precision comparison

In figure 2, graphical comparison illustrated for the performance metrics precision between the proposed ARM-AMO method and the existing ARM method. From this comparison it can be proved that the planned method ARM - AMO is increased in its precision rate which is 80.74% better than the existing ARM method.

Recall: Remember the ratio of positive observations correctly predicted to those in all actual class observations – yes.

Recall = TP / TP + FN

The graphical comparison representation for the above simulation values are illustrated in the following figure 3.

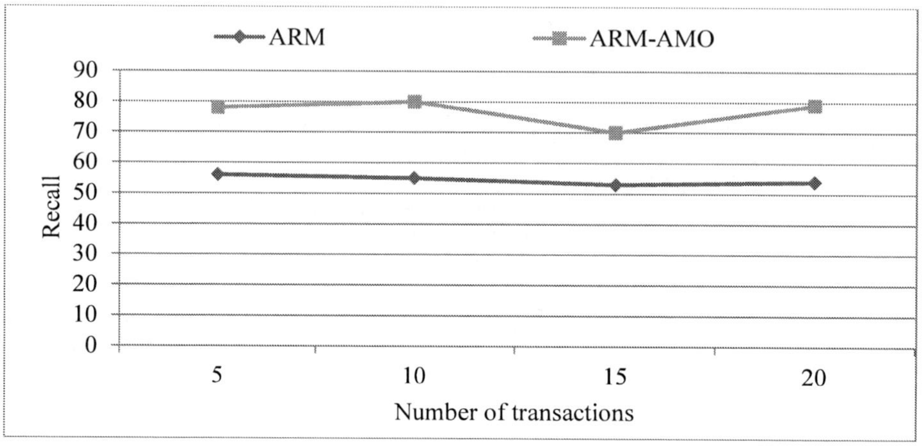

Figure 3. Recall comparison

In figure 3, graphical comparison illustrated for the performance metrics recall between the proposed ARM-AMO method and the existing ARM method. From this comparison it can be proved that the planned method ARM-AMO is increased in its recall rate which is 40.83% better than the existing ARM method.

F-measure: F1 the correct and weighted average reminder value. This ranking thus takes false positives as well as false negatives into account.

F1 Score = 2*(Recall * Precision) / (Recall + Precision)

The graphical comparison representation for the above simulation values are illustrated in the following figure 4 and table representation shown in table 1.

Table 1. comparison of ARM and ARM-AMO value representation

No of transactions	ARM value	ARM-AMO
5	55	78
10	53	81
15	52	70
20	51	80

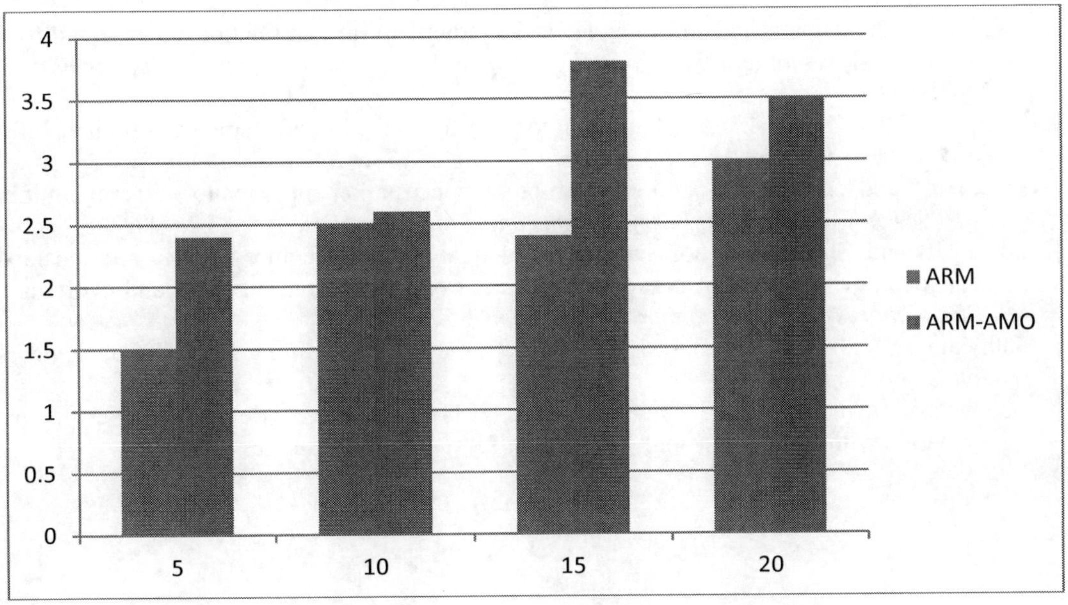

Number of Transactions

Figure 4. F-Measure comparison

In figure 4, graphical comparison illustrated for the performance metrics F-Measure between the proposed ARM-AMO method and the existing ARM method. From this comparison it can be proved that the planned method ARM-AMO is increased in its F-Measure rate which is 10.81% better than the existing ARM method.

5. Conclusion

This work proposes to decrease the number of suggestion rules by proposing a new mining procedure based on the Animal Migration Optix (AM O) known as ARM-AMO. The idea is that rules that are not highly supportive and unnecessary are removed after the data. The first thing to be done is to produce common itemsets as well as association rules using Apriori algorithm. The performance assessment of this research has demonstrated that the technology proposed tends to show better results with an unwanted accurate incident detection rate than the existing method.

References

[1] Edelstein A 2016 Rethinking conceptual definitions of the criminal career and serial criminality *Trauma, Violence, & Abuse* 17 62-71.

[2] Feeley M 2017 The concept of laws in social science: A critique and notes on an expanded view *In Crime, Law and Society* 19-45.

[3] Sevri M, Karacan H and Akcayol MA 2017 Crime analysis based on association rules using apriori algorithm *International Journal of Information and Electronics Engineering* 7 99-102

[4] Adolphs R, Gläscher J and Tranel D 2018 Searching for the neural causes of criminal behavior *Proc. of the National Academy of Sciences* 115 451-2.

[5] Canter D and Alison L 2018 Profiling property crimes. In Profiling property crimes 11-40.

[6] Thongsatapornwatana U 2016 A survey of data mining techniques for analyzing crime patterns *Asian Conf. on Defence Technology* 123-128.

[7] Chau M, Xu JJ and Chen H 2002 Extracting meaningful entities from police narrative reports.

[8] Lin S and Brown DE 2006 An outlier-based data association method for linking criminal incidents *Decision Support Systems* 41 604-615.

[9] Kaza S, Hu D, Atabakhsh H and Chen H 2007 Predicting criminal relationships use multivariate survival analysis *Int. conf. on Digital government research: bridging disciplines & domains* 290-291.

[10] Sharma M 2014 Z-CRIME: A data mining tool for the detection of suspicious criminal activities based on decision tree *Int. Conf. on Data Mining and Intelligent Computing* 1-6.

[11] Li C, Hu Y and Zhong Z 2010 An event ontology construction approach to web crime mining *In 2010 Seventh Int. Conf. on Fuzzy Systems and Knowledge Discovery* **5** 2441-2445.

[12] Fogel GB and Cheung M 2005 Derivation of quantitative structure-toxicity relationships for ecotoxicological effects of organic chemicals: evolving neural networks and evolving rules *IEEE Congress on Evolutionary Computation* 1 274-281.

[13] Sathyadevan S 2014 Crime analysis and prediction using data mining *Int. Conf. on Networks & Soft Computing* 406-412.

[14] Thongsatapornwatana U and Chuenmanus C 2015 Suspect vehicle detection using vehicle reputation with association analysis concept *In Tourism Informatics* 151-164.

IOP Conf. Series: Materials Science and Engineering **925** (2020) 012014 doi:10.1088/1757-899X/925/1/012014

Twitter Spammer Identification using URL based Detection

Yosef Hasan Fayez Jbara[1] and Hyder Ali Segu Mohamed[2]

[1] Department of Cyber Security College of Engineering and Information Technology, SaudiArabia.
[2] Department of Computer Engineering College of Engineering and Information Technology, Saudi Arabia
[1]yosefjbara@yahoo.com, [2]hyderfathima@yahoo.com.

Abstract. Millions of users worldwide participate in social networking sites. The interactions between users and those social sites, as Facebook and Twitter, have an enormous effect as well as sometimes unwanted impact on daily life. Spammers have become a target platform on the prominent social networking sites to scatter huge amounts of damaging and irrelevant information. Twitter has become one of the best-used platforms of all times also thus permits unreasoning spamming. Fake users send unwanted tweets to users to encourage services or sites that affect not just genuine users, but also disrupt the consumption of resources. The main aim of this study is to identify forged users who harm real users, threaten users' identity and therefore their safety and privacy. In this paper, suggest to URL based detection plus machine learning techniques has been developed for the detection of malicious users, celebrities as well as malicious users, based on user characteristics. A crawler is developed for Twitter for the purpose of detecting malicious users, celebrities and non-malicious users and celebrities, and data from around 22 K users were collected from public information. For training and testing purposes, data from around 7,500 users were used in machine learning classifications to classify users. On basis of performance metrics like accuracy, recall, F-measurement also accuracy 5 classifiers were used and compared. With a precision of 99.8 percent, Random Forest is above all classifiers.

Keywords: Celebrities, Malicious users, Non-malicious users, Online Social Networks (OSNs), Twitter.

1. Introduction

Twitter is now a popular media hub where news, jokes, moods and news events can be shared. Users can immediately send tweets to their followers on Twitter. Twitter's real-time search engine may also be used to recover tweets. The ranking of tweets in this search engine varies according to several factors, including the number of users. The popularity of Twitter has made this a popular destination for spammers of all kinds. Spammers have various objectives: to spread ads for sales, phishing or simply to jeopardize the reputation of the system. Fake accounts can be falsified, misleading and spam-powered. The Twitter Rules violate false accounts. They act in an unlawful way. Automatic account connections or attempts for people to mislead, such as posting hurtful links, violent behaviors such as mass tracking, creating many accounts, posting topics or replacing updates, posting links with non-related tweets, plus misusing answering as well as mentioning features, can be used. Real accounts are Twitter rules compliant accounts. Sending email or sending text messages may be used to post tweets. Twitter enables users, directly from smart phones, to publish also exchange 140 messages with various web-based services. Twitter distributes information to a large group of real-time users. Two most popular social networks have been compared in Face book and Twitter as shown in **Table 1.**

Statistics(millions)	Facebook	Twitter
Overall registered users	1380	2000
Fake users	155	42
Compromised accounts	0.6	-

It's the second most popular social website, with a faster increase in the number of malicious users as shown in table 1. We have crawled about 22k users on the Twitter site to detect a large number of user-data. Then we shaped a collection labeled with users classified manually as non-malicious, malicious as well as famous people. We then planned the URL based detection with machine learning techniques attributes of the user behavior to understand their relative discriminatory power in distinguishing malicious users from both types of non-malicious users as genuine and famous. Spammers can use their accounts for different purposes and one of the highest problems in social media. One of these objectives is to disseminate rumors that can have a large impact on a certain business or even on society. In this study we objective to identify the fake profiling accounts of the social network online in order to prevent fake and hurtful news, publicity and fake followers spreading in accordance with the significance of the influence of the social media on society. The related work will be presented, the proposed method will be described, the results analyzed and the research subject will end and our future work will be presented. These characteristics have been utilized for computation of two parameters, i.e. the User Score also the Tweet Score. The feasibility of classifying algorithms was tested using training as well as testing dataset of nearby 7000 users. Machine learning classification used measurement metrics such as accuracy, reminder, exactness and F-meter to compare about 5 classification algorithms. Random Forest classification approach has been found to correctly recognize the majority of malicious users, with only a few legitimate users being misclassified. Figure 1 shows types of spammer detection in twitter.

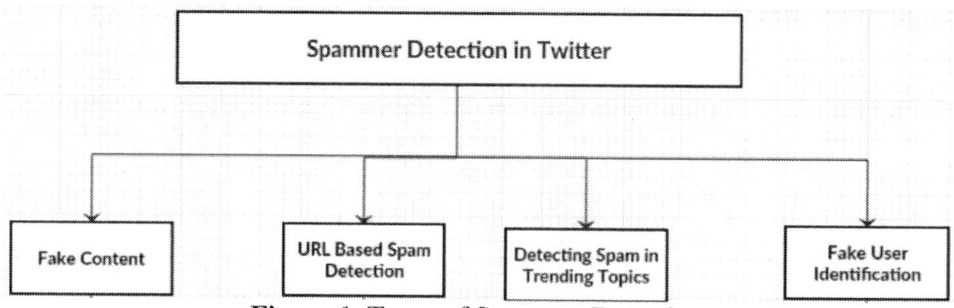

Figure 1. Types of Spammer Detection

1.1. Maliciousurl detection

We introduce for the first time the key concepts of malicious URL identification and then formalize it as a master learning function. A variety of solutions to the harmful URL identification issue have been evaluated. According to the fundamental principles, we categorize them into: (i) Machine Learning approaches, besides (ii) Black listing or Heuristics.

2. Related works

This paper discusses the methods used to identify Twitter spammers. In addition, Twitter spam detection taxonomy reveals [1] that class It is using strategies focused on its capacity to detect: I fake material, (ii) URL-related spam, (iii) spam in trendy subjects and (iv) fake users. In comparison, specific aspects such as user characteristics, material characteristics, design characteristics, structuring and time aspects are the approaches offered. We refer to multiple aspects in this article. Firstly, we are discussing some of the most relevant functionalities and guidelines for an anomalous Twitter account identification (proposed by Academia and the media). Secondly, we are building the checked person and false accounts gold [2] standard. We would then use the Gold Standard to train the updated rules and characteristics of a series of machine-learning classifications. [3] Most Media laws give inadequate success in reporting fake supporters, while Academia's output in spam identification is nice. We refine classifications, also in terms of over fitting elimination, centered on the most promising apps. We present in this paper machine learning algorithms developed for the detection of fake Twitter followers. [4] We manually checked 13,000 purchased fake followers and 5,386 genuine followers on basis of an account created for our study. Then we recognized a number of features that make fake and genuine followers distinctive. We used these features to classify users as fake or frank as attributes for the learning algorithms machine. In our survey, we present a method of classifying fake Twitter accounts. In order to analyze the results of the algorithm Naïve Bayes, [5] we have preprocessed our dataset using a controlled decertification technique called Entropy Minimization Discretization. These false accounts cause several problems, such as consumption of resources and inaccurate results of studies. Often, social network operators determine these counterfeit accounts by employees in full-time. Many researchers have suggested [6] methods to detect fake accounts in social networks online. A new technique that relies on HTML Canvas Fingerprint to identify the accounts of the individual or entity is proposed in this paper. In this post, we present a way to identify fraudulent Twitter accounts. The study defines the minimal set [7] of key factors affecting fake Twitter accounts and uses different classification techniques to then apply the determined factors. The results of these technologies have been compared and according to the precision of the tests, the most accurate algorithm is chosen. Within this paper we built a way of identifying all false followers [8] on a social diagram network based on the centrality of all the nodes in the diagram and of training a classifier centered on a dataset sub-set. The proposed method achieved high precision on fabricated followers identification utilizing only graph-based centrality steps. The approach introduced is standardized in nature and can be utilized independently of the social network site. The scale of the population regulated by an entity, is a vital indicator of the influence of that institution [9] in the online media networks and has significant economic and/or political consequences. These attempts in order to calculate users' interest or manipulate their audience awareness were hampered by the existence of fake profiles. This analysis examined 62 million Twitter user accounts, open to the public, and established a technique for detecting fake profiles created automatically. Mobile Health Observation Systems (MHOS) encounter more difficulties than general health facilities in hospitals when accessing individual patient information [10]. Physicists have also been involved in wireless technology and since it was established this area has been successful. This area has nevertheless received a renewed interest and strong developments over the last ten years, supported by the telecommunications sector in general [11].

3. Proposed method

3.1. Arriving of face project

On 12 December, the False Project started running, The launch of the @TheFakeProject Twitter account in 2012. The following motto is stated in its profile: me only when you're not a fake and that's why The program is aligned with a research project. The IIT-CNR, Pisa-Italy researchers. The biography of the account indicates the web page of the project1. And on this list may find instructions and overall instructions for joining the initiative Motivations and mission goals overview. At a the owners contacted additional researchers in the first stage and The initiative's journalists to advertise [2].

3.2. URL based spam detection

An evaluation of machine learning algorithms for spam tweets has been conducted. For instance, I the spam-not-spam ratio, (ii) the training data collection, (iii) time dependent data, (iv) factor discretization, as well as (v) sampling data, the Authors evaluated the effect on Spam detection efficiency. The following characteristics have been discussed: To test the identification, nearly 600 million public tweets had first been obtained and the writers then used the Pattern Micro web-based credibility scheme to as much as possible classify spam tweets. In addition, a total of 12 lightweight apps were differentiated from this identity data collection from non-spam and spam tweets. The personalities are defined by origins. It is taken into consideration in spam groups focused on machine learning, which later are used for the assessment of spam identification in the trial. To replicate various situations four data sets are pooled. Although no dataset is freely accessible for the mission, in previous researches few datasets were used. 12 apps were obtained after the identification of the spam messages. These tasks are split into two groups, namely user-based apps and tweets. The web-based interface is similar across various elements such as account age plus amount of app likes, lists as well as tweets. JSON specification analyzes the identification of user-based apps. The tweet features, on the other hand, provide the amount of I retweets, (ii) hashtags, (iii) user statistics, as well as (iv) URLs. The assessment results indicate that the evolving distribution of features reduced efficiency, although no changes were seen in the distribution of training datasets. Figure 2 shows URL Based spammer detection.

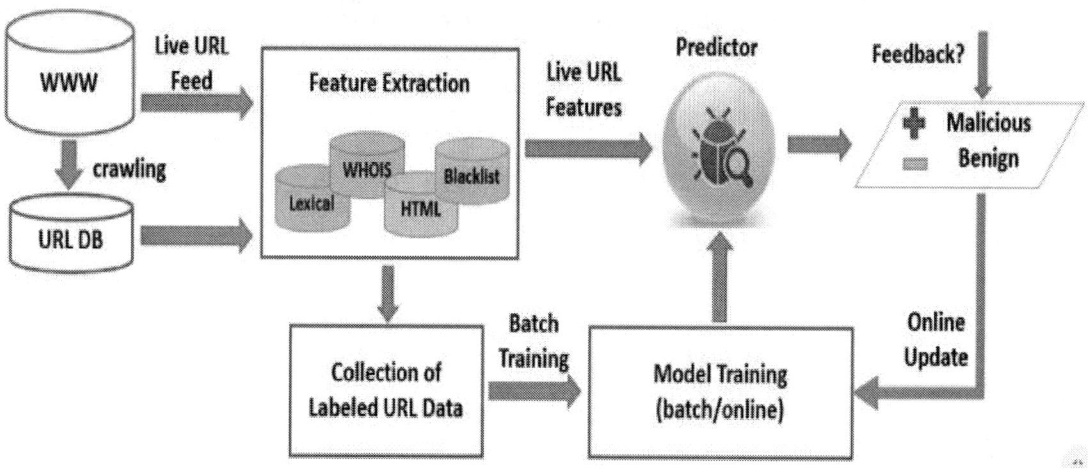

Figure 2. URL Based spammer detection

3.3. Machine learning approaches

Such methods aim to evaluate URLs and their websites by extracting strong URL representations and train a prediction model with hostile and benevolent data processing. Two styles of functions should be considered-static and dynamic characteristics.

3.4. User data collections

We need a research set of users in order to validate the recommended method for detecting abusive, abusive and celebrity individuals on Twitter Social Network. It is important to pre-classify this set into three target groups. As this information is not accessible freely on all of the social media platforms, appropriate stages have also been taken on plan the compilation of user data. Until providing the steps to construct test data, certain words must be described. Under Twitter policy a fraudulent person follows a significant number of people in a short retro, even if his post comprises mostly of links or whether it is used to share random data or to regularly update messages of another person. A non-malicious person is the one who is a true consumer of virtually identical friends and supporters and who tweets mildly. There are users who possess massive numbers of messages. We use the word celebrity for these users to indicate popular people like celebrity / media entities or broad crowds. First, a compilation and evaluation of a subset of such consumers was carefully performed.

3.5. Crawling twitter

A Twitter 4j, Java library for the Twitter API, was designed to acquire a sampling of malicious as well as non-malicious users. Published datasets were obtained utilizing the Twitter REST API, which operates by demanding a particular data form. This implies user information such as IDs, name of the computer, location of the app, information of a relative, followers etc. The JSON was received encoded. A rate limit of 350 host requests per hour for calls to the API is set. Twitter is consistently crawled over a 5-week span of 300 demands an hour for the purpose of avoiding congestion, reaching a minimum of 21,492 people for their most recent 20 posts.

3.6. Classification of users

This portion investigated the viability of implementation for malicious and non-malicious applications of different classification algorithms. Learning algorithms to classify from previously categorized data as well as then gained skills are used to identify unseen users into preferred malicious consumer and celebrity groups.

4. Results

The successful usage of measures such as uncertainty chart, recall, precision, F-message and accuracy has been analyzed. Remembering implies the proportion of users categorized accurately to the amount of users within a group. Precision is specified as the proportion of the number of users correctly categorized to the total class users foreseen. Accuracy is defined as the complete correctness. Confusion matrix has been used to explain these metrics. Recall R, Precision P, as well as accuracy A of the class malicious is calculated as:

$$R= a/ (b +c + a), \quad (1)$$

$$P= a/ (a +g +d), \quad (2)$$

$$A= a/ (c +b +a +d +e +i +g +h +f), \quad (3)$$

F-measure is the harmonic mean between precision as well as recall and is given as

$$F=2PR/ (P+R). \quad (4)$$

7434 users were gathered from the 21,492 samples based on the required user score as well as tweet performance parameters as stated. Of the 7434 applications, 5203 were used for instruction (around 70 per cent), while 2231 (30 per cent) remained for testing. Tests of the different classification algorithms were given. Table 2 shows the outcomes of all 5 machine learning classifiers used with their recall, precision, F-measure as well as accuracy as debated in below.

Table 2. Comparison of machine learning algorithm

No	Machine learning Classifier	Value of Precision	Value of Recall	Value of F-measure	Value of Accuracy
1	BayesNet classifier	99.3	99.3	99.3	96.3
2	Naïve Bayes classifier	95.6	88.5	90.6	95.9
3	SVM classifier	84.5	87.6	85.3	81.9
4	J48 classifier	98.6	99.5	99.6	96.3
5	Random Forest Classifier	99.7	99.7	99.6	99.2

Table 2 explicitly demonstrates that Random Forest provides a maximum precision of 99.2 per cent, which is not unexpected, as this classifier will operate on a data set that does not have a balanced data set. The minimum precision of the SVM rating was 81.9 percent. In Figure 3, plots of precision, accuracy, recall, F-metal and precision is seen.

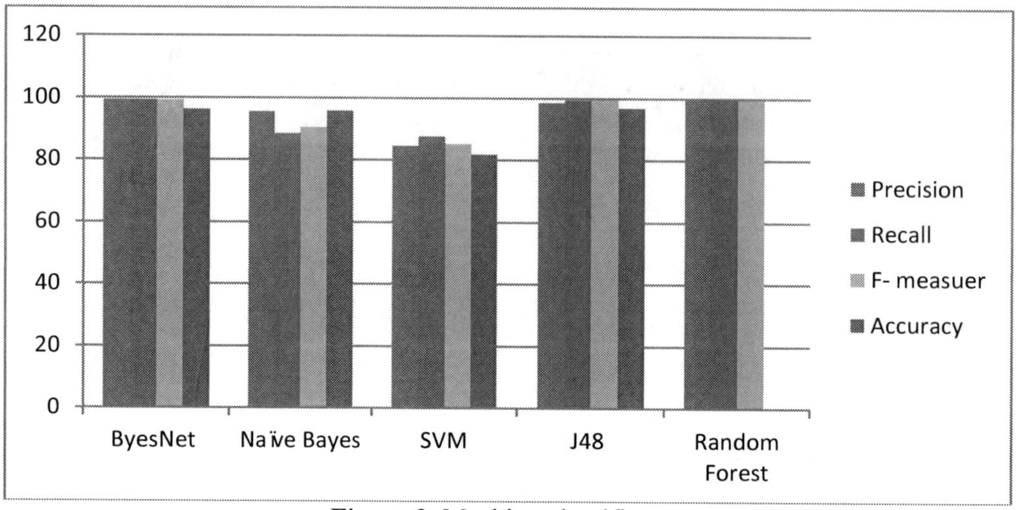

Figure 3. Machine classifiers compared

5. Conclusion

The Virtual Social networks may be explicitly mentioned as a spammer goal. The details exposed on an OSN may be used by an intruder to render the identity of a profile holder impossible, intimidating, impersonating or even harmful. The health and anonymity of social networking users was severely endangered by cyber-attacks. In this article Twitter has been confronted with the topic of finding abusive and offensive users on the most popular website. Crawler has been configured to provide around 22 K user accounts, their messages and follow-up connections on the Twitter website. Some attributes dependent on the users, such as follower count, tweet count, development date, etc., were analyzed, which are helpful for identifying aggressive, negative, and celebrities. The Twitter harassment scheme impacts these characteristics. Our analysis is focused on the framework for spammer identification. In addition, two user performance parameters and a tweet performance were measured using an application model utilizing all our four chosen attributes. A branded sets of users that are categorized as harmful, not harmful and popular have been established on the basis of these two criteria. Celebrities are another non-malicious group of apps. Therefore, a number of approximately 7,500 users were provided, of which 70% were used for instruction and 30% were used for research. 5 classification algorithms have been used and measured using Machine learning classification techniques. The results display the highest accuracy 99.2 percent provided by Random Forest Classifier.

References

[1] Masood F, Almogren A, Abbas A, Khattak HA, Din IU, Guizani M and Zuair M 2019 Spammer detection and fake user identification on social networks *IEEE Access* **7** 68140-68152
[2] Cresci S, Di Pietro R, Petrocchi M, Spognardi A and Tesconi M 2014 A Fake Follower Story: improving fake accounts detection on Twitter *IIT-CNR, Tech. Rep. TR-03*
[3] Khalil A, Hajjdiab H and Al-Qirim N 2017 Detecting fake followers in twitter: A machine learning approach *International Journal of Machine Learning and Computing* **7** 198-202
[4] Erşahin B, Aktaş Ö, Kılınç D and Akyol C 2017 Twitter fake account detection *In 2017 Int. Conf. on Computer Science and Engineering (UBMK)* 388-392
[5] El A, Azab AM, Mahmoud MA and Hefny H 2016 Fake account detection in twitter based on minimum weighted feature set *International Scholarly and Scientific Research & Innovation* **10** 13-18

[6] Abouollo A and Almuhammadi S. 2017 Detecting malicious user accounts using Canvas Fingerprint *In 2017 8th Int. Conf. on Information and Communication Systems (ICICS)* 358-361

[7] Miller Z, Dickinson B, Deitrick W, Hu W and Wang AH 2014 Twitter spammer detection using data stream clustering *Information Sciences* **260** 64-73

[8] Mehrotra A, Sarreddy M and Singh S 2016 Detection of fake twitter followers using graph centrality measures *In 2016 2nd Int. Conf. on Contemporary Computing and Informatics* (IC3I) 499-504

[9] Gurajala S, White JS, Hudson B, Voter BR and Matthews JN 2016 Profile characteristics of fake Twitter accounts *Big Data & Society* **3** 2053951716674236

[10] Aldossary F 2017 Health Observation System Using Cloud Computing *International Journal of MC Square Scientific Research* **9** 8-16

[11] Madhurya VB and Nair GP 2012 Applications of Intelligent Sensor Networks and Wirelss Communication to the instrumentation of Civil Engineering Structures *International Journal of MC Square Scientific Research* **4**.

Twitter Sentiment Analysis on Big Data in Spark Framework

S Nirmala Sugirtha Rajini[1], K Anuradha[2], S Umadevi [3] and E Mercy Beulah[4]

[1,2,3,4] Department of Computer Applications, Dr. MGR. Educational and Research Institute, Chennai, India.
[1]sugirtharaja77@yahoo.com, [2] anu.navine@gmail.com , [3]umakalyani70@gmail.com, [4]mercybeulah@yahoo.co.in

Abstract. Earlier for communication between people is something which is delivered by hand from one person to another. In some words, or letter Social Media an important and also an integral part of everyone's life now days. Twitter is a social media site and its primary purposes are to convert people and allow people to share their thoughts. Twitter an American Microbe Using twitter social media users post their views, opinions and to communicate with messages called as tweets. It is one of the social media that is gaining popularity now days. As of 2014 twitter social media has more than 284 million users are in active in monthly basis and above 500 million messengers sent messages on a daily base. Twitter is created by Jack Dorsey, Noah Class, Biz stone and Evan Willians in March 2006 and it was launched in the month of July in that year. Sentimental analysis means the analysis and classification of the emotions and feelings such as positive thought, negative thought and neutral thought using text analysis techniques within text data. The aim of the proposed analysis is to identify the public opinion using NLP (Natural Language Processing) with n-gram stemming algorithm in Spark framework.

Keywords: Twitter, Sentiment Analysis, Spark, Framework.

1. Introduction

Social media is one of the major platforms to share feelings. Social media contents are the one of the resource issues various services for people to covey opinions about people, services, goods or themselves. From the various social network media twitter is one of the important media where large number of peoples transfer their feeling and ideas. The twitter data is not sufficient to analyze because limited characters are used in the twitter message. By using this twitter data collect the sentiment of the various people. Extraction of proper data is used for business and research purposes. Sentiment of the software peoples mainly controls the quality and productivity of the software.

In this proposed system the tweets which the users tweet can be analyzed using the sentimental analysis the reaction of the users can be known. All the reaction such as positive, negative or neutral using sentimental can be analyzed.

2. Literature survey

Yash Sharma et al., demonstrates a new method of analyzing people opinion available on social network media with the help of Language Processing technique. Here twitter data can be collected from stakeholders of IIT. The output contains the basic emotional values [1]. Richard O. Sinnott et al., studies different analysis methods and estimate various sentiment analysis methods and evaluate their accurateness of the sentiment study on twitter social media in Australia. Here the authors presented a framework based on cloud [2]. Swati Powar et al., proposes architecture for fragment tweet data by using batch processing framework known as TweetSeg. It consists of limited background data with common data for attain enhanced solution for entity recognition. To find the significant data from the twitter data Hadoop architecture can be used here [3].

Yonas Woldemariam introduces the development and combination of a sentiment data analysis into cross media architecture pipeline. This pipeline consists of cleaner for chat, NLP and analyzer for sentiment. Here hadoop architecture with prediction concept based on lexicon and RNTN form can be used to sentiment data analysis. This proposed concept achieves 9.88% exactness of positive, negative and neutral comments on variable length data. But this method shows the better level for classification of positive type comments [4].

Jin Ding et al., conducted a sentiment analysis based on entity point. Initially they develop a dataset with 3000 comments from GitHub. Then the authors developed SentiSW tool for classification and recognize entity. To assess the performance using ten-fold type validation method, it attains 68.71 precision values, 63.98 recall values and 77.19 exactness values [5]. Abdullah Alfarrarjeh et al., proposes a new architecture for analyzing sentiment data by using disaster data value. This new architecture provides the solution for three challenges. The proposed framework can be tested by using Twitter and Flickr dataset values during the time of Napa earth shaking [6].

3. Proposed System Architecture

This proposed system to identify the public opinion using NLP with n-gram stemming algorithm in Spark architecture. This “Trending Topic Analyzer” is developed using theme based sentiment data classification and multi type tweet summarized data. It aims at creating ordered sub summary data, which understanding the topic development of the recent trends. Theme under learning is assumed to contain several unseen sub topics are exposed with the help proposed modelling. This enables a user to completely investigate a trending matter to a better level.

Target data gathering is performed by removing the trending theme on a area basis. Pre-processing steps are used to arrange the end user information. In pre-processing phase, the system also manages non-English tweet that is necessary to avoid disposal of public views about the current trend topics.

Pre-processed data is later analyzed through theme adaptive sentiment data classification to identify the public opinion. Sentiment labelled data will then be processed for sequential summarization. Sub topic detection is achieved using Stream based approach and Semantic based approach. Finally, sub summary candidate selection is executed which gives the highest scored tweets incorporating certain exclusive features of tweets. Redundancy check is performed to take away duplicates from the selected tweet data followed by a threshold check to make sure fair combination of user’s opinions.

Sentimental analysis allows business to find the user sentiment through their tweets. The term sentimental analysis defined as the use of natural language text analysis processing, computational notations and biometric to thoroughly identify, take out quantify a study the subjective data. The following figure 1 demonstrates the proposed framework used to analyze the twitter data.

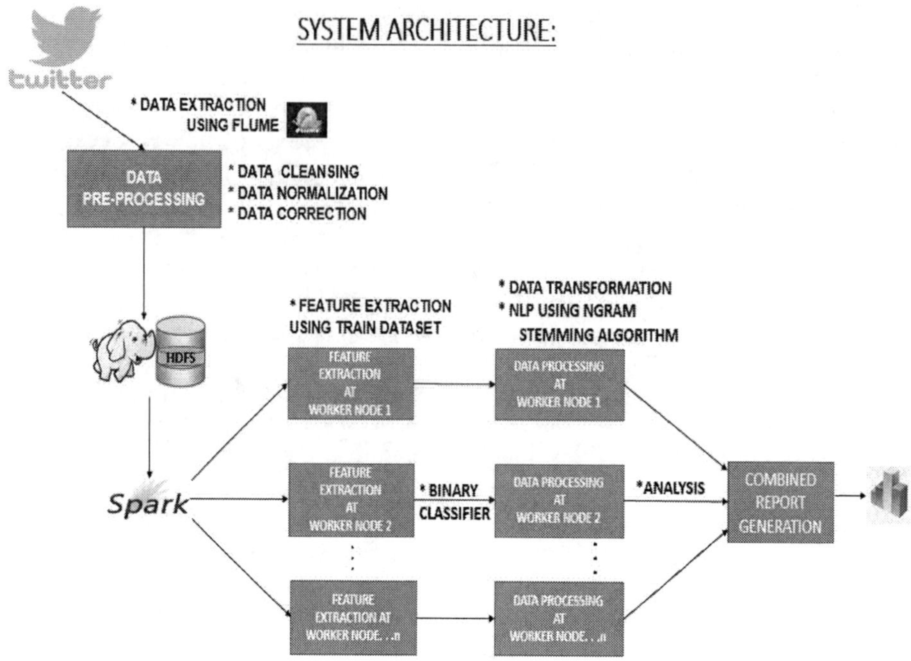

Figure 1. Proposed Framework

Apache Spark is one of the accepted tools used for real-time data processing. Technology is changing from ancient Map-reduce to Apache Spark because of its high speed. Apache Spark is up to 100 times quicker than Map-reduce which become the major cause for various users to move towards Spark. Spark architecture from Apache Foundation is another open source for big data processing which is built specially to overcome the restrictions from the traditional map reduces the jobs. This framework is one of the moving technologies in recent years for the big data growth. Memory abstraction is the most important feature and ability of Spark which enables the allocation of data, it is otherwise called as in-memory data sharing, across the various stages of a map-reduce task.

4. System Implementation

The proposed system can be divided into various phases. The first phase of this system is data extraction. Here the twitter data can be extracted by using flume. It is one of the software used to collect, integrate and transfer large volume of data.

Distinguished characteristics of Flume:

Flume gathers log data from several web servers and stored on centralized area efficiently.
With the help of Flume, collect the data from several servers immediately and transfer into Hadoop.
Flume is used to extract large amount of event data generated by social media.
Flume helps a huge set of sources and destinations types.

Normally tweet information contains noisy data because in twitter the user enters the content with limited letters. So, cleaning and produce meaningful data is important for any type of data analysis. In the pre-processing phase URL link is removed from the tweet data and short terms are also exchanged with proper words.

Detailed pre-processing steps are shown in Figure 2 along with the appropriate algorithm.

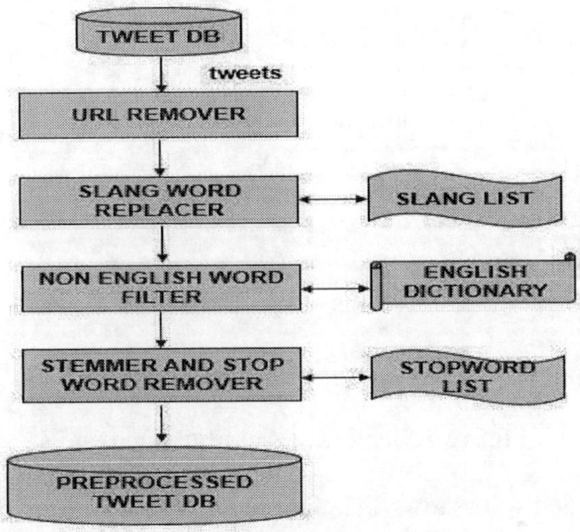

Figure 2. Pre-processing Phase

In feature extraction phase create the major impact of the performance of classify sentiment data. Because tweet data contains @ symbol (refer the user) and emoticons. Feature extraction process improves the feature removal process. By using text feature extraction common sentiment words and adoptive sentiment words are extracted. Emoticons and network based features are extracted by using non text feature extraction method. This proposed system uses semi supervised classification model is used to classify the data with common values and mixed labelled date from different topics. The final phase is sentiment analysis. In this phase twitter text data is processing and tokenizing. Finding accuracy and predicting scores by using NGRAM classification stemming algorithm.

Algorithm

Establish connection with spark and start the session.
Extract data from twitter through flume and load data
Pre-process data
Do tokenize for each partitioning
Feature extraction using train dataset
Classification of data using binary classification algorithm
Nature Language processing using NGRAM analysis and stemming algorithm
Find each token or word counting
Combine all the word count of all the partitions
Predict negative and positive sentiment words
Find prediction accuracy from is count
Show the prediction and analysis report

5. Result and Discussion

Data science world is full of attractive methods and algorithms to take out unseen insights from the data sources. Twitter is one of the famous social network media in this current digital world. It can be used from normal people to famous personalities also. But it contains large volume of data. Analysis twitter data is very difficult because it contains various symbols and nonverbal words. The analysis can be very useful to various marketing people and researchers to predict the result. Here twitter data can be analyzed by using Spark framework with python programming. The following figure 3 shows the sample screen shot of the proposed classification system result. The proposed system is tested by using kaggle dataset.

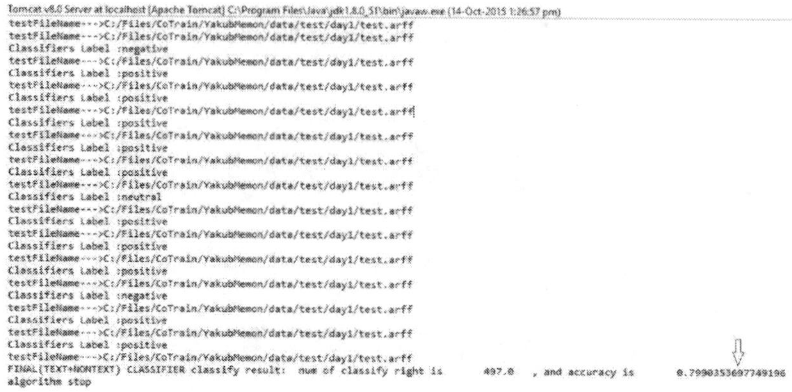

Figure 3. Sentiment classification results

The following figure 4 a) and 4 b) shows the classification result of positive and negative sentiment data.

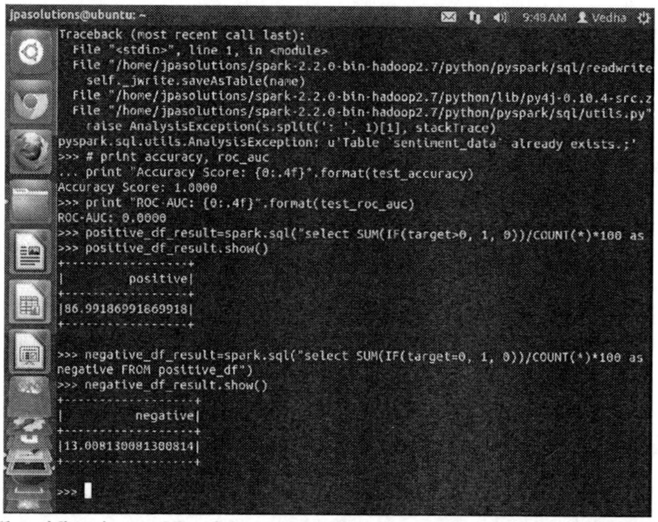

Figure 4. a) Classification of Positive and Negative Sentiment in SPARK environment

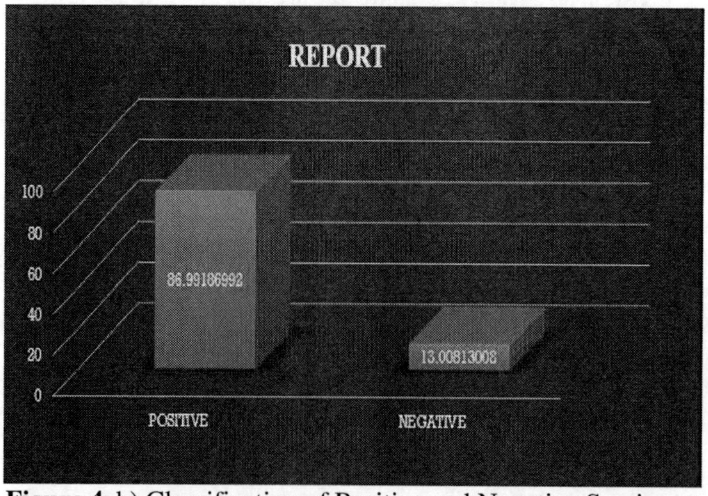

Figure 4. b) Classification of Positive and Negative Sentiment

IOP Conf. Series: Materials Science and Engineering **925** (2020) 012015 doi:10.1088/1757-899X/925/1/012015

6. Conclusion

Social network is playing a major role in human life. Traditionally peoples convey their opinions and emotion through letter for or telephone. But now most of the people share their feelings and ideas through social network. Social network media data analysis is used to predict the future value. Here positive and negative sentiment data are extracted from tweet data. This system is implemented by using spark environment. Sentiment analysis of Twitter social media using big data helped us to analyze large amount of datasets. Sentimental analysis to identify the public opinion using NLP with n-gram stemming algorithm in Spark framework and the report is generated.

References

[1] Sharma Y, Malik I and Jagwani P 2019 Dual Aspect Sentiment Analysis for Opinion Mining. *IEEE Int. Conf. on Electrical, Computer and Communication Technologies (ICECCT)* 1-4

[2] Sinnott RO and Cui S 2016 Benchmarking sentiment analysis approaches on the cloud *IEEE 22nd Int. Conf. on Parallel and Distributed Systems (ICPADS)* 695-704

[3] Powar S and Shinde S 2017 Named entity recognition and tweet sentiment derived from tweet segmentation using hadoop *1st Int. Conf. on Intelligent Systems and Information Management (ICISIM)* 194-198

[4] Woldemariam Y 2016 Sentiment analysis in a cross-media analysis framework *IEEE Int. Conf. on Big Data Analysis (ICBDA)* 1-5

[5] Ding J, Sun H, Wang X and Liu X 2018 Entity-level sentiment analysis of issue comments *Proc. of the 3rd Int. Workshop on Emotion Awareness in Software Engineering* 7-13.

[6] Alfarrarjeh A, Agrawal S, Kim SH and Shahabi C 2017 Geo-spatial multimedia sentiment analysis in disasters *IEEE Int. Conf. on Data Science and Advanced Analytics (DSAA)* 193-202

Analysis and Design of Modular Multilevel Converter (MMC) Based on Full Bridge Switching Modules (FBSM) for Harmonic Elimination using POD PWM Control Method

Varaparla Hari Babu[1] and K Balaji[2]

[1,2] Department of EEE, SPIHER, Avadi, Chennai, India.
[1]varaparlaharibabuphd@gmail.com , [2]kbalaji26@gmail.com

Abstract. Modular Multilevel Converter (MMC) in HVDC for transmitting the power in longer distance because of high in transmission efficiency rate and cost-effective compared to AC system in many industries. The control of voltage and power is earlier and superior method is modular multilevel converter for HVDC transmission. The use of MMC has reduced the filter size and low frequency. Staircase, carrier-based PWM, space vector and selective harmonic elimination are the method using in MMC system control. Unbalanced networks are studied, and strategies are introduced such that in these circumstances MMC contributes to improving the grid efficiency. It involves the key applications in the transmission of high voltage direct current (HVDC) along with other applications of Medium Voltage (MV) and Low Voltage (LV). The MMC with POD-PWM is built using the MATLAB / Simulink method and concludes the advantages and improvements of converter modulation to reduce the voltage ripple and increase the voltage levels.

Keywords: Harmonics, HVDC, MMC, POD-PWM, Unbalanced Load.

1. Introduction

Due to the long distance bulk transmission system, safe link, short circuit level and underground cable system, current (HVDC) transmission system is advanced to AC network. From the high voltage the converters are used to adjust the voltage and transfer the voltage at the receiving end to the long distance and back to ac. Usage of converters have MMC converters to increase harmonics, regulate direct current voltage and rising flowing current. The MMC has the source [3-4] in modular multilevel, close to the cascaded H bridge converter. The multi-terminal DC provides the possibility of interconnecting lattices between regional power systems and generating renewable energies. Reliability in both AC and DC systems, as well as improving stability, energy transmission and efficient use of cables and converters in HVDC systems.

Reduced harmonics and circulation of current are used by multi modular converter. A MMC step leg consists of two upper arm and lower arm muscles. Growing arm has N identical series connected to a cascaded structure with an arm inducer, arm L and arm resistor, arm R. In converter the fault currents are removing by using the Arm inducer. Because of power losses to the converter the arm resistor should be selected as low as possible. Different circuit structures are consisted by MMC of the each. The structure is created a capacitor and two IGBTs with opposing similar diodes based on sub element considering half bridge.

However, full-bridge based sub module structure is consisted of a capacitor and four IGBTs with anti-parallel diodes. Unidirectional cell circuit structure is the other one which is formed a capacitor and one IGBT with anti-parallel diode for the upper side otherwise only one diode without IGBT for the lower side [9]. Sub modules are inserted or bypassed depending on the switching positions in each half-bridge circuit. Two switches work in complementary way otherwise lead to short circuit condition. When the upper switch is ON and the lower switch is OFF, sub module is inserted in the arm. Thus; the terminal voltage of the sub module is equal to the capacitor voltage [10-11]. If the arm current direction is positive, sub module capacitors are charged otherwise capacitor voltages are discharged.

The modern inverters are intended for high-power applications with series connected switching units. Such series switching system connections have a problem with high switching stress on each unit [12, 15]. Using more switching devices, higher voltage rates are found. When the number of switching devices goes up, it would be difficult to produce gate signals for the switching devices. So the MMC is a good topology for converters in the near future. [13-14] Analysis and design of Modular Multilevel Converter (MMC) based on Full Bridge Switching Modules (FBSM) using POD-pulse width modulation (POD-PWM) control technique for reduction of total harmonic diction that occurs in the HVDC line with help of MATLAB/Simulink is proposed in this paper.

2. Proposed system

Figure.1. shows the schematics diagram of MMC in HVDC system. The MMC has several inductor-connected modules. The switching modules (SM) are attached to an arm forming an inductor. The two arms are connected and it is known as one phase inverter. The SM has different forms, for example half bridge and full bridge circuit. The MMC is commonly used in transmission system HVDC, Data, and motor drives with high power. The MMC's advantages are the balance of capacitor voltage, the circulating current power. The capacitor charged when other discharge of the arm current is positive.

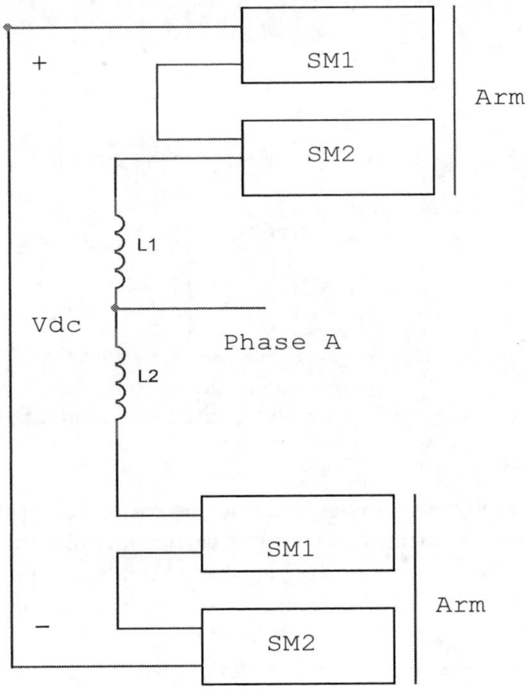

Figure 1. Basic circuit of switching module of MMC

3. Modelling

3.1Full Bridge Switching Modules of MMC

The voltage levels are based on the number of switching modules used in each phase of the inverter. Figure.2 displays the complete Multilevel Converter Modular Bridge. Four switches and switching losses are decreased in the full bridge converter compared to twice the half bridge switching modules which fix the dc fault. Voltage of the switching modules depends on the quality of the bipolar transistor isolated gate. The current flows through the SM capacitor in the arm. The control devices switching frequency is less for decreasing the transfer number. MMC's mathematical modeling is voltage and current in the phase converter's upper and lower arm. In equation (1) and (2), the output voltage of phase an upper and lower arm is given.

$$V_{aupper} = \frac{V_{dc}}{2} - \left(V_{upp} + L\frac{di_{upp}}{dt} \right) \quad (1)$$

$$V_{alower} = -\frac{V_{dc}}{2} + \left(V_{low} + L\frac{di_{low}}{dt} \right) \quad (2)$$

The dc connection current is divided in a steady state equally into three phase inverter legs. Under the current law of Kirchhoff the current and circulating phase current is defined in equation 3 and 4.

$$i_{phase} = (i_{upp} - i_{low}) \quad (3)$$

$$i_{cir\_ct} = \left(\frac{i_{upp} + i_{low}}{2} \right) = \frac{i_{dc}}{3} \quad (4)$$

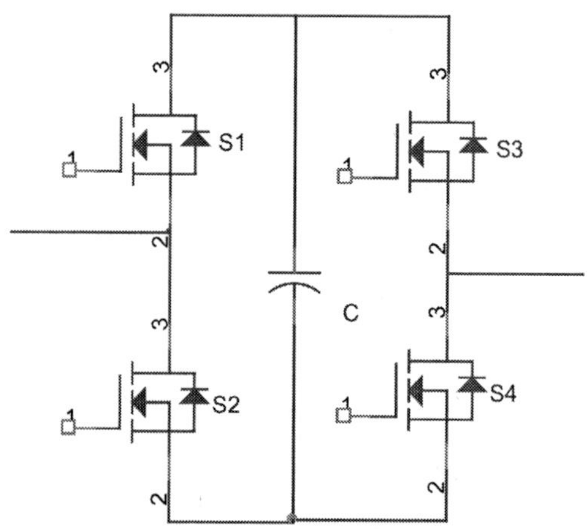

Figure.2 Generalized circuit diagram of FB-MMC

4. Control techniques

The control methods of the modular multilevel converter are carrier based pulse width modulation, space vector and selective harmonic elimination. The multi carrier modulation has been commonly used in MMC and is divided into two forms, such as PWM shifted in phase and stage. In level shifted further classified into Phase Disposition (PD), Alternate Phase Opposition Disposition (APOD), and Phase Opposition Disposition (POD). The modular multilevel converter has dq synchronous reference frame for active and reactive power management. The inverter's active and reactive power is given in equations 5 and 6. The block diagram of active and reactive power control method is shown in figure 3.

IOP Conf. Series: Materials Science and Engineering **925** (2020) 012016 doi:10.1088/1757-899X/925/1/012016

$$P = \frac{3}{2}\left(V_q i_q + V_d i_d\right) \tag{5}$$

$$Q = \frac{3}{2}\left(V_q i_d - V_d i_q\right) \tag{6}$$

The value of direct and quadrature current is represented in Eq. 7 and 8.

$$i_d^* = \frac{2}{3V_d} P \tag{7}$$

$$i_q^* = -\frac{2}{3V_d} Q \tag{8}$$

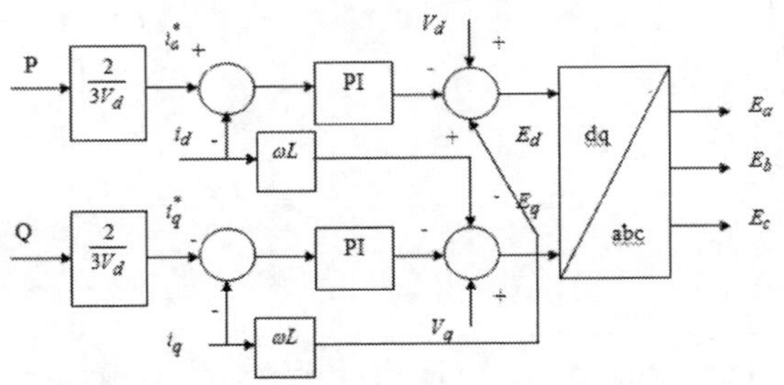

Figure 3. MMC Real and Reactive Power Control

The modulation of the pulse width of the carrier disposition has n equivalent triangular and is symmetrical to zero. Comparing phase carrier voltage with reference, and generating the pulse and feeding into the MMC. The POD-PWM (Phase Opposition Disposition Pulse Width Modulation Technique) is used to reduce the system's THD stage, and is represented in the figure.4. The active power filter based MMC uses the APOD-PWM to reduce the harmonics. The second and other even-order harmonics in the circulating currents are mainly due to the variations in the MMC Sub-Modules (SMs) capacitor voltages.

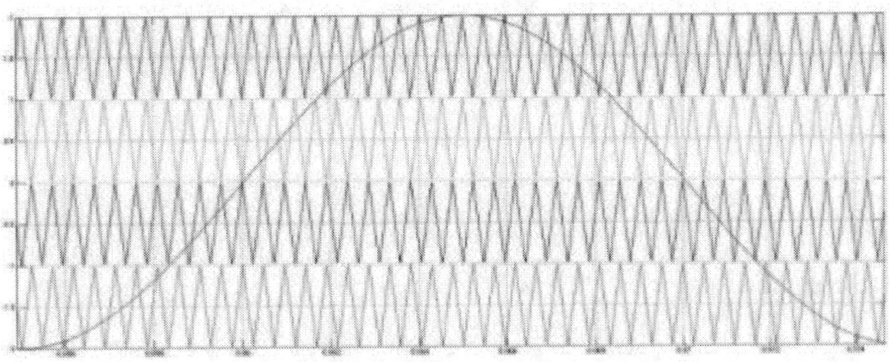

Figure.4 Phase Opposition Disposition PWM

5. Simulink results

The MMC converter has designed and configured to implement the complete POD-PWM control topology configuration of the bridge switching modules. Table 1 reflects the features of a multilevel modular converter based on complete bridge switching modules in the HVDC transmission network. The overall Simulink diagram of the proposed device as well as the active, reactive power with POD-PWM technique are shown in figure.5 and figure. 6.

IOP Conf. Series: Materials Science and Engineering **925** (2020) 012016 doi:10.1088/1757-899X/925/1/012016

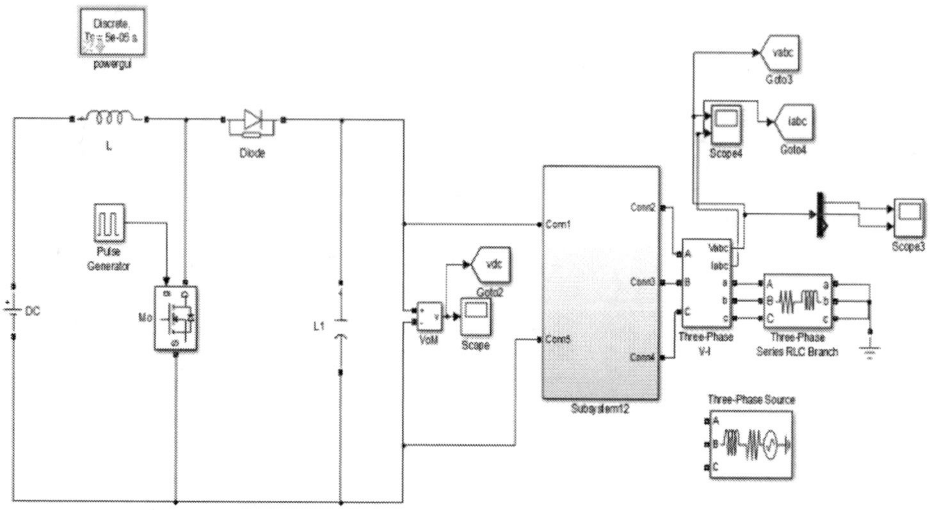

Figure 5. Overall simulink diagram of the proposed DC source fed MMC converter

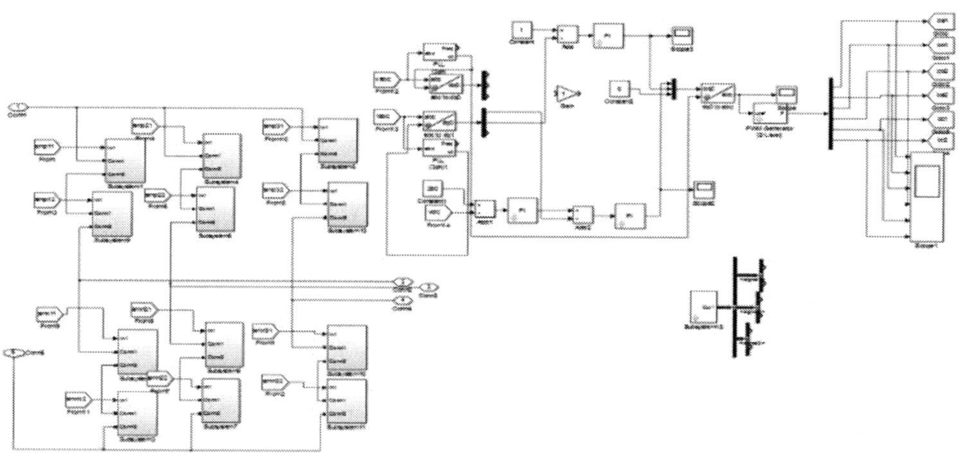

Figure 6. Full bridge switching module with POD-PWM technique

Figure.7 Shows the DC link voltage of the proposed system and it value is 500V and settling time of this DC source boost converter is 0.1s.

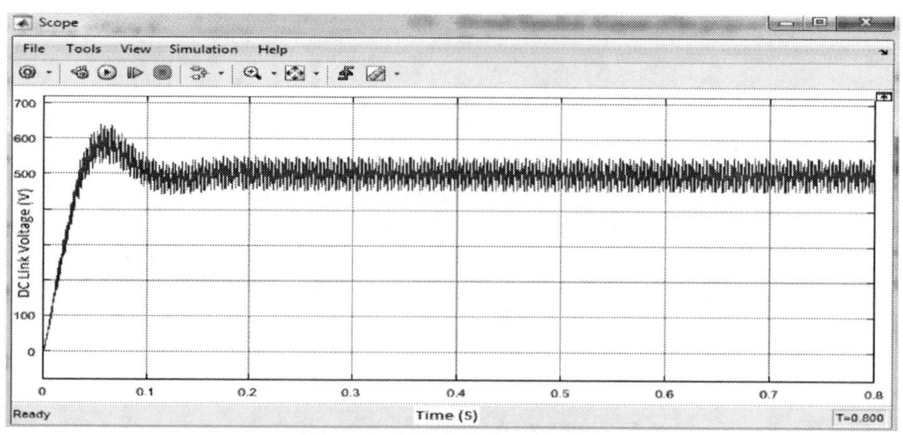

Figure 7. DC link voltage

The generated high DC voltage is fed to the MMC and it produces the 7-level output voltage is shown in figure.8. and its value 350V which is send to the RL load for high power application. Figure.9. shows the %THD of POD-PWM using MMC is 3.98%.

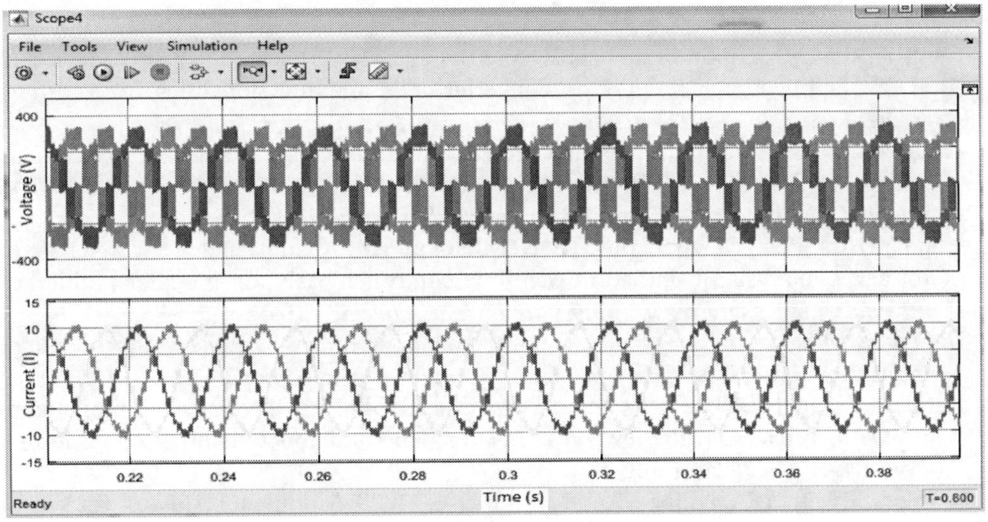

Figure 8. Multi-level's of output voltage and current

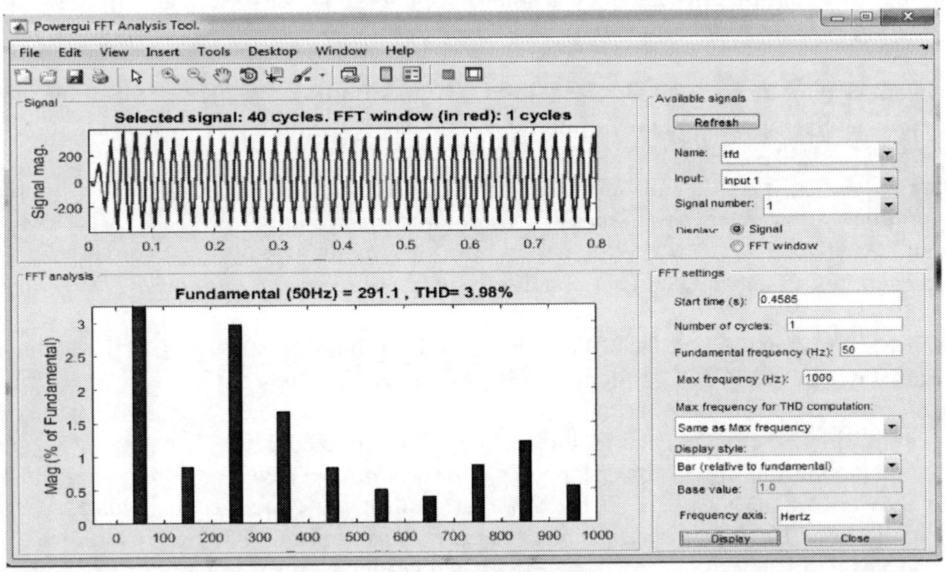

Figure 9. %THD of MMC

Table 1. FBSM in full bridge

S.No	SM Circuit	Voltage levels	Fault handling	Losses
1	Full bridge	0,Vc	Yes	Low

6. Conclusion

The On MMC the concept of modularity was introduced from the switching modules to the system level. For high-power applications for each modular converter type, the modularity principle, complete bridge switching module topologies with POD-PWM control method have been analysed. The MMCs operate in VSM, and operate in CSM for two series of connected MMCs. The system's output voltage is 5level and its THD value is 3.98%. This overall system performance has been analyzed by using MATLAB/Simulink.

References

[1] Nami A, Liang J, Dijkhuizen F and Demetriades G D 2015 Modular multilevel converters for HVDC applications: Review on converter cells and functionalities *IEEE Transactions on Power Electronics* **30**18-36
[2] Zhang L, Zou Y, Yu J, Qin J, Vittal V, Karady GG, Shi D and Wang Z 2017 Modeling, control, and protection of modular multilevel converter-based multi-terminal HVDC systems: A review *CSEE Journal of Power and Energy Systems* **3** 340-352
[3] Yubo YUAN, Peng LI, Xiangping KONG, Jiankun LIU, Qun LI and Ye WANG 2016 Harmonic influence analysis of unified power flow controller based on modular multilevel converter *Journal of Modern Power Systems and Clean Energy* **4** 10-18
[4] Hagiwara M and Akagi H 2009 Control and experiment of pulsewidth-modulated modular multilevel converters *IEEE transactions on power electronics* **24** 1737-1746
[5] Nami A, Liang J, Dijkhuizen F, and Demetriades GD 2014 Modular multilevel converters for HVDC applications: Review on converter cells and functionalities *IEEE Transactions on Power Electronics* **30** 18-36
[6] Rohner S, Bernet S, Hiller M and Sommer R 2009 Modulation, losses, and semiconductor requirements of modular multilevel converters *IEEE transactions on Industrial Electronics* **57** 2633-2642
[7] Vasudevan V and Balaji K 2018 Performance of Cuk-KY Converter Fed Multilevel Inverter for Hybrid Sources *Indonesian Journal of Electrical Engineering and Computer Science* **10** 436-445
[8] Debnath S, Qin J, Bahrani B, Saeedifard M, and Barbosa P 2014 "Operation, control, and applications of the modular multilevel converter: A review *IEEE transaction on power electronics* **30** 37-53
[9] Zeng R, Xu L, Yao L, and Williams BW 2014 Design and operation of a hybrid modular multilevel converter *IEEE Transactions on power electronics* **30** 1137-1146
[10] Guan M and Xu Z 2012 Modeling and control of a modular multilevel converter-based HVDC system under unbalanced grid conditions *IEEE transactions on power electronics* **27** 4858-4867
[11] Guan M, Xu Z and Chen H, 2011 Control and modulation strategies for modular multilevel converter based HVDC system *In IECON 2011-37th Annual Conf. of the IEEE Industrial Electronics Society* 849-854
[12] McGrath BP, Teixeira CA, and Holmes DG 2017 Optimized phase disposition (PD) modulation of a modular multilevel converter *IEEE Transactions on Industry Applications* **53** 4624-4633
[13] Novakovic B and Nasiri A 2017 Modular multilevel converter for wind energy storage applications *IEEE Transactions on Industrial Electronics* **64** 8867-8876
[14] Irshad SM and Ramesh GP 2020 ANFIS-Based MPPT Control in Current-Fed Inverter for AC Load Applications *In Intelligent Computing in Engineering* 459-469
[15] Puttamadappa C and Parameshachari BD 2019 Demand side management of small scale loads in a smart grid using glow-worm swarm optimization technique *Microprocessors and Microsystems* **71** 102886

IOP Conf. Series: Materials Science and Engineering **925** (2020) 012017 doi:10.1088/1757-899X/925/1/012017

A Single Stage 1 Switched Capacitor 7S-7L using Different PWM Topologies for Harmonics Reduction

Shaik Nagulmeeravali[1] **and K Balaji**[2]

[1,2] Department of EEE, SPIHER, Avadi, Chennai, India.
[1]nagulmeeravaliphd@rediff.com, [2]kbalaji26@gmail.com

Abstract. Design and analysis of single-stage 1 Switched Capacitor (SC) boost 7Switch 7-Level (7S-7L) inverter for attaining higher voltage levels with reduced switches count using various PWM techniques is presented. PD, POD, APOD, VF, and CO is different PWM techniques which used to control the switches in the SC circuit. The main objective of this analysis is to reduce the existing drawbacks as increased switches count, high % of Total Harmonic Distortion (THD), and typical PWM control method. Single-stage SC-based boost converter topology in Multi-Level Inverter (MLI) that resolves the existing problem. Single source, two capacitors, and the voltage gain ratio of 4 are attained from the presented topology without the transformer. Furthermore, comparisons are made between the existing parameters such as THD value, voltage levels, and proposed THD, voltage levels with different PWM techniques. SC-based 7L inverter is implemented, in order to compare the performance of proposed better than the existing 5-level MLI and 5-level SC-MLI inverter, and THD is compared with various PWM techniques and 0.9 modulation index. The performance of the proposed method is validated through the simulation results using MATLAB/Simulink.

Keywords: Multi-Level Inverter (MLI), PWM techniques, Switched Capacitor (SC), Total Harmonic Distortion (THD).

1. Introduction

In recent days' solar power system is becoming well-known energy resources by reason of effortless installation, zero environmental pollution, and low continuance, no fuel cost whereas comparing to the wind, hydro, thermal and etc. The energy available from the solar is less at night and cloudy period. For that reason, front end or back end boost converter, and some special converters becomes essential to achieve high voltage but which leads to reduce the system efficiency and increase the circuit complexity. In this paper, 1 SC-based MLI using the LS-PWM technique is presented. It is validated through MATLAB/Simulink, and comparative analysis is accessed for proposed method with/than existing as considered the component count, voltage gain, and %of THD value, controllers. The paper is organized as follow, analysis the survey of the MLI and its control methods are explained in section-2, in section-3 proposed topology and circuit description of SC-MLI, in section-4 modulation control, simulation results, performance and analysis of proposed method in LS-PWM techniques is in section 5, comparative [1] analysis of proposed and existing are presented in section-6, and the last section concludes this analysis.

2. Literature survey

Generally, Multi-Level Inverters (MLI) are familiar as the best results to improve the performance/execution of solar, wind, and other power electronics utilities in high or medium power applications [2-4]. The output of the MLI is stepped voltage waveform at the desired quality. Existing [4-6] have some advantages over than the voltage source inverter, and it has some disadvantages like high switch count, low reliability, higher blocking voltage rating, and extra expenses for designing an immense filter, large capacitor banks required high power supply to attain desired output [7-8]. The developed structure of MLI is SC boost based multi-level inverter is overcome the above-mentioned issues. It reduces the count of the additional element, power supply, and gate driver, high voltage gain with reduced active and passive elements [9, 12]. Sine PWM is used widely, because of its simple operation and provides good results. LS-PWM is studied to establish and applied in that way, enlarging the conversion efficiency [14, 18, 23]. LS-PWM modulation and Carrier overlapping (CO) is used to produce the pulses to the switches in the SC boosting based multi-level inverter [19-22].

3. Proposed switched capacitor boosting 7- level inverter and circuit description

The MLI system consists of 7-level output voltage with a single SC-converter module, one DC source, fewer switches and two capacitors. Block representation of SC-MLI with RL load is shown in figure.1. Charging and discharging of capacitors are utilized to neglect the voltage balancing issues. The proposed boosting has the output of seven-level with reduced switch count. Single unit SC circuit consists of single DC source (Vdc), three switches (S1, S2, and S3) with two diodes (D1, D2), and two capacitors (C1, C2). 7L output voltage obtained from the single-phase inverter is 0, Vdc, 2Vdc, 3Vdc, -Vdc, -2Vdc, -3Vdc. It produces six bipolar (positive polarity & negative polarity) levels and zero state level across the load.

At initial analysis, capacitors are utilized for maintaining the output current (Io) and voltage (Vo) and also eliminates the voltage ripple across the load, input source (Vdc), avoids voltage balancing issues, and switches (S) are kept at ideal and depends on the switches in the switched capacitor boost converter.

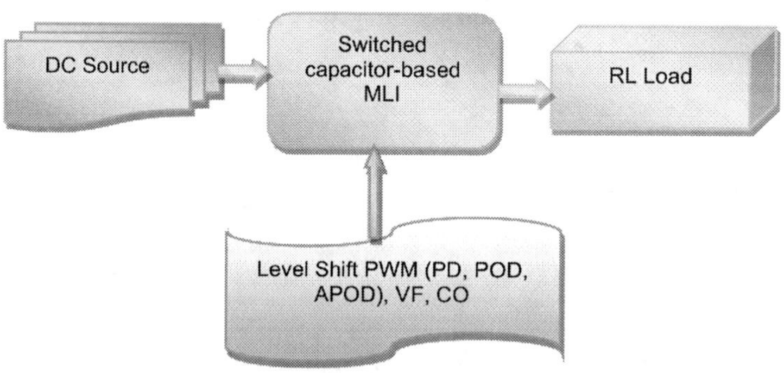

Figure 1. Block representation of SC-MLI

1. Output states of proposed circuit

This paper focuses on the circuit topology of switched-capacitor boost MLI is represented in figure. 2. The 7 states of SC-MLI are represented as 0Vdc to 3Vdc related to the positive half cycle, and 0Vdc to -3Vdc related to the negative half cycle.

IOP Conf. Series: Materials Science and Engineering **925** (2020) 012017 doi:10.1088/1757-899X/925/1/012017

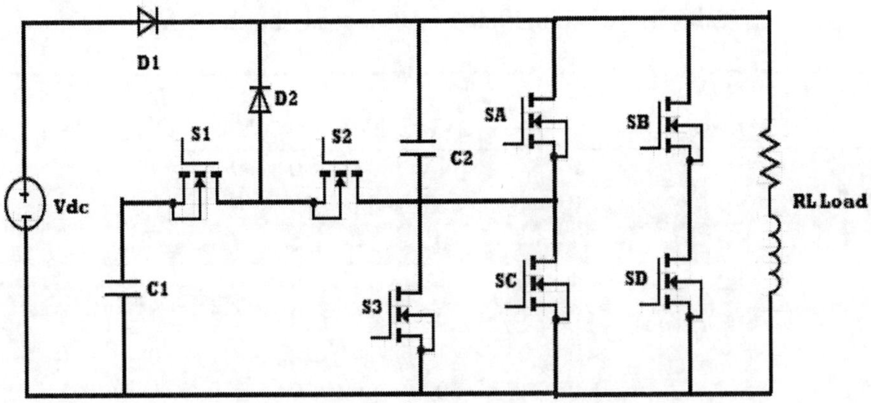

Figure 2 Circuit diagram of 7L-7S SC-MLI

State 0Vdc
During the positive half cycle of 0Vdc state, SA is ON state. Remaining H-bridge inverter is OFF state. The diode in SB switch act as freewheeling. During the negative half cycle of 0Vdc state, SD is ON state. Remaining switches in H-bridge inverter is OFF state. The diode in SC switch act as freewheeling. Thus the DC bus voltage is zero

$$V_{Bus} = 0 \tag{1}$$

State Vdc& -Vdc
In positive mode, S3, SA, and SC are in ON state, DC bus voltage is equal to the input voltage which as capacitor (C2) by turning OFF of S1 and V input supplies power to the load and negative S3, SB, and SD is in ON state. Capacitor C1 is charged to the V input by turn ON of D2 and S3. Then the output voltage level is represented in the below,

$$V_0 = \pm V_{dc} \tag{2}$$

State 2Vdc & -2Vdc
In positive S1, SA, and SC are in ON state, when the capacitor voltage is greater than the input voltage the diode operates as reverse biased, and during the negative half cycle switches S1, SB, and SD are in ON state. Then the output voltage level is represented in the below,

$$V_0 = \pm 2V_{dc} \tag{3}$$

State 3Vdc & -3Vdc
In this the switches S1, S2, SA, and SC are in ON state. Capacitor (C2) is charged. At that time diode is reverse bias. Capacitor (C2) and input source supply the power to the load. The total DC bus voltage is equal to the output voltage, in negative switches S1, S2 SB, and SD are in ON state. The above-mentioned process is repeated in the negative cycle. Then the output voltage level is represented in the below,

$$V_0 = V_{dc} + 2V_{dc} = \pm 3V_{dc} \tag{4}$$

4. PWM strategy
Multicarrier PWM is used when the output voltage levels are 3 or more than three levels. It is generally classified into two methods, such as Phase Shifted (PS) pulse width modulation technique and level shifted (LS) PWM technique. In LS carrier signals are vertically shifted to each carrier signals.

IOP Conf. Series: Materials Science and Engineering **925** (2020) 012017 doi:10.1088/1757-899X/925/1/012017

Table 1. Switching States of Proposed 7L-7S SC-MLI

Output (Vo) / Switches	0	V_{DC}	$2V_{DC}$	$3V_{DC}$	0	$-V_{DC}$	$-2V_{DC}$	$-3V_{DC}$	0
S_1	0	0	1	1	0	0	1	1	0
S_2	0	0	0	1	0	0	0	1	0
S_3	1	1	0	0	1	1	0	0	1
S_A	1	1	1	1	0	0	0	0	0
S_B	0	0	0	0	0	1	1	1	1
S_C	0	1	1	1	0	0	0	0	0
S_D	0	0	0	0	0	1	1	1	1

It is used to generate a pulse to the switches. The performance of the multi-level carrier-based sinusoidal PWM with PD, POD, APOD, CO, and VF is applied for reducing the %of THD. In PD, all triangle signals are in phase. In POD, trapezoidal signals are positive levels are out of phase with triangle signal by negative level at 1800. In APOD, all triangles are 1800 out of phase regarding its nearer triangle signal, In VF, and frequency of the Carrier is varied properly, to balancing the switches. In CO, carriers with the same frequency and amplitude (Amp) are inclined such that the carriers are overlapping each other; overlapping distance between each Carrier is Amp/2.

For 7L, three high-frequency signals (Vdc to 3Vdc) of the same amplitude (Ts – triangular signal), the peak amplitude of the reference sinusoidal signal (Sref). S ref is compared with the triangular signal which generates pulses to turn ON semiconductor switches in SC-MLI. The modulation index of the proposed system is defined below equation (5). The pulse pattern of this proposed circuit is representing in figure. 3.

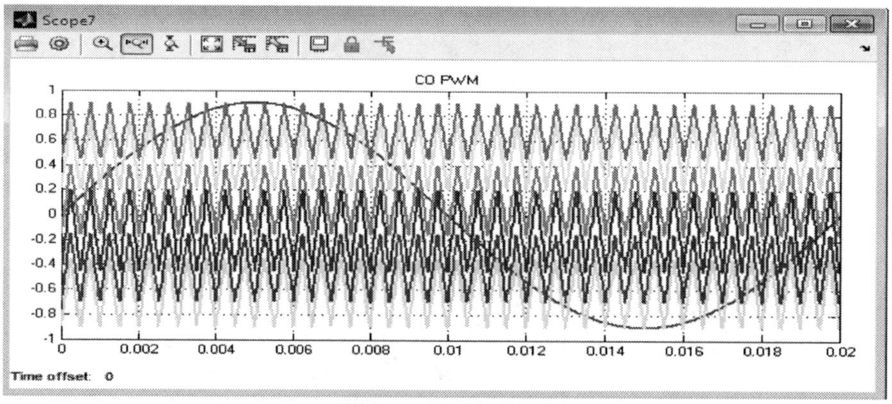

Figure 3. CO-PWM

$$M = \frac{S_{ref}}{3 * S_C} \tag{5}$$

Sref -Peak amplitude of sinusoidal reference voltage
Sc - Amplitude of carrier signal (triangular signal)

Six triangular and one sine wave are used to attain the 7-levels at the output. A reference signal (sine wave) is continuously compared with each triangle signal (the carrier wave). Required pulse is produced when the reference signal is more than or lesser than the triangular signal. Pulse pattern of the 7L-7S inverter with the modulation index of M=0.9 is represented in figure.3

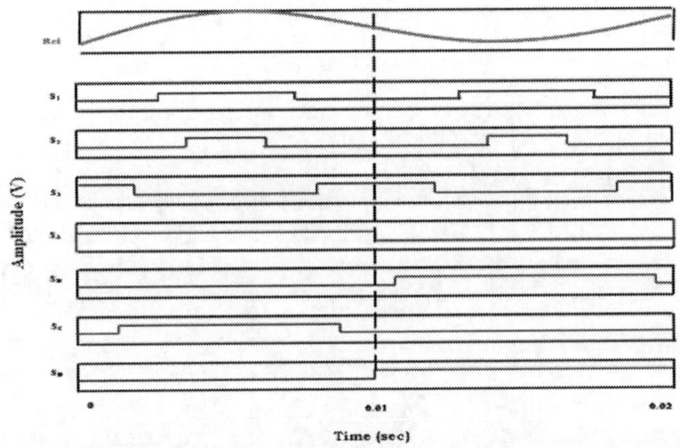

Figure 4. Switching Pulse Pattern of the Proposed Circuit

5. Simulation Analysis

The design of the 7L-7S SC-MLI model has been developed by MATLAB/Simulink software platform. From the designed SC-MLI is shown in figure. 5, maximum 100Vdc is extracted in output 300V0 which a 7-level is attained. For the proposed system, a resistive-inductive load is used, and its specification is represented in table 2.

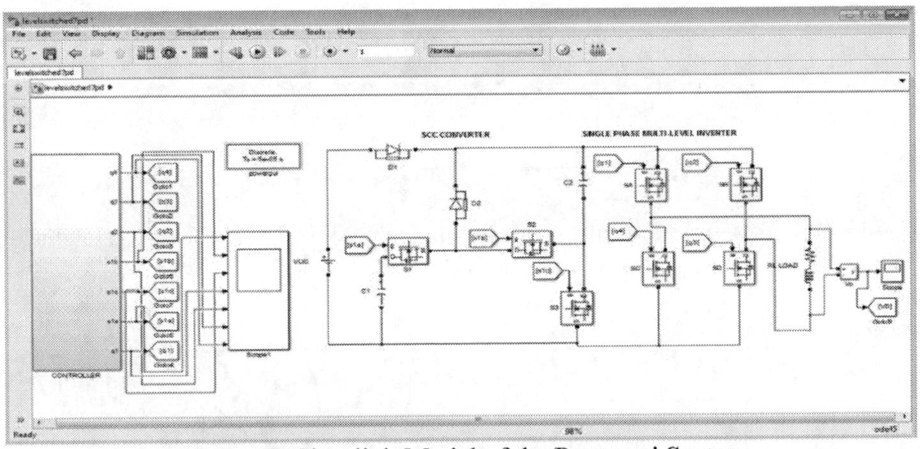

Figure 5. Simulink Model of the Proposed System

Table 2. Simulink model specification

S. No	Parameter Name	Values
1	Source (Vdc)	100V
2	Capacitor (C1, C2)	3000e-6
3	Resistive-Inductive load	100Ω, 100mH
4	Switching Frequency (Fs)	1000
5	Modulation Index	0.9
6	Output voltage (V_0)	280V

Figure. 6 is showing the 7level SC-MLI is 300V, and %THD value is 23.59% in 50Hz with RL load using PD-PWM technique.

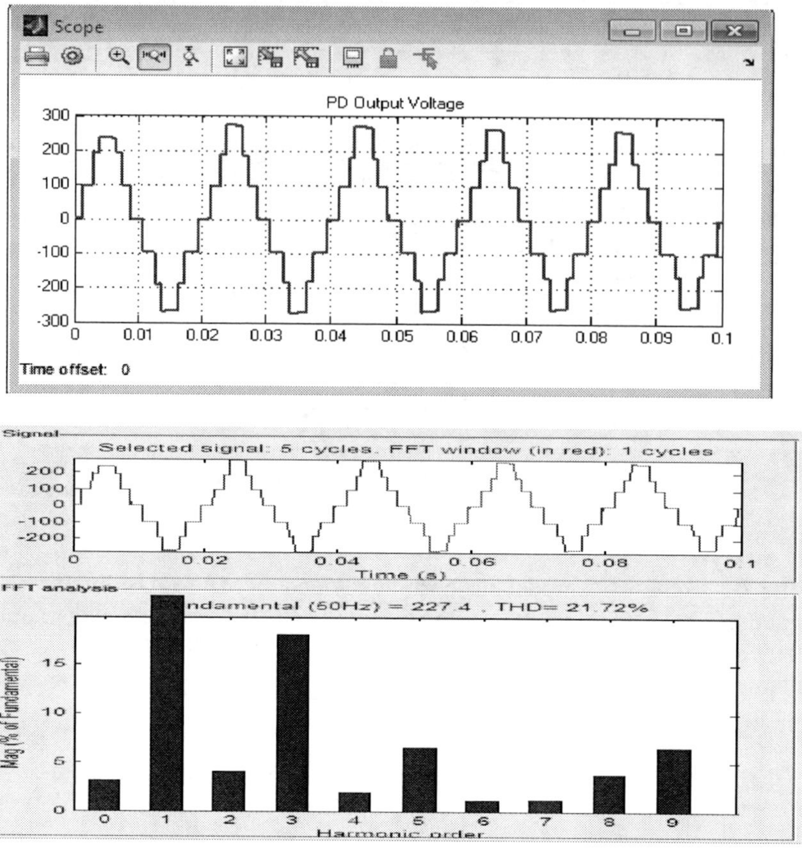

Figure 6. PD and %THD

In figure. 7. 7 level SC-MLI is 300V and % THD value is 7.97% in 50Hz with RL load using POD-PWM technique. THD value of this method is less compared with PD-PWM.

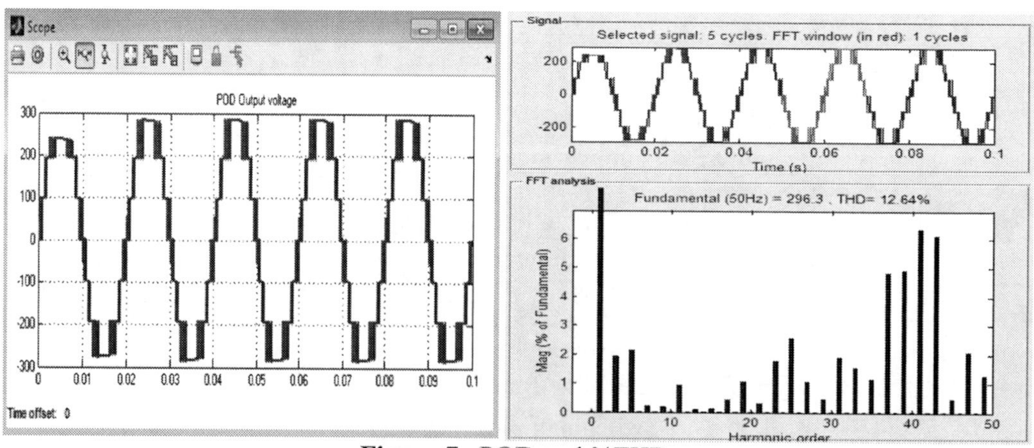

Figure 7. POD and %THD

Figure. 8 shows the output voltage waveform of 7level SC-MLI is 300V, and %THD value is 7.87% in 50Hz with RL load using APOD-PWM technique. THD value of this method is less compared with PD-PWM and POD-PWM.

IOP Conf. Series: Materials Science and Engineering **925** (2020) 012017 doi:10.1088/1757-899X/925/1/012017

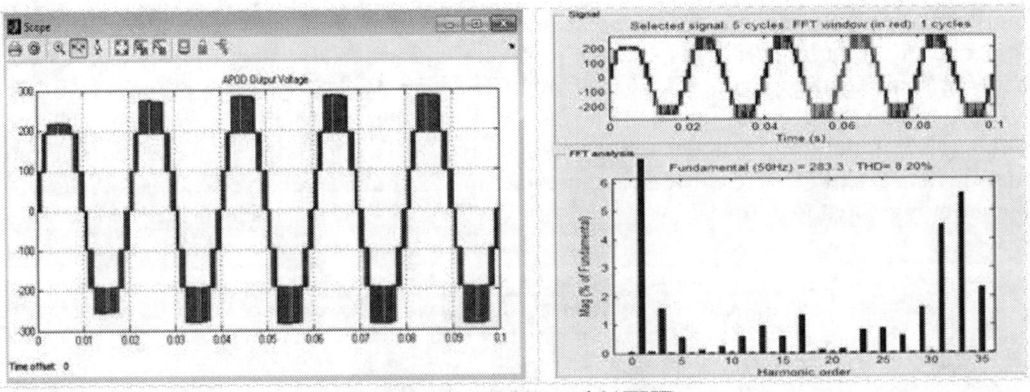

Figure 8. APOD and %THD

Figure. 9 represents 7level SC-MLI is 300V, and %THD value is 15.60% in 50Hz with RL load using VF-PWM technique. THD value of this method is less compared with PD-PWM, POD-PWM, and APOD-PWM.

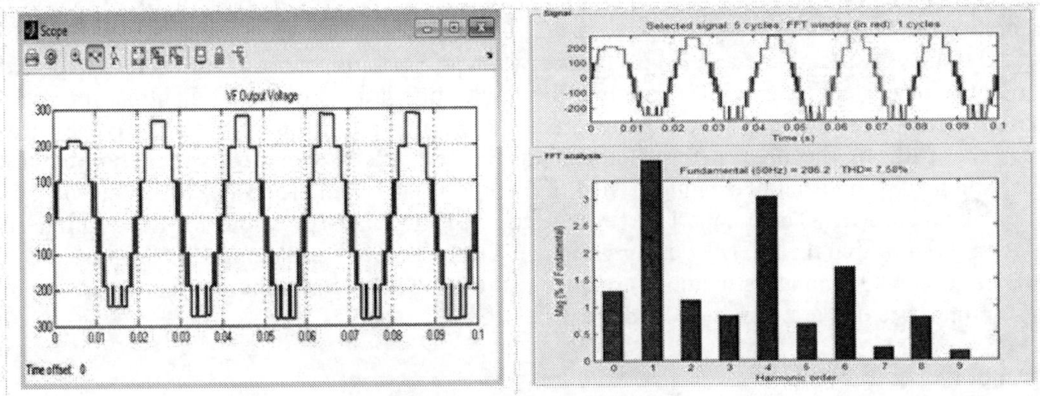

Figure 9. VF and %THD

Figure. 10 shows the output voltage waveform of 7level SC-MLI is 300V, and %THD value is 7.60% in 50Hz with RL load using CO-PWM technique. THD value of this method is less compared with PD-PWM, POD-PWM, APOD-PWM, and VF-PWM.

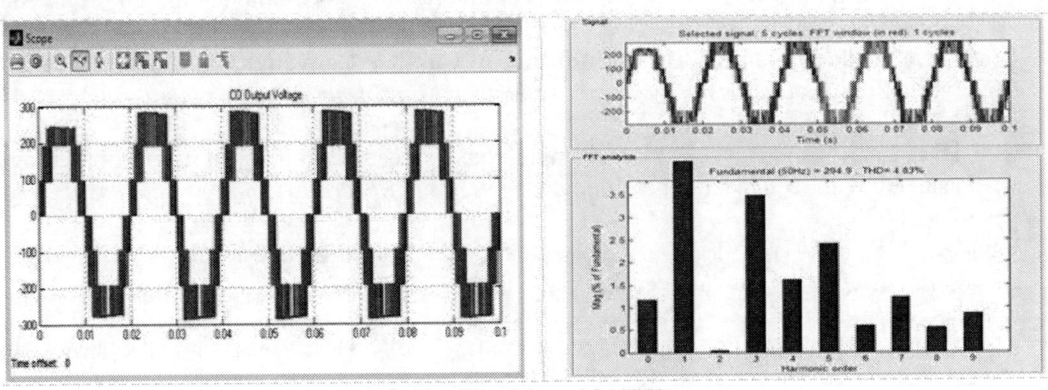

Figure 10. CO and %THD

6. Comparative Analysis

A comparison with various MLI is made for providing a fair decision about presented SC-MLI. For comparison, some of the existing 5L, 5L SC-MLI and proposed SC-MLI is proposed in this paper. Analyses are completed based on result levels (NL), number of MOSFET switches (NS), number capacitors (NCap), number of diodes (ND), input sources (NI), and output voltage boosting ratio. The common advantages of 7L-7S SC-MLI structure over the 5L multi-level inverter, existing 5L SC- multi-level inverter is evaluated in the table. 3

Table 3. Simulink model specification

S. No	PWM Techniques	%THD
1	PD	23.59
2	POD	7.97
3	APOD	7.87
4	VF	15.60
5	CO	7.60

7. Conclusion

This study analyses a single stage single phase switched capacitor boost 7L-7S inverter for attaining high output voltage with a single DC source; the input voltage is boosted efficiently with levels. The proposed method reduces the switch count, reduces the maximum number of source required, and reduces the THD of the presented system using different PWM technique is analyzed. Parameter comparison made between the existing (5MLI, Existing SCMLI) and proposed, which provides a better decision about SC-MLI. The comparison provides the %THD value of various PWM techniques in SC-MLI. Results show that the SC-MLI using various PWM techniques and the CO-PWM provides the less %THD value is 4.83 than other methods are done using MATLAB/Simulink. Further reduction of THD values on the system is complex will be done in future.

References

[1] Solanki VK, Hoang MK, Lu Z (Joan) and Pattnaik PK 2020) Intelligent Computing in Engineering. Advances in Intelligent Systems and Computing *Springer*

[2] Vasudevan V and Balaji K 2018 Modeling and Dynamic Performance of Photovoltaic and Fuel Cell Power Generation for Hybrid Converter System *Medico-Legal Update* 18 539-47

[3] Karasani RR, Borghate VB, Meshram PM, Suryawanshi HM and Sabyasachi S 2016 A three-phase hybrid cascaded modular multilevel inverter for renewable energy environment *IEEE Transactions on Power Electronics* **32** 1070-87

[4] Vasudevan V and Balaji K 2017 Performance of Cuk-KY Converter Fed Multilevel Inverter for Hybrid Sources *Indonesian Journal of Electrical Engineering and Computer Science* **10** 436-445

[5] Babaei E, Laali S and Bayat Z 2015 A single-phase cascaded multi-level inverter based on a new basic unit with reduced number of power switches *IEEE Transactions on industrial electronics* **62** 922-929

[6] Nguyen MK and Tran TT 2017 Quasi cascaded H-bridge five-level boost inverter *IEEE Transactions on industrial electronics* **64** 8525-8533

[7] Vargas RA, Figueroa A, DeLeon SE, Aguayo J, Hernandez L and Rodriguez MA 2015 Analysis of minimum modulation for the 9-level multi-level inverter in asymmetric structure *IEEE Latin America Transactions* **13** 2851-2858

[8] Gupta KK, Ranjan A, Bhatnagar P, Sahu LK and Jain S 2016 Multi-level inverter topologies with reduced device count: a review *IEEE Transactions on Power Electronics* **31** 135-151

[9] Rajeswari R 2015 Power quality improvement of novel PWM based 15 level inverter for modified unified power quality conditioner (UPQC) *International Journal of MC Square scientific research* **7** 23-40

[10] Dhanamjayulu C and Meikandasivam S 2017 Implementation and Comparison of Symmetric and Asymmetric Multi-level Inverters for Dynamic Loads *IEEE* Access *6* 738-746

[11] Khan MN, Siwakoti YP, Li L and Forouzesh M 2018 Single-Phase Switched-Capacitor Integrated-Boost Five-level Inverter *IEEE Region Ten Symp. (Tensymp)* 25-29

[12] Zamiri E, Vosoughi N, Hosseini SH, Barzegarkhoo R and Sabahi M 2015 A new cascaded switched-capacitor multi-level inverter based on improved series–parallel conversion with less number of components *IEEE Transactions on Industrial Electronics* **63** 3582-3594

[13] Barzegarkhoo R, Kojabadi HM, Zamiry E, Vosoughi N and Chang L 2015 Generalized structure for a single-phase switched-capacitor multi-level inverter using a new multiple DC-link producer with reduced number of switches *IEEE Transactions on Power Electronics* **31** 5604-5617

[14] Ali AJ and Ramesh GP 2020 Analysis of Open, Closed Loop PI, PID, FLC and ANN Controllable Wind Energy System Using Γ-ZSI with PMSM *In ICDSMLA 2019* 1143-1155

[15] da Silva RL, Lazzarin TB and Barbi I 2018 Reduced switch count step-up/step-down switched-capacitor three-phase AC–AC converter *IEEE Transactions on Industrial Electronics* **65** 8422-8432.

[16] Khodaparast A, Adabi J and Rezanejad M 2018 A step-up switched-capacitor multilevel inverter based on 5-level T-type modules *IET Power Electronics* **12** 483-91

[17] Liu J, Cheng KWE and Ye Y 2014 A Cascaded Multilevel Inverter Based on Switched-Capacitor for High-Frequency AC Power Distribution System *IEEE Trans. Power Electron.* **29** 4219-4230

[18] Tsunoda A, Hinago Y and H Koizumi H 2014 Level- and Phase-Shifted PWM for Seven-Level Switched-Capacitor Inverter Using Series/Parallel Conversion *IEEE Trans. Ind. Electron.* **61** 4011-4021

[19] Gadalla AS, Yan X, Altahir SY, Hasabelrasul H 2017 Evaluating the capacity of power and energy balance for cascaded H-bridge multi-level inverter using different PWM techniques *The Journal of Engineering* **13** 1713-1718

[20] Jayabalan M, Jeevarathinam B and Sandirasegarane T 2017 Reduced switch count pulse width modulated multi-level inverter *IET Power Electronics* **10** 10-17

[21] Uno M and Kukita A 2018 PWM switched-capacitor converter with switched-capacitor-inductor cell for adjustable high step-down voltage conversion *IEEE Transactions on Power Electronics* **34** 425-437

[22] Kumar KV and Kumar RS 2019 Analysis of Logic Gates for Generation of Switching Sequence in Symmetric and Asymmetric Reduced Switch Multilevel Inverter *IEEE Access* **7** 97719-97731

[23] Puttamadappa C and Parameshachari BD 2019 Demand side management of small scale loads in a smart grid using glow-worm swarm optimization technique *Microprocessors and Microsystems* **71** 102886.

Cascaded Multi-Level Inverter with APOD Based PV System for Induction Motor using DTFC Control

R Murugesan[1] and N C SendhilKumar[2]

[1] Department of EEE, Annamacharya Institute of Technology & Science, Tirupati, AP, India
[2] Department of ECE, Sri Indu College of Engineering and Technology, Hyderabad, India.
[1]rmurugesandr@gmail.com, [2]sendilkumarnc@gmail.com

Abstract. In this paper cascaded multi-level inverter based APOD control is designed to achieve harmonic reduction based on induction motor (IM) application which is fed by switched inductor quasi-Z source inverter (SL-qZSI) is proposed. The control method of proposed induction motor is controlled by direct torque and flux control (DTFC). The proposed converter qZSI is achieving the DC link voltage with improvement and to maintain the constant power supply. The proposed control technique for induction motor directs to reliability improvement and the capability of soft starting. The multi-level inverter with cascaded system is using with APOD control to reduce the total harmonic distortion in the supply voltage. The DTFC achieves the reduced settling time and stator inrush current reductions are achieved. The results and proposed system is analyzed and verified using MATLAB/Simulink environment.

Keywords: APOD PWM, Cascaded MLI, DTFC, Harmonics reduction, IM, Switched inductor quasi-ZSI.

1. Introduction

The Induction motor base drive system has more power density and high efficiency. The induction machine based application is able to operate in constant power and constant torque with wide range under high speed operation condition. In many type of system, the induction motors are preferred [1]. The IM machines for drives are normally adopts for the VSI. The super capacitor or battery is used ad DC power source supply for the system. When the input power supply discharges from capacitor or battery at high current, the voltage is decrease greatly. The control characteristics are affected by voltage which leads to output power reduction. The voltage source inverter the converters are used to improve the bus voltage of the IM based system [2]. This conversion system ensures the constant voltage in DC bus voltage and the mismatch between in the DC bus voltage of inverter and the input dc power supply. At mean time, speed and torque of the machine is operating in wide range since the converter is performs boost state operation. The use of predictive algorithms involves a reasonable estimate of the parameters of the system and, more broadly, greater computing power in comparison to conventional control strategies [3]. Today, multi-phase drive systems are seen as a feasible option and for applications needing redundancies to improve tolerance to failures. In addition, the activity is very close to the well-known Direct Torque Control (DTC) carried out at low switching frequencies. The key benefits of multiphase drives involve that the phase current without growing the phase voltage and the existence of fault tolerance [4]. The Dead-Beat Direct Torque and Flux Control (DBTFC) is another statistical approach, centered on average voltage vectors implemented at constant switching frequency. Industrial drives require soft engine start up. The key downside of CSI fed drive is that because broad inductors are used to realize the current source, they have a slower dynamic reaction compared to that of drives fed by VSI. On IMD applications a cascaded MLI is required [5]. Multilevel inverter has drawn considerable attention many applications due to the without the use of transformer and without the use of complex voltage control circuits [6]. If the number of output stages grows, the output voltage and current harmonics, as well as the EMI decreases. It's because it has basic layout and fast management of this form of MLI. However, in case of cascaded MLI, the need for independent DC-supplies raises the scale, expense and complexity of the multi-winding transformer. In IMD applications even a moving condenser MLI is used. It's because, this MLI requires less independent DC-supplies [7]. However, the MLI flying condenser configuration and control is too complicated. This is because, at the necessary values, the DC-link capacitor voltages and the flying capacitor voltages should be controlled, which needs a very specific and reliable control algorithm. Increasing carrier waveform is out of alignment with its neighbour carrier by 180 degrees in the case of alternate alignment disposition (APOD) modulation [8, 15]. Because in the case of a seven-level inverter, APOD and POD systems are the control methods for MLI and disposition is used to address the APOD System. Using cascaded-inverter with different supplies the multilevel inverter is target the voltage from multiple individual dc voltage sources, which can be solar cells.

In this paper, qZSI converter is utilized to boost power source dc voltage. The high gain is fed to the cascaded multilevel inverter which connected to the induction motor. The proposed system is compared with the [9]. The MLI with cascaded is controlled with the APOD pulse width modulation and harmonics of the MLI is reduced with 11 levels. The voltage can be controlled by the control system of the proposed system using sensors. DTFC is proposed to control the operation of the proposed IM and voltage for the interfaced inverter. The results are analyzed with this article through the MATLAB/Simulink.

2. Literature Survey

The induction motor operation is controlled using the well-known sensor-less based model reference adaptive control method is used to control in the speed and torque. This is based on the lyapunov theory, and stability based control. The improved speed sensor-less vector control algorithm is proposed for the induction motor performance control. The aforementioned control methods are also providing the ripples in speed [10]. Fault diagnosis of single-phase induction motor is analysed and controlled using the acoustic signals. The direct torque flux (DTC) control used in the induction motor as well as the sliding mode controller (SMC) which is giving feedback linearization. It is robustness method and reduces flux ripples. It has low performance during dynamic periods and disturbances [11]. The PV system is fed to the induction motor based water pump. The converter is used to improve the input power supply which is fed to voltage source inverter. The VSI is controlled using field oriented control to achieve the speed and torque control. The interleaved boost converter for achieving high voltage gain based on the fuel cell with controller of MPPT system. The interleaving converter system is increasing the passive elements and enhances the cost and size [12]. The water pumping based induction motor is controlled by using the fuzzy logic control with the adaptive speed estimation which is improve the system response quickly, and ripples free operation. The power quality in the drives are controlled using the fractional order PID control is used to improve the settling time reduction, and rise time minimization. In this [13] the multilevel inverter is used to reduce the total harmonic distortion. The Zeta converter based motor control used for industrial applications. The motor speed is tracked using a hysteresis controller on the back EMF. The proposed method is designed to achieve speed and reduce fluctuation in input voltage. It provides an improved capacity with a high voltage gain rate. It's a zeta and quadratic boost conversion combination [14]. The multilevel inverter with trinary modulation control is used to maximize the voltage levels and which is better while compared with PI controller.

3. Proposed system

In this proposed system, as shown in figure.1. The quasi Z source inverter with switched inductor is interfaced with the PV system to improve the bus voltage of the converter and the system voltage is maintained at stable. Switched inductor qZSI voltage is fed to the MLI with cascaded for the PV power supply to IM with levels voltage as well as reduced harmonics. The cascaded MLI is providing 11 levels to run the induction motor which is controlled by using DTFC. This system consists of photo voltaic system, IM, DTFC method, cascaded multilevel inverter, switched inductor qZSI.

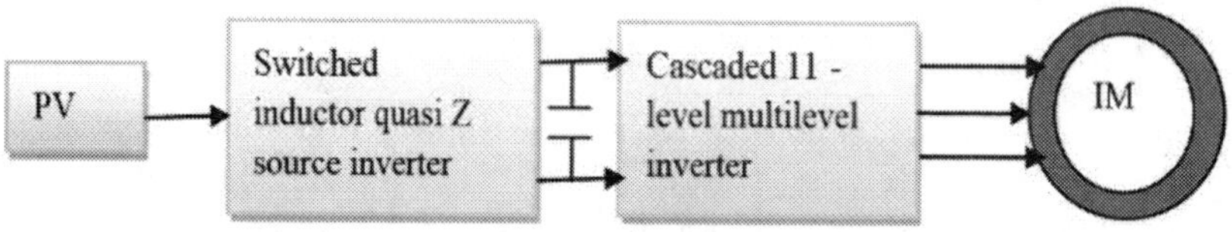

Figure 1. Proposed System Overall Block Diagram

3.1. PV Energy Generation

Photo voltaic system has PV cells in parallel and series combination which is utilized to improve the power respectively. PV produces energy directly from solar system and according to solar irradiances as well as temperature it will be changed. The PV circuit of single diode model is shown in figure.2. In PV cells the sunlight is directly converted into dc power. According to photoelectric effect phenomenon the PV modules are worked.

The following equation shows that PV model's current (I) and voltage (V) (I-V) characteristics:

Where,
I= PV current (A)
K= Boltzmann constant (J/K)
IL= Light generated current (A) in the PV module
V= PV voltage (V)
RsRsh = series and shunt resistance of the single diode PV model
Q = Electron Charge
Io=saturation diode current

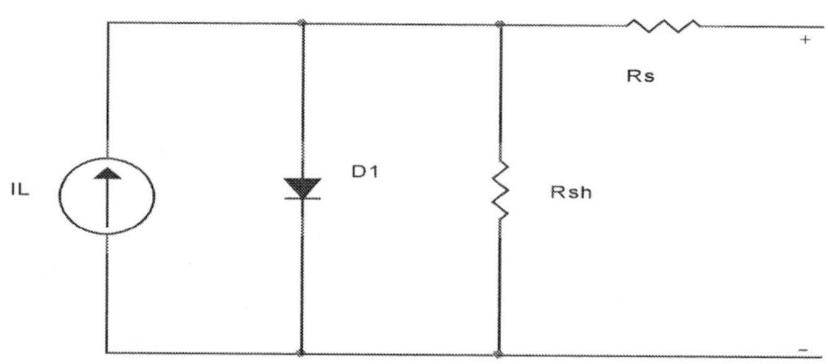

Figure 2. Circuit of single diode PV module

3.2. Proposed qZSI system

The proposed qZSI is used to improve the DC link voltage from the renewable PV system and to maintain the constant voltage for providing input supply of the cascaded multilevel inverter which is developing 11 levels for soft-starting of induction motor (IM). It has low amount of elements to improve the voltage in both modes as illustrated in figure. 3. Normally, the qZSI has two states one is non shoot and other is shoot through mode. Through this, the inrush current is limited and motor soft starting performance is accomplished. Circuit comprises of diodes (Da, Db, Dc, Dn), inductors (La, Lb, Lc), and capacitors (Ca, Cb). Switches in the inverter based conversion legs are turn ON in same time and the relation of the capacitor and inductor voltage is given as,

IOP Conf. Series: Materials Science and Engineering **925** (2020) 012018 doi:10.1088/1757-899X/925/1/012018

$$V_{Lc} = V_{PV} + V_{Ca} \tag{1}$$

$$V_{La} = V_{PV} + V_{Cb} \tag{2}$$

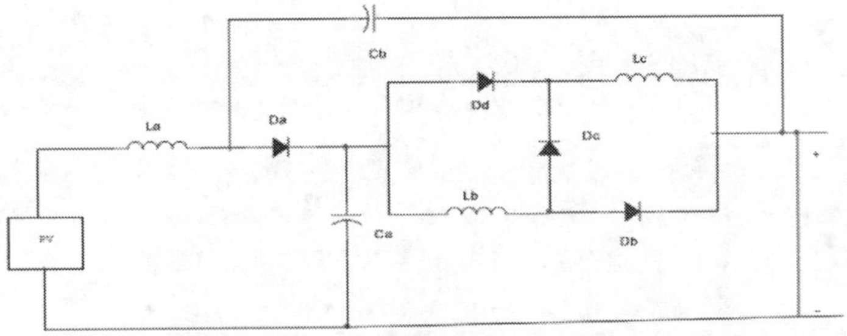

Figure 3. Proposed qZSI Converter Diagram

In non-shoot-through state, the two inductors in the proposed circuit diagram are La which is discharged in this mode as well as the another inductor Lb. The diodes of the circuit will not be conducted in this state named as (Da, Dc) are in OFF condition. The charging of the capacitor is done (Ca, Cb).

4. Proposed system control methods

In this paper, is using the IM based quasi ZSI to achieve the high voltage gain and improve the system stability. The overall proposed control method for the induction motor is shown in figure. 4. It is used to control the response of the IM and improving the exact excitation of the proposed cascaded MLI based induction motor. The output torque of the motor is estimated and fed back to the PI controller. The park and Clarke transformations are implemented to achieve the supply voltage accuracy.

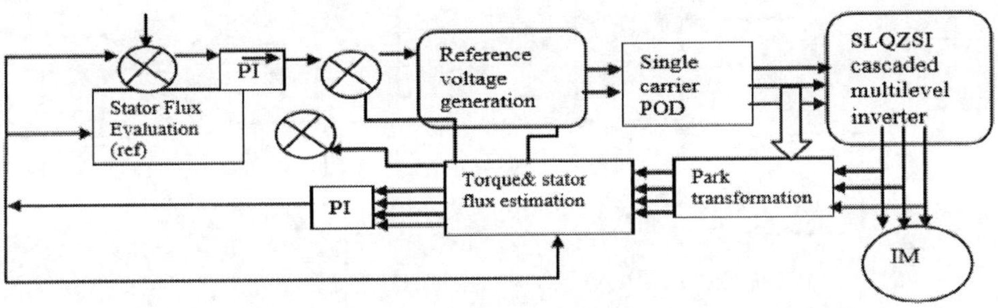

Figure 4. Proposed DTFC Control method

The proposed cascaded MLI is controlled using the APOD based PWM control signals are provided for the power switches of the proposed MLI with cascaded system to achieve the 11 levels in the form of harmonics reduction. Each of the m-1 carrier waveforms are required in this method, as represented in figure.5 for an m-level phase waveform, it is 180 degree disposed from other signals. The control method of the cascaded with MLI is shown in figure 6. In this 180 degree phase out in top two signals each other as well as bottom signal.

IOP Conf. Series: Materials Science and Engineering **925** (2020) 012018 doi:10.1088/1757-899X/925/1/012018

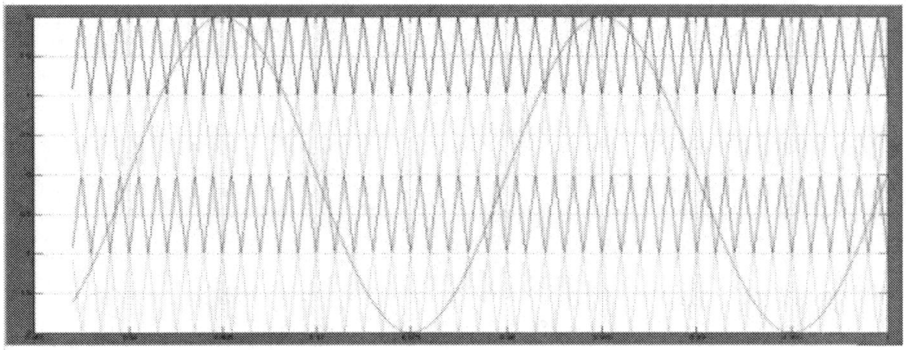

Figure 5. APOD Carrier arrangement

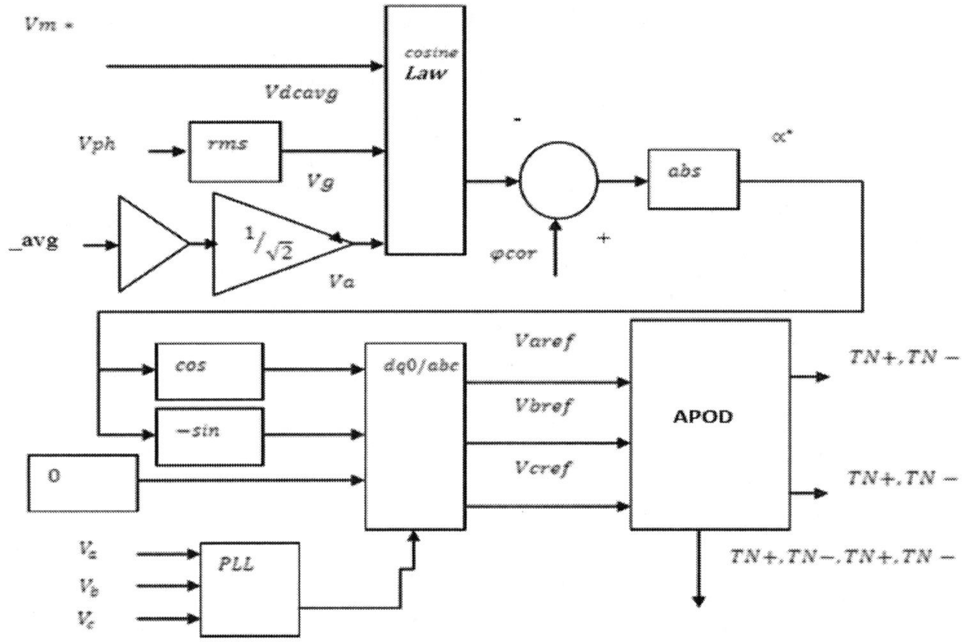

Figure 6. APOD based cascaded MLI

5. Simulation and results

Overall simulink model is shown in figure. 7 the PV renewable energy is supplied to IM through the multilevel inverter with cascaded along with the proposed quasi ZSI to improve the high DC link voltage of the proposed converter and achieve the proposed system stable response. The cascaded multi level inverter is generating 11 levels with reduced total harmonics. The output regulated voltage is fed to the induction motor (IM) which is controlled by using DTFT method. The PV system is generating 100V for the IM fed quasi ZSI for improving the DC link voltage and achieving the reference speed and torque.

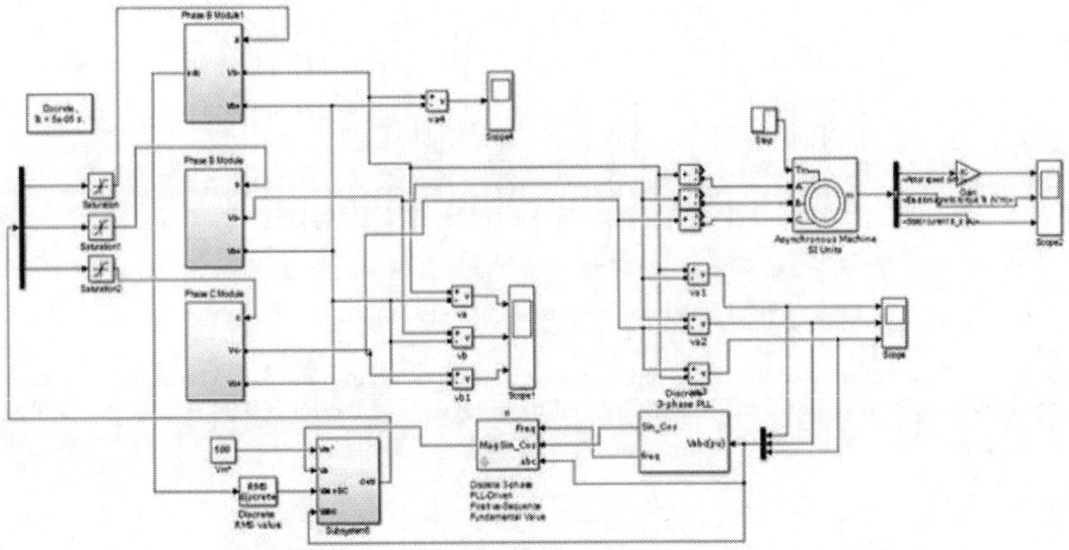

Figure 7. Overall Simulink Model of the DTFC based IM

The DC link voltage is regulated by the proposed quasi Z source inverter which is powered from the solar system generation. The regulated DC voltage is 220V as shown in figure. 8 and it is supplied to the multi-level inverter to improve the output voltage for the proposed DTFC based induction motor.

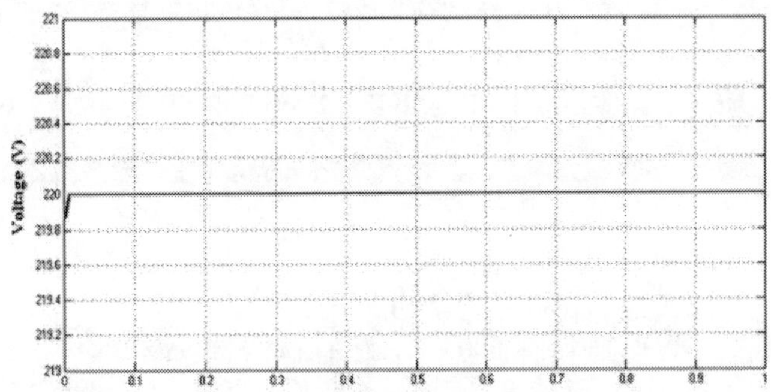

Figure 8. DC link voltage

The cascaded multilevel inverter is used to produce levels for achieving the harmonics reduction in form carrier arrangement with control technique of APOD control for the generating switching signals for the proposed cascaded MLI. The proposed system inverter is achieving 11 levels with 230V for the induction motor as shown in figure. 9 based systems to reduce the inrush current which is done by the control method of the direct torque and flux control. The speed of the proposed DTFC based IM accomplished the speed of 1600rpm which is settled in the time of 0.4 sec as shown in figure 10.

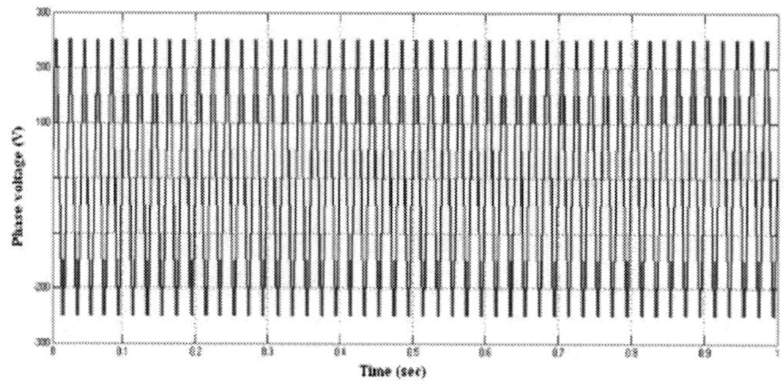

Figure 9. 11 levels of proposed cascaded multi level inverter

The torque and stator current of the proposed system are represented in figure.10. The multilevel inverter with cascaded system is reduced the harmonics in the output of the inverter is 9.04% as shown in figure. 11.

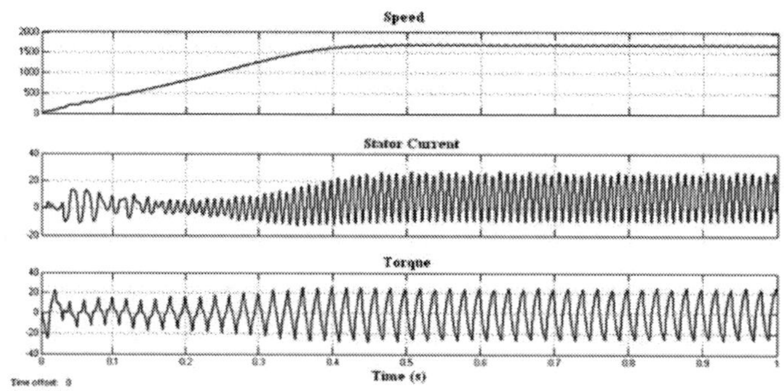

Figure 10. Proposed System induction motor speed, stator current, torque

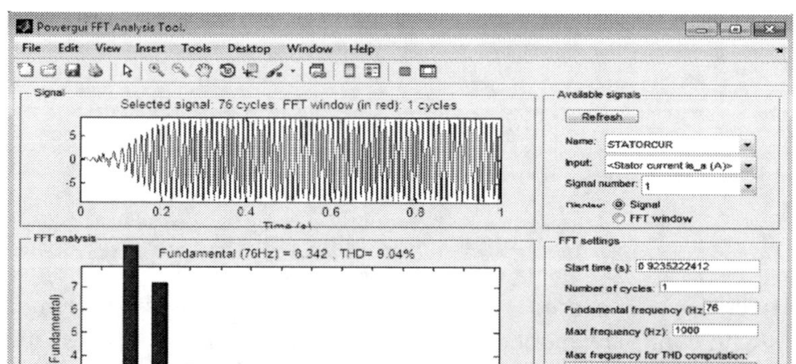

Figure 11. Harmonics reduction cascaded MLI voltage

6. Conclusion

The APOD control based cascaded multi-level inverter is proposed to accomplish the harmonics reduction in the inverter AC output voltage which is fed to induction motor based DTFC control method. The PV generation is fed to propose quasi ZSI for improving the input power supply and it is achieved successfully. The DTFC control is achieved the inrush current reduction as well as limits the current for the stator. In this proposed control method, the settling time of the induction motor speed is reduced and DC link voltage is improved. The MATLAB/Simulink platform is used and verified the results of the proposed system successfully.

References

[1] Korzonek M, Tarchala G and Orlowska-Kowalska T 2019 A review on MRAS-type speed estimators for reliable and efficient induction motor drives *ISA transactions* **93** 1-3

[2] Deng Y, Liang Z, Xia P, and Zuo X 2019 Improved speed sensorless vector control algorithm of induction motor based on long cable *Journal of Electrical Engineering & Technology* **14** 219-29

[3] Glowacz A 2019 Fault diagnosis of single-phase induction motor based on acoustic signals *Mechanical Systems and Signal Processing* **117** 65-80

[4] Lascu C, Jafarzadeh S, Fadali MS, and Blaabjerg F 2016 Direct torque control with feedback linearization for induction motor drives *IEEE Transactions on Power Electronics* **32** 2072-80

[5] Khan K, Shukla S and Singh B 2019 Performance-Based Design of Induction Motor Drive for Single-Stage PV Array Fed Water Pumping *IEEE Transactions on Industry Application* **55** 4286-97

[6] Ramaprabha R, Balaji K, Raj SB and Logeshwaran VD 2013 Comparison of Interleaved Boost Converter Configurations for Solar Photovoltaic System Interface *The Journal of Engineering Research [TJER]* **10** 1726-6742

[7] Rajan S and RK D 2019 Solar PV Fed Induction Motor Drive Using Modified SEPIC Converter *In proc. of the Int. Conf. on Systems, Energy & Environment (ICSEE)*

[8] Shukla S and Singh B 2019 Adaptive speed estimation with fuzzy logic control for PV-grid interactive induction motor drive-based water pumping *IET Power Electronic* **12** 1554-62

[9] Chandrasekaran S and Durairaj S 2019 PV-Based Multilevel Inverter-Fed Three-Phase Induction Motor with Improved Time and Speed of Response *Int. Conf. on Artificial Intelligence, Smart Grid and Smart City Applications* 173-183

[10] Vivert M, Patino D and Diez R 2019 Modulation Strategy and Controlle r for Grid-Tied Trinary Hybrid Multilevel Inverter *IEEE Journal of Emerging and Selected Topics in Power Electronics*

[11] Manie N and Pattanaik B 2019 Zeta DC-DC Converter Based on MPPT Technique for BLDC Application *International Journal of MC Square Scientific Research* **11** 1-2

[12] Umadevi D and Shivakumar EG 2019 Fractional order PID controlled Quadratic-Boost-Converter-Multilevel inverter fed Induction Motor System *IEEE International Conference on Electrical, Computer and Communication Technologies (ICECCT)* 1-6

[13] Lin Y and Hollander MA 2019 Multi-level inverter and method for providing multi-level output voltage by utilizing the multi-level inverter *United States patent US*

[14] Bhuvaneswari P and Ramesh GP 2017 Levenberg-marquardt algorithm to identify the fault analysis for industrial applications *International Journal of Engineering & Technology* **7** 141-150

[15] Puttamadappa C and Parameshachari BD 2019 Demand side management of small scale loads in a smart grid using glow-worm swarm optimization technique *Microprocessors and Microsystems* **71** 102886

Comparison of Artificial Controller Based MRAS Speed Observer for Field Oriented Control of Induction Motor

Meera Shareef Sheik[1] and K Balaji[2]

[1,2] Department of EEE, SPIHER, Avadi Chennai India.
[1]shareefsheikphd@gmail.com, [2]Kbalaji26@gmail.com

Abstract. In many industrial applications, the varying speed of the motor drives has become essential for the fast dynamic load conditions. Therefore, the detailed analysis of Field Oriented Control (FOC) with the sensor-less technique of induction motor is proposed where the reference speed is obtained from the MRAS speed observer, and fuzzy, ANN based controllers control the speed. The FOC not only control the magnitude but also provides the phasor control. The FOC is one of the sensors less technique. The sensor-less invokes the process of estimating the speed; the proposed MRAS speed observer achieves it. The MRAS speed observer gets the inputs such as current and voltage and then it converts to flux to reference speed. The FOC control the torque and flux components independently, and the flux component is obtained by the current. The fuzzy-based control technique is utilized for the control of the speed of the induction motor. In addition to this paper compares the results of fuzzy, and ANN based speed control methods, the results include the settling time, peak overshoot, ripples and accuracy of the speed. The results have been compared in the virtual environment by the MATLAB tool.

Keywords: ANFIS Controller, Field Oriented Controller (FOC), Fuzzy Controller, Induction Motor, MRAS Speed Observer, Settling Time.

1. Introduction

In recent times the application of induction motor with fast varying speed conditions is expanding. The enhancement in speed control requests looking for new control techniques. The typically used controls are scalar and vector control. The scalar control utilizes the Voltage/frequency strategy which invokes the process of controlling the speed by frequency or voltage [1]. The scalar control has the issue of circuit damage when there is confusion in the timing of voltage and frequency control, i.e., voltage and frequency ought to be changed concurrently. The scalar control is mostly used in non-feedback conditions because of its simple structure, financially savvy and error reduction. The feedback condition is also being practicable in scalar control, but it is confounded and produces ambiguity to the control [2]. The feedback condition control only provides control over the torque which is acquired by comparing the motor speed and reference, and they are given through the controller such as PI, PID and advanced controllers to deliver the reference torque [3-4]. The scalar control in the feedback condition has not given authority to control over the flux as well as control over the fast varying speed. To overcome these problems, there is the entry of vector control techniques. The vector control provides the high-quality speed control of induction motor. Both the torque and flux components are controlled individually which builds the intricacy. Thus the FOC and Direct Torque Control (DTC) propelled control schemes are proposed [5]. The FOC and DTC include in controlling the speed by not only taking out the speed value as a reference and converting them into flux but also uses the stator current values as feedback for the reference. The procedure of control for FOC and DTC is in a similar way, i.e., by controlling the flux and torque components however the considerable contrast is DTC utilises the stator flux while FOC uses the rotor flux and there is also need of converting the three-phase current to two-way components(D-Q) in DTC. Therefore, the DTC is quite simple in structure, but it has the disadvantage of less accuracy during quick speed and loads changing conditions [6]. There are two methods of FOC control one is direct control method, and other is an indirect control method. In FOC the rotor sensors either installed or not needed to detect the rotor position and speed. They can be calculated from any estimation method such as MRAS observer, Kalman filter or adaptive methods. Direct FOC and indirect FOC have a difference in calculating the rotor angle. The indirect FOC calculates the rotor angle indirectly by adding the rotor speed with slip speed and then integrating the sum [7]. The usage of indirect FOC includes complexity and ambiguity. This paper utilizes the sensor-less control of direct FOC technique to enhance the simplicity of the system with the comparison of PI and Fuzzy based speed controller. The sensor-less technique includes the usage of estimation methods for detecting the speed of the rotor. The availability of various estimation methods adds the confusedness of selecting the optimal method for speed estimation [8]. The methods are classified into two main types 1) Based on signal injection 2) based on the model. The signal injection method is hard to install in hardware because it requires separate hardware for signal injection as well as it can be used in low-speed applications. The model-based method consists of MRAS speed observer, Kalman filter, sliding mode observer and adaptive methods. These methods provide quick adaption of parameters during dynamic conditions with reduced computing time and easy implementation in both hardware and software [9]. Among these, the MRAS speed observer is used because this method requires fewer motor parameters with two models for estimating the same parameter with different input parameters for each model. This increases the rejection of error to a greater extent than the other estimation methods.

This paper proposes a novel MRAS speed observer to overcome the drawbacks based on the convention MRAS observer. The disadvantage of the conventional method includes not having the capability of quality performance in low-speed applications. The performances involve estimating the speed with varying temperature and resistance values. This paper also utilizes either fuzzy or PI controller for the speed control. Both the fuzzy and ANN based control is implemented and their results are compared for the selection of suitable controller. The proposed system of the induction motor based field oriented control for the speed control is compared with [10, 17] through the settling time reduction and minimization in ripples.

2. Literature survey

To improve the induction motor control with the steady state performance and dynamic response, the model predictive controller used. This control method is explained in [11] the relationship in between the stator voltage and the torque are analyzed. In [12] the two axis stationary reference frame based model reference adaptive system (MRAS) is implemented to estimate the speed of the rotor of an induction motor. The speed is controlled effectively but the settling time is more. In [13] loss minimizing strategy is proposed to achieve maximum efficiency operation for torque control. This control system is using the reference of the optimal stator flux and desired torque. The predictive direct torque controller is used to control the induction motor which combines both the predictive control as well as the direct torque control system. The overall performance of the motor is improved using the kalman filter that used for the estimation of the reliable flux [14-15]. The doubly fed induction motor based speed control method is explained in [16] the author describes the two types of control methods that does not relays on inner current loops and it gives direct voltages for rotor that used for the tracking the reference speed profile.

3. Proposed system

In this proposed system consist of mathematical modelling of induction motor with its control techniques are fuzzy and ANFIS based MRAS speed observer with field oriented controller. DC source fed to the

IOP Conf. Series: Materials Science and Engineering **925** (2020) 012019 doi:10.1088/1757-899X/925/1/012019

three phase inverter which control motor speed and it's settling time by proper control pulse to the inverter switches.

3.1. Mathematical model of IM

The three induction motor consists of three types. These types are classified based on the rotor structure. If the rotor looks like squirrel cage and double squirrel cage then it is called squired cage induction motor & double squirrel cage induction motor respectively and it is called as wound rotor when the rotor is in wound type. In this paper we took a squirrel cage induction motor modelling because it have better efficiency, lesser cost, less maintenance, and there is no need of high starting toque. The modelling can be done in both mechanically and electrically. The induction motor modelling can be done by the use of equations presented in upcoming sections.

3.2. Electrical modeling

The electrical modeling of squirrel induction motor is made through the state space model of 4th order. The three phase parameters of stator and rotor are converter to components of dq.

The q and d axis components is given by,

$$V_{qstator} = R_{stator} * i_{qstator} + \frac{d\psi_{qstator}}{dt} + \omega * \psi_{dstator} \tag{1}$$

$$V_{dstator} = R_{stator} * i_{dstator} + \frac{d\psi_{dstator}}{dt} - \omega * \psi_{qstator} \tag{2}$$

$$\overline{V}_{qrotor} = \overline{R}_{rotor} * \overline{i}_{qrotor} + \frac{d\overline{\psi}_{qrotor}}{dt} + (\omega - \omega_{electrical}) * \overline{\psi}_{drotor} \tag{3}$$

$$\overline{V}_{drotor} = \overline{R}_{rotor} * \overline{i}_{drotor} + \frac{d\overline{\psi}_{drotor}}{dt} + (\omega - \omega_{electrical}) * \overline{\psi}_{qrotor} \tag{4}$$

$$T_{elec\_Mag} = \frac{3}{2} * no.of.pole\left(\psi_{dstator} * i_{qstator} - \psi_{dstator} * i_{dstator}\right) \tag{5}$$

The fluxes are given by,

$$\psi_{qstator} = L_{stator} * i_{qstator} + L_{mag} * \overline{i}_{qrotor} \tag{6}$$

$$\psi_{dstator} = L_{stator} * i_{dstator} + L_{mag} * \overline{i}_{drotor} \tag{7}$$

$$\overline{\psi}_{qrotor} = \overline{L}_{rotor} * \overline{i}_{qrotor} + L_{mag} * i_{qstator} \tag{8}$$

$$\overline{\psi}_{drotor} = \overline{L}_{rotor} * \overline{i}_{drotor} + L_{mag} * i_{dstator} \tag{9}$$

$$L_{stator} = L_{stator\ leakage} + L_{mag} \tag{10}$$

$$\overline{L}_{rotor} = \overline{L}_{rotor\ leakage} + L_{mag} \tag{11}$$

IOP Conf. Series: Materials Science and Engineering **925** (2020) 012019 doi:10.1088/1757-899X/925/1/012019

3.3. Electrical Modeling

$$\frac{d\omega_{mech}}{dt}=\frac{0.5}{J_{rotor-load}}*\left(T_{elec\_Mag}-F_{rotor-load}*\omega_{mech}-T_{mech}\right) \tag{12}$$

$$\frac{d\theta_{mech}}{dt}=\omega_{mech} \tag{13}$$

Where Rstator and is resistance of the stator and rotor , Lstator leakage, is the stator and rotor leakage inductor value, Lmag is magnetizing inductor value, Lstator, is stator and rotor inductance value Vqstator, Vdstator is stator voltage along q axis and d axis respectively, iqstator, idstator is the stator current along q axis and d axis respectively, ωmech is rotor mechanical angular velocity, ωelectical is the rotor electrical angular velocity, i.e., ωmech*number of poles, mech rotor mechanical angular position, Telectromagnetic is electromagnetic torque, Tmechanical is the mechanical torque, Jrotor-load is the inertia coefficient between the rotor and load, Frotor-load is the friction constant between the rotor and load. , is the rotor voltage along q and d axis respectively , is the rotor current along q and d axis respectively, is the stator flux along q and d axis respectively, is the rotor flux along q and d axis respectively.

4. Control methods

4.1. Reference frame

The reference frame has to be chosen to modify the three phase input components i.e., ABC to d-q components and the d-q components are converted back to three phase output components i.e., ABC. Q-axis and D-axis of induction motor is shown in the figure. 1 and figure 2. There are three types of reference frame. They are rotor, stationary and synchronous reference frame. The difference between them are based on the angular position and balancing condition of voltages. In this model rotor reference frame is utilized because the stator voltage is in balance condition and rotor voltages are varying. is taken as rotor is the electrical position of the rotor and β is taken as 0. These may vary for different reference frame. The conversion matrix is given by, Input voltage ABC to d-q:

$$\begin{bmatrix} V_{qstator} \\ V_{stator} \end{bmatrix}=0.3333*\begin{bmatrix} 2*\cos\theta & \cos\theta+1.732*\sin\theta \\ 2*\sin\theta & \sin\theta-1.732*\cos\theta \end{bmatrix}*\begin{bmatrix} V_{abstator} \\ V_{bcstator} \end{bmatrix} \tag{14}$$

$$\begin{bmatrix} \overline{V}_{qrotor} \\ \overline{V}_{drotor} \end{bmatrix}=0.3333*\begin{bmatrix} 2*\cos\beta & \cos\beta+1.732*\sin\beta \\ 2*\sin\beta & \sin\beta-1.732*\cos\beta \end{bmatrix}*\begin{bmatrix} \overline{V}_{abrotor} \\ \overline{V}_{bcrotor} \end{bmatrix} \tag{15}$$

$$\begin{bmatrix} i_{astator} \\ i_{bstator} \end{bmatrix}=\begin{bmatrix} \cos\theta & \sin\theta \\ \dfrac{-\cos\theta+1.732*\sin\theta}{2} & \dfrac{-1.732*\cos\theta-\sin\theta}{2} \end{bmatrix}*\begin{bmatrix} i_{qstator} \\ i_{dstator} \end{bmatrix} \tag{16}$$

$$\begin{bmatrix} \overline{i}_{arotor} \\ \overline{i}_{brotor} \end{bmatrix}=\begin{bmatrix} \cos\beta & \sin\beta \\ \dfrac{-\cos\beta+1.732*\sin\theta}{2} & \dfrac{-1.732\cos\beta-\sin\beta}{2} \end{bmatrix}*\begin{bmatrix} \overline{i}_{qrotor} \\ \overline{i}_{drotor} \end{bmatrix} \tag{17}$$

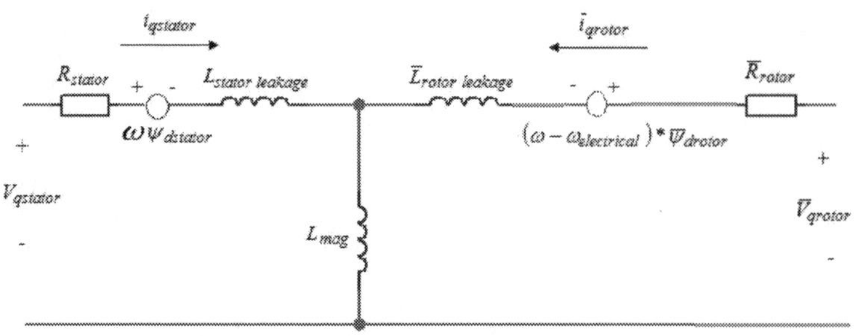

Figure.1 Q-axis of Induction Motor

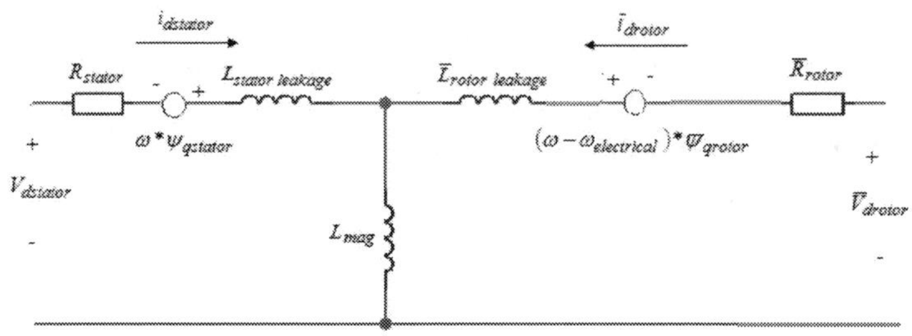

Figure.2 D-axis of induction motor

4.2. Field oriented controller

FOC is one of the vector control strategies which have the ability of controlling the torque and flux separately. The reference torque and flux are generated by the voltage, current and speed parameters from the induction motor. The FOC has two types one is direct FOC and indirect FOC. The direct FOC consists of the sensors fitted to the induction motor which will recognize the speed, position of the rotor as well as flux of the motor. In case of indirect FOC doesn't need flux sensor which is calculated indirectly by the flux calculating equation. The installing of flux sensors will make system complex and it is not attractive one. Therefore, the indirect FOC bring into act to solve the issue. The mathematical equation based on the current and speed is defined to find the flux. There is also existence of another vector control method which is named as DTC. The DTC is simple, lack of sensors but it has the major drawbacks such as ripples production in current, noise production and can't be suitable for low speed conditions. The FOC control scheme has been evaluated as sensor-less technique which is major advantage of DTC through the speed observing techniques. This paper utilizes the advantages of indirect FOC by implementing with sensor-less technique. The proposed system invokes the operation of calculating the speed using the speed observer and they are converted to torque and flux reference and there is also hands of Clarke and Park's transformation. The Indirect FOC block diagram is shown in figure.3.

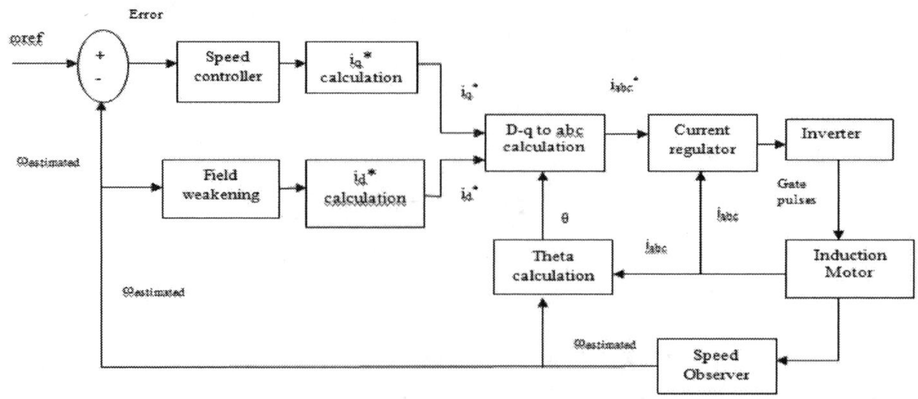

Figure 3. Block Diagram of the IFOC

The indirect FOC process includes the speed is first estimated by using the speed observer. In this paper modified MRAS speed observer is utilized which is discussed in the following section. The estimated speed is then converted as reference torque and flux. The reference torque and flux is converted into dq currents. The q axis current is obtained from torque and d axis current is obtained from flux. Meanwhile is calculated from the estimated speed for the operation of converting the dq components again to abc components. The conversion of dq current to abc current process is necessary to compare the converted abc current with reference abc current obtained from the motor. The comparison is made and the desired gate pulses to the inverter switches are given which is performed by the current regulator.

4.3. MRAS Speed observer

The proposed FOC utilizes the sensor less technique with the help of MRAS speed observer is represents in the figure 4. The sensor-less technique makes the absence of mechanical sensors has the constraints of less sensitivity and specific range. MRAS speed observer is one of the most commonly used speed observer because of its non-complexity, ease of implementation, less calculations and consistency. MRAS speed observer doesn't need more number of machine parameters like other methods such as Extended Kalman Filter, Luenberger observer (ELO) and adaptive observer. These methods also require lot of calculations and memory space. Therefore, the MRAS speed observer is became more attractive. It is classified into different groups namely 1) Based on the estimation of speed using the rotor flux 2) Based on the estimation of speed using back EMF 3) Based on the estimation of speed using reactive power 4) Based on the estimation of speed using electromagnetic torque 5) Based on the estimation of speed using stator current. Among all the methods the speed estimation using the rotor flux is the most popular method. This method has some drawbacks such as poor performance during the temperature and resistance variation, offset problem and less efficient on low speed. The proposed MRAS speed observer overcomes these problems and also increases the accuracy in estimating the speed. The proposed method has the two models one is reference model and adaptive model. These models output are compared and the desired results are given to the adaptive mechanism. Both the models are same for the conventional and proposed MRAS speed observer. Only difference is new adaptive mechanism has been developed. The typical reference model and adaptive model equations modeling are as follows equation 18 and 19 belongs to the reference model.

$$\frac{d\overline{\psi}_{drotor}}{dt} = \sigma L_{stator} * \frac{\overline{L}_{rotor}}{L_{mutual}} * \frac{di_{dstator}}{dt} + \frac{\overline{L}_{rotor}}{L_{mutual}} \left(V_{dstator} - R_{stator} * i_{dstator} \right) \quad (18)$$

$$\frac{d\overline{\psi}_{qrotor}}{dt} = \sigma L_{stator} * \frac{\overline{L}_{rotor}}{L_{mutual}} * \frac{di_{qstator}}{dt} + \frac{\overline{L}_{rotor}}{L_{mutual}} \left(V_{qstator} - R_{stator} * i_{qstator} \right) \quad (19)$$

Equation 20 and 21 belongs to the adaptive model

$$\frac{d\overline{\overline{\psi}}_{drotor}}{dt} = \frac{L_{mutual}}{Rt} * i_{dstator} - \frac{1}{Rt} \overline{\psi}_{drotor} - \overline{\omega}_{rotor} * \overline{\psi}_{qrotor} \quad (20)$$

$$\frac{d\overline{\overline{\psi}}_{qrotor}}{dt} = \frac{L_{mutual}}{Rt} * i_{qstator} - \frac{1}{Rt} \overline{\psi}_{qrotor} - \overline{\omega}_{rotor} * \overline{\psi}_{drotor} \quad (21)$$

The adaptive mechanism is modified by including the torque parameter which is introduced to enhance the stability and response of the system. The adaptive mechanism is given by block diagram. The equation includes,

IOP Conf. Series: Materials Science and Engineering **925** (2020) 012019 doi:10.1088/1757-899X/925/1/012019

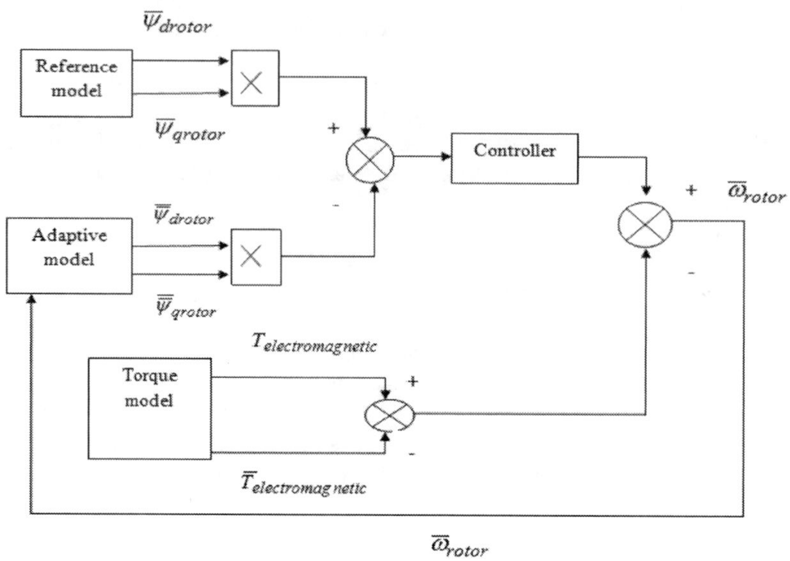

Figure 4. MRAS based IFOC Controller

$$T_{elec\_mag} = 1.5 * \frac{P}{2} * \frac{L_{mutual}}{L_{rotor}} (i_{qstator} * \overline{\psi}_{drotor} - i_{dstator} * \overline{\psi}_{qrotor})$$

$$\overline{T}_{elec\_mag} = 1.5 * \frac{P}{2} * \frac{L_{mutual}}{\overline{L}_{rotor}} (i_{qstator} * \overline{\psi}_{drotor} - i_{dstator} * \overline{\psi}_{qrotor})$$

4.4. Controller

The estimated speed is compared with the reference speed and the output from the comparator is given to the controller part. There are several control techniques available for the control operation. The advanced topologies in control technique have been developed in greater way but they have to meet the following constraints such as they should think, evaluate, develop, improve and organize by their own. The conventional PI and PID controller has the drawbacks of steady state and transient problems. To overcome this advanced controller like Fuzzy logic and ANFIS are proposed.

4.5. Fuzzy Logic Controller

Fuzzy logic controller is designed for regulates and reduce the settling of the speed attained from the motor. The error and change in error is the input of the fuzzy system. Then the output is designed by membership function (MF). It includes the property of IF and THEN condition and used the centroid method. Membership function of the fuzzy controller is shown in figure 5, figure 6 and figure 7.

IOP Conf. Series: Materials Science and Engineering **925** (2020) 012019 doi:10.1088/1757-899X/925/1/012019

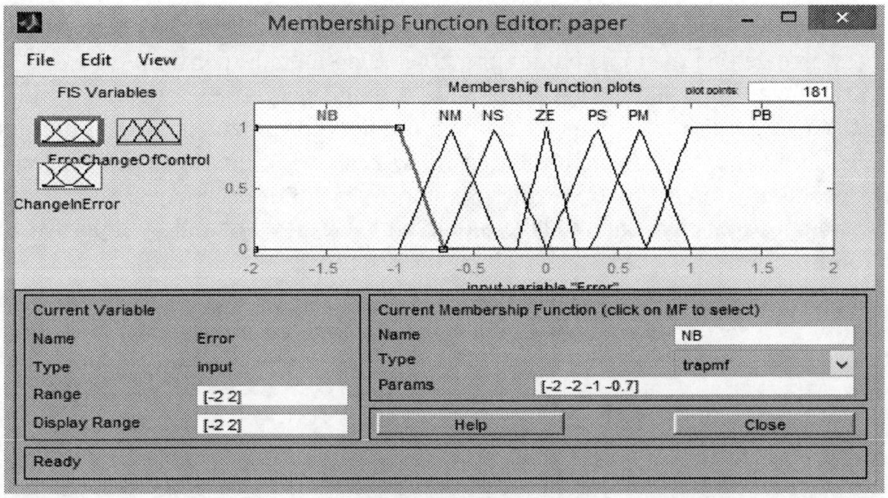

Figure 5. Error

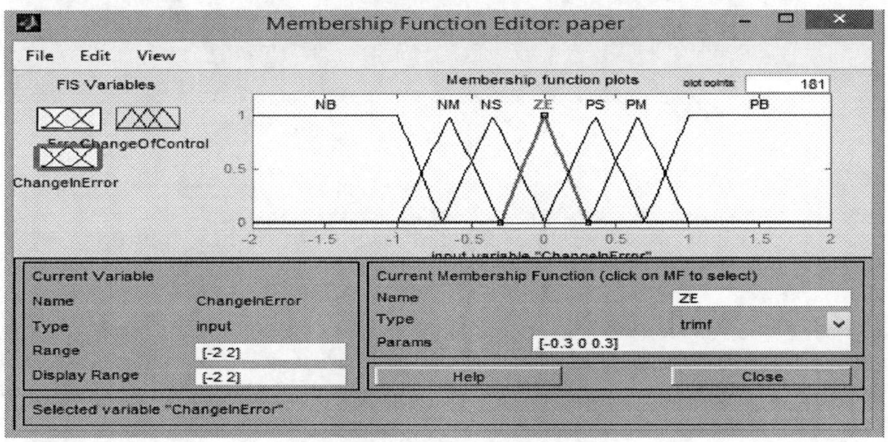

Figure 6. Change in error

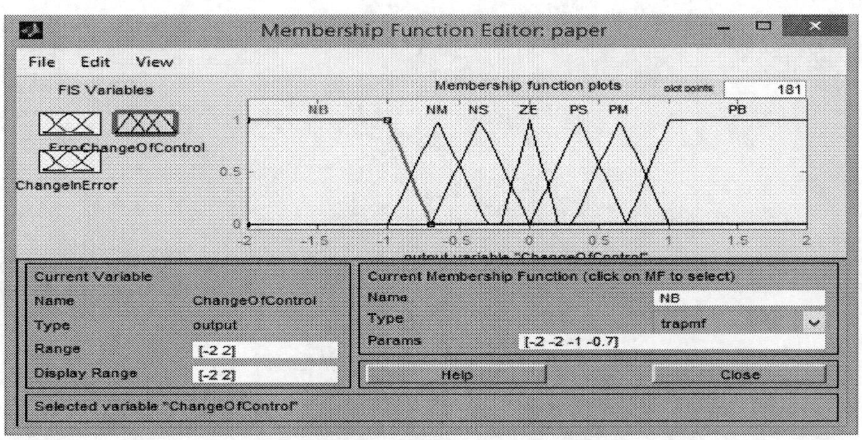

Figure 7. Change of Control

4.6. ANFIS Controller

The Artificial Neural Network Fuzzy Interference system used the property and combines both fuzzy and ANN advantages. The general block diagram for the ANFIS system is shown in Figure 8. It uses the SUGUNO model, which consists of five layers, and this control technique consists of an adaptive node and a fixed node. Nonlinearity mapping is property of ANFIS framework. The ANFIS incorporates the process towards getting the input values and processing them as per fuzzy rules, and the processed output from the fuzzy rules is again sophisticated through neural network define rules to build the accuracy of the system. The layer 1 includes the input data for processing the purpose of the ANFIS. The layer 2 performs the AND operation based fuzzy rules with input data. The layer 3 performs the same AND operation based NN defines rules for improvement of accuracy. The layer 4 is utilized for adjusting the output parameters and the layer5 delivers the desired output by processing the input data with given rules.

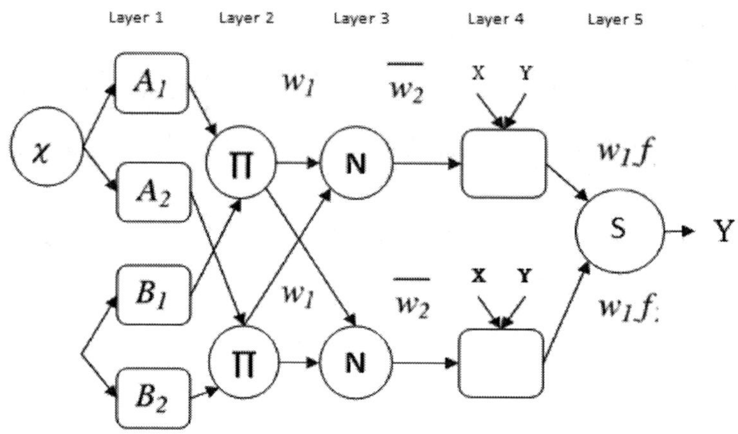

Figure 8. Generalized layered diagram of ANFIS Controller

5. Simulation analysis

The proposed three phase voltage source inverter with Fuzzy and ANFIS technique based MRAS speed observer control strategy is executed and their outcome results are evaluated and advantages are confirmed through MATLAB/Simulink. The diagram of overall Simulink setup and Fuzzy, ANFIS based MRAS speed observer with FOC controller is shown in figure 9.

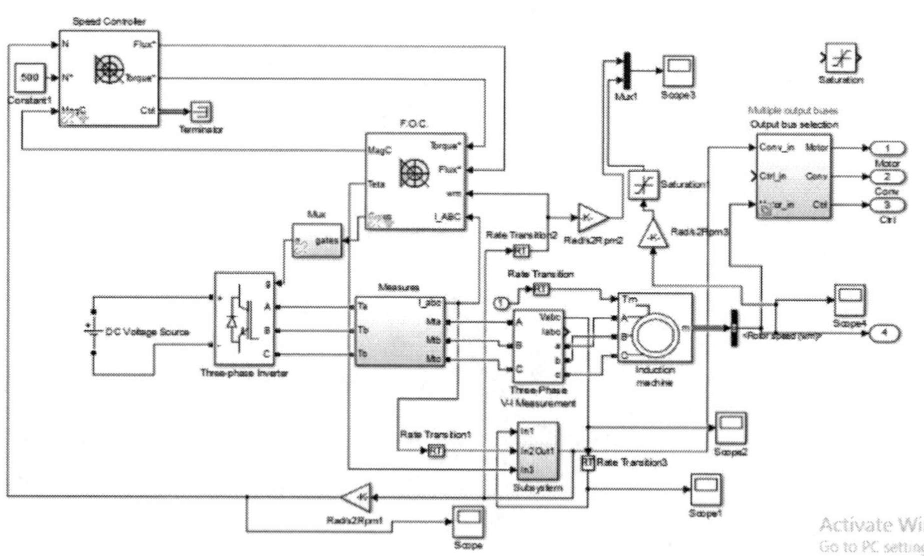

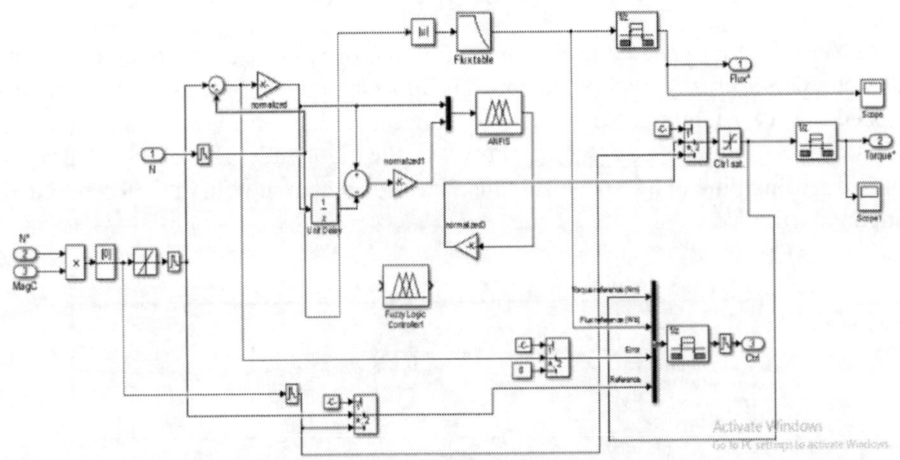

Figure 9. Overall Simulink diagram and Fuzzy, ANFIS with MRAS based IFOC speed controller

The three phase voltage waveform across the load (induction motor) using fuzzy and ANFIS controller is 400Vwithout filter and it represent in figure 10.

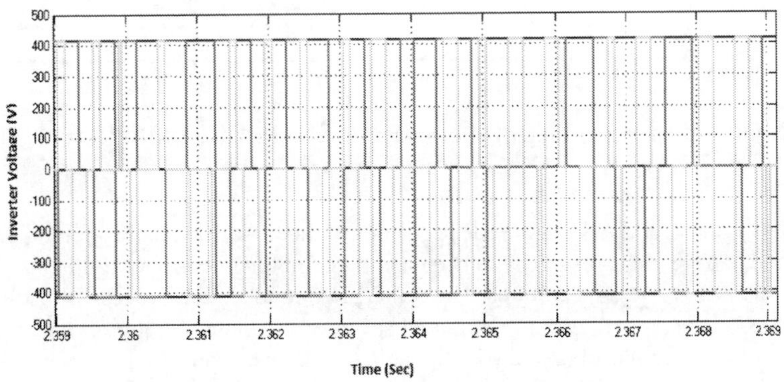

Figure 10. Inverter voltage before without filter

Estimated speed is 500rpm and reference speed is obtained from the MRAS speed observer method. Speed of the proposed induction motor using fuzzy controller is 494rpm with the settling time of 1.7sec shown in the figure 11.

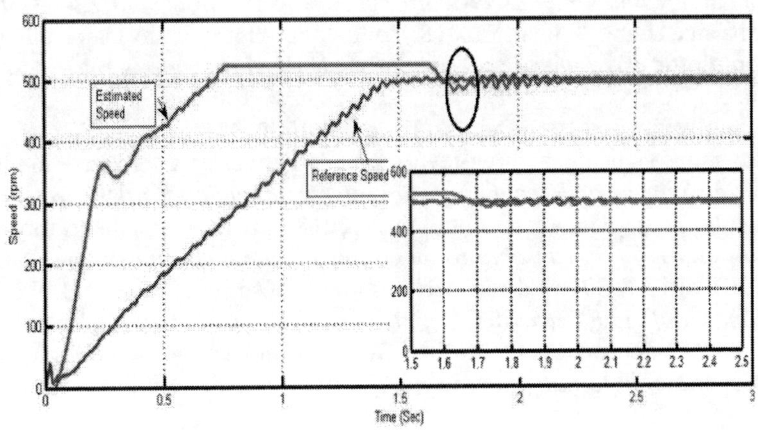

Figure 11. Output speed of motor using Fuzzy based MRAS with FOC controller

ANFIS based control the speed control of induction motor is same as fuzzy control, but its settling time is low compare to fuzzy. Estimated speed is 500rpm and reference speed is obtained from the MRAS speed observer method. Speed of the proposed induction motor using fuzzy controller is 494rpm with the settling time of 1.2sec shown in the figure.12. The obtained proposed control method based settling time of the speed is reduced while compared with the reference analysis which is explained in [10].

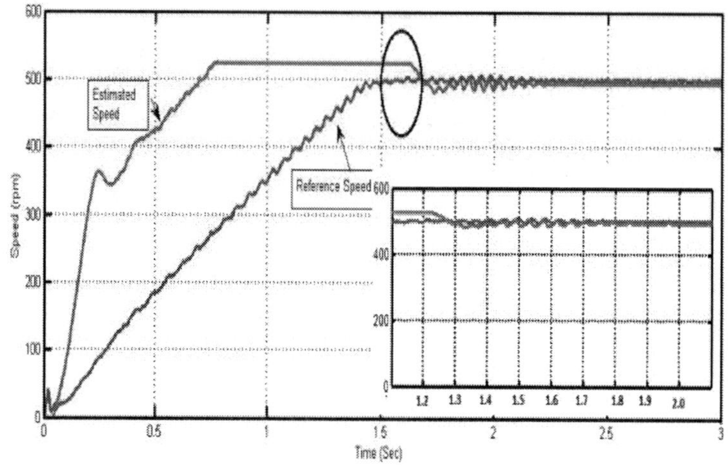

Figure 12. Output speed of the IM motor using ANFIS based MRAS with FOC controller

6. Conclusion

Compares the results of fuzzy, and ANFIS based speed control methods, the results include the settling time of the speed is proposed in this paper. The results have been compared in the virtual environment by the MATLAB tool. The FOC is one of the sensors less technique. The sensor-less invokes the process of estimating the speed; the proposed MRAS speed observer achieves it. Estimated speed of the induction motor is set at 500 rpm and actual speed of the motor is also 500rpm is obtained with the help of proposed controller fuzzy and ANFIS based MRAS with FOC controller. Settling of the induction motor is less at 1.3sec while using the fuzzy based controller. From this analysis and implementation, controls the motor speed and reduces its settling time by using artificial network based MRAS speed observer with FOC controller with the help of MATLAB/Simulink.

References

[1] Singh R, Bajpai SK and Sandhu HS 2015 Comparative study of PWM control and PI control of Induction motor *Bulletin of Electrical Engineering and Informatics* **4** 53-58

[2] Singh B, Sharma U and Kumar S 2018 Standalone photovoltaic water pumping system using induction motor drive with reduced sensors *IEEE transactions on industry applications* **54** 3645-3655

[3] Ramulu C, Sanjeevikumar P, Karampuri R, Jain S, Ertas AH and Fedak V 2015 A solar PV water pumping solution using a three-level cascaded inverter connected induction motor drive *Engineering Science and Technology, an International Journal* **19** 1731-1741

[4] Xu D, Wang B, Zhang G, Wang G and Yu Y 2018 A review of sensorless control methods for AC motor drives *CES Transactions on electrical machines and systems* **2** 104-115

[5] Singh S and Tiwari AN 2017 Various techniques of sensorless speed control of PMSM: A review *Second Int. Conf. on Electrical, Computer and Communication Technologies* (ICECCT) 1-6

[6] Sundaram A 2017 Sensor less Control of BLDC Motor using Fuzzy logic controller for Solar power Generation *International Journal of MC Square Scientific Research* **9** 70-79

[7] Tir Z, Malik OP and Eltamaly AM 2017 Fuzzy logic based speed control of indirect field oriented controlled Double Star Induction Motors connected in parallel to a single six-phase inverter supply *Electric Power Systems Research* **134** 126-133

[8] Ahmed AA, Koh BK and Lee YI 2017 A comparison of finite control set and continuous control set model predictive control schemes for speed control of induction motors *IEEE Transactions on Industrial Informatics* **14** 1334-1346

[9] Thufail MK and Balaji K 2017 Analysis and improvement of Self Tuning Based on Sensor less Control of BLDC Motor *Indian Journal of Public Health Research & Development* **8** 623-629

[10] Hashim A 2019 Speed sensorless vector control of induction motors using rotor flux based model reference adaptive system *Journal of Engineering and Computer Science (JECS)* **17** 1-4

[11] Cho Y, Bak Y and Lee KB 2017 Torque-ripple reduction and fast torque response strategy for predictive torque control of induction motors *IEEE Transactions on Power Electronics* **33** 2458-2470

[12] Pal A, Das S and Chattopadhyay AK 2017 An improved rotor flux space vector based MRAS for field-oriented control of induction motor drives *IEEE Transactions on Power Electronics* **33** 5131-5141

[13] Odhano SA, Bojoi R, Boglietti A, Roşu SG and Griva G 2015 Maximum efficiency per torque direct flux vector control of induction motor drives *IEEE transactions on Industry Applications* **51** 4415-4424

[14] Sankar K 2017 Design of Intelligent Controller for a Hydraulic Solar PV System *International Journal of MC Square Scientific Research* **9** 66-71

[15] Ouhrouche M, Errouissi R, Trzynadlowski AM, Tehrani KA and Benzaioua A 2016 A novel predictive direct torque controller for induction motor drives, *IEEE Transactions on Industrial Electronics* **63** 5221-5230

[16] Ali AJ and Ramesh GP 2020 Analysis of Open, Closed Loop PI, PID, FLC and ANN Controllable Wind Energy System Using Γ-ZSI with PMSM *ICDSMLA 2019* **601**

[17] Puttamadappa C and Parameshachari BD 2019 Demand side management of small scale loads in a smart grid using glow-worm swarm optimization technique *Microprocessors and Microsystems* **71** 102886

A Genetic Algorithm (GA)-PID Controller for Temperature Control in Shell and Tube Heat Exchanger

C Somasundar Reddy[1] and K Balaji[2]

[1,2] Department of EEE, SPIHER, Avadi, Chennai, India.
[1]somasundarphd@gmail.com, [2]balajiphd12@gmail.com

Abstract. In Genetic Algorithm tuned PID method for controls the fluid flow in the heat exchanger is presented in this paper. Shell and tube heat exchanger is the most generally utilized types of heat exchanger for heat transfer in many industrial purposes. The exchanger consists of mechanical part and controlling part. Both the modeled to ensure the efficient operation of shell and tube heat exchanger. The mechanical modeling is completely established on the type of applications. The controller only needs the input fluid and output fluid properties such as temperature and flow rate. Hence the primary objective of the paper is to focus on the controller part for enhancing the heat exchanger performance. This paper proposes the Genetic Algorithm tuned-PID controlling technique to make the settling time and overshoot of set point temperature to be less to a greater extent and the results are compared with the conventional PI method with various tuning algorithms [16]. The performance of proposed GA tuned plant has been compared with existing PID controller outputs with the help of MATLAB/Simulink.

Keywords: ANFIS Controller, BLDC Motor, Back EMF Method, Indirect Field Oriented Controller (FOC)

1. Introduction

Generally, the shell and tube heat exchangers with optimal design and control techniques are used in many industrial applications such as chemical, food and refrigeration industries [1]. In existing, fuzzy-PID has attracted most of the researchers because of its simple structure as well as PID has the advantage of high performance on steady state & fuzzy has the advantage of high performance on the dynamic state [11].GA is one of the methods influenced by nature that generally relies on the basic concept of fittest survival [12]. A detailed methodology for optimal tuning of PID controller has been proposed in this work. Suggested tuning of the operator using the genetic algorithm methodology. The existing method has some drawbacks such as less settling time and high peak overshoot. GA based PID technique has used but there are various mathematical operations are performed between the outputs coming from them. The optimal operation has to be chosen for the desired plant operations. In this paper, the general model of the controller has been developed for the shell and tube heat exchanger using transfer functions. The modeling is based on the flow rate of input fluid & flow rate of fluid coming into the shell, temperature sensor range, valve capacity to increase the flow rate and pressure. The GA-PID structure operation is implemented for obtaining the benefits such as less peak overshoot and settling time has been obtained for the desired set point.

2. Literature survey

Shell and tube are the two type's structure of the heat exchanger. The tube structure belongs to the input fluid or process fluid which is to be heated and shell structure belongs to the fluid which will transfer the heat to the input fluid. The tube structures consist of tubes and shell structure [2]. It also comes under the mechanical part need more to modeling the tube structure and shell structure. The great process also a time of controller in the modeling processes. To understand the complete model of the heat exchanger first is difficult. In existing method was quite difficult to achieve the accurate model [3-5].

The nominal modeling of the heat exchanger should base on following condition such as, open loop condition and closed loop condition. In accordance with the set point is to maintain the output process fluid temperature by the controller. A flow rate of fluid coming to the shell structure and input fluid flow rate are the controlling parameter of the heat exchanger. Different control techniques are presented in practice. P, PI, and PID are the manually tuned controller and Fuzzy, ANFIS are the artificial intelligence controller. The typical PID controller has been attracted by many industries because of easy modeling, tuning of constants flexibility and controllability [6-8]. The constants are normally tuned by the following optimization methods such as Particle Swarm Optimization (PSO), Cuckoo Algorithm, Bat Algorithm, Genetic Algorithm, Ziegler–Nichols method. The other controller such as SMC and MPC has high benefits over dynamic control but it is quite complex to design and hard to implement [9]. Nowadays the combined structure of controllers has been designed. This method gives the nominal values of tuning constants with quick response and acceptableness of changing conditions. In spite of its advantages, it also has disadvantages such as optimization of tuning constants and it can be honorably used for the linear systems rather than the non-linear systems due to its low performance on non-linear systems [10]. The main aim of the design is to utilize both the benefits of the combined controllers.

3. Objective

- To design the shell and tube heat exchanger model in MATLAB/Simulink based on experimental data.
- To regulate the temperature of the heat exchanger output fluid through the implementation of GA based PID control scheme.

4. Proposed model of heat exchanger

The temperature sensor is used in the system to get the feedback from the input fluid coming out from the heat exchanger. The sensor is designed in the range of sensing the temperature 500-1500C. The temperature signal is converted into the current signal by the transducer to the value of 4-20mA.The transfer function of the valve which is used to increase the flow rate of fluid coming into the shell is given by,

$$G_{valve} = \frac{1}{3s+1} * 0.75 * 0.13 \quad (1)$$

To control the temperature coming out of the tube outlet is controlled by the combined structure of Fuzzy-PID controller as well as feed forward controller is also added in this part to diminishes the error caused by the input fluid disturbance coming through the tube inlet. Figure 1 shows the general structure of the shell and tube heat exchanger. The general block diagram in figure.1 is included for easy understanding.

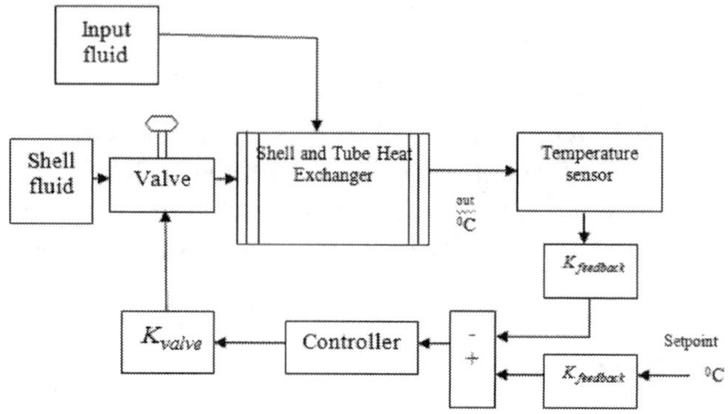

Figure 1. General block diagram

5. Function of Proposed Controller

5.1. PID Controller

The PID controller is one of the most traditional controllers used in the control systems of the industry because of its simplicity and ease of implementation. Even though it has a simple structure the tuning of gain value makes some difficulties in using them. In general, PID consists of three signals such as proportional, integral and derivative signals and they are added each other to generate error correction signal to the plant model. The gains of each signal are a proportional gain (Kd), integral gain (Ki), derivative gain (Kd). The input to the PID controller is error and output coming from the PID is error corrective value. The mathematical equation is given by,

$$PID_{output} = K_p * error + K_i * \int error * dt + K_d * \frac{d_{error}}{dt} \tag{2}$$

There are several tuning methods are experimented for tuning the PID gains. In that GA based PID tuning has been used in this paper because of its fast& adaptable gain process and non-complex structure [13]. GA method consists of following tuning methodology,

5.2. GA Algorithm

It is one of the methods used for optimization. The continuing performance improvement of computational systems has made them attractive for some types of optimization. The genetic algorithm begins with the stages of three evolutionary operators such as replication, fusion, and mutation to arrive at the best possible solution. General process chart of the GA is shown in figure 2

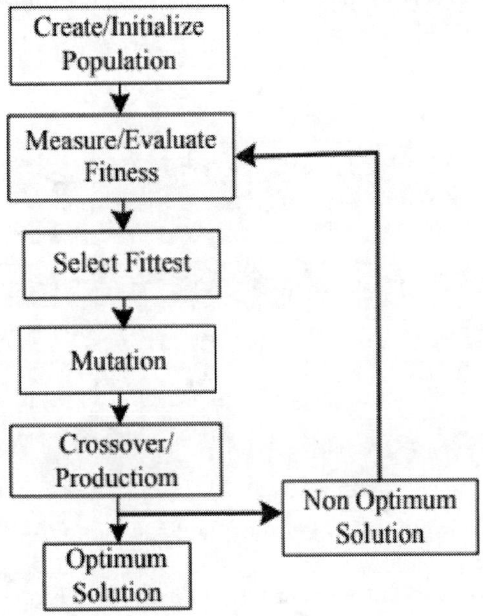

Figure 2. General Block Diagram of GA

5.3. GA based PID Controller

Compared to conventionally tuned PID regulator, GA tuned PID controller have much better output characteristic. The current tuning rule was used based on critical K_{er} gain and critical per time. The integral Ti time is set to infinity in this process, and the derivative Td time to zero. Using this to get the system's initial PID configuration. Using the steepest descent gradient process, this PID configuration will be further optimized. Only the proportional control action will be employed in this system... The Kp will raise to a critical Ker value at which the performance of the device will display repeated oscillations. The PID controller transfer function is given below

$$G_{feedback} = \frac{-18.461s^2 - 6.769s - 0.205}{30s^2 + 31s + 1}$$

6. Simulation Results

The Simulation testing shown in figure. 3 has been done on the transfer function modeled shell and tube heat exchanger. The simulation run time is chosen as 400sec and the set point temperature is kept at 900C, input fluid gain is kept as 20 and Kp, Ki, Kd values are kept as 13.25, 0.36, 56.25 respectively and the results have been obtained for the verification with conventional one using MATLAB/SIMULINK.

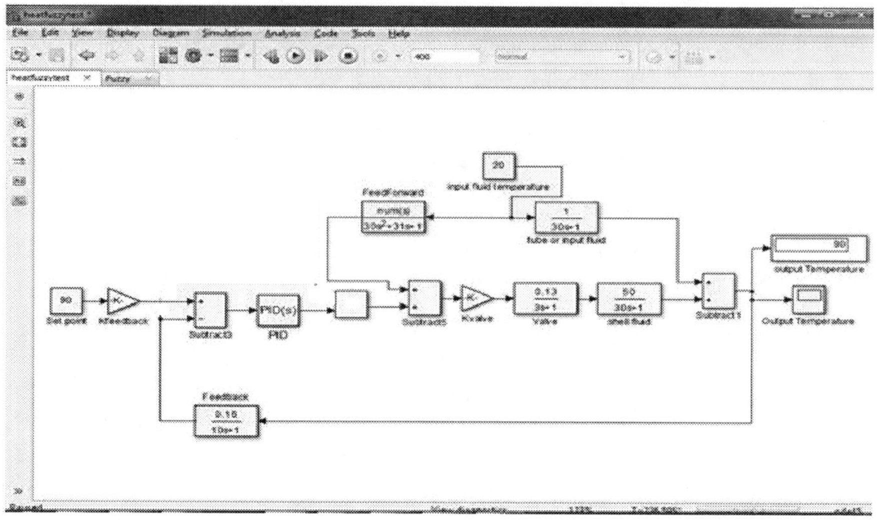

Figure 3. Simulink Diagram of the Proposed GA-PID controller

Figure 4 and 5 shows the input and output set point temperature. Fuzzy-PID controller output is shown in figure 6.

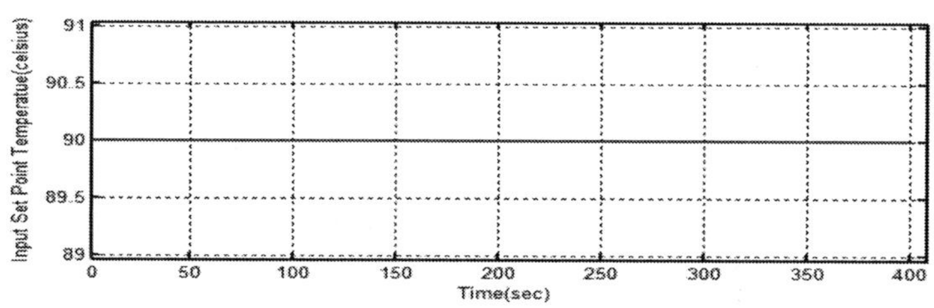

Figure 4. Input set point temperature

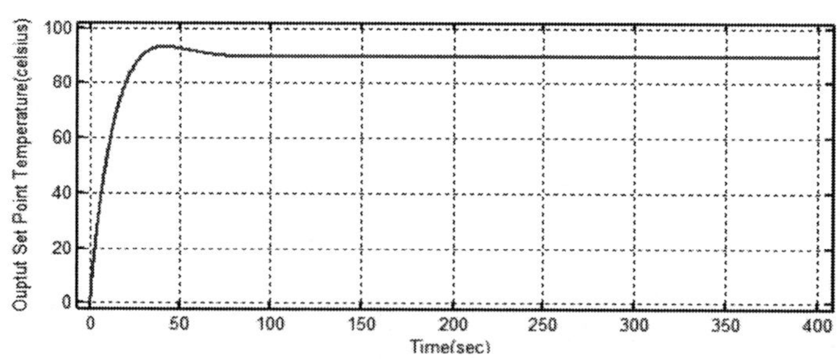

Figure 5. Output set point temperature

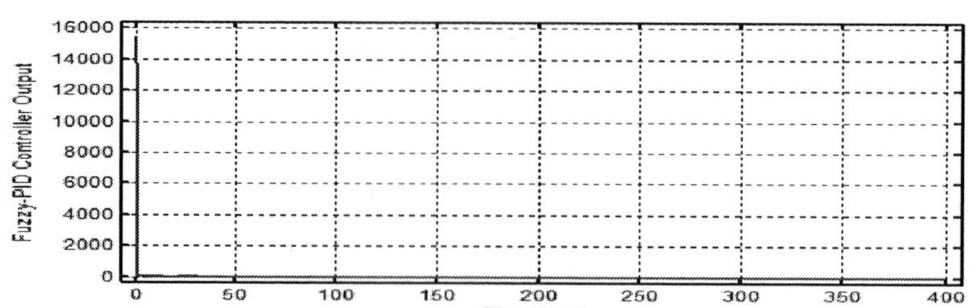

Figure 6. Fuzzy-PID controller output

The settling time and peak overshoot shown in figure.7 and figure .8 has been compared with conventional method [16]. The settling time of conventional method is 74 sec. The settling time of set point temperature using proposed controller is 65 sec.

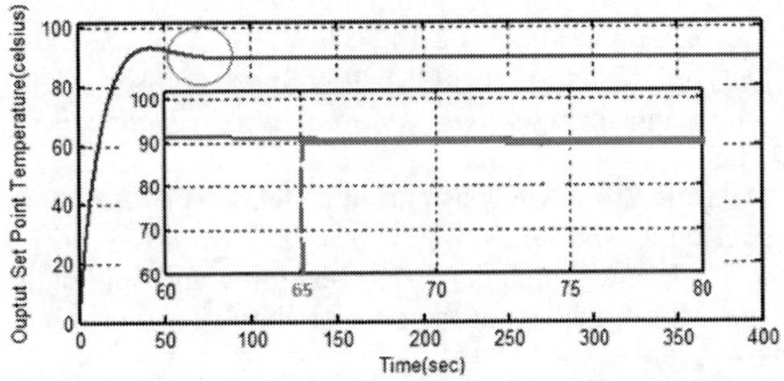

Figure 7. Settling time of set point temperature using proposed controller

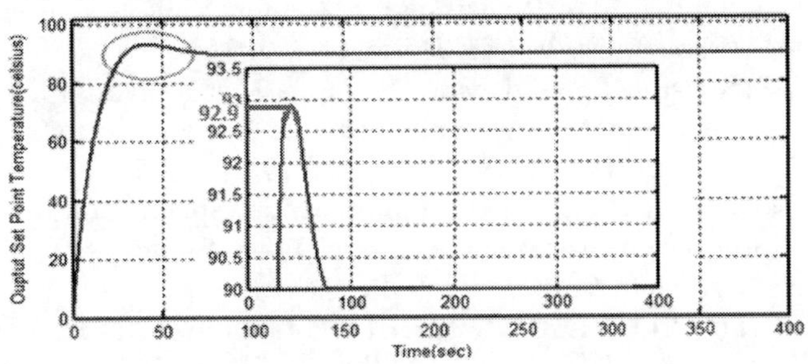

Figure 8. Peak Overshoot of set point temperature using proposed controller

The conventional peak overshoot is 3.54% but in proposed peak overshoot of set point temperature is 3.22%. The peak overshoot percentage is calculated as,

$$\% peak\ overshoot = \frac{92.9-90}{90} * 100 = 3.22\%$$

7. Conclusion

The Simulation has been done on the temperature control on the shell and tube heat exchanger model using Genetic-PID controller. The shell and tube heat exchanger controller part is modelled by the transfer function model. The Genetic-PID with feed forward controller is implemented to reduce the error caused by the input fluid. The genetic-PID which is the combined using multiplication operation has made peak overshoot and settling time of output set point temperature with the least values of 3.54% and 74 sec respectively. The values of peak overshoot and settling time obtained from the proposed control strategy which is less than the PI control method.

References

[1] Costiuc L and Popa V 2009 Simulink Model for a Heat-Exchanger

[2] Nithya S, Sivakumaran N, Balasubramanian T and Anantharaman N 2008 Model based controller design for a spherical tank process in real time *IJSSST* **9** 25-31

[3] Kumar GV 2018 Fuzzy-PID Control Method of Hybrid Derived Boost Converter (HDBC) Using Wind Energy *International Journal of MC Square Scientific Research* **9** 1-10

[4] Vinod S, Balaji M and Prabhakar M 2015 Robust control of parallel buck fed buck converter using hybrid fuzzy PI controller *11th Int. Conf. on Power Electronics and Drive Systems* 347-351
[5] Peng X, Jia M, He L, Yu X and Lv Y 2018 Fuzzy sliding mode control based on longitudinal force estimation for electro-mechanical braking systems using BLDC motor *CES Transactions on Electrical Machines and Systems* **2** 142-51
[6] Al-Bargothi SN, Qaryouti GM and Jaber QM 2019 Speed control of DC motor using conventional and adaptive PID controllers *Indonesian Journal of Electrical Engineering and Computer Science* **16** 1221-8
[7] Thufail MK and Balaji K 2017 Analysis and improvement of Self Tuning Based on Sensor less Control of BLDC Motor *Indian Journal of Public Health Research & Development* **8** 623-9
[8] Jafari SM, Saramnejad F and Dehnad D 2018 Designing and application of a shell and tube heat exchanger for nanofluid thermal processing of liquid food products *Journal of food process engineering* **41** e12658
[9] Gupta S, Gupta R and Padhee S 2018 Parametric system identification and robust controller design for liquid–liquid heat exchanger system *IET Control Theory & Applications* **12** 1474-82
[10] Åström KJ and Hägglund T 2014 Revisiting the Ziegler–Nichols step response method for PID control *Journal of process control* **14** 635-50
[11] Mansor H, Azmi Mat Esa MK, Gunawan TS, Janin Z 2020 Design of travel angle control of quanser bench-top helicopter using mamdani-based fuzzy logic controller *Indonesian Journal of Electrical Engineering and Computer Science* **17** 815-825
[12] Kumar A and Balaji K 2017 PI and Sliding Mode Speed Control of Permanent Magnet Synchronous Motor Fed from Three Phase Four Switch VSI *Journal of Mechanical Engineering Research and Developments* **40** 716-725
[13] Kumar GV 2018 Fuzzy-PID Control Method of Hybrid Derived Boost Converter (HDBC) Using Wind Energy *International Journal of MC Square Scientific Research* **9** 1-10
[14] N C Damasceno and O Gabriel Filho, "PI controller optimization for a heat exchanger through metaheuristic Bat algorithm, particle swarm optimization, flower pollination algorithm and Cuckoo search algorithm *IEEE Latin America Transactions* **15** 1801-1807
[15] Aldossary F 2017 Health Observation System Using Cloud Computing *International Journal of MC Square Scientific Research* **9** 08-16
[16] Ali AJ, Ramesh GP 2020 Analysis of Open, Closed Loop PI, PID, FLC and ANN Controllable Wind Energy System Using Γ-ZSI with PMSM *ICDSMLA 2019* 1143-1155
[17] Puttamadappa C and Parameshachari BD 2019 Demand side management of small scale loads in a smart grid using glow-worm swarm optimization technique *Microprocessors and Microsystems* **71** 102886

IOP Conf. Series: Materials Science and Engineering **925** (2020) 012021 doi:10.1088/1757-899X/925/1/012021

Design of Fuzzy-Pid with Landsman Converter (LC) Fed BLDC Drive Using IFOC Controller

A Sundaram[1] and Peter Romine[2]

[1] Department of EEE, Navajo Technical University, USA.
[2] Department of ECE, Navajo Technical University, USA.
[1]sarumugam@navojotech.edu

Abstract. In this paper, fuzzy - PID control is designed to improve the settling time of DC link voltage of the Photovoltaic (PV) array fed landsman converter (LC). In all platforms such as medicals, automobile the utilization of variable speed drives is increased and which are directs to brushless DC motor demand in increasingly. Indirect Field Oriented Control is used to make precise and robust speed control with ripple minimization of torque. Back-EMF estimation method is contributing with IFOC control for obtaining accuracy speed in sensor-less approach and control the dynamic changes in speed. Dc link voltage of LC is maintained at constant using the Fuzzy-PID control which maintains the BLDC steady state condition while the motor speed at dynamic load as well as set point variation. The performance of IFOC based brushless DC motor is improved with high efficiency and in the MATLAB/Simulink platform the proposed system results are verified.

Keywords: Back EMF, BLDC Motor, Fuzzy-PID control, IFOC, LC converter, MPPT, PV.

1. Introduction

Daily life has become essential in using electric motors that are used in many applications such as electric vehicle, aerospace, air conditioning, and vacuum cleaner. Recently, the brushless direct current motor (BLDC) is highly preferred to employ in more applications due to less maintenance, compactness and high efficiency with low cost. This motor has accurate positions of rotor at high speed drive arrangements. The BLDC motor using electronic commutations technique and it is not requiring brushes for commutation. Normally, a DC motor is combined with friction of mechanical and erosions in electrical system but the brushless DC motor has defeat the aforementioned issues. The mentioned advantages are essential for preferring brushless DC motors in many platforms. In control of BLDC speed is accomplished using PI controller and the switching signals providing PWM. For obtaining better speed control with robust, the PID controller is used and to varied load conditions. In this brushless motors, the commutation error is found as a most concerning problem in high speed range drive system applications which are leads to overall performance of the system will be decreased. Since the commutations precision is impacted by the more non-ideal factors undesirably the elimination of commutation error is so hard. To compensate this issue, the sensor-less control based on 3rd-harmonic compensation method is developed and this control approach is also improving the presentation of drive system. The virtual 3rd harmonic is eliminated the commutation error and increase the effectiveness of the system performance. The output of the motor-fed system should be stable at varying operating conditions. This is done via the motor-connected drive arrangement converter system. The converters of power electronics do the power conversion from the source. The power converters are made of power semiconductor components that can be used in low and high power applications. To improve the energy that fed from source, the power converters are used and doing performance such as dc-dc, ac-dc, ac-ac, etc. The conventional boost, buck-boost converters are used to improve the DC bus voltage due to low cost and simple in structure where the load connected with converter system. However, it does not achieve enough voltage gain for a high step-up application. The power converters are also having the problems such as less efficiency, large capacitor, and high number of passive elements. The landsman converter is better while compared with like SEPIC, ZETA, Cuk converters then it has reduced elements for operation and overcoming the fore mentioned issues. For obtaining high voltage gain from supply voltage, the power switches to be controlled by using control system. The conventional systems are controlled and regulated by using PWM, PI, and PID control to provide gate signals for the power switches with appropriate duty cycle.

The system stability is achieved through Fuzzy-PID control for landsman converter to improve DC link voltage and reduce settling time to maintain the input supply of voltage source converter (VSC) at constant. The Fuzzy-PID approach base implementation feeding power to IFOC control with back EMF estimation based BLDC motor. The power maximization technique of MPPT with Inc& con algorithm is proposed to PV system power extraction for landsman converter (LC) fed brushless BLDC motor. The proposed system is compared with the existing system of [8, 14] in term of DC link voltage improvement and reduced settling time.

2. Literature Review

In [1] the author proposes the simulation model implementation that based on the BLDC motor which operated by four quadrant operation. In this method, the torque ripple of the BLDC drive is reduced. The model predictive controller is used to achieve the effective analysis and monitoring in the developed simulation model that enables control in speed and characteristics in the back EMF and torque. The controller is accomplished the four quadrant operation and there is no power loss. The implemented system is validated with the MPC control scheme and during the regenerative period the energy is conserved. In [2] author suggests a BLDC Zeta converter motor for industrial applications. The motor speed is tracked using a hysteresis controller on the back EMF. The proposed method is designed to achieve speed and reduce fluctuation in input voltage. It provides an improved capacity with a high voltage gain rate. It's a zeta and quadratic boost conversion combination. In [3] the BLDC motor control is accomplished using the PI control as well as the fuzzy logic control system for the control in speed. But in the PI controller based motor speed control has low performance and it has demerit of will not operate under the load disturbances in high degrees. The fuzzy logic controller based speed control has dynamic performance while compared with PI control. In [4] sensor less based BLDC motor is used for the water pump and the BLDC motor is powered from the solar system. The converter is providing DC link voltage with minimum and it is controlled by the fundamental pulse width technique so that the converter not able to provide constant DC supply to load. In [5] proposes the interleaved boost converter for achieving high voltage gain based on the fuel cell with controller of MPPT system. The interleaving converter system is increasing the passive elements and enhances the cost and size. In [6-7] that explain the electro mechanical braking system that based on the utilization of the BLDC motor which is controlled by the method of fuzzy based sliding mode controller. The distribution of the torque is controlled in this approach. In [8] describes the renewable energy based DC-DC converter is interfaced. To improve the input power supply, the MPPT controller used for the converter system. The interleaved boost converter is used for the minimization of the input current ripples in the converter. In [9] author proposes the maximum power tracking controller for the dc-dc converter which is interfaced with the photo voltaic system. The differential MPP is used to tracking the more power but which has more constraints when the low irradiance in solar system. In [10] the model reference adaptive control (MRAC) with compensator for the BLDC motor to achieve high performance. This paper considers the speed control achieving for various speed respect to the different times. But the method is affected by the variations in the parameters and load disturbances. In [11] the BLDC motor drive is used for the

IOP Conf. Series: Materials Science and Engineering **925** (2020) 012021 doi:10.1088/1757-899X/925/1/012021

solar based water pump load and the bidirectional power flow is enabled. To maintain the DC link voltage in constant the energy storage system of grid system is interfaced. In [12] explains the different strategies of the BLDC motor based speed and torque controls are determined. In [13] the both converters of DC-DC buck and boost converters are proposed for the PV system connected voltage source inverter to power the permanent magnet BLDC motor system. To attain the constant speed for the motor, the voltage from the boost converter to be maintained at constant in this paper but also the ripples and settling time is more.

3. Proposed system

In this proposed system, as shown in figure.1. PV source is providing direct current to LC fed brushless DC drive. This paper consists of BLDC motor, single diode PV array, landsman converter, and VSI. LC is regulating high voltage gain which is fed to BLDC connected VSI. By using LC with Fuzzy-PID control method is to accomplish high voltage. In this proposal, the settling time of LC output voltage and high voltage gain are accomplished by the Fuzzy-PID control method. To control the speed and ripple reduction of the BLDC motor, the IFOC with back EMF estimation method is used.

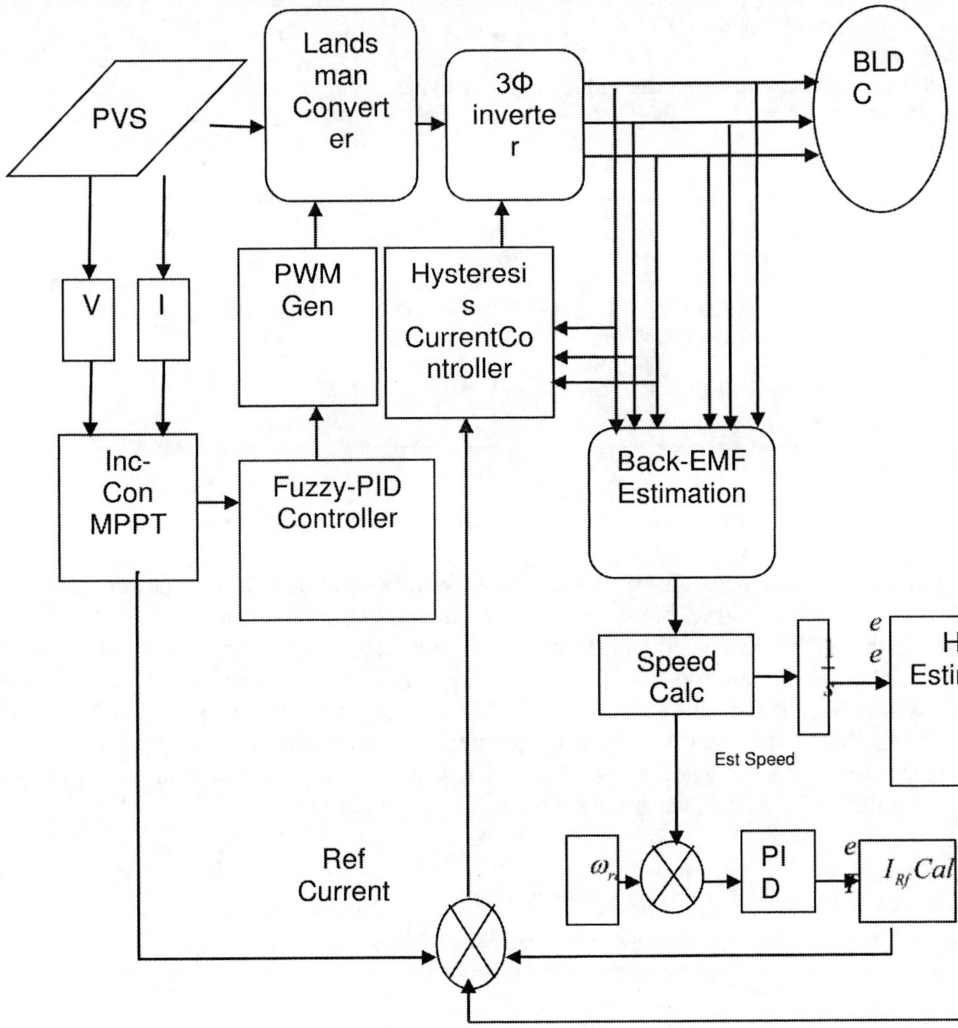

Figure 1. Proposed System Overall Block Diagram

3.1. PV Energy Generation

Photo voltaic system has PV cells in parallel and series combination which is utilized to improve the power respectively. PV produces energy directly from solar system and according to solar irradiances as well as temperature it will be changed. The PV circuit of single diode model is shown in figure.2. In PV cells the sunlight is directly converted into dc power. According to photoelectric effect phenomenon the PV modules are worked. In PV system, the Maximum Power Point Tracking (MPPT) is very essential algorithm during operation which yields high energy from the source and to PV arrays operating point almost reaching the MPP. Generally, the MPPT is used to generate more power that can highly possible to achieve in an electronic system which functions the PV modules. In this paper, incremental and conductance base MPPT is employed.

The following equation shows that PV model's current (I) and voltage (V) (I-V) characteristics: The following equation shows that PV model's current (I) and voltage (V) (I-V) characteristics:

Where,
I= PV current (A)
K= Boltzmann constant (J/K)
IL= Light generated current (A) in the PV module
V= PV voltage (V)
Rs, Rsh = series and shunt resistance of the single diode PV model
Q = Electron Charge
Io=saturation diode current

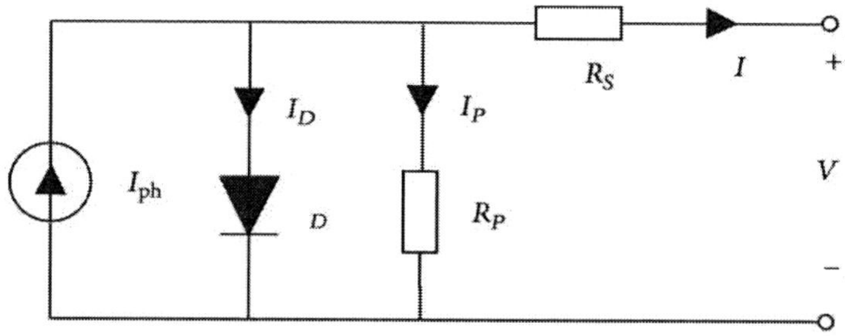

Figure 2. Circuit of single diode PV module

The incremental and conductance (IC) algorithm is overcoming the demerits of perturb and observe method in case of the failure in tracking peak power at atmospheric condition under fast variation. In this method, when the MPP is reached then the operating point disturbing comes to stop. The disturbance of operating point is calculated using the following relationship in between dI/dV and I-V characteristics. While MPP is in right side for MPPT, this is negative and which is from dP/dV and when it is to the left of the MPP, It is positive. This IC has merits over the conventional system because the oscillations are occurred in MPP when the MPPT method is achieving the maximum operating point. The basic incremental and conductance equations are shown in below,

$$dI/dV = -I/V atMPP \quad (1)$$
$$dI/dV > -I/V leftofMPP \quad (2)$$
$$dI/dV < -I/V rightofMPP \quad (3)$$

The MPPT improves the Fuzzy –PID control based PWM signal of landsman converter dc-dc converter. The flowchart of IC technique base MPPT is shown in figure.3.

IOP Conf. Series: Materials Science and Engineering **925** (2020) 012021 doi:10.1088/1757-899X/925/1/012021

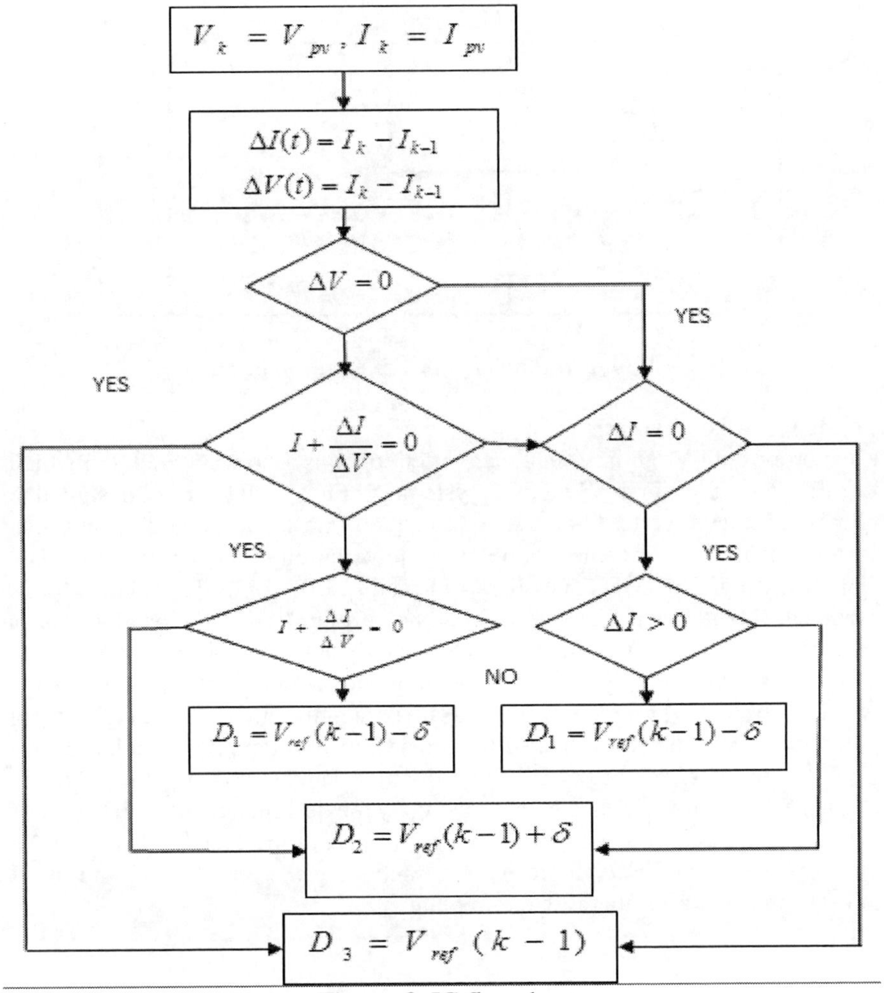

Figure 3. IC flowchart

3.2. Proposed LC system

The LC is dc-dc converter and it is designed to improve the photo voltaic (PV) power and to function in CCM depends on the deviation level of irradiance. Proposed LC is providing improved voltage with less settling time to brushless DC motor. The PV array output voltage is optimized using proposed LC to perform noiseless, safe and better running of the BLDC motor. The DC bus voltage of the proposed BLDC system is regulating and achieving enhanced voltage gain by the function of LC converter. LC comprises of two capacitors (C1, C2), inductors (L1, L2), and diode with the power switch (Sw). The proposed landsman converter is represented in figure.4 generally; LC converter is designed to work in two operation modes such CCM and DCM. In continuous conduction mode, when the power switch is 'ON' state, the voltage in the capacitor C1 is direct to diode (D) gets reverse biased. The inductor current is supplied through the power switch. The voltage across C1 is larger than the result voltage Vdc, here L receiving the energy from the C1 that discharges and the power fed to output. Because of this, the inductor current IL is increased and the voltage across the intermediate C1 decreases. In discontinuous continuous conduction mode, when the switch is in OFF state, the diode operation is biased in forward as well as the IL is flowing through the forward biased diode D. Through the D, the output energy is transferred that stored energy in inductor. At the same time, intermediate capacitor is charged from both input supply and inductor L.

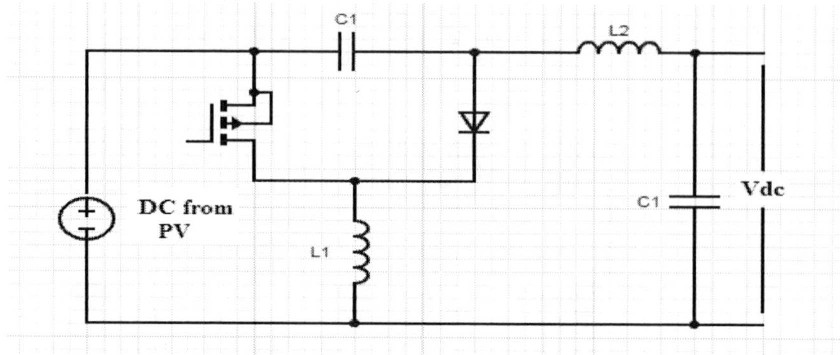

Figure 4. Landsman Converter Diagram

4. Proposed system control methods

In this paper, the proposed PV fed landsman converter contains a power switch (S) which is controlled by using Fuzzy-PID controller. In this proposed system, the Fuzzy-PID control designed to achieve high voltage gain for landsman converter fed brushless DC motor based photo voltaic system. The proposed control is returning the desired reference value even the variation in the set points. This control design is including both fuzzy logic and PID controllers. Normally a PID controller has three parameters values such as proportional, Integral, and Derivative. In this controller design the oscillations can be reduced and the system steady state error is controlled. When the dynamic condition in both load and set point, the proposed Fuzzy-PID controller is used to optimizing and provide switching signals for the proposed converter power switch. In this system, Kp, Ki, and Kd, parameters are determined by using fuzzy logics. The input of the proposed controller Fuzzy-PID consists of two inputs that delta error and error, and three Kp, Ki, and Kd, outputs. The Fuzzy-PID controller used and the settling time is reduced. The input source of the voltage source inverter is increased using by the designed Fuzzy-PID controller. The block diagram of the proposed Fuzzy-PID controller design for proposed dc-dc landsman converter is shown in figure.5. In this system, the Fuzzy logic rules are reducing the error signal and provide three PID parameters to PID controller after that it is fed to plant.

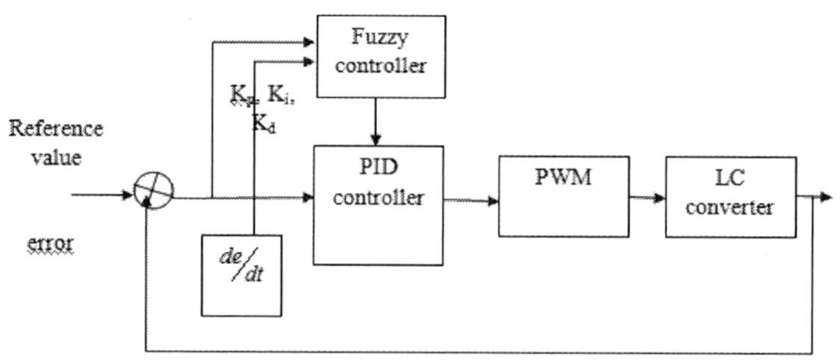

Figure 5. Fuzzy-PID controller Block Diagram

4.1. Back EMF and IFOC system

In The DC link voltage which is improved using landsman converter is fed to three phase voltage source inverter. In this proposed method, the requirement of adding the injection of high frequency signal is not necessary to obtain the position of rotor that achieved using from stator currents and voltages. This sensor-less back EMF is widely applied successfully to more applications which are requiring starting control with simple. To make signals of back EMF, the stator voltages and mathematical model as well as stator command currents are used. The back EMF signals are used to observe the position and speed of the rotor and the sensor-less drive performance is directly influenced by the accuracy of proposed back EMF estimator. From the observed values, the line currents and voltages are can be calculated. By equation (3) the back EMF is calculated that leads to obtain rotor speed. In figure .6 the Back EMF method is shown.

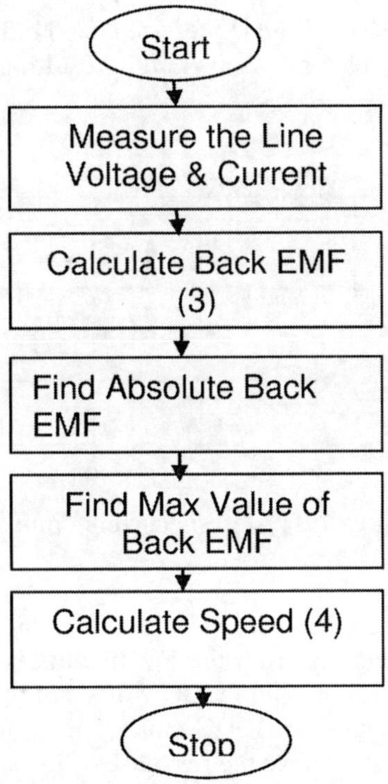

Figure 6. Speed Estimation Method flow chart

In

$$I_{ref} = \frac{T_{ref}}{2 * P * \varphi} \quad (4)$$

Where V_a, V_b, V_c - phase voltages
I_a, I_b, I_c - phase currents
e_a, e_b, e_c - BLDC motor back EMF
ω_r - speed of rotor
P - Number of pole pairs
φ - Flux

The proposed streamlined IFOC scheme is to measure the speed of the BLDC engine, which is calculated using the EMF prediction process behind it. Submit to the PI controller after determining the speed error between approximate speed and reference speed and act as a speed regulator. To produce switching signals by the HC, the hall signal of the BLDC motor is to be computed and is combined with reference current. The estimation of Inc& Con is used to extract the intense power from the source and the path of the period of duty is calculated from the MPPT algorithm balances of the current reference value.

4.2. Hysteresis controller

Present controller function well only because we are correctly characterizing the two party viz. Present error is fed as controller input and produces gate signals when error is below the limit of the hysteresis band or when the pulse reaches the limit of the band it induces gate signals which are low to power switches. As shown in figure.7 if the band limit is preferred band limit will be held at a lower significance because the pulse's high switching frequency is also high, resulting in high power electronic switching losses.

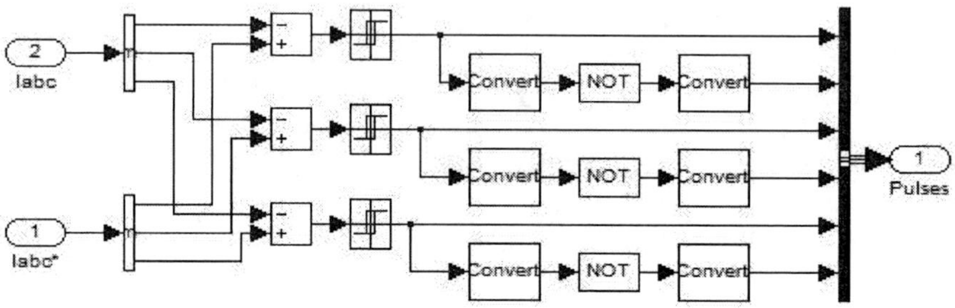

Figure 7. Hysteresis current controller

5. Simulation and Results

Proposed simulink model of the Fuzzy-PID with LC converter which is powered by renewable PV array fed BLDC motor. Low generated energy from the PV module is supplied to landsman converter to achieve high DC link voltage is controlled. Output DC voltage of the PV fed converter is feeding to VSI which is generating three phase voltages for BLDC motor operation. In this proposed system, the main Fuzzy-PID controller has proposed to manage the response of LC converter DC link voltage and reduce the settling time. The proposed system simulink model is shown in figure. 8 The MPPT is used to produce maximum power through the proposed controller which is control the response of the converter. As shown in figure.9.the DC link bus voltage of the proposed LC based on Fuzzy-PID controller is achieving 82 V and settling time is 0.08 sec which is high obtained results while compared with the [8] explains.

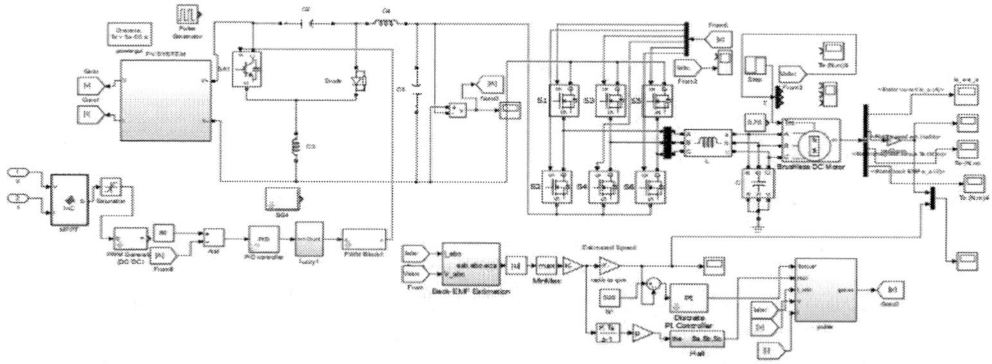

Figure 8. Overall Simulink Model of the Landsman Converter (LC) Fed BLDC motor

The switching signals are generated using the HC which is based on the hall signal valuation from the proposed drive rotor angle. The sensor-less based back EMF method is used to detect the voltage and current of the brushless dc motor stator. From this estimation the proposed system speed and torque can be controlled. The rotor speed as well as the position of the rotor is observed with accuracy using proposed estimation method. The speed of the BLDC has controlled using IFOC control which is performing under the estimation of the back EMF based sensor-less method. In the PI controller the set point or reference torque is generated by reduce the error in the feedback of the proposed system.

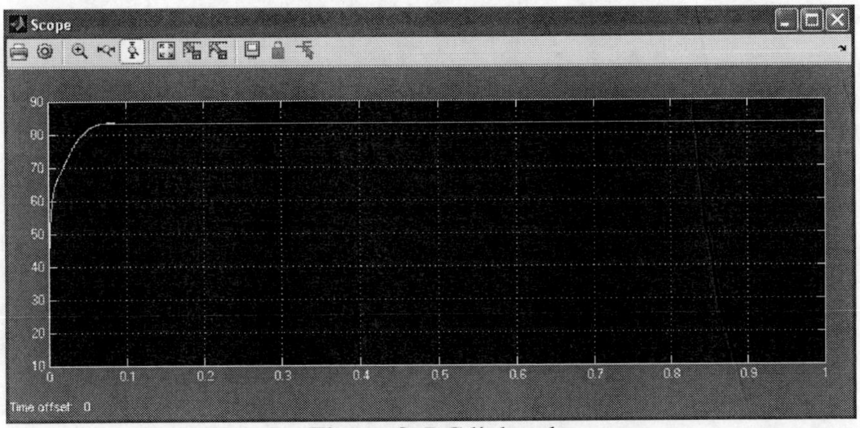

Figure 9. DC link voltage

The Stator current of the proposed BLDC drive is illustrated in figure.10. By using the IFOC and back EMF based sensor-less control. System actual speed is shown in figure.11. The proposed drive system torque is shown in figure.12. Speed that is obtaining using IFOC control method is shown in figure. 13. The actual speed and estimated controlled speed are compared as shown in figure. 14.

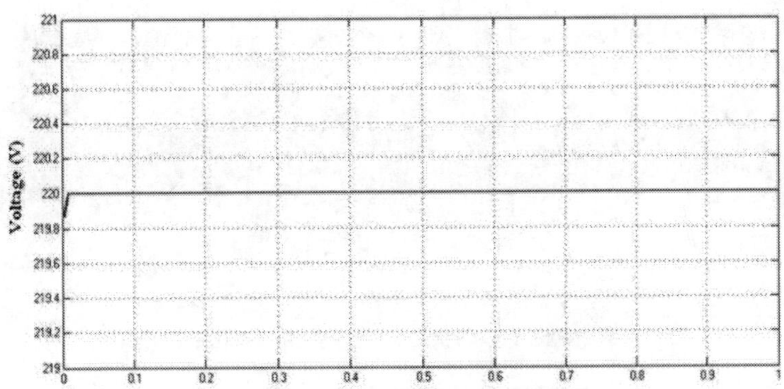

Figure 10. DC link voltage

The cascaded multilevel inverter is used to produce levels for achieving the harmonics reduction in form carrier arrangement with control technique of APOD control for the generating switching signals for the proposed cascaded MLI. The proposed system inverter is achieving 11 levels with 230V for the induction motor as shown in fig based system to reduce the inrush current which is done by the control method of the direct torque and flux control. The speed of the proposed DTFC based IM accomplished the speed of 1600rpm which is settled in the time of 0.4 sec as shown in figure.15.

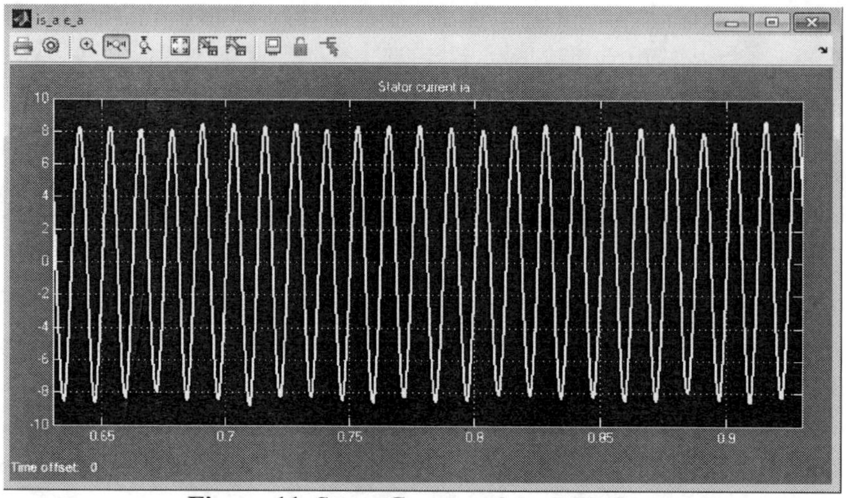

Figure 11. Stator Current of the BLDC motor

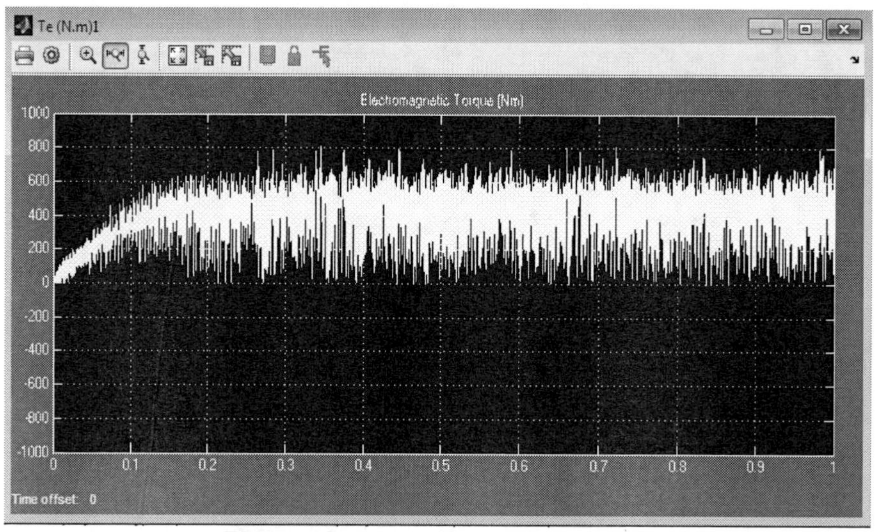

Figure 12. Proposed System Actual speed

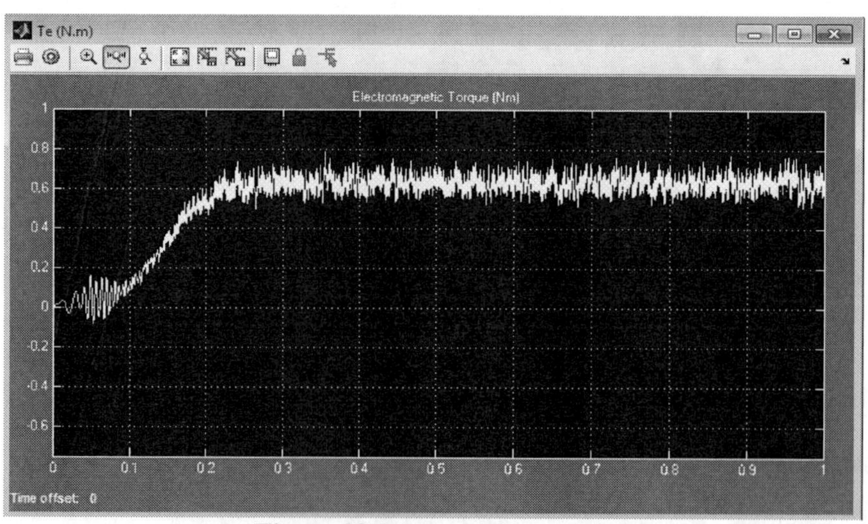

Figure 13. Proposed System torque

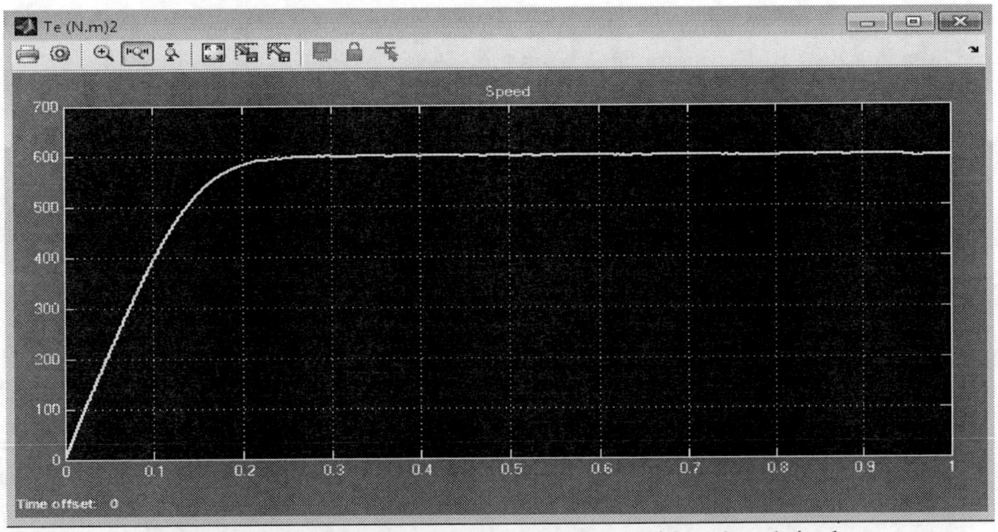

Figure 14. Proposed System speed control with reduced ripple

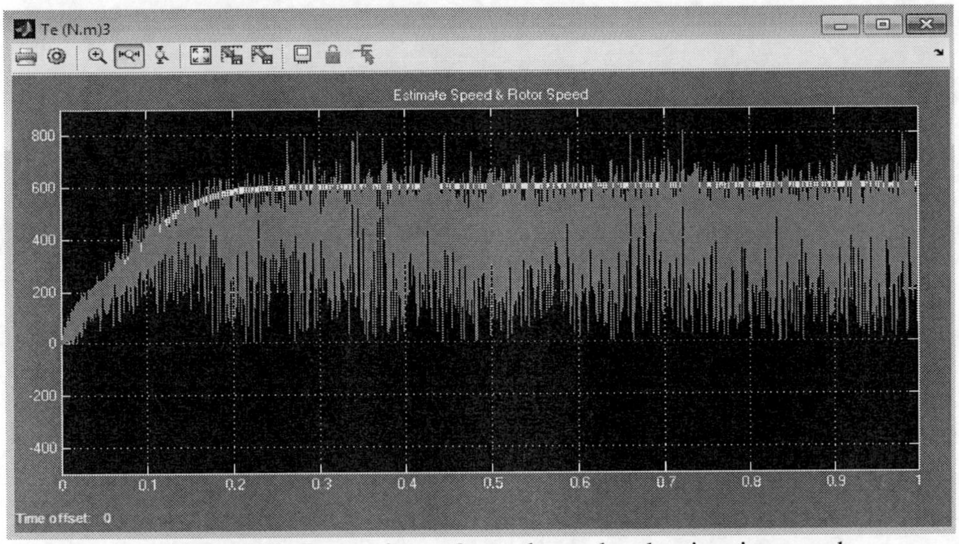

Figure 15. Comparison of actual speed and estimation speed

6. Conclusion

The PV fed brushless DC motor drive is implemented and through the proposed landsman converter (LC) the DC bus voltage is achieved its high voltage gain and the reduced ripples in the output voltage and settling time reduction. These are accomplished by the proposed Fuzzy-PID controller. The stator voltage and currents are observed and the speed of the BLDC drive is controlled by using back EMF estimator and indirect field orient control (IFOC) are respectively. In this paper, the LC is achieved high DC bus voltage and the reduced settling time is 0.07 sec achieved. The ripples in the BLDC drive speed are reduced with efficient control of proposed system.

References

[1] Gopinath M, Yuvaraja T and Jeykumar S 2018 Implementation of four quadrant operation of BLDC motor using model predictive controller *Materials Today: Proc.* **5** 1666-72
[2] Manie N and Pattanaik B 2019 Zeta DC-DC Converter Based on MPPT Technique for BLDC Application *International Journal of MC Square Scientific Research* **11** 1-2

[3] Ahmed AH, Abd El Samie B, Ali AM 2018 Comparison between fuzzy logic and PI control for the speed of BLDC motor *International Journal of Power Electronics and Drive Systems* **9** 1116

[4] Park JS, Lee KD, Lee SG and Kim WH 2018 Unbalanced ZCP compensation method for position sensorless BLDC motor *IEEE Transactions on Power Electronics* **34** 3020-4

[5] Kumar R and Singh B 2018 Solar PV powered-sensorless BLDC motor driven water pump *IET Renewable Power Generation* **13** 389-98

[6] Peng X, Jia M, L He, Yu X and Lv Y 2018 Fuzzy sliding mode control based on longitudinal force estimation for electro-mechanical braking systems using BLDC motor *CES Transactions on Electrical Machines and Systems* **2** 142-51

[7] Reddy KJ and Sudhakar N 2018 High voltage gain interleaved boost converter with neural network based MPPT controller for fuel cell based electric vehicle applications *IEEE Access* **6** 3899-908

[8] Ramaprabha R, Balaji K, Raj SB and Logeshwaran VD 2013 Comparison of Interleaved Boost Converter Configurations for Solar Photovoltaic System Interface," *The Journal of Engineering Research [TJER]* **10** 87-98

[9] Tey KS, Mekhilef S, Seyedmahmoudian M, Horan B, Oo AT and Stojcevski A 2018 Improved differential evolution-based MPPT algorithm using SEPIC for PV systems under partial shading conditions and load variation *IEEE Transactions on Industrial Informatics* **14** 4322-33

[10] El-Samahy AA and Shamseldin MA 2018 Brushless DC motor tracking control using self-tuning fuzzy PID control and model reference adaptive control *Ain Shams Engineering Journal* **9** 341-52

[11] Kumar R and Singh B 2019 Grid Interactive Solar PV-Based Water Pumping Using BLDC Motor Drive *IEEE Transactions on Industry Applications* **55** 5153-65

[12] Verma V, Pal NS and Kumar B 2019 Speed Control of the Sensorless BLDC Motor Drive Through Different Controllers *In Harmony Search and Nature Inspired Optimization Algorithms* 143-152

[13] Arumugam S and Ramesh G P 2017 Sensor less Control of BLDC Motor using Fuzzy logic controller for Solar Power Generation *International Journal of MC Square Scientific Research* **9** 70-79

[14] Puttamadappa C and Parameshachari BD 2019 Demand side management of small scale loads in a smart grid using glow-worm swarm optimization technique *Microprocessors and Microsystems* **71** 102886.

Series Hybrid Power Train for Automotive Application

Shekaina Justin[1], Samia Larguesh[2], Wafaa Shoukry[3], Ghada Naif Alnemer[4] and Shermina J[5]

[1,2,3,4] Department of Electrical Engineering, Princes Nourah Bint Abdulrahman University, KSA.
[5]Department of Computing, Muscat College, Sultanate of Oman

Abstract. In this paper a series hybrid powertrain is presented; factoring in the existing fossil fuel infrastructure established in oil producing economies, environmental impact, drivability characteristics, utilization patterns, advancement in automobile control sub-systems and appropriateness of technology. The prime mover employed for this hybrid are electric hub motors, powered by a super capacitor bank [SCB], which in turn is connected in with a relatively smaller battery pack [BP]. An IC engine is added in the loop to energize the SCB when the charge in the BP falls below the sub-optimal level; and, when the destination or fast charging facility are not reachable with the charge left in the SCB-BP module. A fast charger or a regular charger is plugged in to charge the BP when the vehicle is stationary.

Keywords: Super capacitor bank, Battery pack, Power electronics, Hub motors, Engine-Generator

1. Introduction

About 14% of worldwide greenhouse gas emissions are contributed by the transportation sector and 25% by the electricity and heating sector [1]. It is also true that cities that have high density of automobiles tend to have poor air quality. Although pure plug-in electric vehicles have zero tailpipe emissions, if they are powered by non-renewable sources, their carbon footprint is comparable to that of petroleum powered automobiles. With an overall energy conversion efficiency of fossil fuel power plants being 33% [2]with transmission and distribution losses of about 9% [3]; and after accounting power train, charging and discharging efficiency of about65 to 80%, for plug-in electric vehicle; the actual carbon footprint of a plug-in electric vehicle is comparable to the carbon footprint when petroleum heat energy is converted to vehicle kinetic energy, which is about 16% to 30% for gasoline powered automobiles and 18% to 40% for a diesel powered automobiles [4].However, both IC engine vehicles and electric vehicles have some distinct advantages, some of which are listed below:

Table 1. Comparison of IC Engine and Electric power train

		I.C. ENGINE	ELECTRIC
1.	Technology	Developed✓	Developing
2.	Energy density	Very high ✓	Low
3.	Tail pipe emissions	Yes	No ✓
4.	Efficiency	Poor	High ✓
5.	Maintenance	Moderate	Low ✓
6.	Replacement cost	Moderate✓	Very high
7.	Life span-prime mover	High	Very high✓

8.	Basic cost	Moderate ✓	High
9.	Gross weight	Moderate ✓	High
10.	Energy regeneration	No	Yes ✓
11.	Energy for production	High ✓	Very high

In USA and China, electric cars are gaining popularity at present. But the technology of battery is not mature and appropriate to be adopted for most types of automobiles because of the high cost and mass of the BP. When we factor-in the energy involved in the battery manufacture, toxic materials used for their manufacture and disposal of degraded batteries; the overall advantages of plug-in electrics from the environmental perspective is not positive. Although the cost of running the electric vehicle is much lower when compared to the petroleum fuelled vehicle; when the factors such as initial cost; replacement cost of BP that has an average life span of 5 years, particularly for owners who use their vehicles for limited distances; the cost of ownership and the convenience are not favourable for electrics. And, for oil producing countries like Saudi Arabia, with competency in oil production and dependency of the economy on petroleum commerce; switching over to pure electric vehicles is not a prosperous and prudent preposition for the near future. However, improving the air quality in cities is a pressing need. Moreover, the acceleration and regenerative braking characteristics of the electric automobile cannot be ignored, as the users would appreciate the superior acceleration and lower energy cost of the electric vehicles. Also, the maintenance cost and time intervals of the electric vehicle is bound to be low, when we exclude the replacement cost of the battery pack. As far as utilization is concerned, for example in Saudi Arabia, the average passenger car covers 25000km per year. The pattern of usage varies depending on their profession, lifestyle etc. For a user driving less than 100km daily, the range of 64kWh battery fitted, for example, in HYUNDAI KONA EV is high. The extra mass on account of the battery, costs additional energy to increase the kinetic energy of the car. It also affects the handling and tyre life. But, when the user chooses to engage in a journey beyond 400km, the battery cannot be charged with a regular charger overnight, if he/she wants to return the same day or the next morning, as a regular charger takes 29 hours to charge from zero to full SOC [19]. In that scenario, the user had to utilize a fast charging station. The car cannot be used for long distance travel in roads lacking fast charging infrastructure. The anxiety level of the user will be high when SOC of battery is low and its impractical to use such vehicles during emergency situations. Extensive planning to undertake a trip is inconvenient. The cost to replace the BP after its average life span of 5 years is also very high. The operating cost of the vehicle should match the buyer's purchasing power for vast market acceptability and the configuration of the powertrain should be in alignment with the pattern of usage.

On the other hand, the IC Engine has been around for a very long time. Hence the manufacturing and the fuel infrastructure is well established. Also, there is scope for improvement in efficiency of the IC engine, particularly when operated at steady conditions. The superior life and power density of super capacitors ensure that they could be employed in vehicles, where the variation in the energy demand is very high. As the field of power electronics for automotive energy management has developed considerably, seamless flow of quality power from BP to SCB is made possible; improving the life of the BP significantly and reducing the heat generated in the BP, thereby reducing its cooling requirements.

2. Literature survey

An in-depth study into the efficiency of IC engine powered automobiles show that the fuel energy to kinetic energy conversion efficiency is between 16% to 40% [21], depending on the driving patterns and road conditions, type of vehicle and fuel used, etc. Automobile companies are investing in improving the fuel efficiency mainly because of the carbon footprint targets imposed by various governments [5]. Toyota has developed HCCI with modification in bore-stroke ratio and claims a thermal efficiency of 40%. Mazda has modified the combustion chamber, the inlet manifold and uses a HCCI combined cycle to achieve a thermal efficiency of 38%-41% [7]. The other engine developments having high thermal efficiencies are Achates Power's opposed piston engines of about 50% [8][6][9], RCCI engine of about 60% [10], etc. Emission levels of carbon dioxide, carbon monoxide, particulate matter and nitrous oxide were also reduced. However, these high brake thermal efficiencies don't translate into substantial high mpg/kpl for different drive cycles, as the conditions are of transient nature [17]. The continuously varying torque demand and energy losses on account of braking, idling and transmission are some of the factors that contribute for poor on road efficiency.

In hybrid vehicles with li-ion battery, only part of the energy lost during braking is recovered by converting to electric energy as the degradation of the battery on account of sudden charging [16], [18]imposes a limit on the energy that could be recovered; although the battery chemistry is designed taking into account the quick charging and discharging requirements, compromising on the energy density. Although the fuel consumption of the latest hybrid models is much lower than their gasoline counterparts, scope for improvement using the same configuration without compromising on acceleration characteristics appears to be limited. The transmission systems of these hybrids are complex and, milder the hybrid configuration, the electrical energy regenerated and stored for usage is limited. A hybrid configuration which is efficient in city driving cycle is inefficient for highway.

3. Proposed Drivetrain

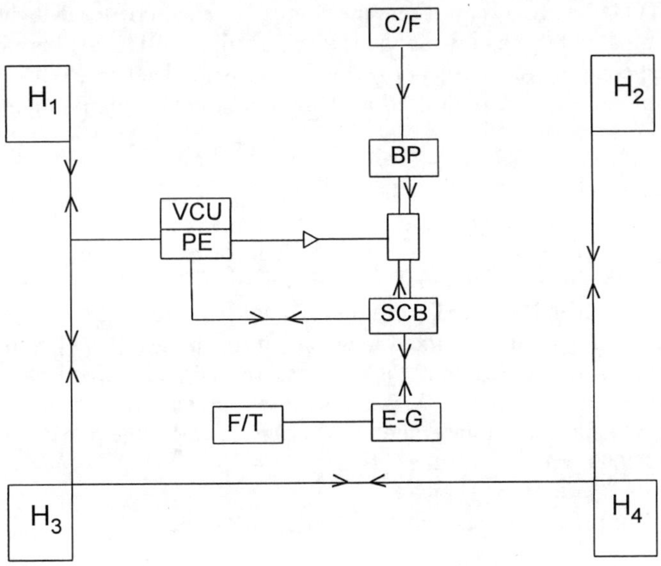

$H_{1,2,3,4}$ ⇒Hub Motors
SCN ⇒Super Capacitor Bank
BP ⇒Battery Pack
E-G ⇒S. Engine-Generator
C/F ⇒Charging Battery
VCU ⇒Vehicle Control Unit
PE ⇒Power Electronics
F/T ⇒Fuel Tank

Figure 1. Block Diagram of the Proposed Drivetrain

HUB MOTORS: H1,2,3,4 are Permanent Magnet Synchronous Machines [PMSM] fitted to the wheels for better response and distribution of mass, to maximize the regeneration during braking and to integrate the functions of anti-lock braking[ABS] and electronic stability program [ESP] within the functions of the hub motors [12]. Mechanical/ magnetic braking is added only for critical conditions.

SUPER CAPACITOR BANK: SCB powers the hub motors. The power electronics and control strategy design support to maximize the energy generated and stored during braking and deceleration [14]. As the heavy charging and discharging is handled by the SCB, and as the life cycle of the SCB is at least ten times that of the battery, the life span of SCB- BP module increases substantially [11, 13]. A thiol-functionalized, nitrogen-doped, reduced grapheme oxide electrode super capacitor bank that has high power/ energy density of 32kW/kg (9.8 kW/L)/ 9.8 Wh/kg at a current density of 50 A/g [15] is utilized.

BATTERY PACK: BP is charged using a fast charger or a regular charger, when the vehicle is stationary. The BP charges the SCB when the state of charge in the SCB reaches the prescribed optimal level and, when the charge in the SCB reaches the minimum set value, and when the charge in the BP also reaches sub-optimal level; the engine-generator [E-G] starts to power the SCB. Based on the destination inputs and SCB charge level, the E-G stops at the appropriate instant.

VEHICLE CONTROL UNIT & POWER ELECTRONICS: VCU along with PE functions as a power rationalization module by varying the current supplied to the hub motors. It also varies the flux of the stator to change the PMSM from motor to generator mode and vice versa. Based on the voltage at the SCB, the torque demand and the contact conditions of wheels with the road, it provides the current required for desirable traction, without compromising stability and eliminating tyre slip. PE module functions as a DC to DC converter, DC to AC converter and as AC to DC converter, depending on the required amplitude and direction of charge flow. When the E-G is generating current, the PE directs the current required by the hub motors and transfers the surplus to the SCB. During instances when the current required by the hub motors are more than the current generated by the E-G, the PE balances the difference by drawing from the SCB.

ENGINE-GENERATOR: The E-G is an IC engine coupled to an alternator that charges the SCB when it reaches the sub-optimal voltage and when the state of charge of BP is also minimum; and when the destination and or fast charging facility is not within reach, with the charge left in the storage module. The operating speed of the engine is optimized for maximum torque and minimum emissions.

4. Novel Engine

The proposed concept is modelled on HYUNDAI KONA EV to compare the different existing models with the proposed models. To reduce the complexity, the PMSM motor with the same specification of KONA EV is employed. The engine selected is a scaled down, striped down version of a 0.9 litre, Achates Power Opposed Piston Gasoline Compression Ignition [OPGCI] two stroke engine that develops 63 kW at 2000 rpm and has a brake thermal efficiency close to 50% [10]; in which the variable geometry turbocharger is replaced by a fixed geometry turbocharger [FGT]; and the supercharger is replaced by an air accumulator, to maintain the air pressure in the inlet manifold high during starting. The displacement of the engine is taken as 374cc, bore as 60mm and stroke as 13mm. The compression ratio is 15.4. Based on the power output of the base engine selected, the power output of the engine is calculated as 20kW @2000rpm.

The block diagram of the proposed steady-state engine is shown below:

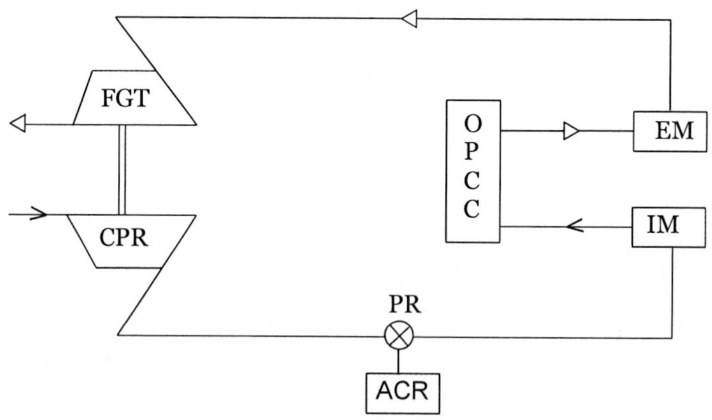

FGT	⇒	Fixer Geometry Turbocharger	ACR	⇒	Accumulator
CPR	⇒	Compressor	EM	⇒	Exhaust Manifold
PR	⇒	Pressure Regulator	IM	⇒	Inlet Manifold
OPCC	⇒	Opposed Piston Combustion Chamber			

Figure 2. Block Diagram of Proposed Steady-State I.C.E

The alternator coupled to the engine is operated at constant load and has a rated output of 20kW.

The BP employed for the model is a 16kWh lithium-ion polymer battery like the one used in KONA EV. The SCB has a capacity of 1.5kWh.

5. Energy Consumption Analysis

Brake thermal efficiency of the base engine [gasoline] at maximum torque = 50%& diesel=55%
Net efficiency of the E-G [gasoline]= 45%& diesel=50%
Average energy required per km for combined drive cycle of KONA EV[20] =0.169kWh
Average energy required per km for the proposed series hybrid is calculated as =0.143kWh

[As the proposed model is lighter by about 250kg when compared to KONA EV (330kg reduction on account of BP and 80kg addition on account of SCB, E-G and fuel tank). This would result in energy savings of about 15% withbetter acceleration and braking characteristics.]

The energy required per kilometre to run the KONA with the proposed drivetrain=0.143kWh

The distance covered by KONA with the proposed drivetrain in E-G mode with: Gasoline = 9.4/0.123*45/100= 29.6km/litre; Diesel=10/0.123*50/100 =35.0km/litre.

Table 2. Comparison of Energy Consumption in Different Configurations of Hyundai Kona

	Model	Gross Wt.	KP/L	kWh/Km
1.	Petrol	1350	12.5	0.713
2.	Diesel	1420	21.0	0.424
3.	Petrol Hybrid	1450	18.7	0.476
4.	EV	160	-	0.169
5.	Concept EV with OPGCI(Gasoline)	1400	29.6	0.143/ 0.318(E-G)
6.	Concept EV with OPDDI(Diesel)	1410	35.0	0.143/ 0.286(E-G)

6. Conclusion

The series hybrid concept presented can be successfully deployed for many automotive applications from small passenger cars to light duty trucks to heavy duty commercial vehicles. The arithmetic is positive for the proposed models with an electric range of about 115km on a 16 kWh battery and an additional 391km on a 13.5litres of gasoline and 472km on a same quantity of diesel. Further studies need to be employed using computational fluid dynamics tools to validate the efficiency of the concept engines. As the size of proposed engine is much smaller than the selected base engine, there could be more heat losses that would reduce the efficiency. However, as the engine is to be operated in steady state, there would be enough room to reduce the cooling losses and modify the combustion chamber to improve upon the efficiency. Simulations need to be performed for various motor combinations to get the optimum acceleration, braking characteristics and energy consumption. Iterations have to be done after detailed simulations for power quality during traction; and optimize the BP, SCB and PE.

Acknowledgment

This research project was funded by the Deanship of Scientific Research, Princess Nourah Bint Abdulrahman University, Riyadh, KSA through the Program of Research Project Funding, grand No.41-PRFA-P-34. I thank the Deanship of Scientific Research of Princess Nourah Bint Abdulrahman University for funding the Research.

References

[1] https://www.epa.gov/ghgemissions/global-greenhouse-gas-emissions-data.
[2] World Bank 2014 Electric power transmission and distribution losses (% of output) *World Development Indicators*
[3] http://www.iea.org/stats/index.asp.
[4] https://www.fueleconomy.gov/feg/atv-hev.shtml

[5] www.nhtsa.gov › lows-regulations.
[6] Hirose I and Hitomi M 2016 Mazda's way to more efficient internal combustion engines *MTZ worldwide* 77 28-33
[7] Barba D 2018 Assessing the Efficiency Potential of Future Gasoline Engines *In SAE 2018 High Efficiency IC Engine Symp.*
[8] Abani N, Chiang M, Thomas I, Nagar N, Zermeno R and Regner G 2016 Developing a 55+ BTE Commercial Heavy-Duty Opposed-Piston Engine without a Waste Heat Recovery System *In Heavy-Duty-, On-und Off-Highway-Motoren* 292-310
[9] Abani N, Chiang M, Thomas I, Nagar N, Zermeno R and Regner G 2016 Developing a 55+ BTE Commercial Heavy-Duty Opposed-Piston Engine without a Waste Heat Recovery System *In Heavy-Duty-, On-und Off-Highway-Motoren* 292-310
[10] Splitter D, Wissink M, DelVescovo D and Reitz RD 2013 RCCI engine operation towards 60% thermal efficiency *SAE Technical Paper*
[11] https://arpa-e.energy.gov/?q=slick-sheet-project/efficient-engine-design
[12] Musolino V, Tironi E and di Milano P 2010 A comparison of supercapacitor and high-power lithium batteries *In Electrical systems for aircraft, railway and ship propulsion Oct 19* 1-6
[13] Zhao Y, Deng W, Wu J and He R 2017 Torque control allocation based on constrained optimization with regenerative braking for electric vehicles *International journal of automotive technology* **18** 685-98
[14] Zhang Q, Deng W, Zhang S and Wu J 2016 A rule based energy management system of experimental battery/supercapacitor hybrid energy storage system for electric vehicles *Journal of Control Science and Engineering* **2016**
[15] Khan MA, Zeb K, Sathishkumar P, Ali MU, Uddin W, Hussain S, Ishfaq M, Khan I, Cho HG and Kim HJ 2018 A novel supercapacitor/lithium-ion hybrid energy system with a fuzzy logic-controlled fast charging and intelligent energy management system *Electronics* **7** 63
[16] Rani J, Thangavel R, Oh SI, Lee YS and Jang JH 2019 An ultra-high-energy density supercapacitor; fabrication based on thiol-functionalized graphene oxide scrolls *Nanomaterials* **9** 148.
[17] Manie N and Pattanaik B 2019 Zeta DC-DC converter based on MPPT technique for BLDC application *International Journal of MC Square Scientific Research* **11** 1-2
[18] Tomaszewska A, Chu Z, Feng X, O'Kane S, Liu X, Chen J, Ji C, Endler E, Li R, Liu L and Li Y 2019 Lithium-ion battery fast charging: a review *ETransportation* **1** 100011
[19] Van Horn C 2006 *Steady state and transient efficiencies of a four cylinder direct injection diesel engine for implementation in a hybrid electric vehicle Doctoral dissertation, University of Akron*
[20] Kouchachvili L, Yaïci W and Entchev E 2018 Hybrid battery/supercapacitor energy storage system for the electric vehicles *Journal of Power Sources* **374** 237-48
[21] https://pod-point.com/guides/vehicles/hyundai/2018/kona-64-kwh
[22] https://www.fueleconomy.gov/feg/Find.do?action=sbs&id=42089
[23] Puttamadappa C and Parameshachari BD 2019 Demand side management of small scale loads in a smart grid using glow-worm swarm optimization technique *Microprocessors and Microsystems* **71** 102886

IOP Conf. Series: Materials Science and Engineering **925** (2020) 012023 doi:10.1088/1757-899X/925/1/012023

Computational Fracture Mechanics: Evaluation of The Structural Integrity in a Penstock Applying the BS7910 Standard and Finite Element Analysis

G O Barrionuevo[1], B A Guerrero[2] and M Walczak[1]

[1] Department of Mechanical and Metallurgical Engineering, School of Engineering, Pontificia Universidad Católica de Chile, Av. Vicuña Mackenna, 4860, Macul, Santiago, Chile.
[2] School of Mechanical Engineering, University of Adelaide, Adelaide, SA 5005, Australia.
[1]gobarrionuevo@uc.cl

Abstract. The present work aims to determine the structural integrity of penstocks, applying the BS7910 standard and finite element analysis (FEA). For the study of fracture mechanics in thin-walled cylinders that have defects inside, the structural integrity of the element is determined through a failure assessment diagram (FAD) where the fracture failure index (Kr) is graphed versus plastic collapse index (Lr). Based on the locus of the initial point of failure, the safety factor is calculated, and it is defined if the element is fit for service or if corrective action must be taken to continue operating; later, crack growth is analyzed, where the critical point of rupture of the penstock and therefore its useful life is determined. To contrast results, a simulation of fracture mechanics is performed in ANSYS, where the stress intensity factor (KI) is determined using the fracture tool, and the useful life of the element is also calculated through a fatigue analysis. The use of tetrahedral elements is recommended for the overall meshing and a cobweb configuration for meshing at crack-tip. Finally, the results obtained are compared, where the mean average percentage error of 3.24% was obtained, denoting the usefulness of the two methods as well as the simplicity of the Paris' law.

1. Introduction

Traditionally, the design of a structural element is carried out under the assumption that the material is continuous, homogeneous, and without defects. Besides that, the supported stress will be less than its yield strength. However, defects in materials are a reality, so when designing, the presence of these defects must be considered to guarantee the safety of people in the first place and ensure the structural integrity of the element. Different failure theories are used in traditional design (Rankine, Saint Venanat, Von Mises, Tresca, Mohr-Coulomb, and so on), depending on if the material is ductile or brittle. However, fracture mechanics improve product design, as well as manufacturing and inspection processes to control the spread of defects that could lead to failure of its components, without the need to use arbitrary safety factors [1].

For the design of pressure vessels, penstocks, storage tanks, and boilers, it is necessary to comply with internationally established standards and codes. The ASME B31.1 [2] code for pressure piping, the API

579/ASME FFS [3] standard that determines the suitability for service of structural elements or the BS7910 [4] standard, which provides methods to evaluate the acceptability of flaws in metallic structures.

Failure of a mechanical element in service occurs due to plastic deformation, fracture, or fatigue. Classical failure theory assumes that a plate subjected to a load will have a uniform distribution of forces along with the plate; however, it is known that at the time of manufacturing or during service, small imperfections are generated in it, which generate stress concentrators [5]. Fracture mechanics is a discipline that links, through analytical expressions, the geometry of the cracked component and the loads to which it is subjected [6]; it is based on the study of displacements, deformations, and stresses in the vicinity of the crack head. Erdogan [7] explored some of the tools that are used today as the stress intensity factor (K_I), where σ: is the applied stress, and a: flaw extension.

$$K_I = \sigma\sqrt{\pi a} \tag{1}$$

It is known that all materials, even the most fragile ones, can develop a certain degree of plasticity at the crack-tip; this improves the strength of brittle fracture [8]. The developed plastic region is known as the crack-tip plastic zone, and according to its extension, it can be approached by three different sub-disciplines [7]:

1. Linear elastic fracture mechanics (LEFM): discipline used when the extension of the plastic deformation is confined to a small area above the head of the flaw, and the deformation of the rest of the body has an elastic behavior.
2. Elastic-plastic fracture mechanics (EPFM): applied when the plastic area on the head of the fissure has a considerable extension, similar to crack length.
3. Plastic fracture mechanics (PFM), used when the fracture is preceded by generalized plastic deformation.

According to Behal & Solodyankin [9], LEFM is enough to characterize the fracture mechanics in 3D objects. LEFM theory assumes that the end of the crack ends in a point, which means infinite stress in that area [10]. In reality, the end has a radius, and the stresses tend to relax due to yielding (figure 1), this radius is known as plasticity radius (r_p) [11], which is given by the following expression, where σ_y is the yield strength.

$$r_p = \frac{1}{6\pi}\left(\frac{K_I}{\sigma_y}\right)^2 \tag{2}$$

The elastic zone is controlled by the evolution of the K_I and must be large enough so that the plastic zone is within it, between 2 to 5% of the length of the crack [5].

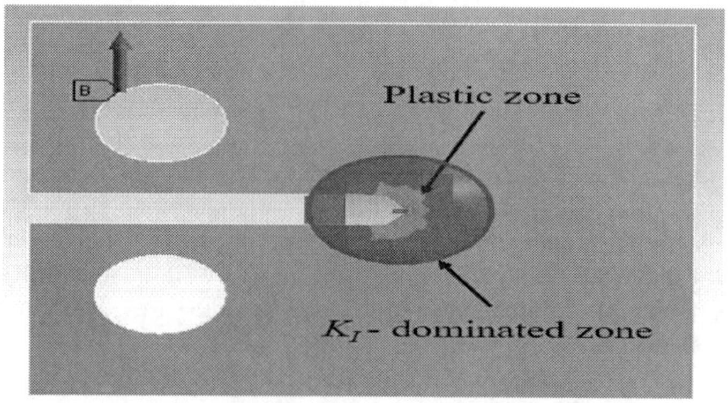

Figure 1. Plastic zone extension in LEFM

In recent years, the term *structural integrity* has emerged, which consists of analyzing an element and determining its suitability for service [12]. Structural integrity analysis contributes to decision-making and maintenance-scheduling, creating an economic and safety culture for mechanical component service [13]. The Welding Institute established a failure assessment diagram (FAD) to evaluate the structural integrity defining the fracture failure index (*Kr*) and plastic collapse index (*Lr*), a complete description of the road map could be found in [14].

Computational Fracture Mechanics
The fundamental objective of computational mechanics is the development and implementation of computational models for the characterization and prediction of the behavior of structural and mechanical components. Computational mechanics is based on the use of numerical methods such as finite differences, finite elements, or boundary elements [15]. The basis of the finite element analysis (FEA) is established in the decomposition of the domain into a finite number of subdomains (elements), these elements are interconnected by a series of points called nodes, which guarantee the continuity of the displacement field, the equations that govern the behavior of the domain also govern the behavior of each element. In this way, it is possible to go from a continuous system with infinite degrees of freedom, to a discrete system with a finite number of degrees of freedom [16]. The displacement (*x*) is calculated applying Hooke's law, where *K* is the stiffness matrix and *F* the global load vector.

$$[K]\{x\} = \{F\} \tag{3}$$

According to Chu & Liu [17], the FEA is effective for predicting the behavior of crack growth due to fatigue. Sun & Zhang [18] developed a method for the evaluation of crack growth using Ansys Mechanical APDL with its software language (UIDL). In addition, Jensen [19] improved this method using the interaction between Matlab and Ansys APDL. Behal & Solodyankin [9], simulated the crack growth and estimated the life of structural elements in 3D, to determine the growth rate of the crack they used a modified Forman's experimental equation and also used FEA in Ansys, where it was determined that a linear elastic analysis is sufficient to characterize fracture mechanics in ductile materials.

In the analysis of crack propagation, quadrilateral and hexahedral elements are used for problems in two and three dimensions, respectively. Zakavi et al. [20] recommend 8 and 9 node Lagrangian quadratic elements for two-dimensional problems, and 20 and 27 node Langrangian three-dimensional problems. The most efficient mesh design for the region around the crack-tip is the cobweb configuration, which consists of concentric rings of elements arranged concentrically around the head of the flaw, the first ring of which consists of singular elements. The design of the cobweb configuration allows a smooth transition from a fine mesh in the vicinity of the crack to a thick mesh in the rest of the component [21]

Regarding the current work, first, BS7910 standard is applied to determine the fracture locus and determine the suitability of the element, this assessment is based on the determination of the fracture and plastic collapse index. Stress intensity factor is compared to fracture toughness for fracture assessment, and the reference stress is compared to yield stress to determine if the element fails by plasticity; then, an FEA is employed to assess K_I, parametrizing the flaw depth (*a*), and flaw length (*c*). Through a fatigue assessment, the fatigue crack growth is evaluated following the Paris law, and the two methodologies are compared.

2. Methodology

2.1. Case study

For the design of a penstock (figure 2), it is regularly considered as thin-walled cylindrical bodies, if there is a significant difference between the wall thickness (*t*) and pipe diameter (*D*) [22].

$$\frac{D}{t} > 10 \tag{4}$$

For thin-walled vessels, the stress state is considered as plane-stress; as a result of internal pressure, mechanical stress called hoop stress (σ_θ) is generated, given by the following expression:

$$\sigma_\theta = \frac{P\,r}{t} \tag{5}$$

Where, P is the internal pressure, and r pipe inner radius, and t wall thickness.

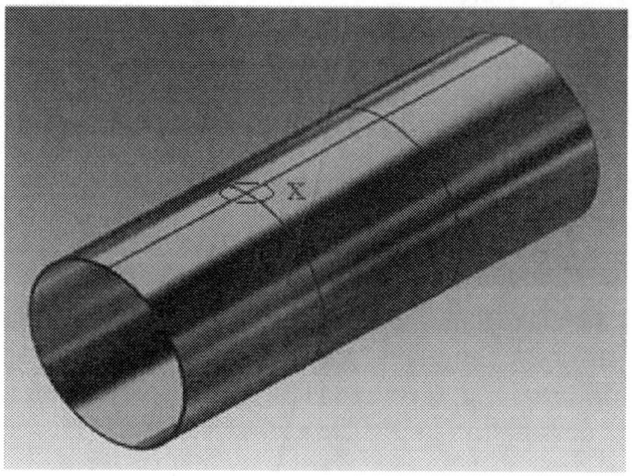

Figure 2. 3-D model of the penstock.

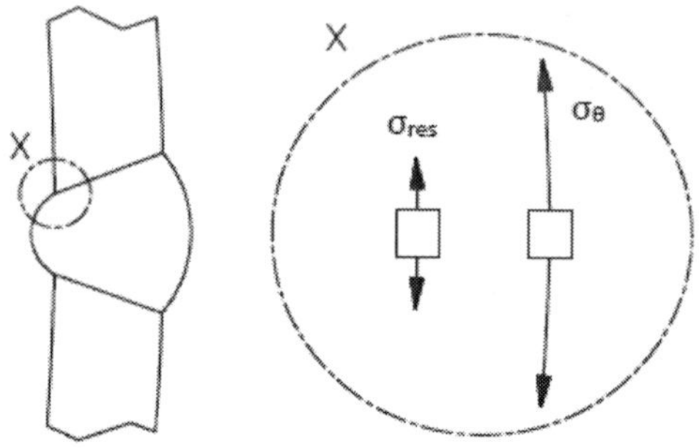

Figure 3. Detail of an element to show the stress state on the penstock wall.

Due to water hammer, cyclic loads must be considered, and therefore the penstock should be designed to avoid fatigue failure. Furthermore, residual stresses remain in a component in the absence of external loads and originate in manufacturing processes such as welding. In a multi-pass weld heat-affected zone (HAZ) is subjected to straining and aging, reducing toughness [23]. BS7910 recommends a magnitude of around 20 % of the yield stress (σ_Y). The maximum stress concentration factor is the sum of the hoop stress and residual stress around the weld (figure 3).

Fracture toughness (K_{IC}) is the maximum value that K_I can reach, and it is related to energy absorption and could be calculated through [4]:

$$K_{IC} = \left[\left(12\sqrt{Cv} - 20\right)\left(\frac{25}{t}\right)^{0.25}\right] + 20 \tag{6}$$

Where Cv is the energy during the impact test.
Nominal parameters for the case study are presented in table 1:

Table 1. Hydraulic data [11]

Parameter	Magnitude	Unit
Pipe inner diameter	2.2	m
Wall thickness	30	mm
Static head of water	4.9	MPa
Water pressure at hammer	6.86	MPa
Crack size	4	mm
Internal pressure change	10000	Times/year

The nominal chemical composition of the material employed for the construction of pressure vessel is ASTM 537 class I, (table 2). The mechanical properties are listed in table 3.

Table 2. Chemical composition of A537 steel alloy [24].

Elements (wt %)								
C	Si	Mn	P	S	Cr	Cu	Mo	Ni
0.24	0.15-0.50	0.7-1.35	0.035	0.035	0.25	0.35	0.08	0.25

Table 3. Mechanical properties of A537 steel alloy [24].

Plate thickness	Yield strength	Tensile strength	Modulus of elasticity	Elongation
mm	MPa	MPa	GPa	%
16 – 40	345	550	200	>18

2.2. Structural integrity: BS7910 procedure

For the evaluation of structural integrity, the British standard BS7910 (Guidance on Methods for Assessing the Acceptability of flaws in metallic structures) has been employed. Which is carried out through a failure evaluation diagram (FAD), where the vertical axis represents the fracture index (K_r), which shows the relationship between K_I and K_{IC}. In contrast, the horizontal axis represents the relationship between the supported stress (σ_{ref}) and σ_Y or plastic collapse index, denoted by (L_r). The location of K_r and L_r index provides the coordinates of the evaluation point or the geometric location of the failure. The positions of the points are compared with the FAD limit curve to determine the acceptability of the structure (figure 4).

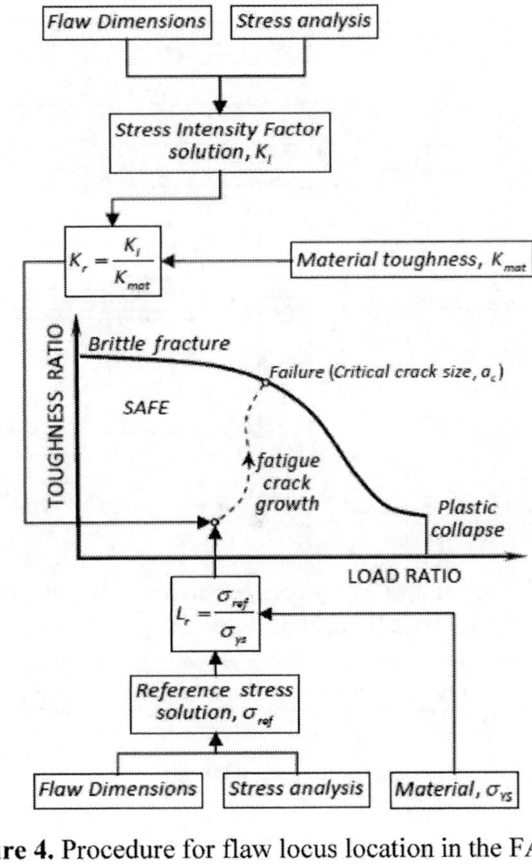

Figure 4. Procedure for flaw locus location in the FAD [3].

BS7910 has three levels of evaluation; at each level, the acceptability of the defect can be evaluated, and the calculation can be iterated to determine the limit value of a parameter such as the crack size, the maximum stress supported, or the fracture toughness. The limit values can be obtained through analytical calculations, or through the location in the FAD to determine which point contacts the FAD limit curve [4].

- Level 1: it is a very conservative evaluation process, requires a simple analysis of failure evaluation diagrams and fracture mechanics.
- Level 2: determines the specific failure of a material, estimates the interaction between fracture and plasticity.
- Level 3: consider the effects of ductile tear and plasticity through the calculation of the integral-J, which implies a direct calculation of the effects of plasticity.

British Standard BS7910, recommends the use of the following equation for the evaluation of materials where welding processes were employed [4]:

$$Kr = \left(1\text{-}0.14\ Lr^2\right)\left[0.3 + 0.7\exp\left(-0.65\ Lr^6\right)\right] \qquad (7)$$

One of the critical aspects to take in a count is the stress intensity correction factor (*Y*) [20], which modifies the magnitude of K_I, *Y* depends on the geometry of the flaw (figure 5).

$$K_I = Y\,\sigma\,\sqrt{\pi a} \qquad (8)$$

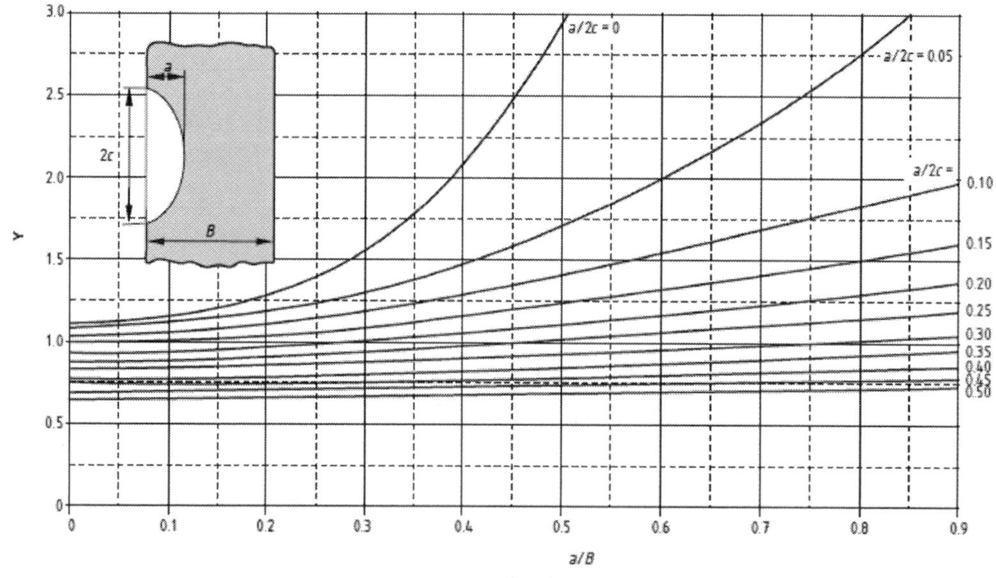

Figure 5. Stress intensity correction factor Y for surface flaws in tension [4].

For the characterization of plastic collapse, it is necessary to determine the referential stress (σ_{ref}), which is a measure of the plasticity at the crack-tip [25].

$$\sigma_{ref} = \frac{3\sigma\alpha + \ [(3\sigma\alpha)^2 + 9\,\sigma^2\,(1-\alpha)^2]^{\,0.5}}{3(1-\alpha)^2} \qquad (9)$$

Where σ is the hoop stress, and α: a geometric constant given by:

$$\alpha = \frac{\frac{a}{B}}{1+\frac{B}{c}} \tag{10}$$

2.3. Computational fracture mechanics: Ansys procedure

A classical finite element analysis (FEA) consists of three steps: pre-processing, solution, and post-process [26][27]. When working with a 3-D element, it is convenient to distinguish between some areas for meshing process (figure 6). In general, two areas of importance are distinguished, the first represents the meshing of the entire element or the global mesh, and the second represents the meshing in the fracture zone, the refined mesh. It is important to distinguish a third zone (transition mesh), where there is a smooth transition between global mesh and refined mesh [28].

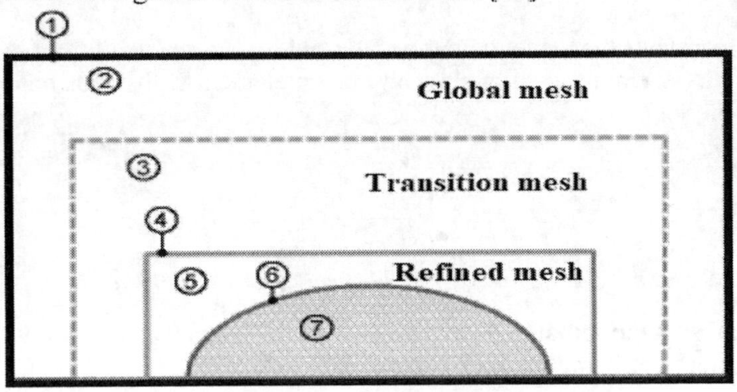

Figure 6. Fracture meshing zones in LEFM [28].

In figure 6, seven zones could be distinguished: 1) the solid, 2) global mesh, 3) buffer zone, 4) fracture limit, 5) fracture zone, 6) front of the flaw, and 7) discontinuity planes. For global meshing (zones 1, 2, 3) tetrahedral elements are recommended to ensure the precision of the results, for zones 4 and 5 a smooth transition mesh is recommended, and in zone 6 and 7 a spider-type mesh is used to represent the singularity of the elements at the crack-tip.

The main objective of the simulation is to determine the K_I, in a penstock that has an axially oriented flaw inside, figure 7 shows the geometry evaluated at the left side and in the right, the flaw location.

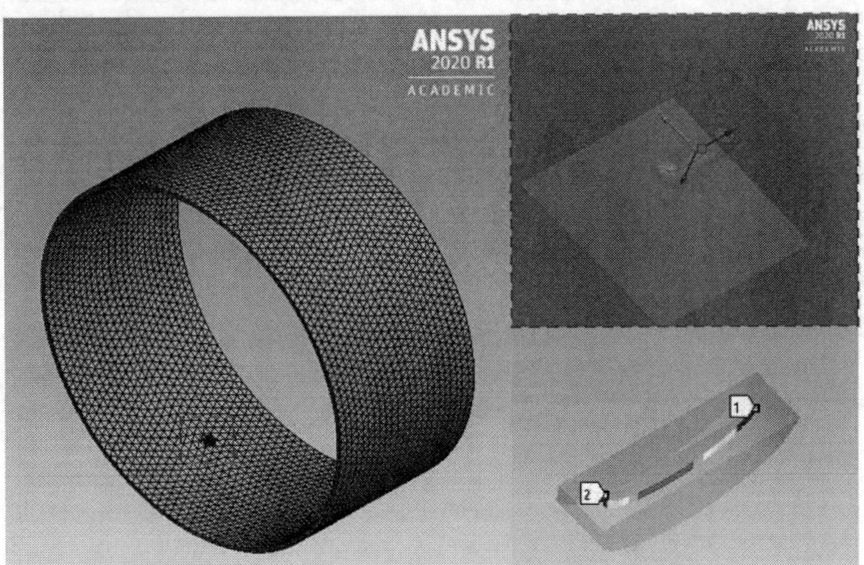

IOP Conf. Series: Materials Science and Engineering **925** (2020) 012023 doi:10.1088/1757-899X/925/1/012023

Figure 7. Simulated geometry and crack details [28].

2.4. Crack propagation

Fatigue is a process of accumulating damage, cracking, and fracturing of components subjected to fluctuating, variable, or cyclic loads [29]. In order to estimate the useful life of a structural element subjected to cyclic stresses, research has shown a correlation between crack propagation and the number of fatigue load cycles called crack growth rate. For a constant amplitude load, the crack growth rate can be described by the Paris' law:

$$\frac{da}{dN} = C\,\Delta K_I^{\,m} \tag{11}$$

Where *C* and *m* are constants that depend on the material [30]. From equation 11, a mathematical model is obtained to determine the number of cycles (*N*) that an element withstands before the fracture, given by the following expression:

$$N = \int_{ai}^{af} \frac{da}{C\left(Y\,\Delta\sigma\,\sqrt{\pi\,a}\right)^m} \tag{12}$$

Where a_i and a_f, represent the initial and final crack length, respectively. Finally, the useful life of the penstock could be calculated using the following equation:

$$\text{life (years)} = \frac{N}{\text{cycles per year}} \tag{13}$$

3. Results and Discussions

3.1 BS7910 results

Table 4 summarizes the results obtained by applying the BS7910 standard; the values used to locate in the FAD are fracture index (*Kr*) and the index for plastic collapse (*Lr*).

Table 4. BS7910 results.

Parameter	Symbol	Magnitude	Unit
Minimum stress	σ_{min}	179.78	MPa
Maximum stress	σ_{max}	251.74	MPa
Alternating stress	σ_a	35.96	MPa
Mean stress	σ_m	217.74	MPa
Residual stress	σ_{res}	69	MPa
Stress intensity correction factor	Y	0.75	-
Primary stress intensity factor	K_{IP}	478.14	$MPa\sqrt{mm}$
Secondary stress intensity factor	K_{IS}	183.41	$MPa\sqrt{mm}$
Fracture toughness	K_{IC}	2717.03	$MPa\sqrt{mm}$
Fracture index	Kr	0.24	-
Reference stress	σ_{ref}	186.87	MPa
Plastic collapse index	Lr	0.54	-

Figure 8 shows the assessment point at no load (A), due to the residual stresses originated in the welding process and the flaw locus (B), which is in the safety zone, and therefore, the structural integrity of the element, is guaranteed. To determine the safety factor, under the penstock is operating, a line is plotted,

starts from the assessment point at no load (A), goes through the point (B) and intersects with the FAD limit curve (C), using the AC/AB ratio the safety factor is obtained.
Through the BS7910 application, two values of K_I were obtained, the first one (K_{IP}) product of the applied stress, and the second one (K_{IS}) product of the residual stresses. The global K_I is the sum of both, with a value of 661.46 MPa$\sqrt{mm}$; this value is well below the K_{IC} (2717.03 MPa$\sqrt{mm}$). Once know K_{IC}, it is possible to determine the maximum stress supported by the penstock, through equation 8 is determined σx = 468.9 MPa.
Finally, knowing K_I to validate the LEFM applicability, the plasticity radius is calculated with equation 2, where r_p = 0.195 mm; equivalent to 4.875 % of the flaw depth, these results are in accordance with [21].

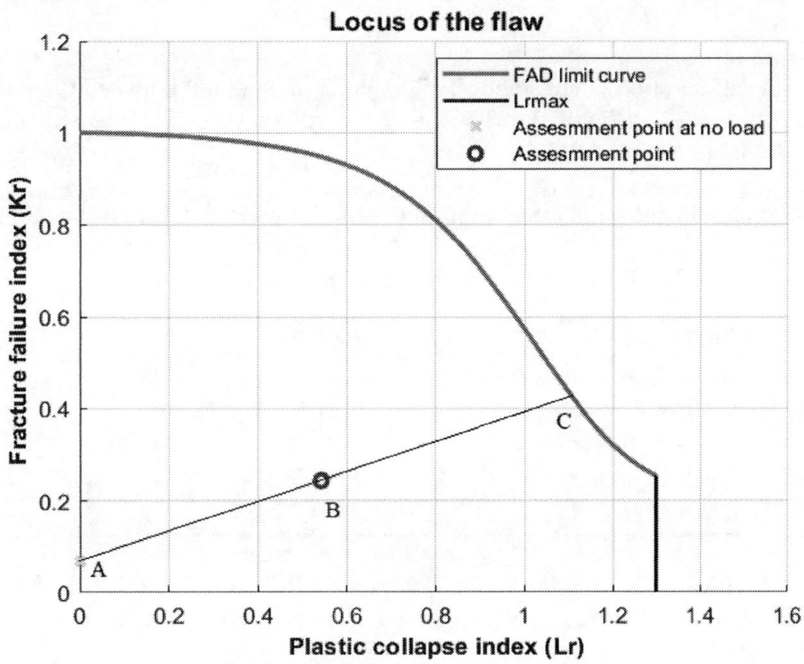

Figure 8. Locus of the flaw applying BS7910.

3.2 FEA results

The parameter that characterizes fracture mechanics is the stress intensity factor, which reaches its maximum value at the crack-tip (figure 9).
The FEA was developed as a linear elastic problem, due to the characteristics of the material employed, with elongation percentage is superior to 18 %, which imply a ductile material. This approach was validated analytically, through the characterization of the plasticity radius, which demonstrates that FEA is appropriate to be applied to fracture mechanics assessment practice of penstocks based on LEFM.

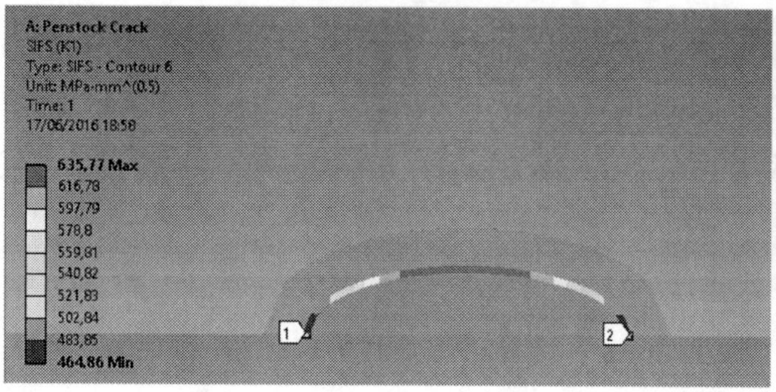

Figure 9. Stress intensity factor (K_I) evaluation in ANSYS.

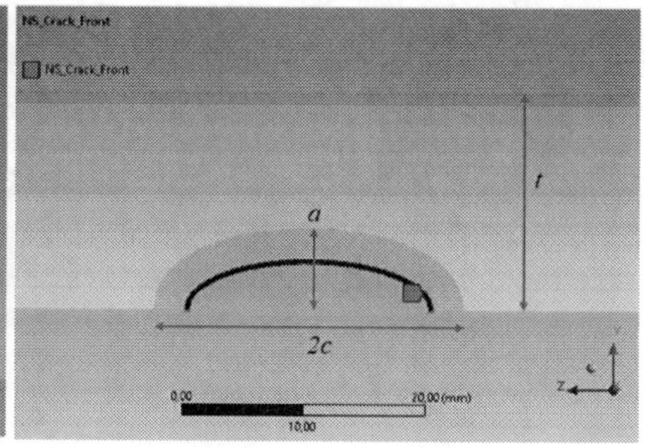

Figure 10. Crack geometry parameters for fatigue evaluation.

3.3 Crack propagation

To determine K_I in each iteration, a constant relationship is maintained between the extension and the depth of the crack $c = a + 1$ (Fig. 10), with which a constant value is obtained for the geometric correction factor Y = 0, 75 (figure 5).

Table 5 shows the evaluation of fatigue life applying Paris' law, with C=1 x 10^{-11}, and m=3.

Table 5. Parameters for crack growth evaluation

Parameter	Magnitude	Unit
Initial size of detected crack	4	mm
Critical flaw size	19	mm
Stress variation	35.96	MPa
Cycles number	495486.8	-
Total life	49.5	years

From figure 11, the critical flaw size is obtained, tending to a failure by a ductile fracture due to the elastic characteristics of the material. When the crack depth exceeds 19 mm, dynamic crack growth begins and its subsequent failure.

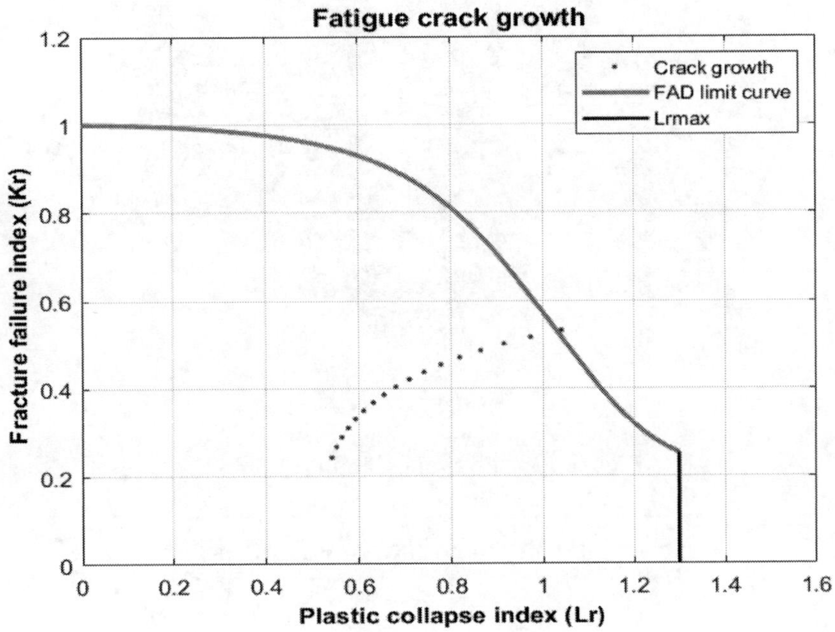

Figure 11. Fatigue crack growth.

ANSYS allows the parameterization of several factors from the model geometry, in this case, the depth (a) and extent (c) of the crack were parameterized to simulate the crack growth, with which the value of the K_I is obtained in each iteration (figure 12).
Through the ANSYS fatigue tool (Fig. 13), and applying the Goodman criterion, with a fluctuating distribution of stresses between a minimum value given by the static head of water and a maximum value given by the effect of water pressure hammer, it is obtained that for the area of interest life fluctuates between 34750 and 535110 cycles.

Outline: No data

	A	B	C	D
1	ID	Parameter Name	Value	Unit
2	⊟ Input Parameters			
3	⊟ Penstock Crack (A1)			
4	P4	Crack --Minor Radius	4	mm
5	P5	Crack --Major Radius	10	mm
*	New input parameter	New name	New expression	
7	⊟ Output Parameters			
8	⊟ Penstock Crack (A1)			
9	P6	SIFS (K1) Maximum	635,77	MPa mm^0.5
*	New output parameter		New expression	
11	Charts			

Name	P4 - Crack --Minor Radius	P5 - Crack --Major Radius	P6 - SIFS (K1) Maximum
Units	mm	mm	MPa mm^0.5
Current	4	10	635,77
DP 1	5	12	702,68
DP 2	6	14	770,35
DP 3	7	16	834,92
DP 4	8	18	897,27
DP 5	9	20	950,35
DP 6	10	22	1019,5
DP 7	11	24	1080,2
DP 8	12	26	1141
DP 9	13	28	1195,7
DP 10	14	30	1260,2
DP 11	15	32	1314,9
DP 12	16	34	1375,9
DP 13	17	36	1432,7
DP 14	18	38	1489,2
DP 15	19	40	1540

Figure 12. K_I evaluation and parametrization in ANSYS.

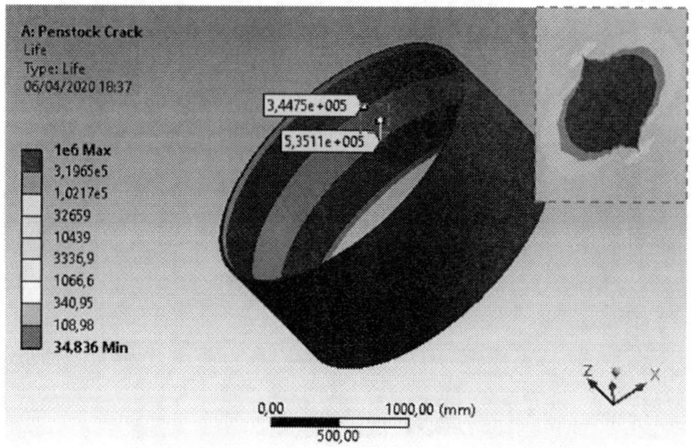

Figure 13. Life evaluation in ANSYS.

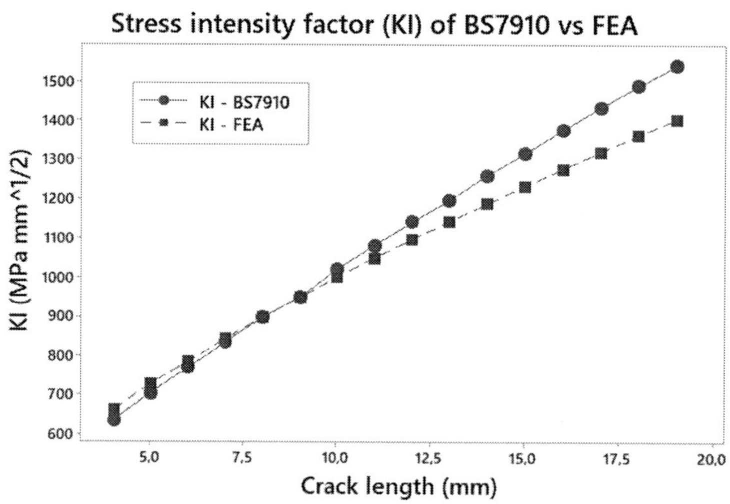

Figure 14. K_I results, BS7910 and FEA results.

3.4 Results comparison

Since the parameter chosen for the characterization of fracture mechanics in the present study was the stress intensity factor (K_I), a comparison of each value is shown in figure 14 as the crack size growths. The FEA of fracture mechanics has generated satisfactory results since a mean average percentage error of 3.24% was obtained concerning the BS7910 standard, with a minimum value of 0.054 and a maximum of 9,726 %. Besides, the life calculated by BS7910 applying the Paris' law gave a result of 49.5 years, and by FEA, an average value of 44 years was obtained, demonstrating the effectiveness of the FEA in the assessment of LEFM problems [31].

4. Conclusions

This paper aims to evaluate the structural integrity in a penstock with a detected flaw inside the pipe. Two methods were applied to determine if the component is suitable to work or if corrective action should be taken to prolong the useful life of the pipe. The following conclusions can be drawn:

(1) BS7910 presents a robust methodology to assess different flaws and taking in count the effect of residual stress around the welded zone; in that sense, the determination of the stress intensity correction factor is vital to determine the fracture ratio and determine the locus of the flaw. Through the FAD, the security factor has been established with a value of around 2.05, which determines the aptitude for service.

(2) Fracture toughness is determined with data provided by the manufacturer on the Charpy test, it relates the impact energy and the thickness of the sheet, obtaining a value of 85.92 $MPa\sqrt{m}$, these data are supported by Welding Research Supplement, where results for fracture toughness of A537-1 steel with a range between 84.6 and 92.3 $MPa\sqrt{m}$ were published.
(3) The fatigue assessment shows that the critical point before the brake is at 19 mm, knowing the fracture toughness was possible to determine the maximum stress that the penstock could hold up, around 470 MPa.
(4) The characterization of the K_I employing FEA has generated satisfactory results. The radius of the plasticity of less than 5% of the crack size has been determined, which ensures that it is a linear elastic problem.

Acknowledgments
This study has been completed under the financial support of the State Secretariat for higher education, science, technology and innovation (SENESCYT) grant number ARSEQ-BEC-000329-2017 and the Research Centre for Nanotechnology and Advanced Materials (CIEN-UC).

References
[1] Pisarski H 2013 Assessment of flaws in pipeline girth welds - A critical review *Weld. World* **57** 933-945
[2] The American Society of Mechanical Engineers 2012 ASME B31.1-2012 Power Piping *ANSI Stand B31. 1*, **1**
[3] American Petroleum Institute 2009 API 579-2/ASME FFS-2 2009- - Fitness-For-Service Example Problem Manual *Society*
[4] BS-7910 2005 Guide to methods for assessing the acceptability of flaws in metallic structures," *BSI Stand. Publ.* **3** 306
[5] Christensen RM 2013 The Theory of Materials Failure *Theory Mater. Fail.*
[6] Anderson TL 2017 Fracture Mechanics, Fundamentals and Applications *CRC press*
[7] Erdogan F 2000 Fracture mechanics *Int. J. Solids Struct.* **37** 171–183
[8] Arnoult XC, Růžičková M, Kunzová K and Materna A 2016 Short review: Potential impact of delamination cracks on fracture toughness of structural materials *Frat. ed Integrita Strutt.* **10** 509–522
[9] Běhal J and Solodyankin K 2012 Crack growth simulation in the course of industrial equipment life extension *20th SVSFEM ANSYS Users' Gr. Meet. Conf.* 1-9
[10] Nowell D and Nowell SC 2019 A comparison of recent models for fatigue crack tip deformation," *Theor. Appl. Fract. Mech.* **103** 102299
[11] Andrade L 2013 Design of penstock pipe for a hydroelectric pumped storage station *J. Chem. Inf. Model.* **53** 1689–1699
[12] Xin W, Yan P, Du J and Cai F 2017 Safety assessment of Cracked K-joint Structure Based on Fracture Mechanics *J. Phys. Conf. Ser.* **843** 012011
[13] Simpson CA, Tonge S, Connolley T, Reinhard C, Marrow TJ and Mostafav M 2019 Validating 3D two-parameter fracture mechanics models for structural integrity assessments *Theor. Appl. Fract. Mech.* **103** 102281
[14] Wintle JB 2003 Which procedures for fitness-for-service assessment: API 579 or BS 7910? *Int. Conf. Press. Vessel Technol.* 1–7
[15] Brighenti R and Carpinteri A 2013 Surface cracks in fatigued structural components: A review *Fatigue Fract. Eng. Mater. Struct.* **36** 1209–1222
[16] Kim JM and Huh NS 2010 On crack interaction effects of in-plane surface cracks using elastic and elastic plastic finite element analyses *Nucl. Eng. Technol.* **42** 680–689
[17] Chu SJ and Liu C 2012 Finite element simulation of fatigue crack growth: Determination of exponent m in paris law *Trans. Korean Soc. Mech. Eng. A* **36** 713–721

[18] Sun Y and Zhang Q 2010 Mix-mode crack growth intelligence analysis and simulation module development based on UIDL and APDL *Proc. - Int. Conf. Electr. Control Eng. ICECE 2010* 1276–1279
[19] Jensen BE 2015 Numerical Analysis of Crack Propagation and Lifetime Estimation
[20] Zakavi B, Kotousov A, Khanna A and Branco R 2019 A new method for analysis of part-elliptical surface cracks in structures subjected to fatigue loading *Theor. Appl. Fract. Mech.* **103** 102258
[21] Cordeiro SGF and Leonel ED 2018 Mechanical modelling of three-dimensional cracked structural components using the isogeometric dual boundary element method *Appl. Math. Model.* **63** 415–444
[22] Kim Y, Hwang IS and Oh YJ 2018 Unified formulae for evaluating load reduction by change in stiffness of circumferential crack considering general piping systems *Int. J. Press. Vessel. Pip.* **165** 68–80
[23] Dong P 2020 Quantitative weld quality acceptance criteria: An enabler for structural lightweighting and additive manufacturin *Weld. J.* **99** 39S-51S
[24] McFarlane B 2016 Boiler & Pressure Vessel Steel Plate
[25] Li Y, Gong B, Lacidogna G, Carpinteri A and Wang D 2019 An improved crack driving force estimation approach for stress-based engineering critical assessment of reeled pipes *Theor. Appl. Fract. Mech.* **103** 102312
[26] Norrie DH 2007 *A first course in the finite element method*
[27] Khennane A 2013 *Introduction to FEA using Matlab and ABAQUS*
[28] ANSYS Inc. 2015 ANSYS Meshing User's Guide **15317** 724–746
[29] Long XY, Liu K, Jiang C, Xiao Y and Wu SC 2019 Uncertainty propagation method for probabilistic fatigue crack growth life prediction *Theor. Appl. Fract. Mech.* **103** 102268
[30] Jacob A, Mehmanparast A, D'Urzo R and Kelleher J 2019 Experimental and numerical investigation of residual stress effects on fatigue crack growth behaviour of S355 steel weldments *International Journal of Fatigue* **128** 105196
[31] Zong L, Shi G and Wang Y 2015 Experimental investigation and numerical simulation on fatigue crack behavior of bridge steel WNQ570 base metal and butt weld *Constr. Build. Mater.* **77** 419–429

Free vibration studies on skew sandwich plates by FEM

Dhotre Pavan Kumar[1] **and C V Srinivasa**[2]

[1,2] Department of Mechanical Engineering, GM Institute of Technology, Davangere, India.

Abstract. Present study mainly deals with the extraction of fundamental frequencies using simply supported and clamped isotropic and laminated faced skew sandwich plates with the orthotropic core. Available literature values were referred to validate the fundamental frequencies obtained using the finite element method. The effects of aspect ratio, skew angle, a ratio of the length-to total thickness of the sandwich plate, and the ratio of a thickness of the core to a thickness of face sheet on the free vibration of skew sandwich plates were examined. The present method is also used to study the effect of fiber orientation angle and laminate stacking sequence on free vibration of antisymmetric cross-ply laminated skew sandwich plates are also extracted. CQUAD8 type of element gives converged, accurate results than the CQUAD4 element in the present study. The variation of frequencies is directly proportional with the skew angle.

Keywords: Fundamental Frequency, Dimensionless frequency parameter, skew sandwich plates, and skew angle, antisymmetric laminated sandwich plates.

Nomenclature:
a: Plate length (mm)
b: Plate width (mm)
h: Total thickness of plate (mm)
t_c: Core thickness (mm)
t_f: Face sheet thickness (mm)
E: Modulus of Elasticity (GPa)
G: Modulus of Rigidity (GPa)
ρ: Density (kg/m^3)
ν: Poisson's Ratio
α: Skew Angle (Degrees)
K_f: Dimensionless Frequency Parameter
ω: Circular Frequency (rad/sec)
C: Core

1. Introduction

Skew sandwich plates now a day frequently used into numerous areas like aeronautical, automobile, civil engineering, and in most structural applications. In skew sandwich plates, the effect of shear deformation is considerably more as compared to laminated composite skew plates, which was the reason behind the widespread applications such plates. Also skew sandwich plates exhibit less weight, more stiffness, more structural efficiency, and more durable.

Much research was made on sandwich plates on the free vibration behaviour for more than two decades. An expression was derived for natural frequency also conducted experiments to validate analytical results [1]. Skew sandwich plates were considered for static and dynamic analysis using finite element displacement model was carried out [2]. Exact solutions were derived for the problems on honeycomb sandwich structures [3]. The p-Ritz method was worked out for free vibration study on Skew sandwich plates with an orthotropic core and laminated facings [4]. A Higher-order shear deformation theory for free vibration of laminated composite and sandwich plates were presented [5]. The vibration parameters of sandwich plates were predicted by a spline finite strip method [6]. Two new C^0 assumed strain finite element [7], facet shell element [8]. To extract the fundamental frequencies of various plates, Reddy's higher-order theory was used. Natural frequencies, displacement and stress eigenvectors were extracted, a semi-analytical [9] and exact analytical solutions based on the propagator matrix method [10] higher-order mixed theory was developed for laminated and sandwich plates. Refined plate theory was proposed on sandwich plates [11].

Free vibration studies on laminated composite and sandwich shells were made by adopting 2D-higher-order shear deformation theories [12]. Free vibration analysis of plates and sandwich plates was discussed using C^o iso-parametric finite element model [13]. Global-local higher-order theory was developed on free vibration of laminated composite and sandwich plates [14].

Structural dynamic analysis was performed on a skew sandwich plate with laminated composite faces based on the high-order shear deformation plate theory [15]. The third-order zigzag theory was used to produce an improved discrete Kirchhoff quadrilateral element on the composite and sandwich plates for static analysis [16] and the free vibration response [17]. C^1 plate bending elements proved to be more efficient than the C^0 plate bending elements to extract exact natural frequencies of sandwich plates [18]. An investigation conducted on the free vibrations of rectangular sandwich plates by applying Hamilton's principle along with the first-order shear deformation theory [19]. C^0 finite element model used for bending and free vibration studies of laminated soft core skew sandwich plates [20].

Various shear deformation theories [21] were considered for the comparison based on the displacement fields [22]. Free vibration analysis of laminated composite and sandwich plates using trigonometric shear deformation theory [23]. The bending behavior [24] and free vibration response [25] of composite sandwich plates with multi-layered face sheets has been investigated using a four nodded rectangular finite element formulation based on a layer-wise theory. The effect of openings and additional mass on free vibration analysis of laminated composite sandwich skew plates were studied [26].

The present effort concerned with the investigation of fundamental frequencies of skew sandwich plates with an orthotropic core. CQUAD4 and CQUAD8 elements are employed in the analysis. The correctness of the results from the elements is confirmed from literature results. The effects of aspect ratio, skew angle, fiber orientation angle, and boundary conditions on the free vibrations of skew sandwich plates are studied.

2. Convergence and Validation

2.1. Convergence

The geometrical representation of the sandwich plate is as shown in Figure 1. The skewed sandwich plate with global and local coordinate systems is as shown in Figure 2. The displacement boundary conditions cannot be applied directly, due to inclination of displacements to the skew edges. To prevail over this, a local coordinate system (x′, y′) normal and tangential to the skew edges is preferred.

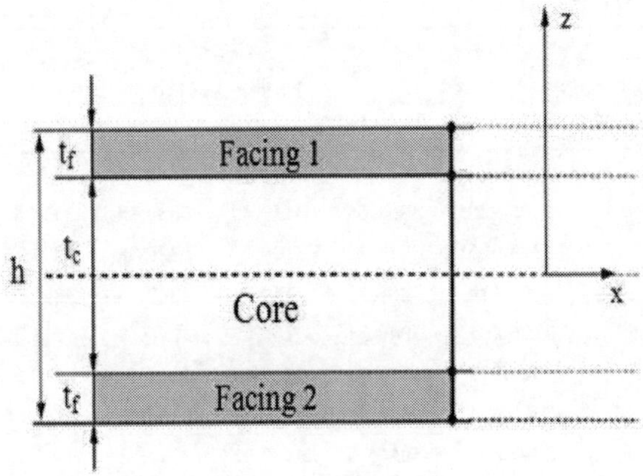

Figure 1. Geometry details of Sandwich Plate

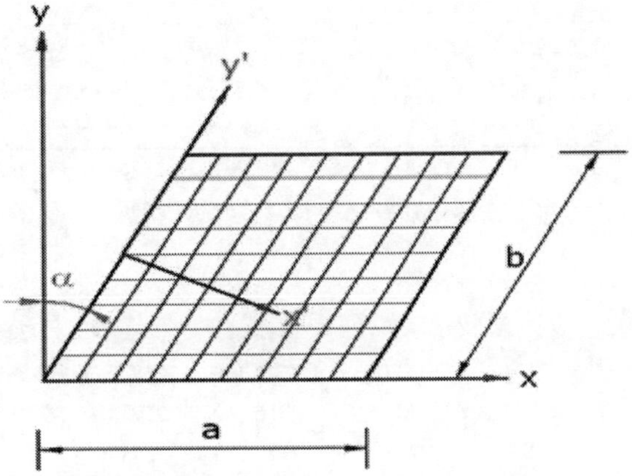

Figure 2. Finite Element Mesh of Skew sandwich plate with a local and global coordinate system.

Total elements in a plate model are optimized to get exact and consistent values. Consequently, it is essential to analyze the convergence of the values. The convergence was made on simply supported and clamped skew sandwich plates having an aspect ratio, length to thickness ratio and the ratio of a thickness of core to facing for skew angles 0°, 15°, 30° and 45° using both elements. The converged detailed results are delivered in Table 1. The material properties used are, for face sheets E= 68.948 GPa, G= 25.924 GPa, ν=0:33, ρ=2768.0 kg/m^3 and for core G_{23}= 0.05171 GPa, G_{13}=0.13445 GPa, ρ=121.83 kg/m^3 [1].

2.2. Validation
Validation of the results from the elements used in the present study is made by matching up to the values for the natural frequency found in the present study to the available literature values. The comparison is shown in Table 2, for a simply supported skew sandwich plate. The material constants employed are similar to those used for the convergence study. The values found in the study are in good harmony with literature results.

In Table 3 the dimensionless frequency parameter (K_f) for skew sandwich plates with graphite-epoxy laminate face sheet and Al honeycomb core are compared with literature results with simply supported edge condition. Face sheets having two layers with anti-symmetric layup sequence are employed. The material properties adopted for face sheet are, E_1= 131 GPa, E_3=3.275 GPa, G_{12} = 3.275 GPa,ν_{12}=0.25 and ρ=1627 kg/m^3 and for core G_{13}=0.1173GPa, G_{23}=0.2415 GPa and ρ=2386 kg/m^3[13]. In Table 3 the results are compared in Dimensionless form using the formula $\mathbf{K_f = 100\omega a\sqrt{(\rho/E_1)_f}}$. *From Table 1 to Table 3 it is observed that CQUAD8 element gives accurate and converge results as than CQUAD4 element. From now CQUAD8 is adopted in the work*

3. Results and discussion

3.1. Sandwich plates with isotropic face sheet and aluminum honeycomb core
The present study considers a variety of parameters such as aspect ratio, ratio of length to thickness of sandwich plates, ration thickness of face sheet to thickness of core, skew angle and boundary conditions.

The results from the numerical methods are presented in the Table 4 for different aspect ratio and skew angle for all sides simply supported and clamped edge condition. The graphical presentation of the results in the Table 4 is as shown in Fig 3 and Fig 4.

Table 1. Convergence for fundamental frequencies (Hz) of simply supported skew sandwich plates ($a/b=1$, $a/h=10$, $t_c/t_f=10$).

Element Density	Element Type	S-S-S-S				C-C-C-C			
		Skew Angle (α)				Skew Angle (α)			
		0^0	15^0	30^0	45^0	0^0	15^0	30^0	45^0
Present (10 x 10)	CQUAD 4	2493.991	2570.462	2827.037	3369.724	3017.56	3081.52	3300.51	3779.85
	CQUAD 8	**2519.097**	**2596.492**	**2856.251**	**3405.989**	**3052.65**	**3117.40**	**3339.14**	**3824.81**
Present (14 x 14)	CQUAD 4	2507.189	2584.146	2842.358	3388.675	3036.29	3100.64	3320.98	3803.53
	CQUAD 8	**2520.018**	**2597.446**	**2857.271**	**3407.130**	**3054.26**	**3119.01**	**3340.76**	**3826.51**
Present (18 x 18)	CQUAD 4	2512.631	2589.787	2848.699	3396.465	3044.04	3108.55	3329.49	3813.30
	CQUAD 8	**2520.395**	**2597.836**	**2857.720**	**3407.607**	**3054.93**	**3119.68**	**3341.47**	**3827.21**
Present (22 x 22)	CQUAD 4	2515.387	2592.644	2851.877	3400.404	3047.97	3112.56	3333.76	3818.26
	CQUAD 8	**2520.586**	**2598.034**	**2857.901**	**3407.851**	**3055.26**	**3120.02**	**3341.78**	**3827.56**
Present (26 x 26)	CQUAD 4	2516.973	2594.288	2853.718	3402.667	3050.23	3114.87	3336.23	3821.11
	CQUAD 8	**2520.696**	**2598.147**	**2858.040**	**3407.993**	**3055.46**	**3120.21**	**3341.98**	**3827.77**
Present (30 x 30)	CQUAD 4	2517.968	2595.319	2854.872	3404.086	3051.65	3116.32	3337.78	3822.89
	CQUAD 8	**2520.765**	**2598.218**	**2858.119**	**3408.082**	**3055.58**	**3120.33**	**3342.10**	**3827.90**
Present (34 x 34)	CQUAD 4	2518.634	2596.009	2855.666	3405.033	3052.60	3117.29	3338.85	3824.09
	CQUAD 8	**2520.811**	**2598.266**	**2858.193**	**3408.143**	**3055.66**	**3120.42**	**3342.21**	**3827.98**
Present (38 x 38)	CQUAD 4	2519.100	2596.492	2856.207	3405.698	3053.27	3117.97	3339.58	3824.93
	CQUAD 8	**2520.843**	**2598.299**	**2858.230**	**3408.186**	**3055.72**	**3120.47**	**3342.27**	**3828.04**

Table 2. Fundamental frequencies (Hz) of skew sandwich plates with simply supported

Authors		Skew Angle (α)			
		0^0	15^0	30^0	45^0
	Raville [1]	23.0000	-	-	-
	Zhou [27]	23.2900	-	-	-
	Bardell [28]	23.0500	-	-	-
	Yuan [6]	23.4100	-	-	-
R. Khare [8]	FOST	23.5994	-	-	-
	HOST	23.4841	-	-	-
	Araújo [30]	23.5000	-	-	-
	Ferreira [29]	23.2600	-	-	-
	Bo Liu [31]	23.2591	-	-	-
	Shenoi [7]	23.6300	-	-	-
		23.4200	-	-	-
	HOST12	23.4258	-	-	-
A. K. Garg [12]	HOST11	23.4268	-	-	-
	HOST9	23.4265	-	-	-
	Voyiadjis [15]	23.3190	24.8550	29.8300	41.4260
A. K. Garg [13]	FOST	23.5279	24.9315	30.0623	43.1677
	HOST	23.4514	24.8438	29.9153	42.7936
Present	**CQUAD4**	**23.0704**	**24.4170**	**29.1940**	**40.9127**
	CQUAD8	**23.0600**	**24.4063**	**29.1799**	**40.9194**

Table 3. Dimensionless frequency parameter (K_f) for simply supported graphite epoxy laminated composite skew sandwich plates with orthotropic core.

AUTHOURS	Skew Angle (α)

		0^0	15^0	30^0	45^0
		9.8130	-	-	-
	ZIGT FE	9.8200	-	-	-
Kulkarni and Kapuria [17]		9.8240	-	-	-
	3D ZIGT	9.8281	-	-	-
	ZIGT	9.8300	-	-	-
		10.0510	-	-	-
Chakrabarti and Sheikh [11]		10.0520	-	-	-
		10.0530	-	-	-
Wang [4]	p-Ritz	10.5550	-	-	-
Kulkarni and Kapuria [17]	TOT	12.0880	-	-	-
Chalak and Chakrabarti [20]	HOZIGT	9.8365	10.2467	11.6056	14.4349
Present	**CQUAD4**	**9.2704**	**9.5546**	**10.5191**	**12.5896**
	CQUAD8	**9.2768**	**9.5612**	**10.5266**	**12.5988**

From the results, keeping the ratio of a/h and t_c/t_f as constant, the following observations were made.

- With increment in the aspect ratio, fundamental frequency diminishes for a given skew angle.
- With increment in the skew angle, fundamental frequency increases for a given aspect ratio.
- The value of natural frequency is high for all sides' clamped edge as compared to the simply supported.

Table 4. Fundamental frequencies [Hz] of skew sandwich plates (a/h=10, t_c/t_f=10)

a/b	All Sides Simply Supported, Angle of Skew (α)				All Sides Clamped, Angle of Skew (α)			
	0	15^0	30^0	45^0	0	15^0	30^0	45^0
1	2532.05	2592.03	2794.67	3232.52	2797.83	2850.8	3032.62	3436.39
1.5	1509.04	2200.22	2269.83	2685.75	1795.36	2408.62	2556.18	2862.47
2	1010.33	1969.91	2108.33	2462.06	1283.64	2145.97	2398.76	2614.63
2.5	724.62	1624.88	1859.73	2345.92	972.97	1787.48	2012.27	2482.03

The results showing the effect of ratio of (t_c/t_f) of the sandwich skew plate on the natural frequency are presented in the Fig 5 and Fig 6. The aspect ratio (a/b=1) and ratio of (a/h=10) of the sandwich skew plate kept constant. From the figures the following observations were made,

- With increment in the ratio of (t_c/t_f), fundamental frequency increases for a given skew angle.
- With increment in the skew angle, fundamental frequency increases for a given ratio of (t_c/t_f).
- The value of natural frequency is high for all sides' clamped edge as compared to the simply supported.

Fig 7 and Fig 8 shows the results of effect of ratio of (a/h) of the skew sandwich plate on natural frequency. The aspect ratio (a/b=1) and ratio of (t_c/t_f=10) of the sandwich skew plate kept constant. From the figures the following observations were made,

- With increment in the proportion of (a/h), fundamental frequency diminishes for a given skew angle.
- With increment in the skew angle, fundamental frequency increases for a given ratio of (a/h).
- The value of natural frequency is high for all sides' clamped edge as compared to the simply supported.

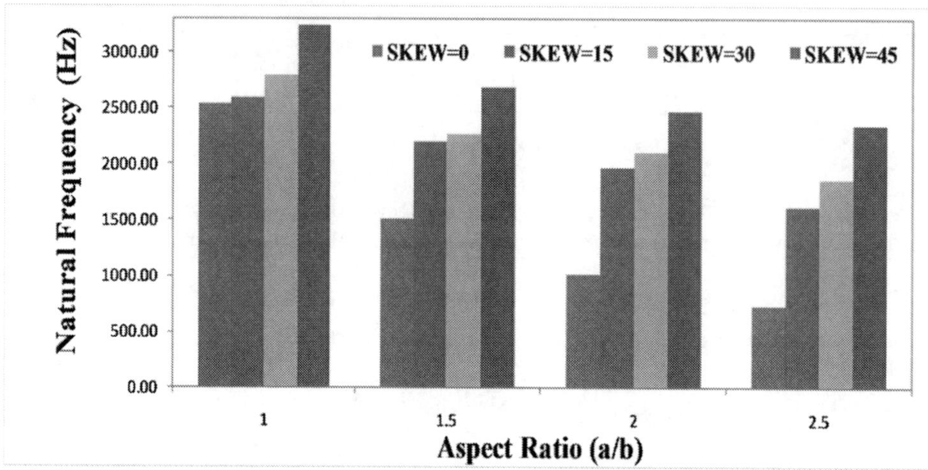

Figure 3. Fundamental Frequency v/s Aspect ratio for simply supported skew sandwich plates.

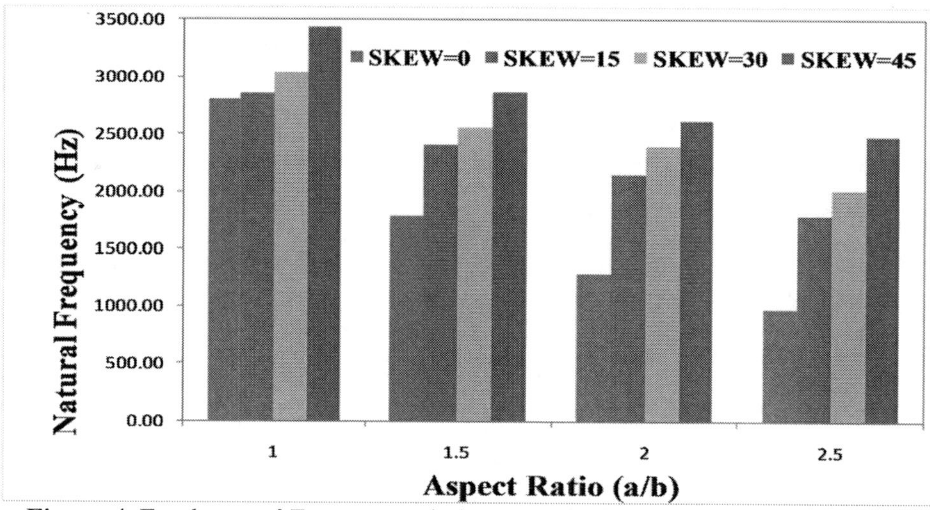

Figure 4. Fundamental Frequency v/s Aspect ratio for clamped skew sandwich plates.

3.2. Sandwich plates with orthotropic face sheet and honeycomb core

The numerical study is made for sandwich skew plates with orthotropic face sheet and aluminium honeycomb core. The material constants employed are, face sheet, E_1=206.84 GPa, E_2= E_3= 5.1711 GPa, ν_{12}=0.25, ν_{23}=0.3 and ρ=1603.1 kg/m^3. And for core, E= 627.9 MPa G_{13}=117.21 MPa, G_{23}=241.32 MPa, ρ= 2351.2 kg/m^3.The results are presented in a Dimensionless form using the formula$\mathbf{K_f} = \mathbf{100\omega a\sqrt{(\rho/E_1)_f}}$. The face sheet is made-up of two layers anti-symmetric laminate with core in the centre. Dimensionless frequency parameters for different aspect ratio and boundary conditions are obtained by varying the skew angle. The values found are as shown in Table 5. The graphical representation of the same is made in Fig 9 and Fig 10.

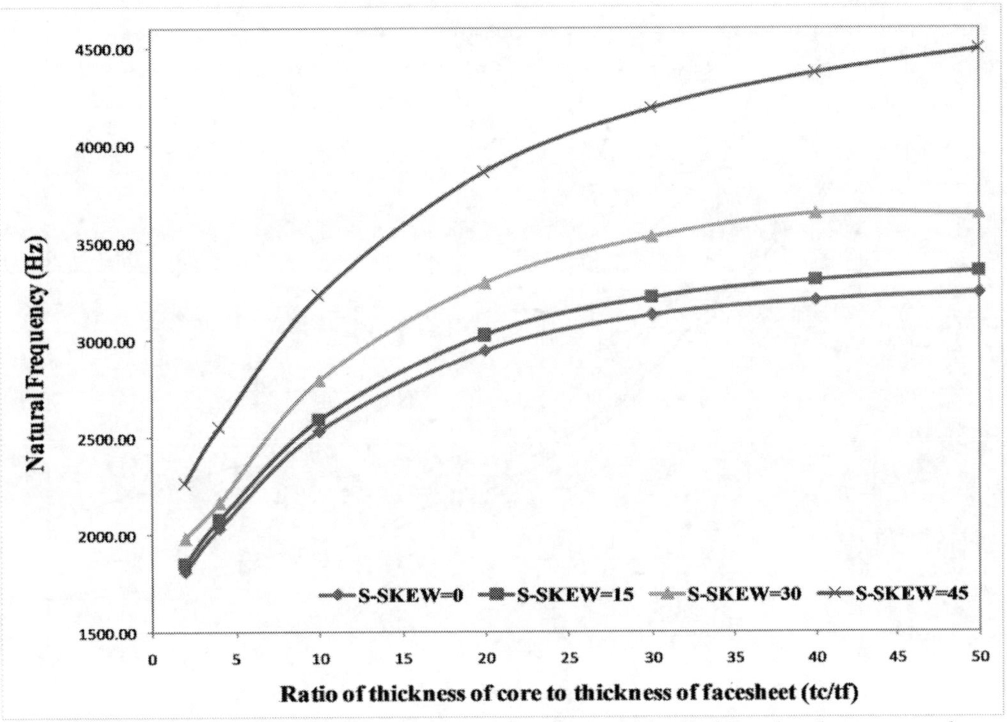

Figure 5. Fundamental Frequency v/s ratio of t_c/t_f with all sides simply supported.

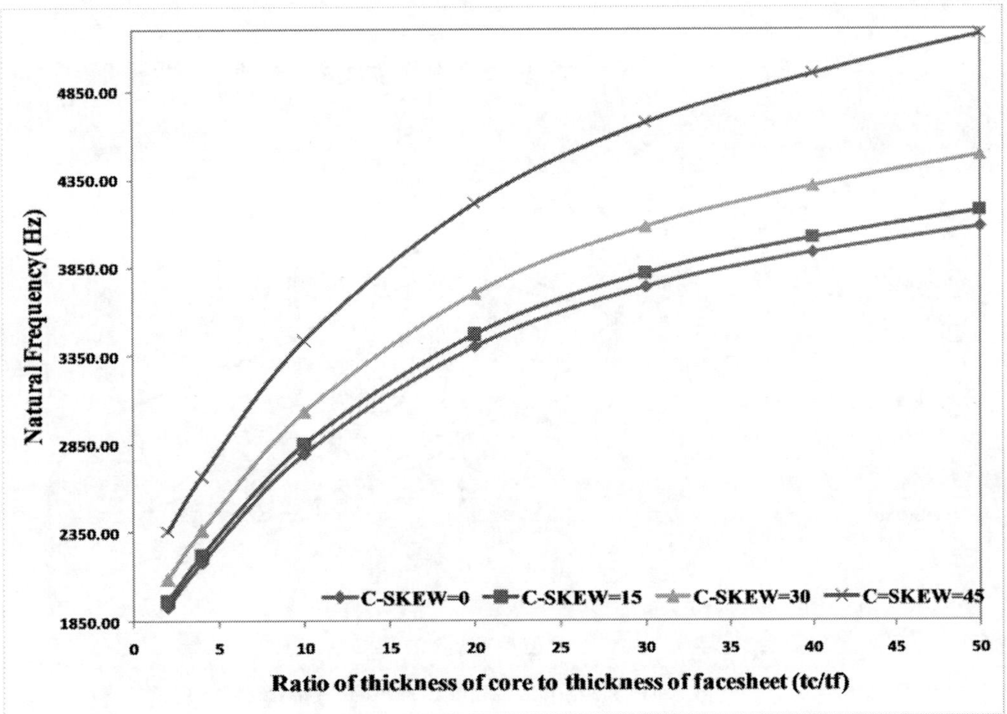

Figure 6. Fundamental Frequency v/s ratio of t_c/t_f with all sides clamped

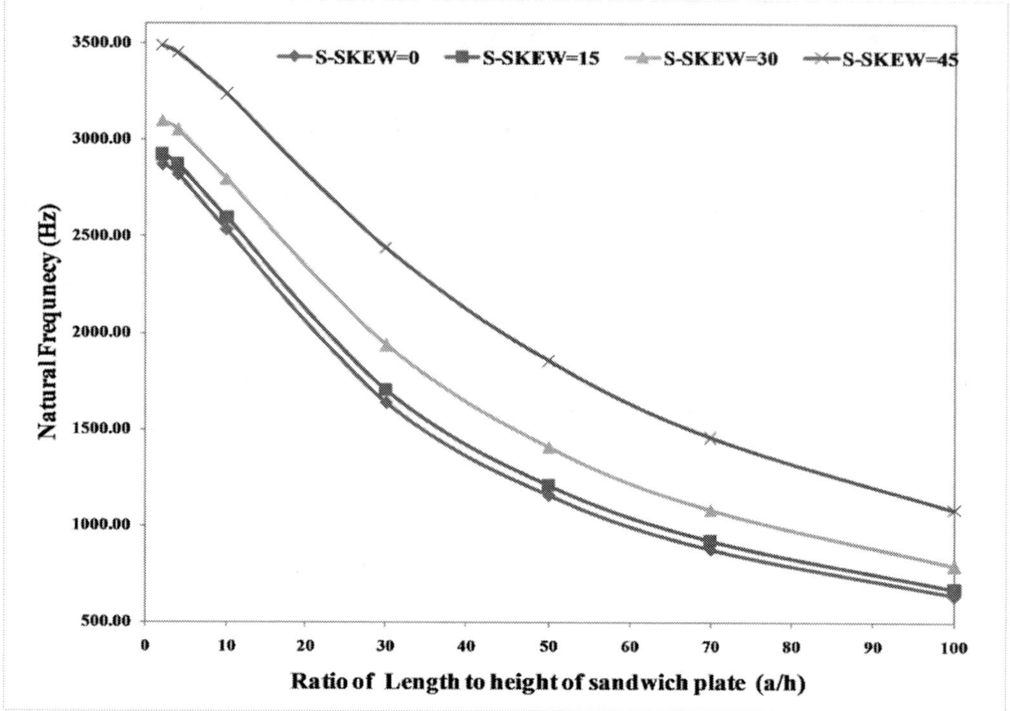

Figure 7. Fundamental Frequency v/s ratio of a/h of skew sandwich plate for all sides simply supported

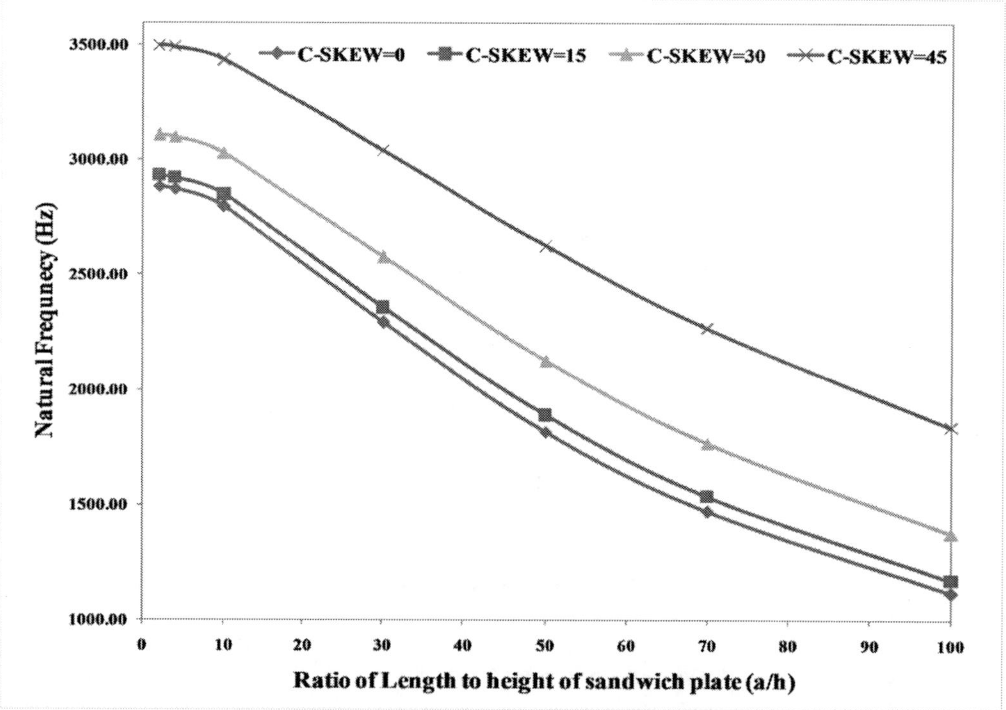

Figure 8. Fundamental Frequency v/s ratio of a/h of skew sandwich plate for all sides clamped edges

IOP Conf. Series: Materials Science and Engineering **925** (2020) 012024 doi:10.1088/1757-899X/925/1/012024

Table 5. Dimensionless frequency parameters of laminated antisymmetric skew sandwich plates ($0^0/90^0/C/0^0/90^0$) ($a/h=10$, $t_c/t_f=10$)

a/b	All Sides Simply Supported Angle of Skew (α)				All Sides Clamped Angle of Skew (α)			
	0^0	15^0	30^0	45^0	0^0	15^0	30^0	45^0
1	9.1174	9.3786	10.2724	12.2037	10.6262	10.8547	11.6391	13.3712
1.5	11.842	12.0111	13.0143	15.7942	14.2666	14.3657	15.1059	17.4253
2	14.8035	14.7184	15.7272	19.4847	18.1454	17.8786	18.3954	21.4101
2.5	17.0208	17.3233	18.4786	23.2834	19.9568	21.2636	21.5831	25.4219

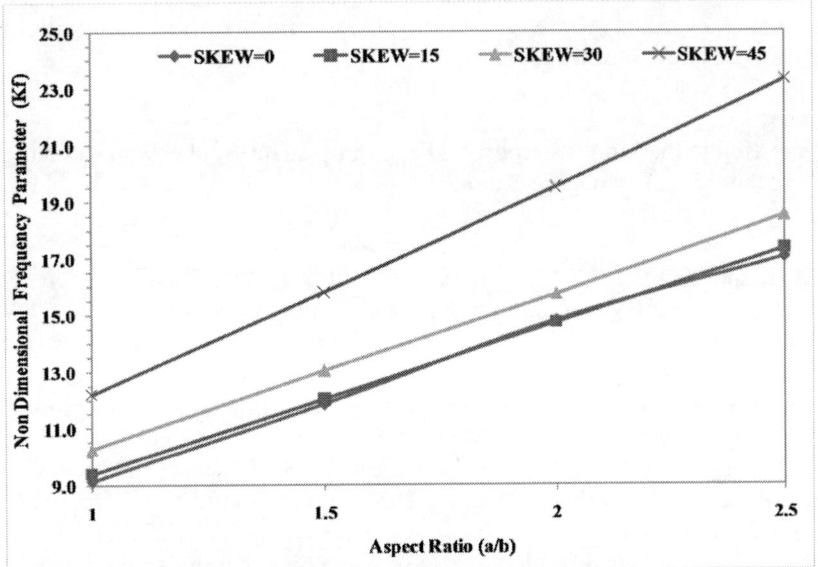

Figure 9. Dimensionless frequency Parameter v/s Aspect ratio for simply supported anti-symmetric cross ply Graphite/Epoxy laminated skew sandwich plates.

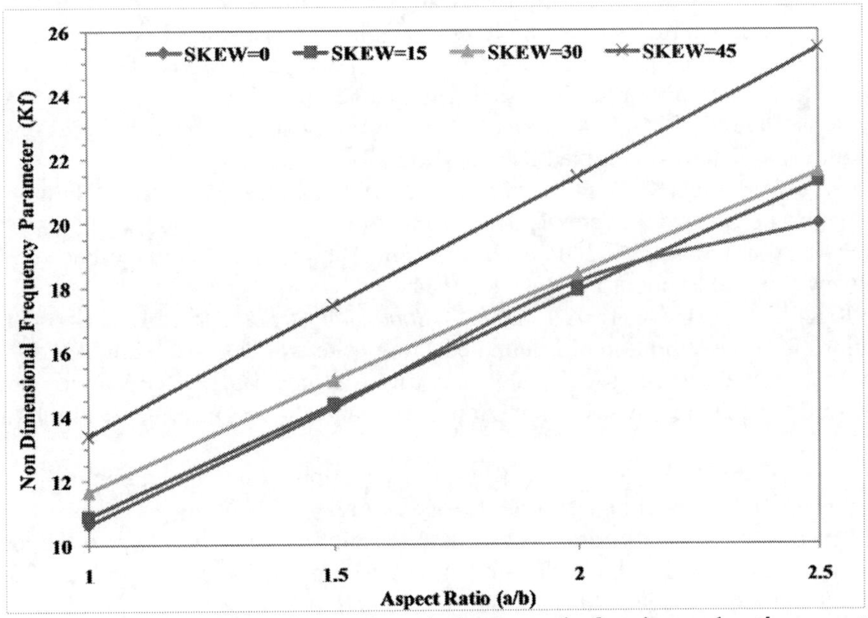

Figure 10. Dimensionless frequency Parameter v/s Aspect ratio for clamped anti-symmetric cross ply Graphite/Epoxy laminated skew sandwich plates.

4. Conclusions

The following points are noticed after carrying out numerical study.

- Both CQUAD4 and CQUAD8 elements have good agreement with the available literature results. But CQUAD8 element yields more accurate results. The same is true for both isotropic and orthotropic skew sandwich plates.
- With increment in aspect ratio, fundamental frequency values diminish in all the cases.
- As skew angle increased the fundamental frequency value also increases in all the cases.
- The ratio (t_c/t_f) has major influence till $t_c/t_f=20$ for simply supported and $t_c/t_f=40$ for clamped boundary condition, there after the influence is negligible.

Natural frequency values decrease considerably for in the influence of ratio of length to total thickness of the skew sandwich plates.

Acknowledgement
The authors acknowledge the kindly support and encouragement by the Management and Principal of GM Institute of Technology, Davangere-Karnataka.

References

[1] Raville M E and Ueng C E S May 16-19 1967 Ottawa *SESA Spring Meeting* Determination of Natural Frequencies of Vibration of a Sandwich Plate pp 490-493
[2] Ng S S F and Lam D K Y 1985 *J. Sound Vib* Dynamic And Static Analysis Of Skew Sandwich Plates **99**(3) pp 393-401
[3] Holt P J and Webber J P H 1982 *J Strain Anal* Exact Solutions To Some Honeycomb Sandwich Beam, Plate, And Shell Problems **17**(1) pp 1-8
[4] Wang C M, Ang K K, Yang L and Watanabe E 2000 *J Sound Vib* Free Vibration Of Skew Sandwich Plates With Laminated Facings **235**(2) pp 317-340
[5] Kant T and Swaminathan K 2001 *Compos Struct* Analytical Solutions for free vibration of laminated composite and sandwich plates based on a higher-order refined theory **53** pp 73-85
[6] Yuan W X and Dawe D J 2002 *Int. J. Numer. Methods Eng* Free vibration of sandwich plates with Laminated faces **54** pp 195-217
[7] Nayak A K, Moy S S J and Shenoi R A 2002 *Composites, Part B* Free vibration analysis of composite sandwich plates based on Reddy's higher-order theory **33** pp 505-519
[8] Khare R K, Kant T and Garg A K, 2004 *Compos Struct* Free vibration of composite and sandwich laminates with a higher-order facet shell element **65** pp 405-418
[9] Rao M K and Desai Y M 2004 *Compos Struct* Analytical solutions for vibrations of laminated and sandwich plates using mixed theory **63** pp 361-373
[10] Rao M K, Scherbatiuk K, Desai Y M and Shah A H 2004 *J. Eng. Mech* Natural Vibrations of Laminated and Sandwich Plates **130** pp 1268-1278
[11] Chakrabarti A and Shah A H 2004 *J. Aerosp. Eng* Vibration of Laminate-Faced Sandwich Plate by a New Refined Element **17**(3) pp 123-134
[12] Khare R K, Kant T and Garg A K 2006 *J. Sandwich Struct. Mater* Higher-order Closed-form Solutions for Free Vibration of Laminated Composite and Sandwich Shells **8** pp 205-235
[13] Khare R K, Kant T and Garg A K 2006 *J. Sandwich Struct. Mater* Free Vibration of Skew Fiber-reinforced Composite and Sandwich Laminates using a Shear Deformable Finite Element Model **8** pp 33-55
[14] Zhen W and Wanji C 2006 *J. Sound Vib* Free vibration of laminated composite and sandwich plates using global–local higher-order theory **298** pp 333-349
[15] Park T, Lee S Y, Seo J W, and Voyiadjis G Z 2008 *Composites Part B* Structural dynamic behavior of skew sandwich plates with laminated composite faces **39** pp 316-326
[16] Kapuria S and Kulkarni S D 2007 *Int. J. Numer. Methods. Eng* An improved discrete Kirchhoff quadrilateral element based on third-order zigzag theory for static analysis of composite and sandwich plates **69** pp 1948-1981

[17] Kapuria S and Kulkarni S D 2008 *Comput. Mech.*Free vibration analysis of composite and sandwich plates using an improved discrete Kirchhoff quadrilateral element based on third-order zigzag theory **42** pp 803-824
[18] Zhen W, Xiaohui R and Wanji C 2010 *Compos. Struct.* An accurate higher-order theory and C^0 finite element for free vibration analysis of laminated composite and sandwich plates **92** pp 1299-1307
[19] Loptin A V and Morozov E V 2011 *Compos. Struct* Fundamental frequency and design of the CFCF composite sandwich plate **93** pp 983-991
[20] Chalak H D, Chakrabarti A, Sheikh A H and Iqbal M A 2014 *Appl. Math. Modell.* C^0 FE model based on HOZT for the analysis of laminated soft core skew sandwich plates: Bending and vibration **38** pp 1211-1223
[21] Pavan Kumar and C V Srinivasa 2018 *J. Thermoplast. Compos. Mater.* On buckling and free vibration studies of sandwich plates and cylindrical shells: A review **33** pp 673-724
[22] Sayyad A S and Ghugal Y M 2015 *Compos. Struct.* On the free vibration analysis of laminated composite and sandwich plates: A review of recent literature with some numerical results **129** pp 177-201
[23] Sayyad A S and Ghugal Y M 2016 *J. Sandwich Struct. Mater.*On the free vibration of angle-ply laminated composite and soft core sandwich plates pp 1-33
[24] Belarbi M O, Tati A, Ounis H and Benchabane A 2016 *Struct. Eng. Mech.* Development of a 2D isoparametric finite element model based on the layerwise approach for the bending analysis of sandwich plates **57**(3) pp 473-506
[25] Belarbi M O, Tati A, Ounis H and Khechai A 2017 *Lat. Am. J. Solids Struct.* On the Free Vibration Analysis of Laminated Composite and Sandwich Plates: A Layerwise Finite Element Formulation **14** pp 2265-2290
[26] Anish, Kumar A and Chakrabarti A 2020 *J. Thermoplast. Compos. Mater* Influence of openings and additional mass on vibration of laminated sandwich rhombic plates using IHSDT **33**(1) pp 3-34
[27] Zhou H B, Li G Y 1996 *Comput. Struct.* Free vibration analysis of sandwich plates with laminated faces using spline finite point method **59** pp 257–263
[28] Bardell N S, Dunsdon J M and Langley R S 1997 *Compos. Struct.* Free vibration analysis of coplanar sandwich panels **38** pp 463–475
[29] Ferreira A J M, Araújo A L, Neves A M A, Rodrigues J D, Carrera E, Cinefra M, et al. 2013 *Compos. Part B Eng.* A finite element model using a unified formulation for the analysis of viscoelastic sandwich laminates **45**(1) pp 1258-1264
[30] Araújo A, Mota Soares C, Mota S C 2010 *Appl. Compos. Mater.* A viscoelastic sandwich finite element model for the analysis of passive, active and hybrid structures **17**(5) pp 529-542
[31] Liu B, Zhao L, Ferreira A J M, Xing Y F, Neves A M A, Wang J 2017 *Compos. Part B* Analysis of viscoelastic sandwich laminates using a unified formulation and a differential quadrature hierarchical finite element method **110** pp 185-192

IOP Conf. Series: Materials Science and Engineering **925** (2020) 012025 doi:10.1088/1757-899X/925/1/012025

Multiple Object detection using Autonomous UAVs based on YOLOv3 algorithm

Murugendrappa N[1], KE Gayathri[1], Harshitha GM[1] and Sony Gupta[1]

[1] Department of Electronics and Communication, GM Institute of Technology, Davangere, Karnataka, India

Abstract. Currently, there has been an enormous increase in interest towards autonomous UAVs and applications with respect to autonomous UAVs such as scientific data collection and security monitoring, infrastructure inspection, search and rescue and much more. Recognizing objects visually is a key step in such applications and is very pivotal to build a complete autonomous system. Nevertheless, the chore of detecting an object is very much challenging and is even more difficult when we have low quality images aboard low cost consumer UAVs. More frequently due to the motion of UAVs, images get blurred, noisy and even onboard cameras have low resolution hence object detection becomes more challenging since identifying objects are quite small and these tasks get even more difficult as there is a necessity of real time detection of objects. This paper focuses on all these aspects and brings an outcome for the object detection and identification.

Key words: Convolution Neural Networks (CNNs), Graphics Processing Units (GPUs), Regions with Convolutional Neural Network (RCNN), Unmanned Aerial Vehicles (UAVs), You Only Look Once (YOLO).

1. Introduction

A number of UAV researches have been approached for the detection and identification of certain sorts of objects like people, landmarks, vehicles and pedestrians for autonomous navigation along with landing in real-time. Out of numerous methodologies, there are only a few that consider recognizing multiple objects at a time, in spite of the very fact that identifying several target objects is clearly a key action for many applications of UAVs.

In our view, the main reasons for the space between technical capabilities and application needs are the below mentioned practical but critical limitations:

(1) Visual perception algorithms often got to be hand-tuned to particular object and context types.

(2) It difficult to create and store a spread of target object models, especially when the objects are diverse in appearance.

(3) Real-time object detection demands high computing power even while detecting single objects, much less when many target objects are involved.
However, in several fields, there is a necessity for object detection and tracking effectively while handling occlusions and other complexities. Numerous researches tried for various approaches in detecting an object which largely depends on the application domain. Few research works that made evolution for the proposed work are depicted as follows.

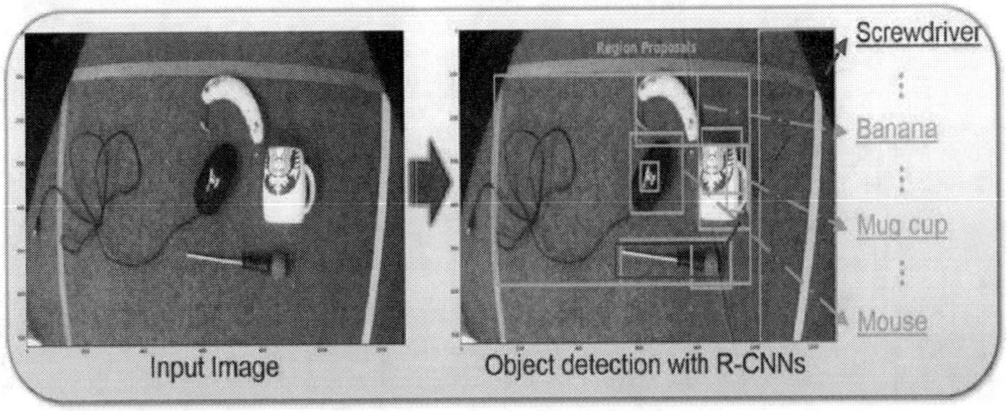

Figure 1. Detection of varied objects using a UAV (drone)

Object detection and recognition using CNN is deliberately demanding, it requires high-end Graphics Processing Units (GPUs) which results in utilizing a lot of power and weight precisely for a low-cost and lightweight drone. In this paper, they propose moving the computation to an off-board computing cloud. They apply Regions with CNNs (R-CNNs), a state-of-the-art algorithm, to detect not one or two but hundreds of object types in near real-time (1). And the UAVs which are popping up as a promising technology for environmental as well as infrastructure monitoring it requires the use of computer vision algorithms to examine the information captured from an on-board camera with the evolution of a single-shot object detector grounded with images which can operate between 5-18 frames-per-second for different platforms with an overall accuracy is taken by UAVs(2), which only recognizes an object to a particular class; object detection also must extract accurate locations of objects. Within the state-of-the-art object detection algorithms, several methods to enhance its performance are proposed. Some unplanned experiments are conducted to prove the effectiveness of the network. Also, applying the network as an auxiliary module to the faster R-CNN algorithm and testing them (3), you only look once (YOLO) at the given image to predict the objects that are present and their location. It predicts multiple bounding boxes and class probabilities for the predicted boxes. YOLO achieves more than twice the mean average precision of other real-time systems (4). Remediate the network structure of YOLO v2 and obtain the YOLO-R network to improve the accuracy of detection; one can intend to integrate more pedestrian context features (5). And this method is more than 6 times faster than the Faster R-CNN at the same detection rate and quantitatively analyzes the relationship between sample scale and detection performance under small scale datasets(6). Computer Vision consists of diversified aspects like image recognition; every object present in the particular image is highlighted with rectangular boxes and tags are assigned to the identified object also with the accuracy of each method used for identifying objects (7).
Two technologies have empowered major tasks like object detection and tracking for traffic vigilance systems due to the increased demand in the features in image for efficient algorithms and detection and tracking of urban vehicle dataset.

2. Object detection model yolov3

YOLOV3 is one of the widely used methods for object detection. Basically, it is the algorithm that explains the working of the code with respect to detecting the objects in the image. Previously, object detection frameworks looked at varied parts of the respective image multiple numbers of times at different scales and processed image classification techniques in order to detect objects. This approach is time consuming and inefficient. It has its own architecture based on CNN and anchor boxes. YOLO in general is an improvement over a region based object detection algorithm. Region based object detection algorithms are much slower and have errors whereas YOLO mainly focuses on speed.

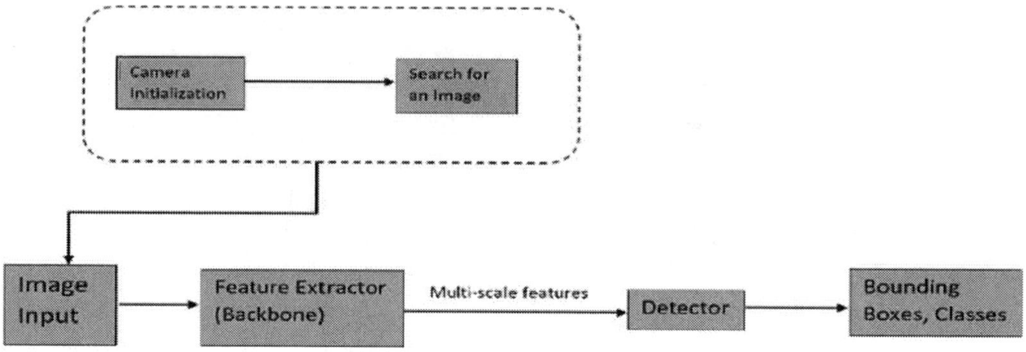

Figure 2. Block Diagram

The above mentioned Figure 2 depicts the block diagram and the steps involved in this process are explained below:

The very first step is camera initialization in the drone.

This in turn, searches for an image in order to input the image to the object detection algorithm.

Feature extraction is a framework that decreases the dimensionality of the given data by removing the redundant data. This in turn increases the accuracy in extracting the features.

YOLOv3 provides an input image to which the output is a particular vector that consists of the class probabilities and the coordinates of the bounding boxes. This is done using a sliding window approach with the stride of zero that gives no overlaps. The whole image gets divided into a grid of particular dimensions. The output vector will have *Pc* which gives the probability of the object being present in that particular image, C_1, C_2...C_n gives the probability of the class that a particular object may belong to, x and y coordinates of the particular image and the height (h) and width (w) of the bounding box.

For multiple objects being present in a single grid, the algorithm uses anchor boxes. Ideally, any number of anchor boxes can be used by the algorithm. With the anchor boxes the output vector will also get modified for a particular grid cell that is the parameters in each output vector that gets multiplied into the number of anchor boxes it has. The YOLOv3 algorithm uses non-maximal suppression to have only one bounding box for each object. The bounding boxes with the confidence less than the threshold will be removed with the non-maximal suppression technique.

3. Result analysis

This is a sample image that is fed to the algorithm and the algorithm is expected to detect and identify the objects present in the image and label them with respect to the particular class assigned. After identification of the objects, each object is assigned with a tag which has the details of the object.

IOP Conf. Series: Materials Science and Engineering **925** (2020) 012025 doi:10.1088/1757-899X/925/1/012025

Figure 3. Image before object detection

Figure 4. Image after object detection

Figure 5. Aerial image before detection

Figure 6. Aerial image after object detection

Figure 7. Real time object detection with class probability and bounding box from a video clip

Figure 8. Image before object detection

Figure 9. Image after object detection

4. Conclusion

This paper explains about various models used in object detection and recognition techniques. Along with, feature descriptors, bounding boxes and segmentation methods based on an image, a video or live stream in python. We have analyzed high accuracy methods to identify the object of same and different classes along with its probability using CNN for feature extraction and pre-defined weights to transform the input data. The result obtained is marked using a bounding box and a tag is assigned to the respective object with its class and probability.

References

[1] Lipeng S, Zhao S, Li G and Liu B 2019 High accuracy object detection via bounding box regression network *Frontiers of Optoelectronics* **12** 324-31
[2] Lan W, Dang J, Wang Y and Wang S 2018 Pedestrian Detection Based on YOLO Network Model *IEEE International Conference on Mechatronics and Automation* Changchun, China, 1
[3] Li G, Song Z and Fu Q 2018"A New Method of ImageDetection for Small Datasets under the Framework of YOLO Network *3rd Advanced Information Technology, Electronic and Automation Control Conference (IEAEC)* 1031-1035
[4] Sri Sai MR, Rella S, Veeravalli V 2019 Object Detection and Identification A Project Report Thesis 14
[5] Mohana and Ravish Aradhya HV 2019 Object Detection and Tracking using Deep Learning and Artificial Intelligence for Video Surveillance Applications *Int. Journal of Advanced Computer Science and Applications (IJACSA)* **10** 4-8
[6] Felicita JJ and Abinaya D 2019 An Efficient Object Detection Model Using Convolution Neural Network *Proc. of the 3rd Int. Conf. on Trends in Electronics and Informatics (ICOEI)* 142-147
[7] Geethapriya S, Duraimurugan N and Chokkalingam SP 2019 Real-Time Object Detection with Yolo *International Journal of Engineering and Advanced Technology (IJEAT)* **8** 578-581
[8] Huang L, Yang Y, Deng Y, Yu YD 2015 Unifying Landmark Localization with End to End Object Detection *arXiv preprint arXiv:1509.04874*
[9] Aradhya HVR 2016 Elegant and efficient algorithms for real time object detection, counting and classification for video surveillance applications from single fixed camera *International Conference on Circuits, Controls, Communications and Computing (I4C)*, Bangalore 1-7
[10] Van Etten A 2018 You Only Look Twice: Rapid multi-scale object detection in satellite imagery *in ArXiv preprint, ArXiv:1805.09512*

Human Action recognition using STIP Evaluation techniques

Latha BM[1], Manjula BK[2], Venkata Sumana CH[3] and Hemalatha KL[4]

[1,2,3] Department of Electronics and Communication Engineering GM Institute of Technology Davangere, India
[4] Department of Information Science, Sri Krishna Institute of Technology, Chikkabanavara, Karnataka, India
[1]lathabm@gmit.ac.in, [2]manjulabk@gmit.ac.in, [3]sumanavch@gmit.ac.in, [4]hema.skit@gmail.com

Abstract. Human action recognition plays a big role in human-to-human interaction and interpersonal relations. As a result, it provides information regarding the identity of someone, their temperament, and condition, it's tough to extract. Recognition of actions within the video isn't a matter for human sensory system. The identification of the actions of the person by the system wants someone special mechanisms. The identification of the actions by the systems are going to be useful in computer vision method. This method is divided into low level action recognition process and high level recognition process. Recognizing the actions victimization the one feature values extracted goes under low level action recognition method. These method is simple to implement and that they don't seem to be reliable all the time. The high level action recognition method needs some special hardware's to discover the actions within the video. The experimental results proved that the proposed methodology achieved better performance in terms of accuracy, sensitivity and specificity.

Keywords: Human Action Recognition; Human Sensory System; Interpersonal relations; Low-level and High-level Action Recognition; Video.

1. Introduction

In recent decades, the human ability to acknowledge another person's activities is one among the most subjects of study of the scientific areas of computer vision and machine learning [1]. As a results, the human action recognition is applied in many applications; human-computer interaction, video surveillance systems and robotics. Where, these applications require a multiple action recognition system. In image and video analysis, human action recognition is a very important analysis direction [2,3]. Within the past a few years, a large range of papers are printed on human action recognition in video and image sequences. Information could also be lost once either intensity or chromatic representations are thought-about in isolation [4]. The method victimization motion trajectories involve following and dense multi-scale optical flow computation, the associated process quality is often over that of STIP-based approaches. Here, we offer a comprehensive survey of the recent development of the

techniques, as well as ways, systems and quantitative analysis of the performance of human action recognition [5].
The videos containing actions performed by persons were collected. The videos were converted into frames and also the frames were pre-processed. Pre-processing is completed by applying median filter [6]. The median filter finds the noises within the frame and replaces the noise by replace the element victimization the median of the neighbor pixels. Options were extracted from the frames. The extracted options are going to be accustomed acknowledge the action of the person within the video [7]. Harris SPIT, Dennis Gabor SPIT and HOG SPIT were used to extract the feature1 values from the video frames. The options were then classified victimization SVM classifier supported the kernel operate [8]. The action is recognized supported the label came back by the classifier. The recognition of actions in the video is helpful in computer vision. This may be useful to develop a system which will acknowledge the actions in the surroundings [9]. This paper has scope in varied fields such as a spread of systems that involve interactions between persons and electronic devices like human-computer interfaces, surveillance systems, patient monitoring systems, Military purpose and Sports. The performance of the planned technique is obtained by activity the performance of the classifier [10, 23]. The Accuracy, Sensitivity and Specificity of the classifier is that the most ordinarily lived performance measure for the classifier. The experimental results show that our technique will considerably improve classification, interpretation and retrieval performance for the video pictures.
Section 2 discuss the studies of existing techniques for human action recognition. Section 3 describes the methodology for each module, where Section 4 provides the validation of proposed methodology with existing studies. Finally, the conclusion of the research study is presented in Section 5.

2. Literature Survey

T. Kalsum, et al, [19] used hybrid features for recognizing the human emotions from the facial expression. In this study a combination of features was utilized for feature extraction like bag of features, speeded up robust transform and scale invariant feature transform. These hybrid features improve the ability of facial expression recognition. While utilizing more number of features, the complexity of the system will increase. Further, Z. Gao, et al, [20] used a new natural inspired algorithm; Adaptive Fusion and Category-level Dictionary Learning (AFCDL) for human action recognition. In this paper, the developed AFCDL model performance was assessed on UCLA dataset. The simulation outcome showed that the developed model achieved 87.8% of accuracy. Still, the undertaken AFCDL algorithm need to focus on feature extraction to achieve better result in action recognition.
M. Liu, et al, [21] developed a new algorithm for human action recognition. Here, the motion and visual enhancement approaches were used on color images to improve the local patterns. Then, convolutional neural network was utilized for extracting the discriminative and robust feature vectors from the color images. The extensive experiment showed that the developed model attained 92.61% of recognition accuracy on UCLA dataset. Additionally, D.K. Vishwakarma, [22] extracted the decisive pose using two fold transformations (ridgelet and Gabor wavelet transform) for better human action recognition. However, these existing models showed low performance in the conditions like poor lighting, head pose variations, etc.

3. Methodology

3.1. Flow Chart

Initially, flow diagram of the proposed model is shown in Figure 1. To get rid of the noise from the videos, the input video frames are preprocessed by median filtering approach. The proposed model performance is improved by reducing the noise within the video. Many forms of noise are present in frames\images. The commonly obtained noise is salt and pepper noise, where it occurs in the white and black pixels.

At first, the unwanted pixels are removed from the video frames by applying median filter. In this research work, the noised pixel in the image is detected by median filter. The identified noisy pixel is replaced by the median value of the neighboring pixel [11]. Figure 2 represents the videos are converted into frames and the Figure 3 states the denoised image.

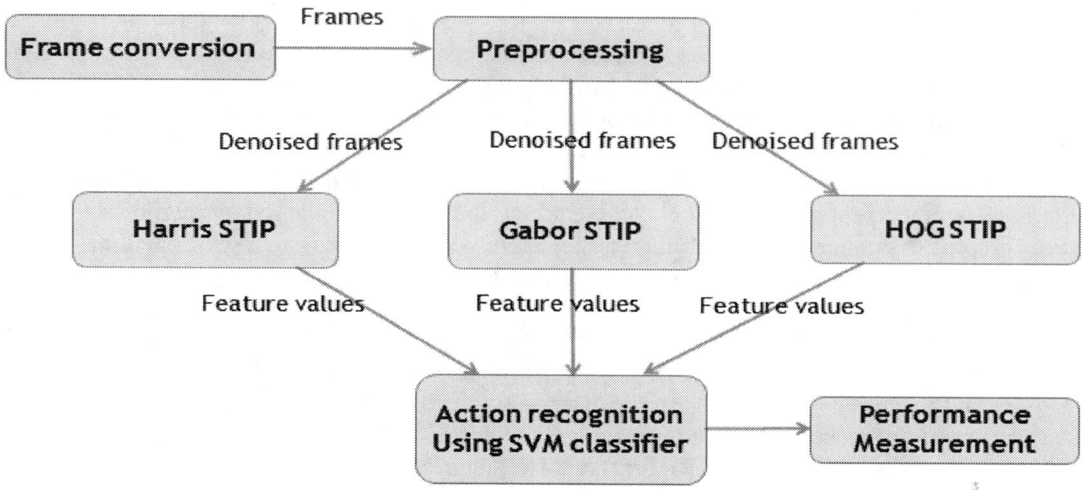

Figure 1. Flow diagram of the proposed model

Figure 2. Graphical illustration of converting video into frames.

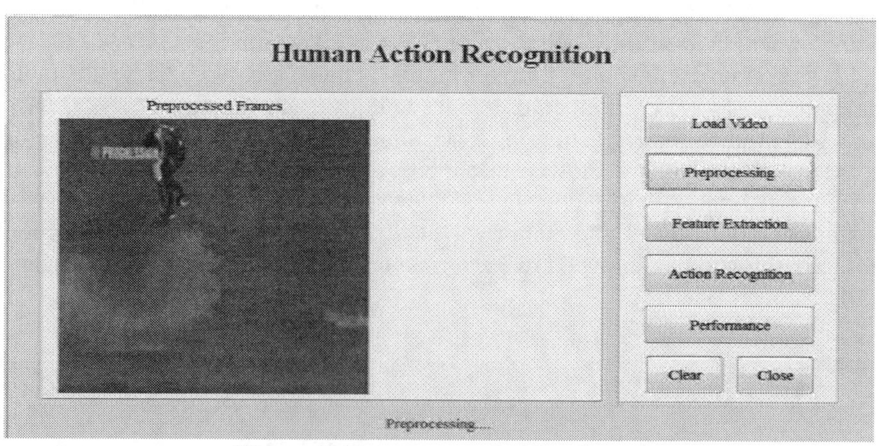

Figure 3. Pre-processed video frame

3.2. Feature extraction

From the preprocessed video frames, the feature values are extracted using STIP descriptors of various sorts. The un-normalized descriptors are extracted from the cuboids around STIP detection within the set of undistorted videos. Hence, the features like Harris STIP, Gabor STIP and HOG STIP are extracted from the frames.

3.2.1. Harris and Gabor STIP

The Harris STIP is used for detecting the corners in the frames [12]. It is used for detecting corner in every pixels of the image by considering the differential of the corner with reference to direction. The nearby edges will look similar, if the pixel in a region of uniform intensity.

In addition, the Gabor STIP is used to find the corners from the exact location of the object by Gabor wavelets. The Gabor function delivers a local spectral energy density for a given position and frequency in a certain direction [13]. The convolution of two perpendicular directions is performed with variously dilated wavelets [14]. The Gabor function provides a local spectral energy density, When s=2 and 4.

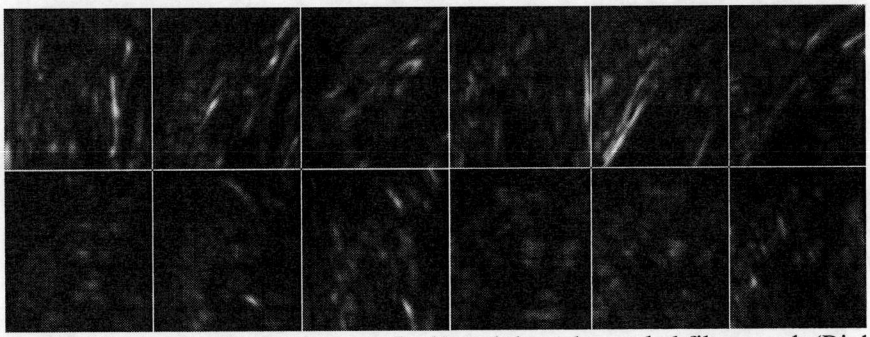

Figure 4. The original input images (Left) and the subsampled filter result (Right)

3.2.2. HOG STIP

In human action recognition, HOG is an effective feature descriptor that is calculated using orientation and magnitude of the denoised images [15]. In HOG feature descriptor, the vertical G_v and horizontal G_h gradients of the images are determined by the equations (1) and (2).

$$G_v = I_N \times [-1,0,1]^T \tag{1}$$

$$G_h = I_N \times [-1,0,1] \tag{2}$$

The determined vertical G_v and horizontal G_h gradient values are utilized for estimating the gradient magnitude $m(x,y)$ and angular orientation $\theta(x,y)$ of the images by the equations (3) and (4).

$$m(x,y) = \sqrt{G_v^2(x,y) + G_h^2(x,y)} \tag{3}$$

$$\theta(x,y) = tan^{-1}(\frac{G_v(x,y)}{G_h(x,y)}) \tag{4}$$

The gradient magnitude $m(x,y)$ and angular orientation $\theta(x,y)$ partition the pre-processed images into several cells. Then, the orientation related to the same cells are quantized and integrated into histogram bins. The obtained histogram bins are merged into final histogram. Total features T_{hog} are determined using equation (5).

$$T_{hog} = B_{img} \times B_s \times N_b \tag{5}$$

Where, B_s is indicated as block size B_{img} is stated as blocks per image, and N_b is denoted as total bins. Figure 5 shows the feature extraction techniques in HAR system.

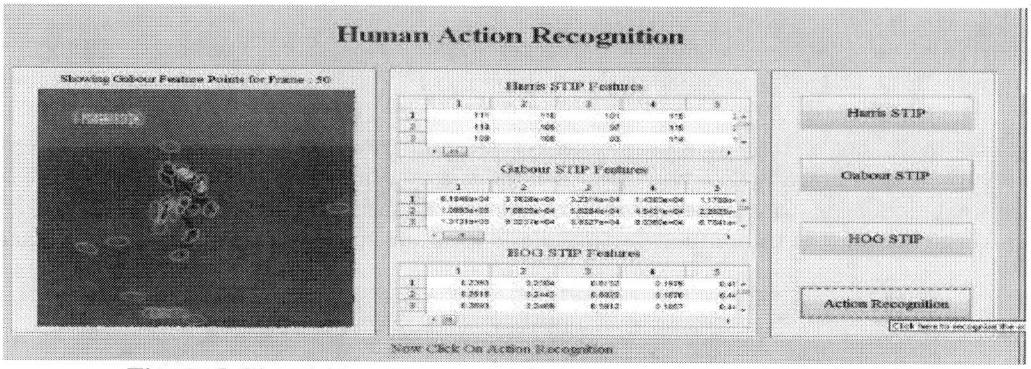

Figure 5. Illustration of feature extraction techniques in HAR system

3.3. Classification using SVM

SVM is a discriminative classifier, which is denoted by a hyper-plane. Recently, SVM classifier is highly utilized in several image processing applications, because it performs effectively in high dimension data. Additionally, the SVM performs well in two class problem, which are related to structure principles and vapnik–Chervonenkis theory [16-17].

The linear discriminant function is calculated using the formula $w.x + b = 0$. However, an optimal hyper plane is utilized in SVM for classification that is denoted in equation (6).

$$pi[w.x + b] - 1 \geq 0, i = 1,2,..N \quad (6)$$

Then, diminish $\|w\|^2$ in equation (6) in order to resolve the optimization issue. Hence the ideal discriminant function is mathematically indicated in equation (7).

$$f(x) = sign\{(w^*x) + b^*\} = sign\{\sum_{i=1}^{N} \alpha_i^* .pi(x_i^* - x) + b^*\} \quad (7)$$

In order to reduce the complexity in high dimensional data, interchange the interior product $(x_i^* - x)$ by linear kernel function$k(x, x')$. So, the ideal discriminant function is re-written as indicated in equation (8). Graphically, the SVM classifier is indicated in Figure 6.

$$f(x) = sign\{\sum_{i=1}^{N} \alpha_i^* .pi. k(x, x_i) + b^*\} \quad (8)$$

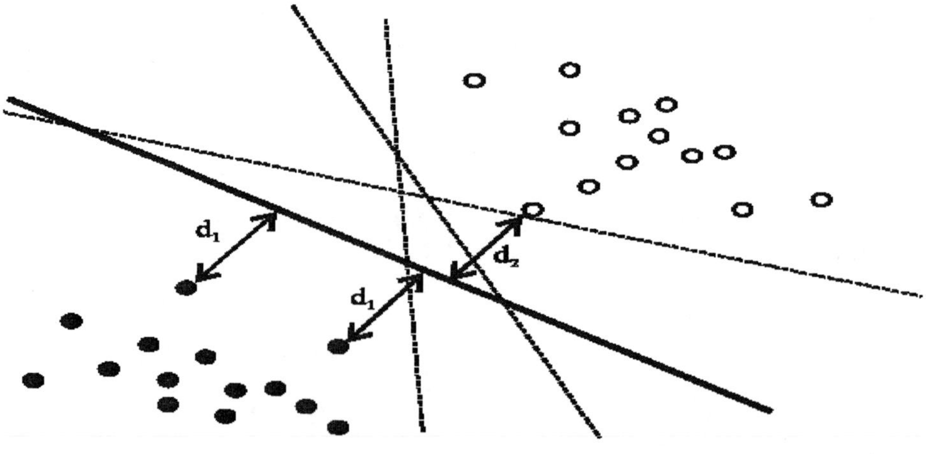

Figure 6. Graphical evaluation of SVM classifier

IOP Conf. Series: Materials Science and Engineering **925** (2020) 012026 doi:10.1088/1757-899X/925/1/012026

4. Results and Discussion

4.1. Performance measure

In this study, MATLAB (2019a) software is applied for experimental simulation with 4 GB RAM, Intel i3 processor and windows 10 operating system. In this research work, the proposed model performance is related with a few previous research models such as Enhanced Skeleton Visualization (ESV) and Adaptive Fusion and Category-level Dictionary Learning (AFCDL). In this scenario, the performance of the proposed model is evaluated in light of specificity, accuracy, and sensitivity. The mathematical expressions of specificity, accuracy, and sensitivity are indicated in the equations (9), (10) and (11).

$$Specificity = \frac{TN}{FP+TN} \tag{9}$$

$$ACC = \frac{(TP+TN)}{(FP+TN)+(TP+FN)} \tag{10}$$

$$Sensitivity = \frac{TP}{TP+FN} \tag{11}$$

Where, TP is indicated as true positive (correctly identified), FP is denoted as false positive (incorrectly identified), TN is denoted as true negative (correctly rejected) and FN is stated as false negative (incorrectly rejected).

4.2. Quantitative analysis

In this study, UCLA dataset is used to validate the proposed and existing models (ESV and AFCDL) performance. Here, 80% of the data are utilized for training and 20% of the data are utilized for testing. In this research, Harris STIP, Gabor STIP and HOG STIP features are used to extract the feature vectors from the pre-processed data. The snapshot shown in Figure 7 is the result of Harris STIP, which identifies the corners by testing each pixel in the image.

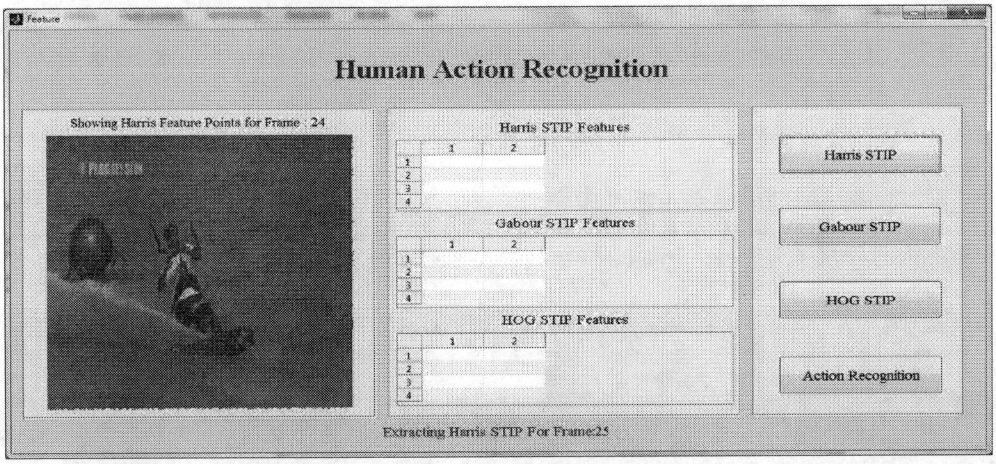

Figure 7. Extracting features using Harris STIP from the video frames

In addition, Figure 8 illustrates Gabor function that gives a local spectral energy density concentrated around a given position and frequency in a certain direction. Additionally, Figure 9 indicates HOG feature extraction that compile a histogram of gradient directions or edge orientations for the pixels within the cell.

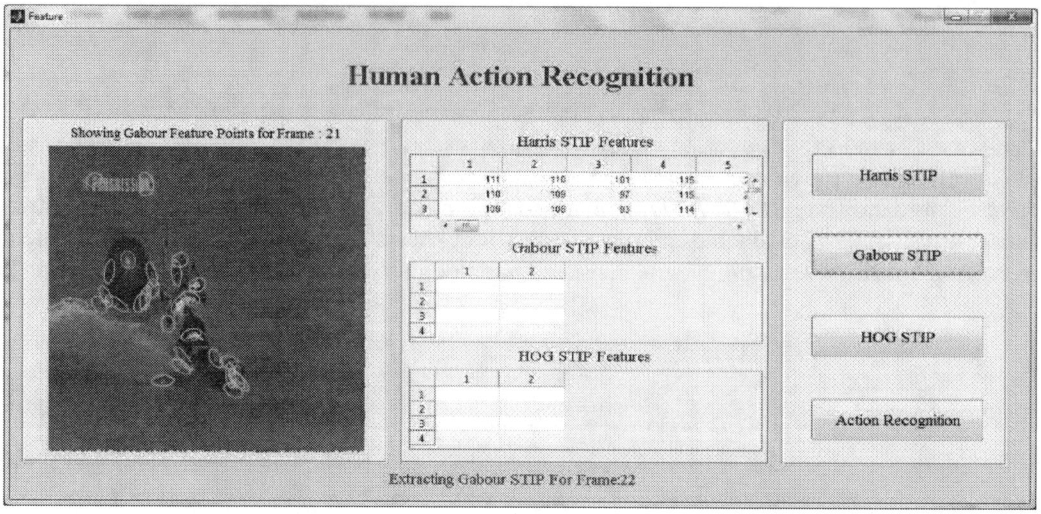

Figure 8. Feature extraction by Gabor STIP from video frames

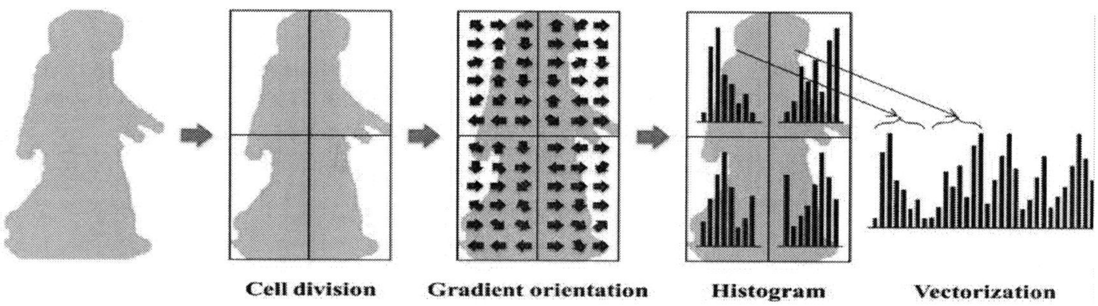

Figure 9. HOG feature points from video frames

Action of the given input video sequence is recognized by the Multi SVM identifier, which is depicted in the Figure 10 that shows surfing is the action identified from the input video processing. Additionally, Figure 11 indicates the GUI performance of the processed video.

Figure 10. "Surfing" is one of the actions identified from the video sequences

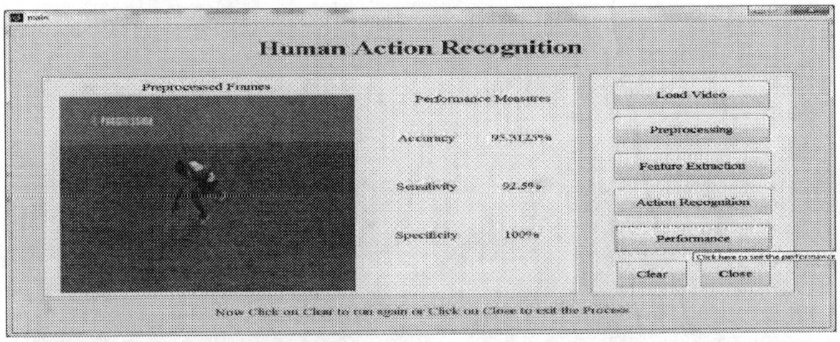

Figure 11. GUI performance of the processed video

4.3. Comparative analysis

Here, the proposed model performance is analyzed in light of specificity, sensitivity and accuracy. The UCLA dataset comprise of twenty actions such as jogging, two hand wave, horizontal arm wave, draw tick, tennis serve, high arm wave, hand clap, forward kick, forward punch, hand catch, pickup, throw, hammer, side-boxing, high throw, draw circle, bend, side kick, draw x, and golf swing. In UCLA database, the proposed model averagely attained 95.3% of accuracy, 100% of specificity and 92.5% of sensitivity that is shown in Table 1. From the experimental outcome, the proposed model showed better performance in human action recognition. The proposed model performance is graphically represented in Figure 12.

Table 1. Performance analysis of the proposed model

Actions	Accuracy (%)	Specificity (%)	Sensitivity (%)
Jogging	96	100	94
Two hand wave	95.4	100	92
Horizontal arm wave	95.9	100	93.5
Draw tick	97	100	93
Tennis serve	94	100	93.45
High arm wave	94	100	92
Hand clap	96.80	100	92
Forward kick	92	100	92.41
Forward punch	95.8	100	90
Hand catch	97	100	95.65
Pickup	97.32	100	93
Throw	95.6	100	92.5
Side-boxing	95.6	100	94
Hammer	95	100	95.31
High throw	96	100	92
Draw circle	96.02	100	92
Bend	95	100	91.63
Side kick	95.21	100	92.5
Draw x	96.28	100	92.25
Golf swing	94.17	100	92.10
Mean	95.3	100	92.5

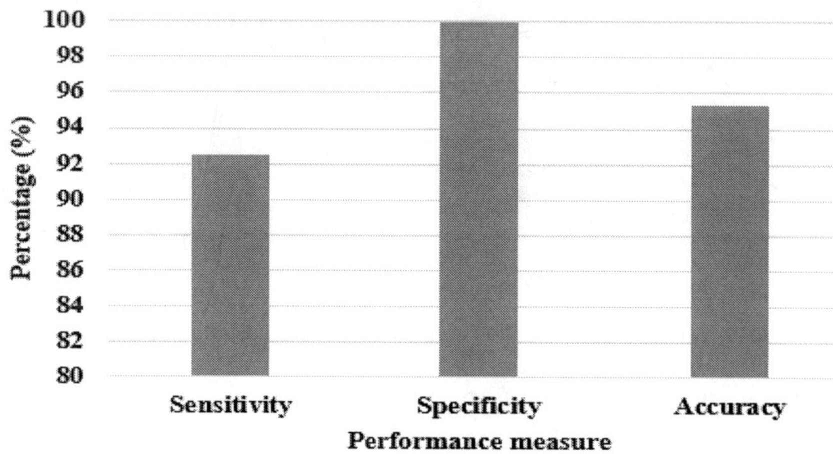

Figure 12. Graphical evaluation of proposed model

In this scenario, Table 2 shows the comparative analysis of proposed and existing models such as AFCDL and ESV in terms of recognition accuracy. Z. Gao, et al, [20] utilized a new natural inspired algorithm; AFCDL for human action recognition. In this literature paper, the developed AFCDL model performance was investigated on UCLA database. In the experimental phase, the developed model achieved 87.8 % of recognition accuracy. Similarly, M. Liu, et al, [21] has utilized motion and visual enhancement approaches on color images for improving the local patterns. Then, convolutional neural network was utilized for extracting the discriminative and robust feature vectors from the color images. Extensive experiment shows that the developed model attained 92.61% of recognition accuracy on UCLA dataset. Compared to the existing research works (AFCDL and ESV), the proposed model achieved 95.3% of accuracy that showed maximum of 7.5% and minimum of 2.69% of improvement in human action recognition. The comparative analysis is graphically stated in Figure 13.

Table 2. Comparative investigation of proposed and existing models

Methodology	Dataset	Accuracy (%)
AFCDL [20]		87.80
ESV [21]	UCLA	92.61
Proposed		95.3

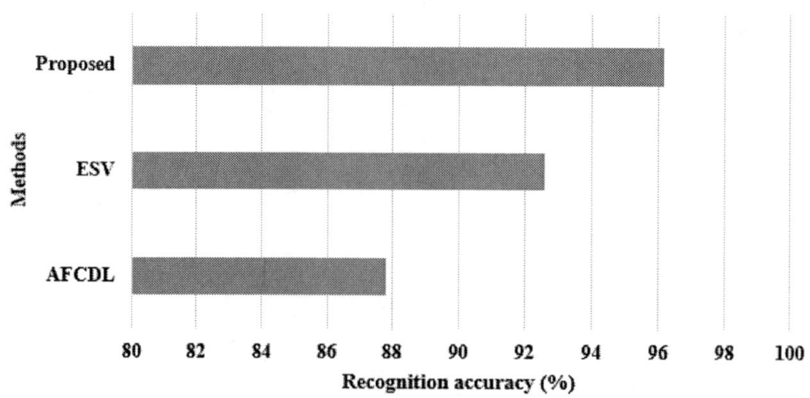

Figure 13. Graphical comparison of proposed and existing models

IOP Conf. Series: Materials Science and Engineering **925** (2020) 012026 doi:10.1088/1757-899X/925/1/012026

5. Conclusion

In this research, the proposed model acknowledges the action of the persons within the video based on the options extracted using color strips. The extracted features area unit based on the STIP that is combined with several of the opposite ideas, so the feature extraction method is simpler. The recognition of the action is completed using the kernel function of SVM classifier. The proposed model provides better accuracy than the existing models, which showed that the misclassification area unit reduced to large extend. STIP descriptors are reformulated into photometric channels that leads to color strips. Here, the proposed model results in an improved balance between discriminative power and photometric invariance, as chromaticity provides a lot of information based on the higher representations. Color strips are effectively evaluated for recognizing human actions on difficult video benchmarks. The proposed model exactly recognizes the action of the person within the video even in the conditions; illumination variations, contrast variations, Abrupt motions and Scaling of the persons in the video.

References

[1] Atiqur R A, Tan J K, Kim H S and Ishikawa S 2008 Human Activity Recognition: Various Paradigms, *International Conference on Control, Automation and Systems*

[2] Lin S C F, Wong C Y, Ren T R and Kwok N M 2012 A Comparison Study on Human Action Recognition from Video Streams *5th Int. Congress on Image and Signal Processing* 1162-1166

[3] Ong C A, and Theng L B 2014 Human Activity Recognition: A Review *IEEE international conference on control system, computing and engineering (ICCSCE 2014)* 389-393

[4] Xue L, Sheng Q Z, Pang C, Zhao X, and Wang S 2012 Effective Approaches in Human Action Recognition *Int. Conf. on Advanced Computer Science and Information Systems (ICACSIS)* 1-7

[5] Luo D, Ekene H K, and Jun O 2012 Human Gesture Analysis using1 Multimodal features *IEEE Int. Conf. on Multimedia and Expo Workshops.*

[6] Chen C, Liu K and Kehtarnavaz N 2013 Real-time human action recognition based on depth1 motion maps *Springer*

[7] Suraj V and Babu R V 2014 Real-time Human Action Recognition from Motion Capture Data, 4th *National Conf. on Computer Vision, Pattern Recognition, Image Processing and Graphics (NCVPRIPG)* 1-4

[8] Nilam N, Sjarif A and Shamsuddin S M 2015 Human Action Invarianceness for Human Action Recognition *7th Int. Conf. on Sciences of Electronics, Technologies of Information and Telecommunications (SETIT)* 263-266

[9] Neziha J, Boujnah N, Htiwich O and Bouhlel MS 2016 Human Action Recognition to Human Behavior Analysis *7th Int. Conf. on Sciences of Electronics, Technologies of Information and Telecommunications (SETIT)* 263-266

[10] Chandni J D, Tushar V R 2016 A Survey on Human Action Recognition from1 Videos, Online *Int. Conf. on Green Engineering and Technologies (IC-GET)* 1-5

[11] Alexandre P, Tabia H, Declercq D and Zanotti A 2016 Feature covariance for human action recognition 6th *Int. Conf. on Image Processing Theory, Tools and Applications (IPTA)* 1-5

[12] Guillermo C, Albuquerque A, Ara´ujo 2009 Harris-SIFT Descriptor for Video Event Detection based on a Machine Learning Approach, *11th IEEE International Symposium on Multimedia.*

[13] Feng Z, Xian-Da Z, Hu Y 2009 Gabor Filter Approach to Joint Feature Extraction and Target Recognition, *IEEE Transactions on Aerospace and Electronic Systems* **45**.

[14] Yuanyuan H, Haomiao Y and Ping H 2012 Action Recognition Using HOG Feature in Different Resolution1 Video Sequences *Int. Conf. on Computer Distributed Control and Intelligent Environmental Monitoring* 85-88

[15] Mohamed I. K, El-Yacoubi M A and Dorizzi B 2012 Human Action Recognition using Continuous Hmms and HOG/HOF Silhouette Representation *ICPRAM* 503-508

[16] Huimin Q, Yaobin M, Wenbo X, Zhiquan W 2012 Recognition of human activities using SVM multi-class classifier *Pattern Recognition Letters* **31** 100-11
[17] Adithyan P, Bhargavi R, and Vaidehi V 2012 Abnormal Human Activity Recognition Using SVM Based Approach *Int. Conf. on Recent Trends in Information Technology* 97-102
[18] Fam B L and Jaward M H 2015 Spatio-Temporal Descriptor for Abnormal Human Activity Detection *14th IAPR Int. Conf. on Machine Vision Applications (MVA)* 471-474
[19] Tehmina K, Anwar S M, Majid M, Khan B, Ali S M 2018 Emotion recognition from facial expressions using hybrid feature descriptors *IET Image Process* **12** 1004-1012
[20] Gao Z, Xuan H Z, Zhang H, Wan S and Choo K K R 2019 Adaptive fusion and category-level dictionary learning model for multiview human action recognition *IEEE Internet of Things Journal* **6** 9280-9293
[21] Liu M, Li, H and Chen C 2017 Enhanced skeleton visualization for view invariant human action recognition *Pattern Recognition* **68** 346-362
[22] Vishwakarma D K 2020 A two-fold transformation model for human action recognition using decisive pose *Cognitive Systems Research* **61** 1-13
[23] Parameshachari BD, Kiran Rashmi P, Supriya MC, Rajashekarappa and Panduranga HT 2019 Controlled partial image encryption based on LSIC and chaotic map. *In ICCSP* 60-63

Colour Image Encryption using Expand-Shrink Operation in Chaos Encryption Algorithm

Rashmi P[1,2], Supriya M C[3] and Kiran[4]

[1]Sri Siddhartha Academy of Higher Education, Tumakuru, India,
[2]Department of BCA, VVFGC, Mysuru, India.
[3] Department of MCA, SSIT, Tumakuru, Sri Siddhartha Academy of Higher Education, Tumakuru, India.
[4] Department of ECE, VVCE, Mysuru, India
[1]rashmibdp@gmail.com, [3]supriya.mc9@gmail.com

Abstract. In recent decades, chaos based encryption algorithm is widely used in image encryption. Most of the chaos based encryption algorithm operate at bit level, which is ineffective in high resolution digital images. In this research paper, a new colour image encryption algorithm is proposed by using expand-shrink operation for confusion and diffusion operation in chaos algorithm. The proposed algorithm consists of confusion-diffusion strategy operated at pixel level. Original colour image is expanded based on intrinsic features of bit planes. Expanded image undergo confusion diffusion operation using Hilbert scan method. In addition, for diffusion operation uses circle map that generates random image. In the experimental section, the proposed algorithm achieved better performance in image encryption on various digital images.

Keywords: Confusion, Diffusion, Hilbert scan, Circle map, Expand, Shrink

1. Introduction

With the growth of computer science and innovative technologies, multi-media data (videos, sounds and pictures) are increasing shared in the internet and also stored in hard disk [1]. However, the unauthorized access of private and personal information are becoming a serious concern in digital world. The issues related to security are vital in many multimedia applications [2-3]. Image encryption is an efficient and classical way for solving the above stated issues. The conventional image encryption standards like AES and DES are utilized for text information. The researchers identified that the conventional image encryption standards are ineffective for multi-media information, because of their properties like high correlation among pixels, bulk data, etc. [4-5]. The basic structure of chaos based image encryption is developed by Fridrich in 1998. Based on this structure, 2 unrelated phases are carried out in chaos based image encryption that are named as diffusion and confusion phases [6].

The image pixels are re-located initially that leads to correlation reduction among the adjacent pixels. Whereas, the confusion operation is carried out in a closed manner [7, 24] and the basic components of the images (pixels values or bits) are not influenced by the external information. The 2D sequence determines the permuting unit (bit or pixel) location. Hence, there is no variations in the statistical information between the confused and the plain images. Self-information is insufficient in the diffusion phase and a few external information need to be introduced. In a few cases, random sequences and one dimensional chaotic sequences are utilized to alter every image pixel values. The bits of pixel values and the basic elements of the images are distributed uniformly. The remaining research draft is organized as: Literature Survey is described in section 2. Related work is discussed in Section 3, and the explanation of the proposed method is illustrated in Section 4. The validation of the proposed method i.e. experimental results is represented in Section 5. Conclusion of the research work is explained in Section 6.

2. Literature survey

In recent decades, secure image transmission gaining more importance, due to the advance of computer vision and multimedia applications. Using chaotic sequence of one dimensional chaotic map, the initial values of the map are predicted easily. Additionally, high dimension chaotic map is undertaken to perform image encryption.

H. Lu, et al, [8] utilized chaotic map that plays a major role to improve Evolutionary Algorithms (EAs) to avoid local optima and speed up the convergence. In addition, dissimilar chaotic maps are utilized in dissimilar phases, where it have dissimilar effects on the EAs. In this paper, authors focused on the concerns in chaotic maps and gave guidance to improve the multi-objective EAs (MOEAs) by implementing several experiments. Additionally, a non-dominated sorting and elitist strategy; NSGA-II algorithm is undertaken to study the effect of chaotic maps. In addition, A. Nair, and K. Kiasaleh, [9] discussed about circle map as an interleaver that creates additional layer of randomness and scrambler effect, and it negates the impact of burst errors due to wireless channel conditions. By effectively using the good correlation and spreading properties of the circle map allows one to achieve this gain. The other critical aspect of the proposed method is the minimization of the size of the side information required for the de-interleaving purposes.

E.S. El-Alfy, and K. Al-Utaibi, [10] developed a new image encryption algorithm on the basis of logistic map for generating 4 chaotic sequences. These sequences are transformed into 4 key streams by utilizing chaotic sequence methodology. The generated key streams are utilized for controlling the multipoint mutation and cross-over operations that results in image diffusion and confusion, correspondingly. K.U. Shahna, and A. Mohamed, [11] implemented a new symmetric cryptography framework for encrypting the images by utilizing Henon map, bit level cyclic shift and Hilbert curve. In this literature study, the image pixel scrambling and key stream generation are carried-out in both bit level and pixel level for ensuring the double scrambling, which is achieved by using cyclic shift and Hilbert curve operations. By using double scrambling methodology, the near pixel correlation values are minimized. Finally, a novel methodology is undertaken for cyclic shift operation. The scrambles image is encrypted utilizing key streams in the diffusion process. Various encryption types discussed using different methods [14-18]. Expand and shrink to enhance the strength of chaos based image encryption technique described in [19]. Performance of the communication for 4x4 indoor MIMO-VLC system are analysed under different noise constraints [21]. Glow-worm Swarm Optimization (GSO) and Support Vector Machine (SVM) is used for decision making process in battery storage to reduce the electricity tariff [22]. In [23] hybrid approach based on chaotic map and henon map presented for encrypting color images.

3. Related work

3.1. Hilbert scan pattern

The Hilbert scan pattern is an effective scan pattern technique in image encryption that scans $2^m \times 2^m$ array points. In this technique, the scan path is constructed from left top, left bottom, right top and right bottom of the square grid [12]. In this technique, Hilbert curve is utilized for shuffling the position of the pixels in the original digital images for obtaining the scrambled image.

In Hilbert scan pattern, the order 0 represents that the Hilbert curve is empty, and the order -1 is attained by using 4 zero ordered curves with 3 straight connector lines. In addition, the higher order curves are also obtained in the similar manner. The Hilbert curve of order 1, 2, 3, and 4 are corresponds to 2×2, 4×4, 8×8, and 16×16 square grids, which are stated in the figure 1 (a), (b), (c), and (d).

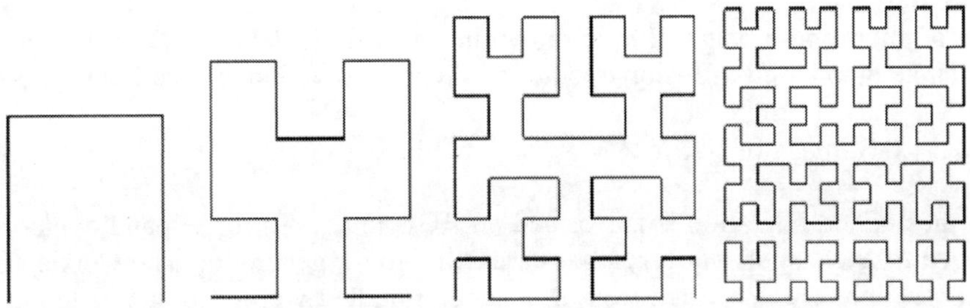

Figure 1. Hilbert curve of order 1, 2, 3, and 4 corresponds to 2 × 2, 4 × 4, 8 × 8, and 16 × 16 square grids, respectively

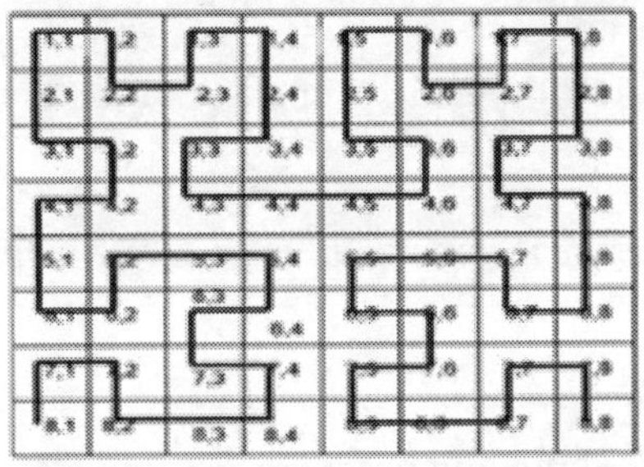

(a) Hilbert

8,8	7,8	7,7	8,7	8,6	8,5	7,5	7,6
6,6	6,5	5,5	5,6	5,7	6,7	6,8	5,8
4,8	4,7	3,7	3,8	2,8	1,8	1,7	2,7
2,6	1,6	1,5	2,5	3,5	3,6	4,6	4,5
4,4	4,3	3,3	3,4	2,4	1,4	1,3	2,3
2,2	1,2	1,1	2,1	31	3,2	4,2	4,1
5,1	6,1	6,2	5,2	5,3	5,4	6,4	6,3
7,3	7,4	8,4	8,3	8,2	7,2	7,1	8,1

(b) Scan coordinate

1	2	3	4	5	6	7	8
9	10	11	12	13	14	15	16
17	18	19	20	21	22	23	24
25	26	27	28	29	30	31	32
33	34	35	36	37	38	39	40
41	42	43	44	45	46	47	48
49	50	51	52	53	54	55	56
57	58	59	60	61	62	63	64

(c) Original image

64	56	55	63	62	61	53	54
46	45	37	38	39	47	48	40
32	31	23	24	16	8	7	15
14	6	5	13	21	22	30	29
28	27	19	20	12	4	3	11
10	2	1	9	17	18	25	24
33	41	42	43	35	36	44	43
51	52	60	59	58	5	49	57

(d) Scrambled image matrix

Figure 2. 8 ∗ 8 matrix permutation using Hilbert scan pattern

3.2. Circle map

Circle map is a dynamical system, which is initially determined by Andrey Kolmogorov. However, the author states that the circle map is a simplified model that is used to driven the mechanical rotors. Especially, a spinning wheel is coupled to spring the motor. Equation (1) denotes the formula to calculate the circle map, where it is a simplified model in electronics [13]. Mathematically, the circle map is represented in equation (1).

$$x_{k+1} = \left\{x_k + b - \left(\frac{a}{2\pi}\right) sin(2\pi x_k)\right\} mod \quad (1)$$

With $a = 0.5$ and $b = 0.2$, and it generates the chaotic sequence $(0, 1)$.

IOP Conf. Series: Materials Science and Engineering **925** (2020) 012027 doi:10.1088/1757-899X/925/1/012027

4. Proposed method

Block diagram of the proposed algorithm is shown in figure 3. This block diagram mainly consists of confusion and circle map based diffusion operation. Steps involved in encryption process.

Step 1: Consider RGB image of size $N \times N \times 3$.

Step 2: Then, the expand operation is performed on RGB image. In the expand operation, the higher two bits of $R, G, and\ B$ channels of every pixels are extorted for generating a new pixel with six bits. In which the lower two bits are generated from R channel, middle two bits are generated from G channel and the higher two bits are generated from B channel. Then, a new image is achieved using these image pixels, which is determined as β channel. Hence, the residual six bits in every pixel of the $R, G, and\ B$ channels are remains unchanged. The height and width of the original image is represented as N, here totally 4 $N \times N$ image with six bits for every pixels are achieved on the basis of $R, G, B, and\ \beta$ channels. The obtained 4 images forms permutation expanded matrix that is graphically indicated in figure 2. Further, the channel β comprises of around 80% of plain image information.

Step 3: Expanded image positions are shuffled randomly by using Hilbert scan pattern as shown in figure 2.

Step 4: Based on the secret key, a random vector sequence is generated using circle map, as shown in the equation (1).

- Update the random vector X by multiplying it with 1014.
- Modulo 256 operations is carried out to get updated random numbers in a range of 0 to 255.

$$\alpha = Modulo(X, 256) \tag{2}$$

- New random finite precision vector α, which is arranged in matrix form to get random image R.

Step 5: Perform Bit XOR operation between confused image and random image R.

Step 6: Resultant image undergo shrink operation to produce final RGB encrypted image.

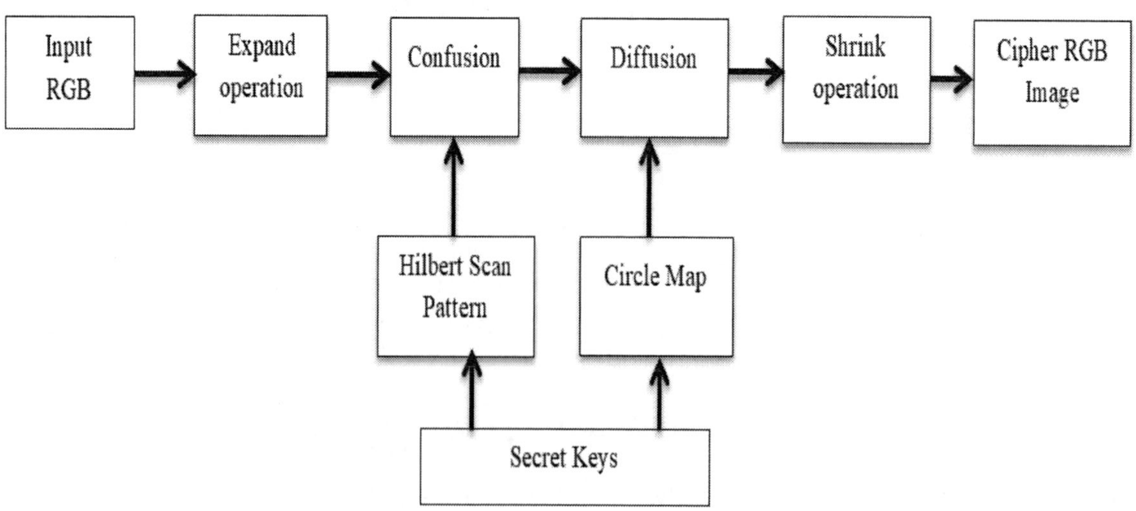

Figure 3. Architecture of proposed colour image encryption

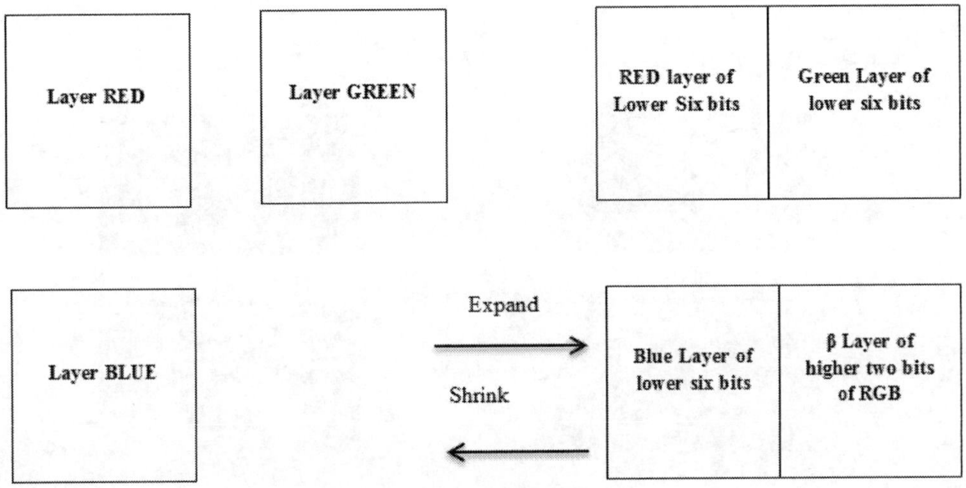

Figure 4. Expand Shrink operation

5. Experimental Results

In this section, the validation of the proposed algorithm is conducted by encrypted analysis and histogram analysis. In this study, MATLAB 2015a is utilized to implement the proposed algorithm along with the configuration of 4 GB RAM, i5 processor of 3.4 GHz Intel Core with 3570 CPU. The test colour images are selected for validating the proposed encryption algorithm. Figures 5 and 6 indicates the encryption results, and it stated that the various size images are also encrypted by the proposed algorithm. The final result images are meaningless and noise-free images that shows the proposed algorithm encrypt the images in an effective way.

Histogram of an image is a graphical representation of pixels distribution based on the intensity levels. For any plain image, an effective encryption algorithm always generates a cipher image of the uniform histogram. Figures 7 and 8 shows the histograms of the various RGB original image and the corresponding cipher images. It is determined that the histogram of encrypted image is close to the uniform distribution.

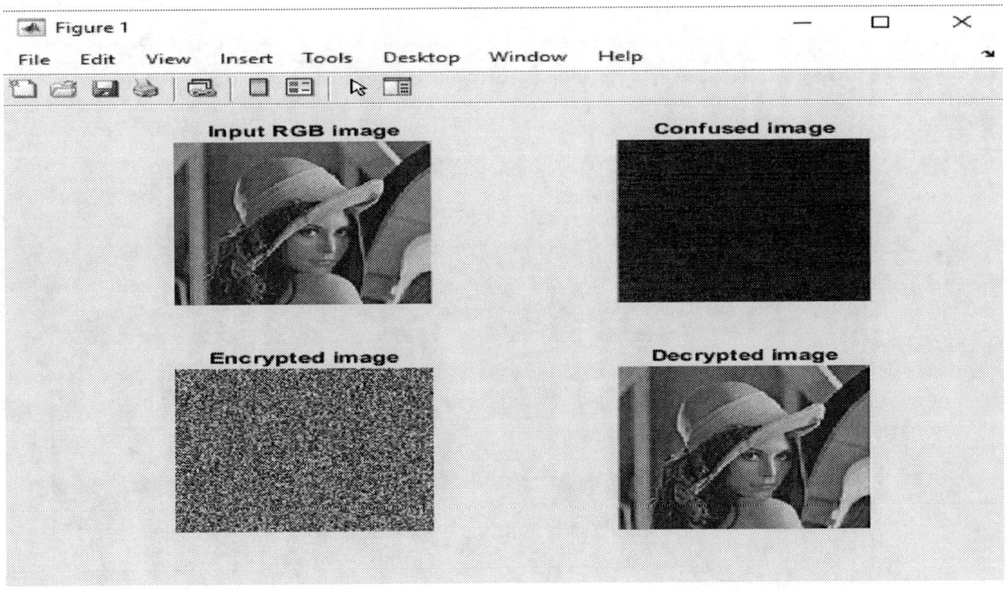

Figure 5. Result of proposed method for Lena input image

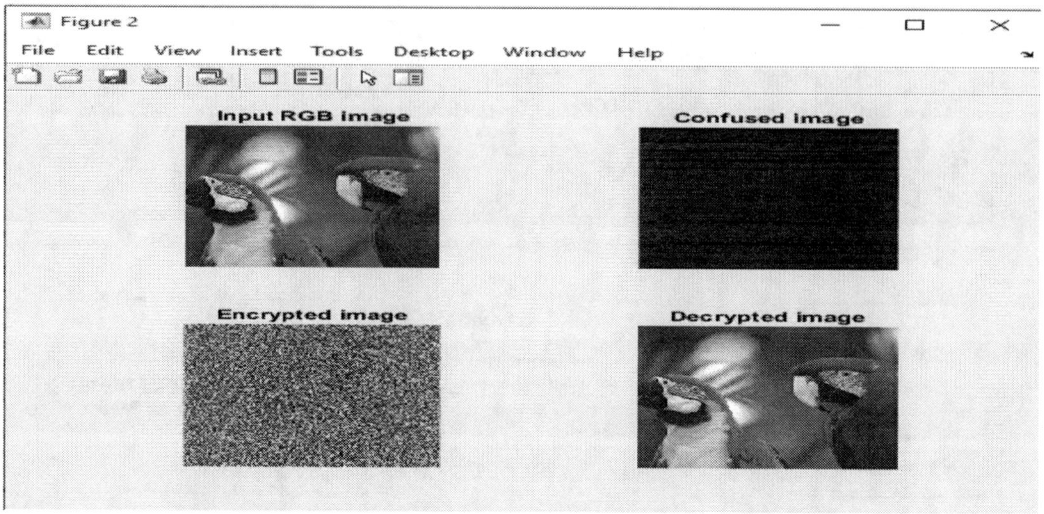

Figure 6. Result of proposed method for Parrot input image

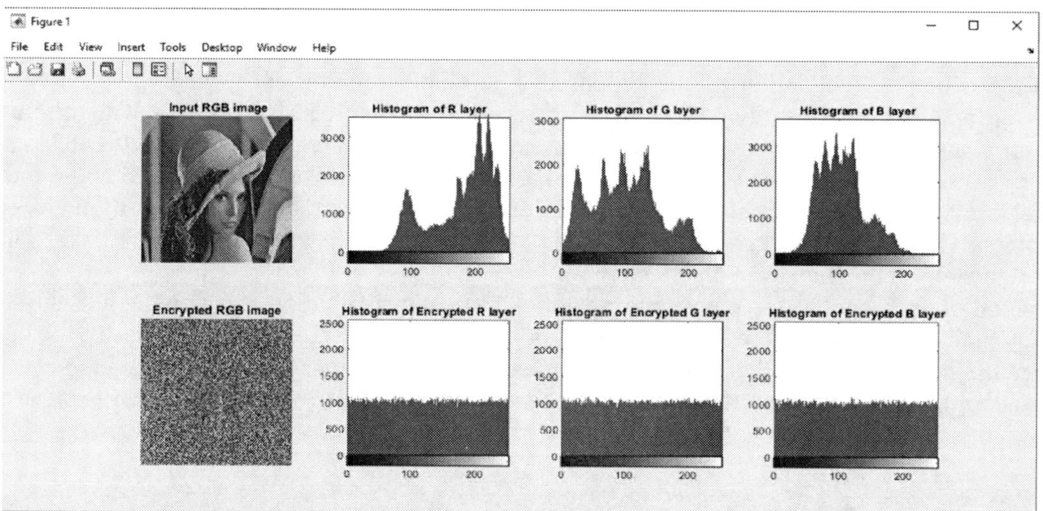

Figure 7. Histogram Analysis of proposed method for Lena input image

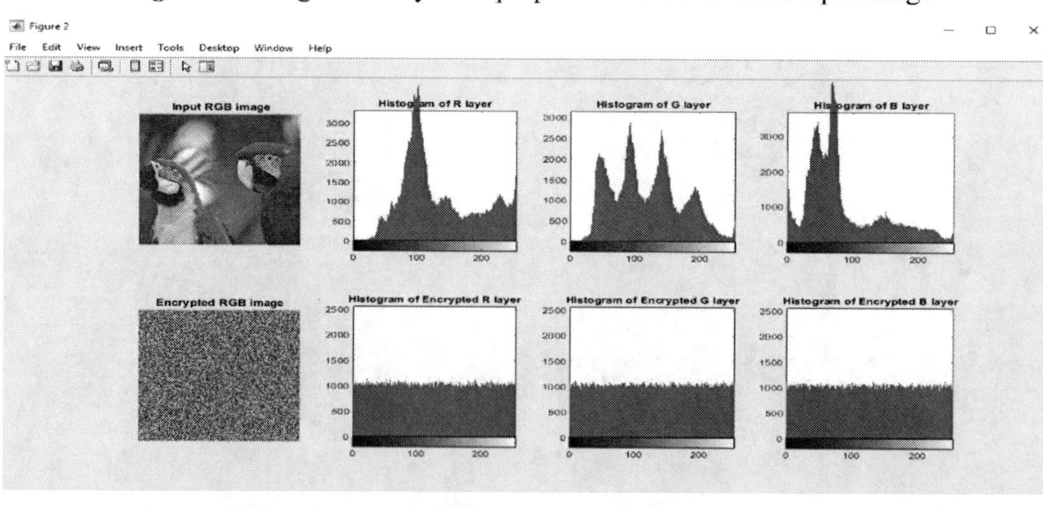

Figure 8. Histogram Analysis of proposed method for Parrot input image

The pixel correlation is the degree of association of the gray values between pixels. Figures 9 and 10 shows the pixel pair distribution of the test images and pixel pair the corresponding cipher images. The formula of correlation coefficient is defined as follows [20]

$$C_{xy} = \frac{cov(x,y)}{\sqrt{\sqrt{D(x)}D(y)}}$$

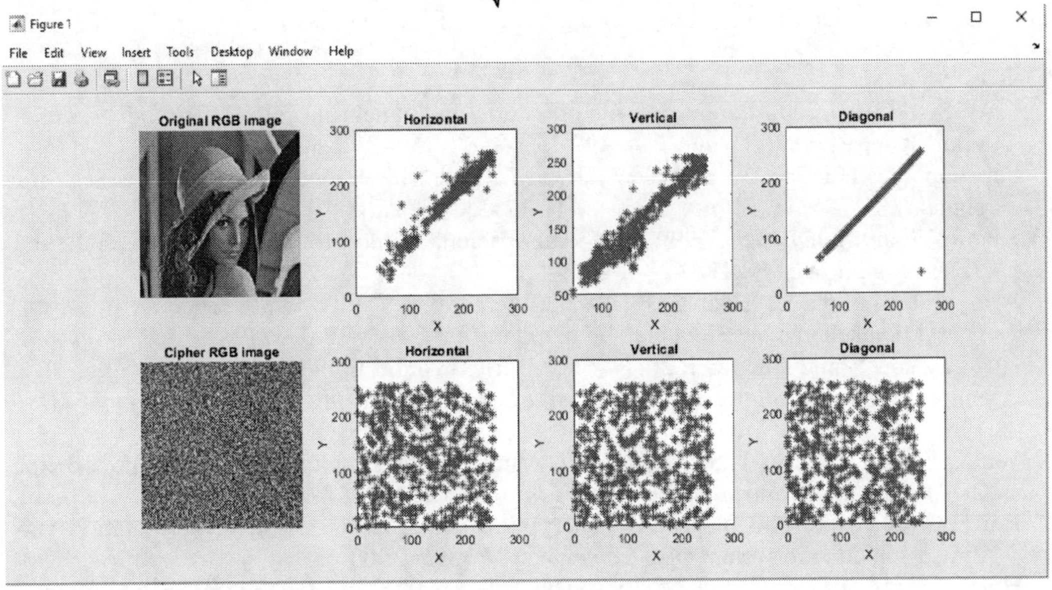

Figure 9. Correlation Analysis of proposed method for Lena input image

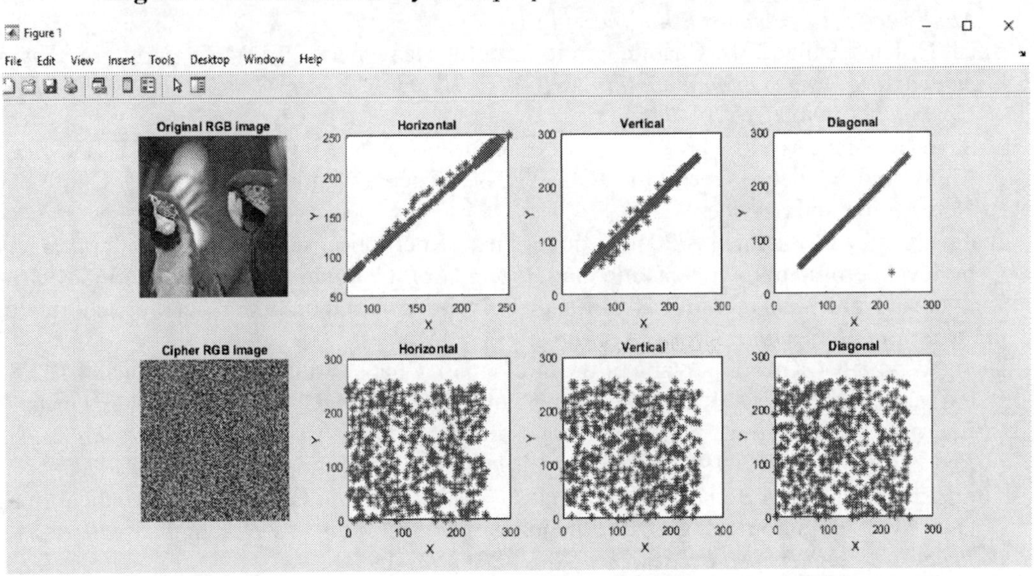

Figure 10. Correlation Analysis of proposed method for Parrot input image

6. Conclusion

In this research, a colour image encryption using expand shrink operation for confusion and circle map based diffusion operation is proposed. This algorithm consists of confusion diffusion strategy, which is operated at bit level. Original colour image is expanded based on intrinsic features of bit planes.

Expanded image undergo confusion-diffusion operation. Here for diffusion operation utilizes circle map. This research article concluded that the proposed algorithm is suitable for real time encryption and also it effectively improves the security of the system. While related to existing algorithms, the proposed algorithm delivers an effective performance in image encryption.

Acknowledgement
The work described in this paper is supported by Sri Siddhartha Academy of Higher Education, Tumakuru, India. Authors would like to thank Hon'ble Vice Chancellor Dr. Balakrishna P Shetty and Registrar Dr. M Z Kurian for providing a strong platform for research work.

References

[1] Chen S and Lü J 2002 Parameters identification and synchronization of chaotic systems based upon adaptive control *Phys. Lett.* **299** 353–358

[2] Papadopoulos H E and Wornell G W, 1995 Maximum-likelihood estimation of a class of chaotic signals *IEEE Trans. Inform. Theory* **41** 312–317

[3] Xiaofu W and Songgeng S 1999 A general efficient method for chaotic signal estimation *IEEE Trans. Signal Process* **47** 1424–1428

[4] Arroyo D, Rhouma R, Alvarez G, Li S and Fernandez V 2008 On the security of a new image encryption scheme based on chaotic map lattices, *Chaos* **18** 033112

[5] Hua Z, Zhou Y and Chen C P 2013 A new series-wound framework for generating 1D chaotic maps *in: Digital Signal Processing and Signal Processing Education Meeting (DSP/SPE)* 118–123

[6] Wu X, Wang D, Kurths J and Kan H 2016 A novel lossless color image encryptionscheme using 2D DWT and 6D hype rchaotic system *Inform. Sci.* **349**137–153

[7] Liu H, Wang X 2011 Color image encryption using spatial bit-level permutation and high-dimension chaotic system *Opt. Commun.* **284** 3895–3903

[8] Hui L, Wang X, Fei Z Qiu M 2014 The Effects of Using Chaotic Map on Improving the Performance of Multiobjective Evolutionary Algorithms Hindawi Publishing Corporation *Mathematical Problems in Engineering* 1-16

[9] Anish N, Kiasaleh K 2016 Chaotic Circle Map Interleaver for OFDM *Int. Conf. on Computing, Networking and Communications (ICNC), Workshop on Computing, Networking and Communications (CNC)* 1-6

[10] El-Sayed M E and Al-Utaibi K A 2011 An Encryption Scheme for Color Images Based on Chaotic Maps and Genetic Operators ICNS 2011: The Seventh International Conference on Networking and Services 92-97

[11] Shahna K U and Mohamed A 2016 A novel image encryption scheme using both pixel level and bit level permutation with chaotic map *Applied Soft Computing Journal* **90** 1568-4946

[12] Sivakumar T and Venkatesan R 2014 Image encryption based on pixel shuffling and random key stream *Int. J. Comput. Inform. Technol.* **3**

[13] Zheng W M 1994 Kneading plane of the circle map, Chaos, Solitons and Fractals, **4** 1221–1233

[14] Parameshachari B D, Kiran, Rashmi P and Supriya M C 2019 Controlled Partial Image Encryption Based on LSIC and Chaotic map *4th Int. Conf. on Multimedia and Image Processing (ICMIP 2019),* University of Malaya, Kuala Lumpur, Malaysia 19-21

[15] Kiran, Parameshachari B D and Panduranga H T 20147 Partial Encryption of Medical Images by Dual DNA Addition using DNA Encoding *Proc. Int. conf. on Recent Innovations is Signal Processing and Embedded Systems (RISE-2017)* 27-29

[16] Parameshachari B D, Rajashekarappa, Sumithar Devi K A and Sunjiv Soyjaudh K M 2014 Partial Image Encryption Algorithm Using Pixel Position Manipulation Technique *4th Int. Conf. on Artificial Intelligence and applications in Engineering and Technology (ICAIET2014)*, Kota Kinabalu, 177-181

[17] Parameshachari B D, Sumithar Devi K A and K M Sunjiv Soyjaudh Partial Image Encryption Approach Using SCAN Mapping and Band Permutation Published at Proceedings of ICSIP 2013, 222–226

[18] Parameshachari B D, Sumithar Devi K A and K M Sunjiv Soyjaudh Secure Partial Image Encryption Scheme Using Scan Based Algorithm, International Journal of advances in Engineering & Technology, **6,** 264-273

[19] Naveenkumar S K, Panduranga H T and Kiran 2014 Chaos Based Image Encryption using Expand-Shrink Concept, International Journal of Informatics and Communication Technology (IJ-ICT) **3** 103-112

[20] Wonyoung J and Lee S. Y 2020 Partial image encryption using format-preserving encryption in image processing systems for Internet of things environment, International Journal of Distributed Sensor Networks **16** 1-17

[21] Subramani P, Rajendran GB, Sengupta J, Pérez de Prado R, Divakarachari PB 2020 A Block Bi-Diagonalization-Based Pre-Coding for Indoor Multiple-Input-Multiple-Output-Visible Light Communication System, Energies, **13** 1-16

[22] Parameshachari B D et al. 2019 Demand side management of small scale loads in a smart grid using glow-worm swarm optimization technique Elsevier Microprocessors and Microsystems **71** 1-9

[23] Rashmi P, Supriya M C and Kiran 2019 Encryption of Color image to enhance security using Permutation and Diffusion Techniques International Journal of Advanced Science and Technology Scopus Indexed Journal **28** 375-384

[24] Narayanan K.L., and Ramesh G.P. 2017 Discrete Wavelet Transform Based Image Compression using Frequency Band Suppression and Throughput Enhancement. International Journal of MC Square Scientific Research, **9** 176-182

IOP Conf. Series: Materials Science and Engineering **925** (2020) 012028 doi:10.1088/1757-899X/925/1/012028

Analysis and Computation of Encryption Technique to Enhance Security of Medical Images

Kiran[1,2], Parameshachari BD[3], Panduranga HT[4] and Silvia liberata Ullo[5]

[1]VTU, Belagaum, India.
[2] Department of ECE, Vidyavardhaka College of Engineering, Mysuru, India.
[3] Department of TCE, GSSSIETW, Mysuru, India.
[4] Department of ECE, Govt. Polytechnic, Turvekere, Tumkur, India.
[5] Engineering Department, University of Sannio, Italy.
[1,2] kiran.mtech12@gmail.com,[3]parameshbkit@gmail.com, [5]silvullo@unisannio.it

Abstract. Recently, there are many advanced techniques used in the medical fields, such as smart health, e-health and telemedicine applications. These techniques rely on open source networks to transmit the digital medical images. These images have very sensitive and confidential information about the patients. Regrettably, most of the regularly used algorithms provide very less security and present a high communication overhead, with high computation costs. Due to these issues, a Region of Interest (ROI) method, based on selective image encryption, is proposed in this paper. An active contour image segmentation is first used to divide the ROI and the Region of Background (ROB). Then, a Hilbert curve, with a Skew Tent map are used to implement the Permutation and diffusion techniques. Namely, the ROI part is permuted according to a Hilbert scan pattern, to decrease the similarities between the inside adjacent pixels. Then pixels in the permuted ROI part are XOR pixel-wise with random numbers, generated based on a predefined threshold value from a Skew Tent map. Finally, encrypted ROI and ROB blocks are combined to generate encrypted images. Given the obtained experimental results, the proposed method is very likely to improve image security, through the employment of correlation checking, key sensitivity, entropy, diffusion characteristic and histogram tools.

1. Introduction

In time critical applications such as internet banking, medical imaging, pay TV, military communication, confidential video conference, satellite security, corporate communication, etc. where the security is utmost important, the selective image encryption plays a vital role. Advanced Encryption Standard [AES] and Triple Data Encryption Standard [DES] are the strong encryption techniques used to secure the images from unauthorized access. The process of encrypting the whole image is a tedious process using these techniques. Partial/Selective Encryption [PE/SE] is the process of encrypting the selected portions from the input image. This process decreases the encrypting and decrypting time. It also increases the robustness of the image.
Zhijuan Dengand and Shaojun Zhong [1] introduced a chaotic mapping based digital image encryption algorithm. In this algorithm the number of iterations used is reduced and the cryptographic space is highly expanded by this algorithm. Akram belazi muhammad Talha et.al [2] explained a new method for encryption for medical images based on chaos, it is the combination of both DNA and chaos, it has two encryption rounds, after which the key generation layer is done. Hossein [3] described image

security algorithm where image encryption is classified into bit-level and pixel-level. Histogram of the image can be changed by bit-level permutation, because it is bit-level computing it is time consuming. M bin younas and Jawad ahmad [4] presented a new technique in wavelet transform domain that is fusion technique. Here the components of high and low frequencies are merged to get improved image content. Four asymmetric keys are generated with respect to each image. Aqeel ur rehman et.al [5] presented two major technologies. They have tested and estimated the scheme of encryption of image using different parameters. This method will more efficient than the previous methods. Lisungu oteko tresor et.al [6] described a new technique for image encryption. This method is suitable only for gray scale images. The speed will be more but the method is restricted only to gray scale images. Ping et.al [7] described a new method known as digit-level permutation, the image is divided and converted into a matrix form, this pixel matrix is decomposed into three digital matrices. Using Henon map this pixel will be shuffled, pixel level permutation is combined with bit level permutation by digit-level permutation. High-speed diffusion operation will be designed. Tresor Oteko Lisungu et.al [8] explained about development of new encryption scheme, where in this method the characteristics of image is changed and compressed. This increases the complexity and makes this method more efficient. Martin Rieger and Andreas Uhl [9] introduced image encryption technique which is restricted to JPEG2000. Uses of this method makes it more efficient and it be used widely for many applications. The security and speed of this method will also increase and give good performance. Xingbin liu and cong liu [10] explained a technique which is used to encrypt image and restricted only for gray images. It considers pixel gray value and information about image blocks position. This scheme is reliable and the security of image is good. Dong xie [11] describe simple method to manipulate image by using matrix multiplication. Crypto system provides great security to the information where it will be hard to crack this image. Hence, this makes it more efficient. Shuqin zhu and congxu zhu [12] explained about 5-D hyper-chaotic map, which is designed by merging 3D Lorenz and logistic map. This scheme includes two types of operations. where in the 1st method which is only related to the pixels of clear text images and the 2nd method undergoes 2 rounds of diffusion and this paper says that theoretical and experimental results ensures the security. Mohammed [13] to overcome the security problem, introduced the privacy preserving data mining. Currently the web users are sharing their confidential data or tasks without risk. Ming li et.al [14] it shows that the newly introduced scheme has higher security due to the improvisation of the 1D chaotic map along with higher performance. They have found few loophole problems in this technique and then they introduced a new attack strategy which uses chosen plain text attack. The newly introduced methods are verified theoretically and practically. Various chaos based encryption techniques were discussed using different methods [15-18, 30]. Dr. Parameshachari B D et.al [19-20] explained about side management and optimization methods in the secure communication system.

The remaining paper is organized as follows: In section II, active contour, Hilbert scan and skew tent maps are briefly explained. Section III shows the proposed ROI based encryption scheme. The experimental analysis and security are presented in section IV and we conclude the paper in Section V.

2. Related Work

2.1 Active Contour Method

For many applications such as motion tracking and image segmentation over the last decade the active contour is the most popularly used method. The deformable contours are used to match the motions and various shapes of the objects. Here we explain how the active contours are set theoretically and also information about the present methods of active contour. Mathematical implementation based approaches: Snakes and setting the levels are used in active contours. Using Energy minimization method, the pre-defined snake points are explicitly shifted by the snakes. The level set approach moves the contour as a function of a particular level completely. Based on the force evolving contours, there are two active contour models in image segmentation: Edge based and region based. Edge detector which uses the image gradient is used in the active contours based on Edge to find the sub-regions boundaries and also to draw the detected boundaries contours. Edge based segmentation is almost similar to the Edge based approach. Instead of searching for geometrical boundaries, the statistical information of image intensity inside each subset is applied in Region based active contours approach. Region based segmentation is almost similar to the region based approach [21-22].

2.2 Hilbert Scan Pattern

Overall, $2^m \times 2^m$ array of points are scanned using Hilbert curve. In a square gird the scan path can be started either from left bottom (LB), left top (LT), right bottom (RB) or right top (RT) [23]. Here the scrambled is obtained by Hilbert curve which shuffles the pixels position in the original image.

The order 0 of the Hilbert Curve is always empty. With the help of three straight line connectors along with four zero order curves, the first order Hilbert curve is obtained. All other higher order curves are obtained similarly. Hilbert curve of order 1, 2, 3 and 4 corresponding to are shown in Fig. 1(a), (b), (c) and (d) shows the first, second, third and fourth order Hilbert curves which corresponds to square grids of 2 × 2, 4 × 4, 8 × 8, 16 × 16.

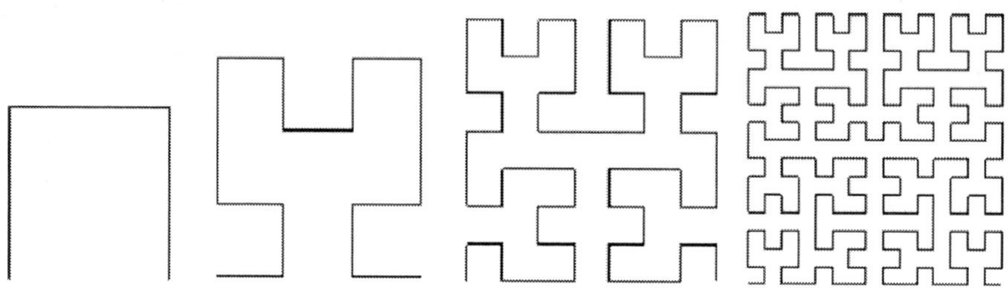

(a) Order1 (b) Order2 (c) Order3 (d) Order4

Figure 1. Hilbert curve of order 1,2,3,4 corresponding to 2x2, 4x4, 8x8, 16x16 square grids respectively

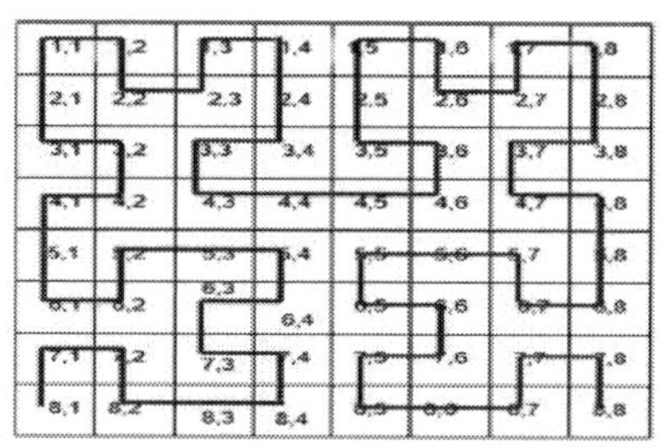

(a) Hilbert

8,8	7,8	7,7	8,7	8,6	8,5	7,5	7,6
6,6	6,5	5,5	5,6	5,7	6,7	6,8	5,8
4,8	4,7	3,7	3,8	2,8	1,8	1,7	2,7
2,6	1,6	1,5	2,5	3,5	3,6	4,6	4,5
4,4	4,3	3,3	3,4	2,4	1,4	1,3	2,3
2,2	1,2	1,1	2,1	31	3,2	4,2	4,1
5,1	6,1	6,2	5,2	5,3	5,4	6,4	6,3
7,3	7,4	8,4	8,3	8,2	7,2	7,1	8,1

(b) Scan coordinate

1	2	3	4	5	6	7	8
9	10	11	12	13	14	15	16
17	18	19	20	21	22	23	24
25	26	27	28	29	30	31	32
33	34	35	36	37	38	39	40
41	42	43	44	45	46	47	48
49	50	51	52	53	54	55	56
57	58	59	60	61	62	63	64

(c) Original image

64	56	55	63	62	61	53	54
46	45	37	38	39	47	48	40
32	31	23	24	16	8	7	15
14	6	5	13	21	22	30	29
28	27	19	20	12	4	3	11
10	2	1	9	17	18	25	24
33	41	42	43	35	36	44	43
51	52	60	59	58	5	49	57

(d) Scrambled image matrix

Figure 2. 8*8 matrix permutation using Hilbert scan pattern

2.3 Skew Tent Map

The skew tent map is one of the popular chaotic maps which is dynamic, simple, nonlinear equation along with a complex chaotic behaviour, is as shown below [24]:

IOP Conf. Series: Materials Science and Engineering **925** (2020) 012028 doi:10.1088/1757-899X/925/1/012028

$$X_{n+1}=\begin{cases}\frac{X_n}{b}, & \text{for } 0 < X_n \leq b\\ \frac{1-X_n}{1-b}, & \text{for } b < X_n < 1\end{cases} \qquad (1)$$

Where $X_{n+1} \in [0,1]$ is the chaotic system state, $b \in [0,0.5] \cup [0.5,1]$ is a control parameter, and $X_n \in [0,1]$ where n is iteration number which generates iterative values.

3. Proposed Technique

Architecture of proposed ROI based advanced encryption scheme is as shown in Figure 2. Throughout this paper, $P \times Q$ represents the size of a medical image. The plain text image pixel value at location i^{th} row and j^{th} column is represented as P (i, j). Three important steps are involved in the proposed scheme. In first step, ROI part of the original image is extracted using active contour method, whereas in the second step, the Hilbert scan pattern employed to change the location of every pixels in the ROI part. In the last step, for diffusion purposes, skew tent map are utilized to get a random matrix R. More details regarding these steps are as mentioned below:

Step 1:

i) The original medical image of size $P \times Q$ is undergo active contour based image segmentation to extract ROI and ROB binary image.

ii) Then ROI part is obtained from multiply original medical image and ROI binary image and size of ROI is $M \times N$.
ROI_Part= Original Medical image.* ROI binary image...............(2)

Step 2:

iii) Starting from the right bottom (RB) position we have to construct a Nth order Hilbert curve. For eg. Fig 2(a) gives the third order Hilbert curve which starts from right bottom (RB) cell in (8×8) square grid. Fig 3(b) gives the scan coordinates of this operation.

iv) The pixels of the original image are scrambled based on this scan pattern.

Step 3:

v) Initial condition for Sine is set i.e., b=0.2838 and X0 = 0.73846.

vi) Iterate the Equation 1 $M \times N$ times to obtain random vector X.

vii) Update the random vector X by multiplying it with 10^{14}.

viii) To obtain a new random number within the range (0-255), Modulo 256 operation is used.
α = Modulo(X, 256).......................... (3)

ix) To get the matrix R we have to arrange the new random finite precision vector α in a matrix form.

x) The ROI block pixels are XORed bitwise with matrix R.
Diff_Block= bitxor (ROI,R)......... (4)

where Diff_Block is the diffused block.

xi) At last combined diffused ROI block and ROB block to form encrypted image.

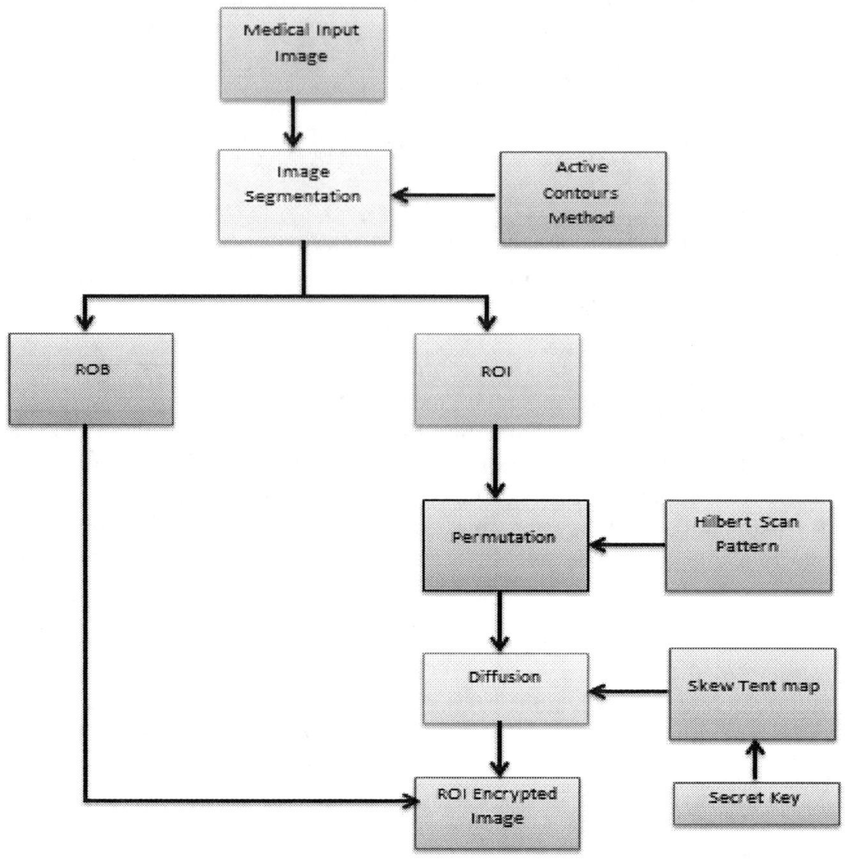

Figure 3. Architecture of proposed ROI based selective image encryption

4. Performance Analysis of Proposed Scheme

The overall performance of the proposed scheme is analysed with the help of different parameters. The following parameters are involved as follows.

4.1 Histogram Analysis

Based on the intensity levels, the graphical representation of pixel distribution is the histogram of an image. A good encryption algorithm for any plain image will generate a cipher image of the uniform histogram. Table 4 gives the various original medical images histograms and their respective cipher images. The histogram of an encrypted image is almost close to uniform distribution.

4.2 Entropy Analysis

Entropy is a measure of degree of randomness in the encryption system. Entropy is calculated using the formula [26]:

$$H(S) = \sum_{i=0}^{2^M-1} P(si)\log_2\frac{1}{P(si)} \quad (5)$$

Where $P(s_i)$ represents the occurrence probability of the ith gray level in an image. The ideal value of entropy is 8 for random image. If it is less , the chance of predictability is more. Table 1 shows entropy of some sample images along with their respective cipher images.

4.3 Mean Square Error (MSE)

Generally MSE is analysed between plain image and cipher image by taking mean of the squared difference between them. More value of MSE leads to higher encryption and more noise in the plain image. Let I1 represents plain image where E1 denotes cipher image after encryption operation. Mathematical equation for MSE [26] given by.

$$MSE = \frac{1}{MXN}\sum_{i=1}^{M}\sum_{j=1}^{N}[X(i,j) - Y(i,j)]^2 \quad (6)$$

4.4 Peak Signal to Noise Ratio (PSNR)

Peak signal-to noise ratio which is always opposite to Mean Square Error (MSE). Cipher image quality generally measured by PSNR quantity. For good security of image, more MSE and Less PSNR. Mathematically PSNR is given as below [26].

$$\mathrm{PSNR} = 10\log_{10}\frac{255}{\mathrm{MSE}} \quad (7)$$

4.5 UACI and NPCR

The sensitivity of proposed encryption technique with respect to secret key and plain image can be checked using two tests. Namely: Unified average changing intensity (UACI) and Number of pixels change rate (NPCR) [25]. The equation 4 is used to calculate UACI.

$$\mathrm{UACI} = \frac{1}{\mathrm{N}}\left[\sum\nolimits_{\mathrm{i,j}}\frac{|\mathrm{C1(i,j)} - \mathrm{C2(i,j)}|}{255}\right] \quad (8)$$

where, 'm' indicates No. of rows, 'n' represents No. of columns. C1(i,j) represents original image and C2(i,j) represents cipher image. Equation 5 shows the calculation of NPCR.

$$NPCR = \frac{\sum_{i,j}D(i,j)}{MXN}X100\% \quad (9)$$

where, 'M' indicates No. of rows, 'N' represents No. of columns and D(i,j) is defined as follows

$$D(i,j) = \begin{cases} 1, C1(i,j) \neq C2(i,j) \\ 0, \quad otherwise \end{cases}$$

where C1(i,j) and C2(i,j) are Original image and Cipher image respectively.

Table 1. Performance parameters for proposed ROI encrypted system

Image Name	Entropy_In (Bit)	Entropy_Enc(Bit)	MSE	PSNR(db)	NPCR(%)	UACI(%)
Baby	2.6682	3.2067	5.6701	40.5949	29.6270	12.3073
Baby Womb	3.5540	6.8615	34.9602	32.6951	75.8653	32.0196
MRI	4.5016	4.7368	36.2388	32.5391	47.5328	15.5345
Hand	4.1229	4.3455	57.6363	30.5238	36.2294	11.7870
Foot	3.7588	3.8080	37.8616	32.3488	28.6262	8.1099

Table 2. Efficiency of proposed ROI encrypted system

Image Name	**Image Size**	**Encryption Time (sec)**	**Time (%) Saving compared to full image encryption**
Baby	761*1024	4.74	70.38
Baby Womb	72*1280	5.31	34.14
MRI	720*960	4.26	52.47
Hand	881*750	3.39	73.78
Foot	750*297	2.70	71.38

Table 3. Timing analysis comparison for existing methods

Image Name	Image Size	Proposed Method	Jan Sher Khan et al. (2019)[29]	Ayoup et al. (2015) [28]	Ullah et al. (2013)[27]
Gray image	256*256	0.43 sec	0.55 sec	7.21 sec	3.12 sec

Table 4. Histogram analysis of proposed ROI Encryption system

Medical Image	Histogram	Encrypted Image	Histogram
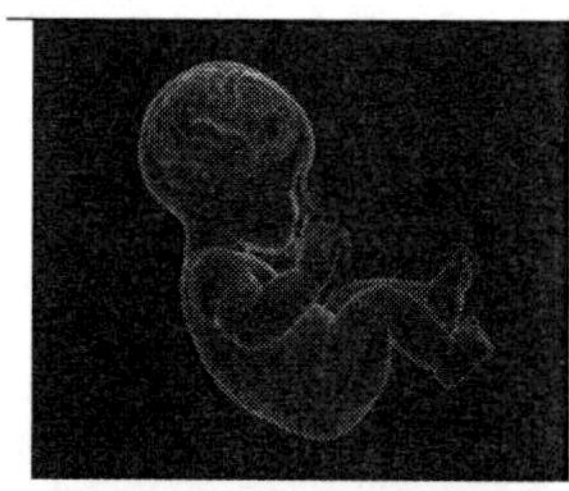	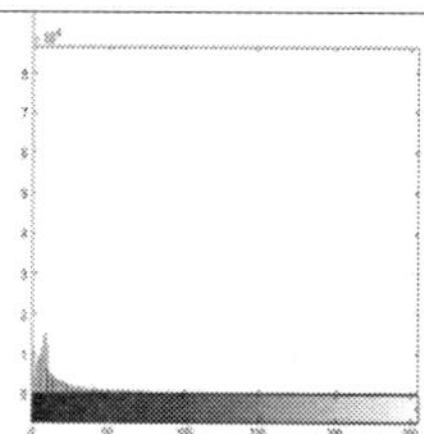	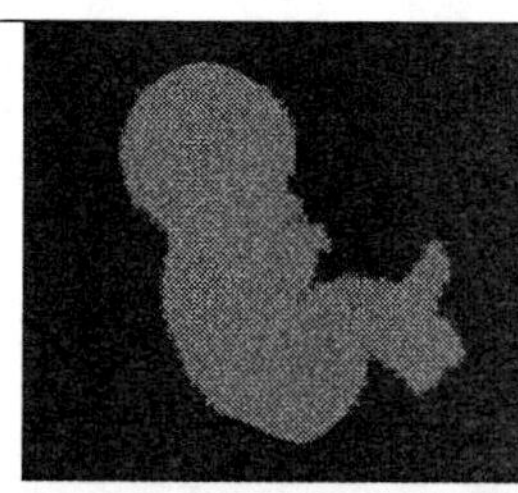	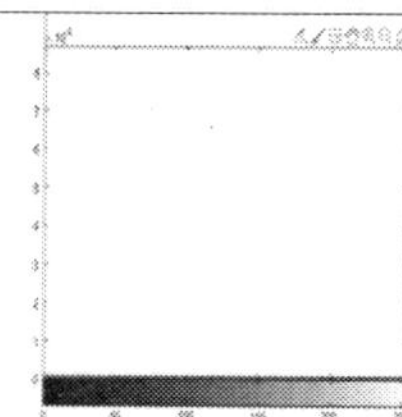
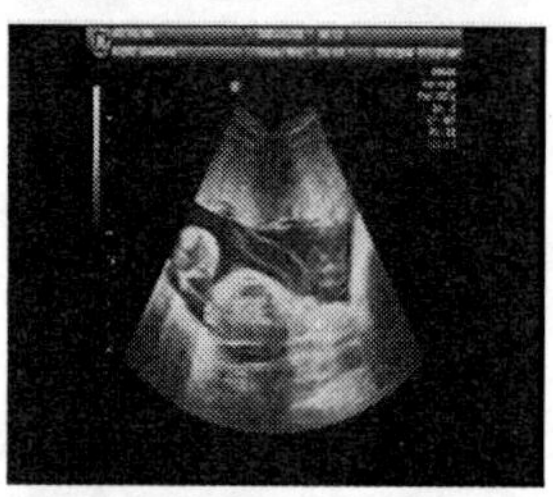	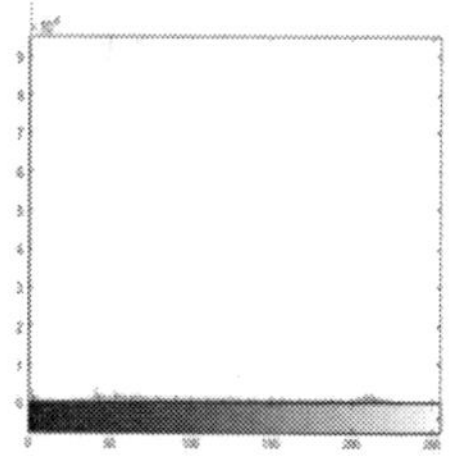		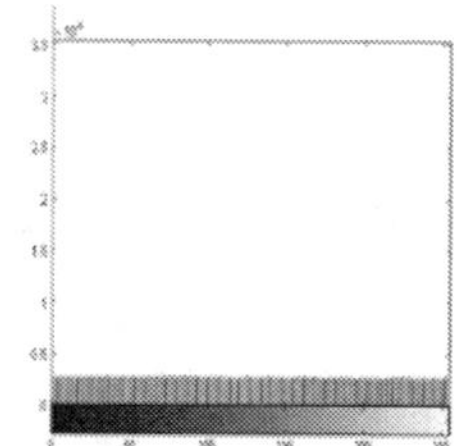
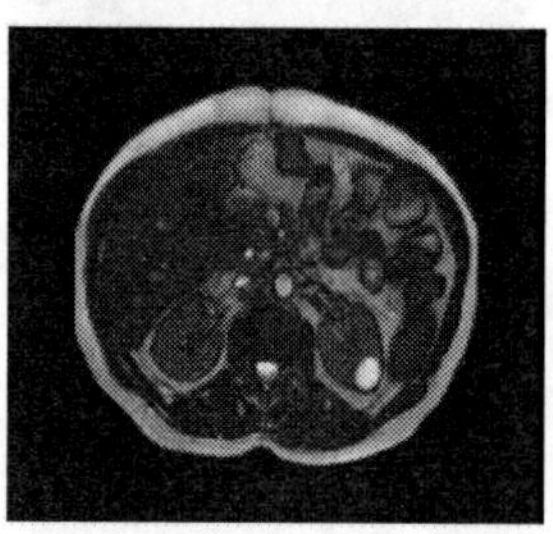	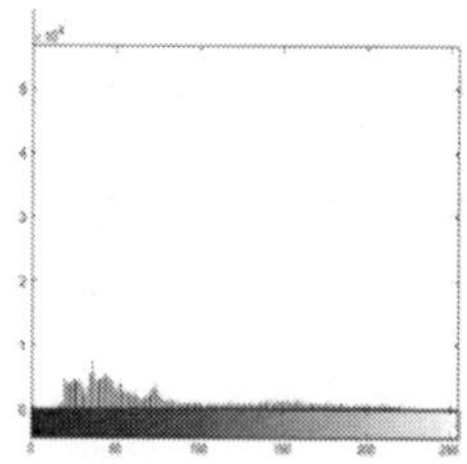	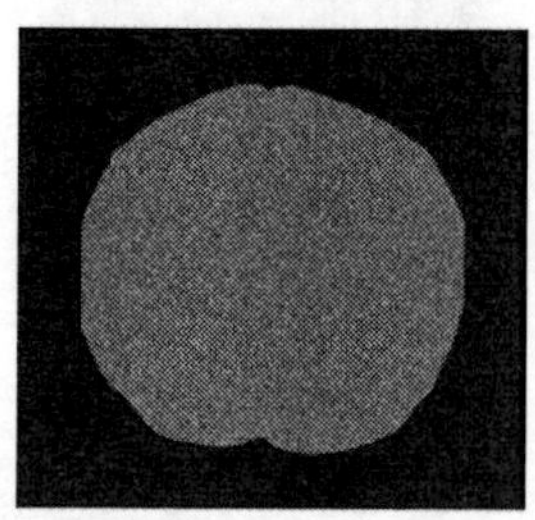	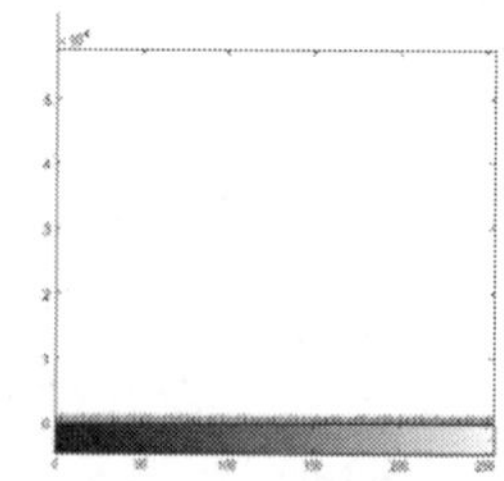

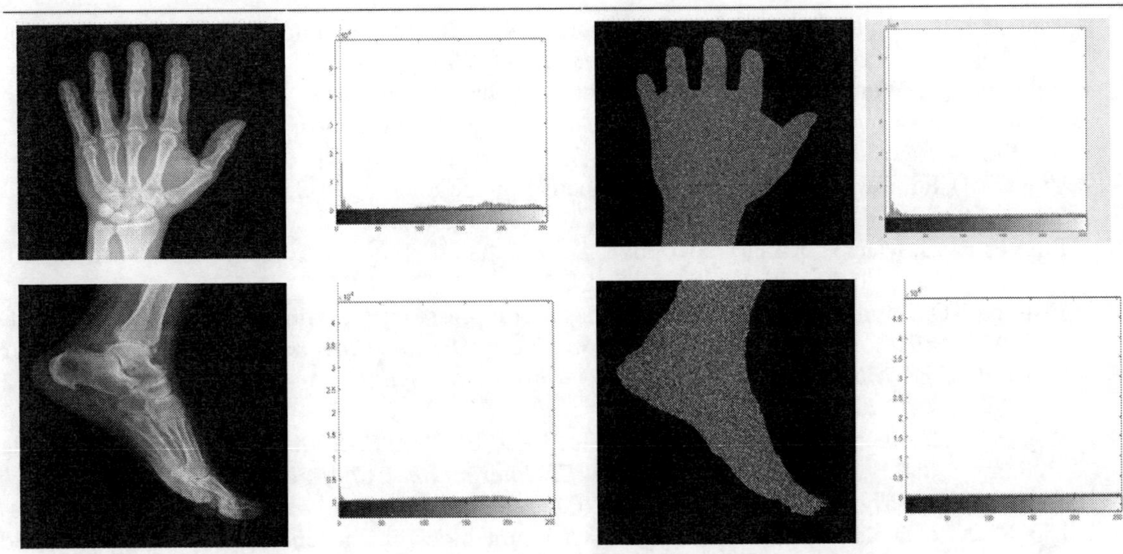

4.6 Time analysis

Matlab 2019*b* is used to test the proposed scheme on gray medical images of various sizes with the system configuration having Intel Core i5 CPU @ 2.53 GHz and 4.0 GB RAM on Windows 10 OS. The encryption time is much lower in the proposed scheme than the other three regularly used encryption techniques; the same can be seen in Table 3 ([25-27]). The proposed encryption technique is a single round image encryption technique which is used for real time, online communication.

From the table 1 we concluded that entropy values of cipher images are more than original plain image. MSE values are increases according to different image that will give the amount of encryption. Because of selective encryption proposed method NPCR values is not much varied that indicates that reduction in computation cost and time.

Table 2 gives the efficiency of proposed method in-terms of execution speed and cost. Compared full image encryption this method saves 70% of computational cost and achieves fast execution time to encrypt image.

5. Conclusion

In this paper, we presented a ROI based selective image cryptographic system to encrypt only ROI part from medical image using the concepts of Hilbert scan and Skew Tent map. Proposed scheme can encrypt any medical image in an effective manner with limited resource utilization and better security. Through active contour method ROI part is extracted. Then ROI part is permuted using Hilbert scan pattern and diffused using skew tent map with predefined threshold values. We can conclude that the proposed technique can be used in real time medical image encryption with limited resource utilization based on the security analysis of the proposed encryption technique.

Acknowledgement

This work is supported by VTU Belagaum, Karnataka and Electronics and communication Research centre, GSSSIETW, Mysuru, Karnataka

References

[1] Zhijuan D and Zhong S 2019 *A digital image encryption algorithm based on chaotic mapping Journal of Algorithms & Computational Technology* **13** 1748302619853470.

[2] Belazi A, Talha M, Kharbech S and Xiang W 2019 Novel medical image encryption scheme based on chaos and DNA encoding *IEEE Access* **7** 36667-36681

[3] Ghadirli HM, Nodehi A, Enayatifar R 2019 An overview of encryption algorithms in color images *Signal Processing* **164** 163-85

[4] Younas MB, Ahmad J 2014 Comparative analysis of chaotic and non-chaotic image encryption schemes *Int. Conf. on Emerging Technologies (ICET)* 81-86
[5] Rehman AU, Wang H, Shahid MM, Iqbal S, Abbas Z, Firdous A 2019 A Selective Cross-Substitution Technique for Encrypting Color Images Using Chaos, DNA Rules and SHA-512 *IEEE Access* **7** 162786-802
[6] Tresor LO, Sumbwanyambe M 2019 A Selective Image Encryption Scheme Based on 2D DWT, Henon Map and 4D Qi Hyper-Chaos *IEEE Access* **7** 103463-72
[7] Ping P, Fan J, Mao Y, Xu F, Gao J 2018 A chaos based image encryption scheme using digit-level permutation and block diffusion *IEEE Access* **6** 67581-93
[8] Lisungu TO, Sumbwanyambe M 2019 Image Compression-Encryption Scheme Based on 2D DWT, SPIHT and Qi Hyper-chaos *Southern African Universities Power Engineering Conference/Robotics and Mechatronics/Pattern Recognition Association of South Africa (SAUPEC/RobMech/PRASA)* 177-182
[9] Rieger M, Hämmerle-Uhl J, Uhl A 2019 Selective JPEG2000 Encryption of Iris Data: Protecting Sample Data vs. Normalised Texture *IEEE International Conference on Acoustics, Speech and Signal Processing (ICASSP)* 2602-2606
[10] Liu X, Xiao D, Huang W, Liu C 2019 Quantum block image encryption based on arnold transform and sine chaotification model *IEEE Access* **7** 57188-99
[11] Xie D 2019 Public Key Image Encryption Based on Compressed Sensing *IEEE Access* **7** (2019): 131672-131680
[12] Zhu S, Zhu C 2019 Plaintext-related image encryption algorithm based on block structure and five-dimensional chaotic map *IEEE Access* **7** 147106-18
[13] Binjubeir M, Ahmed AA, Ismail MA, Sadiq AS, Khan MK 2019 Comprehensive survey on big data privacy protection *IEEE Access* **8** 20067-79
[14] Li M, Wang P, Liu Y, Fan H 2019 Cryptanalysis of a Novel Bit-Level Color Image Encryption Using Improved 1D Chaotic Map *IEEE Access* **7** 145798-806
[15] Kiran, Parameshachari BD, Sahana VS, Apoorva C, Thanuja BS 2018 Auto-Blocking based Image Encryption using ECG Signal *IJCA Proceedings on National Conference on Electronics, Signals and Communication NCESC 2017* **3** 33-35
[16] Parameshachari BD, Panduranga HT, Naveenkumar SK 2017 Partial encryption of medical images by dual DNA addition using DNA encoding *International conference on recent innovations in signal processing and embedded systems (RISE)* 310-314
[17] Parameshachari BD, Kiran, Rashmi P, Supriya MC, Rajashekarappa, Panduranga HT 2019 Controlled partial image encryption based on LSIC and chaotic map *ICCSP* 60-63
[18] Kiran, Rashmi P and Supriya M C 2019 Encryption of Color image to enhance security using Permutation and Diffusion Techniques *International Journal of Advanced Science and Technology* **28** 375-384
[19] Subramani P, Rajendran GB, Sengupta J, Pérez de Prado R and Divakarachari PB 2020 A Block Bi-Diagonalization-Based Pre-Coding for Indoor Multiple-Input-Multiple-Output-Visible Light Communication System *Energies* **13** 3466
[20] Puttamadappa C and Parameshachari BD 2019 Demand side management of small scale loads in a smart grid using glow-worm swarm optimization technique *Microprocessors and Microsystems* **71** 102886
[21] Chan TF and Vese LA 2001 Active contours without edges *IEEE Transactions on image processing* **10** 266-77
[22] Caselles V, Kimmel R, Sapiro G 1997 Geodesic active contours *International journal of computer vision* **22** 61-79
[23] Sivakumar T and Venkatesan R 2014 Image encryption based on pixel shuffling and random key stream *Int. J. Comput. Inform. Technol.* **3**
[24] Kadir A, Hamdulla A, Guo WQ 2014 Color image encryption using skew tent map and hyper chaotic system of 6th-order CNN *Optik.* **125** 1671-5
[25] Wu Y, Noonan JP and Agaian S 2011 NPCR and UACI randomness tests for image encryption. Cyber journals: multidisciplinary journals in science and technology *Journal of Selected Areas in Telecommunications (JSAT)* **1** 31-8

[26] Ahmad J and Ahmed F 2010 Efficiency analysis and security evaluation of image encryption schemes *Computing* **23** 25.

[27] Ullah I, Iqbal W and Masood A 2013 Selective region based images encryption *2nd National conference on information assurance (NCIA)* 125-128

[28] Ayoup AM, Hussein AH and Attia MA 2015 Efficient selective image encryption *Multimedia tools and applications* 1–16

[29] Khan JS, Ahmad J 2019 Chaos based efficient selective image encryption *Multidim Syst Sign Process* **30** 943–961

[30] Ramesh GP and Kumar NM 2018 Radiometric analysis of ankle edema via RZF antenna for biomedical applications *Wireless Personal Communications* **102** 1785-1798

IOP Conf. Series: Materials Science and Engineering **925** (2020) 012029 doi:10.1088/1757-899X/925/1/012029

Traffic Operations and Capacity Analysis in India

Kandukuru Jagan Mohan Reddy[1] **and Paritosh Srivastava**[2]

[1] Noida International University, U.P, India.
[2] Department of Civil Engineering, Noida International University, U.P., India.
[1]jaganmohanreddy120@gmail.com, [2]paritosh.srivastava@niu.edu.in

Abstract. This study highlights the concept of capacity analysis along with traffic operations in urban areas in India. First chapter indicates the introduction of this study through which research aims and objectives can be analysed properly. Moreover, research background and rationale of this study can be also highlighted in this chapter. The second chapter discusses the literature review due to which process of traffic operations and capacity analysis of vehicles can be analysed properly. Using model and theories of traffic flow can help in managing the complexities and issues in traffic management system in urban areas in India. The third chapter defines the research methodology through which tools and techniques regarding this research can be analysed successfully. Using philosophy, approach, design and data collection process, adequate information regarding traffic operations and capacity can be recognised properly.

1. Introduction

The significance of capacity analysis and traffic operations helps in identifying the traffic condition and structure in urban areas in India. The intent of this study is to comprehend the performance structure of vehicles regarding traffic operations in urban areas in India. This study discusses the capacity of vehicles and traffic operations in the context of mixed traffic conditions in India.

1.1. Research background

In India, the status of socio-economic structure and urbanisation have improved through which the capacity of vehicles increased at a rapid pace. The urbanisation level has enhanced from 17.29% in 1951 to 31.6% in 2011 and continued [1]. India is one of the fastest-growing countries all over the world through which the urban population has also increased at a rapid pace. Due to the growth of this population, the mixed traffic behaviour has also increased thus; the traffic management operators can face positive and negative impact due to this traffic situation. Most of the time, these operators have failed to manage the traffic situation that can lead to uncertain accidents. It can negatively impact on the population and violate the structure of mixed traffic operations.

1.2. Aims and objectives

1.2.1 Aims
The foremost aim of this study is to analyse the traffic operations and take adequate measures to handle the mixed traffic to reduce the complexities for the population of urban areas in India.

1.2.2 Objectives

- To analyse the structure of heavy traffic conditions in urban areas in India.
- To recognise the role of traffic management operators during handling mixed traffic situations in India.
- To analyse issues that have a negative impact on the population of urban areas in India.
- To recommend appropriate strategies to overcome these issues and manage heavy traffic situations in India.

1.3 Rationale
Traffic operations are one of the crucial factors for handling heavy traffic situations through which urban people can easily use urban roads without facing any accident or any other complexities. However, sometimes, due to the high capacity of traffic, the traffic management team has failed to maintain the structure of managing traffic conditions. It can negatively impact on the population of urban areas thus uncertain accidents in the street and roads in India can happen that can majorly affect the population. Due to lack of knowledge in understanding the traffic rules and other functions, the traffic management team has failed to handle the heavy traffic that can increase congestion and create delays in traffic.

1.4 Significance of the study
The significance of this study is to acknowledge the operations of traffic and analyse the vehicle capacity in the urban areas in India. Through the aid of several methods, vehicle capacity and other aspects regarding traffic can be analysed successfully that can be beneficial for managing the heavy traffic in India. Moreover, through this study, appropriate strategies for handling the complex situation regarding mixed traffic can be recognised through which uncertain accidents can be avoided. It can be beneficial for local people in urban areas in India.

1.5 Summary
From this chapter, aims and objectives of this study can be analysed properly through which strategies can be incorporated to handle the mixed or heavy traffic condition in urban areas in India. Moreover, issues in the traffic management system and the importance of adequate methods can be recognised to improve the quality of this study. It can help in improving knowledge about traffic condition and situation through which capacity of vehicles can be acknowledged.

IOP Conf. Series: Materials Science and Engineering **925** (2020) 012029 doi:10.1088/1757-899X/925/1/012029

2. Literature Review

The literature review helps in providing adequate information about the traffic operations and conditions of the mixed traffic situation in India. This section also discusses the intervention model and theory to analyse the process of handling the structure of mixed traffic in urban areas in India.

2.1 Conceptual Framework

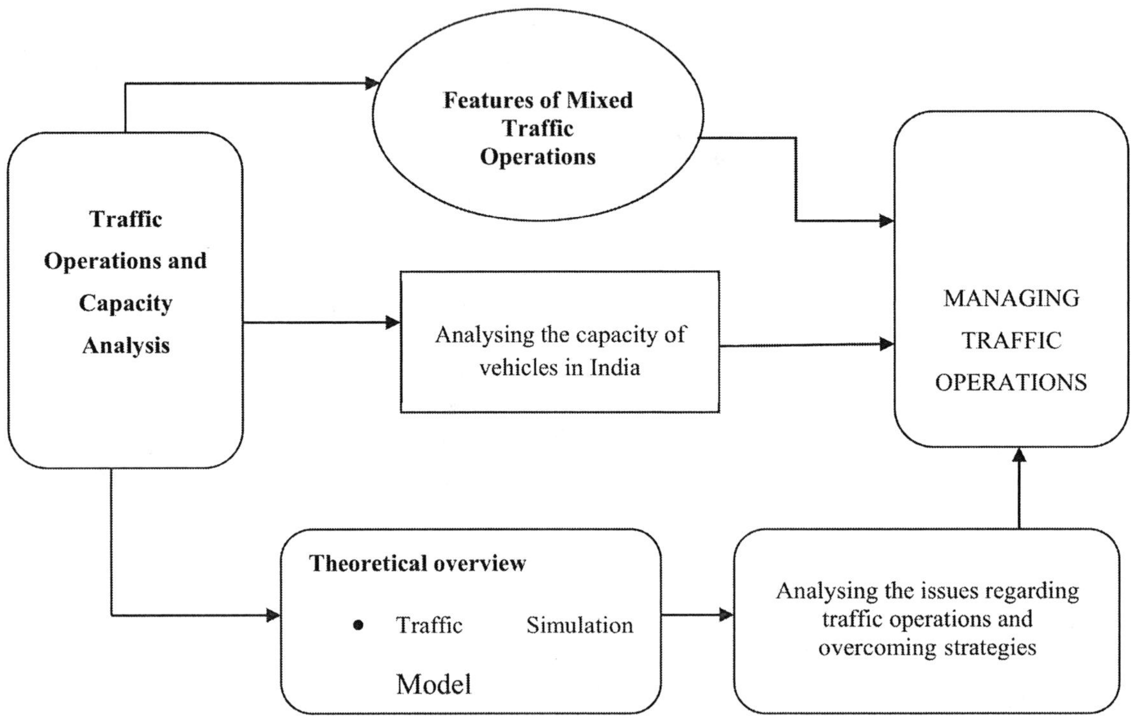

Figure 2.1. Conceptual Framework

(Source: Created by author)

2.2 Features of mixed traffic operations

Mixed traffic operations are very helpful for managing the congestion and heavy traffic situation through which complexities in urban roads can be managed in India. There are several features of mixed traffic operations such as merge, diverse and weaving through which traffic conditions can be successfully maintained. In light of this, using the merging segment, the behaviour of a driver can be analysed properly through which mixed traffic situation can be understood properly [2]. It can be beneficial for urban areas in India through which complexities can be solved and improve the traffic situation. The diverse segment has mainly been used during repairing of roads that can guide the drivers to take a diversion from that repaired road to reduce the complexity natures. Through the aid of this segment, traffic congestion can be avoided

through which efficiency in developing traffic operations can be managed successfully. It can be beneficial for mixed or heavy traffic in the urban roads in India thus, every driver can follow the diversion process to manage the congestion of traffic and improve the capacity of vehicles.
On the other hand, using the weaving segment, vehicles are required to change the lane according to the requirements of the traffic situation. In this way, uncertain accidents and congestion in traffic operations can be successfully managed. It can also be helpful for mixed traffic conditions through which issues and other negative aspects regarding the traffic situation can be managed properly [3]. It can be helpful for the population of urban areas through which traffic congestion and other negative aspects can be avoided. These segments can provide a structural way of organising the mixed traffic and managing the diversity of traffic and maintain the capacity flow of vehicles.

2.3 Analysing the capacity of vehicles in India

Passenger car units play an essential role in understanding the different types of vehicles through which capacity and frequency of vehicles can be understood properly. From this process, it has been acknowledged that there are four types of vehicles used in the urban areas in India. This includes four-wheelers, three-wheelers, two-wheelers and more than four-wheelers [4]. The frequency of these wheelers can be organised successfully through which complexities and other aspects regarding heavy or mixed traffic conditions can be organised properly. It can be helpful for maintaining the structure of traffic operations and reduce the issues regarding mixed traffic.

The traffic management team is needed to make initiative in analysing the traffic situation through which complexities and efficiency of traffic can be organised properly. Additionally, through the aid of PCU, proportionality of vehicles and traffic volume can be organised properly due to which vehicle type on the traffic stream can be identified properly [5]. It can be helpful for managing highway traffic and organising the different LOS services. It would give a significant way of managing the relationship between volume and speed of the four types of vehicles. It can directly impact on the PCU values through which traffic composition on six-lane and four-lane traffic can be managed properly.

IOP Conf. Series: Materials Science and Engineering **925** (2020) 012029 doi:10.1088/1757-899X/925/1/012029

2.4 Intervention model and theory

2.4.1 Traffic Simulation Model

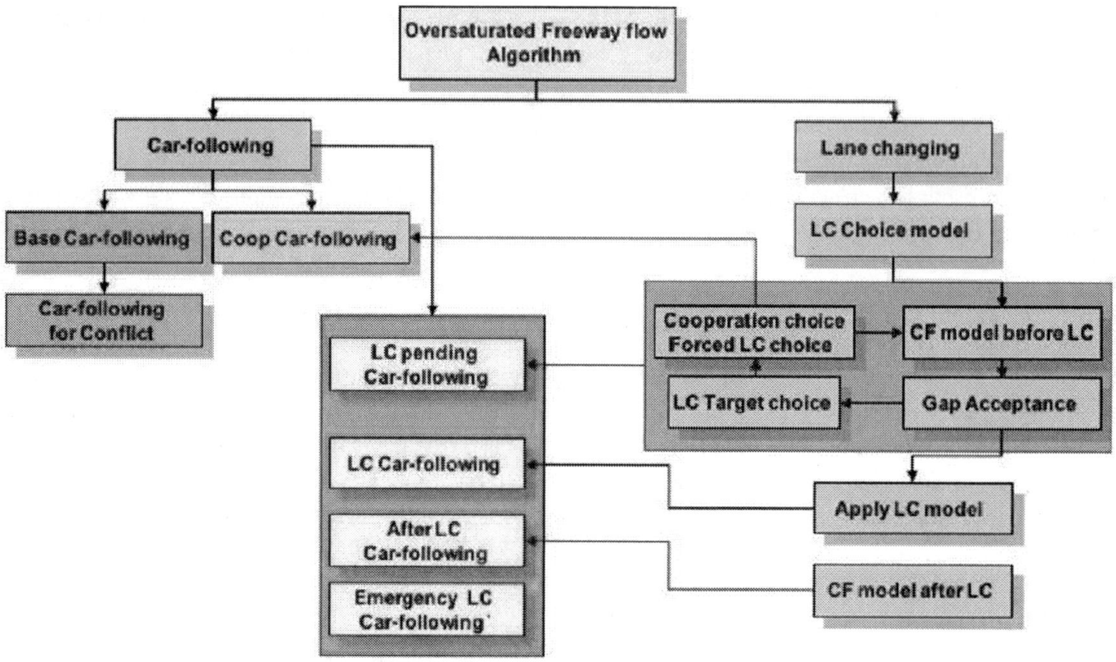

Figure 2.2. Traffic Simulation model
(Source: 6)

Traffic simulation models describe the flow of cars according to the lane through which the capacity of vehicles and structure of lanes can be understood properly. Through the aid of this model, base and coop car following processes can be acknowledged through which conflict among cars within the mixed traffic can be analysed properly [7]. Thus, the traffic management team in the urban areas in India can take crucial measures to handle this traffic situation and manage the issues within the heavy traffic condition. On the other hand, lanes also play a key role in managing the capacity of vehicles using LC choice models. Through this model, drivers can select their individual lanes to decrease the congestion in traffic and manage the flow of traffic conditions. It can increase the gap among the vehicles in the heavy traffic condition that can decrease the rate of uncertain accidents within the urban areas in India.

2.4.2 Traffic flow theory

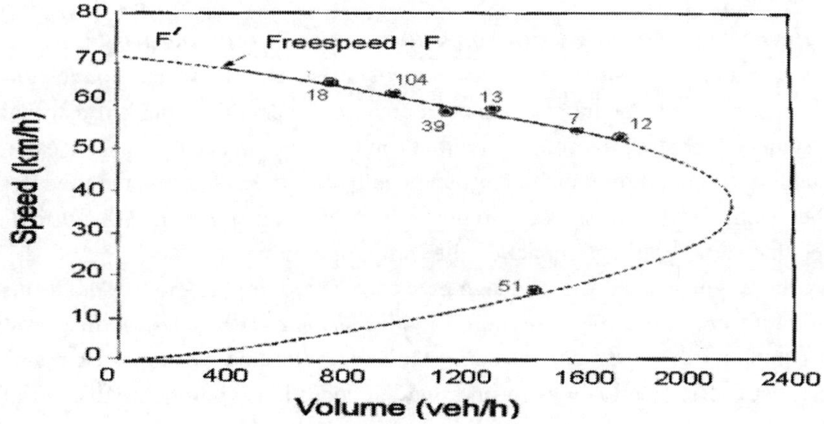

Figure 2.3. Traffic flow theory

(Source: 8)

Traffic flow theory is mainly based on the two factors such as speed and volume through which free speed for vehicles can be successfully managed. If the volume of vehicles is low then the speed and accuracy of that vehicle can be increased. Through the aid of this theory, vehicle capacity and congestion processes in heavy traffic can be analysed properly [9]. Thus, the traffic management team is expected to set rules for limiting the speed in some areas where the population is very high in India. Through the limitation, drivers can follow the speed restriction and reduce the uncertain accidents that can be beneficial for urban people in India. Through the aid of this theory, the traffic management team can guide the vehicles to follow a specific lane and maintain the flow of managing all the segments of the mixed traffic situation. In this way, the traffic management process can be managed successfully and improve the traffic operations within the urban areas in India.

2.5 Impact of intervention model and theory

Intervention model and theory provide a way of handling the heavy traffic situation and improve the structure of the traffic management system. Through the aid of the traffic simulation model, car density and lane capacity can be analysed properly through which mixed traffic conditions can be maintained properly. In this manner, it can be beneficial for organising the structure of handling the traffic and improving the flow of traffic through which vehicle capacity and sustainability can be organised properly [6]. This model also guides the drivers to improve the flow of traffic operations that can give a favourable impact on the urban population in India.

Using traffic flow theory, volume capacity and speed of vehicles can be understood properly that can guide the traffic management team to take suitable measures to deal with those issues regarding traffic. It can provide a positive impact on social lifestyles for urban people due to the complexity of nature. It can help in improving the process of speed circulations according to the volume of vehicles. Thus, drivers can follow a specific route for their driving in the urban areas in India.

2.6 Analysing the issues regarding traffic operations and overcoming strategies

Issues in traffic operations can slow the movement of traffic due to which congestion in vehicles and complexities in drivers in heavy traffic can be increased. Lack of proper infrastructure in the traffic management system can increase the congestion that can lead to fatal accidents. This can negatively impact the urban populations in India through which urban people have failed to cross the road in a secured manner [10]. On the other hand, due to lack of technological advancement process, the structure of performing TMS can be decreased that can negatively impact on the traffic operations.

In order to solve these issues, the traffic management team is required to modify the infrastructure with the help of traffic flow theory and traffic simulation model of their traffic management system. In this way, congestion and complexities among drivers can be solved. Additionally, the traffic management team can incorporate the latest traffic technologies to improve the conditions and operations of traffic due to which efficiency and quality in performance regarding traffic can be organised properly [11]. It can help in providing a favourable impact on the sustainability of traffic capacity and operations as well.

2.7 Gap of Literature

Level of transportation in Indian Traffic system cannot be incorporated within this study due to lack of time management. Moreover, the concept of traffic rules and regulations in Indian Traffic system cannot be included due to lack of adequate resources. Due to these gaps, the quality of this study can be decreased in a certain manner and also slow the progression of this study.

2.8 Summary

From this section, the capacity of the vehicles and traffic operations depending on the lane have been discussed through which structural performance regarding traffic management systems can be analysed. Using traffic flow theory and traffic simulation models, the relationship between speed and volume for a particular vehicle can be understood properly. Moreover, the interconnection between the lane and car simulation can be understood properly in this section.

3. Research Methodology

Research methodology defines the tools and techniques through which the structural way of managing data processing regarding the traffic management system can be organised. Using several methods, the impact of TMS on vehicle capacity and traffic operations can be analysed properly due to which research quality can be increased in a significant manner.

3.1 Research philosophy

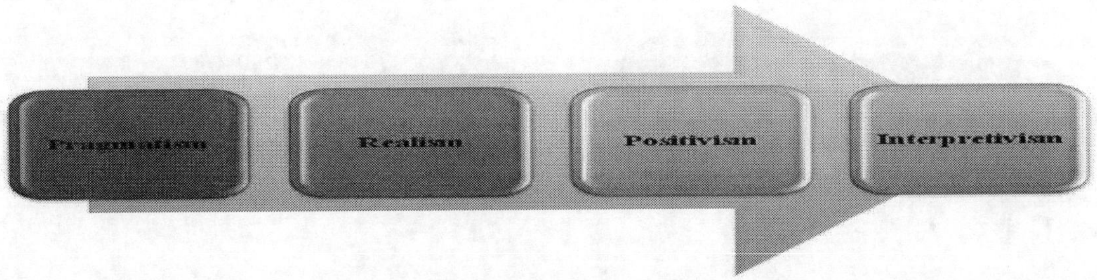

Figure 3.1. Research Philosophy
(Source: 12)

Research philosophy defines the way of including practical information about the content of the research study through which the clarity of the research study can be managed properly. This includes four types like pragmatism, realism, interpretive and positivism [12]. Positivism philosophy has been used in this study to analyse the practical information about the traffic operations due to which mixed traffic conditions can be successfully analysed. This philosophy has also helped in understanding the capacity of vehicles in the urban areas in India through which the quality of this research can be maintained properly.

3.2 Research approach

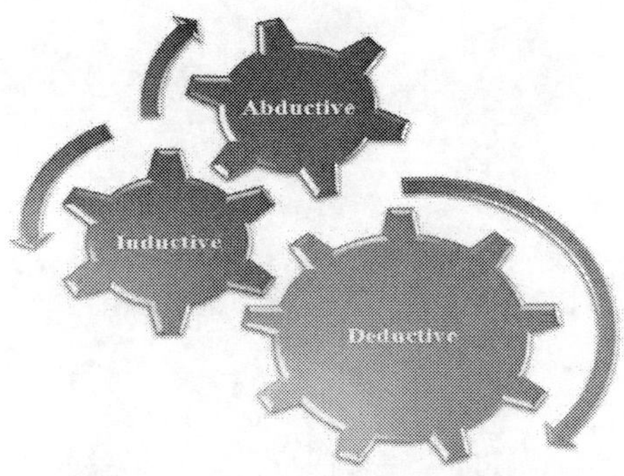

Figure 3.2. Research approach
(Source: 13)

Research approach helps in analysing the validity of the research content due to which specific information can be incorporated within the study. This contains three types such as abductive, deductive and inductive research approaches [13]. The deductive research approach has been incorporated within this study to convert the general information about the traffic management system to specific information about the traffic information system. It can help in improving the process of managing the structure of the traffic condition and deliver adequate information about the traffic rules and regulations regarding the situation in India.

IOP Conf. Series: Materials Science and Engineering **925** (2020) 012029 doi:10.1088/1757-899X/925/1/012029

3.3 Research design

Figure 3.3. Research design

(Source: 14)

Research design assists in establishing a general plan for selecting the appropriate method of data collection process through which essentiality of research can be structured. Primarily, the types of research design can be categorised in two types such as exploratory and conclusive [14]. Conclusive research design has been selected for this research for verifying the existing information about traffic operations and delivering accurate results about the capacity analysis in India. It can give a structural way of handling the heavy traffic situation due to which complexities can be decreased for urban drivers. Thus, they can easily operate their activities without facing any issues and maintain the flow of the traffic management system.

3.4 Data collection method

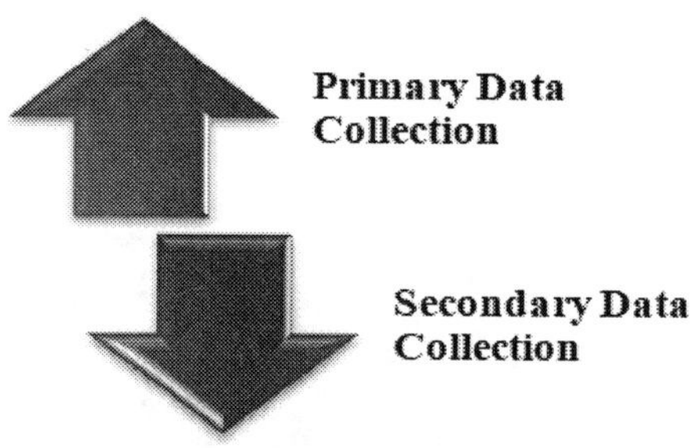

Figure 3.4. Data collection method

(Source: 15)

The data collection method is one of the essential methods in the research methodology through quality data that can be shared using several ways to manage the quality of a research study. This includes primary data collection and secondary data collection methods, depending on the way in which the data has been collected for the research [15]. Primary data collection method has been used in this study in the form of the site selection process through which accurate information about traffic situations can be incorporated. Using this method, this research study is capable of providing the data about vehicle categories, capacity and various others through which complexity nature can be organised properly.

3.5 Data analysis

Figure 3.5. Data analysis

(Source: 16)

Data analysis provides a way of performing data collection methods through which information structure can be successfully synchronised. Depending on the way the data has been collected, the process of analysing data also tends to differ, with the main two categories including qualitative and quantitative data analysis processes [16]. Qualitative data analysis has been used in this study to acknowledge the way of incorporating adequate information about the traffic condition and operations in urban areas in India. Using this method, quality information regarding traffic management systems can be maintained properly due to which the quality and efficiency of this research study can be organised properly.

3.6 Sampling size

Sampling size represents the quantity of research samples from which accurate and quality data can be gathered successfully. For the purpose of conducting this research, using a random probability sampling method, the sample of 51 respondents has been selected from the population of 151. The random simple probability method has been chosen since it allows the researchers to select research data in a random and unbiased manner, making the research outcomes fair and unbiased as well [17]. Thus, the viewpoints of these respondents have been gathered regarding the traffic conditions and operations. It can be helpful for incorporating various kinds of viewpoints through which positive and negative aspects regarding traffic management systems in urban areas in India have been analysed. It would help this research to deliver adequate data about the mixed traffic condition through which clarity and efficiency in research methods have been developed. It would give a structural way of handling more quality essential information to achieve the goals and objectives of this study.

3.7 Ethical consideration

Ethical consideration is required for the research study through which confidentiality and privacy of research participants can be maintained properly. Using the Data Protection Act 2018 (c.12), the security regarding collected data can be developed through which external members cannot access the database of this research. It can be beneficial for this research study by organising the structure of this research. On the other hand, privacy and other essential information regarding respondents can be organised properly. It can be helpful for them that they can execute their operations without facing any issues.

3.8 Summary

From this section, adequate tools and techniques have been used to manage the structure of this research through which the aims and objectives of this study can be reached. This section also delivers appropriate ways of handling the data collection and data processing techniques to manage the clarity and confidentiality of this study.

4. Findings

4.1 Findings 1: Indian Roadways work under mixed Traffic conditions

Based on the data provided in the table, it can be stated that Indian Traffic systems have been observed to work under mixed traffic conditions. However, Apart from this, it can be stated that, movement of too many vehicles have been the main reason for commotion. In this context, it can be stated that vehicles maneuvering with different lane behavior results in traffic conditions that behave homogeneously. Moreover, this has usually been generated due to their static as well as dynamic behavior. Currently, it has been observed that traffic on the road rises rapidly, thus, pressurizing the road conditions. Hence, proper management of traffic has been observed to require better roads along with enhanced traffic management systems. Apart from this, with increase in population the pedestrian count has been observed to increase considerably. This has also resulted in Indian Traffic conditions to be homogeneous in nature. Utilizing public transport would reduce the count of vehicles on road as a result of which Traffic congestion can be reduced comparatively.

4.2 Findings 2: main reasons for traffic congestions

Traffic congestion can be considered as one of the main issues that hinders the construction of Quality roadways in India. This increases pressure on rest road networks as the traffic is diverted in most of the cases. More than 37.50 % of the participants within the research had agreed to the fact that too many cars on Indian Roads are the main reason for increased traffic conditions. Apart from this, an increase in the number of pedestrians on the roads disobeys the traffic signals. This creates commotion as a result of which traffic congestion increases considerably. Roadways can be considered as one of the most commonly used modes of commuting for India after railways. However, the current infrastructure developments have been observed to be comparatively not catering the needs of common Indians. Utilizing public transport would reduce the count of vehicles on road as a result of which Traffic congestion can be reduced comparatively.

4.3 Findings 3: influential factors that impact Free Flow Speeds

Free flow speed can be classified as a term that can be used in order to describe the average speed of a motorist while crusading long distances. However, in case of India, this speed is comparatively low due to the adverse road condition of India. In accordance with the presentations of the table, it can be stated that Vehicle subclasses, Width of the road and quality of the roads have been identified as the main reason for low free flow speed. Maneuvering capacity of a vehicle can also be considered as a major parameter that influences free flow speed. Urban Arterial roadways have been associated with poor maintenance processes and it creates irregularities within roadways. The aforesaid factor can be reflected as the main reason that would initiate reductions in vehicle speed processes. In India, it has been observed that Vehicles change lanes without indication; this can also be considered as one of the major reasons that would help in reducing free flow speed of the vehicle.

4.4 Findings 4: role of urban planners of your city

Data in the table provided above helps in having a glance at the role of the urban planners within a city. In this context, it can be stated that more than 38% of the participants have been observed to be dissatisfied with the role of urban planners. The achievement of rapid growth is expected to be achievable as well as sustainable and this has been observed to present formidable challenges to the urban planners. In most of the cases, it has been observed that the urban planners of a city are associated with corruption. This results in reduction of budgets that have been allocated for construction of Urban Arterial Road network construction. Qualities of roads have been observed to be reduced as a result of this. Besides this, it can be stated that construction of roads has also been associated with certain complications. Instance includes issues like legal conjugations and this hinders the process of creating quality Roadways in the country.

5. Conclusion

This study had aimed at analyzing the performance of urban arterial roads within India. Hence, the concepts of research onion had been deployed in order to accomplish the aims and objectives of the study. Positivism philosophy has been used in this study to analyses the situation of Indian mixed traffic and also helps in improving knowledge about services of urban arterial roads. This has been observed to be aided with the concepts of deductive analysis and descriptive designs. The study had implemented the concepts of mixed methods for framing the contents of the study. One major aspect of this study had been the data validity therefore both primary and secondary analysis has been implemented within the study accordingly. Survey is one of the means of data collection methods that have been used in the study. In order to be specific, Arterial roads form the major aspect of a city's infrastructure that makes the commute process easier as well as effective. However, the type of service provided by these roads has been observed to be problematic. Loads of Obstructions within Roads have resulted in decrease in speed of traffic, apart from this; the issues related to frequent braking has reduced the quality of Roads in India.

5.1 Future Scope

Arterial roads comprise majority of roads in India which connect major town centers in urban areas and have partial access at certain areas like weaving, diverge and merge segments. These roads generally allow movement of smaller vehicles like four wheelers, two wheelers and three wheelers and these roads have specified segments for allowing seamless flow of traffic. However, studies can be carried out within areas through which the performance of Urban Arterial Roads can be increased considerably. Roadways can be considered as one of the most commonly used modes of commuting for India after railways. This has been evident from the fact that most of the roadway transports in India are associated with cheap costs. However, the process of Widening the roads so that there is an increased PCU moving per hour. This can also be considered as one of the major areas of future research. The study also mentions that the Indian Traffic is not homogeneous in nature. However, the study fails in explaining the reason why the Indian Traffic is not Homogenous in Nature. Further Research can be carried out in this field so that the Indian Traffic can be made homogeneous in nature.

5.2 Recommendation

In accordance with the proceedings of the study, it can be stated that lack of maintenance is one of the main reasons that performance of a road is decreased. However, with increase in population has also resulted in increase in road traffic considerations. This in turn have been observed to make the process of maintenance more complex as a result of which the performance of the roadway is decreased considerably. As commented by Gore *et al.* (2019) road closure for maintenance during peak traffic can increase congestion on road as a result of which mobility would be affected. Therefore, closing the road when traffic movement is low can be suggested to the urban planners within Major cities in India. Urban planners have been observed to construct roadways considering the Future expansion capacity of a city. Instances include the construction of Ring roads within Cities like Bangalore and Hyderabad. However, expansion as well as maintenance of Urban Arterial Roads has been observed to be avoided. Lack of Forecasting and Legal complications associated with Roadway construction has been observed to be some of the reason for the same. Hence, the Indian Government can be suggested to look into the matter and ensure proper allocation of funds for the same. Road closure for maintenance during peak traffic can increase congestion on road as a result of which mobility would be affected. However, for maintenance, roads would have to be closed. Road closure during low traffic mobility can be advisable.

Reference

[1] Shriniwas S A 2018 Traffic operations and capacity analysis in India *Transportation Letters* **10** 65-67
[2] Verma A 2016 Review of studies on mixed traffic flow: perspective of developing economies *Transportation in Developing Economies* **2** 5-25
[3] Mankar P U and Khode BV 2016 Capacity Estimation of Urban roads under Mixed Traffic Condition *International Research Journal of Engineering and Technology (IRJET) 3* 2750-2755
[4] Mohan M and Chandra S 2018 Occupancy time-based passenger car equivalents at unsignalized intersections in India *Current Science* **114** 13-46
[5] Raj P, Sivagnanasundaram K, Asaithambi G, and Ravi Shankar A U 2019 Review of methods for estimation of passenger car unit values of vehicles *Journal of Transportation Engineering, Part A: Systems* **145** 401-590
[6] Jin J, Ma X and Kosonen I 2017 A stochastic optimization framework for road traffic controls based on evolutionary algorithms and traffic simulation *Advances in Engineering Software* **114** 348-360
[7] Su G, Li N, Yildiz Y, Girard A, Kolmanovsky I 2019 A traffic simulation model with interactive drivers and high-fidelity car dynamics *IFAC-PapersOnLine* **51** 384-389
[8] Zeng J W, Qian Y S, Yu S B, and Wei X T 2019 Research on critical characteristics of highway traffic flow based on three phase traffic theory *Physica A: Statistical Mechanics and its Applications 53* 121-256
[9] Sainct R, Louis X and Forestier A 2017 Application of averaging techniques to traffic flow theory *Mathematical Methods in the Applied Sciences* **40** 600-625
[10] Hames Y, Kaya K, Baltacioglu E and Turksoy A 2018 Analysis of the control strategies for fuel saving in the hydrogen fuel cell vehicles *International Journal of Hydrogen Energy* **43** 10810-10821
[11] Hosseini S, Ivanov D and Dolgui A 2019 Review of quantitative methods for supply chain resilience analysis *Transportation Research Part E: Logistics and Transportation Review* **125** 285-307
[12] Dougherty MR, Slevc LR and Grand JA 2019 Making research evaluation more transparent: Aligning research philosophy, institutional values, and reporting *Perspectives on Psychological Science* **14** 361-375
[13] Alase A 2017 The Interpretative Phenomenological Analysis (IPA): A Guide to a Good Qualitative Research Approach *International Journal of Education and Literacy Studies* **5** 9-19
[14] Abdulkadiroğlu A, Angrist JD, Narita Y and Pathak PA 2017 Research design meets market design: Using centralized assignment for impact evaluation *Econometrica* **85** 1373-1432

[15] Cyr J 2016 The pitfalls and promise of focus groups as a data collection method *Sociological methods & research* **45** 231-259
[16] Johnston M P 2017 Secondary data analysis: A method of which the time has come *Qualitative and quantitative methods in libraries* **3** 619-626
[17] Etikan I, Alkassim R, Abubakar S 2016 Comparision of snowball sampling and sequential sampling technique *Biometrics and Biostatistics International Journal* **3** 55-84

An Approach for Pose Invariant Face Recognition System Using Log-Gabor Feature

Madhura S Shettar[1] and Poornima Byahatti[2]

[1] School of Computer Science and Engineering, KLE Technological University, Hubballi, India.
[2] Department of Automation & Robotics, KLE Technological University, Hubballi, India.
[1]madhura.shettar@kletech.ac.in, [2]poornima.byahatti@kletech.ac.in

Abstract. Biometrics is the process based on humans' characters like behavioural and physical. Among the biometrics traits, face recognition has attained much importance for the authentication of individuals. Face recognition with pose variation is one of the major challenges among the illumination and expression. The system is widely used as surveillance which can withstand in places like ATMs, Airports, and Streets to track the criminals. The face recognizing mainly depends on the feature extraction from the face. Project aim is to extract those features which are invariant even though the pose changes. In the proposed work face recognition by using Log Gabor features is addressed. Firstly, colour images with variation in poses are captured followed by the extraction of facial features components like eyes, nose, mouth, and chin through Log Gabor features. The identification of the person is done using KNN classifier where the matching is carried out by calculating the Euclidean distance between the training and test pairs. Finally, experimental results show the performance of the system for different number of the users.

Keyword: Pose invariant face, Viola jones, Log gabor features, Knn classifier.

1. Introduction

Biometrics is the field where the human characters are relating to the identification and access controls. Nowadays, biometrics products are widely used in the Colleges, offices and schools to monitor the organization. The face recognition has added significant contribution to the biometrics domain and it is used as video surveillance which can be used in the places like ATMs, Airports, and streets in the city to track the criminals and terrorist's activities. Face recognition process consists of three important phases: face detection, face feature extraction and face identification. In face detection, the face of the person is being detected from the image or scene, followed by face feature extraction which deals with identifying facial features like eyes, length of the nose, width of the mouth, these extracted features are used in the application as input data and finally identification of the face is carried out by comparing with the Created database. Across the globe surroundings several other elements would influence the recognition rate of the face recognition system considerably, like pose, illumination, expression and age. To recognize the frontal views of the faces, various methods of face recognition were developed where the candidate was posing the camera very straight and this type of frontal face recognition system are appropriate for the application like where the person always poses the same way from session to session. But in some other cases the user or person might not look in to camera for reason of being caught, possibly not aware of that face image has been clicked. In these the system has to handle faces

with different poses. Thus, aim of the project is to build efficient pose invariant face recognition system, where the system has to be capable of identifying and / or recognize the person with across poses.
Executions of pose invariant face recognition system as follows. Initially, face is being detected form input image given to the system using face detection method. After this, normalization takes place by removing the noise from the detected face, followed by feature extraction which deals with the facial feature extraction such as length of nose, width of the mouth and eyes etc are extracted and stored it in the database. Finally, recognize the person and labelling the class.
This paper organized as follows: In section 2 survey provided on pose invariant face recognition system. Issues and challenges of face recognition system are discussed in section 3. Problem definition and aim of the work is defined in section 4. In section 5 described the proposed approach with design and implementations. Experimental results have shown in the section 6. Finally, conclusion and future work are given in section VII.

2. Literature Survey

The face recognition systems have developed by many researchers using different approaches. Some of them are discussed below.
The author proposed application for the classroom attendance in order to track the proxy attendance in [1]. Here they make use Radio frequency identification to provide unique identification to student.FEI database used to store the images of the student and Fast Adapter Neural network classifier is used to check whether every student is present in the class or not. If the student couldn't classify then it gives the result as proxy attendance and achieved result is 98%.
The authors explained the feature progressing method in [2]. The facial features will change whenever the pose changes. The model finds the frequently changing features when the pose changes and it plots to that person. To detect the face components, Active appearance model has been used frequently and achieved the result 91.2%.
The authors have used shape descriptor called Affine moment invariants to address the issue of pose variations in [3], where technique divides the entire face in to non-overlapping components. Here, they have taken five images for each person with different like 0 degree,45 degrees,90 degree, 135 and 180 degree and result achieved is 95%
In [4] authors make use of Local Binary Histogram through which every pixel information has obtained by identifying labels in histogram. For the whole image, global binary histogram is developed by concatenating all these local binary histograms & achieved the recognition rate 81.65%
Authors defined reference-based descriptor in [5] for image by taking the resemblance among a face image and a set of reference individuals. Identification of the individual is performed by using the reference-based descriptors of probe and gallery images. The result obtained is 84.7%.
In [6] Hidden Markov Models are fused with Support Vector Machine to detect face very quickly. Authors created frontal face by using Markov Random Fields and high value of Eigen values are extracted as features. By using SVM classifier results are classified and accuracy obtained is 99.617%.
In [7] authors summarize the application of face recognition and feature are extraction based on facial features and holistic features. Facial features are extracted from nose, eyes, mouth and holistic features are extracted from the entire face. Facial Recognition Technology database and the Sheffield face database are used and recognition rate obtained is 91.2%.
In [8] authors proposed method contains three steps: Weber face has been used for image normalization, feature extraction where Daubechies discrete wavelets transform and lower order pseudo Zernike moments are combined and classification is done by using KNN classifier. AT&T and Yale database are used and results obtained are 97% and 98% respectively. Future works includes analysis on performance, such as recognition of ear, palmprint etc.
In [9] face is captured and normalized in to 119x92 pixels. Applying the curvelet features on the normalized face and selecting those sub bands which are invariant to pose. Feature vectors are formed from curvelet invariant features and these feature are given as input curvelet neural network classifier and used FERET, LFW and CMU-PIE database and recognition rate are 89.9%, 63.9%, 90.23% respectively.
In order to calculate the pose a multi view face detection method proposed by authors in [10]. To detect the face image, they have used the combination of Eigen face method and SVM. By making use of irises information input face image is normalized to frontal view. For the face alignment of input pose

image to frontal view Affine Transformation parameters have been used. CAS-PEAL face database has used and 93% recognition rate is obtained for the same.

The NP complete optimization problem proposed by authors in [11]. In a two-level dynamic programming procedure, serially optimizing image columns the complexity can be handled. For each column set of hypothesis computed locally and best hypothesis is computed by globally and achieved recognition accuracy is 94.6%.

In [12] authors presented a method for reconstructing the frontal view virtually from a given non frontal view face image by using Markov random fields. Here Scale invariant feature transform features are extracted and to classify the results the SVM has been used. Used the database like FERET, CMU-PIE and obtained the result 95.5% and 98.5% respectively.

To produce the frontal view virtually from non-frontal view authors in [13] presented locally kernel-based nonlinear regression method. Then the images are converted in to front images and those are classified and recognition rate obtained is 98.5%.

In [14] authors presented method to detect the twelve landmark component which contain pose related characteristics and anatomy. The region of interest is detected by PittPatt face detector. The cascaded classifier used to classify the by making use of Bag of words. Multi-PIE database have been used and result obtained is 82.6% and in [15],authors applied fusion techniques on voice and face by implementing multimodal biometric system to authenticate the person.

In [16] authors used probability distribute functions and to differentiate image from background they used the technique called successive mean equalization transform for preprocessing purposes. Classification is performed by measuring Kullback Leibler distance between the PDFs of a given face and a set of database images. The recognition rate for database FERET =98% is obtained and in [17], authors have discussed methods used to authenticate the face.

3. Issues and Challenges in Face Recognition

After performing literature survey of several papers some issues are identified and described below.

- The main challenge is pose variations of the face, when the user sees camera with different angle or orientation then system can extract information of face region which is visible to the camera and stored in database. If different pose or orientation of the same person comes then system has to be capable of identifying that person by matching the features with the database.
- Illumination is also challenge, if there is no proper illumination or intensity of the light on face region then there would not be proper facial features extraction. The facial feature extraction plays very important role while classifying or identifying the person face and it is difficult get the better accuracy rate.
- Distance of the camera is also plays crucial part, if camera is far away from the user or person then quality of the image will not be good, in which information of the face might lost. So that it will affect the performance of the system.
- Some people or user comes with occlusion, where some regions of the face are occluded with sunglasses, scarf, and hats. In these cases, some parts of face regions cannot be obtained and may fail to extract the facial features like nose, mouth, forehead and eyes.
- Expression is also one of the challenging issue, whenever the expressions changes it's going to affect the information about facial features.

4. Problem Definition

In Face recognition, authentication of the user under the facial pose variation is a major challenge for the research community. Hence there is a scope to develop new algorithm/method to address the issue of pose variation. The proposed system aim is to design and develop the efficient model for Pose invariant face recognition method which helps for identifying the person across the pose.

5. Proposed Approach

5.1. Block Diagram

Fig 1 is the representation of the proposed method. The diagram mainly consists two important phase: Training and Testing phase

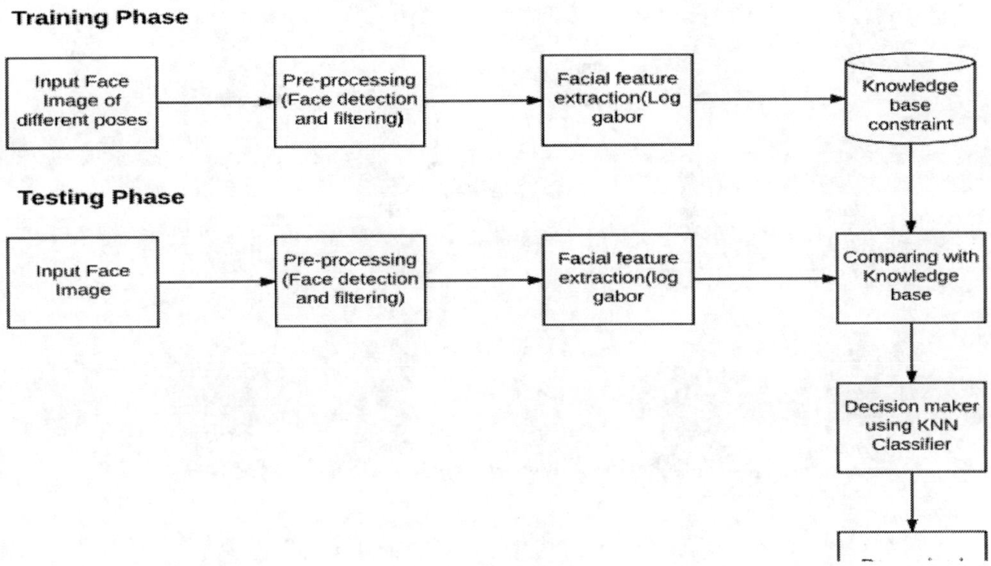

Figure 1. Block diagram of the proposed method

In the training phase, first step is input images with possible pose variation are given to system. Second step is pre-processing which in turn consists of face detection and gray scale image conversion of image. Face detection done by using the Viola Jones method, where it will identify only face from image and it subtracts the rest of the part as a background and the detected face image is converted into gray scale image to remove the noise. Next step is feature extraction, where the log gabor feature are extracted and finally it stored in the database. Once the training is done, testing phase will resume where those images will be given to system as input images which have not trained during the training process. After performing the pre-processing and feature extraction these values will be matched with same person of the database values and classified using the K-NN classifier. Finally, recognition of the person with labelled class name.

5.2. Image acquisition

The database collection plays very important role while during the registration and to store images needed during the identification. In this work, have used the own database and while collection of images of the individual we have used the 14.1 megapixels' camera. Images are taken against homogenous background and each individual the following poses are included: looking left, looking right, looking front, looking up and looking down. The RGB colour space is constructed of 50 users with size of 640x480 pixel resolutions. Fig 2 shows the collection of own database.

Figure 2. Database construction of sample face images with pose variation

5.3. Face Detection and Face Alignment

Object detection is very important in the most of the biometrics applications including the urveillance. Hence object is needed, to identify the face location in the image. In this work, we used the Viola Jones technique to detect the face from image.

Figure 3. Face Detection image

The face image captured from the camera is size of 640x480 pixel resolution. Alignment is process of detecting the region of interest. Here the bounded region by bounding box i.e face is resized into 128x128 pixel resolution. After this resized, image is converted into a gray image to remove the complexity or noise from the image

Cropped Image Gray Scale Image

Figure 4. Face Alignment Process

IOP Conf. Series: Materials Science and Engineering **925** (2020) 012030 doi:10.1088/1757-899X/925/1/012030

5.4. Feature Extraction and Classifier

The feature extraction plays very important role in pose very important. Where, need to select those features which are invariant even though the pose changes. In this work, we used the Log Gabor features. The log Gabor feature has very good in encoding of natural images. In this phase log Gabor feature extraction from gray image is proposed i.e. Gabor wavelet features. Gabor wavelet is function for computing Gabor features of gray- scale image. A one dimensional Log-Gabor function has the frequency response as shown in equation 1.

$$G(f) = \exp\left(\frac{-(\log(f/f_0))^2}{2(\log(\sigma/f_0))^2}\right) \tag{1}$$

Where f_0 f 0 {\displaystyle f_{0}} ffff and σ {\displaystyle \sigma } $\sigma$are the parameters of the filter. f_0f 0 {\displaystyle f_{0}} will give the center frequency of the filter. σ σ {\displaystyle \sigma} affects the bandwidth of the filter. This function calculates the mean squared energy for scale and mean amplitude for orientation, it returns the values as feature vector and the size is 1x160 for each person. The classifier is the one which helps for identification of the person In this work, we are using K-nearest neighbour (KNN) classifier with 1 nearest neighbour to classify the results. This computes pair wise Euclidean distance for each training instances versus testing instance. For classification purpose, some sample images are trained and some other sample images are tested against trained images to authenticate the face of the person under labelled class.

6. Experimental Results

6.1. System Performance Evaluation

The developed method is tested for own created face database, by considering two images of each user with variation in poses. The performance calculated by taking three parameters namely Accuracy, FAR (false acceptance rate) and FRR (false rejection rate).

Accuracy is the ratio of total number of correctly identified users multiplied by 100 to the total number of users attempted and stated as below equation 2.

$$Accuracy = \frac{Number\ of\ correctly\ Identified\ Users}{Total\ Number\ of\ Attempts} \times 100 \tag{2}$$

FAR is the ratio of number of wrongly identified users multiplied by 100 to the total number of users attempted and stated in below equation 3.

$$FAR = \frac{Number\ of\ Wrongly\ Identified\ Users}{Total\ Number\ of\ Attempts} \times 100 \tag{3}$$

FRR is the ratio of number of wrongly rejected users multiplied by 100 to the total number of users attempted and stated in below equation 4.

$$FRR = \frac{Number\ of\ Wrongly\ Rejected\ Users}{Total\ Number\ of\ Attempts} \times 100 \tag{4}$$

6.2. System Performance Analysis

The experiment has been conducted for different number of users to analyze how the system works, as well as how the training phase impacts the accuracy rate. The performance calculated by taking three parameters namely Accuracy, false acceptance rate and false rejection rate. Here, we have tested system for 5, 10,15,20,40 and 50 users.

The performance of the system analyzed as below.

- **3 images for training and 2 images for testing:** The experimental results are tabulated in Table I. The performance analysis of designed system is plotted in fig 5.

Table 1. Performance of Pose Invariant Face Recognition System

Sl. No	Number of persons	Number of persons taken for training	Number of persons taken for testing	Number of correctly identified persons	Number of wrongly identified persons	Accuracy %	FRR %	FAR %
1	5	15	10	10	0	100	0	0
2	10	30	20	20	0	100	0	0
3	15	45	30	27	3	90	10	0.34
4	20	60	40	31	9	78	22	0.57
5	40	120	80	53	27	66.25	33.75	0.42
6	50	150	100	60	40	60	40	0.40

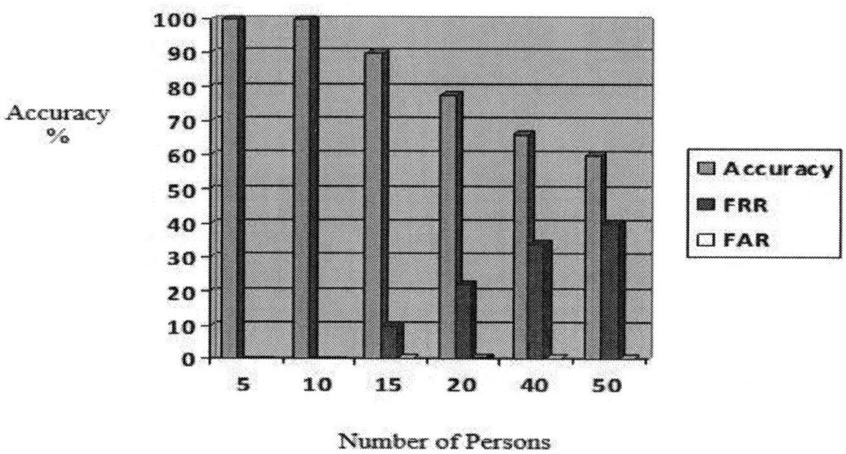

Figure 5. Performance Analysis

- **4 images for training and 5 images for testing:** In this work, tested system for 5,10,15,20 users. The experimental results are tabulated in Table II. The performance analysis of the designed system is plotted in fig 6.

Table 2. Performance of Pose Invariant Face Recognition System

Sl. No	Number of persons	Number of persons taken for training	Number of persons taken for testing	Number of correctly identified persons	Number of wrongly identified persons	Accuracy %	FRR %	FAR %
1	5	20	25	25	0	100	0	0
2	10	40	50	50	0	100	0	0
3	15	60	75	75	0	100	0	0
4	20	80	100	100	0	100	0	0

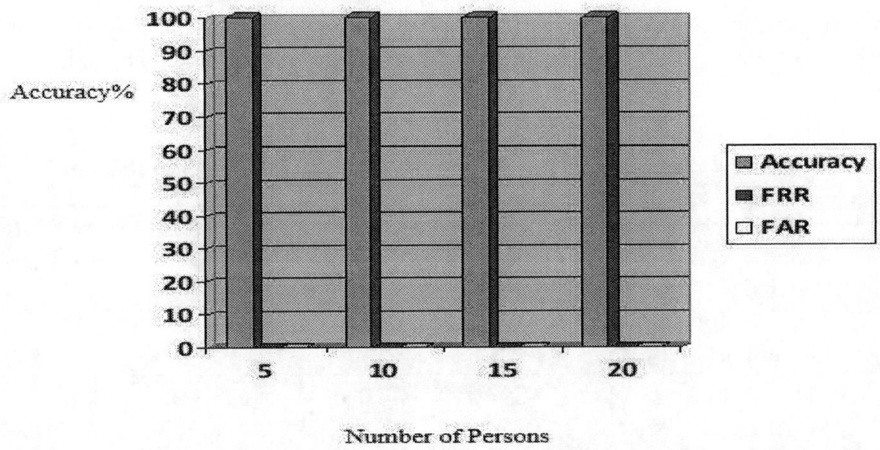

Figure 6. Performance Analysis

- 3 **images for training and 5 images for testing:** In this project, tested system for 5,10,15,20 users. The experimental results are tabulated in Table III and performance analysis is plotted in fig 7.

Table 3. Performance of Pose Invariant Face Recognition System

Sl. No	Number of persons	Number of persons taken for training	Number of persons taken for testing	Number of correctly identified persons	Number of wrongly identified persons	Accuracy %	FRR %	FAR %
1	5	15	25	25	-	100	0	0
2	10	30	50	50	-	100	0	0
3	15	45	75	72	3	96	4	0.05
4	20	60	100	91	9	91	9	0.09

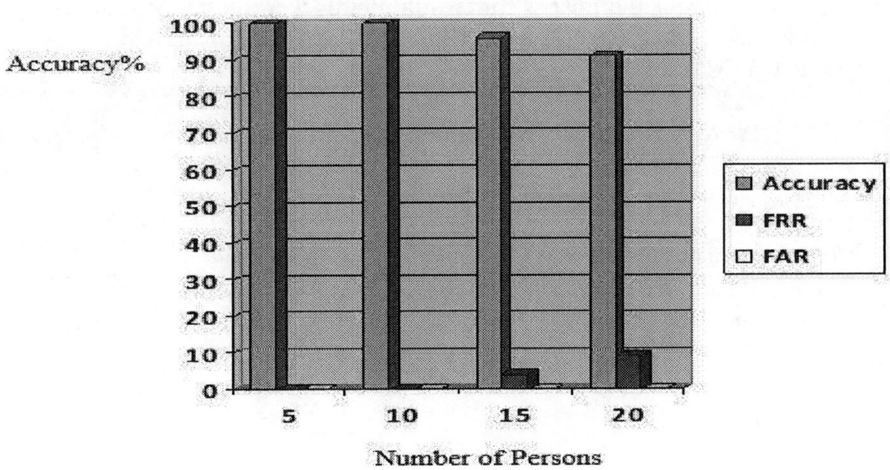

Figure 7. Performance Analysis

- **2 images for training and 5 images for testing**: In this project, tested system for 5,10,15,20 users. Experimental results are tabulated in Table IV and performance analysis is plotted in fig 8.

Table 4. Performance of Pose Invariant Face Recognition System

Sl. No	Number of persons	Number of persons taken for training	Number of persons taken for testing	Number of correctly identified persons	Number of wrongly identified persons	Accuracy %	FRR %	FAR %
1	5	10	25	25	0	100	0	0
2	10	20	50	49	1	98	2	0.04
3	15	30	75	72	3	96	4	0.05
4	20	40	100	89	11	89	11	0.11

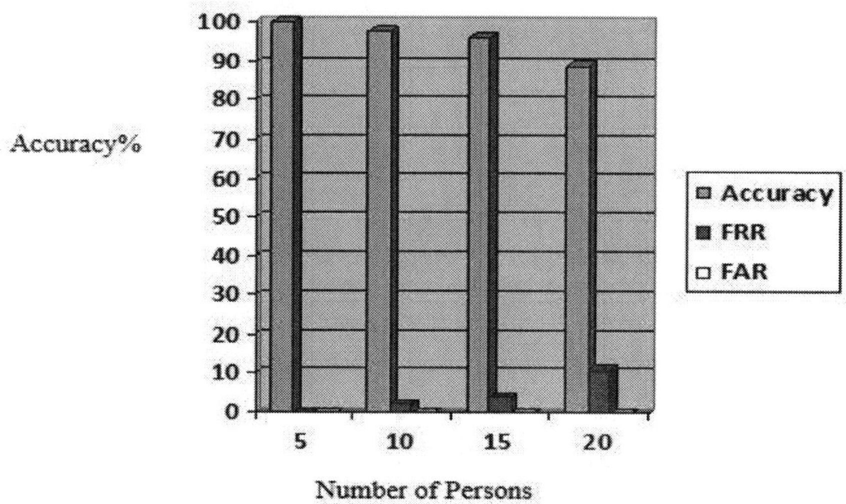

Figure 8. Performance Analysis

- 1 image for training and 5 images for testing: In this work, tested system for 5, 10, 15, 20 users. Experimental results are tabulated in Table V. The performance analysis of the designed system is plotted in fig 9.

Table 5. Performance of Pose Invariant Face Recognition System

Sl. No	Number of persons	Number of persons taken for training	Number of persons taken for testing	Number of correctly identified persons	Number of wrongly identified persons	Accuracy %	FRR %	FAR %
1	5	5	25	21	4	84	16	0.66
2	10	10	50	41	9	82	14	0.36
3	15	15	75	63	12	84	16	0.21
4	20	20	100	76	24	76	24	0.24

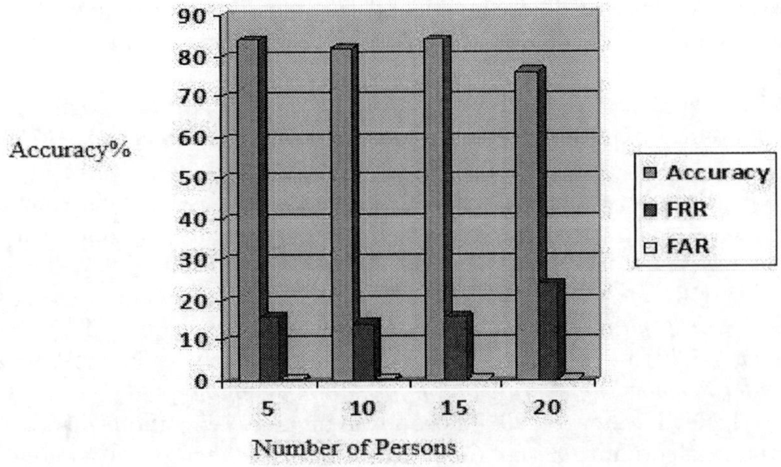

Figure 9. Performance Analysis

7. Conclusion

Face recognition systems are widely used as authentication purposes as it is process of non-contact, which is the important aspect in the securities field. After performing the around seventeen research articles/papers many issues were identified like illumination, expression, occlusion. But, the major challenge is pose variation of the user which has grabbed every ones' attentions in the research community. The proposed system focuses on pose variation problem due to its scope in recognition systems and thus to achieve the better recognition rate by developing efficient model. In the proposed system, have used Viola Jones method to detect the only face of the user and used the image processing techniques for the pre-processing of image. The Log Gabor features are extracted which helps for the improvement in terms of performance of the system and KNN classifier has been used for identification of the person. In future, the system has to handle other challenges like illumination; expression and occlusion are to be address.

Acknowledgments

I would like to express my deepest gratitude to all those who have helped me and special gratitude to my colleague Ms. Poornima Byahatti, my family and KLE Technological University for valuable support.

References

[1] Pss S and Bhaskar M 2016 RFID and pose invariant face verification based automated classroom attendance system *Int. Conf. on Microelectronics, Computing and Communications (MicroCom)* 1-6

[2] Si J, Li W 2016 Pose-invariant face recognition using feature progressing model *IEEE Int. Symp. on Circuits and Systems (ISCAS)* 586-589

[3] Zaeri N 2016 Pose invariant thermal face recognition using AMI moments *UKSim-AMSS 18th Int. Conf. on Computer Modelling and Simulation (UKSim)* 60-64

[4] Panchal P, Patel P, Thakkar V, Gupta R 2015 Pose illumination and expression invariant face recognition using laplacian of Gaussian and Local Binary Pattern *5th Nirma University International Conference on Engineering (NUiCONE)* 1-6

[5] Kafai M, Eshghi K, An L, Bhanu B 2015 A reference-based framework for pose invariant face recognition *11th IEEE International Conference and Workshops on Automatic Face and Gesture Recognition (FG)* **2** 1-8

[6] Indumathi R, Palanivel N 2014 Pose invariant face recognition using HMM and SVM using PCA for dimensionality reduction *Int. Conf. on Information Communication and Embedded Systems (ICICES2014)* 1-6

[7] Oscos GC, Khoshgoftaar TM, Wald R 2014 Rotation invariant face recognition survey *Proc. of the 15th international conference on information reuse and integration (IEEE IRI 2014)* 835-840

[8] Sultana M, Gavrilova M, Yanushkevich S 2014 Expression, pose, and illumination invariant face recognition using lower order pseudo Zernike moments *International Conference on Computer Vision Theory and Applications (VISAPP)* **1** 216-221

[9] Sharma P, Yadav RN, Arya KV 2014 Pose-Invariant Face Recognition Using Curvelet Neural Network *The Institution of Engineering and Technology Biometrics* **3** 128-138

[10] Muruganantham S 2013 Pose invariant face recognition–a new technique *International Journal of Computer Science Engineering and Information Technology Research* **3** 135-40

[11] Hanselmann H, Ney H, Dreuw P 2013 Pose-invariant face recognition with a two-level dynamic programming algorithm *Iberian Conf. on Pattern Recognition and Image Analysis* 11-20 Springer, Berlin, Heidelberg

[12] Ho HT and Chellappa R 2013 Pose-Invariant Face Recognition Using Markov Random Fields *IEEE Transactions on Image Processing* **22** 1573-1584

[13] Arianpour Y, Ghofrani S, Amindavar H 2012 Locally Nonlinear Regression Based on Kernel for Pose-Invariant Face Recognition *The 11th Int. Conf. on Information Sciences, Signal Processing and their Application* 401-406

[14] Efraty B, Papadakis M, Profitt A, Shah S and Kakadiaris IA 2011 Pose Invariant Facial Component-Landmark Detection *18th IEEE Int. Conf. on Image Processing* 569-572

[15] Poornima B and Hatture SM 2017 A Fusion Model for Multimodal Biometric System *International Journal of Engineering Research & Technology (IJERT)* **5**

[16] Demirel H and Anbarjafari G 2008 Pose Invariant Face Recognition Using Probability Distribution Functions in Different Color Channels *IEEE Signal Processing Letters* **15** 537-540

[17] Madhura S and Hatture SM 2017 A Model for Unconstrained Face Recognition System *International Journal of Engineering Research & Technology (IJERT)* **5**

IOP Conf. Series: Materials Science and Engineering **925** (2020) 012031 doi:10.1088/1757-899X/925/1/012031

Fusion Strategies for Multimodal Biometric System Using Face and Voice Cues

Poornima Byahatti[1] and Madhura S Shettar[2]

[1] Department of Automation & Robotics, KLE Technological University, Hubballi, India.
[2] School of Computer Science & Engineering, KLE Technological University, Hubballi, India.
[1]poornima.byahatti@kletech.ac.in, [2]madhura.shettar@kletech.ac.in

Abstract. Deployment of biometric systems in the applications of real world includes the most of unimodal biometric systems. Unimodal biometric system based on the information collected from single source. Sometimes single source of information may not identify the individual correctly because of some limitations such as Non-universality, Noisy data, Intra-class variation, Spoof attacks and Intra-class similarities. Various limitations of unimodal biometric systems are overridden by the multimodal biometric systems which involves multiple sources of information. Multimodal systems can be constructed by fusing of information of multiple modalities. This fusion can take place at various steps of processing such as at image acquisition, extraction of features of the traits, matching of test vectors with trained vectors and during decision taking based on classification. This paper presents a system of multimodal biometrics using face and voice biometric traits by including four fusion methods. Fusion takes place at i) feature level using concatenation of face and voice features, ii) score level using method involving the maximum of mode of scores obtained from two matchers, iii) rank level using borda count & iv) decision level fusion using logical conjunction (AND). Fusing of Log Gabor & Local Binary Pattern (LBP) takes place at the facial feature extraction. The voice features are also fused using Mel Frequency Ceptral coefficients (MFCCs) and Linear Predictive Coefficient features (LPC). Computation of similarity between test feature vectors and training vectors is carried out using Euclidian distance during matching process. KNN Classifier is used during decision making. Performance evaluation of these techniques are also carried out using performance measures such as Accuracy, False Acceptance Rate (FAR), False Rejection Rate (FRR) and ROC curves.

Keywords: Log Gabor, LBP, MFCCs, LPC, Feature fusion, Score fusion, Rank fusion and Decision fusion.

1. Introduction

Implementation of recognition techniques are needed for protection of resources in nowadays as the World of technology evolves. Various ways are available to achieve authentication & authorization techniques. Most of the techniques are being overridden by biometric system because it recognizes the individual identity automatically based on behavioral & anatomical characteristics like signature, voice, face, iris, fingerprint etc. There are two main divisions of biometric systems depending upon the number of modalities used in the biometric system. Those are unimodal biometric systems (make use of single

trait such as face, voice, etc.) & multimodal biometric systems (make use of two or more traits such as (face & iris), (face & voice), etc.)
Some of the limitations embedded in unimodal biometric systems are Non-universality, Noisy data, Intra-class variation, Spoof attacks and Intra-class similarities. Multimodal biometric systems override the several disadvantages of unimodal biometric systems [1]. Unimodal biometric system is not suited for all applications because it makes use of single trait.
There are five ways in which we can achieve multi-biometric system. Those are: 1). Multi-algorithm biometric systems, which involve processing of same biometric information with more than one algorithm. 2). Multi-sensor biometric systems in which acquisition of images takes place using different sensors. 3). Multi-instance biometric systems make use of multiple instances of same biometric trait. 4). Multi-sample biometric systems in which multiple samples of biometric trait are collected from the single sensor. 5). Multi-modal biometric systems make use of the biometric data collected from multiple traits. Only single modality can be used to achieve authentication in first four ways in the above list but in fifth way, multiple modalities can be used to attain authentication. Fusing of the data of multiple traits like fingerprint, iris, voice, palmprint, face, etc. constitutes the multimodal biometric system.
There are two main different paths in which fusion of information can be achieved. Those are fusion before matching and fusion after matching. Attainment of fusion before matching can be again done in two traditions as 1). Sensor fusion technique: In this technique many sensors collect the information from the same biometric trait followed by the integration of such information to construct new information. Features are extracted from the new data then they are used for further processing in recognition. 2). Feature fusion technique: Extraction of features from multiple traits is carried out in this technique at first. Then there is concatenation of features of multiple traits to new feature joint vector. Sometimes these joint vectors are high dimensional vectors so some reduction techniques are needed for selection of useful information for further processing.
Accomplishment of fusion after matching may also be done in three traditions as 1). Score fusion technique: In this technique each individual system calculates its own scores that are nothing but matching similarity measures of input vectors with template vectors. Scores of individual matchers are fused in order to calculate final scores then; these scores will help in taking final decision. 2). Rank fusion technique: Identification and verification are the two types of recognition, where rank fusion technique commonly suits during the identification of the person instead of verification. In this fusion, final decision is being established by determining a new rank obtained from consolidation of multiple ranks associated with an identity. 3). Decision fusion technique: In this technique each biometric system classifies the persons separately and then decisions are fused in order to claim final decision based on various fusion methods available. This fusion method has less information for fusing as compared to other fusion methods.
In this paper, multimodal biometric system is employed in order to evaluate the performance of fusion strategies using face and voice biometric traits in terms of accuracy, FRR and FAR. Fusion of Log Gabor and LBP is carried out for facial feature extraction. MFCCs and LPC features are used during voice feature extraction. Fusion strategies employed in system are feature fusion using concatenation, score fusion using max-min, rank fusion using borda count and decision using AND methods. Classification is done by using KNN classifier with one nearest neighbour. This paper is organized as follows: section 2 summarizes literature survey, section 3 presents the proposed methodology, section 4 demonstrates the experimental results and section 5 illustrates the conclusion.

2. Literature Survey

The literature survey is carried out in order to study the different fusion approaches in multimodal biometric systems and to identify the challenges and issues in multimodal biometric systems. Several research papers have been referred and the outlines of these papers are described below:

Multimodal biometric system evaluates performance of decision fusion with logical operators AND & OR for iris and fingerprint biometric traits using K-nearest neighbour, Neural based classifier and Hidden markov model and proved that neural classifier with AND based fusion attains the highest accuracy of 90.30% in [2]. Multi-algorithmic biometric system is developed by fusing features of face

& iris and multimodal biometric system is employed by using score fusion method of weighted sum in [3]. Daugman's Gabor filters are used for iris. For facial feature extraction Gabor filters and Local binary patterns are used. The value for EER of the developed system is 1.48% that is 85% improvement as compared to unimodal system. Multimodal system involving combination of iris, voice and face using score fusion method was developed. Simple sum rule is used for scores fusion after min-max normalization. Authors achieved accuracy in terms of GAR-92% in [4]. By making use of finger veins, new approach of multi-instance identification system is established based on score fusion method in [5]. EER values of proposed system using three different finger veins (index, middle, ring) in terms of sum weighted, min and max are 0.11415, 0.15714 and 0.08571 respectively. Rank fusion with improvement involving two combinations such as serial and parallel is adapted using in multi-instance, multi-algorithm and multimodal biometric systems in [6]. Values of RR for Multi-instance, multi-algorithm and Multi-modal are 83.6%, 92.63% and 93.4% respectively. Multimodal biometric system is presented using three biometric traits involving iris, voice and face in [7]. Delta Delta Mel Frequency Ceptral Coefficients (DDMFCC) features are used for voice recognition followed by extraction of face features using Scale Invariant Feature Transform (SIFT-based) features then LAB-based features are used in iris recognition. Decision module fusion based on majority voting is performed and attained 90% of recognition rate.

Comparative analysis of performance of various image fusion techniques is carried out and showed that feature fusion is the best method for fusing information than other techniques through results obtained during experimentation. Values of feature level fusion for accuracy, FAR and FRR are 99%, 0.029% and 1.0%. Values for score and decision level in terms of accuracy are 97% and 98% respectively in [8]. System evaluates the performance of unimodal and multimodal (score fusion) systems using face and voice biometric traits. Results varied based on the feature extracted from XM2VTS database and score fusion techniques used and proved multimodal with score fusion achieves the higher accuracy as compared to unimodal system in [9]. Score fusion strategy of weighted score for fusing the palmprint and speech signal is explored in [10]. Palm features are extracted by modified canonical method. Subband based Ceptral coefficients are used for extraction of speech signal features. Clean & degraded databases are applied on system. Evaluation of system is done in terms of EER, FAR and FRR. Results showed that system attains higher accuracy and achieves the EER of 3.54% & 9.17% for clean and degraded dataset. Feature extraction level fusion is carried out based on the information from the two traits of fingerprint and iris in [11]. Decision about recognition is taken for multimodal system and for each unimodal biometric system. The values for FAR and GAR are 1% and 98% respectively. SPR fusion method based on the combination of simple sum and product rule by using face and voice biometric traits in [12]. The double sigmoid normalization technique is used for the normalization of scores before fusion. The sum rule, maximum rule, product rule and minimum rule is also performed in order to compare with the system. Experimental results declared that SPR fusion scheme attains higher accuracy of 89.12% on average. Multimodal system is developed for fusing face and speech using decision fusion techniques with both AND & OR in [13]. Combination of PCA and DCT are used for face identification and for speech text independent mode is used. Results proved that system achieves better results than unimodal systems.

Identification of various issues and challenges in multimodal biometric systems is done through the literature survey such as sensors adjustment and accumulation of dust on sensors in score fusion, non-compatibility of features during feature fusion, normalization of scores due to heterogeneity during score fusion, selection of system design of serial and parallel during rank fusion and limited amount of information during decision fusion techniques.

After carrying out the literature survey, it is observed that various fusion models are available based on the requirements. Numerous systems make use of different feature extraction methods, different fusion techniques at various levels [14]. Choosing among all for application is not an easy task so there is need of evaluation of those systems to know how reliable those are. Hence there is scope for developing system which evaluates the performance of fusion systems and eases way of selection.

3. Proposed Methodology

There is involvement of all processing modules such as image acquisition, extraction of features, creation of template, matching for similarity and concluding the decision in biometric multimodal systems as that of unimodal systems. Additionally, system involves the fusion of information at various levels. The fusion may take place at any of the following phases. 1). at the time of extraction of features of biometric traits. 2). at the time of comparison of input template with stored biometric templates. 3). at the time of concluding of decision based on classification.

Addressing of issues embedded in biometric systems like efficiency, applicability, robustness, accuracy & universality can be done with the implementation of fusion. Fig. 1 shows the block diagram for all levels of fusion. This system consists of two phases such as training & testing phase. In the proposed system, extraction of features from face biometric trait is carried out using both Log Gabor & Local Binary Pattern features. Voice feature's extraction process makes use of Mel-Frequency Ceptral Co-efficient & Linear Predictive Coefficient features from voice biometric trait. Respective features are stored during training according to fusion technique used. Euclidian distance is calculated for testing both the traits during comparison module. Finally, decision is made using K-Nearest Neighbour (KNN) classifier.

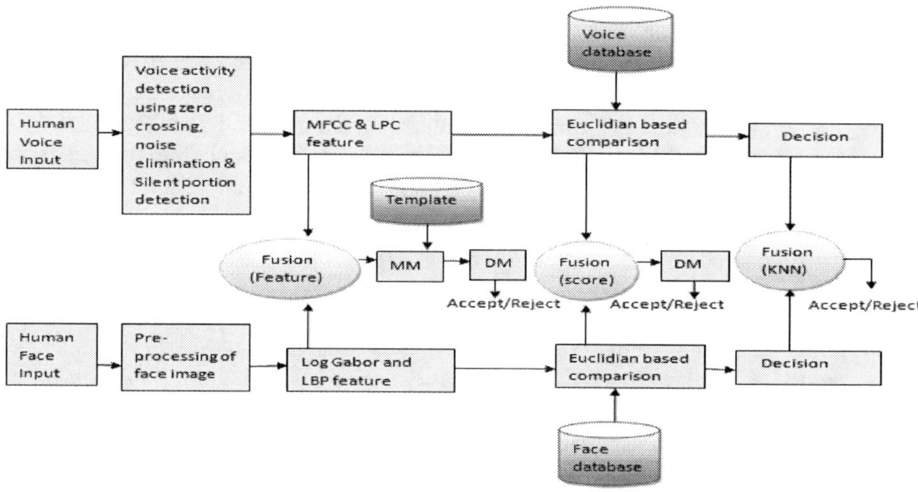

Figure 1. Block diagram of fusion levels

This proposed system involves the four fusion techniques and are described as:

During the process of feature level fusion at first, features are extracted from both face and voice biometric trait. Next, extracted features of both traits are joined together using concatenation fusion technique. Finally, resulting joint vector represents feature set for further processing. Principle component analysis is used in order to reduce the dimension of joint vector.

In this technique scores are calculated using Euclidian distance method for both face and voice trait's features. These scores are fused using maximum score of mode method (max-min). Then final decision is taken based on the fused scores.

Rank fusion is almost similar to score level fusion with slight difference. It is well suited to identification process rather than verification. In proposed system Borda Count rank level fusion technique is used. Borda Count calculates sum of the ranks of individual matchers to obtain final rank.

Decision level technique is carried out using Logical Conjunction (AND) method. Individual matchers calculate scores for trait's features then decisions are taken for both traits. The decisions output by individual matchers are fused using AND technique so when decisions of two matches are same then final decision is same about the matching.

IOP Conf. Series: Materials Science and Engineering **925** (2020) 012031 doi:10.1088/1757-899X/925/1/012031

3.1. Datasets

Image & voice sample acquisition is the first module of proposed system. The system is applied on two datasets. First system makes use synthetic database constructed by combining the AR database for face & voice database .AR database consists of 10 people where each image has size of 576x768 pixels. All frontal views of person are available. For AR database of faces some voice samples are downloaded and they are assigned to people with assumption that those are voice samples of faces of AR database. Fig. 2 shows the samples of face and voice of synthetic database.

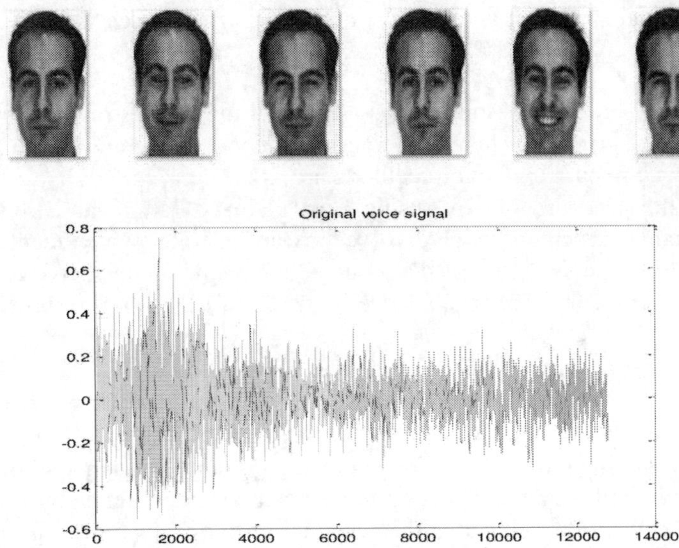

Figure 2. Face and voice Samples of synthetic database

Second database applied on the project is own collected database of 20 persons. Both images and voice samples are collected from our surrounding people through camera and voice recording device respectively. Images are captured by camera with different poses. Voice samples involving utterance of 22 of each person are collected through device. Fig. 3 shows the face and voice samples of own collected database.

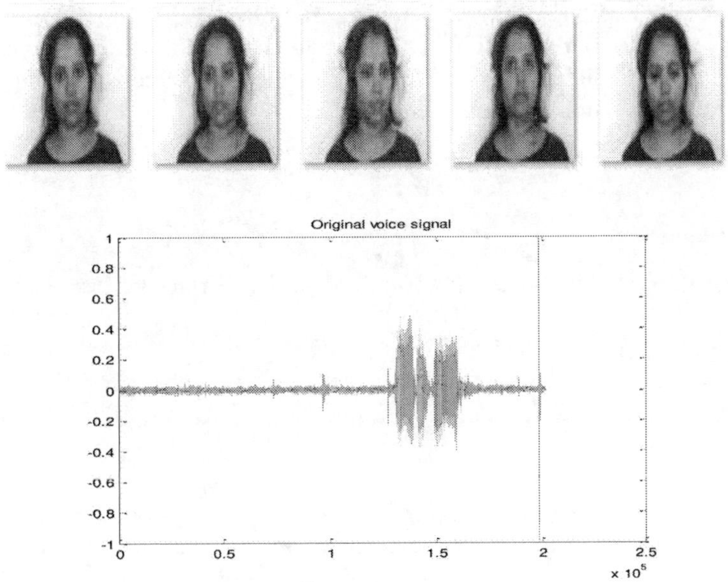

Figure 3. Face and voice samples of own database

3.2. Pre-processing of face and voice biometric traits

Pre-processing of face consists of three steps such as gray scale conversion which is the process of conversion of color RGB image to gray scale image, filtering process which is the process to highlight certain features or exclusion of some features of image. (Here, system makes use of Gaussian filtering) and enhancement of image which is the process of adjustment of digital images so that results are appropriate for display and further analysis of image.

Pre-processing of voice consists of following step of voice activity detection which is the process of separating speech from silent portion or other background noise. Here proposed system makes use of zero crossing rates for voice detection which is the number of times voice signal changes its sign.

3.3. Feature extraction

Features are the building blocks of biometrics so that features are extracted from traits in order to classify the individual through various feature extraction methods or techniques. In the proposed system following facial and voice feature extraction methods are used.

Facial feature extraction is carried out by fusing Local Binary Pattern and Log Gabor. LBP is an effortless and well-organized method which involves texture operator which identifies the image pixel by taking the threshold of all neighbours of that pixel & the result is considered as the binary number. The Equation 1 shows face description with LBP. Histogram of LBP can be represented as

$$Hi = \sum_{x,y} I\{fi(x,y) = i\} \tag{1}$$

Where $i = 0,.....,n-1$ and n is the number of pixels of image output by the LBP operator.

Traditional selection for getting the localized frequency data is done through applying Gabor filters. But it has disadvantage of limited bandwidth so Log Gabor came into existence by Field from which it is possible to create filter with arbitrary bandwidth. Response of frequency for one dimensional Log Gabor filter is given in equation 2:

$$G(f) = \exp\left(\frac{-(\log(f/f_0))^2}{2(\log(\sigma/f_0))^2}\right) \tag{2}$$

Where f_0 and σ are the parameters of the filter.

Voice feature extraction is done by using linear predictor coefficients and mel-frequency Ceptral coefficients. Variance of prediction error is being returned by the LPC function of feature extraction based on the Levinson & Durbin algorithm. This method converts the analog voice signal into digital value in which a value is predicted by linear function of the past values of signal. The most common representation of linear prediction is given in equation 3.

$$\hat{x}(n) = \sum_{i=1}^{p} a_i x(n-i) \tag{3}$$

Where $\hat{x}(n)$ - predicted value, $x(n-i)$ - previous observed values, and a_i the coefficients of prediction. The error generated by this estimate is $e(n) = x(n) - \hat{x}(n)$ where $x(n)$ - true value.

Mel-frequency Ceptral coefficients: MFCCs are nothing but collective coefficients that make up MFC. A type of Ceptral representation of the audio clip derives the MFCC. MFCC is a representation of the real Ceptral of a windowed short-time signal derived from the Fast Fourier Transform.

3.4. Template matching

After feature extraction, those features are stored in database during training phase this is nothing but knowledge base construction. In testing phase features are matched using Euclidian distance in our proposed system. There is use of K-nearest neighbour classifier for matching and for the recognition. KNN classifier is easy to interpret the output and it is instance based learning. This classifier is the simplest among all classifiers. One nearest neighbour classifier is the most suitable classifier and its function is to assign point x to closest neighbour class in feature space. That is illustrated in equation 4:

$$C_n^{\ln n}(x) = Y_{(1)} \quad 4)$$

So in proposed system one nearest neighbour classifier is used for template matching. The proposed system better classifies using KNN classifier than others.

4. Experimental Results

The experimentation is carried out on two databases. First one is the synthetic database involving ten persons where six samples (face & voice) of each person is considered in which four are used for training and other two are used for testing. Second one is the own collected database with pose variation involving twenty persons. Here, also five samples of face & voice of individual is taken for experimentation in that three are used for training and two samples are used for testing. Performance evaluation of fusion techniques is carried out through performance measures such as FAR, FRR & RR. Where RR showed in equation 5: percentage of test sets that are correctly classified among the total number of recognition attempts.

$$RR = \frac{no.of\ correctly\ identified\ users}{total\ no.of\ recogntion\ attempts} * 100 \quad (5)$$

FAR showed in equation 6: sometimes incorrect acceptance of an attempt by invalid user may also be possible. It is required to calculate rate of false acceptances ratio that is of wrongly identified users to total number of recognition attempts.

$$FAR = \frac{no.of\ wrongly\ identified\ users}{total\ no.of\ recogntion\ attempts} * 100 \quad (6)$$

And FRR showed in equation 7: sometimes system may wrongly reject the access attempt of the genuine user hence it is required to calculate the rate of false rejections. Fraction of rejected genuine users to total number of recognition attempts.

$$FRR = \frac{no.of\ wrongly\ rejected\ users}{total\ no.of\ recognition\ attempts} * 100 \quad (7)$$

The results are tabulated in following tables from TABLE 1 to TABLE 2. Experimentation is carried out on all four levels using 10 people from synthetic DB and 20 people from own DB

Table 1. Results of Classification using Synthetic Db

No of users	Fusion level	No of trained samples	No of tested samples	No of correctly identified samples	No of wrongly identified samples	RR in %	FRR in %	FAR in%
10	Feature	40	20	20	00	100	00	00
10	Score	40	20	16	04	80	20	1.05
10	Rank	40	20	20	00	100	00	00
10	Decision	40	20	18	02	90	05	00

The graph of performance analysis on synthetic DB is shown in Fig. 4. In graph FFT means feature fusion technique, SFT means score fusion technique, RFT means rank fusion technique and DFT means decision fusion technique.

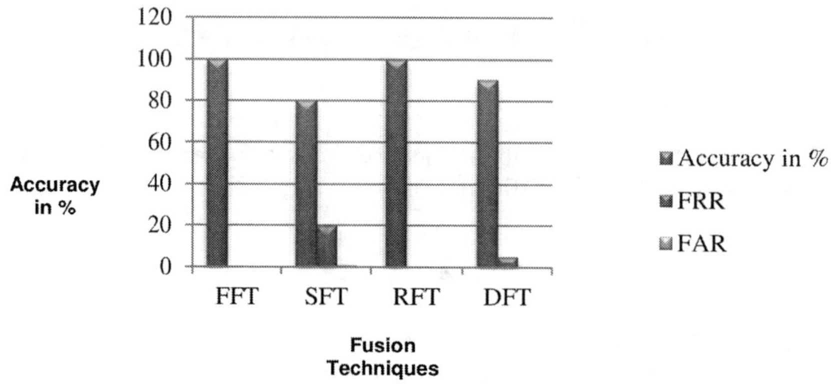

Figure 4. Performance analysis of fusion techniques on synthetic DB

Table 2. Results of Classification using own Db

No of users	Fusion level	No of trained samples	No of tested samples	No of correctly identified samples	No of wrongly identified samples	RR in %	FRR in %	FAR in%
20	Feature	60	40	40	00	100	00	00
20	Score	60	40	06	34	15	85	2.17
20	Rank	60	40	20	20	50	50	1.28
20	Decision	60	40	32	08	80	20	00

The graph of performance analysis on own DB is shown in Fig. 5. Abbreviations for FFT, SFT, RFT and DFT are considered as in previous graph.

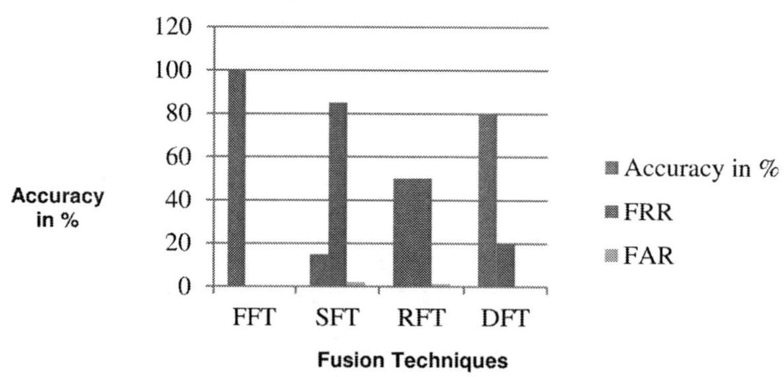

Figure 5. Performance analysis of fusion techniques on own DB

As shown with results system works better on standard databases and achieves higher accuracy. Among all levels feature level provides accuracy of 100% and proved that fusion before matching achieves higher accuracy as compared to other fusion methods which take place after matching

5. Conclusion

Biometric system authorization is designed depend on two biometric traits using face & voice. Proposed system tests 50 users of own collected database involving pose variation and 10 users of AR database involving various expressions. Log Gabor and LBP features are extracted from face. MFCCs and LPC features are used during voice feature extraction. Proposed system implements four fusion levels. Feature fusion fuse features of face and voice using concatenation. Score level strategy implemented using maximum score of mode technique. Borda count technique that takes the sum of scores of

individual matchers is used for rank level fusion. For decision level logical conjunction (AND) is used. K-NN classifier is used for decision making. Experimental results prove that accuracy for feature level fusion is more as compared other fusion levels because fusion at early stage attains better recognition rate. Hence the proposed system is providing the results with better recognition rate. All fusion levels provide better recognition results but score level performs less as compared to other three as demonstrated in experimentation.
In future, it is planned to achieve higher accuracy rate with real time and other databases with more complexity samples.

Acknowledgement
I take this opportunity to express my gratitude to my colleague Ms. Madhura Shettar, my family, A & R Department and KLE Technological University for the valuable support.

References
[1] Madhura S, Hatture SM, 2017 A Model for Unconstrained Face Recognition System *International Journal of Engineering Research & Technology* **5** 1-4
[2] Suneet NG, Vig R, Gupta S 2016 Analysis of Decision Level Fusion in Multimodal Biometrics using Iris and Fingerprint *3rd International Conference on Electrical, Electronics, Engineering Trends, Communication, Optimization and Sciences* 409-416
[3] Kirti VA, Kanade SG, Jadhav DV, Ajmera PK 2015 *A Multi-modal and Multi-algorithmic Biometric System Combining Iris and Face International Conference on Information Processing Vishwakarma Institute of Technology*, 496-501
[4] Sheetal C, Nath R 2015 A New Multimodal Biometric Recognition System Integrating Iris, Face and Voice *International Journal of Advanced Research in Computer Science and Software Engineering*, **5** 145-150
[5] Fateme S, Nasri M 2015 A Multibiometric Finger Vein Verification System Based on Score Level Fusion Strategy *Second International Congress on Technology, Communication and Knowledge, Mashhad Branch, Islamic Azad University, Mashhad, Iran*, 501-507.
[6] Renu S, Das S, Joshi P 2015 Rank Level Fusion in Multibiometric Systems *IEEE Fifth National Conference on Computer Vision, Pattern Recognition, Image Processing and Graphics* 1-6
[7] Barbu T, Ciobanu A, Luca M 2015 Multimodal Biometric authentication based on Voice, Face and Iris. *In 2015 E-Health and Bioengineering Conference (EHB) IASI, Romania* 1-4
[8] Soruba Sree SR, Radha N 2014 A Survey on Fusion Techniques for Multimodal Biometric Identification *International Journal of Innovative Research in Computer and Communication Engineering* **2** 7493-7497
[9] Fawaz A, Zahrani M, Alghamdi T 2013 Score-Level Fusion in Biometric Verification *International Symposium on Biometrics and Security Technologies* 193-197.
[10] Mahesh PK, Shanmukha Swamy MN 2012 Designing an Efficient Multimodal Biometric System using Palmprint and Speech Signal *ACEEE Int. J. on Signal & Image Processing* **3** 76-80.
[11] David MD, Monica B 2012 A Data Fusion Technique Designed for Multimodal Biometric Systems *10th IEEE International Symposium on Electronics and Telecommunications,* 155-158.
[12] Fouda YM 2012 Fusion of Face and Voice: An improvement *International Journal of Computer Science and Network Security* **12** 37-43
[13] Samir A, Yahia B, Messaoud M and Chahir Y 2011 A Multi-Modal Recognition System Using Face and Speech, *IJCSI International Journal of Computer Science Issues* **8** 230-236.
[14] Poornima B, Hatture SM 2017 A Fusion Model for Multimodal Biometric System, *International Journal of Engineering Research & Technology (IJERT)* **5**.

IOP Conf. Series: Materials Science and Engineering **925** (2020) 012032 doi:10.1088/1757-899X/925/1/012032

Universal Fast Marching Method to Identify Liver Image

Sangeeta K Siri[1], Pramod Kumar S[2] and Gavisiddappa[3]

[1] Department of Electronics & Communication Engineering, GAT, Bengaluru, Karnataka, India.
[2] Department of Electronics & Communication Engineering, JNNCE, Shivamogga, Karnataka, India.
[3] Department of Electronics & Communication Engineering, CIT, Gubbi, Karnataka, India.
[1]sangeetamk28@gmail.com, [2]pramod.s86@gmail.com, [3]gavisiddappa@cittumkur.org

Abstract. Liver segmentation is of prime importance in modern liver disease diagnosis and analysis. In our paper, random section of liver image is chosen and its histogram is achieved. From histogram, liver pixel intensity range is obtained. Using the range value, threshold segmentation is carried out which detaches the liver from its adjoining organs. Median filter is employed to curtail the noise. The sigmoidal function is applied to improve anatomical structures of image. Then the image is converted into binary called as speed function. The novel algorithm is designed to locate the start points within speed function without user intervention. These start point evolved outwardly using Fast Marching Method till complete periphery of liver is reached. The proposed algorithm is compared with popularly used segmentation algorithms. The results show that proposed segmentation algorithm is robust in approach.

Keywords: Computed Tomography, Fast Marching Method, Liver segmentation, Magnetic Resonance Imaging, threshold based segmentation.

1. Introduction

Segmentation is a process of dividing the image into meaningful divisions. In medical images, segmentation algorithm is basically used either to separate required organ from its neighbouring organs or put outline for intended organ [1-2].

Liver is an obligatory organ which is accountable for numerous chemical actions that human body needs and it generates chemicals that are used by different parts of human body. For a healthy body, healthy liver is much necessary. The liver is threatened by many diseases like cancer, fatty liver, cirrhosis, etc. The diseases in the liver spread very fast because of its anatomical structures and rich in its biochemical environments.

Liver segmentation is primary requirement for computer based disease analysis and diagnosis. Separating liver from abdominal scanned images is quite tedious task since edges are blurry, huge pixel intensity variation from one patient to another as well as variation of pixel intensity inside the liver image because of diseases or may be due to hereditary problem and close proximity with neighbouring organs. Thus segmentation method based on edge, region, threshold, feature and parameter fails to detect

IOP Conf. Series: Materials Science and Engineering **925** (2020) 012032 doi:10.1088/1757-899X/925/1/012032

the liver image accurately in scanned images. As well as different segment algorithms are used for different modalities like CT, MRI, PET, etc.
The proposed segmentation algorithm first studies the liver pixel intensity dispersion by picking haphazard section of liver. Then liver pixel intensity range is acquired using its histogram. Threshold based segmentation is carried out using this range which precisely isolates liver image from CT/MRI scan images.

2. Related Work

Numerous segmentation algorithms have been suggested by authors which provides good results on particular application. Shenhai Zheng [3] integrates global guassian intensity analysis and local statistical features in level set framework to achieve coarse liver image. The improved Chan-Vese model is used to improve the liver outline. Ankur Biswas [4] propose liver segmentation method in which thresholding is employed to take out liver from abdominal CT scan images. Two geometric contours are initialized which grow using level set method to detach liver from CT scan image accurately. Sangeeta [5] employed Neutrosophic Set (NS) theory to isolate the liver from its neighbouring organs and Chan-Vese (C-V) model to spot liver boundary. Sangeeta [6] used NS theory to segment liver from abdominal CT scan image. A novel step is followed to identify start points inside liver which progress outwardly to spot liver boundary using Fast Marching Method (FMM). Sangeeta [7] used threshold based segmentation to separate liver from CT and MRI scan images and boundary is marked by improved C-V model. L Merg [8] designed liver and liver tumour segmentation algorithm based on 3D dual path multi-scale Convolutional Neural Network. To refine the segmentation results, conditional random field was used which eliminated false segmentation points in the segmentation results. Budak [9] constructed liver and liver tumor segmentation model using two deep Encoder-Decoder Convolutional Neural Network [EDCNN]. An EDCNN separates liver image from CT scan which is input for training of a second EDCNN. The second EDCNN the segment the tomour region within the liver ROI regions as predicted by the first EDCNN. This model produced an average DICE score of 95.22% for test set of CT images.

3. Methodology

Step1: The acquired image is in 1019 X 682-pixel matrix. To speed up the algorithm, the dimensions are reduced to 512 X 512-pixel matrix.

Step2: The liver pixel density distribution varies from one person to another person; it also varies throughout the liver part in medical scan images. There is no definite range of intensity values for liver part. Hence, it is crucial to find pixel range for liver part. In proposed algorithm, this is done by picking random section of liver to learn pixel density dispersion. A histogram of only picked liver part is obtained. From histogram, pixel intensity range is achieved, further highest pixel intensity value T_{max} and smallest pixel intensity value T_{min} is derived.

Step3: Apply threshold segmentation using T_{min} and T_{max}. The threshold function is defined in equation (1).

$$I = \begin{cases} I \ \text{if} \ T_{\min} \leq I \leq T_{\max} \\ 0 \ \text{if} \quad Otherwise \end{cases} \tag{1}$$

Where I is intensity level of image. This threshold function separates liver image from its neighbouring organs like kidney, spleen, heart, ribs, gallbladder and pancreas.

Step4: The noise is reduced by applying median filter of size (3 x 3).

Step5: In the next step, anatomical structures are enhanced by sigmoidal function which is represented in equation (2).

$$S_f = T_{\min} - T_{\max} * \frac{1}{1+e^{-(I-A)/B}} + T_{\min} \quad (2)$$

Where: T_{min} =Minimum pixel intensity value of liver part; T_{max}=Maximum pixel intensity value of liver part; A = Pixel intensity level to be enhanced; B = (T_{min} + T_{max}) /2; I = Intensity level of image.
Step6: Small regions are removed by image erosion morphological operation.

Step7: Convert the obtained image into binary image which is called a speed function.

Step8: Generate the start points automatically within area of the liver.

Step9: The start points and speed function are inputs to FMM. These start points grow outwardly using FMM [10-16] until it detects complete periphery of the liver. This marks liver periphery in CT/MRI scan images.

The proposed model for marking the border of liver image in CT scan and MRI is demonstrated in figure 1.

4. Automatic Start Points Selection

We conducted large-scale investigational experiments to find the impact of start points. The position of start points within liver part and number of start points have great impact on the performance of the FMM. Most of the algorithms identify the start points manually which involve user intervention. We recommend an automatic method which locates the start point automatically within the liver area. To find out start points, subsequent steps are pursued.

Step1: Find out centroid of liver as (cx, cy).
Step2: Choose di=20 which is chosen experimentally.
Step3: Points (cx, cy),(cx, cy+di),(cx+di, cy),(cx+di, cy+di) positioned inside liver are chosen as start points i.e.

5. Experiment and Results

The experiment dataset includes 50 MRI scan images and 50 CT scan images which are offered by M/S "CT scan Center", Hubbali, Karnataka, India. Every scan consists 40-150 slices. Every slice of CT and MRI scan is a 1020 X 682 size colour image.
The outcomes of CT scan and MRI scan images are shown in figure 2 and 3 respectively. Figure 2(a), (b) and (c) present the original CT scan image, picking random part of liver and histogram of picked liver part respectively. Figure 2(d), (e), and (f) threshold based segmentation result, liver periphery marking using FMM and standard image, respectively for CT scan images. Figure 3(a), (b) and (c) present the original MRI scan image, select random part of liver and histogram of selected liver part respectively. Figure 3(d), (e), and (f) threshold based segmentation result, liver periphery marking using FMM and standard image, respectively for MRI scan images.

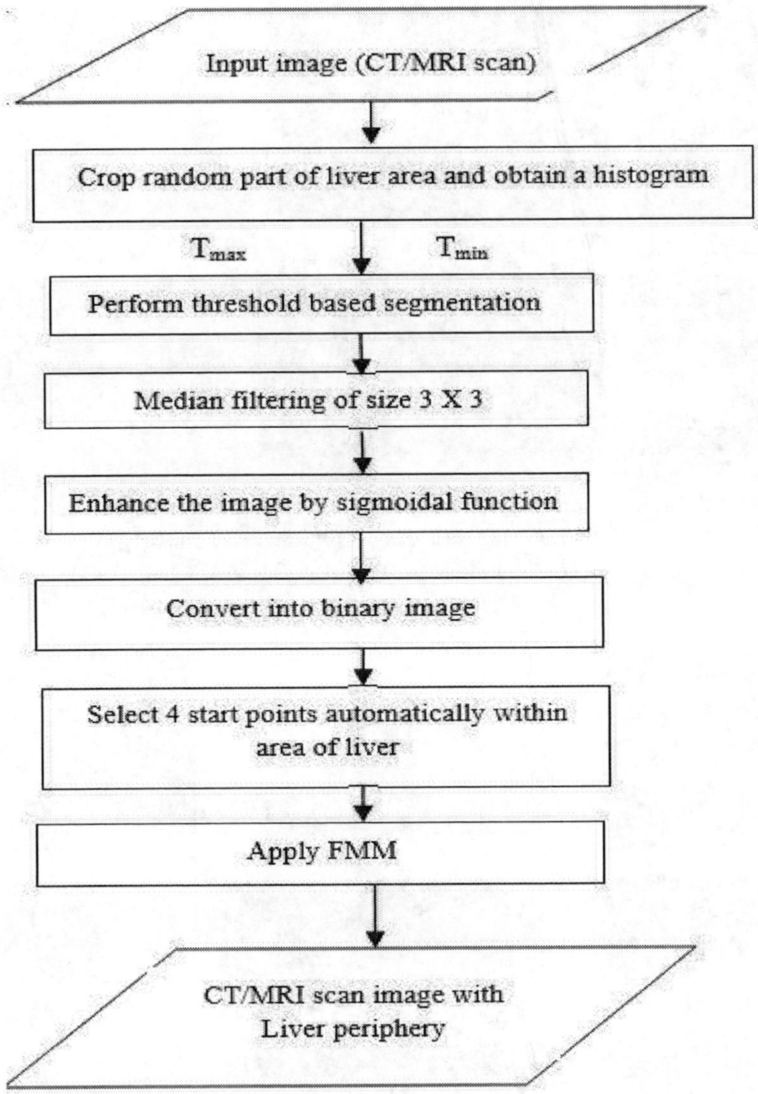

Figure 1. Flowchart of anticipated liver segmentation technique.

6. Comparison with Existing Methods

To evaluate the performance of novel model to identify liver border in abdominal CT and MRI scan images, it is compared with one of the popular region based segmentation method (a). Chan-Vese (C-V) model [17] with specifications: initial contour location= [325,335; 345,335], number of iteration=175 and μ=0.110; (b) Chunming Li et al [18] designed a level set based segmentation method intended for MRI scanned images. This method is also called as LSE (Level Set Evolution) model. The analysis of this method was executed on the medical images with specifications: μ=1.01, ε =1.01, step time=0.101, σ=4, initial contour position= [165,225; 185,235], number of iterations=11; (c) Chunming Li et al. [19] proposed a novel segmentation method based on energy fitting. This technique of segmentation is also called as Region-Scalable Fitting (RSF) model. The analysis was carried out with specifications: σ=3.23, ε =1.01, μ =1.01, step time=0.11 $\lambda 1$=1.01, $\lambda 2$=1.01, number of iterations=24, initial contour position = [165, 210; 185, 210].

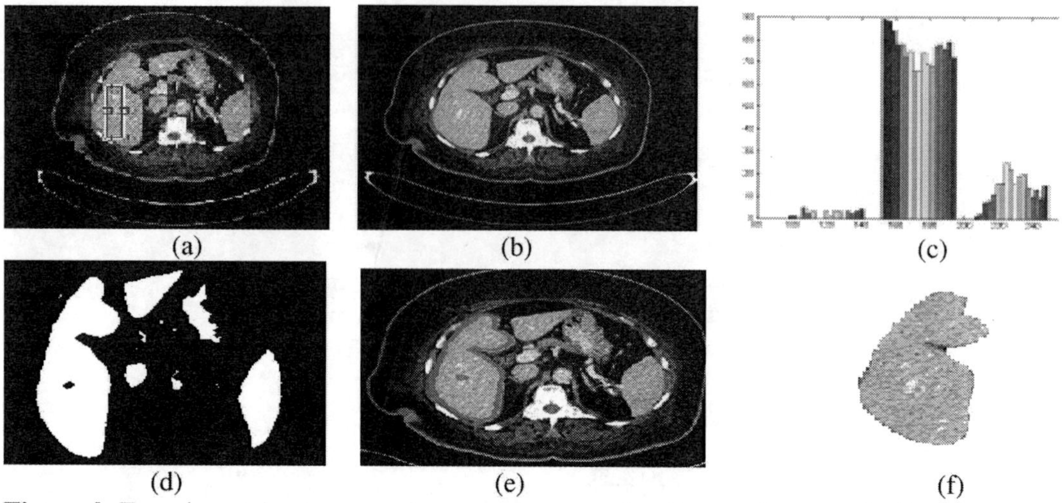

Figure 2. Experimental outcomes of proposed method for CT scan image; (a) 2D CT scan images; (b) pick arbitrary part of liver; (c) histogram of arbitrarily picked liver part; (d) results of threshold based segmentation for CT scan images; (e) periphery of liver marked in CT scan images; (f) reference image of CT scan.

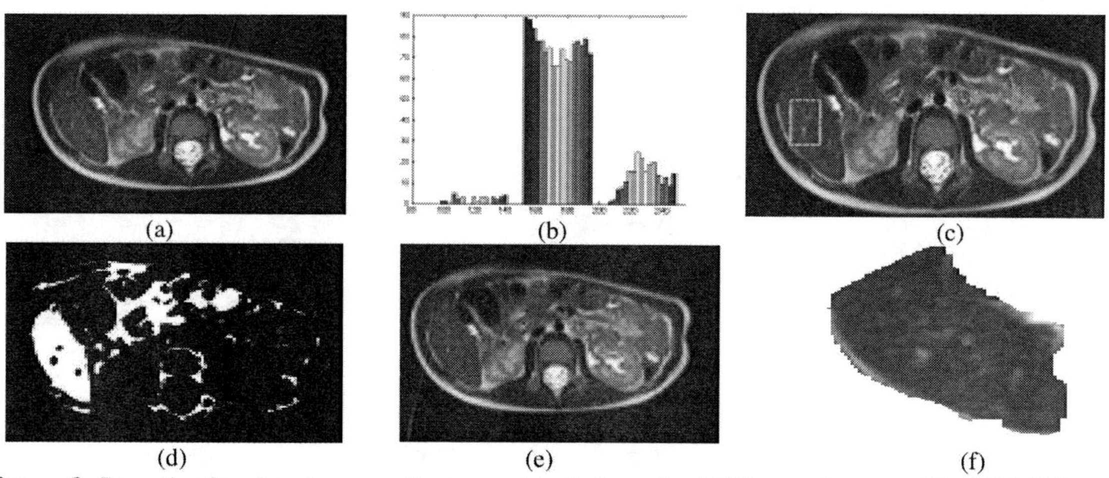

Figure 3. Investigational outcomes of proposed technique for MRI scan images; (a) 2Dl MRI scan images; (b) Pick arbitrary part of liver; (c)) histogram of arbitrarily picked liver part; (d) results of threshold based segmentation for MRI scan images; (e) periphery of liver marked in MRI scan images; (f) reference image of MRI scan.

Figure 4 presents comparison outcomes of proposed technique with existing technique. The figure (a), (b), (c), (d) and (e) show 2D MRI scan, result of C-V model, result of RSF model, LSE model and proposed model for MRI scan respectively. The figure (f), (g), (h), (i) and (j) show 2D CT scan, result of C-V model, result of RSF model, LSE model and proposed model for CT scan respectively
The liver has indistinguishable edge and pixel intensity distribution inside the liver varies from one section of liver to another section and it varies from one patient to another. Hence, all three existing algorithms have limitation, ensuing in vague detection of liver image. In this work, first pixel density variation inside the liver is analyzed by picking random section of liver and hence achieves exact liver separation from neighbouring organ.

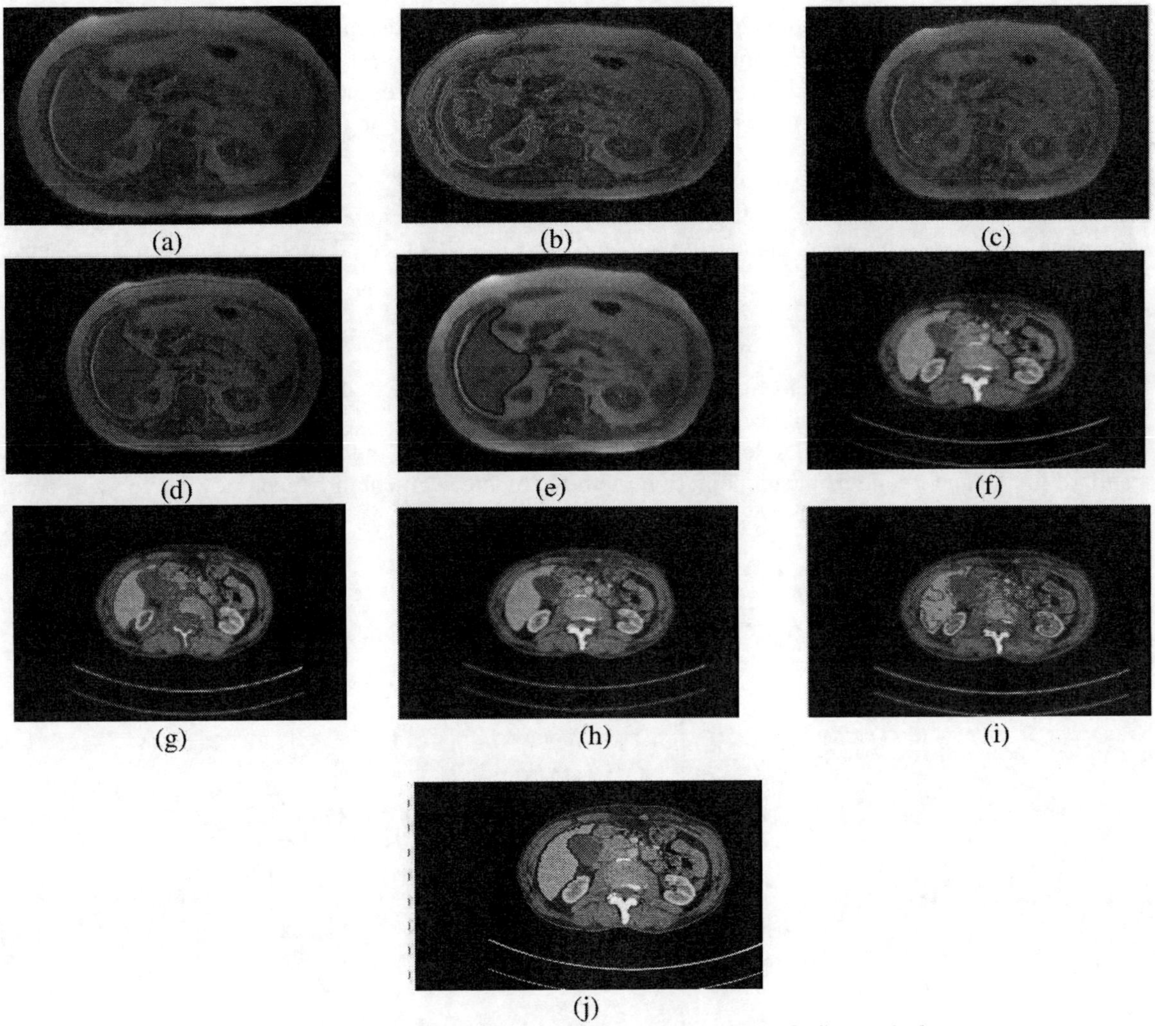

Figure 4. Comparison of proposed technique with existing technique.

The fig (a), (b), (c), (d) and (e) show 2D MRI scan, result of C-V model, result of RSF model, LSE model and proposed model for MRI scan respectively. Fig (f), (g), (h), (i) and (j) show 2D CT scan, result of C-V model, result of RSF model, LSE model and proposed model for CT scan respectively.

In the state-of-art that we have considered for comparison, the authors have to decide the position of initial curve and count of iterations on trial and error bases. These parameters influence the outcome of the segmentation method/ model. Within anticipated technique, there is no human intercession to identify the initial contour position and count of iterations. As soon as liver periphery is recognized, results will be demonstrated on the computer screen.

RSF model, LSE model, C-V model and the anticipated model have been experimented on 50 MRI and 50 CT scan images. The segmentation accuracy for proposed model and state of art has been determined for 50 MRI and 50 CT are illustrated in figure 5 and 6 respectively.

7. Segmentation Performance Evaluation Metric

The segmentation method has many applications like industry, medical images, business, etc. Nowadays, many segmentation methods are available. It is difficult to decide which segmentation algorithm has to be used [20-22]. This is because no universal segmentation method available which provides best result for all types of applications. For different applications, different types of segmentation methods are suitable [23-25]. Selection always depends upon type of data that is provided to segmentation methods. The evaluation metric(s) helps us to select one of the best segmentation algorithms for particular application. This is done by comparing the outcome of proposed model against standard image. The standard image is obtained by manual segmentation which is verified by practicing doctor.

7.1. Accuracy:

The quantity of match between standard image and algorithm segmented image is measured by metric called as accuracy. The accuracy is defined as follows.

$$Seg.Accuracy = \frac{ASP}{AP} \tag{3}$$

Where ASP = Quantity of satisfactorily classified pixels in segmented region and AP = Amount of Pixels in machine segmented image.

The perfect segmentation results segmentation accuracy of 1 i.e. there is 100% similarity between standard image and image from segmentation method. As the segmentation accuracy move away from 1, indicating amount of dissimilarity.

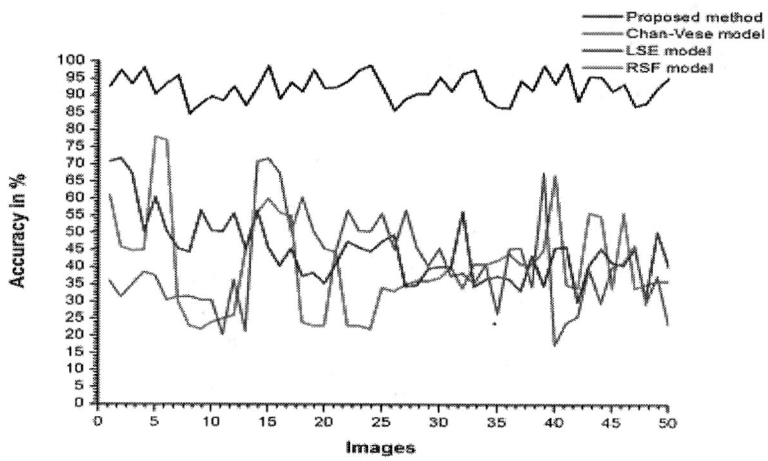

Figure 5. Comparison result using Accuracy for MRI scan images.

Table 1. Average segmentation accuracy for the anticipated method and State-of-art for MRI and CT scan images.

Scanned images	Proposed model (in %)	C-V model (in %)	LSE model (in %)	RSF model (in %)
MRI	74.8	48	51.5	42.3
CT	92.6	40.3	45.6	40.9

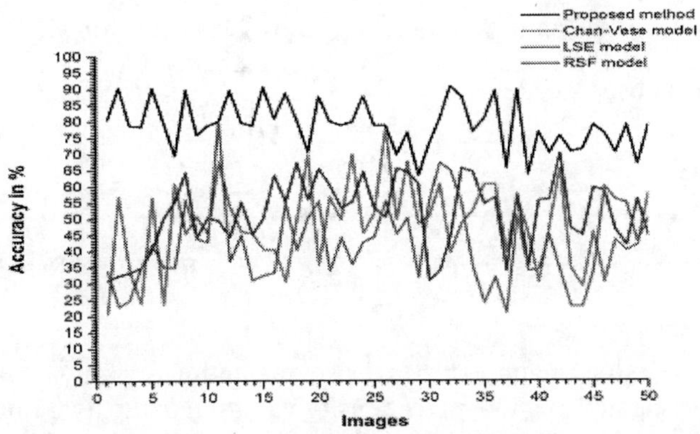

Figure 6. Comparison outcome using Accuracy for CT scan images.

7.2. Volumetric Overlie Error (VOE):

The Volumetric Overlie Error between 2 clusters of pixels SALG and SREF is specified in % and stated as (1-Jaccard Coefficient(JC)) [26, 27]. JC is proportion obtained by intersection and union of algorithm segmented image (SALG) and reference image (SREF). The SREF is acquired by physical separation using paintbrush and verified by doctor. The VOE ranges between 100% to 0%. VOE=100% indicates algorithm segmented image and reference image do not overlap at all and 0% indicates perfect segmentation result by algorithm i.e. algorithm segmented image and reference image overlap on each other. The VOE for proposed model and state of art has been determined for 50 MRI and 50 CT are illustrated in figure 7 and 8 respectively.

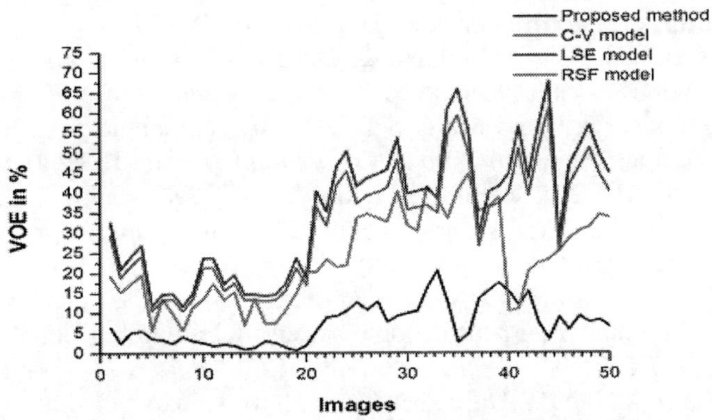

Figure 7. Comparison of results using VOE for MRI scan images.

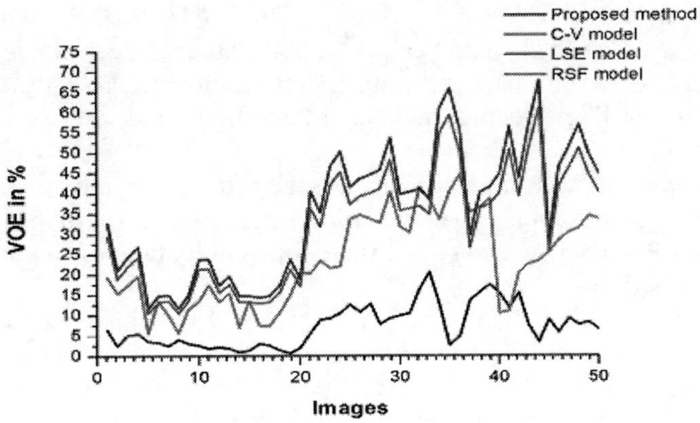

Figure 8. Comparison of results using VOE for CT scan images.

Table 2. Avarage VOE for the anticipated method and state-of-art for MRI and CT scan images.

Scanned images	Proposed model (in %)	C-V model (in %)	LSE model (in %)	RSF model (in %)
MRI	7.4	23.3	35.5	32
CT	5	23.6	27.6	34.4

Table I and II represent average accuracy in % and VOE in % respectively for MRI and CT scan images.

8. Conclusions

In this paper, the authors design the universal liver segmentation technique which is realized on MRI and CT images. In the anticipated method, pixel density variation in the liver image is studied by picking random part of liver image and obtaining the histogram of it. Using this pixel intensity range is achieved. With this, threshold based segmentation is executed which detaches the liver from its neighbouring organs. Subsequently, four starting/seed points are identified automatically which move outwardly using FMM, identifying liver border precisely. This method can be applied to other modalities without any modification. The proposed segmentation algorithm fails to detect liver image accurately in case of diseases exists at the edges. In the future, the authors will focus on segmentation of liver and its tumour.

References

[1] Xiao S, Cheng M, Wang B and Huang S 2013 Automatic Liver Segmentation from CT Images Using Adaptive Fast Marching Method *Seventh International Conference on Image and Graphics* 897-900

[2] Kumar P S and Latte M V 2019 Fully Automated Segmentation of Lung Parenchyma Using Break and Repair Strategy *Journal of Intelligent Systems* **28** 271-289

[3] Zheng S, Fang B, Li L, Gao M and Wang Y 2015 A variational approach to liver segmentation using statistics from multiple sources *Physics in Medicine & Biology* **63** 025024.

[4] Biswas A, Bhattacharya P and Maity S P 2018 3D segmentation of liver and its lesions using optimized geometric contours *Procedia computer science* **133** 240-247.

[5] Siri S K and Latte MV 2017 Combined endeavor of Neutrosophic Set and Chan-Vese model to extract accurate liver image from CT scan *Computer methods and programs in biomedicine* **151** 101-109

[6] Sangeeta S and Latte MV 2019 A novel approach to extract exact liver image boundary from abdominal CT scan using neutrosophic set and fast marching method *Journal of Intelligent Systems* **28** 517-532

[7] Siri SK and Latte MV 2017 Universal Liver Extraction Algorithm: An Improved Chan–Vese Model *Journal of Intelligent Systems* 29 237-50

[8] Meng L, Tian Y and Bu S 2020 Liver tumor segmentation based on 3D convolutional neural network with dual scale *Journal of Applied Clinical Medical Physics* **21** 144-157

[9] Budak Ü, Guo Y, Tanyildizi E and Şengür A 2020 Cascaded deep convolutional encoder-decoder neural networks for efficient liver tumor segmentation *Medical hypotheses* **134** 109431

[10] Sethian J 1996 Level Set Methods and Fast Marching Methods, *Cambridge University Press* 18 89-92.

[11] Sethian JA 1996 A fast marching level set method for monotonically advancing fronts, Proceedings of the National Academy of Sciences, **93,** 1591–1595

[12] Sethian JA and Popovici AM 1999 3-d traveltime computation using the fast marching method *Geophysics* **64** 516–523

[13] Sun H, Sun J G, Sun Z Q, Han F X, Liu Z Q, Liu M C, Gao Z H, and Shi X L 2017 Joint 3d traveltime calculation based on fast marching method and wavefront construction *Applied Geophysics* **14** 56–63

[14] Fomel Y 1998 Fast-marching eikonal solver in the tetragonal coordinates *In SEG Annual Meeting Society of Exploration Geophysicists* 1949-1952

[15] Yatziv L, Bartesaghi A, and Sapiro G 2006 O(n) implementation of the fast marching algorithm *Journal of Computational Physics* **212** 393 – 399

[16] Garrido S, Malfaz M, Blanco D 2013 Application of the fast marching method for outdoor motion planning in robotics *Robotics and Autonomous Systems* **61** 106-114

[17] Chan T, and Vese L 2001 Active Contours Without Edges *IEEE Transactions On Image Processing* **10** 266-77.

[18] Chunming L, Huang R, Ding Z, Chris Gatenby J, Metaxas DN and Gore JCA Level Set Method for Image Segmentation in the Presence of Intensity Inhomogeneities With Application to MRI *IEEE transactions on image processing* **20** 2011.

[19] Chunming L, Kao C Y, Gore J C, and Ding Z 2008 Minimization of Region-Scalable Fitting Energy for Image Segmentation *IEEE transactions on image processing* **17** 1940-9.

[20] Kumar P S and Latte MV 2018 Lung Parenchyma Segmentation: Fully Automated and Accurate Approach for Thoracic CT Scan Images *IETE Journal of Research* **66** 370-83.

[21] Walaa HE, Moftah HM, El-Bendary N and Hassanien AE Performance Evaluation of Computed Tomography Liver Image Segmentation Approaches *International Conference on Hybrid Intelligent Systems (HIS)* 109-114

[22] Li X, Aldridge B, Rees J and Fisher R 2010 Estimating the reference from multiple individual segmentations with application to skin lesion segmentation, *Annual Technical Meeting on Medical Image Understanding and Analysis (MIUA)*

[23] Graaf C, Koster A, Vincken K. and Viergever M 2008 Validation of the interleaved pyramid for the segmentation of 3D vector images in Pattern Recognition Letters, transactions on image processing, **17** 469-75

[24] Manisha S and Chouhan V 2012 Objective Evaluation Parameters of Image Segmentation Algorithms *International Journal of Engineering and Advanced Technology (IJEAT)* **2** 2249 – 8958

[25] Kumar PS and Latte MV 2019 Modified and Optimized Method for Segmenting Pulmonary Parenchyma in CT Lung Images Based on Fractional Calculus and Natural Selection *In Journal of Intelligent Systems* **28** 721–732.

[26] Jaesik M, Powell M and Bowyer KW 2004 Automated Performance Evaluation of Range Image Segmentation Algorithms *IEEE Transactions on Systems, Man, and Cybernetics, Part B (Cybernetics)* **34** 263-71.

[27] Heimann T, Van Ginneken B, Styner MA, Arzhaeva Y, Aurich V, Bauer C, Beck A, Becker C, Beichel R, Bekes G, Bello F, Binnig G, Bischof H, Bornik A, Cashman P, Ying Chi, Cordova A, Dawant B M, Fidrich M, Furst J D, Furukawa D and Grena L Comparison and Evaluation of Methods for Liver Segmentation from CT Datasets *IEEE Transactions on Medical Imaging* **28** 1251-65.

Classification of Diabetic Retinopathy using Residual Neural Network

C Priyadharsini[1], J Jeyapriya[2], R Jagadeesh Kannan[3], G Bharadwaja Kumar[4] and Tulasi Prasad Sakriki[5]

[1]School of Computer Science & Engineering, Vellore Institute of Technology Chennai Campus, India,
[2,3,4,5]School of Computing Science and Engineering, VIT Chennai, India,
[1]priyadharsini.c27@gmail.com, [2]priyacseme@gmail.com, [3]jagadeeshkannan.r@vit.ac.in, [4]bharadwaja.kumar@vit.ac.in, [5]tulasiprasad.sariki@vit.ac.in

Abstract. Diabetic retinopathy (DR) in patients affects retina function owing to chronic elevated excessive blood glucose rates. The DR is a severe medical disorder. Patients with diabetes are easily prone to this medical complication which when not detected and treated at earlier stages, leads to vision loss. Globally, diabetes mellitus is the fifth leading cause of vision loss. So, active research is being conducted in this area to find novel ways in identifying the stages of DR. Specific image recognition methods and computer simulation algorithms were initially used to classify DR, but their usefulness was inadequate in real-time clinical practice. The evolution of deep learning models like convolution neural network performed better in identifying DR and non-referable DR compared to conventional machine learning models. Different variations of CNN architecture are being developed over the period, but more analysis and experimentation needs to be carried out to choose the appropriate architecture for detecting Diabetic Retinopathy. The aim of this research is to apply and understand how the performance of pre-trained Deep learning model – ResNet, a deep layered neural network performs to identify non-referable DR and different types of referable DR.

Keywords: CNN, Deep Learning, Diabetic retinopathy, Neural network, ResNet.

1. Introduction

T as a consequence of hyperglycaemia, a persistent diabetes occurs-elevated blood glucose rates. The presence of hyperglycaemia for a longer period is the starting point of DR development. The numbers of people prone to Diabetes globally are increasing at a very faster pace. In India alone, 69,9 million and 80 million people are projected to cross diabetic by 2025 and 2030. And across the globe, estimated population with diabetes retinal related disease is 382 million and in the year 2025 it might rise up-to 592 million [1]. A diabetes mellitus period in patients was the key predictor for assessing the incidence of diabetic retinopathy. After the duration of 10 years, 20% of Type I and 25% of Type II, 20% of Type I and 60% of Type II, 95% of patients with Type I or Type II acquire this disorder over the span of 30 years. Type I diabetic patients have absolute deficiency of insulin and Type II diabetic patients suffer from defect in progressive insulin secretion [2]. Females are more affected than males in the ratio of 4:3. Pregnancy, obesity, hypertension, hyperlipidaemia, smoking, anaemia are other risk factors associated with progression of DR [3].

Clinical studies show that the progression of DR can be slowed down by bringing blood lipids, blood glucose, blood pressure under optimal control when detected at the earlier stage. But the rate of diabetic population is rising at a rapid pace due to lack of awareness in one-third of the population. And another reason is, as per analysis by experts for 60 million diabetic populations prone to retinal disease, approximate number of Ophthalmologists and Retinal consultants are 12,000 and 3,500 respectively [1]. This is the reason why WHO categorises DR as a major retinal disease and expects attention from professionals and authorities [1, 20]. With advanced machine/ deep learning models and high computational power, computer professionals and engineers can make a significant contribution in identifying DR. This research aims to explore deep learning trends accessible and to explain how ResNet performs in the recognition of various stages in DR in a strong pre-trained model.

Organisation: The remaining research is structured as follows. The following. Part 2 offers a awareness of diabetes retinopathy. The related research articles are discussed in Section 3. Section 4 offers details and the need for ResNet architecture. The ResNet model's study findings are presented in section 5 and potential work is carried out in section 6. And last but not least, Section 7 ends this report.

2. Technologies used

DR is caused when lesions are formed as a reaction to change in structure of blood vessels or due to leakage of fluids like lipid in the retina. As the progression of DR takes place, different symptoms become prominent. This section discusses about features of DR and its classification of different stages [3][4].

2.1. Features of DR

To Major features associated with DR, identified depending upon the shape, colour, texture and size of the symptoms are as follows:

- Microaneurysms
- Retinal haemorrhages
- Hard exudates
- Cotton-wool spots
- Neovascularization

Microaneurysms are identical in nature and dot haemorrhages differ in duration. The microaneurysms, red spherical dots appear small. Blot and flame haemorrhages are big in size. The yellowish white waxy appearance is an attribute associated with hard exudate while cotton wool spots have white fluffy characteristics. If retinal blood vessels alter their shape to obstruct blood supply, there are additional arteries in various places in the eye. This condition of new blood vessel development is called neovascularization.

2.2. Classification of DR

DR is generally graded as a non-proliferative DR (NPDR) and proliferative DR (PDR) in two stages. The development of fresh blood vessels in the eye is correlated with growth here. Non-proliferative DR is a stage where different types of lesions develop in the retina because of micro vascular occlusion which differ in shape, structure, texture but there is no new development of blood vessels. DR phases of NPDR may be mild, moderate and extreme, depending on the level of sensitivity of the pupils. The current step of DR is PDR where blood vessels that contribute to blood leakage contribute to lack of vision as in figure 1. Chart 1. The retinal fundus photos displayed below at different DR levels.

IOP Conf. Series: Materials Science and Engineering **925** (2020) 012033 doi:10.1088/1757-899X/925/1/012033

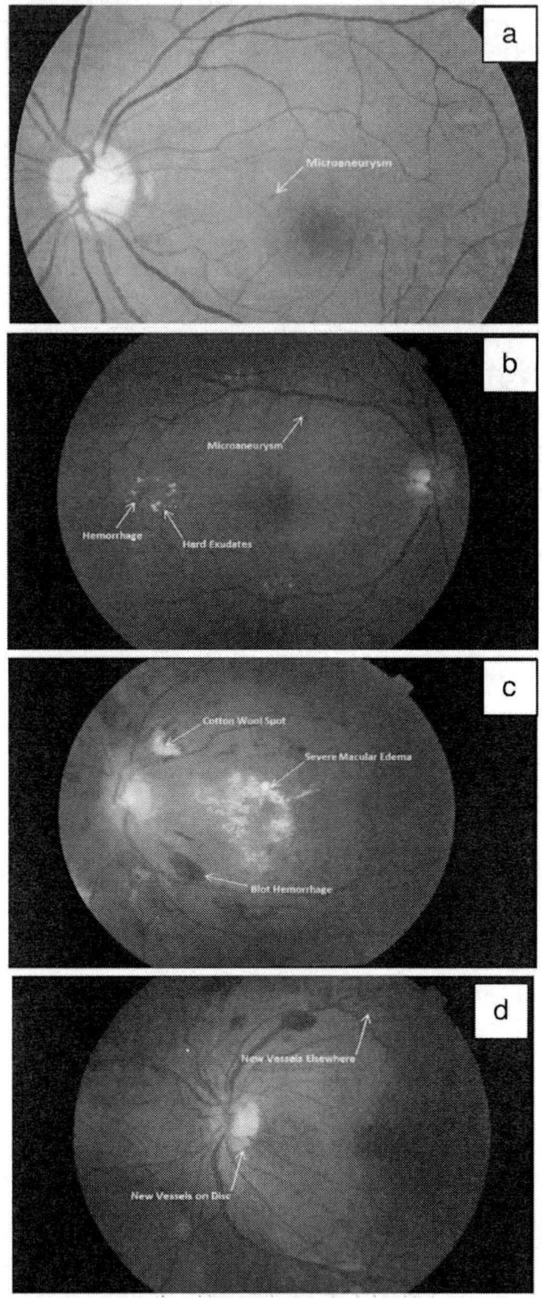

Figure 1. (a) Mild NPDR (b) Moderate NPDR (c) Severe NPDR (d) PDR

According to ETDRS study, non-proliferative DR is classified as follows as in table. 1 based on the severity level [3].

Table 1.non-proliferative DR classification

Grade	Features
R0: No abnormalities	No abnormalities
R1: Mild NPDR	At least one microaneurysm, Retinal hemorrhages, Hard or soft exudates
R2: Moderate NPDR	Strong exudates in at least 1 quadrant (Cotton wool spots) Microaneurysms and / or line and mark haemorrhages
R3: Severe NPDR	Any one of the following 3 features is present (Known as the 4-2-1 rule) • Intraretinal hemorrhages and microaneurysms in all four quadrants. • In 2 or more quadrants Venous Beading • Strong IRMA in 1 quadrant or more
R4: PDR	There are two of the features of the 4 2 1 law

3. Related works

Earlier the use of machine learning models was focussed towards detecting different types of lesions in DR related research works. The paradigm has been changed in the last years to introduce specific profound thinking models to assess the DR gravity point.

In a survey paper published during the year 2014 on ocular disease by Zhuo et al. [5], The author notes that much of his research has focussed on lesion identification relevant to DR. And there were few studies accessible for DR identification that included the DR classification in binary, i.e. reproducible or non-referable DR. Pal et al. [6] studied the implementation of algorithms in machine learning such as the Decision Tree, Classification in Naïve Bayes, Vector Support machines and neighboring K-retinopathy algorithms. The results obtained from applying ML algorithms on Messidor dataset showed SVM had an accuracy of 74.65 % which was highest among the other models.

In Messidor and Eyepacs datasets, Gulshan et al. [7] recorded for the first time the DL for DR recognition. Big picture databases of the fundus were used for monitoring of a deep CNN. The emphasis of the study was only on identification of DRs, but no comparable knowledge on innovative DRs or other DR phases was provided. Abr`amoff et al [8] developed IDx-DR X2.1, a device used for DR detection. The class of classification used in this device are rDR, vtDR and pDR which means referable, vision threatening and proliferative DR respectively. The first designed CNN architecture called AlexNet was used to identify the characteristics of lesions, optic disk and fovea to stage the DR affected retinal images. Lu et al [9]. in his paper on overview of artificial intelligence in ophthalmology, observed that CNN is the most used technique to classify ocular diseases. The author states the performance of DL goes down when more than one disease is included. Therefore, it is important to build systematic AI systems to diagnose general eye disorders in the modern world utilizing multimodal data.

To detect non-DR and DR affected retinal images, Gargeya et al. [10] used CNN for learning deep discriminative features and the result was visualized using heatmap. Features related to pixel height, width and field of view was added with 1024 features obtained after the average pooling layer in CNN and gradient boost classifier on top of it to perform binary classification.

GoogleNet architecture was modified by Takahashi et al. [11] to classify the severity level of DR. Deletion of top 5 layers, reduction in batch size, expansion of crop size are the major modifications applied to GoogleNet architecture. Two different approaches were used by the author in model design. One approach is usage of three coloured retinal images and other retinal image of only one colour for manual staging. The experiment was validated with 20-fold cross validation and compared with ResNet model.

Sequence of pre-processing steps were applied by Garcia et al. [12] before feeding the data on various CNN architectures for detecting DR. The average color of each pixel in the picture is measured and subtracted and the resolution is restated to 256 x 256. After pre-processing data augmentation is carried out as a next step. The results are analysed by applying the resultant dataset on different neural network architecture by varying the learning rates and number of layers.
Though CNN architectures perform better in identifying DR, the number of parameters calculated are huge. So, this paper aims at analysing the performance of ResNet, a neural network architecture where the number of parameter calculation is reduced but the network is deep.

4. Architecture of resnet model

4.1. Description

Krizhevsky et al. [13] shifted the paradigm from conventional machine learning model by proposing a deep conventional neural network architecture called AlexNet. Based on this architecture, the concept of simple design is the number of features that the model increases with rising layers in the neural network. Many models such as ZFNet, VGG16, GoogleNet were born of this idea.
To understand the impact of number of layers in the design of CNN on the performance of the model, He et al. [14] experimented and analysed two CNN models by varying the number of layers- one CNN with 20 layers and other with 56 layers with CIFAR-10 dataset. Figure 2 shows training error plot and test error plot plotted against respective errors vs number of iterations. It is observed that both training error and test error is high for the CNN model with 56 layers. The reason for high error rate is to depict it is difficult for all the models to optimise and not as a result of overfitting or adding more layers.

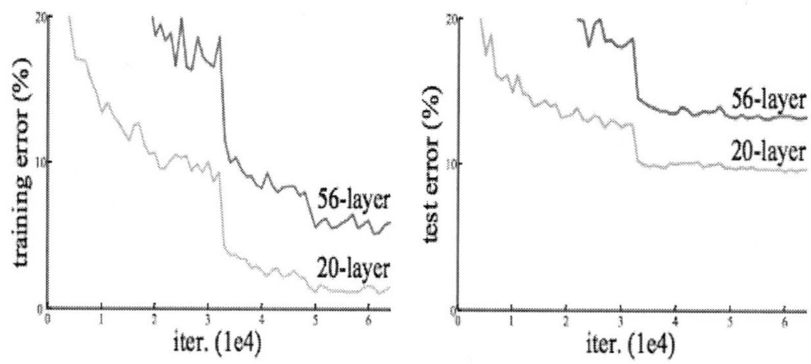

Figure 2. (a) shows the plot between training error percentage vs number of iterations for ResNet 20 and ResNet 56 layer, (b) shows the plot between testing error percentage vs number of iterations for ResNet 20 and ResNet 56 layer

To overcome this issue, residual block was introduced as a new neural network layer which works on the concept of skip or shortcut connections. This architecture was proposed by Microsoft called deep residual learning framework []. Figure.3 shows the structure of residual block.

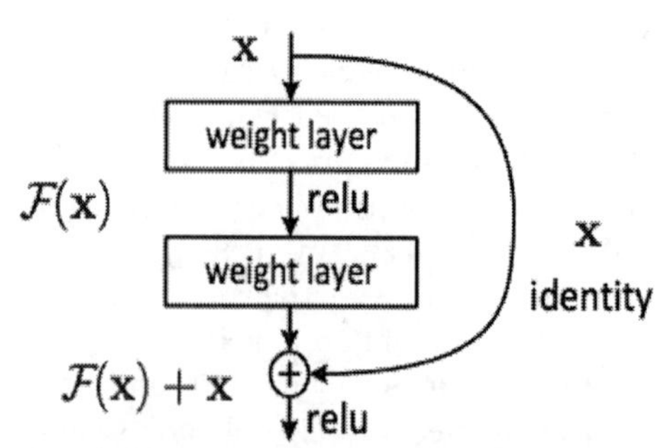

Figure 3. Structure of residual block

IOP Conf. Series: Materials Science and Engineering **925** (2020) 012033 doi:10.1088/1757-899X/925/1/012033

Skip connections improved the performance and mitigated the problem of adding deep layers since vanishing gradient problem is solved by adding the input to the convolution block output. Also, the skip connections ensured that the high layers do not perform less than the low layers by letting the model to learn an identity function.

4.2. Design principle

Different architectures [14] used for ImageNet are shown in figure 4. VGG-19, 34-layered single-scale networks and ResNet-34 are running at a pace of 19.6 billion, 3.6 billion and 3.6 billion per second. It can be noted that 19 layered VGG model has huge number of FLOPs compared to 34 layered ResNet model. The convolution layers in plain baseline model are assigned mostly with 3 x 3 filters. The basic principle of this model is when the output feature map is same or halved, then the number of filters is same or doubled respectively. In ResNet model, shortcut connection is introduced. When the dimensions of input and output is same, identity shortcut F(x{W}+x) is usedAs measurements change, then the projection workaround in ID feature is not used to suit measurements by padding neros for expanded dimensions after the description mapping or by using 1 x 1 convolution.

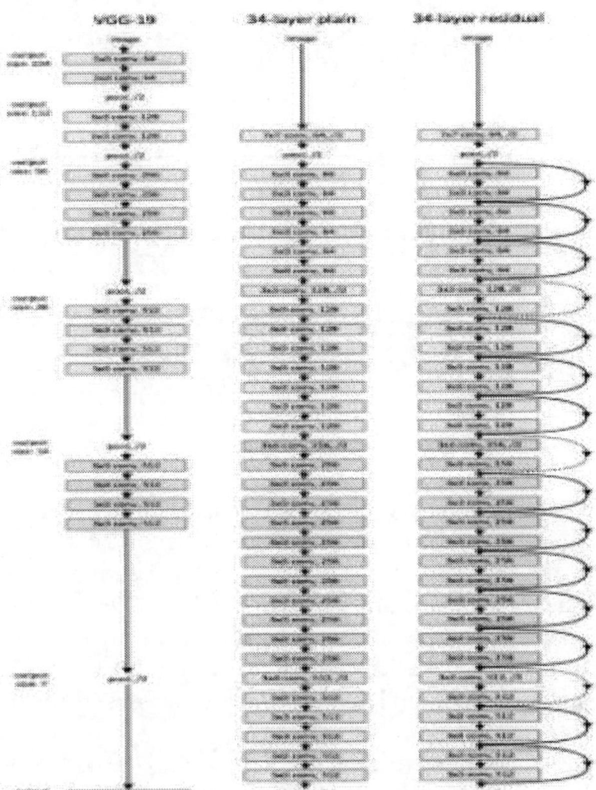

Figure 4. shows architecture of 34-layer residual model, 34-layer plain model and VGG-19 model.

In networks of small size like ResNet 18 and ResNet 34, each ResNet block is of depth 2 layers[15]. In ResNet model of 50 layers figure.6, 101 layers and 152 layers, the 2 layered blocks are replaced by three layered blocks which is shown in Figure .5. The number of FLOPs performed by this model is 3.8 x 10 9.

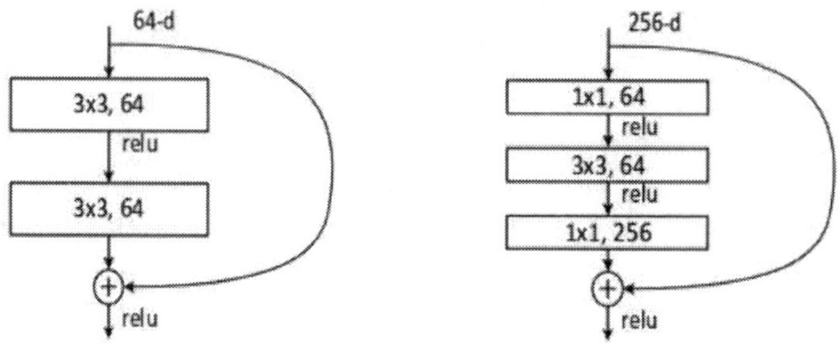

Figure 5. (a) 2 layered residual block and (b) 3 layered residual block

4.3. Specification

Below table tabulates the number of filters, each filter size and output size from convolution layers for ResNet 50 model:

Layer Name	Output size	50-layer
Conv 1	112 x 112	7 x 7, 64, stride 2
Conv 2_x	56 x 56	3 x 3, max pool, stride 2 3 x [1x1,64 3 x 3,64 1 x1,256]
Conv 3_x	28 x 28	3 x [1x1,64 3 x 3,64 1 x1,256]
Conv 4_x	14 x 14	3 x [1x1,64 3 x 3,64 1 x1,256]
Conv 5_x	7 x 7	3 x [1x1,64 3 x 3,64 1 x1,256]
	1 x 1	Average pool, 1000-d fc, softmax
FLOPs		3.8×10^{9}

Figure 6. ResNet 50-layer architecture

4.4. Implementation

In the present work, Fastai deep learning library has been used to retrain the ResNet models. Fastai has many pre-trains: resnet18, densenet121, densenet169, densenet201, densenet161, vgg16 bn, vgg19 bn, alexnet, resnet101, resnet152, squeezenet1 0, squeezezenet1 0. Learning transition is a methodology in which a model trained on a large dataset may be used and the model can then be applied to the dataset from another area. The theory is that the neural network can learn well in very large datasets and that the model would benefit from this experience, particularly if the data set is limited relative to a randomly initialized model. In the last part of the model, the number of classes inside the current dataset has to be modified.

The experiments in the present work have been carried out on NVIDIA K-80 Tesla GPU with 12 GB RAM. In the present work, we have experimented with Resnet34, Resnet50 and Resnet101 architectures. But, when we increase the batch size and use Resnet101, there was a memory limitation on GPU RAM. Resnet34 was not able to produce better results. Hence, we have confined to Resnet50 architecture. We varied the batch size and also optimized the learning rate in the present experiments. The detailed experimentation is given in the next section.

5. Experimental results and analysis

In this research paper, ResNet 50 model is applied on retinal images to analyse the efficiency of the model for two tasks:

- Binary classification: to identify referable and non-referable DR images.

- Multiclass classification: to identify non- referable DR and different stages of DR the retinal image is affected with.

The dataset used for this experiment is collected from Kaggle website as part of the APTOS 2019 Blindness Detection competition. The collection includes a broad variety of high-resolution retina photographs in different imaging conditions. Every topic has a left and right area. Images are labelled both left and right with a subject identifier [16]. For each photograph the clinician has measured with the following scale the occurrence of diabetic retinopathy for 0 to 4: 0- No DR, 1-Make-Up, 2-Mild, 3 – Severe, 4-DR Proliferative. The tests are conducted with 80% of the training photos and 20% of the evaluation photos. The data set's class distribution is seen below Figure .7.

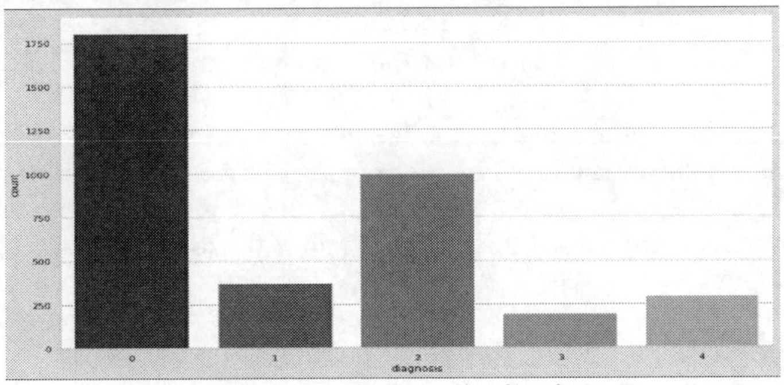

Figure 7. DR class distribution

With repeated experiments on the data, following value for the hyper-parameters are fixed. The number of epochs is set to five, batch size is fixed as 50. Apart from conventional method of tuning the hyper-parameters, for multi-class classification the input image size is changed, and the results are observed. The product efficiency is assessed by means of a uncertainty matrix analysis. Accuracy measure is defined as the ratio of positive and negative observations that are predicted same as the actual value to that of sum of all the possible combinations of negative and positive cases of actual and predicted value. The formula for accuracy is given by:

$$\textbf{Accuracy} = \frac{TP + TN}{TP + TN + FP + FN}$$

5.1. Binary Classification

Image Size = 224,224

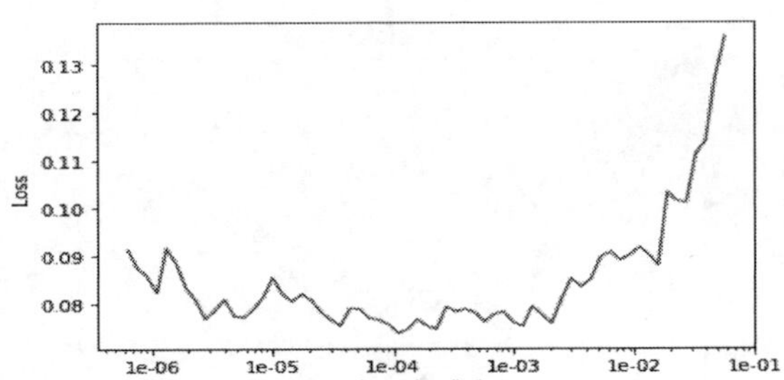

Figure 8. Loss function vs different learning rates

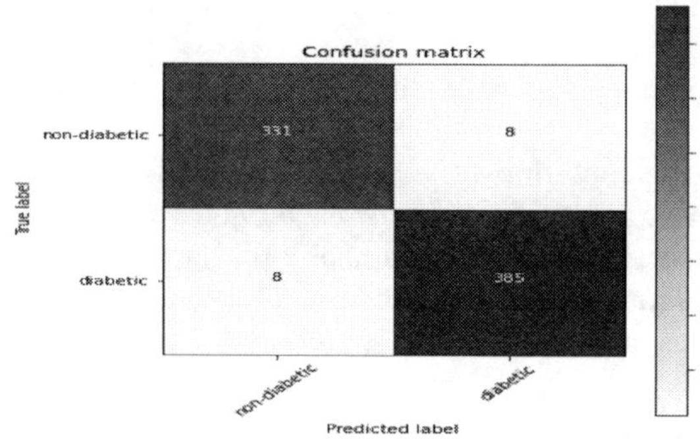

Figure 9. Confusion matrix

To identify referable and non-referable DR, the input image is resized to 224 x 224 and fed into the ResNet 50 model. The results are analysed by plotting a graph between loss function and learning rate for 50 experiments. From the plot Figure .8, we can find that loss is minimum when learning rate is at 5e-6,1e-5. The accuracy of the model at this learning rate is 0.9781420765027322 calculated from the confusion matrix in Figure .9.

5.2. Multiclass Classification

Two types of experiments are carried out by varying the input image size and analysed to check its effect over the performance of the model.

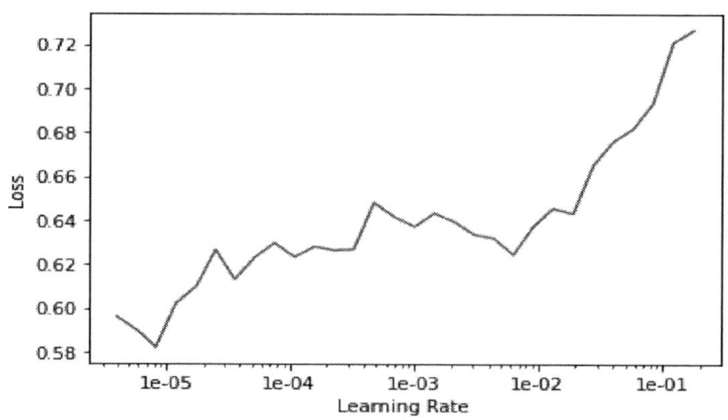

Figure 10. Loss function vs different learning rates

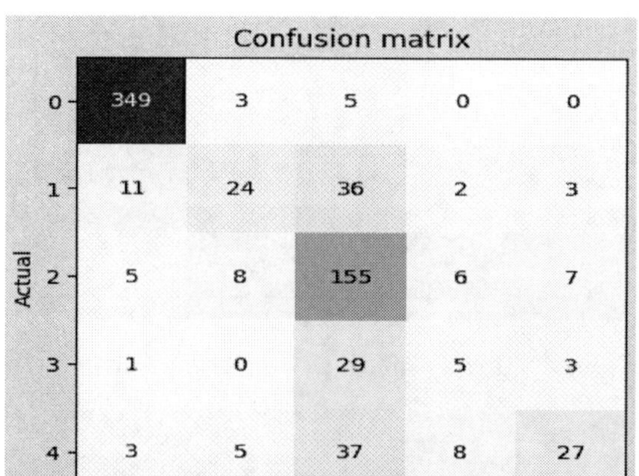

Figure 11. Confusion matrix

To identify DR images belonging to all the 5 categories, the image is resized to 100 x 100 and the results are analyzed for 50 times by varying the learning rates. From the plot figure .10 above, the best-chosen learning rates are 5e-6, 1e-5 and the accuracy estimated using confusion matrix figure. 11 is 0.7650. The reason for less accuracy is the count of images belonging to class one and four but predicted as the images with stage 2 of DR is more

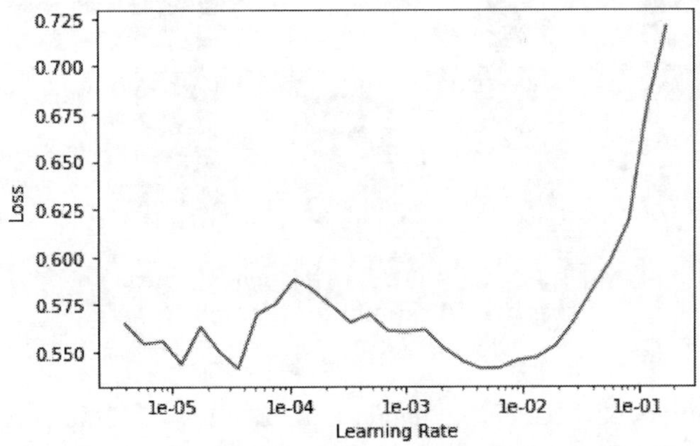

Figure 12. Loss function vs different learning rates

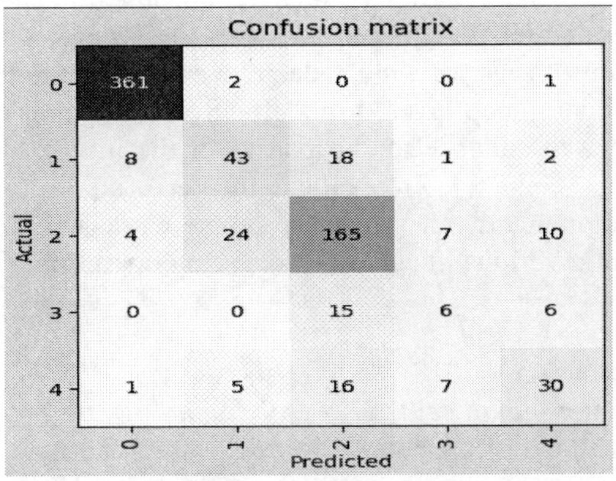

Figure 13. Confusion matrix

The accuracy is found to improve from 0.7650 to 0.8265 when the image size is increased from 100 x 100 to 224 x 244. The reason for this improvement in accuracy could be because of the skip or short connections in ResNet architecture which preserves the features related to DR found at the end of convolution layers as shown in figure 12 and figure 13.

To understand the efficiency of the model on the obtained Kaggle dataset for different classes of DR, additional steps of pre-processing or extraction of features or including hand-crafted features are not included before applying to ResNet model. It is found that ResNet models show promising accuracy of 97.8 % for binary classification. And for multiclass classification the accuracy is 82.65%. Table 2 shows the comparison table. Figure 14 shows the comparison graph.

Table 2. Comparison table

S.No	Methods	Accuracy (%)
[17]	Multiscale amplitude-modulation-frequency-modulation	92
[18]	Vector quantization	90
[19]	Support vector machine	86
Proposed system	**Residual Neural Network**	**97.8**

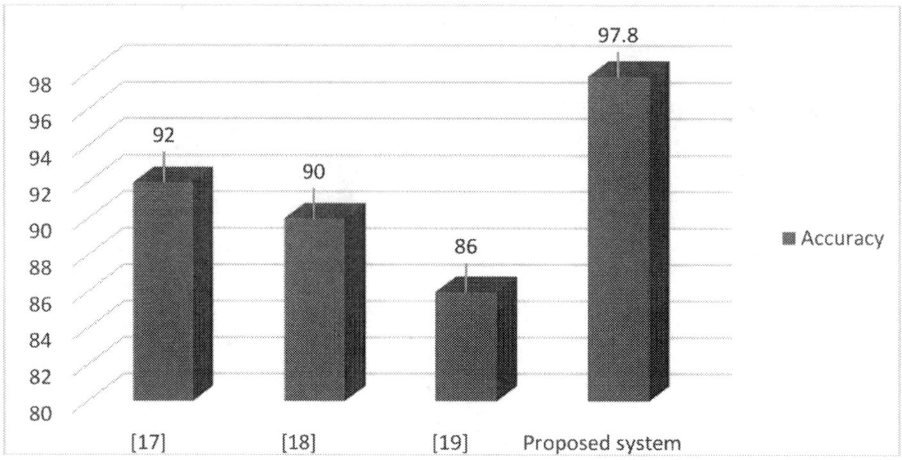

Figure 14. Comparison graph

6. Conclusion

The increase of diabetic retinopathy patients at an alarming rate pose a limitation for all the patients to undergo manual eye examination. Efficient algorithms to detect DR are the need of the hour to detect DR at different stages. So, that proper diagnosis can be planned at the earliest to avoid from vision loss. Deep learning models perform much better compared to conventional machine learning models in identifying DR. The ResNet model is implemented with the Kaggle data set to conduct binary and multiclass classification, in this research article, a deep layered neural network. From the experimental analysis, it can be inferred that a deep layered model – ResNet is able to perform better even on the dataset where additional steps to enhance the properties of the image is not carried out. Future work includes the current work can be extended to evaluate the performance by comparing different ResNet models, comparing ResNet model with other deep layered networks and compare the results of applying ResNet model after performing various image pre-processing operations.

References

[1] Suresh G, Pandey K, Sharma V 2018 World Diabetes Day 2018: Battling the Emerging Epidemic of Diabetic Retinopathy *Indian Journal of Ophthalmology* **66** 1652.

[2] Classification and Diagnosis of Diabetes, Diabetes Care 2015 Jan; **38**, S8-S16, 2015.

[3] Khurana AK Khurana B 2015 Comprehensive Ophthalmology: With Supplementary Book-Review of Ophthalmology *JP Medical Ltd*

[4] Wong TY, Sun J, Kawasaki R, Ruamviboonsuk P, Gupta N, Lansingh VC, Maia M, Mathenge W, Moreker S, Muqit MM, Resnikoff S. 2018 Guidelines on diabetic eye care: the international council of ophthalmology recommendations for screening, follow-up, referral, and treatment based on resource settings *Ophthalmology* **125** 1608-22.

[5] Zhang Z, Srivastava R, Liu H, Chen X, Duan L, Wong DW, Kwoh CK, Wong TY, Liu J 2014 A survey on computer aided diagnosis for ocular diseases *BMC medical informatics and decision making* **14** 80

[6] Pal R, Poray J and Sen M 2017 Application of machine learning algorithms on diabetic retinopathy *IEEE International Conference on Recent Trends in Electronics, Information & Communication Technology (RTEICT)* 2046-2051

[7] Gulshan V, Peng L and Coram M 2016 Development and validation of a deep learning algorithm for detection of diabetic retinopathy in retinal fundus photographs *Journal of American Medical Association* **316** 2402

[8] Abramoff MD, Lou Y, Erginay A, Clarida W, Amelon R, Folk JC and Niemeijer M 2016 Improved automated detection of diabetic retinopathy on a publicly available dataset through integration of deep learning *Investigative Ophthalmology & Visual Science* **57** 5200–06

[9] Lu W, Yan T, Yue Y, Yiqiao X, Changzheng C 2018 Applications of Artificial Intelligence in Ophthalmology: General Overview *Journal of Ophthalmology* 1-15.

[10] Gargeya R and Leng T 2017 Automated identification of diabetic retinopathy using deep learning *Ophthalmology* **124** 962–969

[11] Takahashi H, Tampo H, Arai Y, Inoue Y and Kawashima H 2017 Applying artificial intelligence to disease staging: Deep learning for improved staging of diabetic retinopathy, *Plos One* **12** 6

[12] Garc´ıa G, Gallardo J, Mauricio A, L´opez J and Del Carpio C 2017 Detection of diabetic retinopathy based on a convolutional neural network using retinal fundus images *In International Conference on Artificial Neural Networks ed*Springer **11071** 635–642

[13] Alex K, Sutskever I and Hinton EG 2012 ImageNet Classification with Deep Convolutional Neural Networks *In Advances in neural information processing systems* 1097-1105

[14] He K, Zhang X, Ren S and Sun J 2016 Deep Residual Learning for Image Recognition *In Proceedings of the IEEE conference on computer vision and pattern recognition* 770-778

[15] Rajan A, and Ramesh GP 2016 Glaucoma detection in optical coherence tomography images using undecimated wavelet transform, *Research Journal of Pharmaceutical Biological and Chemical Sciences* 7 878-85

[16] Shahada SAA, Hreiji SM, Atuduand SI, Shamsudheen S 2019 Multilayer Neural Network Based Fall Alert System Using IOT *International Journal of MC Square Scientific Research* **11** 1-15

[17] Agurto C, Murray V, Barriga E, Murillo S, Pattichis M, Davis H, Russell S, Abràmoff M and Soliz, P 2010 Multiscale AM-FM methods for diabetic retinopathy lesion detection *IEEE transactions on medical imaging* **29** 502-512

[18] Satyarthi D, Raju BAN and Dandapat S 2006 Detection of diabetic retinopathy in fundus images using vector quantization technique, *Annual IEEE India Conference* 1-4

[19] Barriga ES, Murray V, Agurto C, Pattichis M, Bauman W, Zamora G and Soliz P 2010 Automatic system for diabetic retinopathy screening based on am-fm, partial least squares, and support vector machines *IEEE International Symposium on Biomedical Imaging: From Nano to Macro* 1349-1352

[20] Prabu S, Balamurugan V, and Vengatesan K 2019 Design of cognitive image filters for suppression of noise level in medical images *Measurement* **141** 296-301

Application of Different Techniques for Underwater Image Processing- A Systematic Review

Prasath R[1] and Kumanan T[2]

[1] Department of Computer Science, Meenakshi Academy of Higher Education and Research, Chennai, India
[2] Faculty of Engineering and Technology, Meenakshi Academy of Higher Education and Research, 600078 Chennai, India
[1]Prasathr05@gmail.com, [2]kumananvetri@gmail.com

Abstract. The area of underwater image processing has gained huge attention among the researchers because of its wide application. Image processing is the approach of developing the input image quality so that it would be understood easily by users in future. Image processing develops the data content of image and changes the images visual impact on the observer. Image processing intensifies the image features and accentuates the features of image like corners, contrast to enhance the photographs display which is much helpful for study and examination. Several methods and techniques have been used by researchers worldwide in order to resolve the underwater image processing issues. This paper reviews different techniques of underwater image processing and also compares the techniques among each other. This paper discusses about the filtering method, histogram-based equalization method and particle swarm optimization techniques and compares them with their advantages.

Keywords: Filtering Method, Histogram Based Method, Particle Swarm Optimization, Underwater Image Processing.

1. Introduction

Specific care is needed for the creation of manuscripts to be replicated by photo-offset. Displayed papers in a technologically unacceptable manner are recalled, whether the amount cannot be completed on schedule, for retyping, or can be celled.

1.1. LNCS Online

In electromagnetic radiation, water is seen as a robust attenuator, thus satellites could connect whole planets from space with ranging of laser and cameras, vehicles in underwater has to bewithin100 meters for best optical sensors. Mechanical waves penetrate within the water, thus some realistic trade-offs are there between frequency, strength of source and distance speed. Sonar's are integral on ship adapt lesser frequencies for reaching the bottom, whereas these adapt lengthy wavelengths that move towards the cost of minimized resolution. For mapping fine-scale types are related to most of practical applications, acoustic and optical imaging stages have to perform relatively near to seafloor. For overcoming these issues, image processing is adopted (Dhanigaivel et al, 2019). There are many researchers and investigators who focused on studying about image processing in underwater (Peng and Cosman, 2017, Galdran et al, 2015, Sowmyashree et al, 2014, and Senthilkumaran and Thimmiaraja, 2014, Lu et al,

2015, Sathya and Bharathi, 2015, Roser et al, 2014, Panetta et al, 2016, Kaur et al, 2014, Wang et al, 2015, Qing et al, 2015, Alharbi et al, 2014 and Galdran et al, 2014). Underwater imaging plays a vital role in research of computer graphics and ocean engineering. Removal techniques for underwater haze have become very famous since adoption of different applications in image underwater. Underwater videos and images are captured for showing underwater to the world. Haze is usual observable fact that minimizes the clear view of underwater images. Additionally, haze minimizes the contrast of scene which would fade the colours. Thus eliminating the haze is challenging and difficult task. There are different methods and techniques developed for removing the haze and quality of the image is improved from underwater images as mentioned by Pati et al (2018).

Underwater images are distorted because of amalgamation and distributed. Captured images from underwater have poor visibility. Due to sea water's dispersion and absorptive nature, it is tough to obtain visible and satisfying images at short or long paths. Noises minimize the details which involve main information. Therefore, best resolving speckled images in underwater are significant for observing in ocean. Such images mostly suffer from distortion of colour and low contrast since minimizing distance of spread light from camera particularly outcome from effects of scattering absorption (Natarajan, 2017).

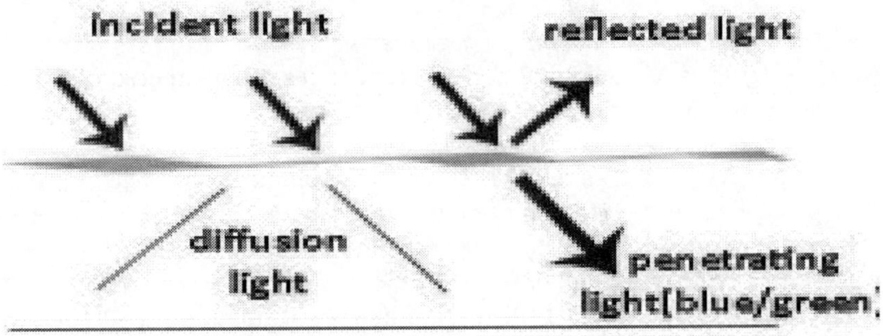

Figure 1. Effects of Water Surface (Source: Raj et al, 2016)

Fig 1 depicts effects of water surface. Several portions of incident lighttravel through the water as well as other portion of light are replicated at surface. Because of choosy absorption characteristics, light component in red colour suffers severe reduction from the surface. Thus light passes profoundly to water mostly involve components of green and blue colour. Diffusion light denotes the light that passes all through the medium of water by scattering (Raj et al, 2016).

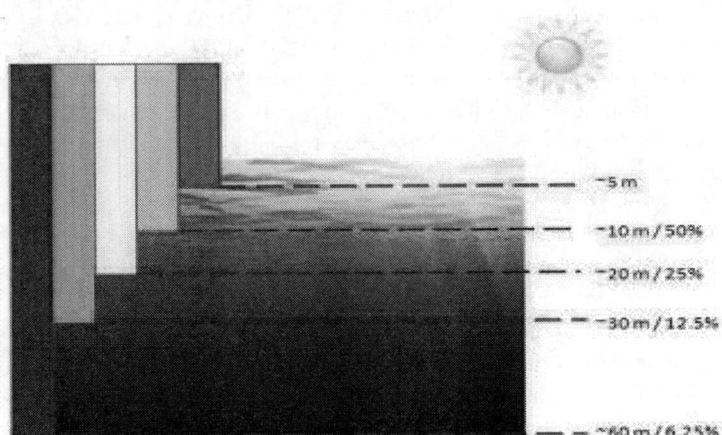

Figure 2. Image about absorption of light in water (Source: Bharal, 2015)

Fig 2 illustrates image about absorption of light in water. In the image, with every 10m intensification of the sunlight, the brightness is reduced by half. More or less all red coloured light minimized to 50 percent from surface. At the same time, blue coloured light continues to go profound in the ocean since blue colour have smallest wavelength thus it passes through the greatest distance in water. Due to this, underwater images are based on green and blue colour (Bharal, 2015). Quantity of sunlight minimizes as it moves profound under water. As an outcome, underwater images seem to be dim. Objects will reach the camera only through part of light and other part of light gets scattered while penetrating through the distance through medium of water. Therefore, underwater images could be captured by camera affect from uneven illumination, visibility degradation, imbalance in the colour, blurring and so on (Raj et al, 2016). Resolution of image and enhancement of image are very important in most of remote sensing applications in image processing like hyper-spectral extraction of data feature (Wang et al, 2015a), segmentation (Zhang et al, 2015 and Shang et al, 2017) and classification (Isikdogan et al, 2017 and Yuan et al, 2015), detecting the cloud (Xie et al, 2017) measurements in remote sensing (Ning et al, 2017 and Martins et al, 2017). Thus, it can be inferred that the enhanced images offer much visibility, interpretability and are good in terms of clarity and colour.

2. Literature review

2.1. Different Techniques of Underwater Image Processing

Sudhakar and Meena (2019) have stated that underwater images are hampered by bad contrast, change in colour, floating and suspended particles. Specifically, with cameras places on AUV and ROV systems has vast spread in military and civil applications. The analysis and processing of underwater images is non-trivial in ocean engineering and several other scientific applications. The main problem in ocean engineering is to capture clear and perfect images in underwater surroundings. Image enhancement is a technique utilized to enhance the underwater image quality by increasing the levels of intensity of the image or image parts so that the output image must be clear compared to the captured underwater image and hence it is utilized for display or for further examination of image. This method will not enhance the intrinsic data of underwater image. This involves manipulation in contrast and grey level, noise removal, sharpening the regions of object, colour correction and filtering and so on. Thakare and Sahu (2014) have stated that the techniques of Underwater Image Enhancement offer a way to enhance the identification of object in underwater surroundings. In underwater surroundings the images get dull due to bad conditions of visibility and impacts like light reflection, light absorption, light bending, light scattering, denser medium, etc. These are the essential factors which causes underwater image degradation. Two techniques are there to address image processing namely restoration of image and enhancement of image (Pati et al, 2018 and Prabu et al, 2019). There are numerous researchers examined about methods for restoring the image (Galdran et al, 2015, Drews et al, 2016, Li et al, 2016a, Peng and Cosman, 2017, Li et al, 2016b, Li et al, 2017 and Peng et al, 2018). Further authors also focused on enhancing the underwater image (Serikawa and Lu, 2014, Fu et al, 2014, Trehan et al, 2016; Sohal and Singh, 2015and Bianco et al, 2015). Enhancing the image is a process adopted for improving the image quality by enhancing its properties. Therefore, it is probable to maximize the objects' visibility and image quality (Kaur and Sidhu, 2015, AbuNaser et al, 2014, Guraksin et al, 2016 and Lakshmi and Loganathan, 2015).

Algorithms for image processing in underwater are categorized as enhancement, restoration or correction in colour and illumination of approaches based on normalization and range from medium computational to high and structural difficulty. Further algorithms based on restoration include de-hazing and de-blurring processes (Dhanigaivel et al, 2019 and Zhao et al, 2015) using DCP (dark channel prior) based techniques or Weiner de-convolution. Many researchers studied about restoration based algorithms. Algorithms based on enhancement do not involve any models which derive from previous picture information or physical phenomena. They adopt histogram or statistical oriented or logarithmic contrast stretching or enhancement and techniques for correcting the colour in their formulation (Ghani and Isa, 2015, Lu et al, 2015). Gradient and entropy optimized algorithms for image processing in

underwater was developed based on partial differential equations and they resulted in automated and efficient enhancement, better outcomes from prior algorithm (Nnolim, 2017 and Nnolim, 2017a).
The below Table 1 shows the various algorithms used for underwater image enhancement:

Table 1. The various algorithms used for underwater

Algorithm	Benefit	Author
WCID Algorithm	Removes artificial lighting and Maintain image quality	Pati et al, 2018
Red Channel Method	Manages artificially illuminated regions gracefully and accomplishes correction of natural colour and equivalent or superior visibility growth compared to other state of art approaches	Galdran et al, 2015
UDCP (Underwater Dark Channel Prior)	Offers better restoration and depth evaluation	Drews et al, 2016; Peng et al, 2018
Underwater image dehazing algorithm and a contrast enhancement algorithm	Dehazing algorithm reduces the loss of information of enhanced UI based on the principle of minimum information loss and underwater imaging optical properties. Contrast enhancement algorithm based on histogram distribution prior increase the brightness and contrast of images	Li and Guo, 2015; Li et al, 2016a; Dhanigaivel et al, 2019
Depth estimation Method	Enhance and restore underwater images and estimates the depth of underwater scene much accurately.	Peng and Cosman, 2017;

2.2. Different Techniques of Underwater Image Processing
Li and Guo (2015) developed an approach based on colour correction and dehazing algorithms for enhancing the underwater image. First and foremost, dehazing algorithm was utilized for removing the impacts of haze in underwater picture. Secondly intensity stretching, saturation, HE and compensation of colour are adopted for improving the visibility, colour, brightness and contrast of underwater picture. Apart from these, this research also adopted bilateral filtering is adopted for addressing the issue of noise occurred due to medium physical properties and Histogram Equalization algorithm. The proposed method maximized brightness, contrast, genuine colour, better visibility and minimized noise level.

Swarna Lakshmi and Loganathan (2015) study presents a comparative analysis of different techniques of image enhancement used for improving underwater images. The underwater images quality is bad because of particular light propagation properties in water. When capturing such images reduce due to several factors namely water ripple, lack of light organic matter availability dissolved in water and also captured images from little distance must be pre-processed before applying any type of operation on these images various techniques of filtering are feasible for pre-processing underwater images. Usually the filters are used to enhance the quality of image, suppress the noise preserve the corner in an image, smoothen and enhance the image properly.

In the study of Banerjee et al (2017) the proposed method processing sequence involves removal of noise using non-linear and linear filters followed by adaptive contrast correction in YCbCr and RGB planes of colour. The proposed method output has been calculated and compared with three major approaches called UCM (UnSupervisedColor correction Process), namely WP (with white patch), GW (Gray World) and APE (Adobe Photoshop Equalization).

The Lu et al (2016) research offers a new solution to optimizing optical photos by utilizing a wavy driven trigonometric filter and the camera's spectral properties in turbid mud. The three major issues of distortion in the optical photography of the underwater include reflection, artificial light and absorption. A new, deep-sea imaging technique that balances variations in attenuation along the direction of diffusion and an effective underwater scenario creation scheme is provided. The input of the report.

In the study of Rai and Bhalla (2016) the pre-processing is undertaken by image denoising. In order to accomplish this the integration of algorithms of image denoising namely KarhunenLoeve Transform (KLT) and Wavelet Transform (WT) and an adaptive filter are used on images to eliminate noise that are either existing in image during injecting or capturing into image during transmission. The wavelet transform has an outstanding performance in the field of denoising while KarhunenLoeve Transform reveals a better performance in signal rebuilt ability and adaptive filter can enhance the image by filtering out the noise of Gaussian.

Thakur and Dutta (2016) developed the pre-processing technique to enhance the quality of degraded pictures in the underwater. Three edge preventing filters considered for the research are wavelet de-noising based on average filter, anisotropic filter and homo-morphic filter. Picture in underwater suffers from transmitting properties of water, lightening disturbance, transmitting limited light range, blurring and low contrast of image, diminishing colour at the time of image capture. Speckle decrease by anisotropic filter enhances the quality of the picture, prevents the edges in picture, restrained the noise, smoothen and enhance the picture. Homomorphic filtering is adopted for correcting non-uniform illumination and for enhancing image contrast.

Pan et al (2018) have proposed a novel approach to enhance underwater images. Firstly, a CNN is trained end to end to evaluate the transmission map. Similarly, the adaptive bilateral filter is used to refine the map of transmission. Secondly a strategy based on white balance is proposed to eliminate the deviation of colour. Laplace pyramid fusion is used to acquire the fusion output of colour corrected and haze free image. Lastly the resultant image is transformed into HWD (Hybrid Wavelets and Directional Filter Banks) domain for edge enhancing and denoising.

The below Table 2 shows the reviews on the filtering techniques used in underwater image processing:

Table 2. Review on the Filtering Techniques Applied for Underwater Image Enhancement

Author	Filtering Technique	Advantages
Liu and Guo, 2015	Bilateral Filtering	Resolves the noise issue occurred due to medium physical properties
Swarnalakshmi and Loganathan, 2015; Madhumatke and Agarwal, 2017; Srividhya and Ramya, 2015	Homomorphic filter, Wavelet filter, Bilateral filter and Anisotropic filter	Improves the quality of image, suppress the noise in image, preserve the image edges, smoothen and enhance the underwater image effectively
Banerjee et al, 2016	Linear and Non-linear filters	Reduces noise of underwater image
Lu et al, 2016;	Guided Trigonometric filter	Refine the map of coarse depth transmission
Rai and Bhalla, 2016	Adaptive filter	Enhance the image by filtering out the noise of Gaussian

2.3. Histogram Based Algorithm

Ashwin et al (2014) combined contrast limited adaptive Histogram Equalization (CLAHE) and dark channel prior techniques for better image quality. Developed method is appropriate to enhance the underwater images quality than existing methods. Jayasree and Thavaseelan (2014) developed HE and algorithm of contrast stretching for improving the quality of underwater picture. Kuber and Dixit (2014) reviewed about modified application in enhancing the image. Enhancing the image is given better input to enhance the perception or interpretability of details in images for person to view with automated techniques in image processing. Different HE techniques such as Global HE, Clipped HE, Bi-histogram equalization, recursive mean-separate HE, dualistic sub image HE and multi-HE techniques are adopted for processing the input image for enhancing its output. This research modifies the brightness preserving dynamic HE (BPDHE) technique for enhancing its abilities in contrast enhancement and brightness preserving whereas it minimizes the computational complexity. There are many modified techniques associated with BPDHE which adopts data of digital images for processing and representation. For representing and processing the images in spatial area allow the method to handle the uncertainty in levels of grey values in a better way and result in enhanced performance.

Brindha and Vijayakumar (2017) stated that images in underwater have poor contrast and colour. Process of enhancing the image is needed for better look. Good enhancement of image for images in underwater is auto level method. Such method maximizes the contrast channel of every colour. This research balances the colour based on each channel distribution namely R, G and B depends on its histogram. Developed method adopts adjusted value for stretching right and left image boundary in histogram relies on distribution. Developed methods illustrate the main enhancement in auto level for image in the underwater. Thus, many objects look clearer in the underwater images. Chen and Wu (2017) developed another HE-based estimation enhance image by distinguishing the suspicious vision of best entropy for keeping up various highlights in quality of image. Exploratory result verified the developed estimation is best to create enhanced images based on subjective human visual estimation and quantitative investigation. Lee and Tseng (2017) introduced an up-gradation technique of colour picture by adapting HE approach without modifying immersion and tint in HSI (hue, saturation and intensity) shading space. Developed method has suitable optical vision over ordinary HE approach because immersion and tint are saved in upgraded method. Picture of backdrop illumination and picture during evening time are adapted for exhibiting the feasibility of developed technique of colour enhancement.

Deperlioglu et al (2018) enhanced the images with HE, HSV (Hue saturation value color space) and V transform algorithm. Shaik et al (2018) applied various histogram based techniques on various images in underwater. This research has provided substitute approach for enhancing the images in the underwater. This strategy develops the principle of white balance and do not need extra data instead of single original image. Apart from these, it was confirmed that method of white balance gives better outcomes when compared with HE (histogram equalization), adaptive HE (AHE) and contrast limited AHE.

The below table 3 shows the reviews on the use of histogram-based algorithm in underwater image processing:

Table 3. Review on the Histogram based algorithm Applied for Underwater Image Enhancement

Author	Histogram Technique	Advantages
Ashwin et al 2014	Contrast Limited Adaptive Histogram Equalization	Improves the intensity and contrast of underwater images
Jayasree and Thavaseelan 2014	Histogram Equalization and Contrast Stretching Algorithm	Enhances the quality of underwater image
Kuber and Dixit 2014	Brightness Preserving Dynamic Histogram Equalization Technique	Improves the contrast enhancement and brightness preserving capabilities while decreasing its complexity of computation
Brindha and Vijayakumar 2017	Histogram Method	Improves the auto level for underwater image
Chen and Wu 2017	Histogram Equalization based Algorithm	Enhances the contrast of underwater image based on maximum entropy to

		manage image quality features
Lee and Tseng 2017	Intensity Histogram Equalization Method	Good visual colourfulness in histogram and the saturation and hue are preserved in the process of enhancement
Deperlioglu 2018	Histogram Equalization Techniques	Improves the R, G and B components of underwater images
Shaik et al 2018	Histogram Equalization, Adaptive Histogram Equalization and Contrast Limited Adaptive Histogram Equalization	Improves the contrast and colour of underwater images

2.4. Particle Swam Optimization

AbuNaser et al (2014) introduced a framework to enhance pictures in underwater image using algorithm of PSO (particle swarm optimization). A pre-processing step was initiated for reducing the scattering and absorbing impacts of water prior adopting a filter based on proposed method for enhancing the picture. Enhanced images quality is numerically measured by adopting proposed method on dataset of underwater pictures. Acquired outcomes illustrate a significant enhancement.

Ye et al (2015) developed a cuckoo-search-PSO based technique for enhancing the images for grey-level pictures is developed and new norm to measure the enhanced picture quality. Developed method improves the image contrast and gives a high-quality base for next image processing. Outcomes of the developed technique are compared with prior techniques of image enhancements and found out developed method stably and rapidly unite to best solution as well as objective function was favourable when compared with other methods.

Chelva et al (2017) devised an algorithm to enhance the image's visual appearance catches by camera in embedded systems based on microcontroller. This study developed a new approach to improve the grey-scale picture quality using PSO algorithm. From the findings of the simulation outcomes, it was found that developed algorithm provide acceptable outcome namely quality of the image and speed than prior methods. Singh et al (2017) adopted grey-levels image technique for enhancing contrast using PSO was found to favourable fit for human interpreter demands.

Zhou et al (2017) found that developed method could efficiently enhance the contrast of pictures using algorithm of multi-scale retinex and nonlinear transformation and their parameters for optimization were chosen automatically with the help of PSO algorithm. When developed method was compared with other classical algorithms for enhancement indicated that developed method efficiently minimizes the impact of uneven and low illumination on next recognition outcomes.

The below table 4 shows the reviews on the use of PSO algorithm in underwater image processing:

Table 4. Review on the PSO techniques Applied for Underwater Image Enhancement

Author	PSO Technique	Advantages
AbuNaser et al; 2014; Alshawatnawi et al, 2015	Particle Swarm Optimization	Increases the visibility, Enhance the brightness and Enhance the contrast of UW image
Ye et al, 2015;	Cuckoo Search Particle Swarm Optimization Algorithm	Robust and Adaptive and Enhance the contrast of underwater image effectively
Chelva et al, 2017;	Particle Swarm Optimization	Reduce the number of particles and reduce the maximum iteration number and also enhances the quality and speed of underwater image
Singh et al, 2017;	Particle Swarm Optimization	Optimize the parameters used in transformation function
Zhou et al, 2017;	Particle Swarm Optimization	minimizes the impact of uneven and low illumination on next recognition outcomes

2.5. Dark Channel Prior Method

Sathya et al (2015) has mentioned in their study that colour change and scattering of light are two major issues in underwater images. The light dispersal deflects the incident light which removes the objects in the atmosphere on several occasions. It rising the underwater image's brightness and clarity. Dark channel approach is used to eliminate haze from the underwater image. This is based on a crucial discovery that some local patches include several pixels of diminished strength on a canal of color in haze-free underwater pictures. For this preceding technique utilizing the haze imaging model the intensity of the haze is measured and a good resolution haze free underwater image can be obtained.

The study of Mallik et al (2016) focused on UW techniques of image enhancement which is needed mainly for navigation and exploration of sea floor, monitoring of underwater environment, predicting possibilities for coral reefs assessment and civil engineering. An approach which comprises an algorithm of haze removal followed by a CLAHE colour structure has been established. This study focuses on enhancement of underwater image through algorithm of haze removal by dark channel prior technique. But it has the capability to darken the underwater image in certain circumstance but it reveals a better outcome by reducing the effect of noise and haze. In order to develop the result of dehazing a HE technique has been considered. CLAHE on the model of RGB has been followed in this study to alter the contrast level and dehaze image intensity.

Guo et al (2017) proposed an ISDCP (Improved Segmentation Dark Channel Prior) defogging method to resolve the fog impacts caused by water physical properties and also resolves the issues of underwater imaging colour cast and fog impact. Due to the mass light refraction in underwater imaging process the impacts of fog could lead to blurring of image. The colour cast is associated closely to varied attenuation ratio while light with varied wavelengths is passing in water. The proposed approach combines the quantitative histogram stretching and Improved Segmentation DCP techniques in to the process of image enhancement.

The UW photos were impaired owing to impacts of dispersion and absorption by Powar and Wagdikar (2017). UW imaging is essential to anticipate artifacts such as tiny organisms, fish and algae in present techniques. Color shift and dispersion of light are two dominant sources of distortion in UW imaging which reduce the contrast and visibility of the images captured. The suggested method thus describes a modern approach to boost UW photos with distance factors and histogram equalization and dehazing algorithms through remote factor estimation. Before the noise particles are eliminated, the distance factor dependent on specific color channel intensities with gamma correction is applied to improve the dimmed picture luminosity. The DCP removes noise and a haze effect and provides good visual quality with adaptive exposure map displaying to change the UW picture areas too light and too dark.

Swetha and Joan (2018) has proposed Dark Channel Prior based on the observation that most local patches in outdoor haze free underwater images comprises certain pixels whose intensity is very small in a colour channel. With Dark Channel Prior essentially superior colour fidelity and enhanced visibility of UW images are obtained and it decreases the complexity of computation and enhances the effectiveness in dehazing effect terms. The below table 5 shows the reviews on the use of DCP algorithm in underwater image processing:

Table 5. Review on the DCP techniques Applied for Underwater Image Enhancement

Author	DCP Technique	Advantages
Sathya et al, 2015;Amer et al, 2018; Venkateswararao and Rao, 2018	Dark Channel Prior Method	Dark Channel Prior method is good compared to image processing using HE
Mallik et al, 2016;Gao et al, 2016; Kaur and Mahajan, 2014; Badhe and Ramteke, 2018	Dark channel Prior and Histogram Equalization	Improves the underwater images quality quantifiably and visually by enhancing the underwater image contrast and reduces artefacts as well as noise in image
Guo et al, 2017;Pramunendar et al, 2018	Improved Segmentation DCP and Quantitative Histogram Stretching	Reduces the complexity of computation and enhances the efficiency in defogging impact terms to meet the real time needs

Powar and Wagdarikar, 2017; Han and Chen, 2016;	Dark Channel Prior with Gamma Correction	Improves the brightness of dimmed images and also improves output by removing the impact of scattering and absorption components avoiding too bright and too dark areas using adaptive exposure map.
Swetha and Joan 2018; Khandelwal et al, 2018; Łuczyński and Birk, 2018	DCP method	Enhances superior fidelity of colour and visibility of underwater image and also reduces the complexity of computation and increases the effectiveness in dehazing effect terms

3. Findings and discussion

3.1. Performance Characteristics

This study focused on the review of various techniques of underwater image processing to improve the underwater images quality. The below Table 6 compares the different techniques available for processing underwater images:

Table 6. Comparison of various underwater Image processing techniques

Author	Year	Technique	Results/.Outcome
2016	Banerjee et al; Lu et al; Rai and Bhalla et al	Linear and Nonlinear filters, Guided Trigonometric filter and Adaptive Filter	Reduces the UW image noise, Refine the coarse depth transmission map and enhance the UW image by filtering out the Gaussian noise
2017	Brindha and Vijayakumar; Chen and Wu; and Lee and Tseng	Histogram Equalization and Intensity Histogram Equalization	Enhances the UW image contrast based on maximum entropy and improves the automatic level of UW image and better visual colourfulness in histogram

2017	Chelva et al; Ye et al; and Zhou et al	Particle Swarm Optimization	Reduce the maximum iteration number and enhance the speed and quality of UW image
2017	Guo et al; Powar and Wagdarikar	Dark Channel Prior Method	Reduces the complexity of computation and improves the brightness of dull images

If external content, e.g. binary files, video clips, or sound recordings, are accessible on your server, simply submit a brief summary of the external material to the volume editors to let them know the URL where it can be located. We can apply the additional content definition to LNCS 'online edition and build a link to your website. If this additional content needs to be modified at some stage, it may also be submitted to the volume editors like all other files directly.

4. Conclusion

Underwater image enhancement is the area of image processing and it is considered as a dynamic sector. Obtaining the objects visibility at short or long distance in underwater scenes is very challenging and a difficult task. The atmospheric light is the main hurdle to process the UW images from bad conditions of visibility under the water, light attenuation and light scattering due to entire reasons which the UW images faces a lot and influence their contrast and visibility which they comprise originally. This review paper has examined different techniques applied for underwater image processing and explained how they enhance the quality of the underwater image. The paper could be extended in future by implementing the reviewed techniques on a real time basis and comparing their performances using appropriate evaluation strategies.

References

1] AbuNaser A, Doush IA, Mansour N and Alshattnawi S 2014 Underwater image enhancement using particle swarm optimization *Journal of Intelligent Systems* **24** 99-115.

[2] Alharbi A, Ammar R, Alhumyani H and Rajasekaran S 2014 Efficient pipeline architectures for underwater big data analytic. *IEEE International Symposium on Signal Processing and Information Technology (ISSPIT),* 000161-000166.

[3] Alshattnawi S, Mansour N, Doush IA and AbuNaser A 2015 Underwater Image Enhancing Using Particle Swarm Optimization *Journal of Intelligent Systems* **24** 99-115.

[4] Amer KO, Elbouz M and Alfalou A 2018 Phase Contrast Enhancement Combined with Dark Channel Prior for Underwater Target Recognition ISRES Publishing **4** 74-81.

[5] Ashwin S, Srinivasan S, Karthik H and Kalimuthu K 2017 Underwater image enhancement: An Integrated Approach, International journal of electrical, electronics and data communication, **5** 102-107

[6] Badhe MMV, and Ramteke PL 2016 Design and Implementation of a Model for Haze Removal Using Image Visibility Restoration Technique International Research Journal of Engineering and Technology **3** 747-751.

[7] Banerjee J, Ray R, Vadali SRK, Shome SN and Nandy S 2016 Real-time underwater image enhancement: An improved approach for imaging with AUV-150. Sadhana **41** 225-238.

[8] Bharal S 2015 Review of underwater image enhancement techniques *International Research Journal of Engineering and Technology* **2** 340-344

[9] Bianco G, Muzzupappa M, Bruno F, Garcia R, and Neumann L 2015 A New Colour Correction Method For Underwater Imaging, The International Archives of the Photogrammetry, Remote Sensing and Spatial Information Sciences Underwater 3D Recording and Modeling, Piano di Sorrento, Italy, 16–17.

[10] Brindha J and Vijayakumar V 2017 Underwater image enhancement using histogram method, *Indonesian Journal of Electrical Engineering and Computer Science*, **8** 687-689.

[11] Chelva MS, Halse SV and Samal AK 2017 A novel approach to quality enhancement of gray-scale image using PSO *Indian journal of science research* **14** 276-283.

[12] Chen BH and Wu YL 2017 An entropy-preserving histogram modification algorithm for image contrast enhancement *International Conference on Applied System Innovation (ICASI)* 1285-1288

[13] Deperlioglu O, Kose U and Guraksin G 2018 Underwater image enhancement with HSV and histogram equalization *7th International conference on advanced technologies* 461-465

[14] Dhanigaivel H, Kannan N and Kumar P 2019 Underwater Image Enhancement by Wavelength Compensation and Dehazing *International Journal of Applied Engineering Research* **14** 37-41.

[15] Drews P, Nascimento E, and Botelho S and Campos M 2016 Underwater depth estimation and image restoration based on single images *IEEE Computer Graphics and Applications* **36** 24-35.

[16] Fu X, Zhuang P, Huang Y, Liao Y, Zhang X and Ding X 2014 A retinex-based enhancing approach for single underwater image *International Conference on Image Processing* 4572-4576

[17] Galdran A, Pardo, D, Picón A and Alvarez-Gila A 2015 Automatic red-channel underwater image restoration, *Journal of Visual Communication and Image Representation* **26** 132-145

[18] Gao Y, Li H and Wen S 2016 Restoration and enhancement of underwater images based on bright channel prior Mathematical Problems in Engineering *Hindawi Publishing Corporation* 1-15.

[19] Ghani ASA and Isa NAM 2015 Underwater image quality enhancement through integrated color model with Rayleigh distribution *Applied Soft Computing* **27** 219-230.

[20] Guo Q, Xue L, Tang R, and Guo, L 2017 Underwater image enhancement based on the dark channel prior and attenuation compensation, *Journal of Ocean University of China* **16** 757-765.

[21] Guraksin GE, Kose Deperlioglu O 2016 Underwater Image Enhancement Based On Contrast Adjustment via Differential Evolution Algorithm *International Symposium on Innovations In Intelligent Systems And Applications* 1-5

[22] Han M, and Chen C 2016. Enhancing underwater image by dark channel prior and color correction. Sixth International Conference on Information Science and Technology (ICIST) 505-510.

[23] Isikdogan F, Bovik AC, and Passalacqua P 2017 Surface water mapping by deep learning *IEEE Journal of Selected Topics in Applied Earth Observations and Remote Sensing* **10** 4909-4918.

[24] Jayasree MS, and Thavaseelan G 2014 Underwater Color Image Enhancement Using Wavelength Compensation and Dehazing *International Journal of Computer Science and Engineering Communications* **2** 389-393

[25] Kaur H and Mahajan R 2014 Improved Dark Channel Prior Using Gabor Filter and CLAHE, *International Journal of Application or Innovation in Engineering and Management* **3** 127-134.

[26] Kaur T and Sidhu RK 2015 Performance Evaluation of Fuzzy and Histogram Based Color Image Enhancement *Procedia Computer Science* **58** 470 – 477, 2015.

[27] Kaur V, Singh A and Dogra AK 2014 A Review on underwater image enhancement *International journal of advanced research in computer and communication engineering,* **3** 7611-7618.

[28] Khandelwal V, Mangal D, and Kumar N 2018 Elimination of Fog in Single Image Using Dark Channel Prior *International Research Journal of Engineering and Technology* **5** 1601-1606

[29] Kuber PSM and Dixit M 2014 A Review on modified enhancements applications *International journal of signal processing, image processing and pattern recognition* **7** 71-78.

[30] Lakshmi RS, Loganathan B 2015 An Efficient Underwater Image Enhancement Using Color Constancy Deskewing Algorithm *International Journal of Innovative Research in Computer and Communication Engineering* **3** 7164-7168.

[31] Lee SL, and Tseng CC 2017 Color image enhancement using histogram equalization method without changing hue and saturation *In 2017 IEEE International Conference on Consumharmaer Electronics-Taiwan (ICCE-TW)* 305-306.

[32] Li and Guo 2015 Underwater Image Enhancement by dehazing and color correction, International *Journal of advanced research in computer and communication engineering* **5** 159-162.

[33] Li C, Quo J, Pang Y, Chen S, and Wang J 2016 Single underwater image restoration by blue-green channels dehazing and red channel correction *In 2016 IEEE International Conference on Acoustics, Speech and Signal Processing (ICASSP)* 1731-1735.

[34] Li C and Guo J 2015 Underwater image enhancement by dehazing and color correction, *Journal of Electronic Imaging* **24** 033023-1-033023-10

[35] Li C, Ji C, Cong R, Pang Y and Wang B 2016 Underwater image enhancement by dehazing with minimum information loss and histogram distribution prior *IEEE Transactions on Image Processing* **25** 5664-5677.

[36] Li J, Skinner K, Eustice R, and Roberson M 2017 WaterGAN: unsupervised generative network to enable real-time color correction of monocular underwater images, *IEEE Robotics and Automation Letters* **3** 387-394.

[37] Lu H, Li Y, Xu X, Li J, Liu Z, Li X, Yang J and Serikawa S 2016 Underwater Image Enhancement Method Using Weighted Guided Trigonometric Filtering and Artificial Light Correction, *J. Vis. Commun. Image R.* **38** 504-16.

[38] Lu H, Li Y, Xu X, Li J, Liu Z, Li X and Serikawa S. 2016 Underwater image enhancement method using weighted guided trigonometric filtering and artificial light correction *Journal of Visual Communication and Image Representation* **38** 504-516.

[39] Lu H, Li Y, Li J, Li X, Kashio Y and Serikawa S 2015 Image Contrast Enhancement for Deep - sea Observation Systems, *The 3rd International Conference on Industrial Application Engineering* 619-624.

[40] Łuczyński T, and Birk A 2018 Underwater Image Haze Removal and Color Correction with an Underwater-ready Dark Channel Prior. *arXiv preprint arXiv:1807.04169*

[41] Madhumatke PP and Agarwal P 2017 Underwater Image Enhancement Techniques: A Review, *International Journal for Research in Applied Science and Engineering Technology* **5** 1574-1580

[42] Mallik S, Khan SS and Pati UC 2016 Underwater Image Enhancement Based on Dark Channel Prior and Histogram Equalization *International Conference on Innovations in Information Embedded and Communication Systems* 139-144

[43] Martins GD, Galo MDLBT and Vieira BS 2017 Detecting and mapping root-knot nematode infection in coffee crop using remote sensing measurements *IEEE Journal of Selected Topics in Applied Earth Observations and Remote Sensing* **10** 5395-5403

[44] Natarajan K 2017 A Review on underwater image enhancement techniques *International research journal of engineering and technology* **4** 780-783

[45] Ning J, Wang R, Deng Y, Li N, Song H and Fei W 2017 AET-IAA: A Novel Approach to Enhance Azimuth Resolution for ScanSAR Interferometry *IEEE Journal of Selected Topics in Applied Earth Observations and Remote Sensing* **10** 5674-5685.

[46] Nnolim UA 2017 Smoothing and enhancement algorithms for underwater images based on partial differential equations *SPIE J. Electron. Imaging* **26** 1–21.

[47] Nnolim UA 2017 Improved partial differential equation (PDE)-based enhancement for underwater images using local-global contrast operators and fuzzy homomorphic processes *IET Image Process* **11** 1059–1067

[48] Pan PW, Yuan F, and Cheng E 2018 Underwater Image De-Scattering and Enhancing Using Dehazenet and HWD *Journal of Marine Science and Technology* **26** 531-540.

[49] Panetta K, Gao C and Agaian S 2015 Human-visual-system-inspired underwater image quality measures *IEEE Journal of Oceanic Engineering* **41** 541-551

[50] Pati V, Patil V and Suryawanshi S 2018 Underwater image enhancement *International Journal of Advance Research in Science and Engineering* **7** 387-394
[51] Peng Y and Cosman P 2017 Underwater image restoration based on image blurriness and light absorption *IEEE Transactions on Image Processing* **26** 1579-1594
[52] Peng Y, Cao K and Cosman P Generalization of the dark channel prior for single image restoration *IEEE Transactions on Image Processing* **27** 2856-2868.
[53] Powar OG and Wagdarikar NM 2017 Underwater Image Enhancement using Dark Channel Prior and Gamma Correction *Proceedings of IEEEFORUM International Conference* 58-62
[54] Pramunendar RA, Wibirama S and Santosa PI 2018 A Novel Approach for Underwater Image Enhancement Based on Improved Dark Channel Prior with Colour Correction *Journal of Engineering Science and Technology* **13** 3220-3237
[55] Qing C, Huang W, Zhu S and Xu X 2015 Underwater image enhancement with an adaptive dehazing framework *In 2015 IEEE International Conference on Digital Signal Processing (DSP)* 338-342
[56] Raj SM, Abhilash S and Supriya MH 2016 A Comparative study of various methods for underwater image enhancement and restoration *IOSR journal of VLSI and signal processing* **6** 30-33
[57] Roser M, Dunbabin M and Geiger A 2014 Simultaneous underwater visibility assessment, enhancement and improved stereo *In 2014 IEEE International Conference on Robotics and Automation (ICRA)* 3840-3847
[58] Sathya R, Bharathi M and Dhivyasri G 2015 Underwater image enhancement by dark channel prior, *International Conference on Electronics and Communication Systems (ICECS)* 1119-1123
[59] Sathya R and Bharathi M 2015 Enhancement of Underwater Images Using Wavelength Compensation Method *International Journal of Innovative Research in Computer and Communication Engineering* 3 1829 – 1835
[60] Seema H, Kaur V and Dhillon N 2016 The study of image detail enhancement techniques *International journal of computer science and mobile computing* **5** 308-313.
[61] Senthilkumaran N and Thimmiaraja J 2014 Histogram equalization for image enhancement using MRI brain images *In 2014 World Congress on Computing and Communication Technologies* 80-83
[62] Serikawa S and Lu H 2014 Underwater image dehazing using joint trilateral filter *Computers in Elecrical Engineering* **40** 41-50.
[63] Shaik M, Meena P, Basha S and Lavanya N 2018 Color Balance for underwater image enhancement *International journal for research in Applied science and engineering technology* **6** 571-581
[64] Shang R, Yuan Y, Jiao L, Hou B, Esfahani AMG and Stolkin R 2017 A fast algorithm for SAR image segmentation based on key pixels *IEEE Journal of Selected Topics in Applied Earth Observations and Remote Sensing* **10** 5657-5673.
[65] Singh N, Kaur M and Singh KVP 2017 Parameter Optimization in Image Enhancement Using PSO *American Journal of Engineering Research* **2** 84-90.
[66] Sohal S and Singh S 2015 Improved particle swarm optimization based image enhancement using fuzzy logic *International journal of engineering and innovative technology* **4** 105-113.
[67] Sowmyashree MS, Bekal SK, Sneha R, and Priyanka N 2014 A Survey on the various underwater image enhancement techniques *International Journal of Engineering Science Invention* **3** 40-45
[68] Srividhya K and Ramya MM 2015 Performance analysis of pre-processing filters for underwater images *In 2015 International Conference on Robotics, Automation, Control and Embedded Systems (RACE)* 1-7
[69] Sudhakar M and Meena MJ 2019 Underwater Image Enhancement using Conventional Techniques with Quality Metrics *International Journal of Innovative Technology and Exploring Engineering* **8** 241-249
[70] Swarnalakshmi R and Loganathan B 2015 A Survey on Under Water Images Enhancement Techniques *International Journal of Computer Science and Engineering Technology* **6** 409-412

[71] Swetha P and Joan FSP 2018 An Underwater Image Dehazing Method Using Dark Channel Prior *International Conference on Advancements in Computing Technologies* **4** 104-109
[72] Lakshmi Narayanan K and Ramesh GP 2017 VLSI Architecture for Multi-Band Wavelet Transform based Image Compression and Image Reconstruction *Journal of Engineering and Applied Sciences* **12** 6281-6285
[73] Thakur S and Dutta P 2016 Enhancement of Underwater Images: A Review *International Journal of Advanced Research in Computer and Communication Engineering* **6** 159-162.
[74] Venkateswararao B and Rao SHK 2018 Underwater Image Enhancement by Dark Channel Prior *SS International Journal of Multidiscplinary Research* **3,** 1-9.
[75] Wang S, Gu K, Ma S, Lin W, Liu X and Gao W 2015 Guided image contrast enhancement based on retrieved images in cloud *IEEE Transactions on Multimedia* **18** 219-232.
[76] Wang L, Dai Q, Xu Q, Zhang Y 2015 Constructing Hierarchical Segmentation Tree for Feature Extraction and Land Cover Classification of High Resolution MS Imagery *Selected Topics in Applied Earth Observations and Remote Sensing IEEE Journal of* **8** 1946-1961
[77] Xie F, Shi M, Shi Z, Yin J and Zhao D 2017 Multilevel cloud detection in remote sensing images based on deep learning *IEEE Journal of Selected Topics in Applied Earth Observations and Remote Sensing* **10** 3631-3640
[78] Ye Z, Wang M, Hu Z, and Liu W 2015 An adaptive image enhancement technique by combining cuckoo search and particle swarm optimization algorithm *Computational intelligence and neuroscience* **2015** 13
[79] Yuan Y and Hu X 2015 Bag-of-words and object-based classification for cloud extraction from satellite imagery *IEEE Journal of Selected Topics in Applied Earth Observations and Remote Sensing* **8** 4197-4205.
[80] Zhang X, Xiao P, Feng X, Feng L, Ye N 2015 Toward evaluating multiscale segmentations of high spatial resolution remote sensing Images *Geoscience and Remote Sensing IEEE Transactions,* **53** 3694-3706
[81] Zhao X, Jin T, and Qu S 2015 Deriving inherent optical properties from background colour and underwater image enhancement, Ocean Engineering **94** 163-172
[82] Zhou C, Yang X, Zhang B, Lin K, Xu D, Guo Q and Sun C 2017 An adaptive image enhancement method for a recirculating aquaculture system *Scientific reports* **7** 6243
[83] Prabu S, Balamurugan V and Vengatesan K 2019 Design of cognitive image filters for suppression of noise level in medical images *Measurement* **141** 296-30153

IOP Conf. Series: Materials Science and Engineering **925** (2020) 012035 doi:10.1088/1757-899X/925/1/012035

Portable Intensive Care Unit Equipment Using IoT

Rakesh Kumar S M[1], Srikanth K V[2], Shubham Kumar[3], Kumar Vaibhav[4] and Madan H T[5]

[1,2,3,4,5]School of ECE, Reva University Bangalore, India
rakesh46kumarsm@gmail.com[1],srikanthrao574@gmail.com[2],ss319saurav@gmail.com[3], vaibhav467@gmail.com[4], madanht@reva.edu.in[5]

Abstract. Many lives have been affected every day in remote areas mainly because the patients are not timely and properly treated by the healthcare centers, that unfortunately resulting in loss of human lives. In most of the rural hospitals, patients are unable to get the timely response because of lack of human resources and facilities. In some cases, it becomes even difficult for hospitals to frequently check patients' conditions and update their record on a regular basis. To deal with these sort of real time problems, we have designed a portable healthcare system, which can be easily adopted in the health care centers. The proposed health care system can be used to measure and monitor various health indicators like temperature, ECG, Blood Pressure, heartbeat etc. The measured parameters of the patient will be recorded using ESP32 module and stored in the cloud server. The Doctors can login to the website, view the parameters and can take the necessary action.

Keywords: Health care system, Human Resource ECG, ICU, ESP32, IoT.

1. Introduction

Non-stop size of affected person parameters along with heart rate and rhythm, respiratory rate, blood stress, and lots of other parameters have grown to be a common function of the care of seriously sick sufferers. While correct and instant choice- making is critical for effective affected person care, electronic video display units often are used to collect and display physiological statistics. An expanding number of, such insights are assembled the utilization of non-intrusive sensors from substantially less seriously sick patients in an emergency clinic's clinical careful units, work and conveyance suites, nursing homes, or sufferers' very own homes to recognize astounding ways of life compromising conditions or to archive standard anyway required records successfully. We commonly consider a patient screen as something that looks for—and cautions against—genuine or presence compromising occasions in sufferers, truly wiped out or something else. Understanding checking might be thoroughly characterized as "rehashed or constant perceptions or estimations of the influenced person [1], their physiological capacity, and the capacity of presence help framework, for the explanation of directing administration choices, together with when to make mending mediations, and evaluation of those intercessions. An influenced individual screen may not best ready guardians to most likely perilous occasions; numerous furthermore offer physiologic enter data used to control straightforwardly associated ways of life bolster gadgets Right now, talk the utilization of PC frameworks to help parental figures inside the arrangement, show, stockpiling, and dynamic, comprising of translation of logical insights, making recuperating suggestions, and disturbing and alarming. Inside the past, most clinical records had been inside the state of coronary heart and breath costs, blood weights, and streams, yet nowadays they envelop incorporating realities from bedside contraptions which degree breath gases, temperature, and haematology notwithstanding coordinating data from numerous assets outside the top to bottom consideration unit (ICU). In spite of the way that we manage sufferers who are in ICUs, the general ideas and techniques likewise are pertinent to other hospitalized sufferers. For example, tolerant checking might be executed for demonstrative capacities inside the crisis room or for restorative purposes inside the working room. Methodologies that solitary a couple of years back had been utilized most straightforward in the ICU right now are mechanically utilized on well-known sanatorium units and in certain circumstances by methods for sufferers at reception.

2. Literature Survey

In the present world, various IOT systems have been adopted for health monitoring systems. Wang et al [1] designed a compatible IOT system for medical devices which was having multiple communication standards. A resource-based data retrieving method (UDA-IOT) was proposed by Xu et al [2] for information-intensive health applications. Peer-to-Peer (P2P) and IoT technologies were combined in a medical system called as a smart box to keep the patients in control. Kolici et al [3] implemented that compared the experimental results for different scenarios. Web Real-Time Communication (Web RTC) was given by Sundholm et al [4] which focused mostly on the secured transmission of data multiple concurrent streams in an efficient manner. By enabling the electronic sphygmomanometer to communicate via Bluetooth, an Android application [5] was developed to record the data such as SBP-Systolic Blood Pressure, DBP - Diastolic Blood Pressure and Heart Rate. That application made it easy to transmit the recorded data using any mobile device and such data is then be recorded, abnormality is found out and message is conveyed to the people. A real-time application [6] was presented with distributed flow environment for the IoT healthcare. When the person under observation moves beyond range, data will be recorded in the local server and communicated later. A Galileo board [7] is a IoT-based device with embedded medical platform for the designed for electrocardiogram (ECG) signal analysis and based on an algorithm, heart function is monitored. In market, few IOT Portable Medical Devices [8] were introduced which upgraded the patient's mobility. But the security threats and few drawbacks were also there while using Portable Medical Devices. When we started to consider light-weight IOT devices, using the existing databases, diseases were predicted. But while such predictions, issues were in storage of databases and analysis using those databases. A new cloud based fine-grained health information access control framework [9] was introduced which addressed the security challenges and the cloud reciprocity issues. A proxy-based approach for end-to- end communication between the IOT enabled living systems [10] was proposed to challenge the real world applications. A portable electric aid device was designed specifically for the blind people in which ultrasonic range finders are mounted on the belt to find the obstacles present in the users way and to direct the blind people through Bluetooth headphone. Another depth sensor based den avigation system for the blind people with high accuracy and to alert user via vibrio tactile feedback in the hand gloves. When the previous works are addressed, there is a limitation of database connectivity between the different cloud environment in monitoring the data in constant time intervals and to analyze data. Considering this limitation, in this paper, we present a cloud-based Internet of Things system that can be implemented in different health monitoring systems.

3. System Architecture

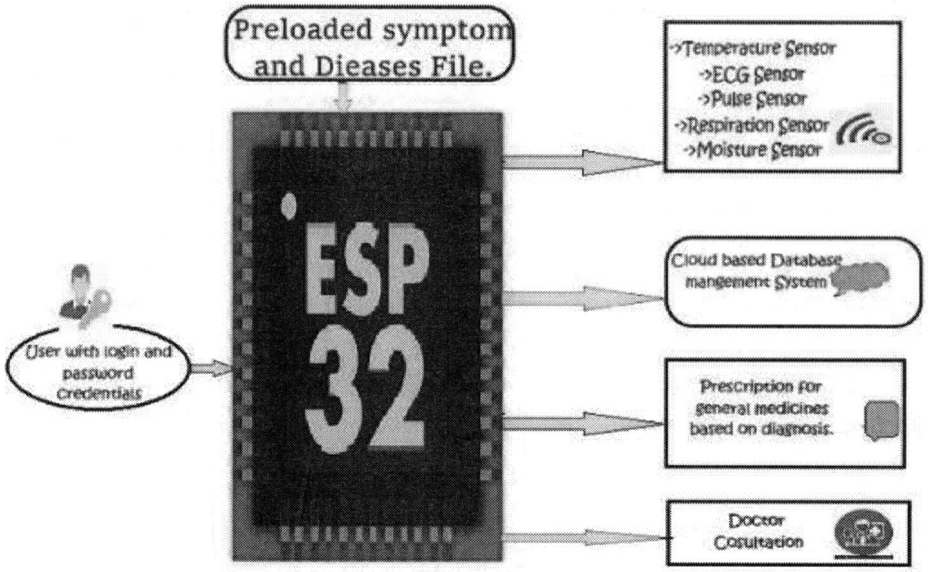

Figure 1. Block Diagram of proposed system

Information about the components:

- Temperature Sensor: Generally, these sensors are structured in a specific way to measure the variations of temperature in an object as well as in the human body. Particularly we have used this temperature sensor for measuring the variations of temperature in the human body. The series that is used for measuring the parameter is ds18b20 the sensor is a wearable gadget and it is made contact with the human body while measuring the parameter and its disconnected after noting down the respective reading. It is provided with pair of metals that produces electrical voltage with respect to variations in temperature.
- ECG Sensor: Normally these days due to pollution humans are in taking the toxic oxygen which is damaging the normal respiration cycle due to this various heart problem are arising firstly when patient consult the doctor about heart-related problem the advice given by the doctor is ECG test which plays vital role in generating the wave for every heartbeat. Since this sensor is also a wearable gadget comes with pair of electrode which arise wet regularly these electrodes are made in contact with human skin at the chest region for generation a wave of electrical activity in a human heart. The series that is used in our proposed work is AD8232.
- Pulse Sensor: Similar to the ECG sensor another sensor is introduced which is used to measure the accurate heartbeat of a person. The sensor is said to be a pulse sensor it comes with led bulb integrated with it which flashes by indicating each heartbeat when a finger is placed on it.

IOP Conf. Series: Materials Science and Engineering **925** (2020) 012035 doi:10.1088/1757-899X/925/1/012035

Figure 1.1. Pulse sensor

- Respiration Sensor: The Sensor innovation gauges the avoidances of the chest and the gut that happen during breathing to straight forwardly quantify the respiratory rate. The single-use sensor changes over and yields the breathing redirections to the projection as a fluctuating low voltage signal when applied to the patient's skin.
- Moisture Sensor: These days the number of old aged people that are getting bedridden are increasing due to various reason and they are unable to go for restroom sometimes due to less consciousness they discharge the urine where they are hence for detecting the moisture content of the bed moisture sensor is used. It makes use of dielectric permittivity of the medium where it is being placed and relates the voltage with dielectric permittivity and detects the wetness of the substance.
- Thingspeak: The above operations performed by sensors are collected and stored in a cloud for reference and by analyzing this reference data the physicians will take the next step for resolving the problem that has occurred. The application used for storing and recovering this information is Thingspeak it's an open- source software it uses the internet for in taking the data that are provided by the sensor using internet standard protocols called HTTP and MQTT.
- ESP32 Wi-Fi module: The above operations performed by sensors are collected and stored in a cloud for reference and by analyzing this reference data the physicians will take the next step for resolving the problem that has occurred. The application used for storing and recovering this information is Thingspeak it's an open-source software it uses the internet for in taking the data that are provided by the sensor using internet standard protocols called HTTP and MQTT. Generally, it is called as Wi-Fi module.

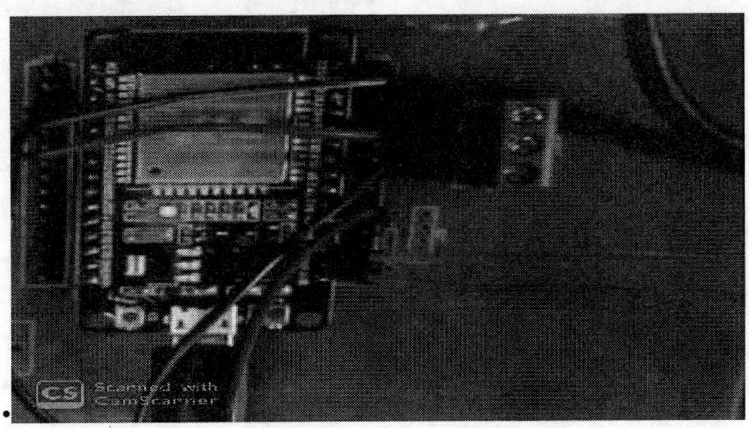

Figure 1.2. ESP32 Wi-fi module

4. Implementation

Overview of the proposed system: The framework is intended to get wellness data of sufferers extraordinarily people who are in a broad consideration unit. By and by the structures are to be had that gathers measurements from different casing portions of the patients. The additional component conveyed in our machine is that it shops the gathered realities inside the cloud, which might be utilized likewise for preparing [7]. Arrangement of measurements is done by methods for individual sensors that are fit for estimating continuous parameters alongside temperature, pulse, electrocardiography, breathing charge, dampness level.

Internal operation of the machine: The machine incorporates particularly 3 layers real layer, network layer, and readiness layer.

Data Fetching: In the physical layer, the centerpiece of the framework ESP32 is associated with wearable gadgets. Sensors are starter dependable to catch constantly patient's wellness records. These sensors are a piece of wearable contraptions. These sensors estimate temperature, pulse, dampness level, and heart yield. Sensors are put in two or three pieces of the casing so right readings are measured.ESP32 is utilized to obtain the concentrating from the sensors and is low-quality machine on-chip with Wi-Fi and double mode Bluetooth.

Data Storing: System layer interconnection organizing grants sensors to transmit sufferer's wellness data to various frameworks like servers.

Cloud Processing:
Wellbeing signs cautions are despatched to cloud carport the use of ThingSpeak. Thingspeak is an open inventory net of factors (IOT) application and API that is utilized to keep and recover records from sensors and utilizations the HTTPS convention over the web or close by place network [8]. The utilization of web this clinical data is made to be needed to Caregivers and clinical specialists for reference. Existing web from household or medical clinic can be utilized for this verbal trade. An alarm message is despatched to overseers and friends and family when readings go the set edge esteem, considering an emergency [9]. When we mindfulness on low force consumption, limit stages ought to be set with the expectation to deal with the crisis conditions. At the point when we mindfulness on low force consumption, limit the indistinguishable time force might be put away by methods for turning off the sensors. At the point when power admission is confined, there will expand the need for low force conventions for conversation. In ESP32 we've worked in a Wi-fi module can transmit records of the use of HTTP protocol, with low vitality consumption.

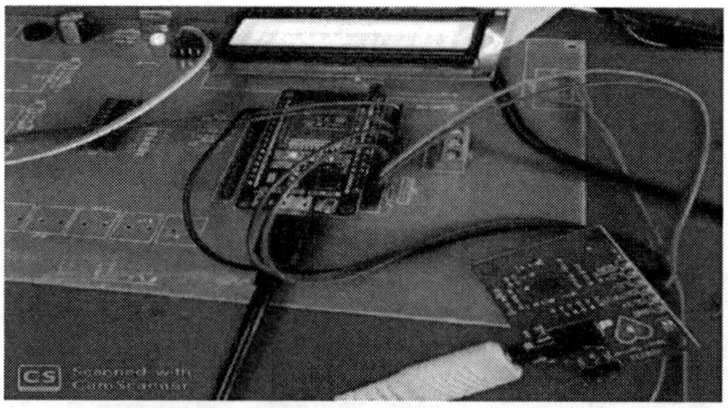

Figure 2.1. Health monitoring system.

Data Receiving: Programming layer the utilization of web this clinical information is made to be needed to specialists and overseers for reference. An alarm message is dispatched to overseers while dissecting surpasses the set limit esteems considering it is a crisis.

IOP Conf. Series: Materials Science and Engineering **925** (2020) 012035 doi:10.1088/1757-899X/925/1/012035

Figure 2.2. LCD Display

A. Algorithm

- Arduino IDE is an open-source programming used to interface the improvement sheets likeESP32.
- The sensors utilized are associated with ESP32 and information is gathered by particular sensors. The processor persistently screens.
- The information is transmitted to the server utilizing thing speak application and put away utilizing one of the database the executive's framework.
- The put-away information can be seen by a client from any point in the world. On the off chance that edge estimations of the sensor surpass quickly the processor will tell the client about a crisis.

B. Flow Diagram

- The product structure has been planned the use of Arduino IDE. The open-source Arduino programming (IDE) is written in highlights from C and C++ which makes it clean to carefully record code and transfer it on progress board like ESP32 which accompanies in-developed Wi-Fi module and Bluetooth.
- Perceive 3 manages a short clarification roughly the product program engineering. The gadgets are imparting among each unique through Serial Port discussion. After the code is dumped into ESP32, the framework holds up till the system is being snared between the Wi-Fi module and sensor station.

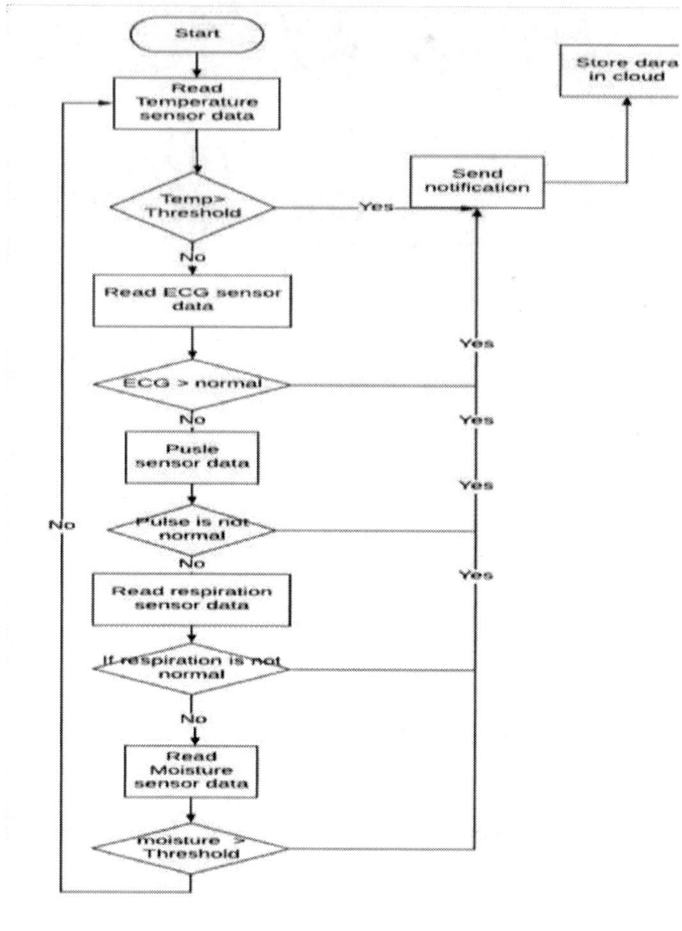

Figure 2.3. Software Architecture

5. Result

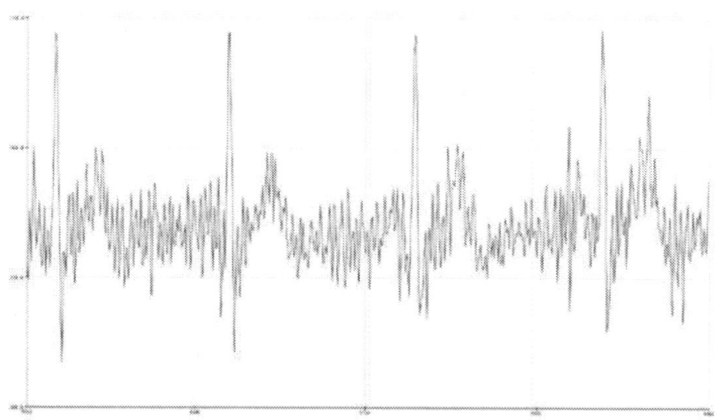

Figure 3.1. Electrocardiograph

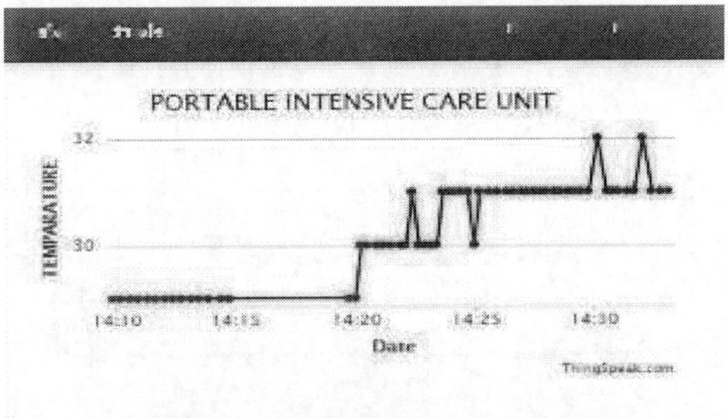

Figure 3.2. Temperature data graph

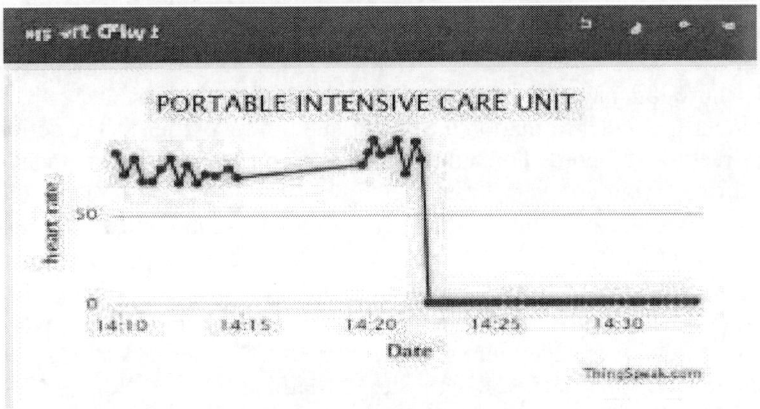

Figure 3.3. Heartbeat Data graph

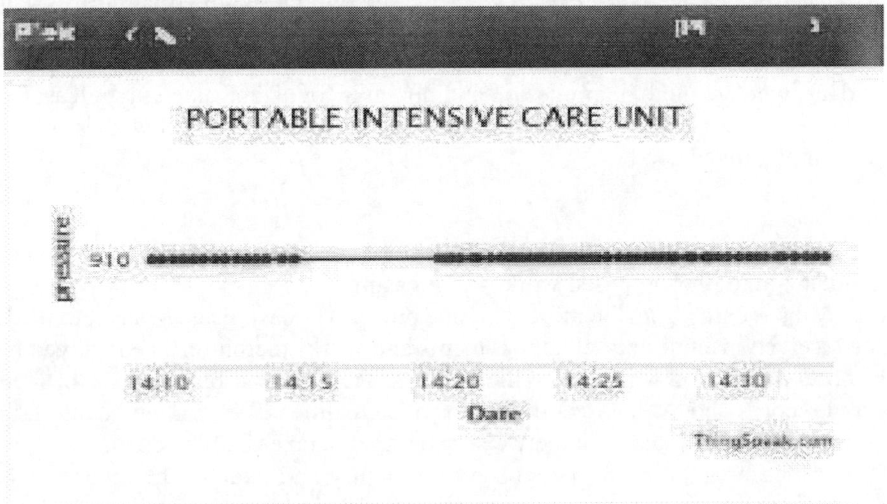

Figure 3.4 Pressure Data graph

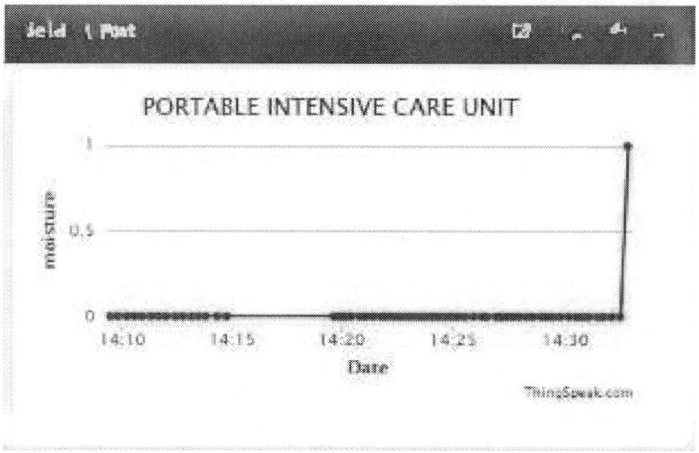

Figure 3.5. Moisture Data graph

- Figure (3.1) shows the perusing of the patient's ECG utilizing present-day hardware. An ECG is a test that identifies and information the power and timing of the electric action on your coronary heart. This records is recorded on a diagram that recommends each segment of the electric sign as it goes through your coronary heart.
- Figure (3.2) shows the perusing of the patient's temperature utilizing present-day-hardware. Human outline temperature is likewise alluded to as euthermia. It can be estimated at numerous areas along the edge of the mouth, ear, armpit and forehead.
- In present-day medicinally time-remembered that common edge levels from 36.5-37.5 confirmation celcius. A temperature of more than 38 recognition Celcius most extreme regularly way you have a fever because of diseases or defilement
- Figure (3.3) shows the perusing of the patient's temperature utilizing present-day-hardware. It can quantify the exchange the volume of blood vessel blood with each heartbeat beat. If the assortment will increase, it shows an exchange the blood volume relationship. The blood float is typically appeared as a waveform utilizing a chart.
- Figure (3.4 and 3.5) shows the identified dampness of the patient's bed. The old matured individuals who are out of commission and not an incognizant state may release pee on the bed itself henceforth we have joined a sensor that can identify measure of wetness of bed. Hence the same is shown in the chart.

6. Conclusion

Considering the real time health care problems in remote areas, we have designed the portable health care unit. The main aim of our proposed work is to measure

The sensors study the records from the more than one pieces of a casing and sends it to the ESP32 where it contrasts the received insights and set edge esteems and works therefore. The received measurements are sent to the observing station which shops the realities. Tracking station contrasts the received records and the acquired records and evaluations if any mistakes realities is gotten the health indicators of the patient and to provide timely interaction between patient, care taker and the doctor.

Our portable unit measures the health parameters frequently and updates the record on regular basis. These are records are made available to care takers and doctor using IoT and cloud computing technologies. We have included the parameters like body temperature, pulse and oxygen level, ECG and blood pressure as health indicators and also set the threshold level for each of these parameters. Every time the parameter is measured, it will be compared with set threshold values. If any parameter is not in the acceptable range, then immediately alarming messages will be sent to care takers and doctors to take up the necessary action.

Both specialist and patients can share their data through portable application and patient get recommendations or treatment data through the application as it were. The exactness of the web application, execution of the framework and circuit are tested and verified. Estimated results are contrasted and standard hardware results and displays the presentation of the proposed circuit with the assistance of the webpage is generally excellent. The proposed portable system can be easily adopted in the remote health care centers to provide the timely response to their patients.

References

[1] Hamim M, Paul S, Hoque SI, Rahman MN and Baqee I 2019 IoT Based Remote Health Monitoring System for Patients and Elderly People *International Conference on Robotics, Electrical and Signal Processing Techniques (ICREST)* 533-538

[2] Sumanta B, Student Member, Prabhu K, SriRam Kumar D 2012 Real-time breath rate monitor based health security system using Noninvasive Biosensor *IEEE Conference* 1-6

[3] Bharat P, Parikh S, Patel J 2018 An Intelligent Real Time IoT Based System(IRTBS) for Monitoring ICU Patient *International Conference on Information and Communication Technology for Intelligent Systems* 390-396

[4] Jagannadha Swamy T, Murthy TN 2019 eSmart: An IoT based Intelligent Health Monitoring and Management System for Mankind *International Conference on Computer Communication and Informatics* 1-5

[5] Makikawa M, Asajima S, Shibuya K, Tokue R, Shinohara H 2002 Portable Physical Activity Monitoring System for the Evaluation of Activity of the Aged in Daily Life *Proceedings of the Second Joint EMBSA3MES Conference* **3**

[6] Qinwu Z, Lin Z, Chen J, Fu H 2011 The Study of Portable Remote Multi-life-parameter Monitoring Network *13th International Conference on e-Health Networking, Applications and Services* 126-129

[7] Duraisamy Rugalendhi S 2017 Secured remote health monitoring system *Healthcare Technology Letters* **4** 228-232

[8] Kamble Bhutad S 2018 IoT based Patient Health Monitoring System with Nested Cloud Security, 2018 *4th International Conference on Computing Communication and Automation (ICCCA),* 1-5.

[9] Mahmud S, Wang H, Esfar-E-Alam AM, Fang H 2017 A Wireless Health Monitoring System Using Mobile Phone Accessories *IEEE Internet of Things Journal* **4** 2009-18

[10] Krishnan DSR, Gupta SC, and Choudhury T 2018 An IoT based Patient Health Monitoring System *International Conference on Advances in Computing and Communication Engineering (ICACCE)* 01- 07

Smart Monitor for Enhancing Security in Smart Home Systems

Bhuvana Janita[1] and Jagadeesh Kannan[2]

[1,2] School of Computer Science & Engineering, Vellore Institute of Technology Chennai Campus,
[1]bhuvana.janita2017@vitstudent.ac.in, [2]jagadeeshkannan.r@vit.ac.in

Abstract. This Smart Home device like smart TV, Refrigerator, Smart Thermostat, Smart bulbs, Smart garage doors are preferred for hassle free life, better ambience and to automate mundane tasks. Smart Home helps in Mindful resource utilization and also in Wellness assisted living. Though there are many advantages, Smart Home devices are basically embedded devices that comes with firmware and limited memory capacity. These devices cannot have an Antivirus installed for preventing Virus, Malware, Worms and other security hazards. The Internet of Things has no standard protocol for communication. This is a serious security threat for Home Automation systems. Proposed Idea is to introduce an edge device that will monitor and manage all the home devices and alert user in the event of their abnormal activity. The proposed idea would mitigate Man-in the –Middle (MITM) attacks and Denial of service (DOS) attacks.

Keywords: Denial of service attacks, Home Security, Home Monitor, Home Automation, Man-in the –Middle attack.

1. Introduction

There are many Benefits of Remote Monitoring Network including locking / unlocking houses by a long range, getting alerted in case of smoke / burn in the house and optimizing the care of gadgets. It is advantageous to support the disabled who are ill and tend to stay at home, instead of at a clinic. The automation system Internet of Things Home fully optimizes the home life and reduces resources. Depending on the context, the client can adjust the atmosphere. Reading the electric and water meter would be highly useful for careful power use. It is also possible to track the corrosion of the units, which ensures their servicing simple. These apps do carry a lot of protection and security concerns, given the benefits. IOT computers have a configuration with less battery and less processing power. They still haven't developed an anti-virus app. These apps carry apparently massive quantities of data and are most susceptible to protection activities, such as Denial of Service (DOS), intrusion threats and Cyber security threats.

2. Related works

Please to understand the limitations of the Smart home devices, Sukhvir et.al performed an experiment that highlighted the devices' contact patterns and attempted to examine the transmissions using wire shark [1]. The study which was carried out on certain smart homes devices like Phillips Hue light-bulb, the Belkin WeMo power switch has showcased the ease with which security and privacy can be compromised in these devices as shown in figure. 1. The case study on Samsung smart things Hub conducted by Earlence Fernandes.et.al Showed that there were access privilege escalations in the design of these systems [2]. The permission that is accessed granted the users with the permission to access all the devices that are connected to the Hub even when permission is not sought.

Devices	Nest smoke Alarm	Philips Light Bulb	Wemo switch
Communication Pattern	**In-direct:**The communication between the User and the IoT device is only through the external server.	**Direct:**IoT device directly communicates with User via the Mobile app and updates the server about the current status	**Direct:**IoT device directly communicates with User via the Mobile app and updates the server about the current status
Message Encryption	Encrypted-OAuth	Message is plain text with status of light(brightness/color/hue/alarm status, etc) in the message	Plain text
concerns	Privacy	Plain text messages showing the status of the devices	No mechanisms to authenticate a legitimate user

Figure 1. Details of devices system

Wei Zhang.et.al exposed Event spoofing done by misbehaving smart apps in Samsung Smart Things. A deceptive SmartApp with the understanding of the host and system descriptors can fake an inappropriate occurrence, which is simple to remember. The Smart Things database will find the event valid and spread it to all Smart Apps that connect to the matching event-related functionality [3]. To detect smartapp mis behavior the Authors have proposed HoMonit that comes close encrypted traffic-induced SmartApps actions with planned actions determined in their files or UI frameworks.
Smart Home devices have heterogeneous protocols and communication pattern. To derive meaningful information from this type of communication will require new techniques. Roberto Morabito et.al have proposed LegIoT (Light weight edge Gateway device) which by integrating the deployment of different constructs with the advantages of container-based scalability, edge gateways are enhanced to allow automated energy management and necessarily show considerations like energy consumption, dual-tenancy and interoperable [4].

3. Smart Home systems

Before A smart home system generally consists of three components, (1) home server, (2) home gateway, and (3) smart home devices. The home server provides storage, integration and distribution function of the information collected from various media in the home. Next, the home gateway interconnects internet and a wired/wireless home network. Home automation systems that illustrated in figure.2 gather data via various sensors and send and receive among systems and gateway products. And it is easily mountable over the environment, and the vendor provide Two specifications: a software development kit (SDK) with specific features and standards to handle larger IoT implementations (e.g., ARM, AVR, PIC) and a tracking system (gateway or web-based platform) to link different instruments to a home automation network. The SDK is not designed specifically for one but for a variety of devices [2] and all the tools do not have a security system.

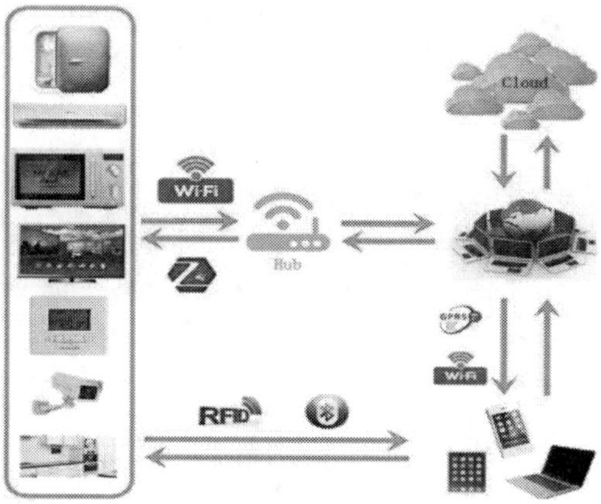

Figure 2. Smart Home system

3.1. Drawbacks in existing Smart Home systems

Although the intention of smart home systems is to automate day to day activities in home or to improve better ambience, currently available systems lack security and centralized monitoring capabilities. Security comes with a cost which the different smart home product vendors hesitate to offer. Multiple products and heterogeneous protocols increase the challenge in providing a security solution. Even the smart home users use different products which require them to handle multiple smart apps individually as in figure.3. Not all Smart Home Users control their smart devices with a Hub. A smart Hub is a device which acts a centralized unit through which all the smart home devices in a home are monitored. The user has the privilege to control all the devices via a single smart hub app which makes it easier for the user.

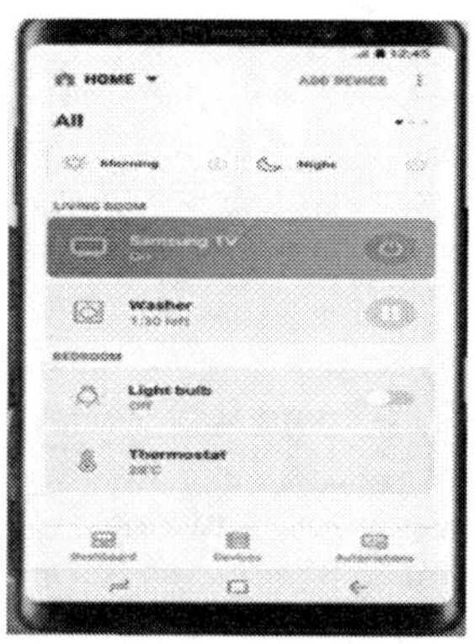

Figure 3. Samsung smart Home Hub UI

3.2. Attacks on Smart Home systems

IoT computers have a firmware with little battery and less processing power as in figure.4. They still haven't activated a bashers-virus app. These apps carry apparently large quantities of data and are most susceptible to protection attempts, such as Denial of Service (DOS), ransomware threats and Ransomware attacks.

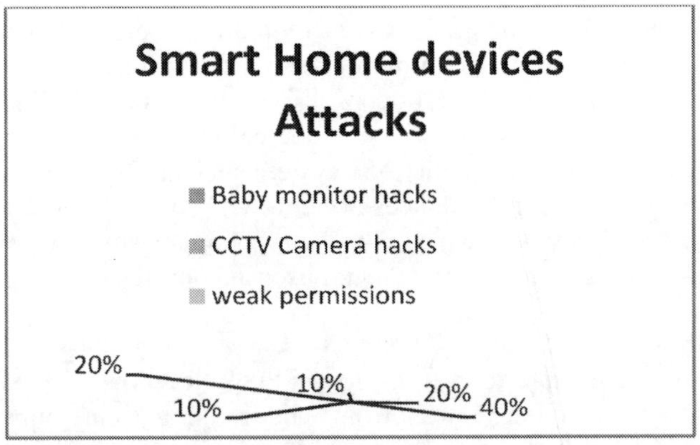

Figure 4. Attacks on smart devices from Media sources

3.2.1. Baby monitor and CCTV Attacks

A person with an intention to hack a house constantly looks for open ports or Wi-Fi hot spots with weak passwords to hack. Unfortunately, there are many websites and apps available in market that will list down the number of open cameras in a particular neighborhood. The number of Baby Monitor and CCTV attacks [6] reported are on constant increase. The users are completely unaware that they are being exposed as there are no systems to alert them.

3.2.2. Mirai/Ransomware Attacks

Figure.5 shows Mirai Attacks that was reported in 2016 is a typical DDOS (Distributed Denial of services attack) on IoT devices like scanners and cameras. This attack is considered a wake-up call that exposed the vulnerabilities of the smart devices that are in day-to-day use and the importance of creating security measures to prevent them.

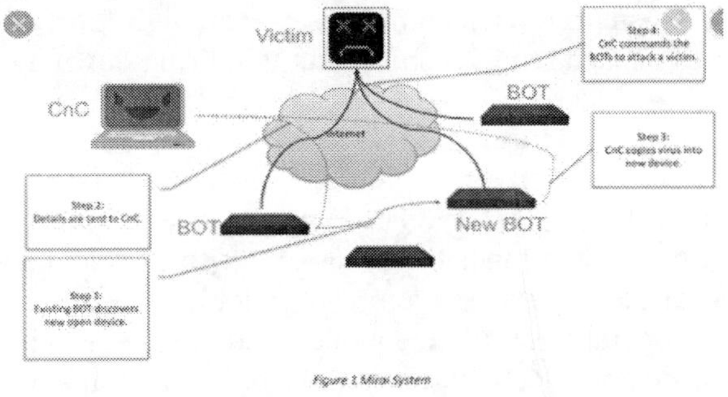

Figure 5. Mirai system

The Hacker/Attacker makes use of a compromised system in a network and installs a BOT in the machine. The compromised device then installs the BOT/Malicious code in the connected devices of the network. This group of affected devices in a network is called a BOTNET which are designed to act on the commands of the Attacker. Some BOTNETS totally affects the targeted device while others silently run in the background parallel sending information about the network devices to the Hacker illegally. The User will not have any clue of the BOTNET in the latter scenario as the devices don't show any abnormality in its daily operations. In case of Ransom Ware Attacks, the malicious code/Malware is installed in the compromised devices and the Attacker demands a ransom amount from the user to return the system to its normal state. In the both the cases early detection is difficult for a normal user and he/she comes to know it only after the damage is done. Hence periodic scanning for any malicious code installed in the devices becomes important.

3.2.3. Weak permission Management

IOP Conf. Series: Materials Science and Engineering **925** (2020) 012036 doi:10.1088/1757-899X/925/1/012036

Fernandes.et.al exposed the lack of fine grained security management system in Samsung smart Home system [2]. Smart Things has a priority isolation process called requirements which specifies the set of operates that a Smart-App can problem to a supported smart home unit. A functionality is consisting of a series of orders and characteristics (properties) (method calls). Orders describe the ways to regulate or actuating a system. Qualities reflect the quality of a system such as On / Off mode. These systems issue permission to the user to access either all devices or none. The author has proposed a solution based on Contextual information from devices using a backend cloud system for rule management and a companion smart app that seeks permission from end user in complicated decision making scenarios.

4. Proposed Smart Monitor

Smart Home systems are vulnerable to man-in-the-middle attack (MITM) attacks, Eaves Dropping, DOS Attacks and Malware injection attacks from Misbehaving smart apps. To protect the Home Network, the devices connected in the network has to be monitored on a regular basis. By introducing an edge device, the latency in the network that is exhausted for decision making can be minimized. The agent software that will be running in the edge device will monitor the devices for the network Traffic parameters and abnormal behaviour of devices can be detected well in advance. The User can be alerted using the companion app based on the analysis. This reduces the latency and also effective in memory and resource restrained devices. The proposed edge device/Hub will be a Raspberry Pi based controller with a sniffer agent installed as shown in figure.6. This device will sniff the packets inside the network and the home network information is sent to a cloud backend service that checks for abnormality in the devices.

To decrease the number of transactions, device profiling is done. Device profiling is the basic data about each device connected in the home network. Based on this the devices inside the home network can be given weight age and only the information about the more vulnerable devices can be analysed. For Instance, a smart camera is more vulnerable than a smart Thermometer. Network information of smart camera can be given more importance for analysis rather than analysing all the devices.

The agent software in the edge device also runs a script to monitor the status of all the device connected in the home network and send the details to the backend cloud service for further analysis. As open ports are the main vulnerabilities of smart device that expose sensitive information, scanning for open ports on a periodic basis is a much needed security check that will bring down number of attacks.

4.1. Working Model of Proposed solution

Cloud is the core of every electronic network which focuses mainly on the delivery of software without physical infrastructure as a central repository. The centrally linked system server sends and evaluates the data produced from our homes and sends the status of our smart telephones. The server manages the whole database and also changes if changes are made at home. The key advantage of this program is that it manages some simple operations with telephone or computers as other forms of devices are built in the home worldwide and are considered automatic or intelligent home everywhere in the world. This system is also very useful in the use of energy. The network introduced comprises of modules and controls of WIFI servers. The Controller board, which runs all operations with our sensors connected to the cloud. Systems = Automatic system access built in the house can be remotely operated from any local machine or mobile phone using an internet link. In WIFI network connectivity, WIFI is used and chosen to link sensors to severs with improved coverage, portability, durability and WIFI requirements, improving device versatility and scalability as a protection framework. Data processing cloud service offers a stronger network for web based applications, because data is sent very quickly, even though very reliable, so all information received by computers because sensors is then processed in a central repository, including data storage and computation. Web service is used to store data. The additional smart phone application also requires the user to control devices and systems without triggering the sensor.

IOP Conf. Series: Materials Science and Engineering **925** (2020) 012036 doi:10.1088/1757-899X/925/1/012036

4.1.1. Device Registration
All smart home devices must be registered and configured in Smart Hub using the smart app of the Hub or web portal. The smart Hub maintains a database about the devices that are connected. The data is streamlined to the home router through the smart Hub.

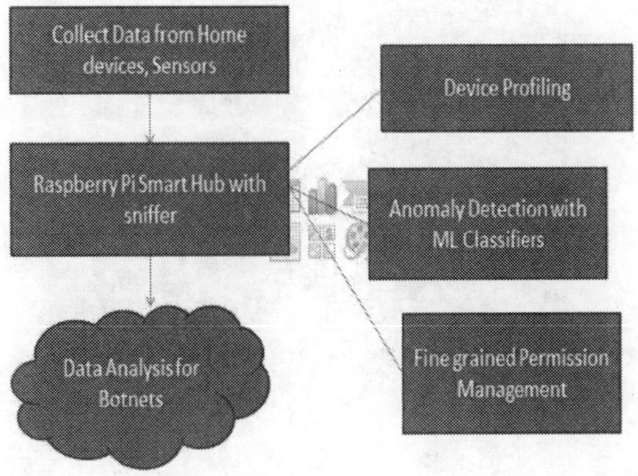

Figure 6. Data analysis from sensors

4.1.2. Device profiling
The Data traffic network is checked against the usual device behavior for abnormal payloads. Device profiling also reduces the time required for analysis as not all smart devices are risky. Every time a new device is registered the device profile is updated in the edge device profile. The network parameters of those devices that have more risk is sent for further analysis.
Port Scanning: A web service is called periodically to check for open ports in all the devices connected to the home and open ports are closed. This will prevent Hackers using open ports a means to infiltrate home devices.

4.1.3. Fine Grained permission management and conflict resolution
This is provided by the smart Hub. Since such decision making is done locally this will avoid latency-issues. Further the data is analysed for botnets based on network traffic on a periodic basis.

5. Simulation and Results
Micro services are used to facilitate each feature of the Smart Monitor Hub. For simulation Azure simulation portal was identified. As the first step, the new devices are registered in the Hub. The device information like device Id, Status (on/off) etc., is stored as xml file. The devices are connected based on REST Principles. The smart devices expose API that should be programmed to connect to the Smart Monitor Hub. The messages are sent in JSON format and the message telemetry is storage in storage endpoint server for further analysis. The IoT hub can be connected to less than 100 devices (in simulation) and handles MQTT, HTTP protocols. The refresh time to update device state is also set in the configuration file. Home status and system design shown in figure 7 and figure 8.

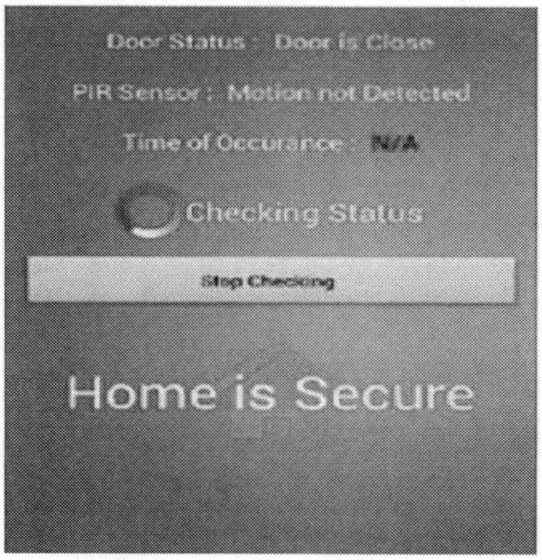

Figure 7. Home status

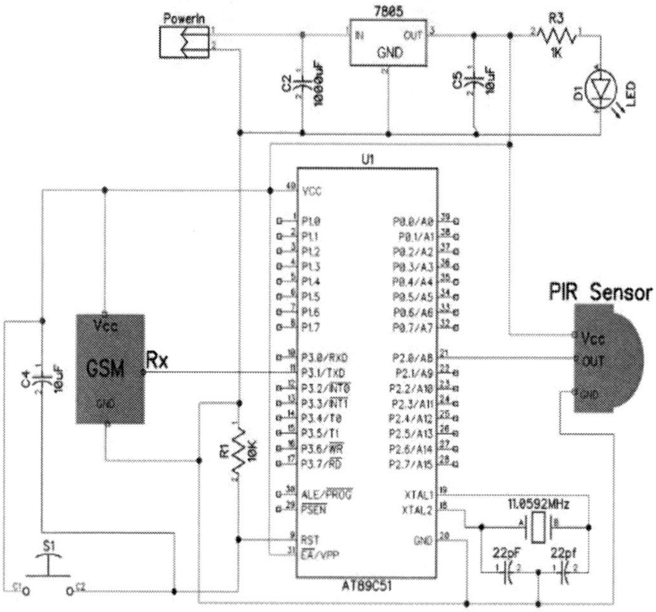

Figure 8. System simulation design

```
private async static Task AddDeviceAsync()
{
    string deviceId = "ParkingLevelOne";
    Device device;
    try
    {
        Console.WriteLine("New device:");
        device = await registryManager.
                    AddDeviceAsync(new Device(deviceId));
    }
    catch (DeviceAlreadyExistsException)
    {
        Console.WriteLine("Already existing device:");
        device = await registryManager.
                          GetDeviceAsync(deviceId);
    }

    Console.WriteLine("Generated device key: {0}",
          device.Authentication.SymmetricKey.PrimaryKey);
}
```

```
private async static Task AddDeviceAsync()
{
    string deviceId = "ParkingLevelOne";
    Device device;
    try
    {
        Console.WriteLine("New device:");
        device = await registryManager.
                 AddDeviceAsync(new Device(deviceId));
    }
    catch (DeviceAlreadyExistsException)
    {
        Console.WriteLine("Already existing device:");
        device = await registryManager.
                         GetDeviceAsync(deviceId);
    }

    Console.WriteLine("Generated device key: {0}",
          device.Authentication.SymmetricKey.PrimaryKey);
}
```

Device probing is done using script run on Kali Linux that looks for all open ports and the log information is periodically sent to the cloud for further analysis.

6. Conclusion and Future works

In this paper the vulnerabilities of Smart Home devices were analyzed. The proposed system can help solve the most obvious issues like eaves dropping, MITM attacks and DOS attacks. Introducing edge device will reduce latency. The existing communication pattern of smart Home devices need not be altered so it is easy to implement in all Homes. Smart Home users with less knowledge on technical details of smart Home systems can also be benefitted with this solution. Based on the data that is communicated between the device app and companion app co further Analysis for malicious activity of misbehaving smart apps can be traced.

References

[1] Sukhvir N, Siddiqiy M, Gharakheiliy HH, Sivaramany V, and Boreli R 2019 An Experimental Study of Security and Privacy Risks with Emerging Household Appliances *IEEE conference on communications and network security* 79-84

[2] Earlence F, Jung J, and Prakash 2016 A Security Analysis of Emerging Smart Home Applications *IEEE Symposium on Security and Privacy* 636-654

[3] Wei Z, Meng Y, Liu Y, Zhang X, Zhang Y, and HoMonit HZ Monitoring Smart Home Apps from Encrypted Traffic CCS *In Proceedings of the 2018 ACM SIGSAC Conference on Computer and Communications Security* 15-19

[4] Roberto M, Petrolo R, Loscrì V, Mitton N LEGIoT: A Lightweight Edge Gateway for the Internet of Things *Future Generation Computer Systems*, **81,** 2018.

[5] Isa S Implementing Smart Home using Cisco Packet Tracer Simulator *International Journal of Engineering Science Invention Research & Development* **4**

[6] abc NEWS. 2013 Baby Monitor Hacking Alarms Houston Parents.http://goo.gl/LpuJzg

[7] Tatikayala SG, Meerolla M, Jyostna G, Reddy Lakshmi Eswari P and Magesh E Mitigating Mirai Malware Spreading in IoT Environment *Centre for Development of Advanced Computing(C-DAC),*

[8] Narayanan KL, Ramesh GP and Divya V Robust and brittle secured video for IOT *International Journal of Engineering and Technology* **7** 93

[9] Shahada S, Ahmad A, Mohammed Hreiji S and Shamsudheen S 2019 IoT Based Garbage Clearance Alert System with GPS Location using Arduino *International Journal of MC Square Scientific Research,* **11** 1-8

[10] Badawi I and Ali W 2019 Underground Pipeline Water Leakage Monitoring Based on IoT *International Journal of MC Square Scientific Research,* **11** 01-08

IOP Conf. Series: Materials Science and Engineering **925** (2020) 012037 doi:10.1088/1757-899X/925/1/012037

Low Cost Autonomous Driving Architecture using Ultrasonic Sensors and Camera

S Manivannan[1] and R Jagadeesh Kannan[2]

[1]Senior Embedded Software Developer
[2]School of Computer Science & Engineering, Vellore Institute of Technology Chennai Campus
[1]sivan.manivannan@gmail.com, [2]jagadeeshkannan.r@vit.ac.in

Abstract. The existing autonomous vehicles cost are very high due to inclusive of various range of LIDAR, RADAR and multiple cameras for autonomous features. The proposed autonomous vehicles can operate at low speed with limited sensors which will cut the cost by half. The purpose of this research is to make an auto driven platform, which can be ported in any electric vehicles. The prototype also has features including a safety system, control panel, route recognition, android device assistance, Wi-Fi, & battery updates. Our study findings are carefully analysed. This leads to less emissions of the electric vehicle powered by man.

Keywords: Autonomous Driving, Beacon, Electric Vehicle, Open CV, Raspberry Pi 3

1. Introduction

ICE-powered vehicles are the biggest challenge to green gas emissions in emerging transport technology. The recent autonomous vehicles are also too costly to afford in public transportation. Our research in autonomous driving architecture leads to low cost autonomous vehicle which can operate in closed environment with limited speed 25Kmph (Ref ARAI Certification Specification for EV). The planned stand-alone vehicle includes an 8Mega pixel monitor, User Interface touchscreen display, Ultra Sonic Sensor Design, Beacon module, distance speedometer, Motor driver & DC motor. The proposed architecture has been implemented in smaller vehicle with operating voltage of 12Volt with UI, Array of ultrasonic sensor, HC-06 module. This entire architecture can be ported in existing electric vehicle is added advantage. The Raspberry Pi 3 is used as vehicle controller unit and atmgega328 is used as secondary controller. The secondary controller receives data from array of ultra-sonic sensors and process the data and feedback the signal information to vehicle control unit.

2. Literature review

The primary autonomous robot development and collision avoidance is route planning. Mobile Robot Road Planning is to find a collision-free route from which to divide the goal into global route and local route maps [1]. The global path management is focused on knowledge of the work situation of the robot. Local path scheduling is useful for applications that do not know or slightly know the mobile robot's work atmosphere [2-3]. The able to sense the real reality and to respond appropriately relies on independent behaviour in urban environments [4]. That involves making sure in compliance with traffic regulations and viability constantly for some pleasure-optimized operations [5]. This paper presents a new, longitudinal framework for planning developed as a discreet plan. In the near future, independent driving will become a truth. In recent years, driver assistance systems for manufacturing vehicles have increased steadily their automation standard [6-7].

3. Background of technologies behind autonomous vehicle in closed environment

The closed environment will comprise of speed limit to 25Kmph (For all vehicles), perfect lane marking throughout the region, traffic signs shall be properly installed as shown in figure.1. For example, consider the environment like amusement parks, Universities. In OpenCV, the road lanes are detected using Hough Line Transform. The edges and thresholds are defined to detect turning lanes and differentiate dotted partition lines in the mid of the road. The figure.2 illustrates the lane detection and action to be taken for left, right curve. The raspberry Pi 3 with Pi camera V2 is used in autonomous vehicle to detect the lanes.

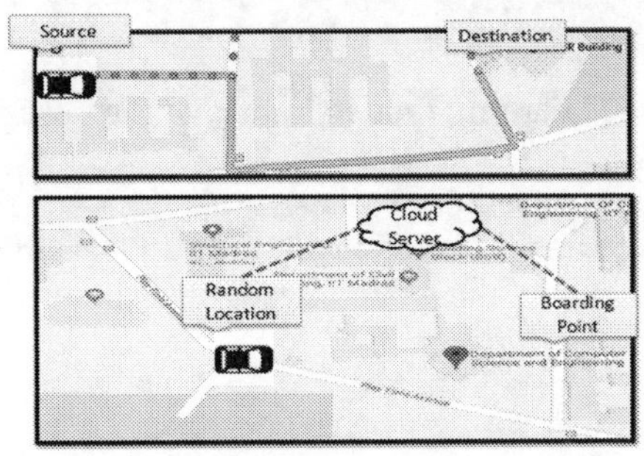

Figure 1. Autonomous vehicle in closed environment map

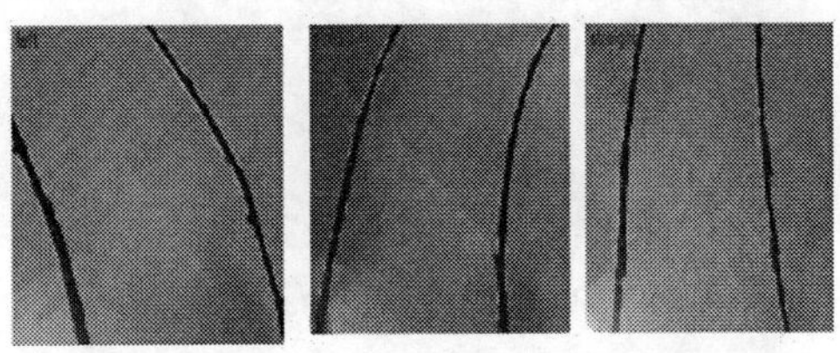

Figure 2. Lane detection using OpenCv

3.1. Motor Controller

Generally, the motor controller can be controlled by two different signals. To control the speed, PWM signal is passed to PWM pin and to control the direction in forward or reverse, 5volt signal is passed to Direction pin. In our controller, a single PWM signal can control both speed and direction. When the PWM signal is 50% duty cycle, the motor stops running. If the PWM has less than 50% duty cycle, the motor will rotate in Clockwise direction. If the PWM signal has more than 50% duty cycle, motor will rotate in Counter clockwise direction.

3.2. Path following accuracy using beacon

To improve the accuracy in corner detection and landmark detection, the beacons are placed in every corner, landmarks to emit unique identification number. In figure.3 indicates the location of beacon modules in every corner.

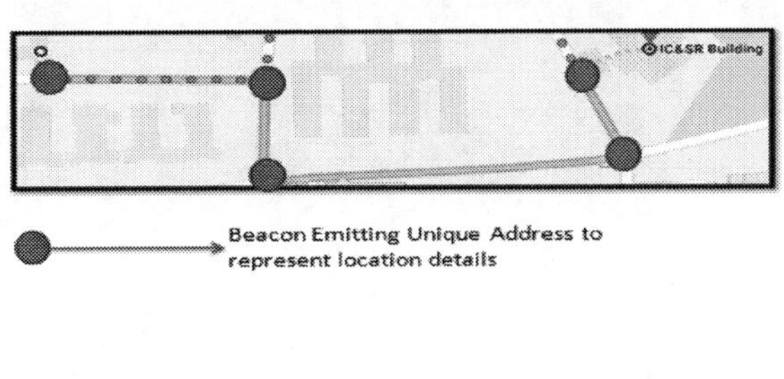

Figure 3. Location of beacons in every corner

3.3. Path following accuracy using beacon

Beacon modules are placed in every corner which emits unique value. Initially the vehicle is operated in learning mode by manual driving, where the controller observes the unique number of beacon in each corner. Once the complete is mapping is done, the vehicle is changed to auto driven mode.

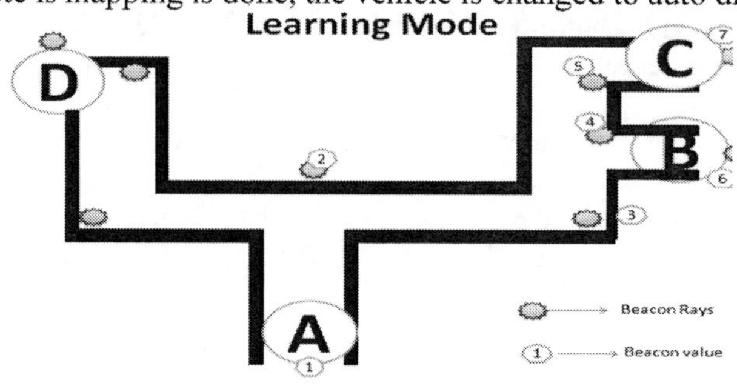

Figure 4. Learning mode to observe the beacon values

3.4. Path following accuracy using beacon

From figure.4 if a vehicle wants to travel from point A to C. In learning mode, it recognizes all the turns from A to C. Then in auto driven mode, the vehicle recognizes the beacon value and takes action which was taken in manual mode as shown in figure.5.

A location to C Location

Beacon Rays	1	2	3	4	5	6	7
Action	Straight	Right	Left	Straight	Right	-	Stop

C location to A Location

Beacon Rays	1	2	3	4	5	6	7
Action	Straight	Left	Right	Straight	Left	-	Stop

Figure 5. Data mapping algorithm

4. Architecture overview

In proposed architecture as shown in figure.6 the ultrasonic sensors used to detect obstacle in 10ms time delay and cameras are used for lane detection and object identification like differentiate between cat and wooden block. GPS and beacon module used for accurate localization. The three objectives of sensor fusion in architecture is Lane detection, object classification and localization. The controller programmed to have different call thread for each sensor input, so it can be run in parallel.

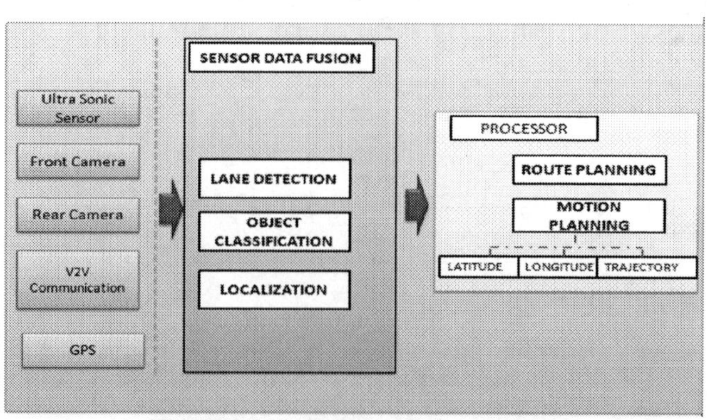

Figure 6. Architecture Overview

IOP Conf. Series: Materials Science and Engineering **925** (2020) 012037 doi:10.1088/1757-899X/925/1/012037

The MCU is designed with a route search algorithm, obstacle avoidance, & a route after an algorithm is configured for the system. The figure 7 denotes the prototype of autonomous vehicle. As per the architecture, it is fitted with Camera in front for lane detection and ultra-sonic sensor for object detection.

Figure 7. Prototype of Autonomous vehicle

In the sealed environment, prototype consists of Beacon, GPS with 32-bit microcontroller. The EV database monitors the GPS coordinates of the vehicle at different station boxes located in closed environments. With the aid of the Button, the car is pushed in automatic mode.

5. Discussion

The system is sponsored by the built-in solar panel for more study that can run for up to six hours. The performance of the vehicle with a solar panel is evident from Table.1 and Table. 2. From figure 8 we can able to refer the simulation results for the proposed work.

Table 1. Power consumption of autonomous vehicle

Vehicle operative data	**Value**
Power intake by vehicle / hour	410 W
Total Operating time	6 Hours
Total Energy consumption	2460 Watt Hour

Table 2. Solar power generation in autonomous vehicle

Solar power generation data	**value**
Power generated by solar panel	500 W
Average fall of direct sunlight per day	5 Hours
Total Energy	2500 Watt Hour

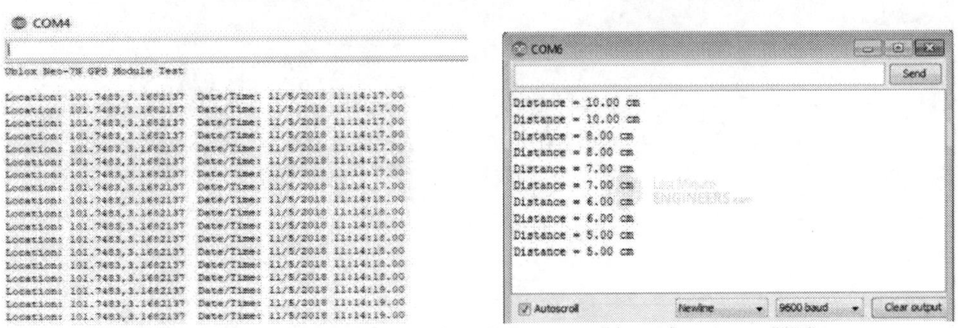

Figure 8. Simulation results of range and location over GPS

IOP Conf. Series: Materials Science and Engineering **925** (2020) 012037 doi:10.1088/1757-899X/925/1/012037

6. Conclusion and Future works

Autonomous driving architecture provides pollution less transportation and cost effective auto driven electric vehicle. End users like business, university campus and entertainment parks will benefit from this design. This innovation used in vehicles helps renewable transport. Physically disabled people will benefit from automated mode.

References

[1] Yalong Z, Zhenghua L and Le C 2017 A new adaptive artificial potential field and rolling window method for mobile robot path planning *29th Chinese Control and Decision Conference (CCDC)* 7144-7148

[2] Constantin H, Michael A and Christoph S 2016 A generic driving strategy for urban environments, *Intelligent Transportation Systems (ITSC) 2016 IEEE 19th International Conference* 1010-1016 2016

[3] Gaku T, Makoto O and Uto K 2017 Posture-to-posture trajectory planning for an autonomous car using triple clothoid segments, *Decision and Control (CDC) IEEE 56th Annual Conference* 299-306

[4] Sundaram A, Ramesh GP 2015 Investigation of Solar Based SL-QZSI Fed Sensorless Control of BLDC Motor *Intelligent Computing in Engineering. Advances in Intelligent Systems and Computing* 1125

[5] Shermin S and Mubarakali A 2019 Smart agriculture using IoT *International Journal of MC Square Scientific Research* **11** 543-545

[6] Biruk Z and Demissie M IoT based lawn cutter *International Journal of MC Square Scientific Research* **11** 2019

[7] Waleed AB 2019 Underground pipeline water leakage monitoring based on IOT *International Journal of MC Square Scientific Research* **11**

IOP Conf. Series: Materials Science and Engineering **925** (2020) 012038 doi:10.1088/1757-899X/925/1/012038

Doctor Finder: Find doctors on the Go

R Jagadeesh Kannan[1], Harsh Tamakuwala[2], Siddharth Kale[3] and Harsh Rohan Bhowmick[4]

[1]School of Computer Science & Engineering, Vellore Institute of Technology Chennai Campus, India
[2,3,4] School of Computing Science and Engineering, VIT Chennai, India,
[1]jagadeeshkannan.r@vit.ac.in,[2]tamakuwalaharsh42774@outlook.com,
[3]siddharthkale97@gmail.com

Abstract. Currently most people don't have access to world-class health care. People tend to go to their local doctors and get their treatment delayed in case of serious diseases. These cases require a specialist to tend to the needs of the patients. A system is created that helps a person find a doctor based on the symptoms the user is facing. The system helps the user to get redirected to the nearest specialist that is appropriate to the special case of each user. This article discusses how the system was designed and implemented and what exactly was the need to implement such a system.

Keywords: Doctor Recommendation System, Disease Prediction, Intelligent Healthcare Management System, Applications of technology in Healthcare

1. Introduction

The world is ridden with various diseases which can be hard to get diagnosed. And what makes it even worse is that it even harder to get these diseases treated even if the disease is correctly diagnosed. In many third world countries, an unprecedented number of people die due to undetected diseases. The most prominent problem rattling the healthcare sector is to identify accurately a disease that a patient might be contacted with as early as possible. It is even more crucial in the case of terminal diseases like Cancer, HIV, COVID-19, etc.

People observe various cases that the families and friends were a late diagnosis of a terminal disease ended up in the demise of our loved ones. Only if there was any way of knowing that we can test ourselves at home just by seeing initial symptoms, that could have saved some of our loved one's life. That gave us the initial idea of creating such a system, our main inspiration is drawn from such cases only.

Various studies and articles have pointed out and further explained in their sources [35][36] that how serious can a late diagnosis be. The source in [36] states about 10000 people die from a late cancer diagnosis. This study depicts that we urgently need a system to readily diagnose our patients.

At the time of writing this paper, there is an ongoing corona pandemic in the world. As of now, South Korea has done a laudable job of controlling the spread of the virus. As per the source [37], it was all possible due to the massive scale testing that the Koreans were able to conduct. In the article [37] from

the University of New South Wales, Sydney, Raina MacIntyre, an infectious disease scholar stated, that the key to controlling an epidemic is to diagnosis at scale.
All of the above sources point out the need for early diagnosis in the case of terminal diseases. And more often than not people do not approach specialists and wait for the ailment to get worse before consulting a specialist. This poses a serious threat to any patient.
The other problem that came to our mind that most countries like India face a lack of awareness of different kinds of specialists. People tend to visit local doctors or just general doctors. This mentality of not approaching specialists results in even more delayed diagnosis and ultimately late treatment. It is very crucial in many cases that the patient visits a competent medical professional that can properly detect the disease and provide efficient and cost-effective treatment.
The source in [33], states that there is less than one doctor per 1000 people in India, which is well below the WHO standard. This even results in even fewer specialists per 1000 people. Just by looking at these statistics, one can assume that people surely struggle with finding an appropriate competent specialist in India. Even a recent UK based study [34], suggested that in almost 50% of the hospitals there have at least one place vacant for respiratory consultant post.
The above studies and statistics suggest that how hard can be it for a normal person to just get an appointment with a local doctor. The empty posts in hospitals might also translate to increased workloads of doctor and that could stall the early and rapid testing and diagnosis of the patients.
Surely government and health care professionals can create laws and systems to ensure that people get more easily diagnosed or there is enough medical staff available for rapid treatment. But this can take years to achieve. And this is not possible instantly for many countries. That's why there is a need for a system that can correctly guide patients to the nearest available specialist that can treat their ailment and is perfectly reliable. The new system discussed here will also decrease the load on the hospitals as it will guide the patients to the correct specialist. Thus, balancing the load from the general doctors and other specialists whom the patients might have wrongly concerned.
With this new system, a pipeline is created that enables anyone to just put what they are suffering through in just a sentence to the app. From there the analysis of the symptoms and conditions takes place. Based on the symptoms, a probabilistic model predicts the disease that is more like to happen with the particular case of the patient. Information about the disease is also provided along with possible treatment. Various doctors are listed that can effectively treat the possible disease and are near to the user's location. There is also a method to book an appointment with doctors and have the prescription noted to the customer through the app.

2. Technologies used

2.1. REST API

To create an actual system that intends to serve the customers, there was a need to bring the functionality of the models to them. For creating such a pipeline, there had to be a system in place that can communicate between the client and the model. That's why it was decided to use and create APIs to handle communications.
System has specifically incorporated the REST API. APIs are as described in [1], Application Programming Interface, facilitate communication between the client and the webserver. Through APIs, various programs and devices can transfer data, which is made possible by exposing the devices to a set of data and functions. And REST is a common architectural style that is applied to the system to create API for our webservers.
As proposed by Roy Fielding [2] REST is essentially a subset of WWW and based on HTTP. The messages in the REST are not dependent on the state of the conversation going on between various clients and web services, essential it is termed as stateless. The REST agents give uniform interface semantics - basically make, recover, update and delete - as opposed to discretionary or application-explicit interfaces, and control assets just by the trading of representations. [2]
The method of creating the REST API can be subdivided into the following phases in which we tacked different problems, found solutions, and implemented as best suited to the system's needs.

2.2. Selecting the framework and language
Initially, it was decided to go with NodeJS, but as things moved on, it was observed that the core machine learning pipeline was being developed in python. Moreover, NodeJS being single-threaded will not a currently processing CPU the oncoming requests even for a second. That's results in creating an asynchronous process, which is hard to get used to. [3]
Keeping the above points in mind, it was decided to go with a framework based in python. Which helped in two ways-
To have a single language for our system that will make it much easier to design and manage.
The ability to host the API as well as the system handling the ML model on the same server, thus saving resources and the hassle of communicating between different resources.
As for the framework it was decided to go with Flask. Flask is a popular framework designed for creating web services and is written in python. Flask is developed in such a way that it doesn't require any special libraries or tools making it a microframework. Flask is considered as a quick and easy to use framework, which was decided to use as rapid prototyping was used to create this system. [4][5].

2.3. Development of Routes for Website and App
Here API helped to create various methods with which any device can communicate.
Routes, as defined by Flask documentation [5], is a meaningful URI to help locate users to a page. Here we have created several routes with specific URI to handle various oncoming requests.
These routes also include the methods with which the oncoming data is manipulated or data is retrieved from a Database server and send back to a client. The system is incorporated to use both of these functionalities in various routes.

2.4. Handling the website
It was noticed that a website interface is certainly important to reach out to as many people as possible. The Website Interface doesn't have a separate backend, instead, configuration was need so that the API can handle and serve the website as well. There are specific API calls that are specifically modeled to handle the website and its various functionalities.

2.5. Database
The reasons and arguments why a special kind of database is needed to suit the needs of this system have to be mentioned. As to quote the source [10], NoSQL is the current emerging alternative to the traditional database system which chiefly includes the RDBMS. NoSQL is considered to be a database that does not completely replace SQL but indeed compliments the SQL to a point that there is a coexisting scenario in place. [10]
As evident from various sources [6][7], it is very evident that the data in the Healthcare sectors tend to be much more unstructured and unformatted. It is very difficult to handle unstructured data through the traditional method of RDBMS.
As there is a lot of data in Healthcare that is just arbitrary and cannot be structured [7] and modeled specifically to suit a rigid relational schema [8], that's why a system like NoSQL is needed for our system. Another reason is that the NoSQL database is needed for the system is that it was expected that if this system is implemented in the healthcare sector, it would see that the data will continue to grow, as mentioned above it will be unstructured, but more specifically their would-be occurrence of problems of RDBMS that are related to huge data size. That will include lower management and operational costs, low writing and reading speeds, low scalability (various DBMS tends to get slower after a certain data amount is reached). [9]
Hence, it was decided that RDBMS would stand in the way of data management of unstructured data and we would very well like our system to be scalable, that's why, the system has used Firebase, a NoSQL database for the project.
Firebase Database is a Google product, that is a very fast, NoSQL based, cloud-hosted database. It let its users store and sync data in real-time all across the globe. [11]

The database comprises of the doctors that the system is suggesting to our user. Every doctor is categorized and stored based on the specialty of the doctor.
Other than the private information of the doctors and the users the database also holds all the patient appointments, records of the app queries of the user, user medical history, and user test reports. Out of this data, only Doctor information and appointment detail are very well structured. Rest can be very unstructured and sometimes even some records may be present for one user or one not.
For example, a user with the treatment of a broken leg might have an X-Ray report to his name in the database, and the other patient could have the sonographic images of her baby. In this way, there is very ambiguity about what may be uploaded to the database.

2.6. ML Models
Below are brief discussions and a bit of explanation of the ML models that have been tested out or used in the system at any point in time in the development of the system.

2.7. Decision Tree
A decision tree has a binary tree-like structure where each node represents a question kind of information upon which the output at the leaf nodes is reached. [22]

2.8. Additive decision tree
A classical decision tree has some weak branches and some strong branches. These divisions are created due to the information obtained from the data. Some test cases may be less of input; some may have very diverse symptoms that the model gets confused about what the disease might be. Hence, to overcome these weaknesses, one such method devised was additive decision trees. In this method, the weak classifiers were trained recursively to convert them to strong classifiers [23].

2.9. Incremental decision tree
This is an algorithm that gives a decision tree as an output. It trains the current model with the newly obtained data, without re-training the model on the previous data. Besides the increasing data problem, this model had no such major change inaccuracy [24].

2.10. Neural Network for multi-label classification
Neural networks are often used by data scientists as an approach to possibly define a simple model into a complex model to gain more accuracy. Neural networks are inspired by the working of neurons in the brain. It is a circuit of artificial neurons meant to solve much more complex problems which moreover has a huge amount of data. [25] Neural nets are generally used for computer vision purposes or Natural language processing where the data to be used as an input is much more complex. [26][27] Various industries have devised their libraries for the implementation of neural networks.

2.11. Incremental decision tree
This is a classical probabilistic classification approach. This method calculates the probability of occurrence of an event given the input. The formula for calculating the probability is shown in figure.1.

Likelihood — Class Prior Probability

$$P(c\mid x)=\frac{P(x\mid c)P(c)}{P(x)}$$

Posterior Probability — Predictor Prior Probability

$$P(c\mid X) = P(x_1\mid c)\times P(x_2\mid c)\times \cdots \times P(x_n\mid c)\times P(c)$$

Figure 1. Formula for calculating posterior Probability

IOP Conf. Series: Materials Science and Engineering **925** (2020) 012038 doi:10.1088/1757-899X/925/1/012038

Naïve Bayes classification naively assumes that each class is independent of one another and predicts one output. [30]

2.12. Multinomial Naïve Bayes Classifier

This classifier is used when the data is multinomially distributed. It uses the same approach as naïve Bayes but for each of the instances of features. This way probability for class can be calculated without any independence and the top three outputs can be used by our system with probability for each of them to give the prediction some meaning [31].

2.13. Cloud

Cloud computing can be simply defined as the delivery of a wide range of services through the internet. These services or resources include data servers, storage of data, networking services, databases as well as software [12]. The main aspect of cloud computing is to remove the need for the local hard drive for storage and instead, use cloud-based storage solutions. As long as the device is connected to the web, it can be used to access the data stored in the database or the applications that can be used to do the same [13].

Companies that use this technology enable the users to store applications and data on remote databases and also give them the ability to retrieve the files back via the web [14]. Cloud computing does all the heavy work lifting which involves a lot of processing and crunching of data and allows the user to access them from a remote location from anywhere in the world.

Cloud computing services can be deployed in various ways, also termed as models. The deployment models are as follows [15]:

2.13.1. Public cloud

It is a type of cloud hosting model where all the services are deployed over a network that is open for public usage.

2.13.2. Private Cloud

it is also known as an internal cloud where the platform is implemented or deployed on a secure network only accessible by the client who has the permission to access it

On the technical side, both cloud types, Public and Private, may have little to no difference at all in the structural side of the cloud but are much different in terms of security offered by the service providers to the client [16].

2.13.3. Hybrid Cloud

It is a type of cloud hosting that is integrated i.e., the arrangement of one or more cloud servers, public, private, or community cloud services bound together but function as separate entities. It helps to move workloads between private and public clouds according to computing and cost needs which also gives the businesses greater flexibility and more data deployment options too.

After the selection of the right type of cloud, the type of services has to be selected. The services of the cloud include SaaS- Software as a service, PaaS- Platform as a Service, IaaS- Infrastructure as a Service. The cloud service model SaaS is a type of software distribution model where a third-party provider hosts the services and makes them available to the clients over the network. PaaS is a computing model where a third-party provider provides or delivers necessary hardware and software tools that are generally required for the development process, to the client deployed over the network. Here, the PaaS provider hosts the software and hardware on its infrastructure which helps the users avoiding stalling hardware and software to develop or run a new application. IaaS is a form of model that provisions virtualized computing resources deployed over the network. Here, a cloud provider hosts the infrastructure components present in an on-premises data center like storage, networking hardware, and also servers. The customers using IaaS access the resources or the infrastructure services through the internet and use the provider's services to install the elements required of an application stack.

In the project model, we deploy our SaaS software application on the public cloud using GCP Firebase cloud services [17].

3. System architecture

The main objective of the application is to process the patient's input and display the details of the respective doctors and hospitals nearby with their previous statistics and location marked on the map. The patient inputs the symptoms and the model detects the type of disease corresponding to the input and the user henceforth, can consult the required doctor and track the details and location of the doctor or the hospital. The interface also provides the user with information about different types of diseases and gets required details regarding their symptoms, required treatments, and steps for prevention.

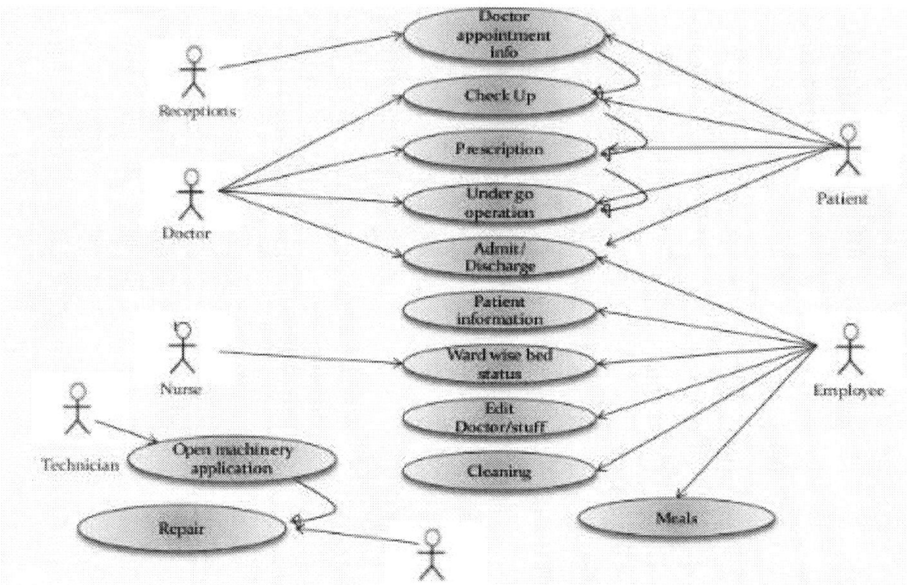

Figure 2. Use Case for the whole system

The model system application as shown in figure.2 is a two-part system containing the front-end part which acts as an interface for the user and the back-end part which stores the details of the users and connects with the ML model hosted in a web server using required APIs. Specifically, in order to serve this purpose, the REST API [18] is used for this system. To avoid a separate backend process requiring to store the inputs of the user and to preserve the privacy of the users, the API was configured to handle and serve the application by using specific API calls that are specifically modeled to handle the application and its various functionalities [19].

For the database part, GCP cloud storage service namely, Firebase storage [20] is used to store the details of the user as well as the doctors. Apart from individual details, it also includes the location of the hospitals and doctors as well as treatment information and prevention details too. The flow chart of the system is represented in figure.3.

- The architecture of the application involves four modules namely, collection, organization, recommendation, and updation.
- Collection phase: In this phase, the data from the user is collected in the form of symptoms and then submitted to the server.
- Organization phase: In this phase, the input data is then converted into respective JSON format or HTTP format to make it easier to process it in the server [21]. The server here, hosting the ML model is used to process the data of the user. After the data organization, the data is sent to the server using the REST APIs.

- Recommendation phase: The data from the application after the organization phase, goes to the web server where the data is processed and provides the details required from the database.

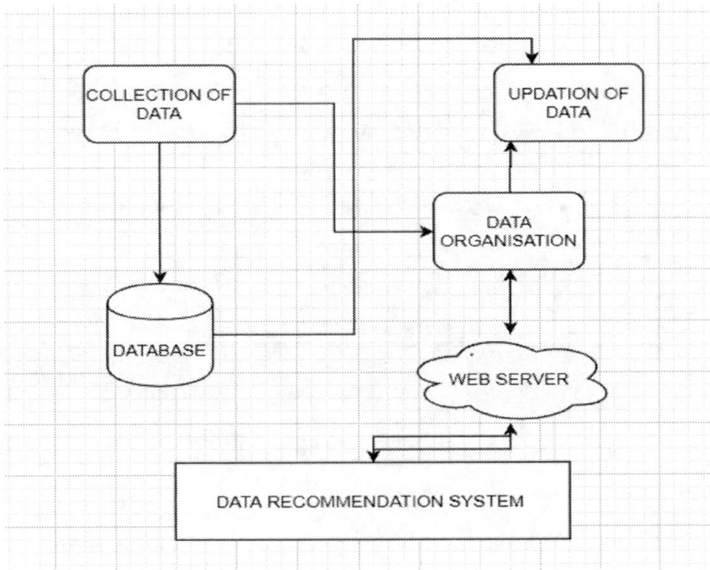

Figure 3. System Architecture explained by the flowchart

The architecture of the system is described as follows,

3.1. Data Collection phase
According to the system architecture given above, apart from the application having the option of storing various details of the user and doctors including the location, the application also enables the user to input symptoms to detect the corresponding disease. The data is given by the user in the form of symptoms. The user can either manually input the diseases in the text box given in the application or can use the search box and select the diseases from the list.

3.2. Data Organization Phase
After the selection of the symptoms, the input is then organized in a form acceptable by the web server and then sent to the server for further processing. This is the data organization phase. After this phase, the data is sent to the web server using required APIs, here, REST APIs. As it is the core process of the system, a specific route is created that handles this process [11]. The route is supposed to get Symptom data as an HTTP post method from the client-side device. The data is then sent to the ML model pipeline handling process hosted in the Heroku web server.

3.3. Data Recommendation System phase
Beacon, next comes the data recommendation phase, where the ML model hosted in the web server is used to process the input and detect the corresponding diseases according to the probabilities and the related various information and treatment info is then fetched from the database and then sent to the mobile application.
The architecture of the data recommendation phase is as follows

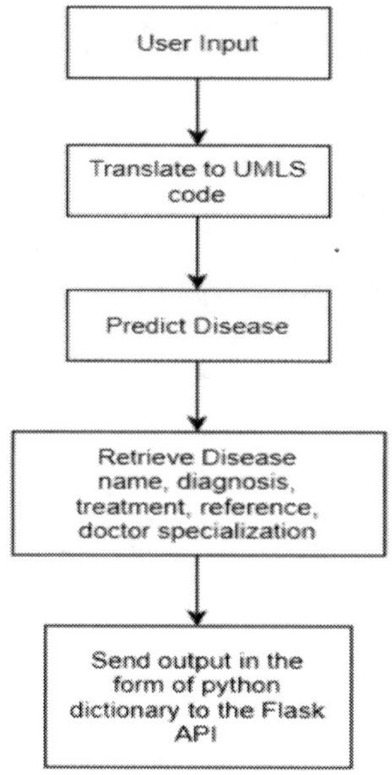

Figure 4. Flowchart of the data recommendation phase

According to the architecture figure.4 given above, it is divided into 5 modules,

3.4. User input
In this module, the input from the application via the APIs is sent to the server where the ML pipeline is hosted.

3.5. Input translation and output format
The UMLS defined codes were used as input and output of the machine learning model. Hence, after receiving the symptoms as input from the user, the basic English terminology was converted to a UMLS code and fed into the model after appropriate pre-processing which includes pivoting the whole input to convert it into a two-dimensional matrix containing binary values. The output given by the model was the UMLS code for the disease which was then to be translated into the English language so that it is understandable to the user. But, before converting it to the English language, the same UMLS code was used to retrieve disease information/diagnosis, treatment info, reference links, and doctor specialization info. These UMLS codes were used so that no syntactical errors may occur.

3.6. Disease Prediction
Various methodologies were tried for the disease prediction from the symptoms. Each of them along with its pros and cons is mentioned below:

3.6.1. Decision Tree
For classification on a small amount of data, decision trees are very accurate. These models were implemented which gave pretty accurate output if enough input was fed to them. A basic decision tree was used since the decision tree formed by it was not complex enough and accuracy change between different variations was not noticeable. The decision tree was implemented using the sklearn library. [22]

$$E(S) = \sum_{i=1}^{c} -p_1 log_2 p_1 \tag{1}$$

3.6.2. Additive Decision tree

IOP Conf. Series: Materials Science and Engineering **925** (2020) 012038 doi:10.1088/1757-899X/925/1/012038

A classical decision tree has some weak branches and some strong branches. These divisions are created due to the information obtained from the data. Some test cases may have less input, some may have very diverse symptoms that the model gets confused about what the disease might be. Hence, to overcome these weaknesses in the models, various methods are used. One such method devised was additive decision trees. In this method, the weak classifiers were trained recursively to convert them to strong classifiers. Various pre-processing methods like interpolation can also be used. Since the dataset had an equal number of cases, this method did not affect the accuracy that much but this was more of a future endeavor when the data might become more varied. [23] This variation of the decision tree was implemented from scratch on a small amount of data which was then compared with the standard decision tree approach on the same data. The main concern at the moment was for accuracy which this method did not solve.

3.6.3. Incremental decision tree

This was also a future endeavor for the model. When new diseases are added to the dataset, currently the model has to be trained completely from scratch to accommodate the newly obtained data which can be very resource-consuming and the system has to be down for maintenance more often to update the machine learning model. To tackle this problem an incremental decision tree was implemented. This is an algorithm that gives a decision tree as an output. It trains the current model with the newly obtained data, without re-training the model on the previous data. Upon researching more, this model might be dangerous in the medical domain. There might occur an instance where a disease's symptoms are completely changed due to some previous human error or even due to the adoption of the microorganisms. In such cases, the whole model needs to be trained again else, the model might predict the old outputs which are false. Besides the increasing data problem, this model had no such major change inaccuracy. [24] This variation of the decision tree was also not focused on accuracy since its main task was to reduce the workload required to accommodate the new data into the current environment. Hence, this approach also met the dead end.

3.6.4. Neural Network for multi-label classification

Neural networks are often used by data scientists as an approach to possibly define a simple model into a complex model to gain more accuracy. Neural networks are inspired by the working of neurons in the brain. It is a circuit of artificial neurons meant to solve much more complex problems which moreover has a huge amount of data. [25] Neural nets are generally used for computer vision purposes or Natural language processing where the data to be used as an input is much more complex.[26][27] Various industries have devised their libraries for the implementation of neural networks. The accuracy of the model could still be increased; hence, this method was tried. Also, there was one more problem in the current approach. The current model predicted only one disease. Hence, in some scenarios, when two or more diseases have a quite similar probability of occurring, the current approach might fail to give correct predictions. Hence, after that, approaches were more probability-based. Since having some actual number related to the disease the prediction might make more meaning. One such approach was using a neural network. Upon testing the model, scores for each possibility were calculated and the top three were predicted as output. This approach was very resource-consuming for practical purposes. And the approach was even more resource consuming during training, hence when some new data is obtained, retraining the whole model could end up being more costly. Also, concurrent access to the model could drastically increase the waiting time for the users since the computation of the output is more time-consuming. [28]

3.6.5. Hybrid model

For the multi-label classification problem, various custom prototypes were tried. One approach was based on the theory of random forest implementation. Random forest is a collection of various decision trees which are fed different data upon input, and it gives an output based on the most voted output by the trees. [29] This approach was modified by giving each tree the same input but in an incremental manner. After the prediction from the first tree, that case could be hypothetically scraped out from the input and the rest of the input could be fed into the second possible tree. This way 3 diseases could be predicted but without any actual numbers to rely on. This method was also implemented from scratch on a small amount of data but no appropriate scoring mechanism was found to get concrete proof about the result. Another method was tried using the same decision tree but gradually feeding the input. A buffer of the outputs could be created at various intervals to give the top three outputs. A score-based mechanism has to be accommodated to get the top three outputs based on the score. This method is not too reliable since it might give various false results since the score calculated is relative to the other inputs fed at that interval. This method might fail also when the symptoms of the first disease in the dataset are matched completely with the input. Hence, for every iteration, it will give the same output. This can be solved by running the above method more times, each time feeding randomness into the input. This method could be more time-consuming.

3.6.6. Naïve Bayes Classifier

Since the main focus currently was the accuracy and then a proper grading mechanism of the output, Naïve Bayes Theorem was used.

Naïve Bayes classification naively assumes that each class is independent of one another and predicts one output. [30]

$$P(A|B)=p(B|A)P(A)/P(B) \tag{2}$$

3.6.7. Multinomial Naïve Bayes Classifier

This classifier is used when the data is multinomial distributed. It uses the same approach as naïve Bayes but for each of the instances of features. This way probability for class can be calculated without any independence and the top three outputs can be used by the system with probability for each of them to give the prediction some meaning. [31] The Naïve Bayes and Multinomial Naïve Bayes approach was implemented using the sklearn library.[32] Instead of directly getting the output, the probabilities were taken as output and they were mapped to each disease in the dataset. The top three disease names were then sent to the flask API along with the disease information.

3.6.8. Final Takeover

Currently, Multinomial Naïve Bayes classifier is used in the system due to its simplicity, less resource-consuming approach. Research is still ongoing on reducing the resource consumption by the neural networks to make it more practical since the outputs obtained by that approach were promising too and can be more scalable and reliable on huge amounts of data. Some hypothetical hybrid models could also be created to reduce the maintenance cost during the training with new data once the scaling problem is solved.

3.7. Retrieving other disease info

After the model was prepared, the next task was to provide some disease information and mapping it to the doctors. No APIs were found for getting the disease-related information. No datasets were also found of the mapping between disease and the disease information. There were two future possibilities: redirecting the user to some other search website/engine using the search query as the disease name but that could be less reliable due to some differences in results among the websites. Another option was to create a dataset with the disease information and the mapping with the doctors. This method was found more reliable since the disease information can be manually checked for information so that correct and relevant information is visible to the users.

A csv file was created containing the mapping between disease name and disease information, treatment information, reference links & specialist doctors for more reliable information for the users. The mapping of the disease with the specialization of the doctors was necessary to search for relevant doctors in the local area. There were several mappings in the project.

3.8. Sending Output to the Flask API

Using the UMLS code, disease name, disease diagnosis, disease treatment, reference links, specialization of doctor required for worse cases were retrieved by the python script and sent to the flask API in a python dictionary format.

3.9. Data Updation phase

Finally, in the data updating phase, the diseases predicted with necessary information and doctor details are displayed in the application for the user to view. Here, the result includes the name of the detected diseases displayed based on probabilities and also the information of the doctors and treatment details. The user can also track the location of the doctor from his present location by using the map tracking feature present in the application.

4. Results

Most of the algorithm approach changes were due to logical reasons and not due to metrics. In the system architecture section, we have discussed in detail what was the reason for choosing each algorithm. Here we have tried to elucidate what exactly were results we had obtained that lead us to our final model. Naïve Bayes (NB) and Neural Networks were used instead of the Decision Tree because of its multiple output prediction capabilities while consuming fewer resources and maintaining the accuracy of prediction. But some key metrics which were also useful like execution time, accuracy were noted and are shown as figure. 5.

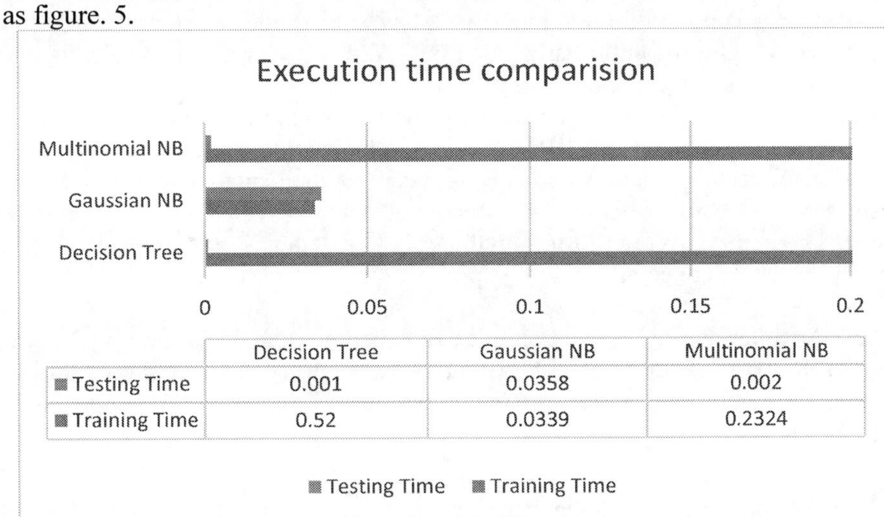

Figure 5. Execution time comparison of ML models

The precision and recall are 1 for 2 of the algorithms which means that this is a hypothetically perfect model which is practically not possible. Here, we get those values due to insufficient training data. For training data, we had 1 test case for each of the disease, and for each disease, symptoms were almost different. Also, during testing, the 6 diseases chosen were exactly predicted with 0.99 probability because they might have almost distinct symptoms compared to other diseases. For testing, all symptoms for that disease except for a random single symptom were used. On further testing, it was also found that the minimum number of symptoms required to predict the disease was different for every disease. Hence, as a common barrier, a threshold of 0.05 probability was set upon which meeting the disease output is shown else the user is asked to enter more symptoms. This way more accurate output can be shown to a user.

As suggested from the above results we need to get more training data which covers more cases and diseases.

Neural Networks seemed promising but due to high resources consumption and inability to accommodate varying inputs puts its hold on use in this case. For now, based on the results and testing a Bayesian approach seems to be the most suitable for this system.

IOP Conf. Series: Materials Science and Engineering **925** (2020) 012038 doi:10.1088/1757-899X/925/1/012038

5. Future work

The ability to improve upon ourselves is the key that has made the human race come so far. And the same applied to our system. We know it is not perfect yet and even in the future will be needing constant changes. Below are some key features and improvements that we are working on bringing in the version of our system or we hope to include these features in our system or similar systems someday.

Neural network models can be created to predict output with more probable outputs. Currently, the probability of the user having a disease is shown as per the number of symptoms he possesses. But this cannot be a proper metric as the disease output may also depend on the severity and also the occurrence of the disease. A neural network model can be created by taking the number of people getting that disease metric into account as well. This way location-based disease output can be predicted which could be more accurate than the current model. Also, the hybrid model could be accommodated for future updating of the models to be less resource consuming and less time-consuming.

As we would like to reach out to more and more people, a voice operation model is what we feel that this system needs. As our main aim is to help people with no knowledge about Healthcare, we notice that many people are unable to read and write, and there are also people with disabilities who cannot operate our system in traditional ways. A voice model can mitigate this problem to some extent.

Moreover, our models will need to adapt to accommodate various local diseases. There are certain diseases common in one are, and most of the time that some symptoms translate to a specific local disease rather than an uncommon foreign one. But it has to be made sure that the uncommon disease that might also have a high probability isn't ruled out. And that can be easily accommodated by our current system as we predict multiple possible diseases.

Currently, we cover over 200 commonly occurring diseases, and in the future, we would like to expand on that as well. We plan to cover at least 1000 diseases in the next version of the system.

Our ultimate aim is to create a system where a patient can get medical assistance on the go. And along with that, there is a vast pool of disease and treatment data is created so that doctors and medical professionals have something to refer to for anything. If this data is stored, managed, and shared with all the data privacy laws in concern, it could help boost the research in the medical field and can save millions of lives.

Lastly, we want our interface to be even simpler. We believe that incorporating Natural Language Processing into our interface can make our system easier to use. One example could be that patient instead of entering the symptoms, can just say in one-two lines, what exactly is happening to them. And our system can handle the rest.

6. Conclusion

In this work, an application of machine learning, cloud, Web-based APIs were studied and successfully implemented for the cause. The machine learning model used was Naïve Bayes classifier on multinomial distribution.

From the results, it can be concluded that the application can be deployed successfully for the benefit of the common crowd to get pre-treatment information and contact a suitable specialist doctor.

Besides, the current implementation, various future aspects and different use cases were also discussed in future works to make this implementation even more accurate.

References

[1] REST API Design Rulebook: Designing Consistent RESTful Web Service Interfaces
[2] Mark M Web Services Architecture *World Wide Web Consortium*. 2016.
[3] Shah H and Soomro TR 2017 Node Challenges in Implementation *Global Journal of Computer Science and Technology*
[4] Flask Foreword. Archived from the original on 2017, pp. 11-17.
[5] Welcome to Flask. https://flask.palletsprojects.com/en/1.1.x/
[6] Amir G and Haider M 2015 Beyond the hype: Big data concepts, methods, and analytics *International Journal of Information Management* **35** 137-144

[7] Kale S, Tamakuwala H, Vijayakumar V, Yang L, and Rawal Kshatriya B 2020 Big Data in Healthcare: Challenges and Promise *Proceedings of 6th International Conference on Big Data and Cloud Computing Challenges. Smart Innovation, Systems and Technologies* **164**
[8] Michael S 2010 SQL databases v. NoSQL databases. *Commun. ACM* **53** 10–11.
[9] Jing H, Haihong E, Guan LE and Du J Survey on NoSQL database *International Conference on Pervasive Computing and Applications, Port Elizabeth* 363-366.
[10] Nayak AP, Poojary A and Dikshay 2013 Type of NoSQL databases and its comparison with relational databases *International Journal of Applied Information Systems* **5** 16-19
[11] Firebase: store and sync data in real-time. https://firebase.google.com/products/realtime-database
[12] Gantz J and Reinsel D The digital universe in 2020: Big data, bigger digital shadows, biggest growth in the far east *IDC iView: IDC Analyze the Future.*
[13] Pankaj S 2013 Cloud Computing: Types, Architecture, Applications, Concerns, Virtualization and Role of IT Governance in Cloud International Journal of Advanced Research in Computer Science and Software Engineering **3** 2013
[14] Buyya R, Chee Y S and Srikumar V 2008 Market-Oriented Cloud Computing: Vision, Hype, and Reality for Delivering IT Services as Computing Utilities, *Department of Computer Science and Software Engineering, University of Melbourne* **9**
[15] The Revolution of Cloud Computing https://www.trianz.com/insights/revolution-that-is-cloud-computing
[16] "Private vs. Public Cloud: What's the Difference?" https://www.expedient.com/knowledgebase/blog/2014-06-05-private-vs-public-cloud-whats-difference/. Date accessed: 15 Jan. 2020.
[17] "Firebase Cloud Storage How-to and Use Case" https://expertise.jetruby.com/firebase-cloud-storage-how-to-and-use-case-dd9810fe041b Date accessed: 15 Jan. 2020.
[18] Mark M 2016 Web Services Architecturem, *World Wide Web Consortium Relationship to the World Wide Web and REST Architectures*
[19] Neumann A, Laranjeiro N, and Jorge B 2020 An Analysis of Public REST Web Service APIs. *IEEE Transactions on Services Computing* 1-1
[20] Chatterjee N, Chakraborty S, Aakash D and Asoke N 2020 Real-time Communication Application Based on Android Using Google Firebase *IJARCSMS,* **6**
[21] "JSONObject by Android developers" https://developer.android.com/reference/org/json/JSONObject. Date accessed: 5 Feb. 2020.
[22] "Decision trees in machine learning" https://towardsdatascience.com/decision-trees-in-machine-learning-641b9c4e8052
[23] Luna JM, Gennatas ED, Ungar LH, Eaton E, Diffenderfer ES, Jensen ST, Simone CB, Friedman JH, Solberg TD, Valdes G 2019 Building more accurate decision trees with the additive tree *Proceedings of the National Academy of Sciences* **116** 19887-19893
[24] Chao S and Fai W 2009 An incremental decision tree learning methodology regarding attributes in medical data mining *International Conference on Machine Learning and Cybernetics* **3** 1694-1699
[25] Hecht N and Robert 1992 Theory of the backpropagation neural network *Neural networks for perception. Academic Press* 65-93.
[26] Kulkarni AD 2001 Computer vision and fuzzy-neural systems *Upper Saddle River NJ: Prentice Hall PTR* 2001.
[27] Goldberg Y 2017 Neural network methods for natural language processing *Synthesis Lectures on Human Language Technologies* **10** 1-309
[28] Cerri R, Barros RC and De Carvalho A 2014 Hierarchical multi-label classification using local neural networks *Journal of Computer and System Sciences* **80** 39-56
[29] Liaw A and Wiener M 2002 Classification and regression by random Forest *R news* ***2*** 18-22
[30] Rish I 2001 An empirical study of the naive Bayes classifier *IJCAI 2001 workshop on empirical methods in artificial intelligence* **3** 22
[31] Schneider KM 2004 On word frequency information and negative evidence in Naive Bayes text classification I*nternational Conference on Natural Language Processing* 474-485

[32] "Naive Bayes- Scikit Learn" https://scikit-learn.org/stable/modules/naive_bayes.html
[33] Less than one doctor for 1000 population in India: Government to Lok Sabha https://economictimes.indiatimes.com/industry/healthcare/biotech/healthcare/less-than-one-doctor-for-1000-population-in-india-government-to-lok-sabha/articleshow/59697608.cms?utm_source=contentofinterest&utm_medium=text&utm_campaign=cppst. Article accessed 15 April 2020.
[34] A critical shortage of lung specialists could leave the NHS struggling to cope this winter - new study. https://www.brit-thoracic.org.uk/about-us/pressmedia/2019/a-critical-shortage-of-lung-specialists-could-leave-the-nhs-struggling-to-cope-this-winter-new-study/. Article accessed 10 January 2020.
[35] Driemeier M, de Oliveira PA, Druzian AF, Lopes Brum LF, Pontes ER, Dorval ME, Paniago AM Late diagnosis: a factor associated with death from visceral leishmaniasis in elderly patients," Pathogens and global health **109** 283-9
[36] Late cancer diagnosis kills 10,000 a year according to government tsar, https://www.theguardian.com/society/2009/nov/29/late-cancer-diagnosis-kills-thousands, Article Accessed 28 March 2020.
[37] Dennis Normile Coronavirus cases have dropped sharply in South Korea. What's the secret to its success? Article accessed 2020.
[38] Shahada S, Abu A, Mohammed Hreiji S and Shamsudheen S 2019 IOT Based Garbage Clearance Alert System with GPS Location Using Arduino *International Journal of MC Square Scientific Research* **11** 1-8
[39] Badawi WA 2019 Underground Pipeline Water Leakage Monitoring Based On IoT *International Journal of MC Square Scientific Research* **11** 01-8.

Machine Learning Techniques to Improve the Success Rate in In-Vitro Fertilization (IVF) Procedure

Patil N Sujata[1], S M Madiwalar[2] and V M Aparanji[3]

[1,2] Department of Electronics & Communication Engineering, KLE Dr M S Sheshgiri College of Engineering & Technology, Belgaum, Karnataka, India
[3] Department of Electronics & Communication Engineering, Siddaganga Institute of Technology, Tumukur, Karnataka, India
[1]sujata.patil@klescet.ac.in, [2]shwetamadiwalar85@gmail.com

Abstract. In Vitro Fertilization (IVF) usually assesses the embryo quality by visual morphological methods to transfer the potential embryo. But the success rate of IVF still remains low because of variations in selection process. The main objective is to improve the implantation rate by predicting the quality of embryos transferred from Day-2 to Day-3. Here using the Machine Learning techniques, thousands of the images trained together for the Day-2, the selection of embryos to come for the further assessment i.e. for Day-3. This will assist the doctors to check for the quality embryo without human intervention. We have also compared the results obtained by our Artificial Intelligence methods precision of >0.98 and also generalizes the method for potential embryo selection. Around 3000 plus embryo images are trained by CNN based Azure model and the results were validated using the Machine Learning techniques. Potentially viable embryo will help improve the implantation and success rate.

Keywords: Artificial Intelligence (AI), Blastocyst, Convolutional Neural Network (CNN), Embryo, In Vitro Fertilization (IVF), Machine Learning.

1. Introduction

In recent years, deep learning has changed the way images are handled by computers. In fact, it has brought in quite a revolution in computer science itself. Deep learning systems mimic working of biological brain. These systems are not programmed but trained to understand and interpret data, much like biological structures. The implementation of the paper work is to improve the success rate by selecting a potentially good embryo. This will be done if a growth of Day-2 embryo is better and the morphological features are defined precisely. The percentage of embryos go to Day-3 depends on the False positive rate of Day-2. Machine Learning technique is used to assess the rate of True Positive and False positive rate [6]. The total number of fragments detected can be neglected when they are considered for Day-3 growth. Here the Deep Learning will help us to predict these good quality embryos

[5]. Combinations of these basic elements in certain specific ways tend to enable them to understand images, predict events, etc. Here we have discussed more on how to improve the success rate by knowing the false positive rate. If the number of embryo images from Day-2 is more than the rejected then the success rate will be improved. Deep learning has changed that. In this research, I have used such systems to understand and interpret embryo images. I have trained these systems with embryo images collected during this research and tested the accuracy of evaluation. Surprisingly the results are exceptionally good [5].
Machine learning and deep learning are related but not same. Experts differ in their definition. So I will record my opinion based on the historic developments in this field. In data science, machine learning refers to certain statistical operations on data to understand complex and big data [6]. These methods include linear regression, random forests, k-nearest neighbours etc., which are essentially statistical procedures. These have been used in data mining, recommendation systems, etc. which have been very effective on the internet. On the other hand, artificial neural networks use brain like structures to interpret data, similar to the way brain interprets data. They have been around for more than 50 years but there was no supporting hardware platform to prove their capabilities. Mobile revolution brought in the change. Mobile devices needed highly number crunching capacity, multi-core parallel execution, and had to work with extremely low electric power. This made large computing capacities to be built at data centres for relatively low cost. This combined with developments like Convolution Neural Networks (CNN) made image recognition [13] an easy task for them. Currently, compute power required to process data on artificial neural networks is available over Internet. Web services have evolved to support image learning capabilities. Currently we have used the web services provided by Microsoft Azure [6].

2. Literature Review

Pegha Khorsavi, Eshan Kazemi April 2019, suggested that the use of Artificial Intelligence (AI) method being trained with thousands of embryo images will precisely predict the quality of embryo image without actually needs to look into the development stages. Here they have implemented an AI approach based on Deep Neural Networks (DNNs) to choose the most viable and good quality embryos using a huge collection of human embryo time-lapse images (about 50,000 images) from a high-volume fertility center in the United States. They developed a framework (STORK) based on Google's Inception model. Here the STORK predicts blastocyst grade with an AUC of >0.98 and generalizes well to images from other clinics outside the US and outperforms individual embryologists [18].
Chavez Badiola, AFS Farias, September 2020 have developed a model comprising of various morphological feature extraction algorithms using digital micrographs, along with other parameters for assessment to predict the viable or the highest grade embryo to predict the implantation rate. Here the system is evaluated using five dissimilar classifiers: Probabilistic Bayesian, Support Vector Machines (SVM), Deep Neural Network (DNN), Decision Tree, and Random Forest (RF), using a K-fold cross validation to assess the model's generalization competencies. In the database A, the SVM classifier achieved an F1 score of 0.74, and AUC of 0.77. In the database B the RF classifier obtained a F1 score of 0.71, and AUC of 0.75. Their results suggest that the system is able to predict a positive pregnancy test from a single digital image [19].
Human Embryo Image Generator Based on Generative Adversarial Networks; here the authors D Dirvanauskas, R Maskeliūnas, V Raudonis Sensors, 2019 used human embryo images obtained during cell development processes for training a deep neural network (DNN). The proposed algorithm used generative adversarial network (GAN) to generate one-, two-, and four-cell stage images [20]. We achieved a misclassification rate of 12.3% for the generated images, while the expert evaluation showed the true recognition rate (TRR) of 80.00% (for four-cell images), 86.8% (for two-cell images), and 96.2% (for one-cell images).
Also CL Curchoe, CL Bormann in January 2019 have discussed about the use of AI & ML methods will help to predict the most viable embryo image which will lead to implantation rate [21]. In this work each and every aspect of patient carefulness was scrutinized, including sperm morphology, sperm identification, identification of empty or oocyte containing follicles, predicting embryo cell stages,

IOP Conf. Series: Materials Science and Engineering **925** (2020) 012039 doi:10.1088/1757-899X/925/1/012039

predicting blastocyst formation from oocytes, assessing human blastocyst quality, predicting live birth from blastocysts, improving embryo selection, and for developing optimal IVF stimulation protocols [5].

Here our methods discuss about the how the machine learning approach will help to assess the quality of embryo images from Day-1 through Day-5. The selection of viable embryo in Day-1 will lead to healthy development in Day-2 and Day-2 to Day-3 so on. Further the growth will be till the blastocyst stage. The percentage of rejection will be more in Day-1 and as it goes further the rejection rate reduces, the blastocyst stage will have more potential embryo selection. This will help to improve the success rate [5]. Most of the literature review discussed above use the Blastocyst stage as the main criteria for assessing the implantation rate by using machine learning techniques. But here in this work we have considered the assessment analysis from Day-1 to Day-5. Here the false positive rate of the respective day embryo images are calculated prior feeding it to the machine learning system for day-2 training process. This procedure will help to achieve the good success rate.

Cleavage Stage

Morphological analysis of the embryo images at the cleavage stage is multivariate and also complex so it is important to characterise the different features of the Day-2 or Day-3 human embryo parameters. In this stage the percentage of fragments will be considered as the number of cell counts so the fragments need to be segmented by morphological assessment. The cleavage stage will include other categorical parameters such as shape of the blastomere cells, cell division and number of cells in a radius of Zona Pellucida (ZP) (i.e. the outer radius of the cell which includes the blastocysts). The rate at which the blastomere cells divide can be observed in association with the morphological cleavage of embryo growth. The rate of true positive cells identified with Day-2 will carry forward to the next stage i.e. Day-3, this will help to improve the success rate of implantation [13]. Also the Day-3 embryo images will have 8-cells and positive true values at this stage will lead to Day-5. Machine learning technique is applied with the attributes considered in all respects of its assessment parameters so that the correct classification is done. As indicated in the figure 1 each of the embryo parameters for the respective days is shown with cleavage and the number of blastomere cells from (a) to (e).

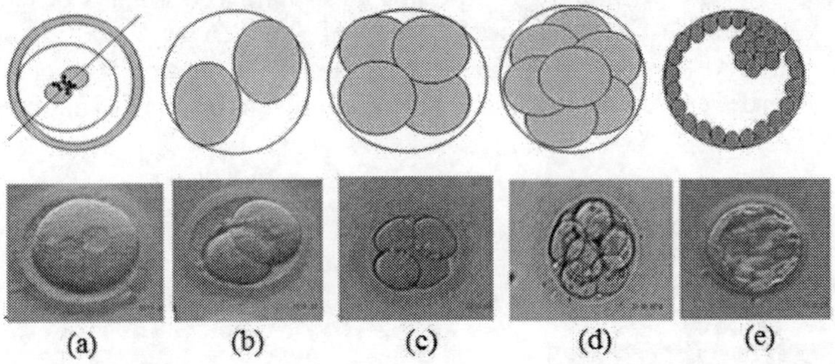

Figure 1: (a) Grading based on presence and symmetry of nucleoli (b) Number of cleaved cells (c) 4-cells regular even shaped sized embryos. (d) Based on the number of fragments in a cell (e) Blastocyst grading on ICM and TE mass.

Implantation Issues

Implantation of the embryo at the women's womb will involve the various physical parameters of the subject where it will be placed after fertilization. Figure 2 shows the embryo implantation procedure where the embryo will be placed into the women's womb using a catheter. Usually the embryo is transferred in the blastocyst stage, but the observation for its development is done for all the five days. Figure 2 shows the stage wise process of transfer. In order to have at least one embryo to be implanted, generally the clinicians will transfer more than two embryos into the women's womb. Usually the transferring of the embryo and how many healthy eggs to be transferred will be decided by the medical experts [6]. These decisions are based on patient's phase and other factors may be related to the fertility centre where the patient is of concern, the country, religious problems etc.

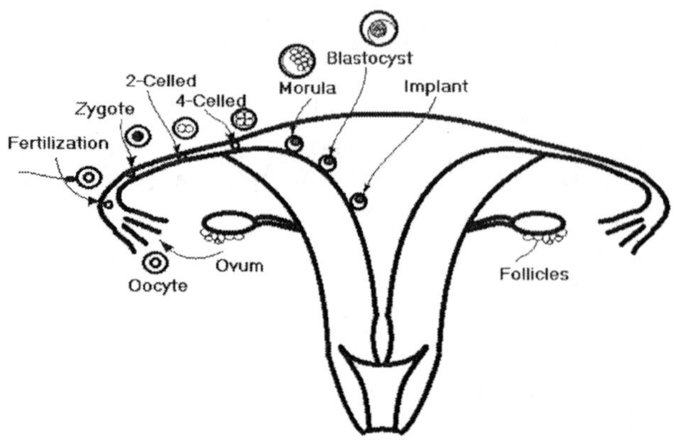

Figure 2 Embryos development and transfer stages

The transfer of multiple embryos will lead to complications for the mother as well as the baby. Sometimes it may present a greater risk for the women's uterus. The health of the mother and baby may be seriously affected this causes the abnormal child, serious injury to the mother etc. So effective measures have to be taken by the infertility clinics to reduce the multiple transfers of the embryos. Automating the embryo selection process is essentially an image processing and image recognition task. It involves capturing a digital copy of the subject (fertilized egg, in our case) and applying various types of filters and feature extraction algorithms to understand the image. For example, we can use edge detection algorithms to highlight boundaries of organs in a biomedical image. Variation in the intensity and color of parts of image can help recognize organs. For example, the light intensity varies as a second order difference across a blood vessel but remains almost constant along the blood vessel. Algorithms to detect circles in an image are useful to detect organs with spherical, ovoid and ellipsoid shaped organs. In recent times, Machine Learning and Artificial Intelligence are changing the way we use computers for image processing.

IVF centre has supported with almost 500 embryo images comprising of around 100 images of each day. All the possible combination of the images is taken for the study of Day-2, Day-3 and Day-5. Normally the microscopic images have different magnification, illumination and also contain overlapping cells [5, 6]. These images were prepared for analysis by resampling and resizing to maintain some uniformity in the input data.

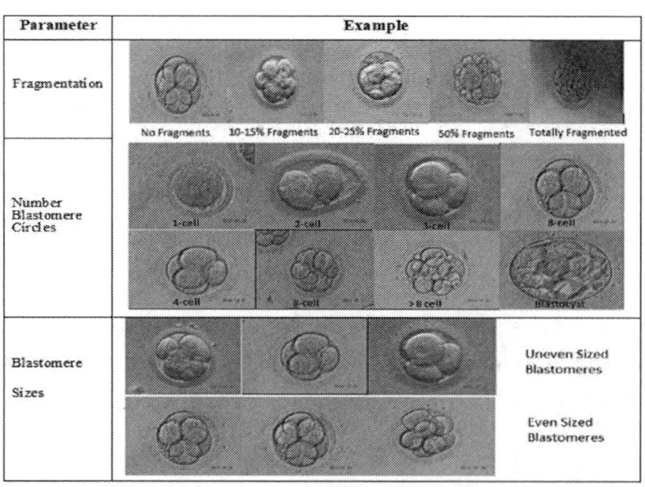

Figure 3: selecting the quality of embryo based on features

Images were divided into two datasets, one of which contained training dataset images. These images were chosen from all the class of sets containing the different category of images such as grades, fragments, sizes of blastomere, etc. The other set included the test dataset images which are used for validating the software results. Around 70-80 % of the dataset is taken for training the system and remaining 20-30% are for the test dataset.

Medical Image Analysis by using Machine Learning techniques are comparatively modern methods for the accurate grading of the embryos. To estimate the grading of the embryos basically these deep learning techniques make use of statistical properties of the data. The prediction rate achieved is more than 85% compared with existing methods of image classification [13]. We will get better result with the use of these statistical properties though there is a very minute difference in the cell boundaries. In many of the techniques discussed we find some error between expected output and applied input. The error is used to tune the machine learning system. Some of the methods popular methods include Regression, Support Vector Machine, Auto encoder, etc. Linear regression is useful for prediction as well as classification. The procedure involves identification of a geometric hyper-plane that separates the data into separate classes. Support vector machines also work on a similar principal but they calculate an exponent of the error so that the separation between the classes can be stretched for easy recognition. Auto encoders find a match between the applied input and existing templates. Implementation of a machine learning algorithm requires study of statistical methods for image processing.

Most of the classification algorithms are processed using the neural networks which help to improve the accuracy of the classifier. ANNs are mainly a pool of layered arrangements used as a scheme of consistent processing elements, intermittently called nodes, which resemble just as that of biological nerves or neurons by function [7]. The figure 4 shows the Neural Network system for classification of images. Here the embryo images are taken from the microscope, pre-processed to for magnification compensation, the training the dataset (i.e. 70% of total samples) with required number of samples then evaluation is done with training iteration. If the model is working for all the dataset, then freeze the network. The weights are being denoted by numerical values and by adjusting the values of these weights we are able to approximate the network with desired function. The weights can be adjusted systematically to draw the specific function [13]. The layers will be fully connected and the weights will be continued further to add on to the final result of the classifier.

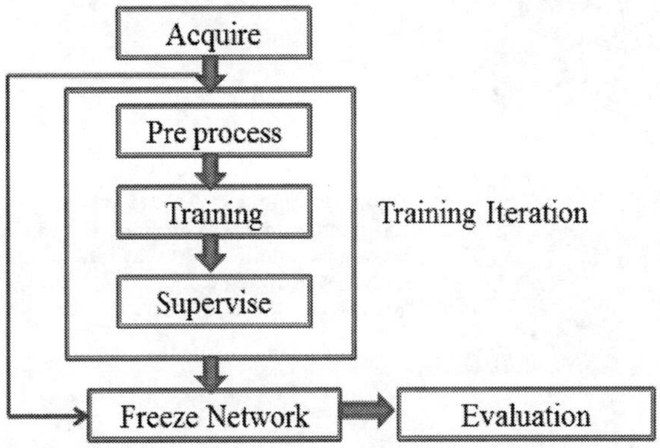

Figure 4: Neural network training system

Here every node connection in the fully connected network proceeds with several different access points and then estimates the result based on a single neuron. Here the output of this node is fed to the next neuron for processing and repeating this procedure will give the final output. The participation layer obtains the estimated inputs and the output layer yields a desired output. Appropriately there is a distinction between each layers in the neural network, the middle layer are called hidden layers.

The Neural network system used to train here involves acquiring the microscopic images; pre-process those using the techniques described earlier [17]. Training process involves uploading of pre-processed images and trains them for little iteration. Then supervise and test it for the obtained dataset. Generally, the deep learning models will directly act on the unprocessed data so the convolutional neural networks help for feature extraction. The Machine Learning techniques will help to predict the grading and proving to be a highly accurate validation process for the prediction of the implantation rate. Figure 3 shows how the how the neural network training can be used as a multilayer perceptron (MLP) model which are designed to have minimum pre-processing. After the convolutional layer the pooling layer which is the maximum pooling layer, the fully connected will have all the activations connected to the previous layer [8-10].

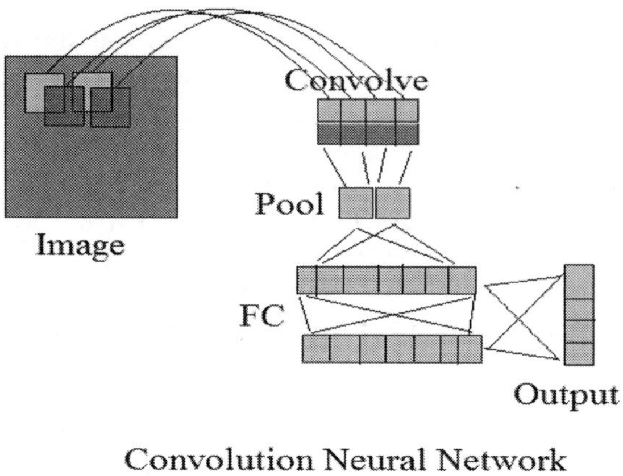

Figure 5: Convolutional Neural Network system

Figure 5 shows the embryo images are given as a input to the CNN model that convolve in an effective manner to produce the relevant data set for the training as well as testing from the 525 embryo images. The result show how these CNN based classifiers will help to improve the success rate of implantation. Our model is able to predict the grading for the respective days of the embryo images with an accuracy precision of 88% and the network to be able to recall at the rate of 87% for the all the features of the embryo images considered.

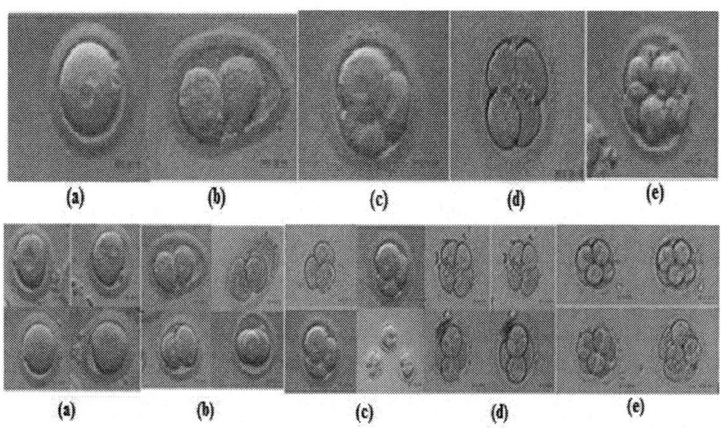

Figure 6: Input to CNN at various stages of embryo development Top row shows one zygote. Second row shows several zygotes in a batch (a) initial stage, (b) Day-1, (c) Day-2, (d) Day-3, and (e). Day-4

As indicated in the figure 6 the position of the embryo cell is not regular it may be placed any corner of the zygote and also we observe that there exist undistinguishable intensity variations. In (b) as seen from the figure 6 sometimes there will be three cells instead of two for day 1.5 thus it may lead to misclassification. So in order to clearly distinguish between the viability of the embryo the false positive cells have to be eliminated at each stage so that it helps to reduce the chances of misclassification.

3. Application of AZURE API

Microsoft also provides an API for the analysis of images with our own attributes; here we can give our features of respective day embryo images for the CNN model. This API helps us to develop the system according to our requirement, test the model for all the available parameters and also we can validate the result with our Machine Learning algorithms developed. Microsoft AZURE uses the Representational State Transfer (REST) API to train the CNN model. Here we can train the model with

the embryo images day wise and validate the result. Before feeding it into the system the images are pre-processed and resized to 255x255 pixel size so as to increase the efficiency of the model. The quality of the embryo images are retained so that there is a clear distinction between the features.
The features of the embryo images are provided with attributes for the day-1 it is the nucleoli and alignment of nucleoli at the pronuclear junction, day-2 or day-3 it is shapes, sizes and the number of fragments with respect to cleavage stage, for day-4 it is morula stage and finally for day-5 it is Trophectoderm (TE) membrane, Inner Cell Mass (ICM). Allowing these to assess the parameters in a appropriate way will help to improve the success rate of implantation.

4. Results & Discussion

Here the results show that refer Table 1, in case of Day-3 embryo images the acceptance percentage is 80.85% which is more than the traditional algorithmic approach. Also the embryo images considered from Day-2 to Day-3 is 76% which indicates that the embryo will be implanted in the blastocyst stage is 80.85%. This will help to assess the quality of embryo to be implanted in Day-5. Even the success rate is comparatively better with Machine Learning Technique. The system involves training with CNNs and Deep Neural Networks (DNN), as a result of which the building of the network starts with the interconnection of neurons as nodes. Training of the network is done rigorously by uploading the embryo images, assign the tags to them with given features, then look for the iteration to mark. Once the iteration is marked the network is ready to use for the validation of results. Work described here focuses on new techniques for human embryo image classification based on a training sequence for the convolutional neural network system.
The results clearly depict as to how the machine learning method for the assessment of the human embryo images helps to predict the grade of the embryo. The AI based CNN model is used to classify these embryos and is encouraging, in particular considering that they have been obtained using a small training set with very few positive samples. For pro-nucleated (initial stage of embryo development) oocytes and embryos, it is probable that other type of descriptors not specifically designed for prevalent textural images might be used. In fact, the alignment and the number of nucleoli or the number and size of blastomeres are not textural features. The two possibilities (dynamic and static observation) might be used together and integrated in a more thorough analysis for practical aims in a normal IVF clinical setting.
In a recent paper discussed by Sarah Armstrong, Priya Bhide that the Time Lapse Imaging Systems will assist the embryologists to assess the quality in a better manner [1]. The Time Lapse Imaging Systems will give the thousands of embryo images with each day assessment as a measuring parameter. But they are much expensive and prone to error as there is no control on assessment. Hence it's better to select the viable embryo by predicting the growth of Day-2 embryo images by discarding the false positive cells.

5. Conclusion

Here the machine learning techniques used for the embryo image classification to eliminate the false positive samples of each day is evaluated and it is seen that it helps to select only the most viable embryo images for the further days. Hence the overall efficiency of the classifier is improved with proposed method. In this paper the machine learning algorithm is trained with more than 100 plus images for each day and the rejection rate for day-1 was 28%, for day-2 it was 17%, day-3 12% and for the final stage blastocyst stage is was found to be less than 5%. Measurement of the accuracy is evaluated by the ration of number of true cells identified to the total number of cells fed to the machine learning model. In the literature review mentioned in section 2 most of the research is done on the blastocyst stage where the embryo features are clearly defined but whereas in the proposed method we have discussed the machine learning algorithm which will assess the quality of the embryo images for the next day to be evaluated. Here the accuracy of the system is increased and also the success rate of the embryo images by more than 85%.

Time points	Day 2				Mc. Nemar	p-value
	Accepted	Rejected	Total	%		
Day 3						
Accepted	63	9	72	76.60	3.5714	0.0587
Rejected	19	3	22	23.40		
Total	82	12	94	100.00		
%	87.23	12.77	100.00			
Day 5						
Accepted	66	10	76	80.85	1.3846	0.3268
Rejected	16	2	18	19.15		
Total	82	12	94	100.00		
%	87.23	12.77	100.00			

Day 5	Day 3				Mc. Nemar	p-value
	Accepted	Rejected	Total	%		
Accepted	60	16	76	80.85	0.5714	0.4497
Rejected	12	6	18	19.15		
Total	72	22	94	100.00		
%	76.60	23.40	100.00			

Table 1: Percentage of Acceptance and Rejection with respect to days

The above table 1 shows percentage of the embryo images accepted and rejected according to the respective days. Above results shows that in case of Day-1 embryo images the acceptance rate is high because the features are not defined clearly. As it goes for the further development the acceptance rate reduced as number of fragments with respect to the abnormal embryo images will be more. Here each fragment is considered as a cell and it will be counted for further assessment. The sample result for the Day-3 and Day-5 embryo images is as shown in the table above. The blastocyst stage is completely developed and it gives a better acceptance rate. Cleavage stage of the embryo will also be considered for the success rate as it gives us the total number of healthy embryos transferred to Day-3. As shown in the graph below it gives the overall assessment of the single embryo selection or a potential embryo implantation [5].
The comparison of these individual days result shows that if a single viable embryo is selected and transferred to women's womb which leads to the implantation definitely the success rate is going to improve. These Machine learning techniques [6] helps us to do these in an easier way. The selection criteria are more flexible and once the training is done it's very easy to implement it on the setup.

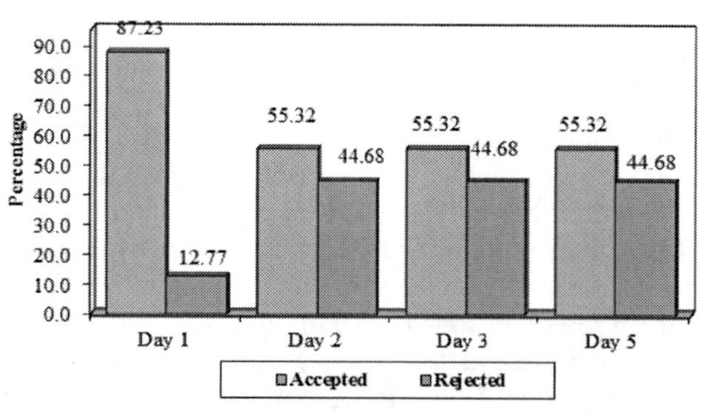

Figure 7: Percentage of acceptance and rejection

Using these statistics, the precision, recall and F-measure can be obtained from the following formula. Here, precision represents the correction rate in detected boundary pixels. Recall represents the detected boundary pixels in the ground truth pixels. F-measure is the harmonic mean of precision and recall and is used as the overall measure of the detector performance. We express the precision, recall and overall measure as:

$$\text{Precision} = \frac{\text{TP}}{\text{TP+FP}} \tag{1}$$

$$\text{Recall} = \frac{\text{TP}}{\text{TP+FN}} \tag{2}$$

$$\text{F\_Measure} = (1+\beta 2)\frac{\text{Recall}*\text{Precision}}{\beta 2\ \text{Recall}*\text{Precision}} \tag{3}$$

$$\text{F-Measure} = \frac{\text{2TP}}{\text{TP+FP+FN}} \tag{4}$$

For β=1, the accuracy of the blastomere cell is calculated as:

$$\text{Accuracy} = \frac{\text{TP+FP}}{\text{Total number of cells detected}} \tag{5}$$

Blastomere ratio is calculated by the number of true cells detected with the actual input cells in blastomere.

$$\text{Blastomere ratio} = \frac{\text{True cells detected}}{\text{Total cells in Blastomere}} \tag{6}$$

The use of Machine Learning algorithm gave excellent results in the analysis of the healthy development of the embryo selection from Day-2 to Day-3 based on the embryo features. Considering the example where all positive inputs are given, learning the concept of "fragments, nucleoli, grade, fertilization count, and number of cells in blastomere, also embryo features: embryo cell, radius, fragments specified, fertilization count and finally the grade.

References

[1] Sarah A, Priya B, Vanessa J 2019 Time-Lapse systems for Embryo incubation and assessment in assisted Reproduction *Cochrane Database of Systematic Reviews* **5**

[2] Meyers S, Burruel V 2019 Equine non-invasive time-lapse imaging and embryo development *Reproduction, Fertility and Development* 31 1874-84

[3] McLennan HJ and Saini A 2020 Embryo evaluation by AI and multi-spectral auto-fluorescence imaging: livestock embryology needs to catch-up to cilinical practice *Theriogenology* **150** 255-262

[4] Bovin J, Bunting L, Collins JA, Nygren KG 2007 Human Reproduction (Oxford, London): International estimates of infertility prevalence and treatment-seeking: potential need and demand for infertility medical care *PubMed Journal* **22** 2800.

[5] Sujata NP, Wali UV, Swamy MK 2016 Automation Technique for classification of Human In-Vitro Fertilized (IVF) Embryos using Digital Image Processing Techniques *International Journal of Technology and Science* **3** 46-49

[6] Sujata NP, Wali UV, Swamy MK 2018 Deep Learning Techniques for Automatic Classification and Analysis of Human In Vitro Fertilized Embryos Journal of Emerging Technologies and Innovative Research (JETIR)

[7] Johansson M, Berg M 2005 Women's experiences of childlessness 2 years after the end of in vitro fertilization treatment *PubMed Journal* **19** 58-63

[8] Emre B, Shevlin M, Sutcliffe A 2010 Reproductive Biomedicine Online: Growth of children conceived by IVF and ICSI up to 12 years of age. *Pediatrics* **126** 270

[9] Robert GE, Johnson MH 2011 Reproductive Biomedicine Online: Robert Edwards: The path to *IVF* **23** 245–262

[10] Malchoff CD and Kobolt U 1991Clinical Chemistry, Principles, Techniques and Correlations *China: Springer* **55** 625
[11] Callum FR, Hall MI and Heesy CP 2006 Were Basal Primates Nocturnal? Evidence from Eye and Orbit Shape. Ross Lab University of Chicago **67** 695-707.
[12] Psaty BM, Heckbert SR, Koepsell TD, Siscovick DS 1995 The risk of myocardial infarction associated with antihypertensive drug therapies *University of Washington, Seattle* **274** 620-5.
[13] Sujata NP, Uday VW, Swamy MK 2016 Application of Vessel Enhancement Filtering for Automated Classification of Human In-Vitro Fertilized (IVF) Images. *International Conference on Electrical, Electronics, Communication, Computer and Optimization Techniques (ICEECCOT),* 27978-1
[14] Saritha KR, Bongso A 2001 Comparative Evaluation of Fresh and Washed Human Sperm Cryopreserved in Vapor and Liquid Phases of Liquid Nitrogen *Journal of Andrology* **22** 857-862
[15] Abdel-Salam AH, Abdel-Bakey NF 2001 Life table and biological studies of Harmonia axyridis Pallas (Col., Coccinellidae) reared on the grain moth eggs of Sitotroga cerealella Olivier (Lep, Gelechiidae) *Journal of Applied Entomology* **125** 455-462
[16] Hussain J, Salam A and Gohar 2001 A Study on the Cryopreservation of Stallion Semen with Alpha Lipoic Acid. *International Journal of Pharmaceuticals* **01**
[17] Samaneh H, Farsi MM, Khafri S 2016 Should We Perform Semen Analysis, DNA Fragmentation, and Hypo-osmotic Swelling Tests together? *International Journal of Molecular and Cellular Medicine* **5** 246–254
[18] Khosravi P, Kazemi E, Zhan Q, Malmsten JE 2019 Deep learning enables robust assessment and selection of human blastocysts after in vitro fertilization *NPJ digital medicine* **2** 1-9
[19] Chavez-Badiola AFS, Farias G, Ruiz M 2020 Predicting pregnancy test results after embryo transfer by image feature extraction and analysis using machine learning *Scientific Reports* **10** 1-6
[20] Dirvanauskas D, Maskeliūnas R, Raudonis V 2019 Hemigen: human embryo image generator based on generative adversarial networks *Sensors* **19** 3578
[21] Curchoe CL, Bormann CL 2019 Artificial intelligence and machine learning for human reproduction and embryology presented at ASRM and ESHRE *Journal of Assisted Reproduction and Genetics* **36** 591-600

IOP Conf. Series: Materials Science and Engineering **925** (2020) 012040 doi:10.1088/1757-899X/925/1/012040

Manufacturing Strategy in the Era of Industry 4.0

PM Kulkarni[1], Rohit R Mutkekar[2], Sanjeev S Ingalagi[3] and Govindraj Mane[4]

[1,3,4] Department of MBA, KLS, Gogte Institute of Technology, Udyambag, Belagavi, Karnataka, India
[2] Department of MBA, Goa Institute of Management, Sattari, Goa, India
[1]pmkulkarni@git.edu, [2]rohitrm@gim.ac.in, [3]ssingalagi@git.edu, [4]grmane@git.edu

Abstract. The evolution of Industry 4.0 has influenced the manufacturing sector with the latest technology for improving the manufacturing process. This study is carried out to understand and upgrade information with regards to small and medium enterprises (SMEs). In the context of the above study, the manufacturing strategies of small and medium enterprises (SMEs) are evaluated for the adoption of Industry 4.0. The main purpose of this study is to identify the opinions of the small and medium enterprises (SMEs) in Belagavi District of Karnataka state on the manufacturing strategy for the adoption of Industry 4.0 technology for improving manufacturing. This research used a structured questionnaire with a sampling method of random sampling with a nominal and Likert Scale of 67 small and medium enterprises (SMEs) in the manufacturing of auto-component products. The data was analyzed through advanced statistical tool Structural Equation Modeling for understanding the results. A major contribution from this study has highlighted the challenges influencing the manufacturing strategy of small and medium enterprises (SMEs) for the adoption of Industry 4.0 and has suggested measures to overcome these challenges through actual application of Industry 4.0 manufacturing strategy

Keywords: Industry 4.0, Strategic Management, Manufacturing Strategy, Automobile Sector, Structural Equation Modeling.

1. Introduction

Seventeenth-century witnessed the first industrial revolution, during this revolution steam engines, water power, and mechanization were witnessed in the various manufacturing sector. While the second industrial revolution witnessed assembly line-based manufacturing. The third industrial revolution was influenced by computers and automation in the manufacturing process. The fourth industrial revolution also called Industry 4.0 which was introduced in the year 2011 at Hannover Fair, Germany. This technology is based on Cyber-Physical Systems which would support the production process of the firms[1]. However, to transform a small and medium scale organization into digitalization, change in manufacturing strategy is essential for the firms. Further, the influence of the business environment influences changes in the acceptance of new technologies, this affects short-term and long-term growth and performance of the firms. Hence, firms need to develop technology-driven manufacturing strategies

for acceptance of new manufacturing technology such as Industry 4.0 [19]. Hence, the manufacturing strategies that are required in Industry 4.0 needs a detailed and inclusive manufacturing roadmap [20]. However, small and medium enterprises (SMEs) have to evaluate their readiness with regards to factors associated with the acceptance of industry 4.0[2]. From the manufacturing strategic perspective, a comprehensive roadmap is required to implement Industry 4.0 in small and medium enterprises (SMEs) [3].The present research aims to offer a framework to understand the readiness of small and medium enterprises (SMEs) for acceptance of Industry 4.0 and its influence on the manufacturing strategy of these firms. This study is unique as it explores potential technical and economic benefits that small and medium enterprises (SMEs) derive through acceptance of Industry 4.0.
This study provides an insight into the challenges that small and medium enterprises (SMEs) experience while transitioning into Industry 4.0. The study also investigates factors that are associated with regards to manufacturing strategies such as investment in machinery, supply chain management, the skill of the workforce, and the role of management in the implementation of the technology in the SMEs. Hence, this study intends to answer the following research questions.

RQ1: What are the challenges for manufacturing strategy for implementation of Industry 4.0 in small and medium enterprises (SMEs)
R*Q2*: What should be the future course action for successful implementation of Industry 4.0 in small and medium enterprises (SMEs)

The research questions are investigated through the application of well-defined structural equation modeling (SEM) method. This method of investigation through SEM provides a relationship with regards to factors associated with manufacturing strategy and readiness of small and medium enterprises (SMEs) for acceptance of Industry 4.0. In general, we developed and analyzed the factors influencing manufacturing strategy for acceptance of Industry 4.0 and the readiness of small and medium enterprises (SMEs) towards Industry 4.0. The remainder of the paper is organized as follows. In the next sections of this paper, we have tried to explain manufacturing strategy, challenges for acceptance of Industry 4.0 from the perspective of SMEs and research design, and methodology with regards to manufacturing strategy through Industry 4.0. An effort has been made in this study to analyze the reliability of the factors selected for the study. This study also provides the profile of the respondents and results are presented through structural equation modeling. In the last section of this paper, the main issues and challenges of acceptance of this technology are presented with a conclusion and further development of manufacturing strategy in the light of Industry 4.0.

2. Study on Manufacturing Strategy and industry 4.0

Manufacturing strategy: "Manufacturing strategy is a competitive tool for enhancing manufacturing productivity". Skinner (1969) defined manufacturing strategy as "exploiting certain properties of the manufacturing function as a competitive weapon" [4]. Likewise, Hayes & Wheelwright (1985) define "manufacturing strategy as a consistent pattern of decision making in the manufacturing function which is linked to the business strategy" [5]. The above definitions mention that manufacturing strategy plays a very important role in the business and provides a competitive advantage to the manufacturing process of the firms. Competitive advantage to firms is derived through the adoption of the latest technology-driven tools of manufacturing. Today's, firms are adopting Industry 4.0 tools to become competitive in all respect. Industry 4.0 can be defined as a technology that is based on design principles and techniques. These design values or principles enables manufactures to forecast the right procedures and solutions and the application of technology provides effectiveness in the process of manufacturing [6]. Therefore, the above discussion indicates that manufacturing strategy would be effective through Industry 4.0. Hence, firm's competitive advantage would be effective through the application of tools of Industry 4.0

2.1 Challenges of adoption of Industry 4.0 in Small and Medium Scale Industries in India

Industry 4.0 leads to significant changes in the manufacturing process of firms. This would entirely change and transform the traditional manufacturing process and operation of the firms [7]. Industry 4.0 adoption in small and medium enterprises (SMEs) firms are influenced by factors like manufacturing strategy, employee skill-sets, cost of implementation in small and medium enterprises (SMEs), level of digitization in the firm, level of digitization of supply chain of the firm, and commitment of top management towards Industry 4.0. These above factors were well documented in the literature of the large and medium scale firms [8]. However, from the perspective of small and medium-scale firms of the Indian industry needs more detailed study to understand the influence of these factors in the adoption of Industry 4.0. Hence, this study has considered these factors and has attempted to investigate the interrelationship between manufacturing strategy and factors influencing small and medium scale firms in developing a manufacturing strategy for implementing Industry 4.0. Table.1 presents the review of the literature on challenges influencing small and medium scale firms in the implementation of industry 4.0. Table.1 presents challenges of adoption of Industry 4.0 in small and medium enterprises (SMEs).

Table 1. Literature Review on Challenges of Implementation of Industry 4.0 in SMEs

Sr.No.	Author and Year	Details	Outcomes
1	Müller, Julian; Maier, Lukas; Veile, Johannes; Voigt, Kai-Ingo,2017	The research focused on collaborating measure between the large, small and medium firms for implementation of Industry 4.0	Most of the Small and medium scale firms are in the early phase of implementation, hence the collaboration with large firms needs more understanding of Industry 4.0.
2	Egon Müller and Hendrik Hopf,2017	The development of competencies would add value in implementing Industry 4.0 in SMEs.	There is a resource constraint with regards to finance, workforce for application of Industry 4.0 in SMEs.
3	Mohammed Bakkari and Abdellah Khatory,2017	Market forces need to be considered while implementation of Industry 4.0 in the SME sector.	Opportunity to identify markets for products produced by SME need to be identified and develop strategic manufacturing roadmap for production to sales of the product produced using Industry 4.0
4	David Grube, Ali A. Malik & Arne Bilberg, 2017	SME sector long duration of implementation due to constraints related to finance and manpower in SMEs.	There is partial research to address the challenges of the implementation of Industry 4.0 in the SMEs sector.
5	Sameer Mittal, Muztoba Ahmad Khan, David Romero, Thorsten Wuest,2018	There are very few SMEs who have reached the maturity of implementation of Industry 4.0	The majority of the research outcomes were focused on large manufacturing units, there is little research with regards to SMEs implementation of Industry4.0.

6	Morteza Ghobakhloo, 2018	Industry 4.0 is strategic for manufacturing companies and therefore firms have to take careful steps for implementation of industry 4.0 in manufacturing.	A strategic roadmap may be influenced by various factors that might be significantly different from large manufacturing firms in comparison to small firms.
7	AlexandreMoeuf, Robert Pellerin, Samir Lamouri, Simon Tamayo-Giraldo& Rodolphe Barbaray,2018	SMEs are fully equipped with regards to the implementation of Industry 4.0 as it influences the production, planning, and control of these firms.	The role of strategic manufacturing might support the development of production, planning, and control measures for successful implementation of Industry 4.0.
8	David Goerzig, Thomas Bauernhansl,2018	The present digital transformation measures undertaken by SMEs are enough due to the lack of clarity with regards to the production process with Industry 4.0	Manufacturing strategy provides a key for the identification of key digital manufacturing transformation and provides support to implement the same in SMEs.
9	Leos Safar, JakubSopko, Slavomir Bednar, Robert Poklemba, 2018	Implementation of Industry 4.0 needs a higher level of digitalization in the manufacturing process of SMEs.	Role of manufacturing strategy provides an opportunity to design and develop methods of digitalization of manufacturing process for effective implementation of Industry 4.0
10	DóraHorváth, Roland Zs. Szabó,2019	Employee skill-sets are an significant feature of the implementation of Industry 4.0 in SMEs.	Strategic manufacturing allows designing, implementation of the strategy to involve in developing skill-sets of employees and prepare them for implementation of Industry 4.0 in SMEs.

3. Methods

Research design is the blueprint for collection, measurement, and analysis of data. Exploratory research design has been used to get more insights on the role of Industry 4.0 in manufacturing strategy of SMEs. The data was collected through the structured questionnaires and the data was collected from small and medium enterprises (SMEs), operating in the foundry industry at Belagavi, Karnataka, India. The structured questionnaire was tested for reliability analysis and results of reliability analysis showed that all six factors that is manufacturing strategy, employee skill-sets, cost of implementation, digitization of small and medium scale firms, digitization of supply chain in small and medium enterprises (SMEs) and top management commitment towards Industry 4.0 showed the reliability of more than 0.89. Reliability results are the summated scores on the scale as predictor components in the study factors. Since summated scales are assembled for interrelated variables level of importance. The test results are considered when they give stable and reliable responses over the test outcomes. The test outcomes are calculated through the method of Cronbach's alpha; this is calculated through a formula as under;

The formula for calculation of Cronbach's alpha for reliability analysis	$\alpha = \frac{N.\bar{c}}{\bar{v}+(N-1).C}$ Where: N = the number of items. $\bar{c}$ = average covariance between item-pairs. $\bar{v}$ = average variance.

Variables derived from test instruments are stated to be reliable only when they offer stable and reliable responses over a repetitive administration of the test. The details with regards to reliability analysis are presented in Table.2. The reliability results of more than 0.80 and above are acceptable for future analysis of the study [9]. The sample for the study was collected from 67 foundries unit's operation in small and medium scale sector at Belagavi, Karnataka. The profile of foundry units is presented in Table.3. The data collected is analyzed through structural equation modelling, which is based on regression analysis. The statistical intervention applied for the study includes structural equation modeling (SEM). This method is a multivariate analysis that is applied to understand the relationships between the variables. In this method factor analysis and multiple regression method is applied to evaluate the relationship between measured variables and latent variables. This method is applied as it provides estimates of manufacturing strategy for Industry 4.0, were in manufacturing strategy technique are interrelated to dependent factor that is manufacturing strategy. The variables under consideration for structural equation modeling (SEM) are based on the fit indices in path coefficient that includes p-value and standard error and model through the root mean square error of approximation and chi-square. The Chi-Square provides results with regards to any discrepancy between variables in the model, any results with less than P-value of 0.05 are accepted indicating that variables in the study are consistent and any value higher than the P-value of 0.05 is rejected indicating that the variables in the study are not consistent. Further, Root Mean Square Error of Approximation (RMSEA) fitness of the study model, value of 0 shows that the model of fit, while more than 0 shows the lack of fitness of the model. The results from RMSEA and SRMR show that the study results are consistent with the study model. Standardized Root Mean Square Residual (SRMR) results of less than 0.09 shows a good model fit. In the structural equation modeling, a comparative fit index is presented covariance matrix the results range from 0.0 to 1.0, higher the value of the comparative fit index better is the model fit. The goodness of fit index indicates sample size fit for the study, results in the range of 0-1.0 is a better fit for the study. Tucker Lewis Index indicates sample size, results from this test need to more than 0.90, which is considered fit for the study results [31,32].

Table 2. Reliability Analysis of Study Variables

Sr.No	Reliability Analysis	
	Variables of the Study	Reliability Scores
1	Manufacturing strategy	0.88
2	Employee skill-sets	0.91
3	Cost of Implementation of Industry 4.0	0.95
4	Digitization of small and medium enterprises (SMEs)	0.92
5	Digitization of supply chain in small and medium enterprises (SMEs)	0.83
6	Top management commitment towards Industry 4.0	0.93

Table 3. Profile of the Study Respondents based on Number of Employees

Sr.No	Profile of the Respondents		
	Profile of the Respondents	N	Percentage
1	100 to 150 employees	15	22.38%
2	70 to 100 employees	28	41.73%
3	50 to 70 employees	7	10.44%
4	Less than 50	17	25.37%
	Total	67	100%

4. Results

The analysis of the study is presented in section four of the study. Table.4 presents a descriptive analysis of the study, while figure: 1 shows the structural equation relationship between the independent and dependent variables of the study. Table.5 shows structural equation fit analysis results and Table. 6 show relationship results with regards to the independent and dependent variables of the study.

Table 4. Descriptive Analysis

Factors	N	Mean	Std. Deviation
Manufacturing Strategy	67	2.52	1.283
Employee Skill-Sets	67	3.13	1.266
Cost Structure	67	3.76	1.195
Digitization of SME'S	67	2.73	1.286
Digitization of Supply Chain of SME's	67	3.10	1.339
Top Management Commitment towards Industry 4.0	67	2.58	1.437

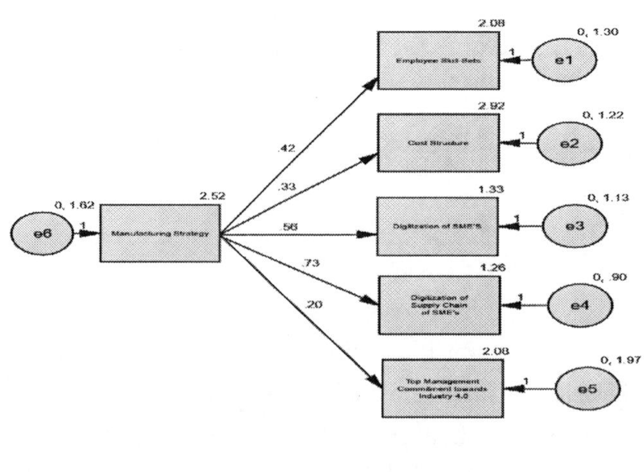

Figure 1. Model Fit Analysis Results

Table 5. Results of Structural Equation Modeling (SEM)

Model Fit Analysis	Obtained Values
CMIN	1182
Df	341
Significance	.000
CMIN DF	1.469
GFI	0.95
NFI	0.92
RFI	0.96
TLI	0.97
CFI	0.95
RMSEA	0.052

The study results show that mean value of (2.52) for manufacturing strategy, employee skill sets is having a mean value of (3.13), cost of implementation has a mean value of (3.76), level of digitization of small and medium scale has a mean value of (2.73), digitization of supply chain management has a mean value of (3.10) and top management commitment has a mean value of (2.58). The structural equation modeling analysis presented in Table.4 shows that CMIN is obtained at (1182), while Degree of freedom is at (341). GFI is ($0.95 > 0.90$) than, NFI is at ($0.92 > 0.90$), RFI is at ($0.96 > 0.90$), TLI is at ($0.97 > 0.90$), CFI is at ($0.95 > 0.90$) and RMSEA is ($0.052 < 0.05$) all the variables are in the acceptable range; hence the model fit is found in the study results.

Table 6. Results of Regression Analysis

Independent		Dependent	Estim ate	S. E	C. R.	P	Label
Employee Skill-Sets	=	Manufacturing Strategy	.417	.110	3.78	000	Supported
Cost Structure	=	Manufacturing Strategy	.334	.107	3.12	.007	Not Supported
Digitization of SME'S	=	Manufacturing Strategy	.556	.103	5.41	.008	Not Supported
Digitization of Supply Chain of SME's	=	Manufacturing Strategy	.730	.092	7.95	000	Supported
Top Management Commitment towards Industry 4.0	=	Manufacturing Strategy	.199	.136	1.46	.142	Not Supported

Results from regression analysis shows that small and medium enterprise (SMEs) have taken effort to develop skill-sets of employees for Industry 4.0, the P value of (000) shows that employee skill-sets development has been included in the manufacturing strategy. While regression results with regards to cost and investment by SMEs has been supported with P value is (0.007) which is more than (0.005) that is confidence level. Further, results with regards to digitization in SMEs have also shown negative results with P-value of (0.008) which is higher than (0.005) of confidence level. There is an acceptance with regards to supply chain management, as most of the study SMEs are the vendor for large automobile manufacturing firms, hence, they have taken effort for partial digitalization of supply chain management, therefore, p-value shows (0.000) which is lower than confidence level of (0.005). Preparedness of top management of SMEs towards Industry 4.0 has shown an P-value of (0.142) which is higher than (0.005), this indicates that top management of these firms need more time to accept and implement the concept of Industry 4.0 in SMEs. The overall analysis of results shows that relationship between dependent variable that is manufacturing strategy and independent regression analysis shows that manufacturing strategy is supported with employee skill-sets, while cost structure, digitization of small and medium scale industries and top management commitment towards industry 4.0 in SMEs is not support. However, level of digitization of supply chain management is supported in the study results.

5. Discussion

The discussion in the study is focused on the research questions developed for the study, which are developed based on challenges of small and medium scale sector for acceptance of Industry 4.0. In this section, the first research question is about the challenges for manufacturing strategy for implementation of Industry 4.0 in small and medium scale firms.
RQ1: What are the challenges for manufacturing strategy for implementation of Industry 4.0 in SMEs

The smart factory is the main theme of Industry 4.0, these smart factories operate manufacturing on Cyber-Physical Systems[10]. Hence, the traditional methodology of manufacturing would change the SMEs operation and manufacturing. This would influence directly on the workforce with regards to skill-sets [11]. These skill-sets include critical thinking and advanced technological skills such as Big data, robotics, machine learning, cyber security[12]. Therefore, the traditional nature of employee engagement and responsibility assessment would change in the organization's security [12]. The study results show that with regards to SMEs employees skill-sets are in the development stage for the implementation of Industry 4.0. Therefore, manufacturing strategy implementation would be supported in this sector. Further, challenge with regards to cost-benefit analysis is an important aspect in the implementation of these technological solutions in the SMEs [13]. Industry 4.0 is expected to large with regards to personnel and hardware. However, the expected benefit of this investment would be in reduced variable costs and enhanced quality and productivity [14]. The study results with regards to cost on implementation showed that negative relationship with regards to manufacturing strategy and cost of implementation of Industry 4.0 in SMEs. The result with regards to digitization in SMEs needs to be evaluated for application of Industry 4.0[15]. In the present situation of study, the level of digitization has not penetrated much in the SMEs. Hence, manufacturing strategy implement for digitization in SMEs would require more time for implementation of industry 4.0. Supply chain operations in Industry 4.0 would be based on Cyber-Physical Systems. This would influence supply chain operations [16]. The traditional supply chain management would change significantly [15]. However, with digitalization in SMEs, there is partial and, in few cases, full digitization of operations in supply chain management. Therefore, the study outcomes have shown that SMEs would be able to implement digitization in supply chain management. However, for complete implementation of the digitization of supply chain management new procedures are needed to manage supply chain management. The role of top management is critical for the success of Industry 4.0[17]. Since the implementation of Industry 4.0 requires extensive firm's consequences and heavy investment support from top management of SMEs [18], the study results have indicated that top management evaluates manufacturing strategy with regards to cost-benefit analysis and investment in industry 4.0 technologies.

R*Q2*: What should be future course actions for successful implementation of Industry 4.0 in SMEs

This paper identified five independent factors and one dependent factor in the use of Industry 4.0 tools and technologies in manufacturing. The study focused on understanding variables concerning manufacturing strategy and its influence on Industry 4.0 in SMEs. The most important factor that influenced the SMEs for implementation of Industry 4.0 is the cost structure, even though long term benefits are witnessed with high-quality products and enhanced productivity, but the investment is much higher for small and medium. This factor of cost has influenced the top management commitment towards Industry 4.0 and thereby influencing the reduction in the digitization of the firms and operations in the firms. Apart, from manufacturing strategy, digitization of the firms brings a higher level of success of Industry 4.0, also employee adaptability to enhance skill-sets to match the needs of Industry 4.0. Therefore, firms need to design strategy based on the predicted changes in the firms in the implementation of Industry 4.0 in the firms.

6. Conclusion

Academic interest in Industry 4.0 has gained a lot of interest among the academicians. The findings from this study would support the future direction of research. Influence of manufacturing strategy to manufacturing units needs deeper studies to understand the influence on Industry 4.0, such studies would support in understanding the success factors of Industry 4.0 on different sectors of the manufacturing industry. Further, the influence of Industry 4.0 would also influence the relationship between business environment and organization and have a direct impact on the value chains, hence future studies need directions on the influence of Industry 4.0 on the value chain of the firms, especially with regards to small and medium-scale firms. The present study was confined and limited with regards to SMEs at Belagavi, Karnataka operating in automobile component sector, further studies in other sectors of SMEs and understand implications on manufacturing strategy in implementing Industry 4.0 would provide broad perspective on challenges of implementation of Industry 4.0 in SMEs.

References

[1] Posada J, Toro C, Barandiaran I, Oyarzun D, Stricker D, de Amicis R, Pinto EB, Eisert P, Döllner J and Vallarino I 2015 Visual computing as a key enabling technology for industrie 4.0 and industrial internet *IEEE Computer Graphics and Applications* **35** 26-40.

[2] Gokalp E, Şener U and Eren PE 2017 Development of an assessment model for Industry 4.0: Industry 4.0-MM *International Conference on Software Process Improvement and Capability Determination* 128-142

[3] Sarvari PA, Ustundag A, Cevikcan E, Kaya I, Cebi S 2018 Technology roadmap for Industry 4.0 *In Industry 4.0: Managing the digital transformation* 95-103

[4] Skinner W 1996 Manufacturing strategy on the "S" curve *Production and operations management* **5** 3-14

[5] Wheelwright SC 1985 Restoring the competitive edge in US manufacturing *California Management Review* **27** 26-42

[6] Liao Y, Deschamps F, Loures EDFR and Ramos LFP 2017 Past, present and future of Industry 4.0-a systematic literature review and research agenda proposal *International journal of production research* 55 3609-3629.

[7] Müller JM and Voigt KI 2018 Sustainable industrial value creation in SMEs: A comparison between industry 4.0 and made in China 2025 *International Journal of Precision Engineering and Manufacturing-Green Technology* **5** 659-670.

[8] Ninan N, Roy JC and Thomas MR 2019 Training the workforce for industry 4.0. *International Journal of Research in Social Sciences* 9 782-790.

[9] Bagozzi RP 1994 Measurement in marketing research: Basic principles of questionnaire design *Principles of marketing research* 1 1-49

[10] Palazzeschi L, Bucci O and Di Fabio A 2018 Re-thinking innovation in organizations in the industry 4.0 scenario: new challenges in a primary prevention perspective *Frontiers in Psychology* 9 30-42

[11] Pinzone M, Fantini P, Perini S, Garavaglia S, Taisch M and Miragliotta G 2017 Jobs and skills in Industry 4.0: an exploratory research *IFIP International Conference on Advances in Production Management Systems* 282-288.
[12] Wolf M, Kleindienst M, Ramsauer C, Zierler C and Winter E 2018 Current and Future Industrial Challenges: Demographic Change and Measures for Elderly Workers in Industry 4.0 *Annals of the Faculty of Engineering Hunedoara-international journal of engineering* 16
[13] Schröder M, Indorf M and Kersten W 2014 Industry 4.0 and its impact on supply chain risk management *In Proceedings of the 14th International Conference Reliability and Statistics in Transportation and Communication* 114-125.
[14] Bauernhansl T, Ten Hompel M, and Vogel-Heuser B 2014 Industrie 4.0 in Produktion, Automatisierung und Logistik: Anwendung-Technologien-Migration *Wiesbaden: Springer Vieweg* **1** 648
[15] Lichtblau K, Stich V, Bertenrath R, Blum M, Bleider M, Millack A, Schmitt K, Schmitz E and Schröter M 2015 Industrie 4.0-readiness *IMPULS-Stiftung, Cologne*
[16] Tan JSK, Ang AK, Lu L, Gan SWQ and Corral MG 2016 Quality analytics in a big data supply chain: commodity data analytics for quality engineering IEEE Region 10 Conference (TENCON), 3455-3463.
[17] Sila I 2013 Factors affecting the adoption of B2B e-commerce technologies *Electronic commerce research* **13** 199-236.
[18] Kagermann H, Wahlster W and Helbig J 2013 Recommendations for implementing the strategic initiative Industrie 4.0 Final report of the Industrie 4.0 Working Group. *Communication Promoters Group of the Industry-Science Research Alliance, Acatech, Frankfurt am Mai*n.
[19] Ivanov, D., Dolgui, A., Sokolov, B., Werner, F. and Ivanova, M. (2016), "A dynamic model and analgorithm for short-term supply chain scheduling in the smart factory Industry 4.0", International Journal of Production Research, 54, 386-402.
[20] Sarvari PA, Ustundag A, Cevikcan E, Kaya I and Cebi S 2018 *Technology roadmap for Industry 4.0", in Ustundag, A. and Cevikcan, E. (Eds), Industry 4.0: Managing the Digital Transformation*, 95-103
[21] Müller J, Maier Lukas, Veile J, Voigt KI 2017 Cooperation strategies among SMEs for mplementing industry 4.0, In: Kersten, WolfgangBlecker, *International Conference of Logistics (HICL)* **23** 301-318
[22] Egon M, Hendrik H 2007 Competence Center for the Digital Transformation in Small and Medium-Sized Enterprises *Procedia Manufacturing* **11** 1495-1500
[23] Bakkari M and Khatory A 2017 Industry 4.0: Strategy for more sustainable industrial development in SMEs. *In Proceedings of the IEOM 7th International Conference on Industrial Engineering and Operations Management* 11-13.
[24] Grube H, Malik D, Ahmad A, Arne B 2017 Generic Challenges and Automation Solutions in Manufacturing SMEs *Proceedings of the 28th DAAAM International Symposium, pp.1161-1169, B. Katalinic (Ed.)* 11-2
[25] Sameer M, Muztoba AK, David R, Thorsten W 2013 A critical review of smart manufacturing & Industry 4.0 maturity models: Implications for small and medium-sized enterprises (SMEs) *Journal of Manufacturing Systems* **49** 194-214
[26] Ghobakhloo M 2018 The future of manufacturing industry: a strategic roadmap toward Industry 4.0 *Journal of Manufacturing Technology Management* **29** 910-936
[27] Alexandre M, Robert P, Samir L, Simon TG and Barbaray R 2018 The industrial management of SMEs in the eraof Industry 4.0 *International Journal of Production Research* **56** 1118-1136
[28] David G, Thomas B, 2018 Enterprise Architectures for the Digital Transformation in Small and Medium-sized Enterprises *Procedia CIRP* **67** 540-545
[29] Safar L, Sopko J, Bednar S and Poklemba R 2018 Concept of SME business model for industry 4.0 environment *TEM Journal* **7** 626

IOP Conf. Series: Materials Science and Engineering **925** (2020) 012040 doi:10.1088/1757-899X/925/1/012040

[30] Dóra H, Roland ZS 2019 Driving forces and barriers of Industry 4.0: Do multinational and small and medium-sized companies have equal opportunities *Technological Forecasting and Social Change* **146** 119-132
[31] Hu LT and Bentler PM 1999 Cutoff criteria for fit indexes in covariance structure analysis: Conventional criteria versus new alternatives *Structural equation modeling: a multidisciplinary journal* **6** 1-55
[32] Bentler PM and Bonett DG 1980 Significance tests and goodness of fit in the analysis of covariance structures *Psychological bulletin* **88** 588

Plant design, optimization, and ethanol production from *Zea maize* L.

S Keerthi[1], D M Gurumurthy[1], N M Sneha[1], S Bhavana[1], A S Shwetha[1], N G Abhishek[1] and S P Kaushik[1]

[1] Department of Biotechnology, GM Institute of Technology, Davanagere, India
E: mail: keerthis@gmit.ac.in

Abstract. Ethanol production from corn grain is taken a point of convergence as an alternative fuel development recently. Herein, we studied the production of ethanol from Zea maize L with designing the production plant and optimization process, respectively. As an additional source, the maize grains are processed to a fine flour and then exposed to liquefaction, where the temperature was maintained at 85 ° C for 4 hours. Furthermore, it was exposed to saccharification for 4 hours at 50°C with subsequent refining. The ethanol yield acquired from the process was observed to be 10.6, with the purity of 95%. The distinctive ethanol properties are likewise evaluated with the specific gravity, relative thickness and the refractive index were 0.79, 0.85, and 1.4, respectively.

Keywords: Amylase, fermentation, glucoamylase, saccharification, Saccharomyces cerevisiae.

1. Introduction

Energy is humans fundamental prerequisite, and it is a definitive gift for all essentials. Fuels from food serve as a great deal of vitality for regular day to day existences in various manners. The more significant part of our vitality prerequisites is met from non-inexhaustible sources such as non-renewable energy sources [1]. As a result of the wasteful usage of the world's oil assets, we have confronted a few emergencies before and are confronting a comparative circumstance soon. Bioethanol is viewed as one of the most reasonable fuel where numerous nations, including the United States, Brazil, and China, have permitted the mixing of bioethanol into fuel [2-4]. At present, most bioethanol is created from starch or sugar-based feedstock, for example, corn grain (US and China) and sugar stick (Brazil) [5].

Maize is not just utilized as food and feed, besides as feedstocks for sustainable fuel ethanol creation [6]. Fuel ethanol production through organic aging of sugars removed from sugar-rich yields, (for example, sugar stick) and boring harvests, (for example, maize, and wheat) is effective in practice today. Ethanol from lignocellulosic materials is still in the early formative/trail stages [6, 7]. Maize production in India accounts for the third-largest harvest after rice and wheat. Maize that gives food, feed, and grub and fills in as a wellspring of essential crude materials for a few mechanical items, principally starch, maize oil, maize syrup, mixed drinks, beauty care products, and biofuel [8, 9]. Harvest yields held pace with quickly expanding worldwide populace. The available biomass can be used for improved cellulosic innovations like plant rearing that guarantee to improve both the quality and amount of grain. Because of the profitability, India now the giant producer of the maize. This pattern is probably going to proceed, and the overabundance from the food gracefully can be utilized to make sure about us on the vitality stage and to expand the future fuel request of the nation as an individual and overall [1].

2. Methodology

Ethanol is created by fermenting 6-carbon sugars (principally glucose) through the glycolytic pathway. Corn starch is utilized as a feedstock for hydrolysis, where glucan is changed over to glucose by catalyst hydrolysis and glucose aging to ethanol by yeast [6]. The detailed process was described in Figure 1 with slight modification described previously [5, 6, 10].The mature pound is isolated into ethanol and deposits (utilized for feed creation) by refining. Fine corn flour is mixed with water to form slurry called "mash." The liquefaction of corn flour is carried out in a 1-liter flask with 500 g total mixture. α-amylase is added, then the temperature of cornflour slurry is reduced to 85 °C and is maintained for 4 hours at a waterbath. The saccharification by the glucoamylase enzymes, the glucose syrups were produced by the hydrolysis of remaining oligosaccharides.

Furthermore, the conditions were maintained at the temperature of 55–60°C and pH 4.2–4.5 for glucoamylase activity. After the saccharification of the mash, the Mash was cooled and transferred to a fermenter. *Saccharomyces cerevisiae* was added to the fermentation process, where the whole process requires 48-72 hours, with the concentrate rate up to 10-12% of ethanol. Further, separation of the solid and liquid phase of the mash, the distillation process was performed. The remaining liquid content in the Mesh was removed by dehydration [5], and the available DDGS can be used as animal feed (Figure 2).

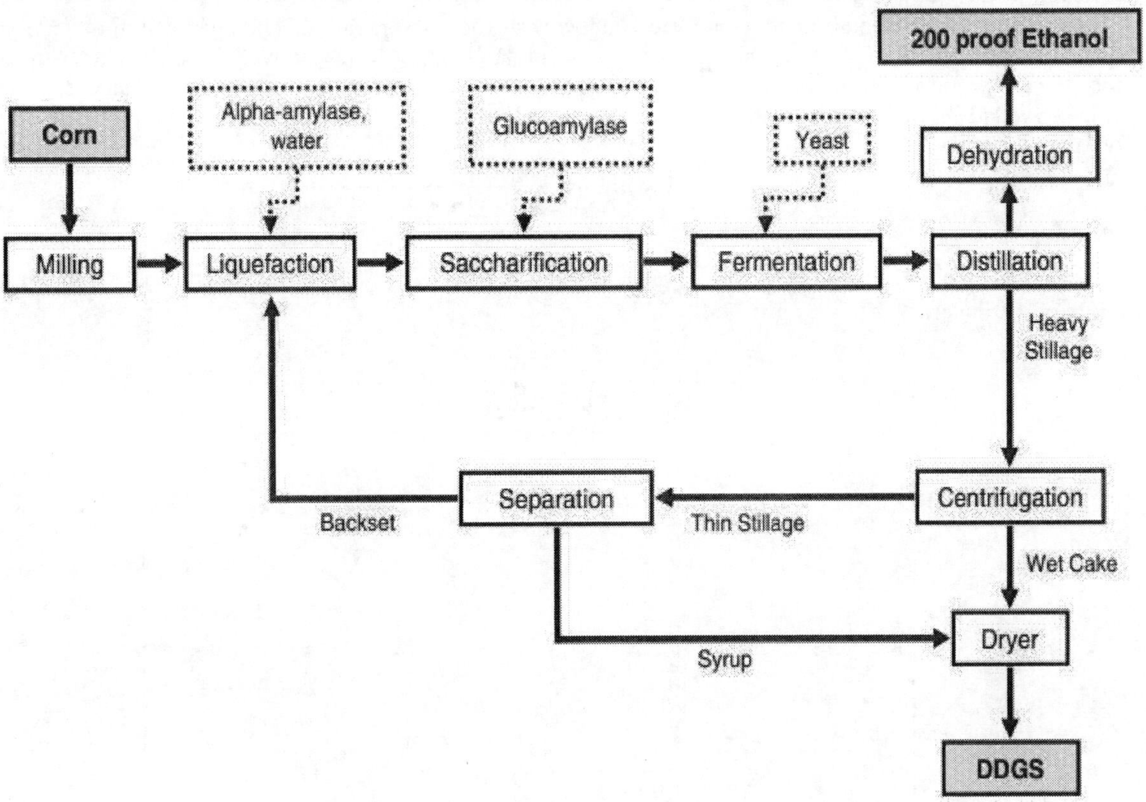

Figure 1. Process followed for ethanol production

3. Plant Design

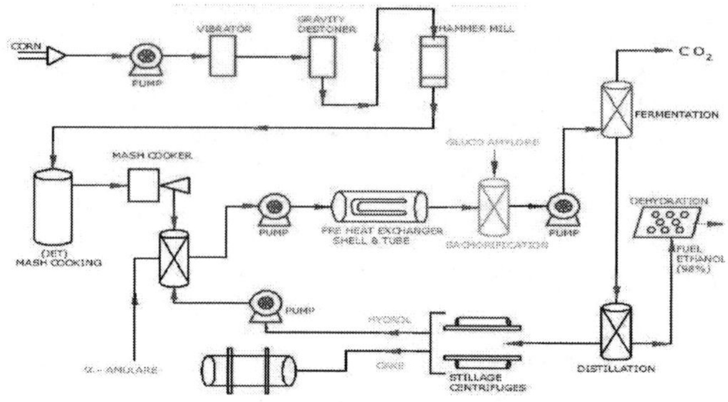

Figure 2. Plant design for ethanol production

4. Result and Discussion

All our critical experimental procedures were conducted at room temperature. The level of bio-ethanol yield (%) and specific gravity (SG) was calculated by the formula (1) and (2), respectively. The parameter for each process was carefully considered, as recorded in figure 3. The values of the quantitive test for alcohol product was analyzed (Table 1). A comparative analysis with standard alcohol with the current alcohol production was presented the table 2.

$$\%\text{yeild} = \frac{\text{weight of sample (corn)used}}{\text{weight of distillate (ethanol)}} \quad (1)$$

Where the weight of sample: 885g; the weight of beaker:55.1g; the weight of distillate beaker: 129.72g; the weight of distillate:129.72 – 55.1g = 83.49g

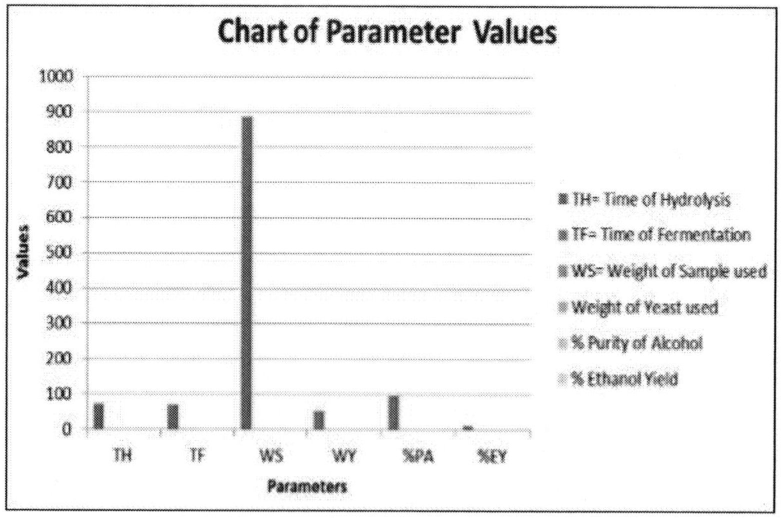

Figure 3. Process scaled on the parameters against achieved values

Specific gravity

$$S.G\ of\ ehanol = \frac{weight\ of\ ethanol}{weight\ of\ water} \quad (2)$$

IOP Conf. Series: Materials Science and Engineering **925** (2020) 012041 doi:10.1088/1757-899X/925/1/012041

Weight of SG bottle: 7.9g; weight of S.G bottle +water : 29.87g; weight of S.G bottle + ethanol: 23.6g S.G of ethanol : 23.6/29.87 : 0.79

Table1. Values for the quantitative test of alcohol

Parameters	Values
Temperature (°C)	29
pH	9.2
Refractive index	1.35
Specific gravity	0.85
Relative density	0.73
Boiling point (°C)	783.7
Melting point (°C)	-1114°C
State at 25°C	Liquid
Color	Colorless
Odor	Characteristics

Table 2. Comparative quantitative analysis of standard and experimental alcohol produced from Maize

Parameters	Standard value	Experimental
Temperature	27	29
pH	9.2	7.30
Refractive index	1.40	1.35
Specific gravity	0.79	0.85
Relative density	0.8	0.73
Boiling point (°C)	780-787	783.70
Melting point (°C)	-1150 °C	-1114°C
Colour	Colorless	Colorless

The production of ethanol from corn by utilizing the straightforward refining process was adopted. The available corn grains are gathered as a crude material, which was initially hydrolyzed for 72 hours to convert starch to maltose completely. At the aging time frame of about 96 hours, the complete conversion of all the glucose, fructose absolutely, and sucrose into cell vitality and in this way to deliver ethanol and carbon dioxide as metabolic waste pro-weight. The total amount of corn utilized for the production of ethanol was 885 grams. This critical measure of corn test was utilized to guarantee that a considerable amount of ethanol was extricated. The amount of *Saccharomyces cerevisiae* was added is 52.2g to help advance the pace of aging of corn. The level of ethanol produced was evaluated to be 10.6 percent. The low yield of ethanol shows that the corn has less ethanol content in contrast with potatoes, where the ethanol yield was observed at 38 to 45% [11]. Moreover, corn as a crude material for ethanol production could not be affordable because of its low rate yield. The level of virtue of liquor production was resolved to be 95% [8]. Furthermore, the ethanol from potato has a parentage virtue of 95%, which is similar to corn ethanol. This observation determined that ethanol from corn has a similar caliber to that of potato. The quantities analysis it was revealed that the temperature of the liquor was 29 ° C, in contrast to RT of 27 ° C. The pH was esteemed at 9.2, showing that ethanol is exceptionally basic. With the specific gravity (S.G) of 0.79, ethanol from corn is slightly heavier than water (0.85). Other analysis, such as boiling point (78°C), and melting point (1114°C) shows that the liquor is mostly unpredictable. At the temperature of 25°C, the ethanol was found to be at a liquid state.

5. Conclusion

India is a developing economy and is the fourth largest client for petroleum product on the planet, with the imports of around 70%. The mission for ecologically economical and sustainable power source alternatives like ethanol can be mixed with gas that could be utilized as biofuel. India, on the other hand, the largest in the world for maize production. This reason India developing alternative fuels form the plant product that is grown. Besides, comparative production from different harvests which contains more starch also makes their place as a source for alternative fuel production. Maize, herein we produced, can also be used for the production of ethanol, which can benefit economically as well as for the value-added products.

References

[1] Pallavi SDCSM, RS SAKRK 2011 Corn to Ethanol: Retrospect's and Prospects Maize: Leading to a new Paradigm
[2] Gnansounou E 2016 Production and use of lignocellulosic bioethanol in Europe: Current situation and perspectives *Bioresource technology* **101** 4842-4850
[3] Moreira JR, Pacca SA and Parente V 2014 The future of oil and bioethanol in Brazil *Energy Policy* **65** 7-15
[4] Sheehan J 2001 The road to bioethanol: a strategic perspective of the US Department of Energy's national ethanol program *ACS Publications* 2-25
[5] Yu J, Xu Z, Liu L, Chen S. Wang S, Jin M 2019 Process integration for ethanol production from corn and corn stover as mixed substrates *Bioresource technology* **279** 10-16.
[6] Schwietzke S, Kim Y, Ximenes E, Mosier N, Ladisch M, Ethanol Production from Maize, 347-364.
[7] Zaldivar J, Nielsen J, Olsson L 2001 Fuel ethanol production from lignocellulose: a challenge for metabolic engineering and process integration, *Applied microbiology and biotechnology* **56** 17-34
[8] Mbonu O, Udeozor P, Umoru G, Uti D 2016 Production of Bio-fuel from sweet corn (food to fuel), Journal of Pharmacognosy and Phytochemistry **5** 43
[9] Persson T, Garcia AG, Paz J, Jones J, Hoogenboom G 2009 Maize ethanol feedstock production and net energy value as affected by climate variability and crop management practices Agricultural Systems 11-21.
[10] Cardona CA, Sánchez ÓJ 2007 Fuel ethanol production: process design trends and integration opportunities *Bioresource technology* **98** 2415-2457
[11] Lareo C, Ferrari MD, Guigou M, Fajardo L, Larnaudie V, Ramírez MB, Martínez-Garreiro J Evaluation of sweet potato for fuel bioethanol production: hydrolysis and fermentation *SpringerPlus* **2** 493.

Fatigue and Facture Behaviour of Al6061-Al_2O_3 Metal Matrix Composite: Effect of Heat Treatment

Mahendra H M[1], G S Prakash[2], Keerthi Prasad K S[3] and Rajanna S[4]

[1,2] Department of Industrial Engineering Management, MSRIT, Bangalore, India.
[3] Department of Mechanical Engineering, VVIET, Mysore, India.
[4] Department of Mechanical Engineering, GECK, Kushalanagara, India
[1]Mahendra.hun@gmail.com

Abstract. Metal matrix composites of 6061Al - Al_2O_3 matrix are associated as standard materials inside the space of flying, car and marine applications to move forward the physical properties. Particulates of Al_2O_3 with 40 µm, 60 µm &100 µm were reinforced utilizing Die casting method. Next stage of fortress is being changed from 0-16wt% in steps of 4wt%. Microstructural characterization has been conducted for the composites to check the homogeneous movement of particles. Fatigue and Facture Toughness properties of 6061Al - Al_2O_3 composite have been inspected. Effect of fortress on the Disintegration, Shortcoming and Facture toughness properties was considered carefully. A considerable enhancement in Fatigue and Facture Toughness properties was observed by the incorporation of particulates. Formation uniformly dispersed particulates in the matrix was observed for the composite up to 12% filler composition.

Keywords: Al6061, Al_2O_3; Die- Casting; Mechanical Properties; Corrosion test; Fatigue test; Facture Test.

1. Introduction

Immaculate metals of aluminum have slight quality and we cannot utilize with specific reason wherever extreme to twisting and burst is essential extra components are settled to base metals for surge control. The prime mechanical properties of aluminium6061 are achieved in alloying trappings and warm .medicines. This invigorates the establishment of small warm rushes through line by signal of relocation and improve its mechanical properties. Commonly utilized aluminum combination for mechanical application is Al6061 due to its ravishing wide-ranging properties those are small thickness, incredible quality, ductility sturdiness and strife to fatigue. It takes remained broadly expended in airplane essential components and extra endlessly focused on basic application.

Al6061 combination is as well defencelessness to push oxidization angry. It is happened by homogeneity of the 6061alloy and natural persevering stresses associated by its innovation approaches. The point of the work are to look at the gear of warm conduct on the microstructure, unbending nature of aluminum 6061metal and aluminum 6061/Al2O3 MMC's. A blend of Al2O3including to aluminium6061 amalgam and extra reinforcing is accepted by the way of possible reserves of assistant advances the mechanical properties of this 6061alloy primarily for temperature tradition. So, presently this examination on the properties of heat treatment is utilized as a assets of developing the mechanical properties of aluminum 8% Al_2O_3 alloy. Aluminum 6061 is utilized in a kind of cast and fashioned frameworks and circumstance of heat treatment.

Die casting procedure is a reasonable and low-cost technique to convey MMCs. Other than being essential, versatile, and appealing, as compared with other strategies, it licenses the era of components in broad sum. Other than, this sort of planning is by commercial utilize for the course of action of particulate Al-based composites Al_2O_3alloy.Aluminium 6061 are utilized in a kind of a cast and made systems and circumstance of heat treatment.

The Metal matrix composites ensure a wide choice of orchestrating applications. They have been broadly organized for mechanical devices ask to headway the mechanical, chemical and warm presentation of mechanical components. [1] But, its small break quality [2] limits this one utilization in various applications. Copious effort has been truly in instruction to move forward the facture sturdiness of Al_2O_3. A curiously approach is to upgrade another fortifying organize they are particles, hairs and strands to theAl6061- Al_2O_3system [3–6].

In show disdain toward of the reality that, these managing with offerings certain challenges with troublesome course of action organize, more costs, difficult to urge a typical dissemination of the other organize. The break strength of Al6061- Al_2O_3 composite materials can be made strides by the combination of metal components by a plastic execution sharp on the Al6061- Al_2O_3organize [7-11]. In formerinvestigationAl6061- Al_2O_3composites was organized to recognize extraordinary mechanical properties and awesome break strength. Mishra et al. observed the microstructure of break sturdiness Al6061- Al_2O_3 arrange composites by divergent extents of alumina as the diluents [12].

Mixed mode fatigue crack growth experiments were performed using the compact tension shear specimens made of Al 6061-T6 alloy for mode mixity angles of 30°, 45°, and 60° [13]. Based on the overall performance of the models during the entire crack growth regime, the results of the present investigation clearly show that Irwin's model and one of the Tanaka's model were consistently found to predict the mixed mode fatigue life close to the experimental data. Whereas, Richard's and Yan's models, again based on the overall performance, are found to be conservative models consistently for the prediction of the mixed mode fatigue life.

The effect of building orientation on the very-high-cycle fatigue (VHCF) response of Ti-6Al-4V specimens produced through selective laser melting (SLM) process with three different building orientations (0°, 45° and 90°) was experimentally assessed [14]. Size of defects induced fatigue failures and the stress intensity factor range decrease with the number of cycles to failure. The building orientation significantly influences both the defect size and the resulting VHCF response.

Crack closure induced by laser peening (LP) and its effect on fatigue life extension of 2024-T351 aluminum alloy with initial fatigue crack was studied [15]. The fatigue life of the LP-treated pre-cracked specimen was 412% and 44.7% higher than the untreated specimen and LP-treated specimen without pre-crack, respectively. The crack closure around the pre-crack resulting from the LP-induced plastic deformation is mainly responsible for fatigue life extension.

The high-temperature fatigue behaviour of a heat treated AlSi7Cu3Mg alloy was reported [16]. Fatigue testing has been performed at room and higher temperatures. The fatigue strength decreases by increasing the testing temperature from 473 to 573 K. The uniform distribution of fine strengthening precipitates in the α-Al matrix is affected by higher test temperature, which promotes coarsening phenomena of the precipitates. The failure mode changes from brittle to ductile at 573 K.

Ceramic coatings were produced on AA7075-T6 alloy by micro-arc oxidation (MAO) [17]. Effects of duty cycle on microstructure and fatigue behavior of the coated samples were investigated. The influence of the residual stress relaxation on the fatigue life at high and low cyclic stresses was explored. Defects on coating surface including pores and cracks decreased the fatigue life of the substrate.

IOP Conf. Series: Materials Science and Engineering **925** (2020) 012042 doi:10.1088/1757-899X/925/1/012042

In the present investigation focused on the fatigue properties of Alumina reinforced Al6061 alloy.

2. Material

In Table 1 as shown Composition of matrix metal 6061Al alloy. In Table 2 and 3 are represented the mechanical and physical properties of matrix metal 6061Al alloy respectively. The alumina properties are depicted in Table 4.

Table 1. Chemical composition of 6061

Mg	Si	Fe	Cu	Cr	Mn	Zn	Ti	*Al*
0.9	*.62*	*0.33*	*0.28*	*0.17*	*0.06*	*0.02*	*0.02*	*Balance*

Table 2. Mechanical Properties of Al6061

Property	**Values**
Yield strength(MPa)	110
Ultimate strength (MPa)	207
Elongation (%)	16
Hardness (BHN)	75
No of cycle	2000/min

Table 3. Mechanical Properties of Al6061

Property	**Values**
Density	2.7 g/cm^3
Melting Point	580°C
Modulus of Elasticity	70-80 GPa
Poisson's Ratio	0.33

Table 4. Properties of alumina

Property	**Values**
Density (Gm/cc)	3.89
Porosity (%)	0
Elastic Modulus (GPa)	375
Shear Modulus (GPa)	52
Bulk Modulus (GPa)	228
Poisson's ratio	0.22
Fracture Toughness(MPa)	4
Hardness	1440
Compressive strength(MPa)	2600

3. Experimental

3.1 Die Casting

Die casting method, the stable die and it is constructed with steel. The molten metal pouring into mould cavity with high pressure and high speed. It takes duplet upright platens on which boosts are situated

IOP Conf. Series: Materials Science and Engineering **925** (2020) 012042 doi:10.1088/1757-899X/925/1/012042

which add the die shares. In duel platen, one die platen is fixed and additional die platen is employed for released and locked die. Inserted the necessary metal into the shot sleeve and then presented into the casting die by a hydraulically-driven piston. When the metal has hardened, the mould cavity is opened, and removes the casted metal. Thus, Al6061/Al_2O_3 composite with frequent dissimilar compositions (0, 4, 8, 12 and 16wt %) was attained rectangular solid shaped bars with dimension of 25mm and a length of 150mm.

3.2 Facture Toughness Test

The facture toughness testing of the composite material models were prepared according to ASTM E8 standard, through the using of three point bending machine in figure 2. The dimension of the models was 10mm width and 4mm thickness and length 80mm as exposed in below figure 1. Facture test were examined in Raghavendra Metallurgical centre peenya, Bangalore.

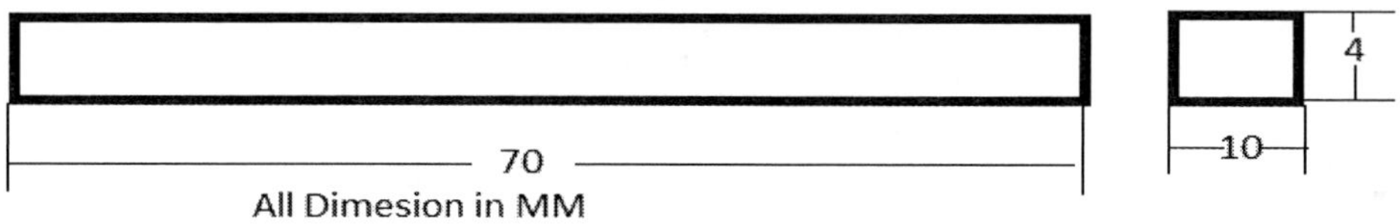

Figure 1. Facture test specimen according to ASTM standard.

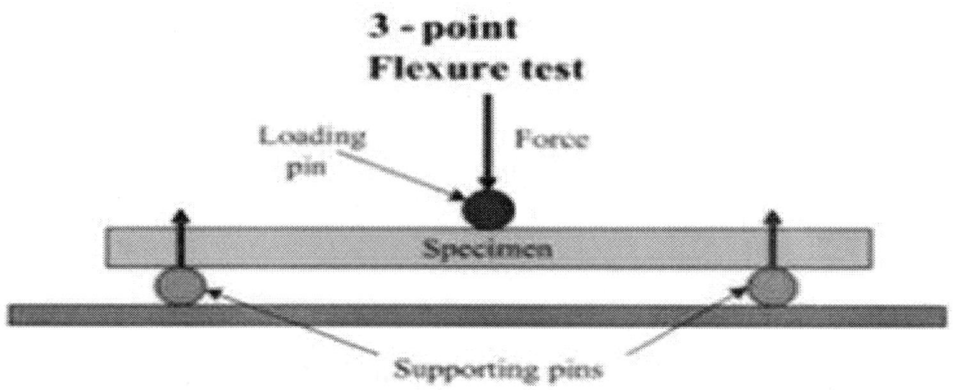

Figure 2. 3-Point Bending test.

3.3 Fatigue Test

The fatigue testing of the MMC samples were prepared as per the ASTM E446 standard, with the use of digital testing rotating bending fatigue test machine. The samples with dimension of diameter 10mm and 6mm and measure length 100mm are shown in Figures 3. The samples were introduced for test with a speed of 1500rpm and temperature range from 100 to 300. Fatigue test was investigated in Dr. Ambedkar Institute of Technology, Bangalore.

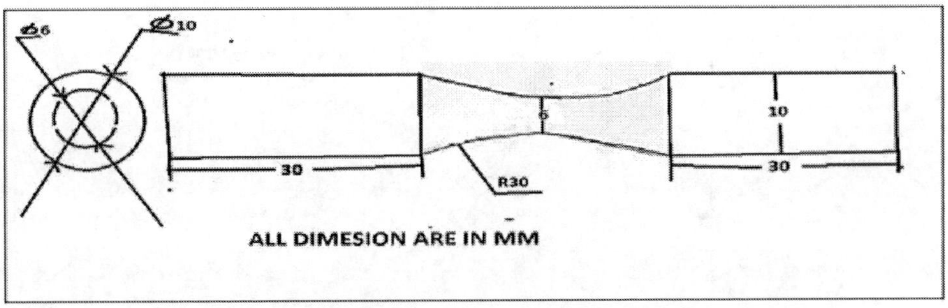

Figure 3. Fatigue test specimen according to ASTM standard.

4. Results and Discussion

4.1 Without Heat Treatment of Facture Test

Table 5. Facture Test Result of Al6061/Al_2O_3 composite.

Microns	Reinforcement (%)	Facture stress in MPa	% of Elongation
40	0	118.25	12
	4	146.45	23
	8	173.28	62
	12	198.32	125
	16	182.46	112
60	4	132.26	17
	8	158.45	53
	12	179.21	118
	16	167.52	103
100	4	118.85	13
	8	142.12	46
	12	169.23	112
	16	150.82	97

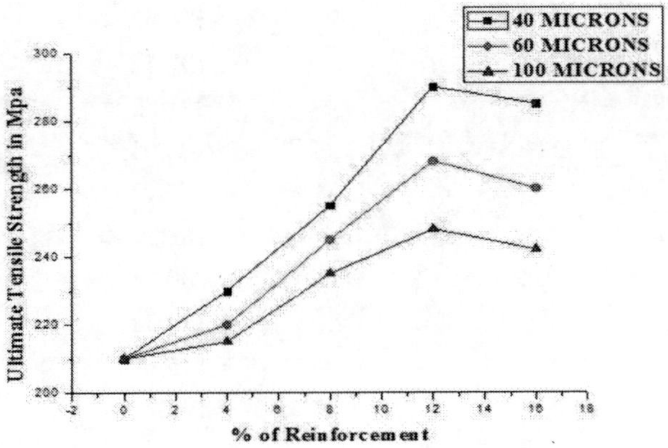

Figure 4. Ultimate Tensile Strength V/S Reinforcement of Al6061/Al_2O_3 composite.

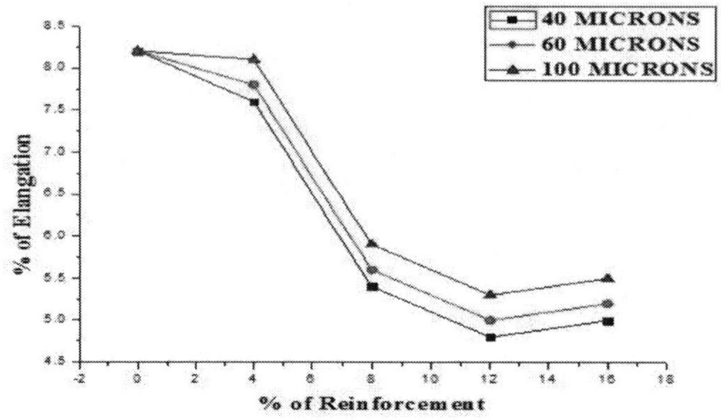

Figure 5. % of Elongation V/S Reinforcement of Al6061/Al_2O_3 composite

Above plot provides the details about the Facture toughness of prepared composite material. There is substantial modify in the graph for composite materials differentiate to pure base material. Increase in all the fatigue properties can be obviously looking in the figure 4&5. Adding of particulates enhance the stability of base material. A reduction in elongation at break is observed by the adding of particulates. This is due to the interaction developed between matrix and filler.

Facture toughness properties such as shown in Table 5. The values are improved step by step up to 12 % of particulates in the aluminum matrix. Further, the values found to be reduced. Agglomeration of particulates or reduction in the consistency of dispersal reduces the Facture toughness properties of the composites at higher loading stage. From result we can conclude 12% of 40 microns particle has a good facture toughness properties compare to other two particles such as 60 and 100. Here we can select 12% of 40 microns particles aircraft and automobile application.

4.2 With Heat Treatment of Facture Test

Table 6. Facture Test Result of Al6061/Al_2O_3 composite.

Microns	Reinforcement (%)	Facture stress in MPa	% of Elongation
40	0	142.45	23
	4	170.52	35
	8	197.52	73
	12	221.82	138
	16	203.65	121
60	4	155.42	29
	8	182.24	68
	12	200.54	132
	16	189.32	115
100	4	145.32	25
	8	167.28	62
	12	188.44	124
	16	172.38	110

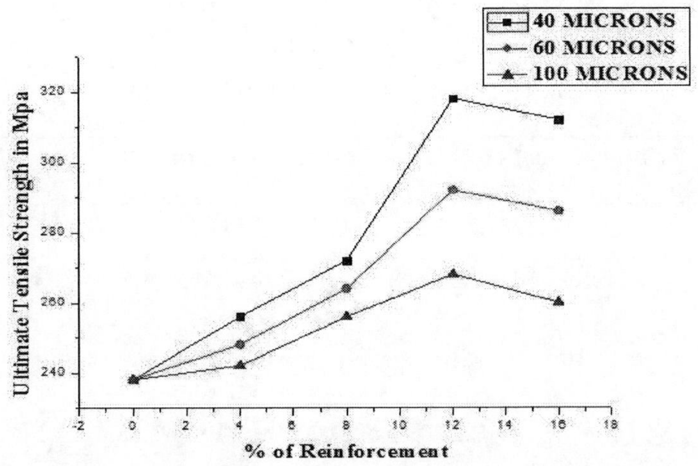

Figure 6. Ultimate Tensile Strength V/S Reinforcement of Al6061/Al_2O_3 composite

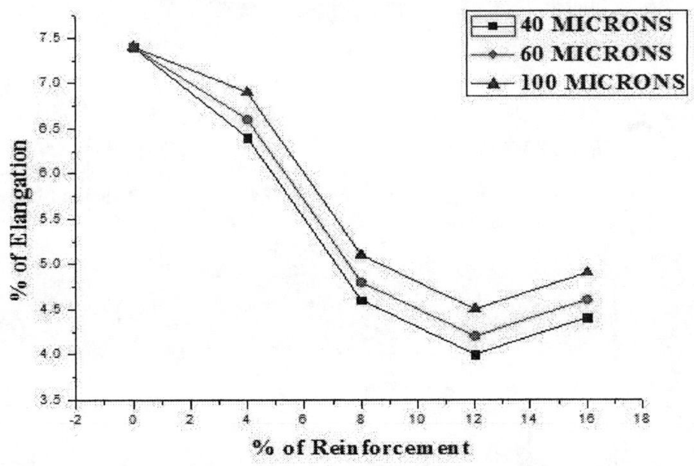

Figure 7. % of Elongation V/S Reinforcement of Al6061/Al_2O_3 composite

After Hear treatment above plot provides the details about the Facture toughness of prepared composite material. There is substantial modify in the graph for composite materials differentiate to without heat treatment of the Al6061/Al_2O_3 Composite. Increase in all the fatigue properties can be obviously looking in the figure 6 &7. Addition of Heat increases the stability of Al6061/Al_2O_3 Composite. A reduction in elongation at break is observed by the addition of heat. Addition of heat increases the bonding between matrix and filler.

Facture properties such as shown in Table 6. From result we can conclude 12% of 40 microns particle has a good Facture toughness properties compare to other two particles such as 60 and 100. Here we can select 12% of 40 microns particles aircraft and automobile application.

4.3 Without Heat Treatment of Fatigue Test

Table 7. Fatigue Test Result of Al6061/Al_2O_3 composite.

Microns	Reinforcement (%)	Fatigue stress in MPa	Number of Failure Cycle X 10^4
40	0	118.25	12
	4	146.45	23
	8	173.28	62
	12	198.32	125
	16	182.46	112
60	4	132.26	17
	8	158.45	53
	12	179.21	118
	16	167.52	103
100	4	118.85	13
	8	142.12	46
	12	169.23	112
	16	150.82	97

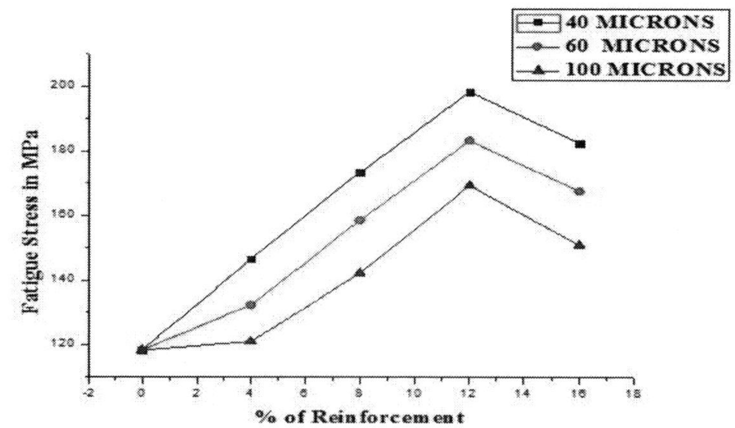

Figure 8. Fatigue stress V/S Reinforcement of Al6061/Al_2O_3 composite

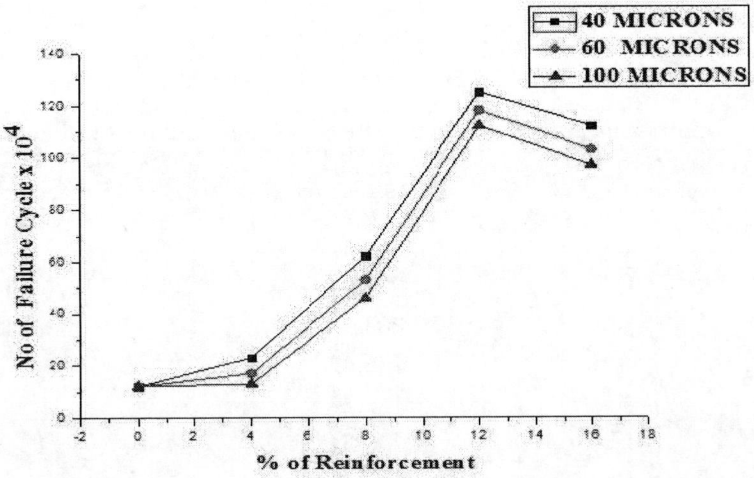

Figure 9: No of failure cycle V/S Reinforcement of Al6061/Al_2O_3 composite

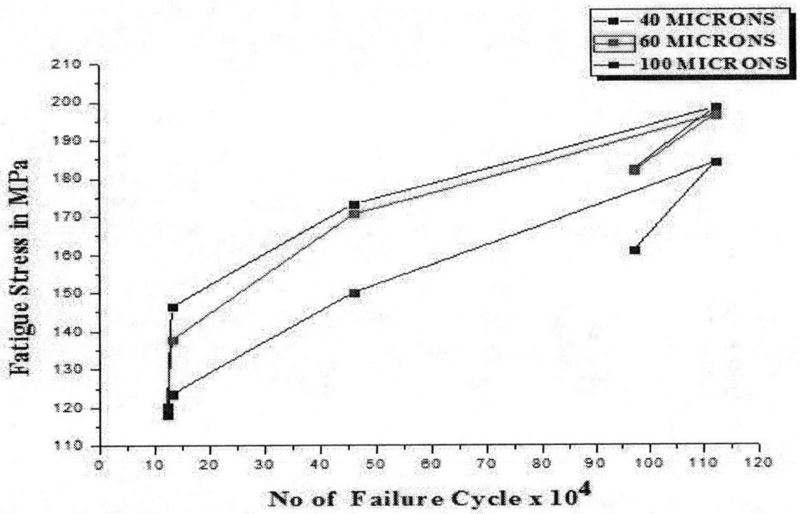

Figure 10: Fatigue Stress V/S No of failure cycle of Al6061/Al_2O_3 composite

Above plot provides the details about the fatigue properties of prepared composite material. There is substantial modify in the graph for composite materials differentiate to pure base material. Increase in all the fatigue properties can be obviously looking in the figure 8, 9 &10. Adding of particulates enhance the stability of base material. A reduction in elongation at break is observed by the adding of particulates. This is due to the interaction developed between matrix and filler.

Fatigue properties such as shown in Table 6. The values are improved step by step up to 12 % of particulates in the aluminum matrix. Further, the values found to be reduced. Agglomeration of particulates or reduction in the consistency of dispersal reduces the Fatigue properties of the composites at higher loading stage. From result we can conclude 12% of 40 microns particle has a good facture toughness properties compare to other two particles such as 60 and 100. Here we can select 12% of 40 microns particles aircraft and automobile application.

IOP Conf. Series: Materials Science and Engineering **925** (2020) 012042 doi:10.1088/1757-899X/925/1/012042

4.4 With Heat Treatment of Fatigue Test

Table 8. Fatigue Test Result / of Al6061Al_2O_3 composite

Microns	Reinforcement (%)	Fatigue stress in MPa	Number of Failure Cycle X 10^4
40	0	142.45	23
	4	170.52	35
	8	197.52	73
	12	221.82	138
	16	203.65	121
60	4	155.42	29
	8	182.24	68
	12	200.54	132
	16	189.32	115
100	4	145.32	25
	8	167.28	62
	12	188.44	124
	16	172.38	110

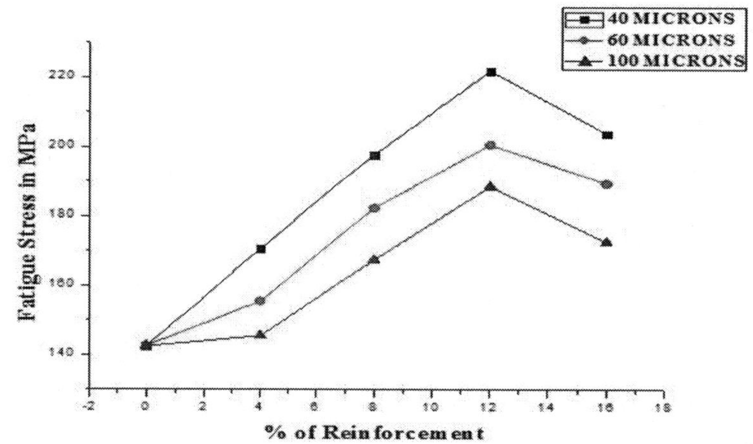

Figure 11. Fatigue Stress V/S No Reinforcement of Al6061/Al_2O_3 composite

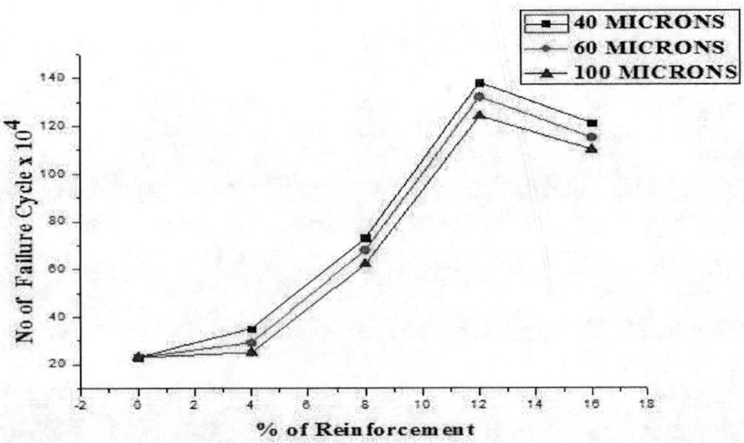

Figure 12. No of failure cycle V/S Reinforcement of Al6061/Al_2O_3 composite

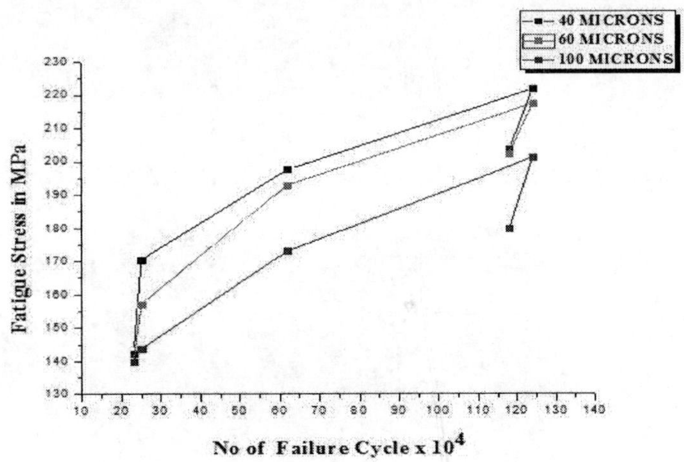

Figure 13. Fatigue Stress V/S No of failure cycle of Al6061/Al_2o_3 composite

After Hear treatment above plot provides the details about the fatigue properties of prepared composite material. There is substantial modify in the graph for composite materials differentiate to pure base material. Increase in all the fatigue properties can be obviously looking in the figure 11, 12 & 13. Addition of Heat increases the stability of Al6061/Al_2O_3 Composite. A reduction in elongation at break is observed by the addition of heat. Addition of heat increases the bonding between matrix and filler.

Fatigue properties such as shown in Table 8. From result we can conclude 12% of 40 microns particle has a good Fatigue properties compare to other two particles such as 60 and100 Here we can select 12% of 40 microns particles aircraft and automobile application.

4.5 Microstructure studies by SEM

Figure 14 displays the Fatigue & Facture Test SEM picture of Al60601/Al_2O_3metal matrix composite. A well microstructure is seeing in all Al60601/Al_2O_3 metal matrixes composite and it shows a better scattering of particulates during the matrix.

IOP Conf. Series: Materials Science and Engineering **925** (2020) 012042 doi:10.1088/1757-899X/925/1/012042

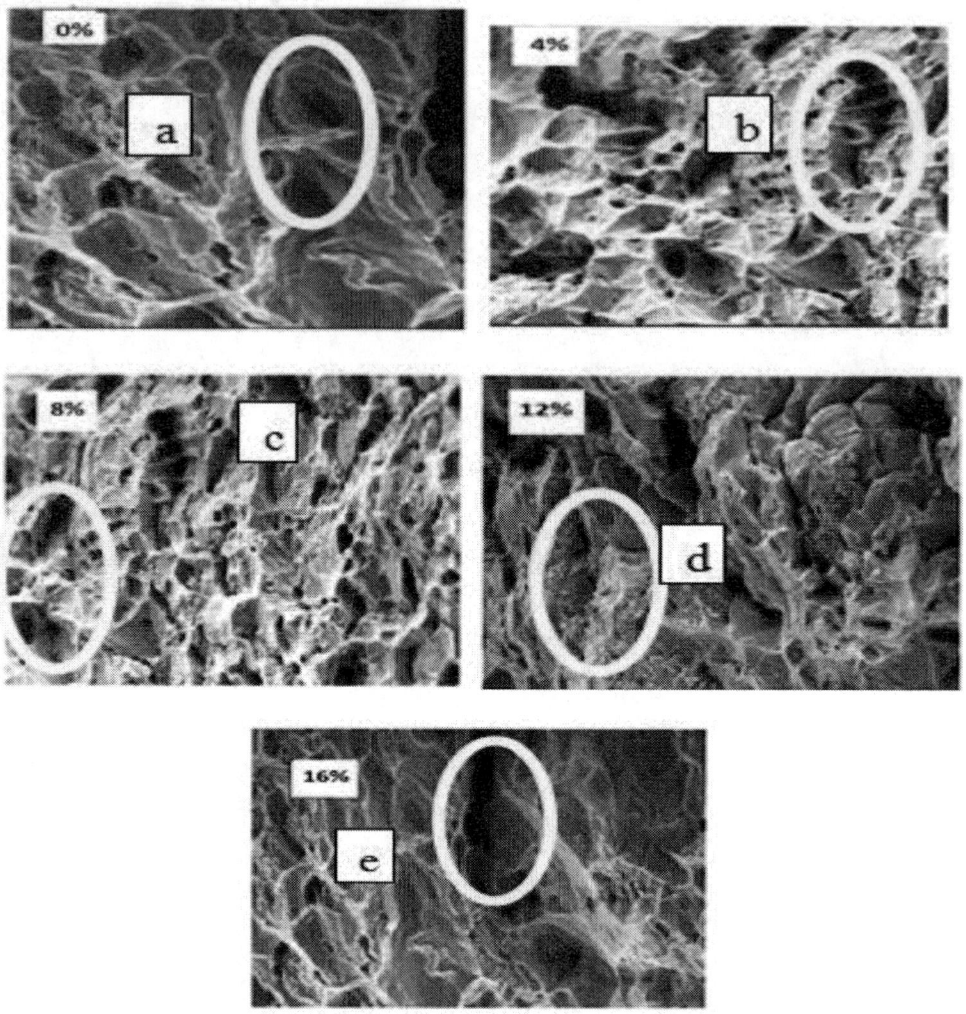

Figure 14. Fatigue & Facture Test SEM Images (a - e) of Al6061/Al_2O_3 composite

a) Cup cone pattern b)Trans granular facture c)Trans granular facture d)Inter granular facture e) River pattern

5. Conclusion

Three-point bending test of an aluminum 6061 reinforced particulate Composite impregnated with Al_2O_3 are studied to determine the flexural stress, flexural strain and maximum bending strength. In the present work, aluminum 6061 and Al_2O_3 particles are mixed in different compositions. It is observed that the composite with a composition of 4%Al_2O_3enhances the Facture and Fatigue properties. The Fatigue property and Facture stress is greatly influences by Al_2O_3. Aluminum composite material having good mechanical properties on increasing the percentage of alumina. Composites with Al2O3particles up to 12% increase the tensile strength 290MPa and elongation 4.8% in aluminum alloy matrix containing Cu-Zn -Mg. Finally, it can be concluded that the composite material developed in the present investigation in found to be good in terms of mechanical properties. SEM microphotographs clearly show the nature of dispersion and it reflects in the mechanical properties. Heat treatment increases the stability of the resulting composites due to the enhanced interaction between filler and matrix phases.

IOP Conf. Series: Materials Science and Engineering **925** (2020) 012042 doi:10.1088/1757-899X/925/1/012042

References

[1] Chan HM 1997 Layered ceramics: processing and mechanical behaviour *Annu. Rev. Mater. Sci.* **27**, 249–282

[2] Lugovy M, Slyunyayev V, Orlovskaya N, Blugan G, Kuebler J and Lewis, M., Apparent *fracture toughness of Si3N4-based laminates with residual compressive or tensile stresses in surface layers. Act. Mater,* 2005,**53**, 289– 296.

[3] Wang XM, La PQ, Wang BJ and Yang GL 2016 Toughening effect of ZrB2 in Al2O3-ZrB2 nanocomposite ceramics *Rare Metal. Mater. Eng.* **45** 1714–1718.

[4] Trejo-Arroyo D, Zárate-Medina J, Alvarado-Orozco JM, Contreras-Garcia ME, Boldrick MS, Muñoz-Saldaña J 2013 Microstructure and mechanical properties of Al2O3-YSZ spherical polycrystalline composites *J. Eur. Ceram. Soc.* **33** 1907–1916.

[5] Backhaus-Ricoult M, Eveno P 1993 Creep properties of an alumina-zirconia composite reinforced with silicon carbide whiskers, *J. Eur. Ceram. Soc.* **11** 51–62

[6] Qu XY, Wang FC, Shi CS, Zhao NQ, Liu EZ, He CN, He F 2018 In situ synthesis of a gamma-Al_2O_3 whisker reinforced aluminium matrix composite by cold pressing and sintering *Mater. Sci. Eng. A* **709** 223– 231

[7] Masanta M, Shariff SM, Choudhury AR Evaluation of modulus of elasticity, nano-hardness and fracture toughness of TiB2–TiC–Al2O3 composite coating developed by SHS and laser cladding *Material Science and Engineering: A* 528 5327

[8] Park SJ, Jang YS 2001 Effect of Chemical Treatment of Kevlar Fibers on Mechanical Interfacial Properties of Composites *Journal of Colloid and Interface Science* **237** 91.

[9] Prabu SB, Karunamoorthy L, Kathiresan S, Mohan B 2006 Influence of Stirring speed and Stirring Time on Distribution of Particles in Cast Metal Matrix Compopsite *Journal of Materials Processing Technology* **171** 268-273

[10] Donthamsetty S, Rao N 2010 Investigation on Mechanical Properties of A356 Assisted Cavitation, *Journal ofMechanical Engineering* **41** 2010

[11] Zamani MH, Baharvandi H 2011 Fabrication and studying the mechanical properties of A356 alloy reinforced with Al203-10% ZrO2nano particles through stir casting *Advances in Materials Physics and Chemistry* 26-30

[12] Kheder ARI, Marahleh GS, Al-Jamea DMK 2011 Strengthening of Al by SiC, Al2O3 and MgO *Jordan Journal of Mechanical and Industrial Engineering* **5** 533-541.

[13] Sajith S, Murthy KSRK, Robi PS 2020 Experimental and numerical investigation of mixed mode fatigue crack growth models in aluminum 6061-T6 *International Journal of Fatigue* **130** 105285.

[14] Guian Q, Li Y, Paolino DS, Tridello A, Berto F and Hong Y Very-high-cycle fatigue behavior of Ti 6Al-4V manufactured by selective laser melting: Effect of build orientation *International Journal of Fatigue* **136** 105628.

[15] Hu Y, Cheng H, Yu J, Yao Z 2020 An experimental study on crack closure induced by laser peening in precracked aluminum alloy 2024-T351 and fatigue life extension *International Journal of Fatigue* **130** 105232

[16] De Mori A, Timelli G, Berto F, Fabrizi 2020 A High temperature fatigue of heat treated secondary AlSi7Cu3Mg alloys *International Journal of Fatigue* **138** 105685

[17] Da W, Hao J, Li C, He D, Jia D, Zhang Y, Tan Z 2020 Residual stress relaxation and duty cycle on high cycle fatigue life of microarcoxidation coated AA7075-T6 alloy *International Journal of Fatigue* **130** 105283

[18] Krishna LR, Madhavi Y, Sahithi T, Wasekar NP, Chavan NM, Rao DS. 2018 Influence of prior shot peening variables on the fatigue life of micro arc oxidation coated 6061–T6 Al alloy. I*nt J Fatigue* **106** 165–74

[19] Madhavi Y, Krishna LR, Narasaiah N 2019 Influence of micro arc oxidation coating thickness and prior shot peening on the fatigue behavior of 6061–T6 Al alloy *Int J Fatigue* **126** 297–305

[20] Krishna LR, Madhavi Y, Sahithi T, Rao DS, Ijeri VS and Prakash O 2019 Enhancing the high cycle fatigue life of high strength aluminum alloys for aerospace applications *Fatigue Fract Eng Mater Struct* **42** 698–709
[21] Ye Z, Liu D, Zhang X, Wu Z, Long F 2019 Influence of combined shot peening and PEO treatment on corrosion fatigue behavior of 7A85 aluminum alloy *Appl Surf Sci* **486** 486:72–9
[22] Wang Q, Liu X, Wang W, Yang C, Xiong X, Fang H. Mixed mode fatigue crack growth behavior of Ni-Cr-Mo-V high strength steel weldments *Int J Fatigue* **102** 79–91
[23] Demir O, Ayhan AO, Iric S, Lekesiz H. Evaluation of mixed mode-I/II criteria for fatigue crack propagation using experiments and modeling Chin J Aeronaut **31** 1525–34
[24] Sajith S, Murthy KSRK, Robi PS 2019 Prediction of accurate mixed mode fatigue crack growth curves using the Paris' law *J Inst Eng Ser C* **100** 165–74
[25] Sajith S, Murthy KSRK, Robi PS 2018 A simple technique for estimation of mixed mode (I/II) stress intensity factors *J Mech Mater Struct* **13** 141–54
[26] Qian G, Jian Z, Pan X and Berto F 2019 In-situ investigation on fatigue behaviors of Ti-6Al- 4V manufactured by selective laser melting *Int J Fatigue* **133** 105424
[27] Xu S, Zhu S, Hao Y and Qian G 2018 A new critical plane-energy model for multiaxial fatigue life prediction of turbine disc alloys *Eng Fail Anal* **93** 93:55–63
[28] Hu Y, Wu S, Wu Z, Zhong X, Ahmed S, and Karabal S 2020 A new approach to correlate the defect population with the fatigue life of selective laser melted Ti-6Al-4V. *Int J Fatigue* **136** 105584
[29] Liu X, Chen E, Zeng F, Cong T and Domblesky J 2019 Mechanisms of interior crack initiation in very-high-cycle fatigue of high-strength alloys *Eng Fract Mech* **212** 153–63.
[30] Pan X and Hong Y 2019 High-cycle and very-high-cycle fatigue behaviour of a titanium alloy with equiaxed microstructure under different mean stresses *Fatigue Fract Eng Mater Struct* **42** 1950-64.

IOP Conf. Series: Materials Science and Engineering **925** (2020) 012043 doi:10.1088/1757-899X/925/1/012043

Computer Simulation of Electrochemical Studies of Al6061-Al_2O_3 Metal Matrix Composite: Effect of Heat Treatment

Mahendra H M [1], G S Prakash[1], Keerthi Prasad K S[2], Rajanna S[3] and Jobhish Johns[4]

[1] Department of Industrial Engineering Management, MSRIT, Bangalore, India.
[2] Department of Mechanical Engineering, VVIET, Mysore, India.
[3] Department of Mechanical Engineering, GECK, Kushalanagara, India.
[4] Department of Physics, RRCE, Bangalore, India.
Email: Mahendra.hun@gmail.com

Abstract. The aim of this work is to investigate the corrosion Studies of Aluminum Oxide (Al_2O_3) reinforced Aluminum 6061 metal matrix composites in Different Concentrations of HCl & NaCl solutions using weight loss method, Potentio-dynamic & EIS Technique. The composites have been prepared by Die- casting method. Aluminum 6061 / Al_2O_3 MMCs containing different sizes such as 40, 60, and 100 microns and 0, 4, and 8 weight percentages of Al_2O_3 particulates are casted. The corrosion characteristics of Aluminum 6061 / Al_2O_3 composite and the unreinforced alloy were experimentally assessed. The results indicated that corrosion rate of metal matrix composites is lower than that of matrix material Al 6061 under the corrosive atmosphere irrespective of exposure time and concentration of corrodent. Aluminum 6061 / Al_2O_3 composite become more corrosion resistant as the Al_2O_3 content is increased. This is because of the formation of stable oxide layer over the specimens. Scanning Electron Microscopy (SEM) shows the degree of attack of HCl & NaCl solution on the surface of the investigated material.

Keywords: Al6061, corrosion, Heat treatment, Metal matrix composites, SEM.

1. Introduction

Pure metals of aluminium have slight strength and we cannot employ with particular purpose wherever tough to distortion and rupture is necessary additional components are fixed to base metals for surge power. The prime mechanical properties of aluminium6061 are attained in alloying trappings and heat treatments. This stimulates the foundation of little heat hastens through line by gesture of displacement and enhance its mechanical properties [1, 2]. Commonly employed aluminium alloy for mechanical application is Al6061 due to its gorgeous wide-ranging properties those are little density, great strength, ductility toughness and conflict to fatigue, it takes stayed widely consumed in aircraft basic components and additional vastly stressed structural application [3].

Al6061 alloy is too vulnerability to stress oxidization furious. It is happening by in homogeneity of the 6061alloy and intrinsic enduring stresses connected by its invention approaches. The aims of the job are to examine the paraphernalia of heat behaviour on the microstructure, rigidity of aluminium 6061metal and aluminium 6061/ Al_2O_3 MMC's. A mixture of Al_2O_3 adding to aluminium6061 alloy and additional strengthening is assumed by way of imaginable funds of auxiliary progresses the mechanical properties of this 6061alloy mainly for temperature convention. So, now this investigation the properties of heat treatment is used as a resources of cultivating the mechanical properties of aluminium 8% Al_2O_3alloy.Aluminium 6061 are employed in a kind of cast and wrought systems and situation of heat treatment [4, 5].

Corrosion study of aluminium alloy AA6063-T5 was investigated in molybdate-containing NaCl solutions [6]. It was suggested that the inhibition initiation preferentially occurred over Fe-rich cathodic IMPs. The formed layer provided inhibition with an efficiency of ~90% after 4h of exposure. High efficacy of ~70% was achieved even after one week of exposure. A two-step oxidation-reduction mechanism of corrosion inhibition by aqueous moly dates was noticed.

The effect of different precipitates on the corrosion behaviour of an Al-Cu-Li alloy through quasi-in-situ transmission electron microscopy observation was studied [7]. The corrosion mode was directly influenced by the aging time, evolving from IGC to the less detrimental intra-granular corrosion. The maximum and average corrosion depths first increased and then decreased with prolonged aging time. The severe corrosion striations within the grain interior were due to the corrosion of anodic T1 phase, while the corrosion cavities were caused by cubic Al5Cu6Li2 phase.

Al-Cu-Fe alloys in the vicinity of $Al_{65}Cu_{20}Fe_{15}$ have been prepared. Among these, one type of particles of a spherical shape was considered as pure cubic β phase in previous reports [8]. Additionally, there are places where the τ phase appears to have out-most layer enriched in Cu. The surface of the particles is faceted and the facets can be classified to three types according to their detailed structures. A close examination and analysis led to the conclusion that these three types are parallel to the atomic planes of {1 1 0}, {1 1 1}, and {1 0 0}, respectively.

Chemical interactions between aqueous vanadium species and aluminium alloy AA6063-T5 were investigated in vanadate-containing NaCl solutions [9]. The study suggests that a two-step mechanism of corrosion inhibition in which V+5 species are first reduced to V+4 or V+3 species above cathodic IMPs, and then oxidized to mixed-valence V+5/V+4 polymerized compounds.

Corrosion inhibition of aluminum alloy AA6063-T5 by vanadates ($NaVO_3$) in 0.05 M NaCl solution was investigated by electrochemical and weight loss measurements, and associated with microstructure and Volta potential data [10]. Electrochemical potentio-dynamic polarization measurements indicated that the vanadates provided mixed corrosion inhibition effects, mitigating both oxygen reduction, occurring on cathodic IMPs, and anodic metal dissolution reaction, occurring on anodic sites, such as Mg_2Si and interphase boundary regions. The weight loss measurements show that the inhibition efficiency decreases after long term exposure.

Cu-MoO_3/g-$C3N_4$ hybrid nano composite was developed by two step strategy of one-pot pyrolysis followed by impregnation method. The structural, phase, morphology and electronic environment of MoO_3, g-$C3N_4$ and Cu in the composite has been determined by various characterization methods [11]. The remarkably enhanced photo catalytic performance was attributed to the combined effect of the double Z-scheme mechanism and defective MoO_3. In situ formation process of MoO_3/g-C_3N_4 hybrid follows a direct Z-scheme charge transfer by generating a great number of defects at the solid-solid interface, similar to that of a conductor and offer low electric resistance.

In the present investigation focused on the corrosion properties of Alumina reinforced Al6061 alloy.

2. Methodology

2.1 Specimen Preparation

Al 6061 samples for electrochemical analysis were prepared with the dimension of 2 cm × 1 cm. The surfaces of the samples are thoroughly polished by different grade of emery papers. After getting smooth surface, samples should wash with distilled water and then with acetone and allowed to dry in air for further use. Solution preparation Solutions of hydrochloric acid medium was prepared by using

IOP Conf. Series: Materials Science and Engineering **925** (2020) 012043 doi:10.1088/1757-899X/925/1/012043

analytical grade hydrochloric acid and double distilled water. It was standardized by potentiometric method. Hydrochloric acid of concentrations 1N was prepared by appropriate dilution.

2.2 Die Casting

Die casting method, the stable die and it is constructed with steel. The molten metal pouring into mould cavity with high pressure and high speed. It takes duplet upright platens on which boosts are located which add the die shares. In duel platen, one die platen is fixed and additional die platen is employed for released and locked die. Inserted the necessary metal into the shot sleeve and then presented into the casting die by a hydraulically-driven piston. When the metal has hardened, the mould cavity is opened, and removes the casted metal. Thus, composite materials with numerous different compositions (0, 4, and 8wt %) were attained rectangular solid shaped bars with dimension of 25mm and a length of 150mm.

3. Result and Discussion

3.1. Corrosion Test (without Heat Treatment)

Table 1. Corrosion Test Results of 6061 Aluminum alloy metal matrix reinforced with alumina.

Microns	Reinforcement	Corrosion current	Corrosion Rate in mpy
	0	1.462×10^{-3}	2.912
40	4	2.152×10^{-3}	2.413
	8	2.325×10^{-3}	1.961
60	4	1.854×10^{-3}	2.523
	8	2.123×10^{-3}	2.124
100	4	1.673×10^{-3}	2.652
	8	1.912×10^{-3}	2.356

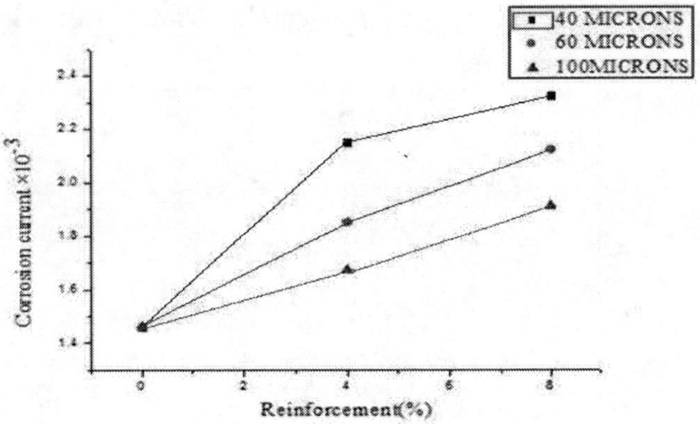

Figure 1. Difference of Corrosion Current for Al6061 with Different Particle of A_2O_3 (Without Heat Treatment)

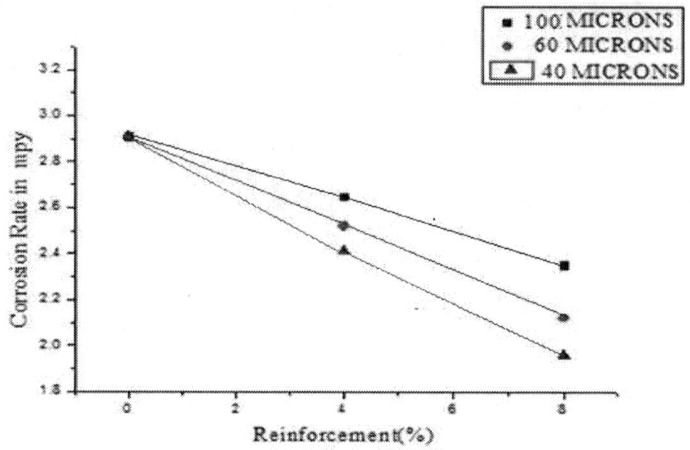

Figure 2. Difference of Corrosion Rate for Al6061 with Different Particle of A_2O_3 (Without Heat Treatment)

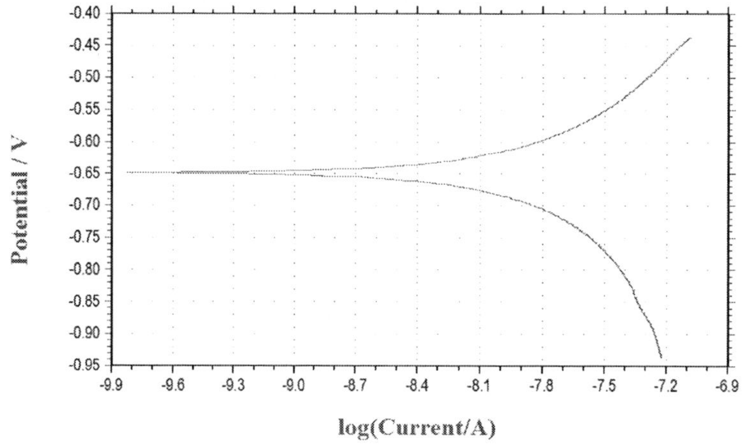

Figure 3. Potential/V v/s Log (Current/A) (Without Heat Treatment)

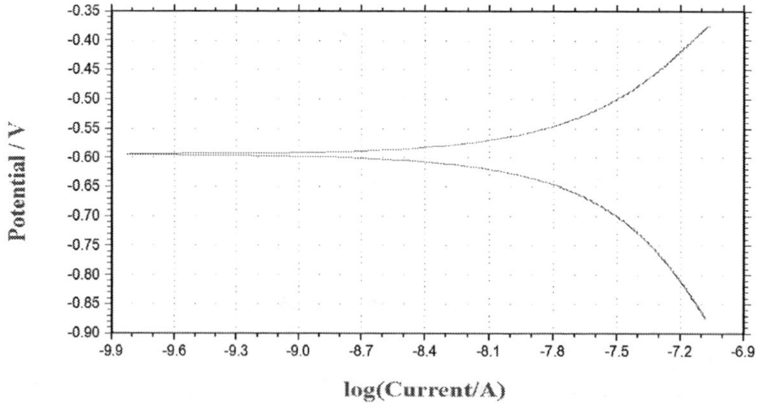

Figure 4. Potential/V v/s Log (Current/A) (Without Heat Treatment)

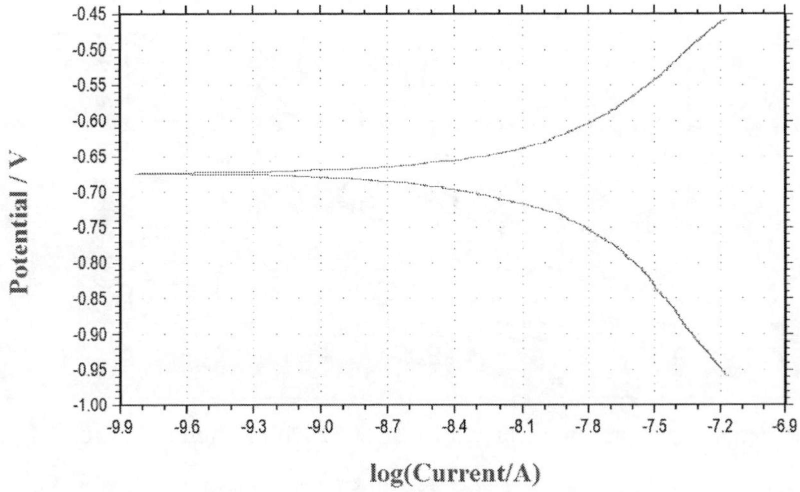

Figure 5. Potential/V v/s Log (Current/A) (Without Heat Treatment)

The influence of hydrochloric acid standard taking place the corrosion amount of $Al6061/Al_2O_3$ alloy was deliberate by technique of tafel polarization. Potentio-dynamic division curvatures of $Al6061/Al_2O_3$aluminium alloy now dissimilar deliberation of HCl standard at laboratory temperature. Corrosion constraints such by way of corrosion potential (E_{corr}), density of corrosion current (i_{corr}), slope of anodic and cathodic are gained since curves of tafel polarization technique. Outcomes are listed in table 1. The corrosion amount sprightly attained since CH software. The outcome shows the diminution in the corrosion amount with a cut in the concentration of HCl standard. The Figure 1 and 2 displays the dissimilarity of corrosion current and corrosion rate. In above graph says reinforcement increases corrosion current is increases and corrosion rate decreases. From result we can conclude 8% of 40 microns particle has a good wear properties compare to other two particles such as 60 and 100. Here we can select 8% of 40 microns particles aircraft and automobile application.

3.2. Corrosion Test (without Heat Treatment)

Table 2. Corrosion Test Results of 6061 Aluminum alloy metal matrix reinforced with alumina. (With heat treatment)

Microns	reinforcement	Corrosion current	Corrosion Rate in mpy
40	0	1.912×10^{-3}	3.162
	4	2.578×10^{-3}	2.762
	8	3.323×10^{-3}	2.562
60	4	2.345×10^{-3}	2.962
	8	3.023×10^{-3}	2.753
100	4	2.162×10^{-3}	3.155
	8	2.872×10^{-3}	3.012

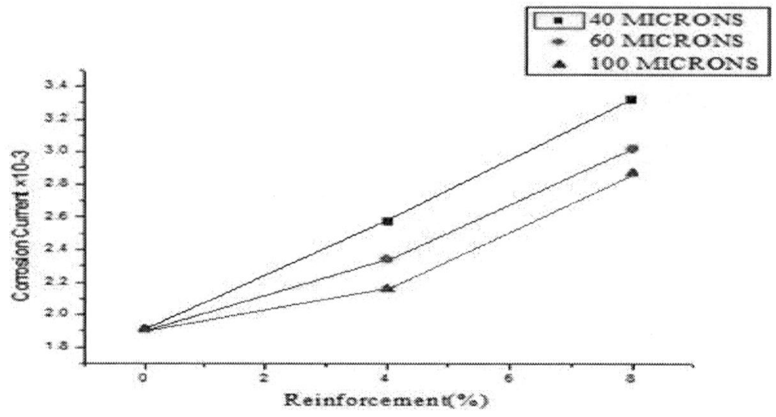

Figure 6. Difference of Corrosion Current for Al6061 with Different Particle of A_2O_3 (With Heat Treatment)

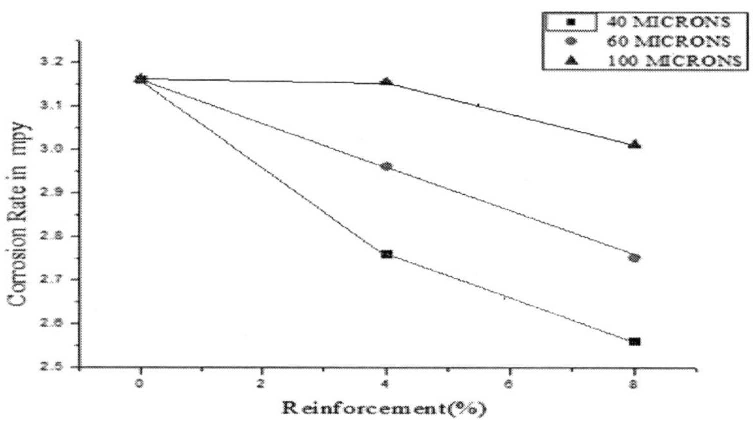

Figure 7. Difference of Corrosion Rate for Al6061 with Different Particle of A2O3 (With Heat Treatment)

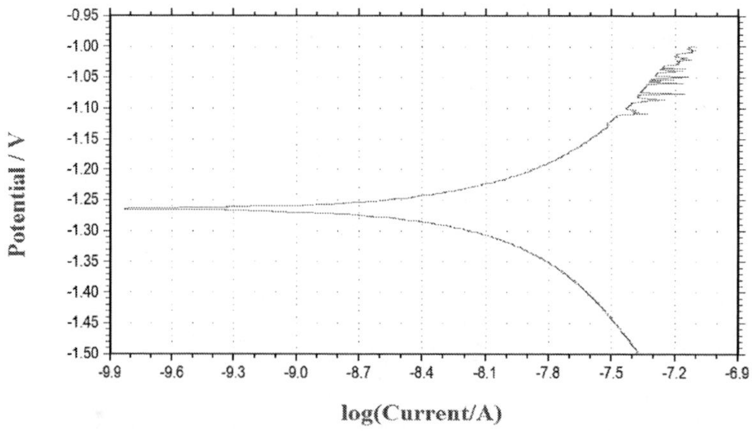

Figure 8. Potential/V v/s Log (Current/A) (With Heat Treatment)

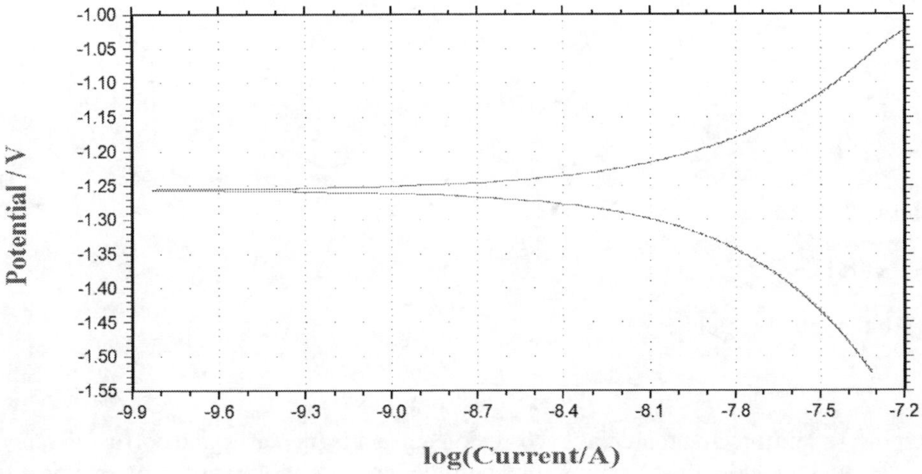

Figure 9. Potential/V v/s Log (Current/A) (With Heat Treatment)

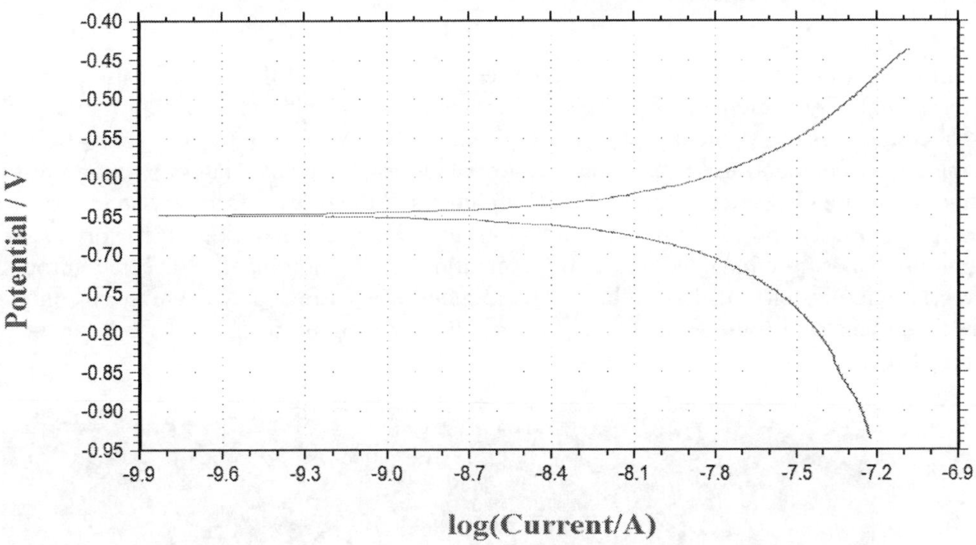

Figure 10. Potential/V v/s Log (Current/A) (With Heat Treatment)

After heat treatment the influence of hydrochloric acid standard taking place the corrosion amount of Al6061/Al_2O_3 alloy was deliberate by technique of tafel polarization. Potentiodynamic division curvatures of Al6061/Al_2O_3aluminium alloy now dissimilar deliberation of HCl standard at laboratory temperature. Corrosion constraints such by way of corrosion potential (E_{corr}), density of corrosion current (i_{corr}), slope of anodic and cathodic are gained since curves of tafel polarization technique Figure 8, 9 and 10. Outcomes are listed in table 2. The corrosion amount sprightly attained since CH software. The outcome shows the diminution in the corrosion amount with a cut in the concentration of HCl standard. The Figure 6 and 7 displays the disparity of corrosion current and corrosion rate. By way of a percentage of reinforcement increases the corrosion current also increases, as the size of the reinforcement increases the corrosion current decreases. Then, as a percentage of reinforcement growths the corrosion amount decreases as the size of the particulate increases the corrosion rate increases, this is happened due to creation of oxide film going on the apparent of the components.

Oxide layer Formula,

$$Al+ (OH)_2= Al (OH)_2$$

Where,

Al=Aluminum

$(OH)_2$=Hydracids ions

$Al (OH)_2$= Aluminum Hydracids

In above graph says reinforcement increases corrosion current is increases and corrosion rate decreases. From result we can conclude 8% of 40 microns particle has a good wear properties compare to other two particles such as 60 and 100. Here we can select 8% of 40 microns particles aircraft and automobile application.

3.3 SEM Analysis

Scanning electron microscopy (SEM) pictures of the Al6061alloy previously corrosion trial and reinforcement Al_2O_3 elements are exposed in picture respectively. The Corrosive performance of Al6061composites SEM pictures of Al6061 matrix alloy with Al_2O_3 (0, 4 and 8wt %) composites occupied later polarization trainings in decinormal chloride standard, later common pre-treatment, are existing in pictures. A relationship of the SEM pictures of the models formerly and later the polarization trainings obviously shows simple exterior worsening because of pitting corrosion in all the models. Pitting predisposition intended for aluminum alloys is predominantly affected through the micro structural heterogeneity of these alloys. The reduction in corrosion worsening of composites can be accredited to matrix-reinforcement boundary grit development of micro or sub-micron local cubicles on the metal surface.

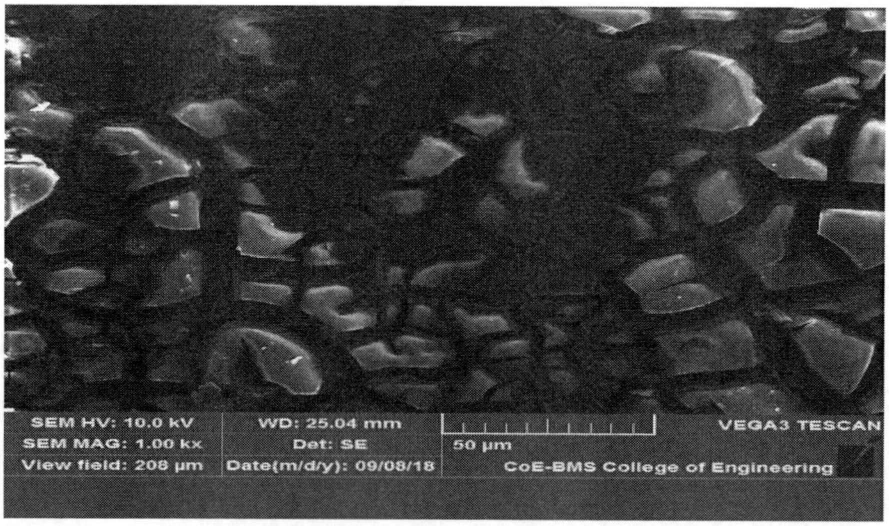

a) Without Heat Treatment (8wt %)

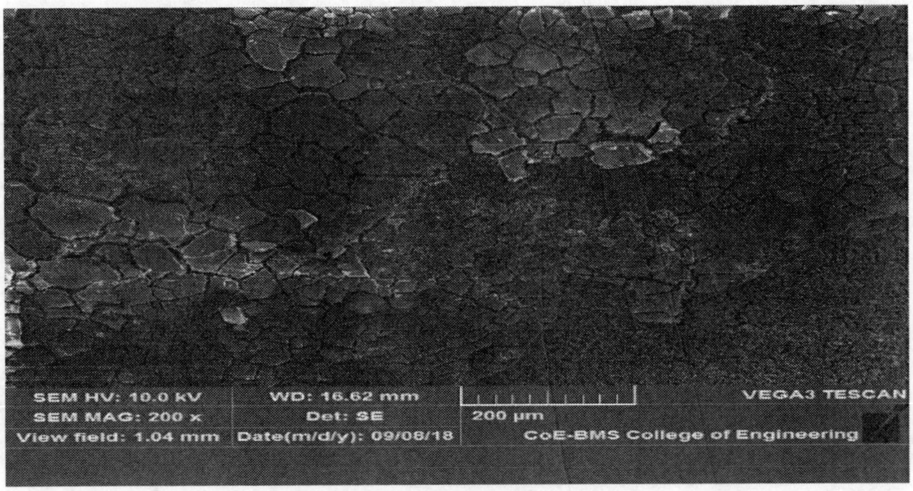

b) With Heat Treatment (8wt %)

Figure 11. SEM Photographs of Al6061 -Al_2O_3 Matrix

4. Conclusion

The corrosion performance of Al6061/Al_2O_3 alloy in dissimilar concentrations of HCl solutions was researched by potentio-dynamic technique. Al6061/ Al_2O_3 alloy undergoes corrosion in HCl medium. The corrosion rate reduces with reduces in the concentration of HCl solution. Corrosion current enhances with enhance in concentration of hydrochloric acid. SEM images shows that as the concentration of HCl solution enhances, formation of pits on the Al6061/ Al_2O_3 sample also enhances. A reduction in the corrosion rate by reducing the formation of pits on the surface was noticed upon heat treatment. Interaction between both the phases increases the corrosion resistance of the composite material. The material developed in the present investigation exhibits better corrosion resistance compared to other composites reported in literature.

References

[1] Stansbury EE and Buchanan RA Fundamentals of Electro-chemical Corrosion, *ASM Int Mater Park* 2000.

[2] Umoren SA 2009 The Inhibition of aluminum corrosion in hydrochloric acid solution by exudate gum from Raphiahookeri *Desalination* **247** 561-572

[3] Obi-Egbedi NO, Spondiasmombin L 2012 As a green corrosion inhibitor for aluminium in sulphuric acid:Correlationbetween inhibitive effect and electronic properties ofextracts major constituents using density functional theory Arabian J Chem **5** 361-373

[4] Nnanna LA 2008 Comparative study of corrosion inhibition of aluminium alloy of type AA3003 in acidic and alkalinemedia by Euphorbia hirta extract *Afr J Pure Appl Chem* **5** 265-271

[5] Abdel-Gaber AM 2008 Inhibition of aluminum corrosion in alkaline solutions usingnatural compound *Mater Chem Phys* **109** 297-305.

[6] Dmitry SK, Illia D, Sefer B, Ryl J, Wrzesińska A, Makarova IV, Bobowska I, Kurilo II, and Claesson PM 2020 Surface and Corrosion Properties of AA6063-T5 Aluminium Alloy in Molybdate-containing Sodium Chloride Solutions *CorrosionScience* 105686

[7] Bo J, Haisheng W, Tian Y, Yi D, Liu H, Hu Z Effects of aging time on corrosion behavior of an Al-Cu-Li alloy *Corrosion Science* 108759

[8] Li C, Carey C, Li D, Caputo M, Bouch R, Hampikian H 2018 A study on spherical particles in Al65Cu20Fe15 alloy prepared by arc melting *Materials Characterization* **140** 162–171

[9] Dmitry SK, Sommertune, Cem Ornek J, Jacek R, Kurilo II, Claesson PM and Pan J 2018

Corrosion Inhibition of Aluminium Alloy AA6063-T5 by Vanadates: Local SurfaceChemical Events Elucidated by Confocal Raman Micro-Spectroscopy *Corrosion Science* https://doi.org/10.1016/j.corsci.2018.12.011

[10] Dmitry SK, O¨ rnek C, Claesson PM, Sommertune J, Zharskii IM, Kurilo II and Pan J 2018 Corrosion Inhibition of Aluminum Alloy AA6063-T5 by Vanadates: Microstructure Characterization and Corrosion Analysis *Journal of The Electrochemical Society* **165** C116-C126

[11] Sulagna P, Swain G, and Parida KM 2018 Highly efficient charge transfer through double Z-scheme mechanism by Cu promoted MoO3/g-C3N4 hybrid Nano composite with superior electrochemical and photo catalytic performance *Nanoscale* **10** 1039

[12] Kharitonov DS, Dobryden IB, Sefer B, Zharskii IM, Claesson PM and Kurilo II 2007 Corrosion of AD31 (AA6063) Alloy in Chloride-Containing Solutions, *Prot. Met. Phys.Chem. Surfaces.* **54** 291–300

[13] Huang Y, Mu S, Guan Q and Du J. 2019 Corrosion Resistance and Formation Analysis of a Molybdate Conversion Coating Prepared by Alkaline Treatment on Aluminum Alloy 6063 a b c d, J. *Electrochem. Soc.* **166** 2019 C224–C230

[14] Dychtoń K, Kwolek P 2018 The replacement of chromate by molybdate in phosphoric acid based etch solutions for aluminium alloys *Corros. Eng. Sci. Technol.* **53** 234–240

[15] Gharbi O, Thomas S, Smith C and Birbilis N 2018 Chromate replacement: what does the futurehold?, *Npj Mater. Degrad.* **2** 12

[16] Zhu Y, Sun K, Frankel GS 2018 Intermetallic phases in aluminum alloys and their roles in localized corrosion *J. Electrochem. Soc.* **165** C807–C820

[17] Kharitonov DS, Ornek C, Claesson PM, J. Sommertune, I.M. Zharskii, I.I. Kurilo, J. Pan, Corrosion Inhibition of Aluminum Alloy AA6063-T5 by Vanadates:Microstructure Characterization and Corrosion Analysis *J. Electrochem. Soc.* **165** C116–C126

[18] Zhu D, Yu LG, Zhang RH 2018 Dissolution Effects with Different Microstructures of Inconel 718 on Surface Integrity in Electrochemical Machining *J. Electrochem. Soc.* **165** 872-878

[19] Wang XD, Qu NS, Guo PF, Fang XL, and Lin X. 2017 Electrochemical Machining Properties of theLaser Rapid Formed Inconel 718 Alloy in NaNO3 Solution *J. Electrochem. Soc.* **164** 548-559

[20] Guo PF, Lin X, Xu JJ, Li JQ, Liu JR and Huang WD Electrochemical removal of different phases from Laser Solid Formed Inconel718 *J. Electrochem. Soc.* **164** 151-157

[21] Sui S, Chen J, Fan EX, Yang HO, Lin X, Huang WD The influence of Laves phases on the high-cycle fatigue behavior of laser additive manufactured Inconel *J. Mater. Sci. Eng. A.* **695** 6-13

[22] Zhang X, Zhou X, Hashimoto T, Liu B, Luo C, Sun Z, Tang Z, Lu F 2018 Ma Y Corrosion behaviour of 2A97-T6 Al-Cu-Li alloy: The influence of non-uniform precipitation *Corros. Sci.* **132** 1–8

[23] Jiang B, Cao F, Wang H, Yi D, Jiang Y, Shen F, Wang B and Liu H 2019 Effect ofaging time on the microstructure evolution and mechanical property in an Al-Cu-Li alloy sheet *Mater. Sci. Eng. A* 740–741

[24] Huang JL, Li JF, Liu DY, Zhang RF, Chen YL, Zhang XH, Ma PC, Gupta RK and Birbilis N 2018 Correlation of intergranular corrosion behaviour with microstructure in Al-Cu-Li alloy *Corros. Sci.* **139** 215–226

[25] Liu RH, Zhu DY, Liu Y, Xu JF, Li YL, Chen XH and Zhang ZQ Correlation between artificial aging and intergranular corrosion sensitivity of a new Al-Cu-Li alloy sheet *Mater. Corros.* **68** 65–76

[26] Mock M, and Albe K 2017 Diffusion of yttrium in bcc-iron studied by kinetic Monte Carlo simulations *J. Nucl. Mater.* **494** 157-164

[27] Pereira Falcón JC, Echeverría A, Afonso CRM, Zambrano Carrullo JC, Amigó Borrás V Microstructure assessment at high temperature in NiCoCrAlY overlay coating obtained by

IOP Conf. Series: Materials Science and Engineering **925** (2020) 012043 doi:10.1088/1757-899X/925/1/012043

laser metal deposition, J. Mater. Res. Technol. **8** 1761-1772
[28] Lu J, Chen Y, Zhang H, Zhao C, Zhao X, Guo F, Xiao P 2019 Superior oxidation and spallation resistant NiCoCrAlYbond coat via homogenizing the yttrium distribution *Corros. Sci.* **159** 108145
[29] Song R and Kumai S 2019 Solidification structure and secondary particles in vertical-type high-speed twin-roll cast 3003aluminum alloy strip, Mater. Trans. **60** 2019 322-329
[30] Lu J, Chen Y, Zhao C, Zhang H, Luo L, Xu B, Zhao X, Guo F, Xiao P. significantly improving the oxidation and spallation resistance of a MCrAlY alloy by controlling the distribution of yttrium *Corros. Sci.* **153** 178-190
[31] Xueyun G, Huiping R, Yiming L, Haiyan W 2017 First-Principles study of phase stability and solubility in Fe-RE (Y,La, Ce) Alloys *Rare Met. Mater. Eng.* **46** 3188-3192
[32] Saenko I, Fabrichnaya O and Udovsky A 2017 New thermodynamic assessment of the Fe-Y system *Journal of Phase Equilibria and Diffusion* **38** 684-99.

Study on High Frequency Vibrations using Statistical Energy Analysis for Different Composite Materials

Sridhar N Shastry[1] and S Subrahmanya Swamy[2]

[1] Department of Mechanical Engineering, RNS Institute of Technology, Bangalore, India.
[2] Global Academy of Technology, Bangalore, India.
[1]nagaraj.sridharshastry@gmail.com

Abstract. Statistical Energy Analysis developed in early 1960s is used to determine the vibration response and energy levels in the structures subjected to high frequency vibrations by using statistical approach. In this work, Statistical Energy Analysis parameters like coupling loss factor and velocity response for composite plate like structures are analytically determined. Velocity response and coupling loss factor for L-shaped composite plate like structure subjected to point load have been computed for three different composite materials. Velocity response and coupling loss factors were determined and compared for the materials considered. The results show that composite materials having high specific modulus had low velocities and high coupling loss factor.

Keywords: Coupling loss factor, High frequency vibration, Velocity response.

1. Introduction

Materials with high specific stiffness find wide application in aerospace structures where minimum weight is required Specific stiffness or Stiffness to weight ratio is the ratio of elastic modulus to mass density of a material. In the field of automotive and aerospace industries, advanced composite laminates having high-strength-weight ratio are widely used. A composite laminate is a combination of fibrous materials that are bonded together layer by layer to obtain the required engineering properties- bending stiffness, strength, coefficient of thermal expansion, and in-plane stiffness. The individual layers consist of high-modulus, high-strength fibres in a polymeric, metallic, or ceramic matrix material.

In the design of aerospace structures and machineries operating at high speeds, study of vibration response at high frequencies is of paramount importance to arrive at methods to control vibrations in high frequency range. Finite Element Analysis (FEA) become less practical in the area of high frequency vibrations. Statistical Energy Analysis (SEA) was developed by Lyon [1,2] in early 1960's to determine the vibration response and energy levels in the structures subjected to high frequency vibrations by using statistical approach.

SEA predicts vibration response in structures by dividing it into subsystems and measuring the energy level and energy transmitted across the subsystems. Vibrations encountered in engineering applications are characterised based on the frequency range- low, mid and high frequency region [3]. Low frequency region can be defined as an area where resonance peak dominates. In this region one can clearly observe the effect of end conditions and change in geometrical parameters. In high frequency region, unlike in

IOP Conf. Series: Materials Science and Engineering **925** (2020) 012044 doi:10.1088/1757-899X/925/1/012044

low frequency region, study of individual modes and frequencies becomes difficult due to high modal overlap. The mid frequency region exhibits transient effect of low and high frequencies.
SEA involves predicting the vibration response of a complex structure by dividing it into a number of subsystems and is characterized by mean energy per mode and the change in energy level between subsystems. Experimental SEA helps in analysing the plate and beam like structures with different configurations subjected to bending wave transmission. SEA is characterized by loss factor comprising of internal loss factor and coupling loss factor. Internal loss factor corresponds to damping in the subsystem itself and coupling loss factor (CLF) corresponds to the energy dissipation during flow across the subsystems. The change in energy level between subsystems is defined by the loss factor.
SEA has been widely applied for high frequency vibration analysis of isotropic materials. Different configurations like L-shaped, T-shaped, ribbed plates, have been analysed [4]. The coupling loss varies with the configuration employed. The parameters for the SEA have been computed using analytical, experimental, FEA [5], TSEA [6] and other hybrid methods [7-8]. The main drawback in experimental technique lies in measuring the response at source and receiver at a large number of points [9] for composite materials which are heterogeneous. FEA, used as an auxiliary method for SEA, has limitations in high frequency range [15,17].
In this work, SEA parameters have been computed for L- shaped configuration using classical wave approach. Classical wave approach predicts reliably the response in high frequency domain when compared with other experimental and numerical techniques. Analysis was carried out using three different composite materials for L-shaped configuration. The influence of specific modulus on CLF and velocity response was investigated. CLF varied linearly with the increase in frequency. The results obtained using classical wave approach employed in this work are in close agreement with the results obtained by Abdullah [15].

2. Statistical Energy Analysis

In SEA, a complex structure is divided into number of coupled subsystems. Exchange of energy takes place between the coupled subsystems along the coupling junctions, where each subsystem represents mode type. During the transfer of energy between coupled subsystems, loss of energy level is due to coupling loss factor and internal loss factor. The relation for energy flow between coupled subsystems (Figure 1) is established by using power balance equation.

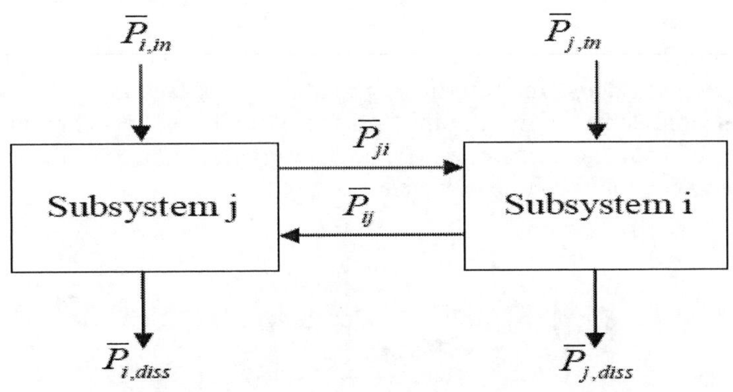

Figure 1. Energy flow across two Subsystems

The power balance equation is established on the assumption that the energy flow between two coupled subsystems is proportional to the difference in the subsystem modal energies. Assuming that power loss is due to internal damping factor η only, the input power to a subsystem becomes equal to the dissipated power given by:

$$\overline{P}_{i,in} = \overline{P}_{i,diss} = \omega\eta\left\langle\overline{E}\right\rangle \tag{1}$$

where, $\overline{P}_{i,in}$ - power injected in subsystem i,

IOP Conf. Series: Materials Science and Engineering **925** (2020) 012044 doi:10.1088/1757-899X/925/1/012044

$\langle \overline{E} \rangle$ - Frequency averaged energy in subsystem *i*,

η-Structural damping loss factor,

$\langle \ \rangle$ - indicates spatial averaging, and bar indicates frequency averaging.

The power balance equation is given by [10, 11],

$$\overline{P}_{i,in} = \overline{P}_{i,diss} + \overline{P}_{ij} \tag{2}$$

$$\overline{P}_{j,in} = \overline{P}_{j,diss} + \overline{P}_{ji} \tag{3}$$

where, the power transmitted between subsystem i and j is given by,

$$\begin{aligned} \overline{P}_{ij} &= \omega\eta_{ij}\langle \overline{E}_i \rangle - \omega\eta_{ji}\langle \overline{E}_j \rangle, \\ \overline{P}_{ji} &= \omega\eta_{ji}\langle \overline{E}_j \rangle - \omega\eta_{ij}\langle \overline{E}_i \rangle. \end{aligned} \tag{4}$$

The power balance equation for n systems is given by [12, 13],

$$\omega \begin{bmatrix} \left(\eta_1 + \sum_{i\neq 1}^{N} \eta_{1i}\right) n_1 & -\eta_{12} n_1 & \cdots & -\eta_{1N} n_1 \\ -\eta_{21} n_2 & \left(\eta_2 + \sum_{i\neq 2}^{N} \eta_{2i}\right) n_2 & \cdots & -\eta_{2N} n_2 \\ \vdots & \vdots & \ddots & \vdots \\ -\eta_{N1} n_N & \cdots & \cdots & \left(\eta_N + \sum_{i\neq N}^{N-1} \eta_{Ni}\right) n_N \end{bmatrix} \times \begin{bmatrix} \frac{\langle \overline{E}_1 \rangle}{n_1} \\ \frac{\langle \overline{E}_2 \rangle}{n_2} \\ \vdots \\ \frac{\langle \overline{E}_N \rangle}{n_N} \end{bmatrix} = \begin{bmatrix} \overline{P}_{i,1} \\ \overline{P}_{i,2} \\ \vdots \\ \overline{P}_{i,N} \end{bmatrix} \tag{5}$$

3. Analysis and Procedure

SEA was carried on L-Plate like structure made of composite materials. Glass/Epoxy, Boron/Epoxy and Graphite/Epoxy were considered for determining coupling loss factors and velocity response for the configuration considered subjected to a point load. Two laminates of a composite having four layers oriented at zero degree with L-configuration (Figure 2) were considered for the analysis. Classical SEA technique was used to determine SEA parameters.

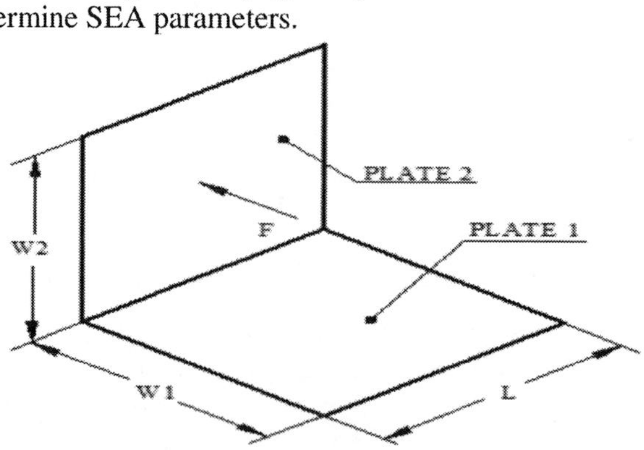

Figure 2. L-Plate Configuration

IOP Conf. Series: Materials Science and Engineering **925** (2020) 012044 doi:10.1088/1757-899X/925/1/012044

The stacking sequence of the laminate is given in table 1. Length of coupling 'L' and the width of plate 'W' were considered as 1m. A unit transverse harmonic load with a frequency range from 1000 Hz to 13000 Hz was applied on one of the plates and responses on both the plates are calculated. The CLF and velocity responses have been computed using the analytical wave approach through a program developed using MATLAB. The properties of the different composite plates are given in Table 1.

Table 1. Mechanical and Physical properties of composite materials

Material & Physical properties		Glass/ Epoxy	Boron/ Epoxy	Graphite/ Epoxy
Elasticity modulus of plies	E_x	39 GPa	204 GPa	181 GPa
	E_y	8.6 GPa	18.5 GPa	10.3 GPa
Poisson's ratio of plies	ν_{xy}	0.28	0.23	0.28
	ν_{yx}	0.0617	0.0208	0.0159
Average density	ρ	2100 kg/m^3	2500 kg/m^3	2200 kg/m^3
Specific Modulus	(E_x/ρ)	18.57 MPa/kg/m^3	81.6 MPa/kg/m^3	82.27 MPa/kg/m^3
Number of plies	n	4	4	4
Orientation angle	θ	0	0	0
Plate size		L = 1 m W1 = W2 = 1m	L = 1 m W1 = W2 = 1m	L = 1 m W1 = W2 = 1m
Total thickness	t	4X10^{-3} m	4X10^{-3} m	4X10^{-3} m

3.1. Analytical formulation for orthotropic plate

SEA for orthotropic plates can be applied only when modal density is high. Analytical classical wave approach for determining CLF and input power for semi-infinite orthotropic plates subjected to force impedance are discussed in the following section. The CLF for a line junction is given by [14]:

$$\eta_{ij} = \frac{\overline{\delta f}_i \tau_{ij}}{\pi f \left(2-\tau_{ij}\right)} \tag{6}$$

where, $\overline{\delta f}_i$ is average modal frequency spacing, τ_{ij} is the transmission co-efficient.

Average modal frequency spacing is given by,

$$\overline{\delta f}_i = \frac{h}{\sqrt{3}A}\sqrt{c'_{Lx} c'_{Ly}} \tag{7}$$

where, h is thickness of the plate, A is the surface area of the plate, c'_{Lx} and c'_{Ly} are the longitudinal wave speed on X and Y directions respectively.

Longitudinal wave speed for orthotropic plate is given by

$$c_{Lx} = \sqrt{E_x / \rho\left(1-\upsilon_{xy}^{2}\right)} \quad \text{and} \quad c_{Ly} = \sqrt{E_y / \rho\left(1-\upsilon_{yx}^{2}\right)} \tag{8}$$

where, E_x and E_y are Elastic modulus in X and Y directions respectively, ρ is the density of the orthotropic material considered, and υ_{xy}, υ_{yx} are the Poisson's ratios.Transmission co-efficient for orthotropic plate is given by,

$$\tau_{ij} = \frac{4\,Re\left(Z_i\right)Re\left(Z_j\right)}{\left|\sum_{s=1}^{S} Z_s\right|^2} \tag{9}$$

where Z_i is the force impedance in Z-direction and here, S is the number of subsystems connected at a common joint.

Force impedance is given by,

$$Z = \frac{4}{\sqrt{3}}\rho h^2 \sqrt{c'_{Lx} c'_{Ly}} \tag{10}$$

4. Results and discussion

The variation of CLF and velocity response against the variation of frequency for different composite materials considered has been plotted. The CLF of the composite material plotted for analytical approach seems to be good agreement with results given by Abdullah [15].

Figure 3 and figure 4 show the variation of velocity response in plate 1 and plate 2 with respect to increase in frequency respectively. A harmonic load was applied on plate 1 and velocity response was computed on both the plates and plotted. The energies of both the plates were computed and inversed to determine the velocities of plates using inverse method [16]. The velocity of plate 1 and plate 2 against frequency in the range of 1000 Hz to 13000 Hz has been plotted for three materials Glass/Epoxy, Boron/Epoxy and Graphite/Epoxy.

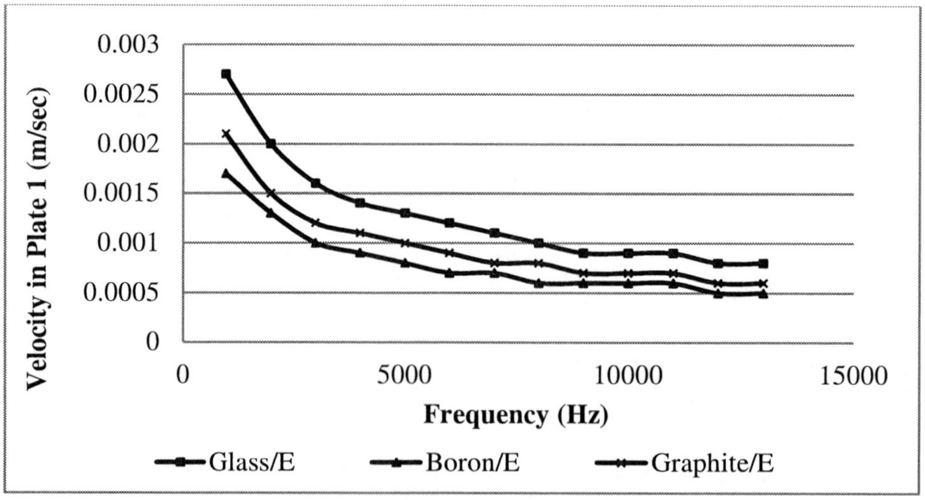

Figure 3. Velocity vs Frequency (Plate 1)

The velocity of plate1 and plate 2 decreases with the increase in frequency. It can be seen that the Glass/Epoxy had higher velocity response compared to that of Boron/Epoxy and Graphite/Epoxy. The velocity response of Glass/Epoxy was higher because of lower longitudinal wave speed. Longitudinal wave speed was highest in Boron/Epoxy among the three materials. Velocity of plate 1 was higher compared to that of plate 2 since load was applied on plate 1.

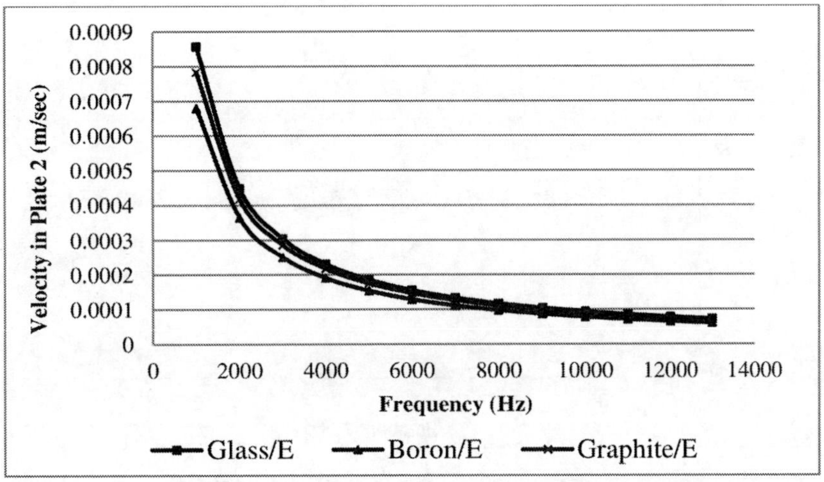

Figure 4. Velocity vs Frequency (Plate 2)

CLF was plotted against frequency for all the three materials as shown in figure 5. Figure shows a semi log plot of CLF vs frequencies. It can be seen that the CLF in Glass/Epoxy was least among the three materials. CLF can also be viewed as function of specific modulus from equation 7. The specific modulus of Glass/Epoxy was least among the three materials while the magnitude of specific moduliof Boron/Epoxy and Graphite/Epoxy werenearly equal. The CLF is independent of internal loss factor at high frequencies and depends more on specific modulus. The CLF of Boron/Epoxy and Graphite/Epoxy is in close match since the specific moduli of both the materials are nearly same. Coupling loss is observed to vary linearly with increase in frequency.

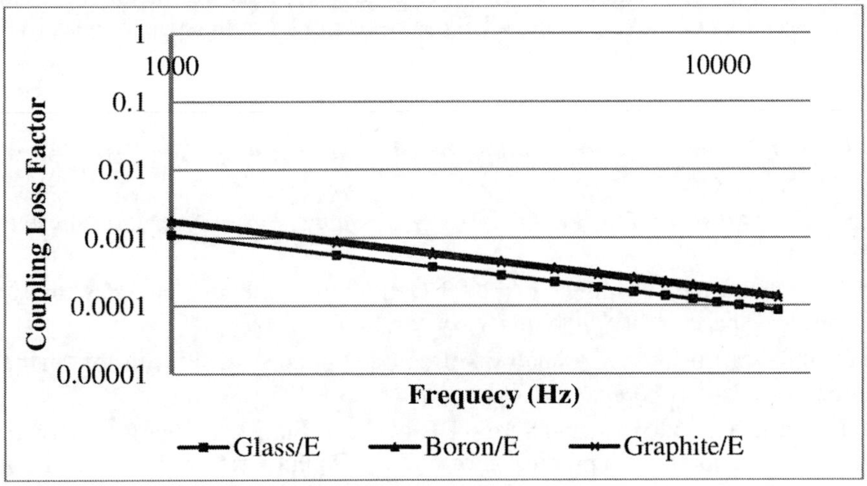

Figure 5. Coupling loss factor vs Frequency

SEA parameters were determined by a modal based method for symmetrically laminated composite plates by Abdullah [15]. The figure 6 shows the effect of CLF on fibre orientation with frequency. The results obtained by modal based approach showed considerable enhancement of CLF with change in fibre orientation while the conventional SEA was independent of fibre orientation. The enhanced result of CLF was observed only in mid-frequency region while both the methods had close results in high frequency region.

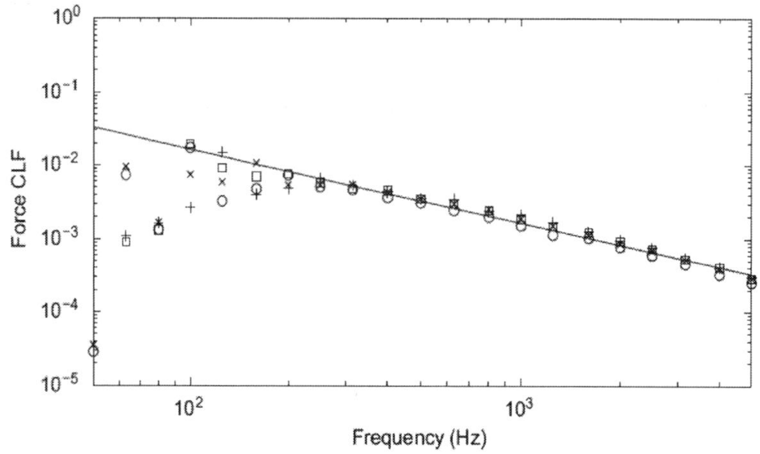

Figure 6. Coupling loss factor vs Frequency: o: θ=0⁰, x: θ=15⁰, □: θ=30⁰, + θ=45⁰, ___: Analytical

Comparing figure 5 and 6, it can be inferred that there is a good correlation between the results obtained by modal based method and classical wave approach in the high frequency domain.

5. Conclusion

The SEA studies on different composite materials were carried out and velocity responses and coupling loss factors were compared. The velocity response and coupling loss factors were computed using analytical classical wave approach. The coupling loss factor, which is independent of internal damping, was found varying linearly with frequency. It can be seen that the material having high specific modulus has high coupling loss factor. This work demonstrates that for heterogeneous materials the CLF depends on specific modulus. It is observed that higher the specific modulus lower will be the velocity response in the plates. A novel way of looking at the CLF has been shown in this paper specially for composite materials.

References

[1] Lyon RH 1995 *Theory and Application of Statistical Energy Analysis* 2nded(Butterworth-Heinemann, Boston)

[2] Lyon RH 1975 *Statistical Energy Analysis of Dynamical Systems*(Massachusetts Institute of Technology)

[3] RabbioloG, Bernhard RJ and Milner FA 2004 Definition of a high-frequency threshold for plates and acoustical spaces *J. of Sound and Vibration* **277** 647–67

[4] Hopkins C 2002 Statistical energy analysis of coupled plate systems with low modal density and low modal overlap *J. of Sound and Vibration* **251**193-214

[5] Giuseppe P, Melill G, Laudiero A and Sergio DER 2019 A Statistical Energy Analysis (SEA) model of a fuselage section for the prediction of internal Sound Pressure Level (SPL) at cruise flight conditions *Aerospace Scienece and Technology* **88** 340-49.

[6] LangleyRS, Hawes DH, Butlin T and Ishii Y 2019 A derivation of the Transient Statistical Energy Analysis (TSEA) equation with benchmark application to plate systems *J. of Sound and Vibration* **445** 88-102

[7] Edwin Reynders PB, Robin Langley S 2018 Cross-frequency and band-averaged responsevariance prediction in hybrid deterministic statistical energy analysis method *J. of Sound and Vibration* **428** 119-46

[8] Hawes DH, Langley RS, Butlin T and Ishii Y 2019 A hybrid Finite Element – Statistical Energy Analysis method for impulsive and transient loading *J. of Sound and Vibration* **459** 1-19

[9] Robinson M and Hopkins C 2014 Prediction of maximum time-weighted sound and vibration levels using transient statistical energy analysis, Part 1: theory and numerical implementation *Acta Acust., United with Acust.* **100** 46-56

[10] Fahy FJ 1998 An alternative to the SEA coupling loss factor: rationale and method for experimental determination *J. of Sound and Vibration* **214** 261-67

[11] Shankar K Keane AJ 1997 Vibrational energy flow analysis using a substructure approach: the application of receptance theory to FEA and SEA *J. of Sound and Vibration* **201** 491-513

[12] Hopkins C 2009 Experimental statistical energy analysis of coupled plates with wave conversion at junction *J. of Sound and Vibration* **322** 155-66

[13] Hodge CH, Nash P and Woodhouse J 1987 Measurement of coupling loss factors by matrix fitting an investigation of numerical procedures *Applied Acoustics* **22** 47-69

[14] Lyon RH and De Jong RG 1998 Theory and Application of Statistical Energy Analysis 2nd ed., (R H Lyon Corp)

[15] Abdullah S 2013 Numerical determination of statistical energy analysis parameters of directly coupled composite plates using a modal-based approach *J. of Sound and Vibration* **332** 361-377

[16] D A Bies and S Hamid 1980 Insitu determination of coupling loss factors by the power injection method *J. of Sound and Vibration* **70** 187-204

[17] Nagaraja KC, Rajanna S, Prakash GS and Rajeshkumar G. 2020. Improvement of mechanical and thermal properties of hybrid composites through addition of halloysite nanoclay for light weight structural applications *Journal of Industrial Textiles* 1-19.

IOP Conf. Series: Materials Science and Engineering **925** (2020) 012045 doi:10.1088/1757-899X/925/1/012045

Study of Quasi-Static Compression Behaviour of Hollow Tubes Made of Glass/Epoxy Composite

Sampath H P[1], CV Srinivasa[2] and Veeresh Naik[3]

[1,3] Department of Mechanical Engineering, City Engineering College, Bangalore, Karnataka, India.
[2] Department of Mechanical Engineering, GM Institute of Technology, Davanagere, Karnataka, India.
[1] hpsampath037@gmail.com, [2]srinivasacv@gmit.ac.in

Abstract. Composite materials have been used widely in various engineering fields. Composite materials possess low weight, corrosion resistance property, and specific strength, and due to these facts, composite materials are becoming popular among researches and scientists. Now a day, the glass fibers offer the potential to acts as a reinforcing material for composites alternative to the use of glass fiber, carbon fiber and other man-made fibers. In present work hallow tube composites of epoxy – glass fiber was prepared by hand lay-up technique. Mechanical characterizations of prepared hollow tube composites were carried out. Mechanical results were improved; we considered the hollow tubes of glass fiber and epoxy in different proportions such as 50% of glass fiber and 50% of epoxy, 60% of glass fiber and 40% of epoxy, 66% of glass fiber and 34% of epoxy. The maximum compression strength was obtained in the ratio of 50:50 that is 57.69N/mm^2, the maximum load at the peak was obtained in the ratio of 60:40, that is 14.61kN and the maximum deformation at the peak was obtained in the ratio of 60:40 that is 8.03mm.

1. Introduction

Fibre-reinforced polymers receive more attention in various structural applications. Conventional materials used in fibre-reinforced polymer composites, such as carbon fibres and glass fibres have great features in high tensile strength, chemical resistance, dimensional stability as well as excellent insulation properties. Today, the use of fiber reinforced composite materials are more selective material rather than metal as seen in many applications of technology, present knowledge has reached a good level in some of the field related to composite. The composite material is a combination of two or more materials which is known as a structural material that consists of two or more combined constituents that are combined at a microscopic level and are not soluble in each other [1]. The reinforcement phase is one of the important constituents in composite material whereas another phase is matrix material in which the first constituent is added. Fibers, flakes, and particles are used as material for reinforcement in composite material [2]. In the fabrication of a composite continuous matrix phase may be considered. The matrix acts as a protection layer for fiber reinforcement from environmental harms, fibers which are the main load barring agents in composites and this matrix phase even acts as stress transferring agent to fibers and makes the orientation and location of fibers for desired specification. N. Movahedi et al. [3] studied on Quasi-static compressive behavior of the ex-situ aluminum-alloy foam-filled tubes under elevated temperature and concluded that the ex-situ ordered aluminium cellular structure filled tubes gives excellen.t compressive properties compared with the empty tubes. Foam filled tubes gives excellent energy absorption, axial & radial compression properties [4]. The axial crushing tests of empty and partially foam-filled thin-walled circular and square columns were performed by Altin et al. [5]. The impact of physical measurement on the crashworthiness of these structures was investigated.

Measurements of the internal cylinder were found to have huge effect on the basic crashworthiness of froth filled twofold cylinders. The arrived at the midpoint of pound power, explicit vitality ingestion, vitality assimilation per stroke and vitality engrossing viability factor of flimsy walled roundabout structures are higher than those of dainty walled square structures, individually. Froth filled single and twofold round cylinder structures are prescribed as crashworthy structures because of their high pulverize power effectiveness and vitality engrossing proficiency [6]. Many researchers have investigated compressive strength of different types of thin walled tubes which are honeycomb type [7], natural/glass hybrid [8], metal-composite [9] and natural fiber. Analysis of crashworthiness performance of thin-walled cylindrical tubes was carried out by Rezvani M J et al. [10]. Many composites are prepared by reinforcing epoxy and steel embedded with natural fibers and synthetic fibers in vinyl ester etc. The study of Albahash and Ansari [11] on natural jute fiber with Kevlar fiber in the application of energy absorption hollow tube, from their investigation it's shows that the hybrid natural/synthetic fiber-reinforced polymer have a better result in crashworthiness efficiency compared with single natural jute fiber. They determined that foam-filled square columns show highest crash performance. These types of models were validated experimentally by Abramowicz and Jones [12]. Alexander [13], Pugsley [14], Abramowicz [15] and Abramowicz and Jones [16], Singace et al. [17] and Wierzbicki and Bhat [18] to predict the crush behavior of the circular tubes. The common main observation of these studies was that the energy absorption capacity can be increased by using metallic foams. The compression properties can be improved or can be enhanced by using some best metal tubes based on the application. Hybrid composites can be made using composition with carbon/aramid tubes [19] with glass.

Glass fiber reinforced plastic (GFRP) thin-walled tubes have excellent mechanical properties and highly feasible design ability, and thus have been shown many applications in oil and construction industrials. The strength of the composite is enhanced due to existence of fiber reinforcement Synthetic fibers such as E-glass fibers are used in various applications from centuries by mankind. Many possibilities of using E-glass fibers have been explored, by utilizing the useful mechanical properties. Production of hard boards, particle boards, sandwich panels and hollow tubes were done by using most of the E-glass fibers. Researchers have developed new dimensions of using synthetic fibers for the production of polymer Synthetic fiber based composites has become important materials in day today uses and also in other applications like air craft, spacecraft etc. due to the light weight, high strength to weight ratio, corrosion resistance and other advantages.

2. Materials and Fabrication

To fulfil the objective of this research work, the compression behaviour of glass fiber reinforced polymer composite tubes manufactured with three different compositions were analysed.

2.1 Materials

2.1.1 E-Glass Fiber

E-glass or electrical grade glass was originally developed for standoff insulators for electrical wiring. It was later found to have excellent fiber forming capabilities and is now used almost exclusively as the reinforcing phase in the material commonly known as fiber glass.

2.1.2 Epoxy resin

Epoxy resin has high strength and mechanical adhesiveness properties due to which these are commonly widely used in industrial applications. It has good temperature and chemical resistant as it is a good solvent. Atul Ltd. Lapox L–12 is used for our composite material.

2.1.3 Hardener

Lapox K-6 is a light yellow aliphatic polyamine hardener suitable to cure epoxy resin at ambient Temperature. When cured with epoxy resin it provides short pot life and faster cures speed. Due to very low viscosity and low dosage, it finds suitable for applications like adhesive, motor and casting of small electrical component. E-Glass fiber, Epoxy resin and Hardener materials are shown in 'Figure.1'.

IOP Conf. Series: Materials Science and Engineering **925** (2020) 012045 doi:10.1088/1757-899X/925/1/012045

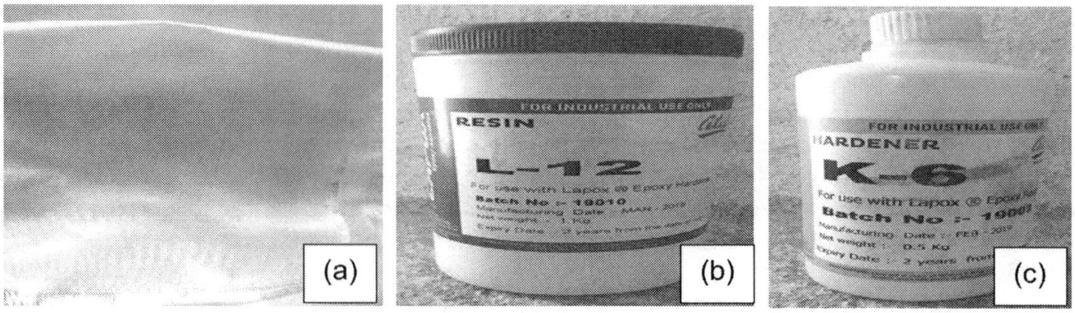

Figure 1. Materials used for Fabrication, (a) E-Glass, (b) Epoxy Resin and (c) Hardener K-6

2.2. Fabrication

In the present work E-glass fiber is taken as reinforcement in composite where as matrix resin used is epoxy for preparation of composite hollow tubes. The E-glass fiber mat cut around 350mm to 450mm and also specimens are prepared according to the ASTM standards. In the present work specimens are prepared for three different compositions. The table 1 shows the compositions used for fabrication of composite hallow tube. The moulds from selected materials are prepared to required dimensions. The Prepared moulds are then cut to specified dimension by using hacksaw blade for the fabrication of required composite material.

2.2.1 Hand layup method

The prepared mould is then fixed to lathe machine running at low speed for hand layup method. Outer surface of the mould is wrapped with polythene sheet for better finish and easy removal, a thin layer of epoxy resin is applied. One layer E-glass fiber mat is wrapped on the epoxy applied followed by a layer of epoxy resin. The same process is continued until the required thickness of the composite tubes for the application is met, then cured it for 12 to 24 hours. The prepared composite is cut to the required dimension for testing using hacksaw blade. The hacksaw blade cutting process doesn't give perfect dimension for the specimen so the cut specimen is then filed for getting exact dimension as per ASTM standards. The 'Figure.2'shows the fabrication of a glass fiber composite hollow tube.

Table 1. Composition of Testing Specimens

Configuration	Composition	
	Reinforcement (E-Glass)	Matrix (Epoxy Resin)
1	50	50
2	60	40
3	66	34

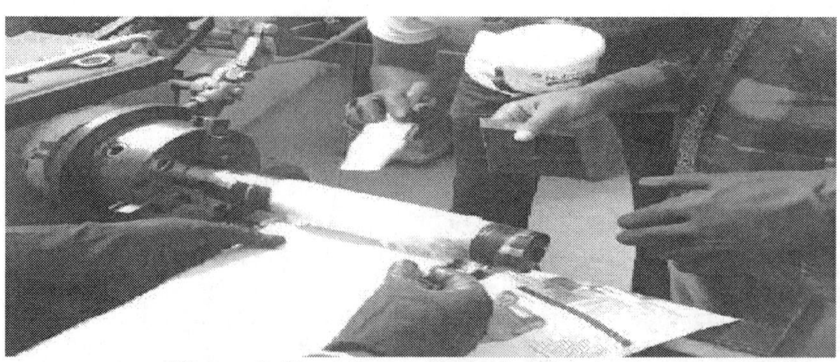

Figure 2. Hand Layup method of fabrication

3. Experimentation

This test procedure covers the determination of compressive strength of hollow tubes made of glass-epoxy composite. ASTM-D3410 standard is used for preparing specimens in this present work. The specimen dimension used is 250mm length and 50mm diameter is as shown in 'Figure.3'. The standard used entails the compressive properties determination in the form of circular test specimens. The test is conducted for specimens by considering three configuration specimens and the results are discussed in next section.

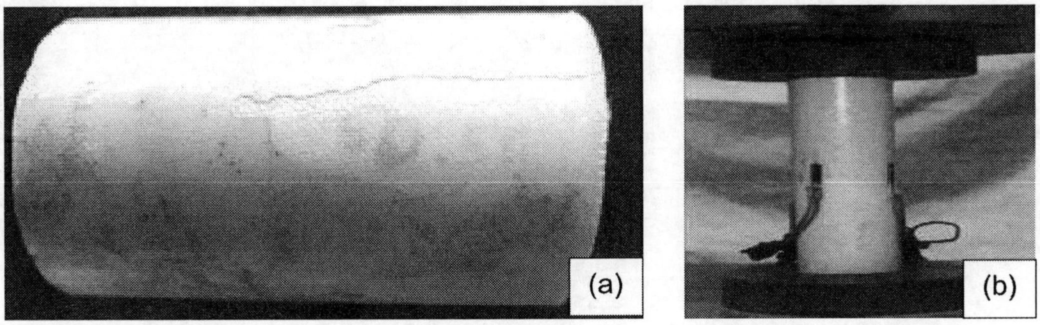

Figure 3. a) Compression Test Specimen, b) Experimental setup of compression test

The compression test is conducted according to the ASTM standards. Each test specimens of hollow tube are 250mm length, 50mm inside diameter and 2mm thickness were prepared and for this test UTM of capacity 1000kN used. Testing machine is calibrated before use. After testing of specimens, the readings are taken and calculated the required mechanical properties and tabulated. 'Figure.4' shows the different configuration specimens after compression testing.

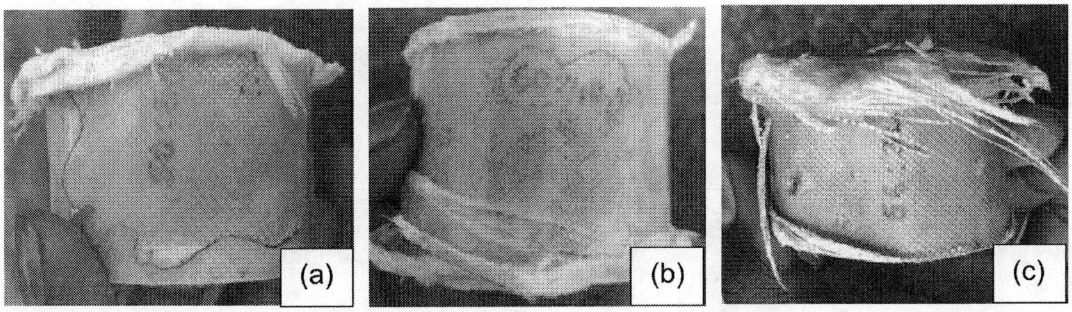

Figure 4. Specimens after testing a) *configuration 1,* b) *configuration 2,* c) *configuration 3*

4. Results and Discussion

It is very important to know the mechanical properties of the polymer composites before any application; it is one of the most important aspects to determine the mechanical properties like stiffness, strength, elastic modulus etc. Mechanical properties of fiber reinforced composites depend on the nature of matrix material, the distribution and orientation of the reinforcing fibers, the nature of the fiber-matrix interfaces. A very small change in the physical nature of the reinforcement for a given matrix may result in prominent changes in the overall mechanical properties of composites. The results obtained for compression test are shown and discussed individually.

4.1. Compression test results for different compositions

Figure.5 shows load v/s Deformation curve for a composite hollow tube made up of E-glass fiber and epoxy resin. The maximum load of the tested machine is 1000kN and specimen cross-section area is 233.67mm2. Output data of the configuration 1, configuration 2 and configuration 3 are, load at peak is 13.9kN, 14.6kN and 12.67kN, deformation at peak load is 7.45mm, 8.15mm and 7.29mm respectively.

Observations can be clearly noted from that; peak load and deformation was highest for 60-40 composition as compare to other compositions.

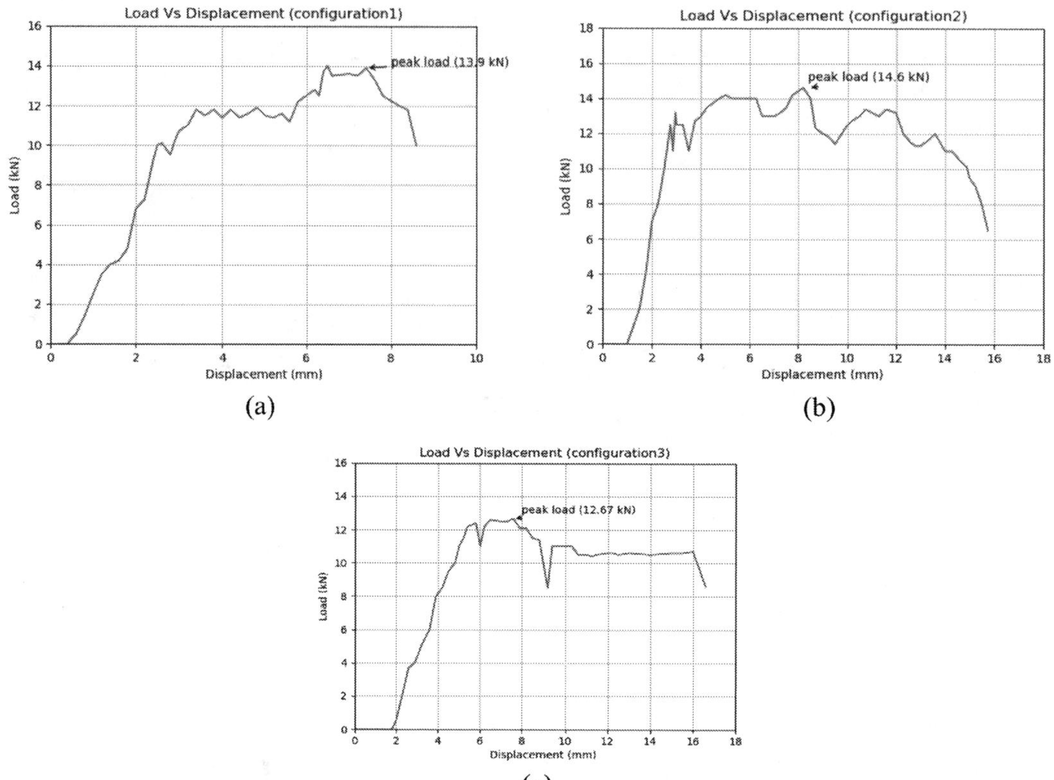

Figure 5. Load v/s Deformation curve for a composite hollow tube Made up of E-glass fiber and epoxy resin (a) Configuration 1, (b) Configuration 2 (c) Configuration 3

4.2. Comparison between different configuration

The comparisons of load at the peak, deformation at peak load and compression strength for three configurations are listed in Table II. These comparisons are illustrated graphically in 'Figure.6' as shown below.

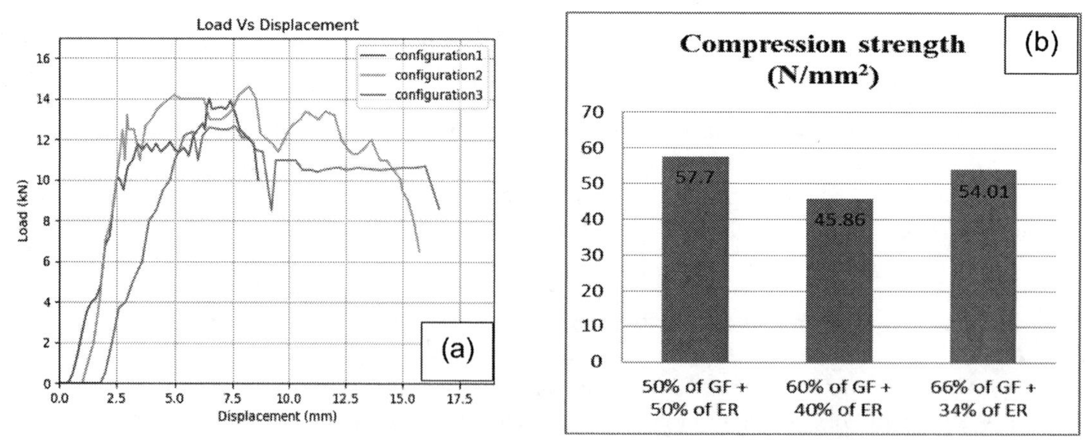

Figure 6. (a) Load v/s Deformation of different configurations, (b) Compression Strength of different configurations

The compressive properties of models like square tube made of metals and composites were compared from the work done by Ma Quanjin et al. [19] Chamila Batuwitage et al. [20] Daniel Paul et al. [21] H.R. Zarei et al. [22] and J.M. Babbage et al. [23] and aluminum composite [23] showed a peak load between a range of 24kN to 47kN. Foam filled aluminum tube [22] showed a range of 65kN to 71kN. The results were compared with glass-epoxy composites with addition of hollow glass tubes which showed a range of 8kN to 14.5kN with different composition where as in our work we got a range of 12.67kN to 14.61kN as peak load for crushing.

5. Conclusions

In this work, the comparison between three different compositions of hollow tubes was made up of glass fibre and epoxy polymer composite. The experiments are conducted to evaluate the mechanical properties such as the quasi-static compression test of prepared polymer hollow composite tubes. The compositions used are 50% of matrix material and 50% of fibre fraction, 40% of matrix material and 60% of fibre fraction, and 34% of matrix material and 66% of fibre fraction. The following observations are made and conclusions are drawn based on the limited test results.

It is observed that matrix material has a significant influence on the load-displacement relationship of the glass fibre epoxy polymer hollow composite tube. Load carrying capacity increased when there is a good adhesion property between fibre and matrix. From the results, we can conclude that glass fibre of 60% and epoxy of 40% have higher yielding capacity then compared to the other two ratios of the composite. Compression strength observed for the first composition of 50% fibre and 50% epoxy was 57.69N/mm^2 is higher compared to other two compositions. Deformation at the peak for 60% of fibre and 40% of matrix composite is 8.15mm which is high compared to other two compositions.

By comparing the quasi-static compression test, it can be concluded that for forecasting the average crushing force of compression test of hallow tubes, 9 layers of samples with a depth of 50mm and thickness of 2mm with a fiber layout, it can be used to the mean force of the quasi-static test. It is also possible to predict that the mean force values and degree of destruction of these specimens under axial dynamic test through the gradient of displacement force.

References

[1] Najafzase asl O, Pol MH, and Golshan NR Experimental investigation of parameter affect on behavior of composite tubes under dynamic and quasistatic axis loading *In Persian, Journal of Science and Technology of Composites* **34**

[2] Venkatesha Prasanna G and Venkata Subbaiah K 2013 Modification, flexural, impact, compressive properties & chemical resistance of natural fibers reinforced blend composites, *Malaysian Polymer Journal* **8** 38-44

[3] Movahedi N and Linul E 2017 Quasi-static compressive behavior of the ex-situ aluminum-alloy foam-filled tubes under elevated temperature conditions *Materials Letters* **206** 182-184

[4] Su M, Wang H and Hao H 2019 Axial and radial compressive properties of alumina-aluminum matrix syntactic foam filled thin-walled tubes *Composite Structures* **226** 111197

[5] Altin M, Acar E and Güler MA 2018 Foam filling options for crashworthiness optimization of thin-walled multi-tubular circular columns *Thin-Walled Structures* **131** 309-323

[6] Li Z, Chen R and Lu F 2018 Comparative analysis of crashworthiness of empty and foamfilled thin-walled tubes *Thin-Walled Structures* **124** 343-349

[7] Wang Z, Yao S, Lu Z, Hui D and Feo L 2016 Matching effect of honeycomb-filled thinwalled square tube-Experiment and simulation *Composite Structures* **157** 494- 505

[8] Sanjay M and Yogesha B 2017 Studies on natural/glass fiber reinforced polymer hybrid composites: an evolution *Mater Today Proc* **4** 2739–47

[9] Hua G, Su Zhun G, Liu Q, Li G and Li Q 2017 On crushing characteristics of different configurations of metal–composites hybrid tubes *Compos Struct* **175** 58–69

[10] Rezvani MJ and Jahan A 2015 Effect of initiator, design, and material on crashworthiness performance of thin-walled cylindrical tubes: a primary multi-criteria analysis in lightweight design *Thin Wall Struct* **96** 169–82

IOP Conf. Series: Materials Science and Engineering **925** (2020) 012045 doi:10.1088/1757-899X/925/1/012045

[11] Alkbir MFM, Sapuan SM, Nuraini AA and Ishak MR 2014 Effect of geometry on crashworthiness parameters of natural kenaf fibre reinforced composite hexagonal tubes *Mater Des* **60** 85–93
[12] Abramowicz W and Jones N 1984 Dynamic axial crashing of square tubes *Int J Impact Eng* **2** 263–81
[13] Alexander JM 1960 An approximate analysis of the collapse of thin cylindrical shells under axial loading *Q J Mech Appl Math* **13** 10–5
[14] Pugsley AG 1979 On the crumpling of thin tubular structures *Q J Mech Appl Math* **32** 1–7
[15] Abramowicz W 1983 The effective crushing distance in axially compressed thin-walled metal columns *Int J Impact Eng* **1** 309–17
[16] Abramowicz W and Jones N 1984 Dynamic axial crushing of circular tubes *Int J Impact Eng* **2** 179–281
[17] Singace AA and Elsobky H 1996 Further experimental investigation on the eccentricity factor in the progressive crushing of tubes *Int J Solids Struct* **32** 589–602
[18] Wierzbicki T and Bhat SU 1986 A moving hinge solution for axisymmetric crushing of tubes *Int J Mech Sci* **28** 135–51
[19] Ma Quanjin MSA, Salim MRM Rejab, Bernhardi OE and Yunus Nasution A 2020 Quasi-static crushing response of square hybrid carbon/aramid tube for automotive crash box application *materials today proceedings* **27**
[20] Chamila B, Fawzia S, Thambiratnam D, Liu X, Al-Mahaidi R and Elchalakani M 2018 Impact behaviour of carbon fibre reinforced polymer (CFRP) strengthened square hollow steel tubes A numerical simulation *Thin-Walled Structures* **131** 245-257
[21] Daniel P, Velmurugan R and Gupta NK 2019 Improvements in the crushing behaviour of glass fibre-epoxy composite tubes by the addition of hollow glass particles *Thin-Walled Structures* **141** 111-118
[22] Zarei HR and Kro ger M 2008 Optimization of the foam-filled aluminum tubes for crush box application *Thin-Walled Structures* **46** 214–221
[23] Babbage JM and Mallick PK 2005 Static axial crush performance of unfilled and foam-filled aluminum–composite hybrid tubes *Composite Structures* **70** 177–184

IOP Conf. Series: Materials Science and Engineering **925** (2020) 012046 doi:10.1088/1757-899X/925/1/012046

Inelastic Behavior of En 31 Steel Metallic Surface Subjected to Hard Body

Darshan C J[1], Mahendra H S[2], Hemaraju[3] and Rashmi S N[4]

[1] Production Engineering and System Technology, Department of IPE, The NIE, Mysore.
[3] Department of IPE, The NIE, Mysore, India.
[2,4] Department of Mechanical Engineering, BGS Institute of Technology, Adhichunchanagiri University, BG Nagara, India.
[1]darshancj17@gmail.com, [2]mahen041@gmail.com, [4]rashmi.ip@nie.ac.in, [3]hemarajucrp@gmail.com

Abstract. In machineries of thermal power plants elements such as coal conveying systems, grinding balls, slurry carrying systems involves transfer of load and displacements. In all these machineries every element is subjected to the hard body such as particles from coal and slurry cause continuous wear results with abrasive wear. Due to this abrasive wear there is inelastic and damage of both the objects was takes place. Since many elements of all these machineries were made from En category steels this study was focused on En 31 steel. In the present study rubber wheel abrader was used to conduct the experiments. 106 microns, 212 microns and 425 microns graded abrader was used for conducting the test. En category steel i.e. En 31 was used as target material for the studies. The sand flow rate of 95 grams/minute was maintained. The weight loss was calculated and found that 0.004, 0.012 and 0.021, for an abrader size of 106 microns following with 0.005, 0.012 and 0.020 & 0.007, 0.022 and 0.023 for an abrader size 425 microns. For all these three tests a normal load of 11.37 N, 35.31 N, and 58.85 N was applied. Scanning electron microscope (SEM) was used to study the morphological analysis. There is light influence on volume loss due to the influence of abrader. Weight loss of the material does not solely depend on increasing in abrader size. Inelastic deformation and wear modes are seen as subject to mechanical properties of the target material. Disfigurement highlights describe the volume loss of the wear.

1. Introduction

In thermal power plants the machineries like coal conveying systems, coal handling systems, fluid and slurry handling systems, ash handling systems, many elements were made from En category steels. EN category steel offer excellent properties like wear resistance, toughness properties at low temperatures, good elastic modulus. High strength properties enable them to use in shafts, punches, drills, dies, mining equipments, ploughing elements of agriculture. In all these applications the common feature is two rubbing surfaces. Due to continuous operations involved there will be the transfer of displacements and motions. Transfer of displacements and loads causes the failure of machine elements leading to permanent machine breakdowns and efficiencies. During continuous rubbing a third body i.e. dirt or sand particles are present between the contact surfaces. As continuous relative motion takes place the third body will rubbing over the surface tends to abrade the surface, so that causing ploughing the surface. Due to this material was displaced from contact surface, so that weight losing will take place leads to element failure. Hence to design the elements of the machineries, it is the strong thrust needed to understand the mechanisms of wear in all the directions. It is also strong necessity required to understand the nature of wear as how it behaves with respect to load and nature i.e abrasive, adhesive and so on [1, 12].

Abrasive wear: whenever two contact surfaces interlocks due to their surface morphology, ploughing takes place which leads to abrasive wear. The mechanism is shown in figure 1(a)

Figure 1. Schematic image adhesive wear mode.

Adhesive wear: If the contact surfaces are very smooth and enough adhesive bonding stresses exists between two pairs which could be sufficient enough to resist the relative sliding displacement then adhesive wear takes place. The mechanism is shown in figure 1 (b). Understanding of the wear is only limited theoretical based on the process like micro-cutting, micro-fatigue and micro-cracking [2]. Many of the researchers found theoretical model for understanding the volume loss and degree of penetration. **Hokkarigawa and kato** introduced a parameter called degree of penetration D_p which was used to predict various modes of wear phenomenon and given by,

$$D_p = R\sqrt{\frac{\pi H}{2W}} - \sqrt{\left(\frac{\pi H}{2W}\right)R^2 - 1}$$

Where D_p = Degree of penetration
R = Pin tip radius
H = Hardness
W = Normal load

From the model developed they determined that in three types of abrasive wear modes the grooves were formed as a result of abrasive particle and there is a plastic flow of the material by creating the groove having ridges on both sides. They found experimental and theoretical estimates of morphology of deformation in ploughing, cutting and wedge formation [3]. **Kayaba and Kato** [4] developed a model for predicting adhesive wear proposing various models of wear. The size of the wear particle predicted does not simply equal to the size of contact. The complexity of wear marvel was evaluated by characterizing boundaries like wear modes and wear rates. The complication of wear phenomenon was quantified by defining parameters like wear modes and wear rates. **Jonas Allebert** et al [5] tried to understand the influence of overlay welded high chromium white iron and quenched, tempered low alloy carbon steels. The test results showed that the wear of the material was dependent on abrasive size. **Seth** et al [6] made an attempt to understand the wear behavior of steel which were coated by thermally sprayed aluminum-titanium alloy coatings. The test results showed that wear performance of coating with aluminum and titanium was comparable within the weight percentage. **Beata Bialobrzeska & Piotr Kostencki** [7] conducted laboratory and field experiments to study the wear behavior of selected low alloy boron steels. They identified the two pre dominative wear modes which were micro-cutting and micro-ploughing. **Victor Gomez** et al [8] attempted in identifying effect of size of abrasive particles on wear. The test results indicated that the mass fraction of two different sized powders had a bearing on wear modes of the micro-scale level & wear rate was not found to have a direct relation with mass fraction. **Vytenis jankauskas** et al [9] made an attempt to characterize wear behavior of manually arc welded hard facings with low carbon and stainless steel matrix. They found the experimental results that improvement in wear resistance when tungsten carbide content was an order of 42 to 48%. **Hernandez** et al [10] made an attempt to understand role of temperature on abrasive of boron steel and hot forming tool steels. The wear rate was found to remain either a constant or improved as the temperature increases up to 400°C.. The wear rate was found to increase with increase in temperature above 400°C. **Bin Hwang** et al [11] studied the role of tensile properties on abrasive wear of steel saw wires. The two body wear was found to be influenced by tensile strength and micro-hardness of test materials. Whereas wear loss of three body wear was found to be influenced by elongation or ductility. The point of the work introduced in this paper was to give the tool to foreseeing versatile and inelastic characteristic of En 31 steel when exposed to abrasion wear.

IOP Conf. Series: Materials Science and Engineering **925** (2020) 012046 doi:10.1088/1757-899X/925/1/012046

2. Experimentation

Experimentation was done by using dry sand abrader test rig. Dry sand wheel abrader confirms ASTM G-65 standards. The schematic view of dry sand abrader was shown in figure 2.

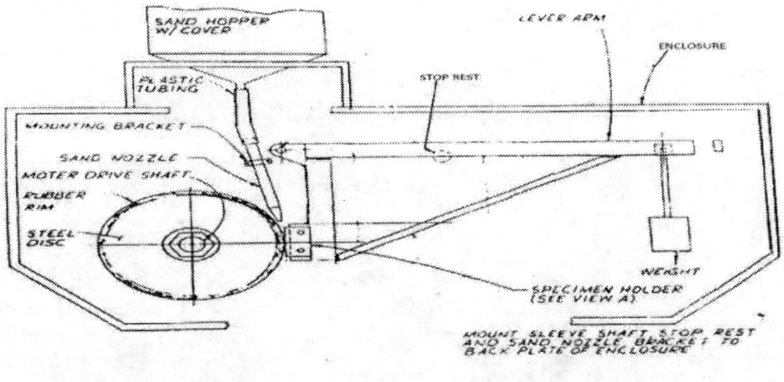

Figure 2. Schematic view of dry sand abrasion test rig.

Three levels of different loads, varying size of abraders, constant speed, constant speed and flow rate was followed during entire test. The test parameters followed during the experimentation was tabulated and shown in table 1.

Table 1. Test Inputs

Sl no	Abrader type	Abrader size	Normal Load in N	Wheel velocity in rpm	Test time in minutes
01	Graded Sand	106 microns	12	200	8
02	Graded sand	212 microns	35	200	8
03	Graded Sand	425 microns	59	200	8

The test includes loading the target specimen in the slot provided in the test rig. Calculating the leverage loss the normal load was found for three different levels. The normal load was applied so that the chlorobutyle rubber wheel having contact with the target specimen. The wheel has been allowed to rotate at a speed of 200revolutions per minute. The standard size sand abrader was sieved and filled to the hopper of the test rig. The sand flow rate of 100 grams/minute was maintained during the test. The test has been conducted for 8 minutes. After the test the target specimen was weighed and test result was recorded. Before the experiment also the target specimen was weighed. Weight loss of the target specimen was recorded by taking initial and final weight of the specimen.

3. Results and Discussions

The weight loss for target material was calculated by weighing before and after experimentation and weight difference of target test sample with normal load and abrader sizes was tabulated table 2.

Table 2. Volume loss of target material for different normal loads and abrader sizes.

Sl. no	Abrader size in microns	Load in N	Volume loss in mm^3
		12	0.004
1	106	35	0.012
		59	0.021
		12	0.005
2	212	35	0.012
		59	0.020
		12	0.007
3	425	35	0.022
		59	0.023

Weight loss the specimen was calculating by taking initial weight and final weight of the specimen by before and after conduction of experiments. The weight loss calculated was tabulated in the table 2. The weight loss of 0.004,

0.0021 and 0.021 was found for the normal load of 11.77 N, 35.31 N and 58.85 N and abrader size of 106 microns. Similarly, weight loss of 0.005, 0.012 and 0.020 was found for the same normal load and abrader size of 212 microns. And the weight loss of 0.007, 0.022 and 0.023 was found for the same normal load and abrader size of 425 microns.
The variation of volume loss of the target specimen subjected to three different sand abraders was plotted in the graph and as shown in the figure.3.

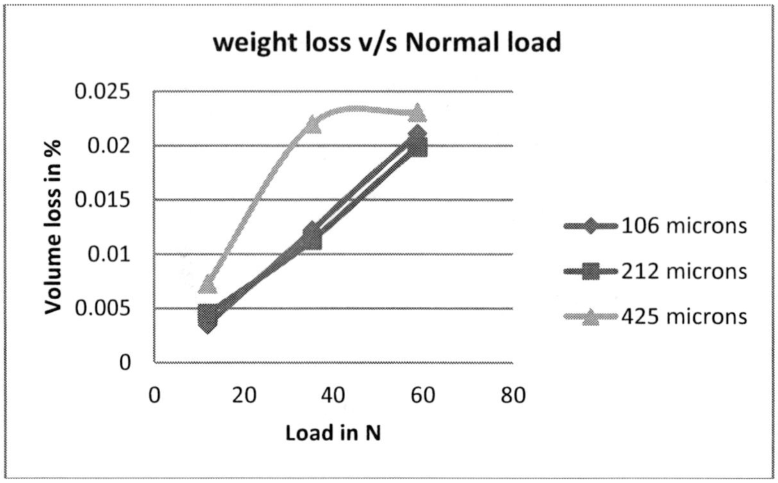

Figure 3. Variation of volume loss for different normal loads.

Figure 3 shows the dependency of volume loss of target material for three different sized abraders. The curve with blue color belongs to volume loss of target specimen when the grain size of the abrader is 106 microns. Weight loss of 0.004, 0.012 and 0.021was takes place when the normal loads 12 N, 35 N and 59 N respectively. The curve with red color belongs to volume loss of target specimen when the grain size of the abrader is 212 microns. Weight loss of 0.005, 0.012 and 0.020 was takes place when the normal loads 12 N, 35 N and 59 N respectively. The curve with green color belongs to volume loss of target specimen when the grain size of the abrader is 425 microns. Weight loss of 0.007, 0.022 and 0.023 was takes place when the normal loads 12 N, 35 N and 59 N respectively. From the results it was found that there is no much influence of weight loss at the selected normal load levels except the weight loss takes place at a normal load 35.31 N and abrader size 425 microns.

The variation of weight loss of the target specimen subjected to three different normal load levels was plotted in the graph and as shown in the figure.3.

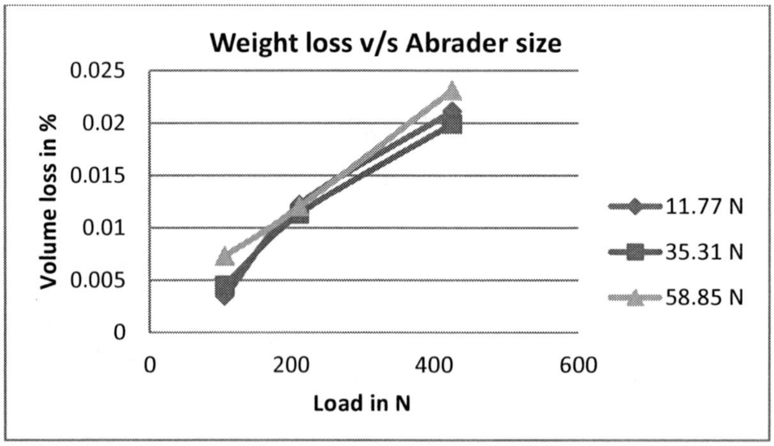

Figure 4. Variation of volume loss for different sized abraders.

From the figure the blue curve shows the weight loss of the target specimen when the normal load was 12 N. The weight loss takes place for this normal load was 0.004, 0.012 and 0.021subjected to three abrader sizes of 106 microns, 212 microns and 425 microns. The red curve shows the weight loss of the target specimen when the

normal load was 35 N. The weight loss takes place for this normal load was 0.005, 0.012 and 0.020 subjected to three abrader sizes of 106 microns, 212 microns and 425 microns. The green curve shows the weight loss of the target specimen when the normal load was 59 N. The weight loss takes place for this normal load was 0.005, 0.012 and 0.020 subjected to three abrader sizes of 106 microns, 212 microns and 425 microns.
In order to understand the deformation morphology of the abraded specimens a scanning electron micrographic study was carried out and discussed below.
Figure 5 shows scanning electron micrographs when the target specimen was subjected to a normal load of 11.77 N and three abrader sizes.

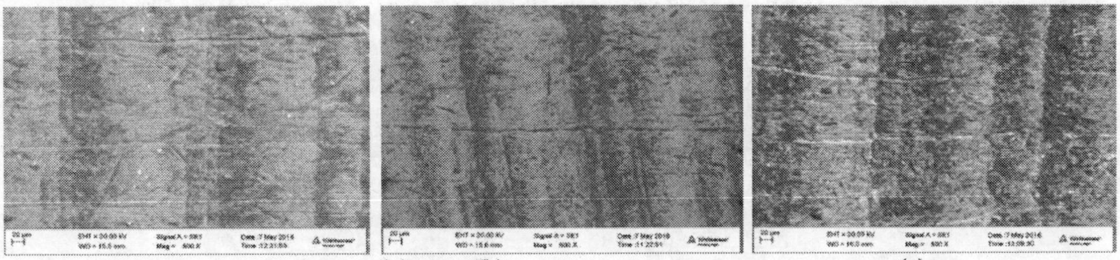

(a) (b) (c)

Figure 5. SEM micrographs of wear scar- for normal load of 11.77 N, abrader size 106 microns(a), 212 microns (b) & 425 microns(c) with a maginification of 500 X

Micrographs shown in figure 5 (a) & figure 5 (b) have one groove but these grooves were not well defined. The scratches were found in both micrographs 5 (a) and 5 (b). But theses scratches are also not defined clearly. Many number of grooves were found in micrograph shown in figure 5 (c), but not well defined.
Figure 6 shows micrographs scanned by scanning electron microscope (SEM), when the target specimen was subjected to a normal load of 35.31 N and three different abrader sizes.

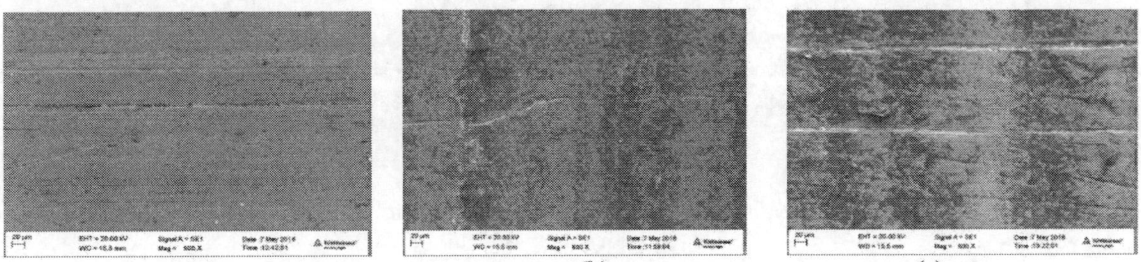

(a) (b) (c)

Figure 6. SEM micrographs of wear scar- for normal load of 35.31 N, abrader size 106 microns(a), 212 microns (b) & 425 microns(c) with a maginification of 500 X

Two number of grooves were found in micrographs shown in figure 6 (a) and figure 6 (c). No well defined grooves were found but many scratches were found in micrograph in figure 6 (b).
Figure 7 shows micrographs scanned by scanning electron microscope (SEM), when the target specimen was subjected to a normal load of 58.85 N and three different abrader sizes.

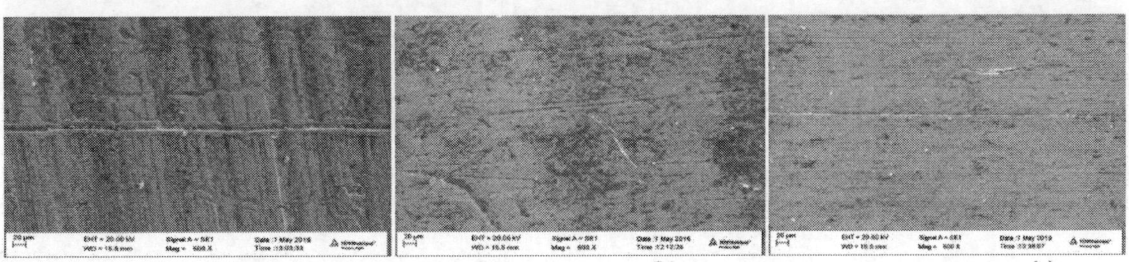

(a) (b) (c)

Figure 7. SEM micrographs of wear scar- for normal load of 58.85 N, abrader size 106 microns(a), 212 microns (b) & 425 microns(c) with a maginification of 500 X

There are two grooves each one at micrograph shown in figure 7 (a) and figure 7 (c) along with little no scratches. In micrograph shown in figure 7 (b) no well-defined grooves were found but more number of well-defined scratches were found.

In order to better understanding the variation of weight loss with different normal loads has been shown in figure 8 to figure 13. The wear scars were scanned by scanning electron microscope.

Micrographic study of wear scars when target specimen was subjected to different normal load levels and abrader size of 106 microns has been shown in figure 8.

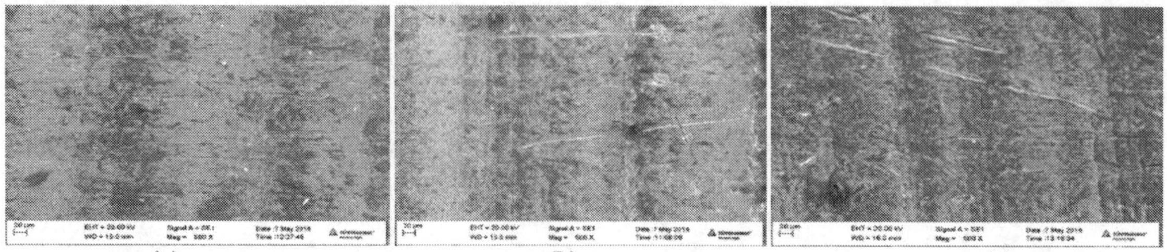

(a) (b) (c)

Figure 8. SEM micrographs of wear scar- for abrader size of 106 microns, normal load 11.77 N (a) 35.31 N (b) & 58.85 N (c) with a maginification of 500 X

Micrographs shown in figure 8 (b) and figure 8(c) have more density of scratches without any grooves where as micrographs shown in figure 8 (a) has very less density of scratches without any grooves.

Micrographic study of wear scars when target specimen was subjected to different normal load levels and abrader size of 212 microns has been shown in figure 9.

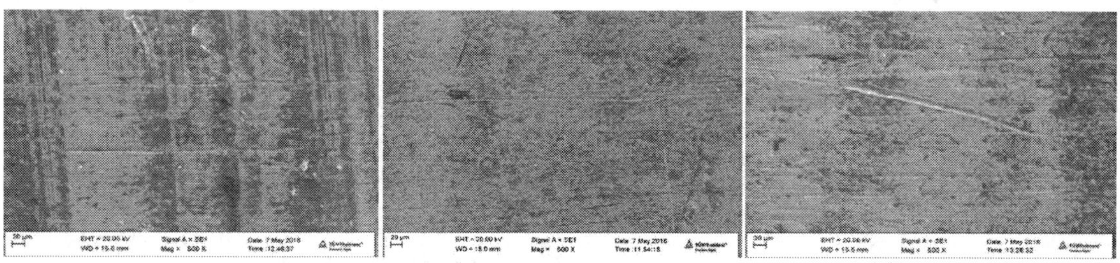

(a) (b) (c)

Figure 9. SEM micrographs of wear scar- for abrader size of 212 microns, normal load 11.77 N (a) 35.31 N (b) & 58.85 N (c) with a maginification of 500 X

All the micrographs shown in figure 9 (a), 9 (b) and 9 (c) do not have any grooves but many scratches are found.

Micrographic study of wear scars when target specimen was subjected to different normal load levels and abrader size of 425 microns has been shown in figure 10.

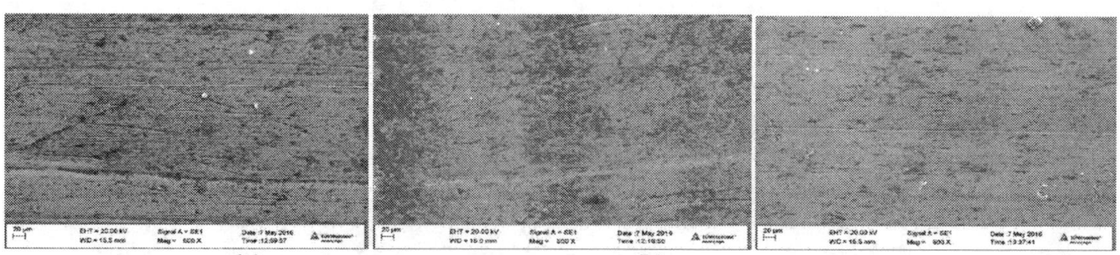

(a) (b) (c)

Figure 10. SEM micrographs of wear scar- for abrader size of 425 microns, normal load 11.77 N (a) 35.31 N (b) & 58.85 N (c) with a maginification of 500 X

Micrographs shown in figure 10 (a) shows grooves of different intensities running across the length. Micrographs shown in figure 10 (b) & 10 (c) does not shows any cleared grooves but micrographs shown in figure 10 (b) & 10 (c) are comparable.

Micrographic study at lower magnification which is at 500 x could not reveal the feature which could explain the weight loss on variations in abrader sizes. A further study was carried out at magnification level of 1500 X to understand the in detail inelastic deformation of the target specimen.

IOP Conf. Series: Materials Science and Engineering **925** (2020) 012046 doi:10.1088/1757-899X/925/1/012046

Micrographic study of wear scars when target specimen was subjected to a normal load of 11.77 N and three different abraders has been shown in figure 11.

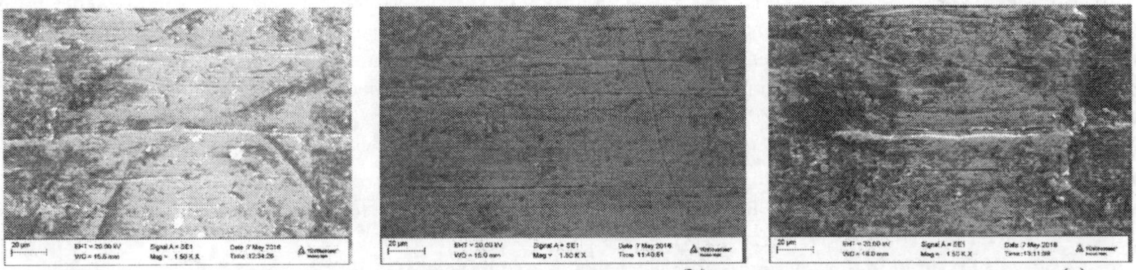

(a) (b) (c)

Figure 11. SEM micrographs of wear scar- for normal load of 11.77 N, abrader size 106 microns (a), 212 microns (b) & 425 microns(c) with a maginification of 1500 X

Micrograph shown in figure 11 (a) and figure 11 (b) show very few not well defined and shorter in size grooves. These two micrographs are comparable. Micrograph showed in figure 11 (c) shows well defined bigger grooves with mild cut ridges. Micrograph shown in figure 11 (c) is different from micrograph shown in figure 11 (a) and shown in figure 11 (b).

Figure 12 shows the scanning electron micrographs study of wear track on EN 31 at a normal load 35.31 N and different grades of abraders.

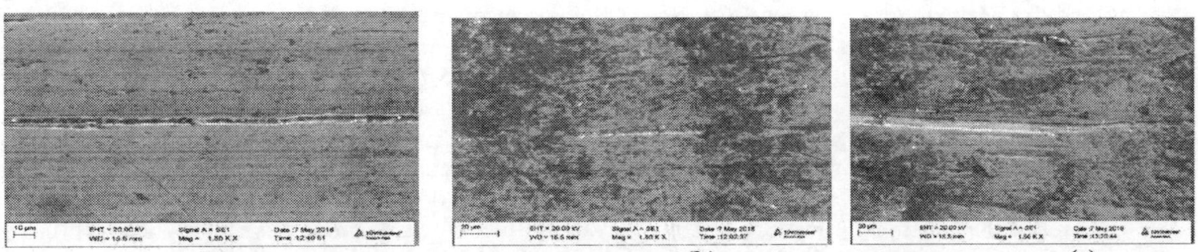

(a) (b) (c)

Figure 12. SEM micrographs of wear scar- for normal load of 35.31 N, abrader size 106 microns (a), 212 microns (b) & 425 microns(c) with a maginification of 1500 X.

Micrographs shown in figure 12 (a) shows a narrow and discontinuous groove running across the length. Micrographs shown in figure 12 (b) shows non continuous, shallow and number of grooves compared to features observed in micrograph shown in figure 12 (a). Micrograph shown in figure 12 (c) shows groove which is wider, deeper with not well defined ridges running across the length. At few places partially chipped metallic material is observed.

Figure 13 shows the scanning electron micrographs study of wear track on EN 31 at a normal load 58.85 N and different grades of abraders.

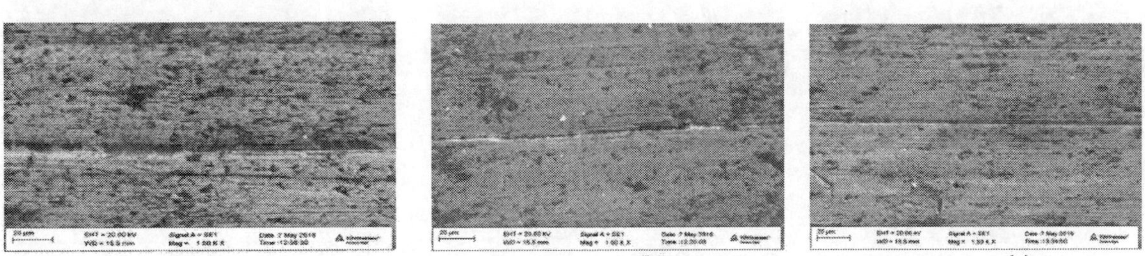

(a) (b) (c)

Figure 13. SEM micrographs of wear scar- for normal load of 58.85 N, abrader size 106 microns (a), 212 microns (b) & 425 microns(c) with a maginification of 1500 X.

Micrographs shown in figure 13 (a), 13 (b) and 13 (c) shows grooves which are running across the length. The groove features in figure 13 (b) and figure 13 (c) are comparable whereas grooves in micrograph shown in figure 13 (a) is different from grooves found in micrograph 13 (b) and micrograph 13 (c). The groove found in micrograph 13 (a) is shallow but little wider.

4. Conclusions

The wear was found to be consequence of deformation. No such inelastic deformation was found due to the influence of abrader size. Significant deformation was found by the impact of normal load. The wear was found with different modes and well defined mode was not found. The deformation features characterize the wear volume losses. The grain sizes of the abrader are found to have little influence on volume loss when compared to the wear loss with normal load. It is required to carry out the study for the materials which exhibit different micro-structural features. It is necessary to correlate the study with the developed theoretical model with different instant of loads.

References

[1] Hemaraju Ranganatha S and Shashidhara KN 2016 Studies on wear loss and deformation morphology in three body abrasion *IJITEE* **5** 15-21
[2] Burwell JT 1957/58 Survey of possible wear mechanisms *Wear* **1** 119-141
[3] Hokkirigawa K and Kato K 1988 An experimental and theoretical investigation of ploughing, cutting and wedge formation during abrasive wear *Tribology Int.* **21** 51-57
[4] Kayaba T and Suzuki S 1976 An investigation of surface damages by rolling contact, Technology Report *Tohoku Univ.* **41** 21-4
[5] Jonas A, Jungedal M and Waara P 2013 wear on overlay welded HCWI vs quenched and tempered low alloyed carbon steels evaluated with granite in a laboratory drum test machine *Wear* 330-331
[6] Seth S, Jones AH and Lewis OD 2013 Wear performance of thermally sprayed Al-Ti alloy measured by three body micro-scale abrasive wear test *Wear* **302** 972-980
[7] Beata B and Kostencki P 2015 Abrasive wear characteristics of selected alloy boron steel as measured in both field experiments and laboratory tests *Wear* 328-329
[8] Victor AO Marcelo G, de Macedo CS, Souza RM, Scanadian C 2015 Effect of abrasive particle size distribution on the wear rate and wear mode in micro scale abrasive tests *Wear* 328-329
[9] Vytenisjankauskas, Antonov A, Varnauskas V, Skirkus R 2015 Effect of WC grain size content on low stress abrasive wear of arc welded hard facings, with low-carbon or stainless steel matrix *Wear* 328-329
[10] Hernandez S, Hardel J, Winklemnn H, Rodriguez Ripoll M, Prakash B 2012 influence of tempareture on abrasive wear of boron steel and hot forming steels *Wear* 290-291
[11] Bin H, Park K, Hoi-Bong K, Su Bae D, Cho YR 2012 Effect of tensile properties on the abrasive wear of steel saw wires used for silicon ingot slicing *Wear* 290-291
[12] Nagaraja KC, Rajanna S, Prakash GS, and Rajeshkumar G 2020. Improvement of mechanical and thermal properties of hybrid composites through addition of halloysite nanoclay for light weight structural applications *Journal of Industrial Textiles* 1-19

IOP Conf. Series: Materials Science and Engineering **925** (2020) 012047 doi:10.1088/1757-899X/925/1/012047

Comparative Study on Mechanical Properties of SiC/Gr &Al_2O_3/Gr reinforced AL6061 hybrid metal matrix composites

P Vignesh[1], R Sureshkumar[1], A Arun[1] and S Rethinavel[1]

[1] Department of Mechanical Engineering, Arasu Engineering College, Kumbakonam, Tamilnadu, India

Email: vigneshpackkirisamy@gmail.com

Abstract. Silicon carbide / Graphite and Alumina / Graphite reinforced AL6061 Hybrid metal matrix composites are fabricated by stir casting (liquid metallurgy) route. Four samples A,B,C,D with varying proportions in both matrix and reinforcements by fixing graphite proportion (5%) constant for all sampled are prepared. Mechanical properties of all the samples are compared with matrix material (AL6061). Scanning electron microscope is used to examine the microstructural characteristics of the composite samples. Mechanical test results exhibit 3.5% (sample C) increase in the hardness number than the base matrix. But, yield and ultimate tensile strength are reduced with all the reinforcements. Microstructural characterisation clearly depicts the presence of cracks, Agglomeration of reinforcements, cast defects on the surface of prepared composites which leads to poor yield and ultimate tensile strength.

Key words: AL6061 Hybrid metal matrix composites, Alumina, Graphite, Silicon Carbide, Stir Casting,

1. Introduction

In recent times, the engineering world is in requisites of newer materials to meet their demands. Such newer materials are also called as composite materials, their usage in military, automobile, aviation, defence sectors, keeps on increasing owing to their less weight, high thermal resistance, high strength to weight ratio, wear resistance, high hardness, and stiffness when compared to conventional engineering materials. In this way it is witnessed that hybrid metal matrix composites have overcome the limitations of other composite materials from recent researches. A hybrid metal matrix composites consist of a minimum of two distinct phases for reinforcement with the matrix phase. Aluminium based hybrid metal matrix composites find way in recent times to replace older materials. Researchers fabricated different grades of aluminium (AL6061, AL7075, AA6351, AL6082, AL2024) based hybrid MMCs with ceramic reinforcements like Alumina (Al_2O_3), Silicon Carbide (SiC), Titanium Boride (TiB_2), Titanium Oxide (TiO_2), Boron Carbide (B4C), etc. These reinforcements enhance the mechanical properties of composites to meet industrial demands. B.Jayendra,D.Sumanth,G.Dinesh, Dr. M.Venkateswara Rao had observed that reinforcement of B4C and Graphite in AL7075 enhanced the hardness, impact, and tensile strength of the composites considerably [1]. M.Satheesh, M.Pugazhvadivu found 8% reinforcement of

IOP Conf. Series: Materials Science and Engineering **925** (2020) 012047 doi:10.1088/1757-899X/925/1/012047

coconut shell ash (CSA) in AL6061-Sic matrix phase has increased the hardness and tensile strength by 46% & 47.31% when compared to 10% CSA mixed sample [2]. V.Jaya Prasad, K.Narasimha Rao, N.Kishore Babu R reinforced ceramics (TiB_2 /SiC) and observed the addition of ceramics in aluminium increases the mechanical properties of the composites [3]. Abhishek Sharma,Vyas mani sharma,Jinu paul has found reinforcement of graphene and carbon nanotubes (CNT) in AL6061-SiC matrix increases the nano hardness of composites by 27% and microhardness values by 36% than alone AL6061 [4]. V. Anirudh, M.Vigneshwaran,E.Vijay,R.Pramod, GB Veeresh Kumar studied the TiB2 and graphite-reinforced AL6061 alloy and concluded increasing the proportion of reinforcements increases the hardness and UTS of composites than alone AL6061 [5]. V.Mohanavel, K.Rajan, P.V.Senthil, S.Arul in another study informed after the dispersion of Al_2O_3 and Graphite in AA6351 alloy increased the mechanical properties of composites than pure AA6351 [6]. B.Ramgopal Reddy, C.Srinivas has studied the reinforcement of SiC and fly ash in AL6082 matrix as a base and found a considerable enhancement in UTS, Hardness, and wear resistance in the composites [7]. Cheng-jin Hu, Hong-ge YAN, Ji-hva CHEN.Bin SU reinforced Graphite and SiC in AL2024 matrix and concluded tensile strength and elongation of composites were reduced with reinforcements [8, 9]. From this literature survey it was observed that aluminium based hybrid metal matrix composites are prepared by various processes such as Squeeze casting, Vacuum hot pressing, Friction stir processing, Stir casting, and Powder metallurgy. Among the above methods stir casting is mostly preferred owing to some advantages over other methods, one of the important phenomena where stir casting stands prior to other methods is the preparation of a wide range of shapes with larger sizes is possible and this process is economically suited. There is a lack of research concentration in AL6061 based hybrid metal matrix composites. So, this work is aimed to fabricate a hybrid metal matrix composite of AL6061 reinforced with SiC/ Al_2O_3 (varying proportions) with constant proportion (5%) of Graphite using stir casting process and to compare their mechanical properties with the base matrix (AL6061).

2. Methodology

2.1. Material Selection

AL6061 was chosen as matrix material because of its cast ability and being utilized in wide range of applications such as in aircraft wings, fuselages etc., table 1 and table 2 shows the chemical composition and mechanical properties of AL6061.

Table 1. Chemical Composition of AL 6061

Component	Wt %
Mg	0.9
Si	0.62
Fe	0.33
Cu	0.28
Cr	0.17
Mn	0.06
Zn	0.02
Ti	0.02
Al	97.6

IOP Conf. Series: Materials Science and Engineering **925** (2020) 012047 doi:10.1088/1757-899X/925/1/012047

Table 2. Mechanical Properties of AL 6061

Mechanical Properties	Value
Hardness Rockwell B scale	60 RHN
Ultimate Tensile Strength	310 MPa
Tensile Yield Strength	276 MPa

SiC was chosen as one of the reinforcements during fabrication of composites. Silicon in SiC enhances the fluidity property of AL6061 during casting, meanwhile SiC possessing capacity to withstand in high temperatures with melting point of 2830°C.Owing to its high hardness and strength $Al_2O_3$was chosen as other reinforcement. Due to its high thermal resistance capability graphite was chosen as reinforcement with a constant proportion in all samples.

2.2. Stir Casting

AL6061 aluminium hybrid metal matrix composite samples A, B, C, D were prepared by stir casting process (under argon atmosphere) by adding SiC/Graphite and Al_2O_3/Graphite with varying proportions as shown in table 3. Stir casting has been selected for fabrication of composites as the reinforcements having uniform distribution in the matrix phase in this process. Al6061(CP) was taken in form of ingots and melted in the graphite crucible (750°C) and then followed by mixing SiC with stirring speed 350 rpm for 15 minutes and graphite is mixed to the molten mixture to prepare samples A&B. Then above procedure is followed by replacing SiC with Al_2O_3 to prepare samples C&D respectively. SiC and Al_2O_3 were used in the form of powders with 20μm particle size. Mechanical type stirrer is used in the stir casting setup. Stir casting setup and work flow chart is shown in the figures 1 and 2 as a pictorial illustration. Molten mixture was poured into the die to get circular specimen of 225mm length and 20mm diameter.

Table 3. Proportion of Reinforcements (in %)

Sample	Al 6061	Sic	Al_2O_3	Graphite
A	85	10	-	5
B	80	15	-	5
C	85	-	10	5
D	80	-	15	5

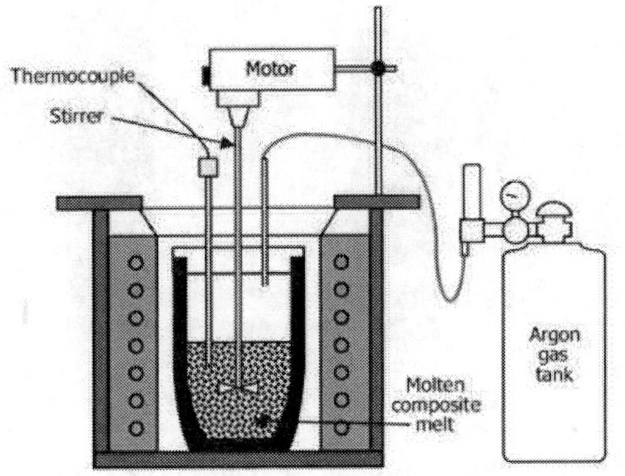

Figure 1. Stir Casting Setup

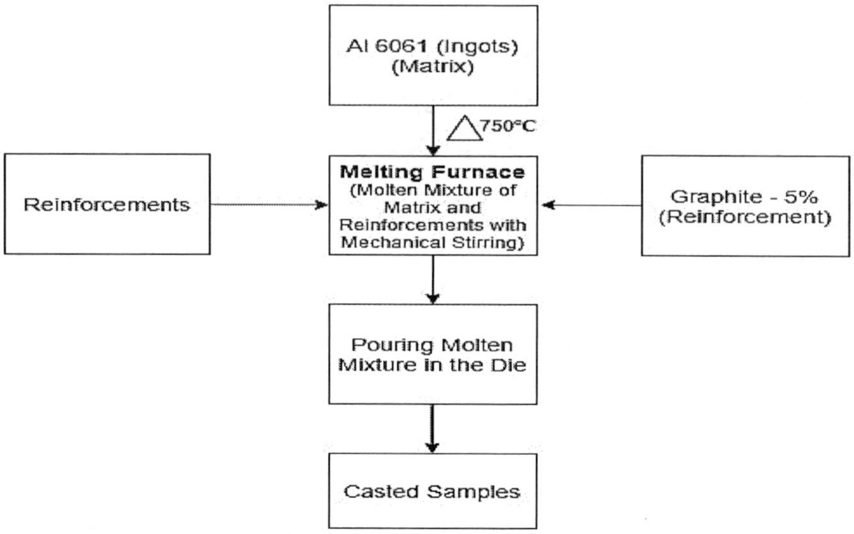

Figure 2. Process Flow Chart of Stir Casting

3. Results and Discussion

3.1. Hardness test

Rockwell hardness test machine (B-Scale) was used to measure hardness values in the fabricated samples. Hardness test specimens were prepared as per ASTM standards and the test results are shown in the below chart 1. Pre-treatment was given to the test specimens; its surfaces were degreased and polished to get even surface. Load of 450kgs was applied for 10 seconds. Sample "C" exhibits higher RHN value than the base matrix.

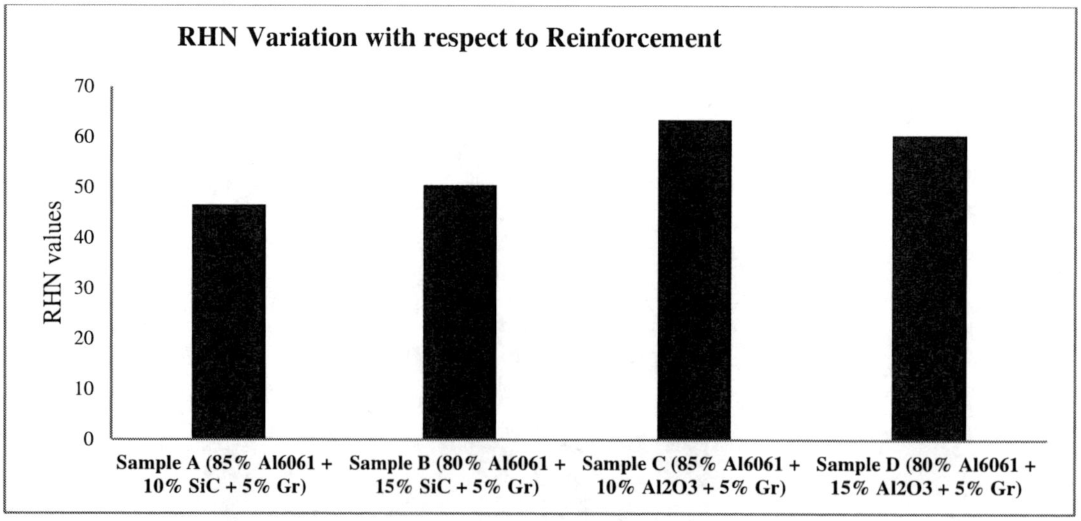

Chart 1. RHN Values Comparison Graph of Composite Samples

3.2. Tensile test

As per ASTM SA370 standard tensile test specimens were prepared like the design shown in the figure 3. Tests were carried on the prepared specimens in universal testing machine. The ultimate tensile strength (UTS) and yield strength values are obtained for all samples. The values are given in the below charts 2 & 3. Test results reveals that the yield strength and the ultimate tensile strength of all samples are decreased when compared with the AL6061 base matrix.

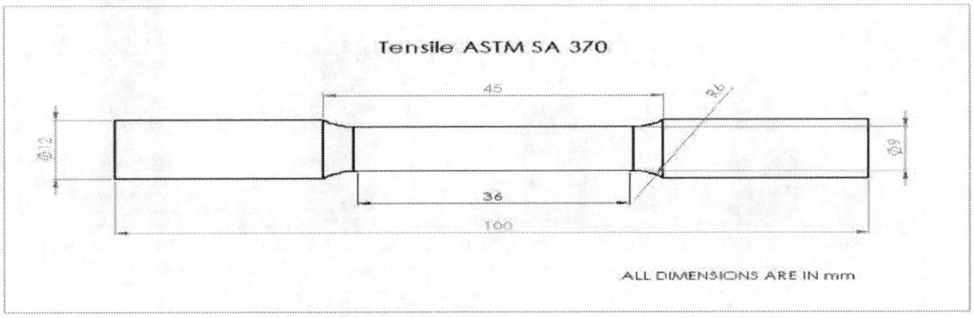

Figure 3. Design of Tensile Test Specimen

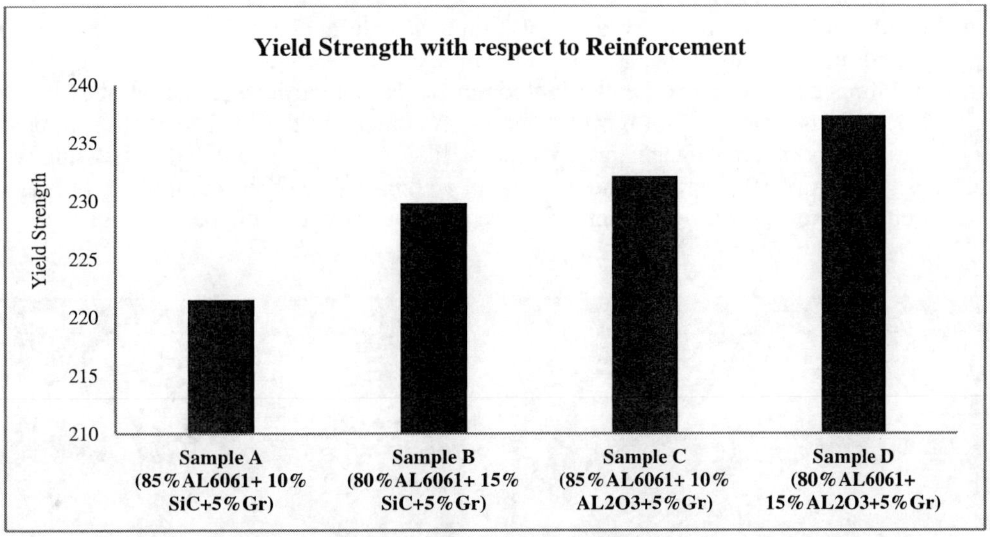

Chart 2. Yield Strength Results

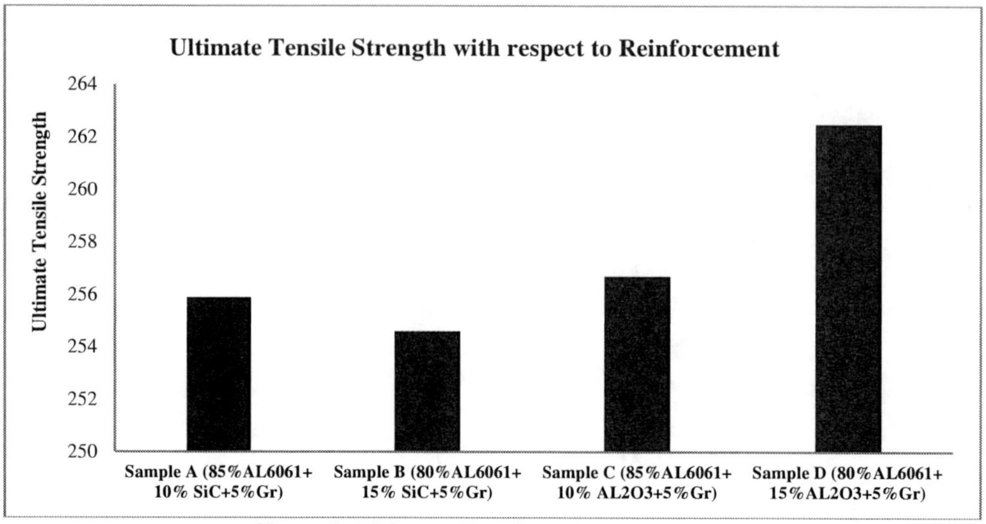

Chart 3. Ultimate Tensile Strength Results

3.3. Micro structural characterization

Scanning electron microscope was used to observe micro structural changes on the stir casted specimen surfaces with high magnification. Micro structural images of different samples were given in below figures 4,5,6&7. Image results show the distributed reinforcement particles in the AL6061 matrix phase with higher level of agglomeration of reinforcements and cracks in the surface of the composites. For the sake of getting good quality images pre-treatments like mirror polishing and degreasing were done on the surface each sample. The cracks and cast defect on the surface of the composites fails to improve the yield and ultimate tensile strength when compared with as cast AL6061 base matrix.

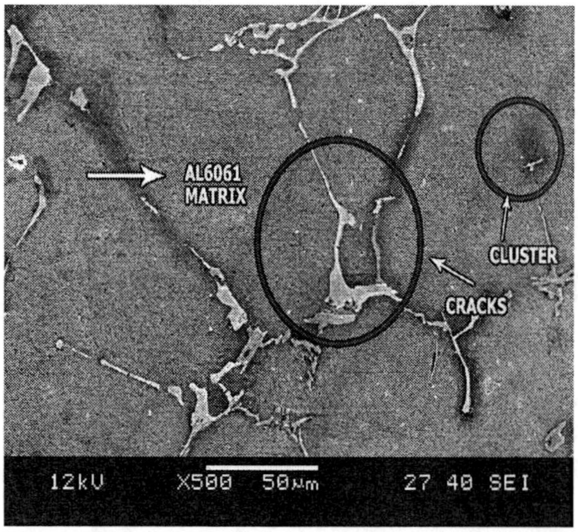

Figure 4. Sample A
(85% Al6061 + 10% SiC + 5% Gr)

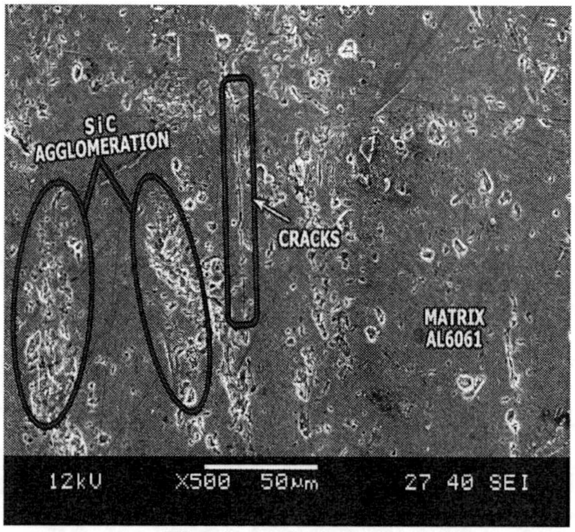

Figure 5. Sample B
(80% Al6061 + 15% SiC + 5% Gr)

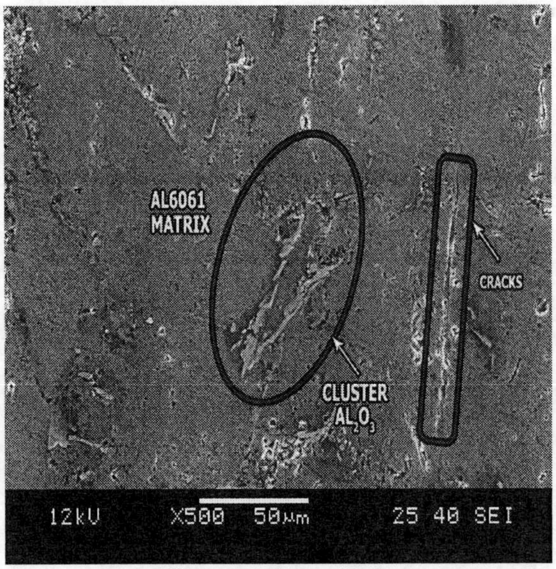

Figure 6. Sample C
(85% Al6061 + 10% Al_2O_3 + 5% Gr)

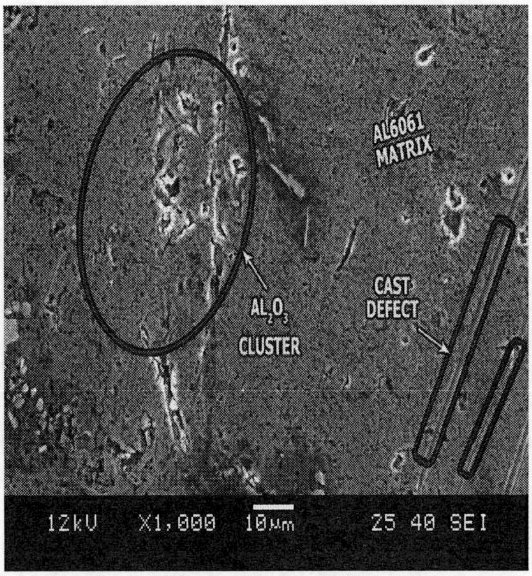

Figure 7. Sample D
(80% Al6061 + 15% Al_2O_3 + 5% Gr)

4. Conclusion

AL6061 based hybrid metal matrix composites reinforced with Al_2O_3/Graphite and SiC/Graphite samples with varying proportions with fixed proportion of graphite (5%) were fabricated using stir casting process. From the experimental results the following comparison statements can be drawn. There is an increase in the hardness value with increasing the proportion of the reinforcements, Sample "C" exhibits 3.5% increment in RHN number then the base matrix when compared with other samples.
Micro structural characterization images of all the samples A, B, C, D clearly shows the distribution of reinforcement particulates in the matrix phase with clusters of reinforcements in some places. Also, cast defects and surface cracks are present in the samples, which reduce the mechanical properties of the composites drastically. Cast defects are the major limitations found in the study.
This work will be extended by reducing the clustering of reinforcement particles in the stir casted composite by performing further heat treatment process. Secondly, optimization of process parameters could be done with optimization tools.

References

[1] Jayendra B, Sumanth D, Dinesh G, VenkateswaraRao M 2020 Mechanical Characterizationof stir cast AL-7075/B4C/Graphite reinforced hybrid metal matrix composites *Materials proceedings* **21**1104-1110.

[2] Satheesh M, Pugazhvadivu M 2019 Investigation of physical & mechanical properties of Al-6061-SiC/CSA (Coconut Shell Ash) hybrid composites *Physica B :Condensed Matter* **572** 70-75

[3] Jaya Prasad V, Narasimha Rao K, Kishore Babu RN 2020 Mechanical and tribological characterization of AL metal matrix composites reinforced with micro ceramics particles(TiB2/SiC) *Materials Proceedings* **23** 637-641

[4] Abhishek S, sharmaV, Paul J 2019 A comparitive study on microstructural evolution & surface properties of graphene/CNT reinforced AL-6061-SiC hybrid surface composites fabricated via friction stir processing *Transactions of non-ferrous metals* **29** 2005-2026

[5] Anirudh V, Vigneshwaran M, Vijay E, Pramod R, Veeresh Kumar GB 2018 Influence of TiB2 & Graphite reinforcement on AL6061 *alloy Materials Proceedings* 25341-25349.

[6] Mohanavel V, Rajan K, Senthil PV, Arul S 2017 Mechanical behaviour of hybridcomposites (AA6351+Al2O3+Graphite) fabricated by stir casting method *Materials today* **4** 3093-3101

[7] Ramgopal Reddy B, Srinivas C 2018 Fabrication & characterization of Silicon carbide and fly ash reinforced aluminium maetal matrix hybrid composites *Materials today proceedings* **5** 8374-8381

[8] Hu C, Yan H, Chen V and Bin SU 2016 Microstructures and mechanical properties of 2024AL/Gr/SiC hybrid composites fabricated by vacuum hot pressing *Transactions Of Non Ferrous Metals Society Of China* **26** 1259-1268.

[9] Nagaraja KC, Rajanna S, Prakash GS, and Rajeshkumar G 2020 Improvement of mechanical and thermal properties of hybrid composites through addition of halloysite nanoclay for light weight structural applications *Journal of Industrial Textiles* 1-19

IOP Conf. Series: Materials Science and Engineering **925** (2020) 012048 doi:10.1088/1757-899X/925/1/012048

Protective Coatings for Bio-Composites – A Review

Srinivasa C V[1], Allamaprabhu S Ani[2] and Jyothi Prasad B M[3]

[1,2,3] Department of Mechanical Engineering, GM Institute of Technology, Davangere, India.
srinivasacv@gmit.ac.in[1], allamaprabhu_16me008@gmit.ac.in[2], jyothi_16me024@gmit.ac.in[3]

Abstract. Even though Bio-based composites are eco-friendly, they are susceptible to degradation owing to both mechanical as well as biological factors. While it is evident that moisture affects the composites considerably, there are other factors such as heat and abrasion that could further accelerate the degradation. This article aims to encapsulate a few techniques for protection of the natural fibre-based composites. The reasons of thermal degradation and efforts to minimize the combustibility of substrates by application of aqueous, organic as well as intumescent coatings along with chemical treatments is discussed in brief. An account on the methods of applying a barrier coat and its subsequent curing is given. The need of gel coats, their advantages and several treatments to enhance the fibres to combat against degradation are outlined.

1. Introduction

Composites made of natural fibres have several advantages such as lower density, better thermal insulation and mechanical properties when compared to composites reinforced with synthetic fibres. Further, unlike synthetic fibres, natural fibres are broken down by bacteria once they are no longer in use. To enhance the life of the natural fibre composites a top layer called coating is applied on them that acts as a barrier against the external factors. The use of natural fibres in composites is beneficial because of its relatively high strength, hardness, and low density. They are eco-friendly, less abrasive to the tool and less irritating to the skin and respiratory system of the people who use them. Also, their properties vary significantly based on the chemical composition and structure of the fibres.

Natural fibre reinforced composites (NFRC) are gaining a lot of attention and importance due to the concept of sustainable development. Also, a life cycle assessment study by S. Joshi shows that NFRCs are superior than Glass Fibre Reinforced composites [1]. Although the sensitivity of such composites towards moisture is a cause of concern. This paper intends to review the different kinds of coatings for protection and enhancement of properties of composites.

Coating is the process of covering the surface of the object by applying a protective top film on it. The purpose of applying the coating may be decorative, functional, or both. From a surface engineering point of view, a coat is a layer of material deposited onto a substrate to enhance the surface properties and protect the substrate from external attacks. There is a wide range of coating processes ranging from just a few microns, up to several millimetres depositing upon different types of material. Even though there are a few wood preservatives, ultraviolet radiation absorbers, water repellents, fire retardants, paints and coatings, that are developed to shield bio-based composites from environmental degradation, all these solutions still pollute the environment in some way; therefore, it is essential that novel technologies are developed.

1.1. Reasons for degradation of Natural Fibre based composites: [2]

- Biological Degradation due to natural Enzymatic and Chemical Reactions by Fungi, Bacteria, Insects, Termites.
- Mechanical and Thermal Degradation due to factors such as abrasion and external heat leading to cracks and fracture of the top layer which in turn expose interior of composite.
- Weather Degradation in the form of shrinking, cracking, swelling due to the Sun, Rain and cycling temperature changes.

These fibres have several advantages over synthetic fibre composites in terms of mechanical properties along with lower densities, and biodegradability. However, their hydrophilic nature is a major disadvantage. Natural weathering is often a cause of changes in the properties of bio composites. Generally, shallow cracks appear on the surface after a few months of natural exposure. Several different formulations of coating as well as their application methods, along with their outcomes have been outlined in this paper.

2. Hydrophobic Coatings

The best of inventions accredited to humans are simply through manifestation and thorough observation of nature and its essence. Therefore, it seems quite logical to draw inspiration from nature itself even for development of protective coatings. The epidermis of a lotus leaf acts as a super-hydrophobic layer due to its unique surface roughness. Several efforts have been made to emulate this property. A surface is often considered as super hydrophobic when it exhibits a contact angle above 150° [3]. Fluor-containing compounds work as excellent hydrophilizing agents due to their low surface energy. Their high reactivity also enables a relatively high bonding strength of the coat towards the sample surface [4]. Also, it was seen that moisture sorption is reduced significantly with the increase in cross-linking of the fibre cell wall.

A mixture of epoxy resin and silica nano particles could be used to coat over glass slides using dip coating. Upon curing, when the polymer coating is exposed to oxygen plasma, it imparts a distinct surface roughness that mimics the roughness on a lotus leaf. Similarly, Ke et al. [5] have fabricated a super hydrophobic surface by sequential drop coating of silane-modified-silica solutions onto glass substrates. Thin super hydrophobic films were developed by assembling silica nanoparticles on glass substrates by C. Sun, L. Ge and Z. Gu. They employed two methods to create the lotus effect. In the first method, a template of 8 μm sized silica was coated with smaller particles by electrostatic deposition. The other method was by casting a Poly-dimethyl-silohexane mould on the template particles array.

The fibre-based composites must be subjected to a pre-treatment called silanization to form silane groups that act as a fibre-matrix coupling agent. Silanol (Si-OH) groups react with -OH groups of the fibres and the matrix functional groups which facilitates the adhesion of the coating on to the composite substrate. Zhidong Ma et al. developed a composite organic coating using Polyvinyl butyral, polymethyl-hydrosiloxane (PHMS) and chitosan, for use as a protective barrier for fibre-based composites [6]. However, these methods were not suitable and scalable for industrial applications. An industrially applicable, robust system for creating super hydrophobic surfaces on biobased composites has been developed at Chalmers University. A silica nanoparticle / polymer binder dispersion technique was devised for simplifying the application of hydrophobic barrier coat by spraying onto the composite surface like paint.

3. UV Curable Coatings

UV curable epoxy films were developed by M M. Aung using Jatropha seed oil. The acrylation process was reported to be slower and during the curing a thin film would be formed which further deterred the curing process as well as decreased the gel content [7]. Liang et al. developed an organic UV treatable coating using Tung oil-based derivative - Tung Maleic Anhydride. Challenges experienced in the curing of bio-based UV treatable coatings such as insufficient reactivity to UV radiation (which is paramount in achieving high crosslinking) and volume shrinkage were overcome. The high crosslinking density lead to improvement in thermal stability and enhanced the hydrophobicity and gel content.

Polyurethane coatings are majorly applied in the areas of rigid foams, elastomers, and sealants, but majorly used as adhesives. The surface properties such as pencil hardness, adhesion and flexibility of the coatings are especially important. Renewable cardanol can be used to prepare highly branched UV-curable oligomers that have a pencil hardness of 3H. Coatings based on Cardanol exhibited excellent adhesion, glass transition temperatures and hardness. Gaikwad et al. developed polyurethane coatings from Karanja and Cotton seed oils and their resulting mar hardness and pencil hardness was comparable to that of synthetically produced acrylic polyurethane 4H [8]. In addition, a coating with a hardness of 5H was developed by the use of hyperbranched polyurethane acrylate.

The hydrophobic property of PU coating was further improved upon by Zhang et al. by mixing the PU with Poly-dimethyl-silohexane (PDMS) and Aluminium flakes. The coating also had self-cleaning properties owing to its high Water Contact Angle (WCA) of about 117°. Upon further modification with nano SiO_2 they achieved super hydrophobicity (WCA 151.5°) and lowered glossiness. This makes the method suitable for stealth applications. Acrylated epoxidized soybean oil (AESO) is also good barrier against moisture when polymerized by UV light initiator. The physical, mechanical and water resistance properties of coated composites were found to improve in comparison with untreated samples [9].

4. Fire Retardant Coatings

Coatings have been developed to reduce fire hazards, prevent combustion of materials and reduce the heat evolved or emitted during their combustion, while also meeting regulations of fire-safety. They do so by interfering with the polymer combustion process [10]. Mineral flame retardants such as aluminium tri-hydroxide are very effective in reducing the risk of fire.

Treatment of cellulose fibres with a strong acid produces cellulose nanocrystals. The regions rich in cellulose are crystallized, while the amorphous regions of low order are easily penetrated by the acid and are hydrolysed. This method has been reported to be an effective way to encourage the formation of char during thermal degradation. It is interesting to note that chemical modifications like these can enhance properties and improve fire proofing. Hemicellulose, when subjected to thermal degradation, releases the water adsorbed by it. Further, there is rupture of its weak side branches, along with a decrease in mass. Notable causes of weight loss are dehydration and decarboxylation reactions that lead to release of gases like Carbon monoxide, Carbon dioxide, water vapour, furfural and acetic acid. Thermal degradation of hemicellulose is less affected by other biomass because its decomposition takes place at lower temperatures. The presence of minerals favours ring scission and produces volatile products due to the prevailing dehydration reactions and formation of furans. However, a notable disadvantage of hemicellulose is that it has a low initial thermal decomposition temperature which endangers its use as a filler in polymeric matrices.

Owing to the fact that the properties of substances are a result of their structure and composition, the stability and thermal decomposition rate of starch also depends on the same factors. Thermal stability of corn starch decreases as the ratio of amylose increases. Parallelly, due to presence of hydroxyl groups in Chitosan, the possibility of its usage in flame proofing applications is examined. The presence of carbohydrates promotes char formation [11]. Clay combined with chitosan has been used as a barrier coating on poly lactic acid (PLA) and poly urethane (PU) for high-performance insulation and packaging applications. Resistance to inflammation and burn spread on composite fibre can be induced by a coat of alumina trihydrate which is aqueous in nature, calcium carbonate and antimony trioxide in combination with Vinylidene-chloride acrylate which acts as a binder [12]. Another method of reducing the net heat evolved is by addition of low porosity and low melting inorganic Ceepree glass in combination with epoxy composites, where an inorganic carbonaceous layer is formed that acts as a flame inhibitor. This additive coating technology finds application in protection of the layers of composites from intense flames. The coated matter burns down, thus protecting the internal parts. Protection of metals from corrosion and fire is achieved by a fine coating of ethylene-chloro-trifluoro-ethylene copolymer (Solvay Solexis Halar) which has also been found to be effective with very thin coating [13].

Another method of barrier protection is intumescent coatings. Intumescent coating is a fire retardant and fire resistant material, which swells when heated, thus protecting the material underneath. Alcoa Corporation has patented an intrusive powder blend that contains fibrous or vitreous calcium magnesium silicate which finds applications as a mastic when mixed with a binder like polyvinyl acetate or acrylic resin. Additionally, an intumescent firestop or sealant can be made using the Expantrol 4BW commercialised by 3M along with hydrated sodium silicate-borate and a terpolymer of acrylate-vinyl acetate-ethylene, zinc borate, polyol, organic phosphate plasticizer and glass fibre. Dispersed carbon nanotubes are employed in silicon coatings in order to make surfaces such as textiles, foams, cables, metals and wood to be flame proof [14]. The barrier coat on the surface can be as thin as 10 mm. The coatings are made from unsaturated polyester resins and are subsequently cured. Glass reinforced polyester laminates can be coated with this material with a thickness of as little as 0.4 to 0.5 mm, the applied coat also provides a smooth surface. Commonly, formulations of ordinary unsaturated polyester resins and curing catalysts like peroxides are alternatively combined with thixotropic agents in order to control the flow properties. Flame retardancy of composites is quite useful in construction, marine and railway applications.

Different chemical treatments alkalization, salinization, and combination of both (alkalization & salinization) done on jute, ramie, sisal, and curauá fibres exhibit an improvement in the durability of the composites that are subjected to heat. Furthermore, the treatments also improved the adhesion of the natural fibres to the polymer matrix. An US patent avows that, including highly water-resistant resin components in an emulsion of polyvinylidene fluoride (like Arkema's Kynar Aquatec), enhances the water resistance of intumescent coatings that are otherwise hydrophilic. Although, intumescent coatings swell up to protect the substrate when heated, some intumescent coatings applied on wooden panels are susceptible to cracking upon exposure to flames and burst open fully, thereby losing ability to protect the inner substrate. A study by N Amir [15] shows that the addition of rockwool fibres to an ammonium poly phosphate – melamine – pentaerythritol formulation, inclination to cracking is controlled. Another intumescent coating was developed using Chicken Egg Shells, ammonium polyphosphate, pentaerythritol and melamine [16]. The coating was found to enhance the barrier properties by forming a multicellular insulating foam reducing the conduction of heat into the underlying substrate. It was observed that the coating had acquired a thermal stability superior adhesion along with water repellence.

5. Gel Coats

Multiple properties like resistance to thermal degradation, conductivity to heat and electricity can be effectively brought out through polymer coatings, as it requires the application of different additives. While there is always a possibility of negatively affecting other properties of the composite structure, the additives can be applied in separate layers on the substrate. However, by the means of gel coats, this can be done in a single layer. Gel coats prove to be functional barriers that have versatile properties along with a superior surface finish. M Landowski developed a technique for the protection of glass fibre reinforced composites. Studies on both Macro and microstructural damages a typical Glass Fibre

Reinforced Polymer laminates was done using gel coat based on isophthalic acid and urethane modified vinyl ester [17]. They proposed that protection against blistering and degradation of the laminate due to water absorption could be done using a combination of nanoparticles reinforced gel coat (10 % nano particles). This is due to tortuosity of diffusion paths and resin rich boundary coating that prevents water diffusion inside laminate and subsequently its mechanical degradation. Multifunctional gelcoats are discussed extensively in a study by Pomázi et al. based on their functionalities; primarily flame retardancy, electric conductivity and hydrophilicity. Several methods of gelcoat preparation, application by spraying, rolling, in-mould coating, curing by exposure to ultraviolet radiation are discussed. Also, common defects occurring over the course of gel coating and testing methods are outlined [18].

6. Organic Coatings

Several eco-friendly and non-polluting plastic products have been developed by utilization of polymers made from spent sugarcane, apple, and oranges. Composite blends of poly vinyl alcohol and lignocellulosic fibres are derived from these spent fruits. Hexa-methoxy-methylamine which is a crosslinking agent, could be used for reducing water sorption on composites as they show a promising water repellent characteristic along with good binding properties. Bio-based coatings were used by T.H Mokhothu et al. for reducing the amount of water absorbed by composites containing phenolic resin. Polyurethane (PU) and Polyfurfuryl alcohol (PFA) resin were the two kinds of coatings examined. Results showed that samples coated with PFA had better mechanical performance and higher shielding against moisture in comparison with other bio-based coatings [19]. It was also observed that high crosslinking density caused higher hydrophobicity. It is evident that moisture diffusion is the reason for hydrolysis and the consequential plasticization of the matrix. Le Duigou et al. tried using Poly (L-Lactic Acid) as an extra layer over the flax based composites to protect them from degradation from saline sea water in order to overcome the degraded mechanical properties of the composite due to wet ageing. The outcomes show that PFA serves as a good bio-based coating for fire proofing of treated natural fibre reinforced composites [20]. Sodium silicate extracted from rice husk also has the potential to be used as thermal insulator. Water absorption increases when a coat of Cellulose Nano Fibrils (CNF) was applied onto a cardboard. The hydrophilicity of CNF film and coatings can be reduced by incorporating alkyd resins. Higher water content results in increased drying costs, which is unacceptable unless it can be compensated by improved product properties.

7. Fibre Treatments to Combat Degradation

A super hydrophobic cotton fabric was made by low pressure, plasma enhanced, chemical vapor enhanced deposition technology. Lauryl methacrylate was used as a functional monomer for treating the fabric. The properties like washing stability, hydrophobicity, permeability of water vapor and mechanical tensile strength of the resulting fabric are given in the study by L Xu et al. A stable and highly breathable, hydrophobic cotton fabric was successfully prepared without affecting its tensile strength. Due to the nano structure cells and grooves that are formed after the plasma treatment, the fraction of wetting area was diminished. Generally, water drops rapidly wet the surface of untreated cotton samples, but it was seen that water drops rounded on the treated surfaces and were non-wetting. To give a quantitative account, the contact angles improved from 87.73° to 163.65°, which explains the non-wettable property [21]. Increasing the temperature of water accelerates the rate of degradation, this is because of the combined effect of both increased absorption and accelerated enzymatic reactions due to the hot water. E. Rodríguez and G. Francucci discovered a novel method to treat plant fibres and they analysed its influence on the water absorption and mechanical properties of composites made of vinyl ester matrix. Jute fibres were treated with an alkali and polyhydroxy butyrate and the treatment enhanced the resistance against water absorption [22].

Protective coat of TiO_2 nanoparticles in combination with a fatty acid, can increase hydrophobic property of nonwovens prepared from recycled jute. It was seen that the samples that were coated showed superior hydrophilicity in response to the TiO_2 nanoparticle pre-treatment. Subsequent treatment with steric acid (even at a low concentration), further improved the water repellence [23]. Further another sol–gel treatment was employed that helped in minimizing the fibre swelling. Seepage of water and its constituents corrupts the dimensional stability and also degrades mechanical properties and eventually creates a suitable environment for fungus attack. The reduction in swelling of the fibres is another aspect of importance so as to not deteriorate the original thickness and mechanical performance. Another noticeable feature of this coating is that the recycled fibres do not need initial cleaning; hence fibre strength is not affected.

Epoxy coating is another effective way to simultaneously increase the resistance of bamboo composite material against acid attack and improve its bond strength with concrete for the application of concrete reinforcement [24]. This paves a way to use coatings for not only external protection of the composites but also as a protective layer over reinforcements. Addition of Sodium silicate into PLA retards the fire propagation to half the rate in comparison to that of neat PLA and about 10% compared to fibre mat composites without sodium silicate coating [25]. When applied to fibre composite made of abaca, the rate of fire propagation was found to decrease by about 40%. Among abaca, jute and sisal fibres, abaca absorbs sodium silicate highly, due to the high degree of network structure found in abaca fibres. The silicate coating applied on the Abaca mat also increases the Elastic modulus of the composites, meaning that the silicate coat boosts the overall stiffness.

8. Conclusions and Scope for Further Research

It is conspicuous that interfacial adhesion between natural fibres and matrix resin decreases over time, which weakens the capacity of the composite to carry loads. The compatibility between fibre and the polymer matrix could be improved. Adhesion to substrates, especially in the case of natural fibres, requires a primer coat or a pre-treatment, but this can be avoided by self-priming formulas. Adhesion of the barrier coats to fibre or elastomer surfaces provides an R&D opportunity. Furthermore, there is a need for single coat barriers that are feasible in terms of cost and are easier in application. Some intumescent coatings on wooden panels are susceptible to cracking when exposed to flames and they open at the full extent of their expansion, thereby losing ability to protect the inner substrate. Flame retardant coatings are usually prone to water damage, whereas coatings that impart hydrophobic properties are advertently susceptible to flame damage. Materials that can heal themselves when damaged can be designed to sense failure and as a result, respond automatically to restore structural function. Possibly, nanocomposites can be used in coatings to impart self-healing properties.

References

[1] Joshi S, Drzal L, Mohanty A and Arora S 2004 Are natural fibre composites environmentally superior to glass fibre reinforced composites? *Composites Part A: Applied Science and Manufacturing* **35** 371–376
[2] Rowell R 2012 *Handbook Of Wood Chemistry And Wood Composites* (Boca Raton: CRC Press)
[3] Nosonovsky M and Bhushan B 2009 Superhydrophobic surfaces and emerging applications: Non-adhesion energy green engineering *Current Opinion in Colloid & Interface Science* **14** 270–280
[4] Ma M and Hill R 2006 Superhydrophobic surfaces *Current Opinion in Colloid & Interface Science* **11** , 193-202 Available: 101016/jcocis200606002
[5] Ke Q, Fu W, Jin H, Zhang L, Tang T and Zhang J 2011 Fabrication of mechanically robust superhydrophobic surfaces based on silica micro-nanoparticles and polydimethylsiloxane *Surface and Coatings Technology* **205** 4910–4914
[6] Ma Z, Sun M, Li A, Zhu G and Zhang Y 2020 Anticorrosion behavior of polyvinyl butyral (PVB) / polymethylhydrosiloxane (PMHS) / chitosan (Ch) environment-friendly assembled coatings *Progress in Organic Coatings* **144** 105662

[7] Aung MM, Yaakob Z, Abdullah LC, Rayung M and Li WJ 2015 A comparative study of acrylate oligomer on Jatropha and Palm oil-based UV-curable surface coating *Industrial Crops and Products* **77** 1047–1052
[8] Gaikwad MS, Gite VV, Mahulikar PP, Hundiwale DG and Yemul OS 2015 Eco-friendly polyurethane coatings from cottonseed and karanja oil *Progress in Organic Coatings* **86** 164–172
[9] Zhang W, Jiang S and Lv D 2020 Fabrication and characterization of a PDMS modified polyurethane/Al composite coating with super-hydrophobicity and low infrared emissivity *Progress in Organic Coatings* **143** 105622
[10] Laoutid F, Bonnaud L, Alexandre M, Lopez-Cuesta JM and Dubois P 2009 New prospects in flame retardant polymer materials: From fundamentals to nanocomposites *Materials Science and Engineering: R: Reports* **63** 100–125
[11] Laufer G, Kirkland C, Cain A A and Grunlan JC 2012 Clay–chitosan nanobrick walls: completely renewable gas barrier and flame-retardant nanocoatings *ACS Applied Materials & Interfaces,* **4** 1643–1649
[12] Laufer G, Kirkland C, Cain AA and Grunlan JC 2012 Clay–chitosan nanobrick walls: completely renewable gas barrier and flame-retardant nanocoatings *ACS Applied Materials & Interfaces* **4** 1643–1649
[13] Chandrasekaran S, Kundel N K, Garg B and Chin H B 1990 *US Patent No 4957961* Washington DC: US Patent and Trademark Office
[14] Weil ED 2011 Fire-Protective and Flame-Retardant Coatings - A State-of-the-Art Review *Journal of Fire Sciences* **29** 259–296
[15] Amir N, Othman SW and Ahmad F 2015 Fire resistance properties of ceramic wool fibre reinforced intumescent coatings
[16] Yew MC, Sulong NHR, Yew MK, Amalina MA and Johan MR 2014 Fire Propagation Performance of Intumescent Fire Protective Coatings Using Eggshells as a Novel Biofiller *The Scientific World Journal* **2014** 1–9
[17] Landowski M, Budzik M and Imielińska K 2013 Water absorption and blistering of glass fibre-reinforced polymer marine laminates with nanoparticle-modified coatings *Journal of Composite Materials* **48** 2805–2813
[18] Pomázi Á and Toldy A 2019 Multifunctional Gelcoats for Fibre Reinforced Composites *Coatings* **9** 173
[19] Mokhothu TH and John MJ 2017 Bio-based coatings for reducing water sorption in natural fibre reinforced composites *Scientific Reports* **7**
[20] Duigou AL, Deux JM, Davies P and Baley C 2010 PLLA/Flax Mat/Balsa Bio-Sandwich Manufacture and Mechanical Properties *Applied Composite Materials* **18** 421–438
[21] Xu L, Deng J, Guo Y, Wang W, Zhang R and Yu J 2018 Fabrication of super-hydrophobic cotton fabric by low-pressure plasma-enhanced chemical vapor deposition *Textile Research Journal* **89** 1853–1862
[22] Rodríguez E and Francucci G 2015 PHB coating on jute fibres and its effect on natural fibre composites performance *Journal of Composite Materials* **50** 2047–2058
[23] Arfaoui MA, Dolez PI, Dubé M and David É 2018 Preparation of a hydrophobic recycled jute-based nonwoven using a titanium dioxide/stearic acid coating *The Journal of The Textile Institute* **110** 16–25
[24] Rahman N, Shing LW, Simon L, Philipp M, Alireza J, Ling CS, Wuan LH and Nee SS Enhanced bamboo composite with protective coating for structural concrete application *Energy Procedia* **143** 167–172 2017
[25] Thongpin C, Srimuk J, Hipkam N and Wachirapong P 2015 Effect of natural fibre types and sodium silicate coated on natural fibre mat/PLA composites: Tensile properties and rate of fire propagation *IOP Conference Series: Materials Science and Engineering* **87** 012078

Experimental Investigation of Mechanical Properties of Al6061 Reinforced with Iron oxide

S P Shivakumar[1], A S Sharan[2] and K Sadashivappa[3]

[1] Department of Mechanical Engineering, SJM Institute of Technology, Chitradurga, India
[2,3] Department of Mechanical Engineering, Bapuji Institute of Engineering and Technology, Davangere, India
[1]shiv180cta@gmail.com

Abstract. The present work aimed at investigating the mechanical properties of the aluminum matrix composites. The aluminum 6061 alloy as a matrix reinforced with the weight fraction 2, 4 and 6% of particles of iron oxide (Fe_2O_3) of size 40μmis prepared using the stir casting technique. The prepared specimens are tested as per the standards for tensile, bending, impact and hardness properties of the composite. From the results it is observed that the mechanical properties of the said composite have increased with the increment in the weight percentage of iron oxide up to 4% of reinforcement. The significant reaction between the reinforcement and the matrix forms the better bonding between them. The enhanced mechanical properties of the aluminum-iron oxide composites are due to the fine, well bonded and homogeneous distribution of the iron oxide particles in the aluminum matrix. The maximum improvements were observed at 4 wt% of Fe_2O_3.

Keywords: Aluminium alloy, Iron oxide, mechanical properties, MMCs, Stir casting.

1. Introduction

Aluminum matrix composites referred as AMCs are now a day's attractive for their high strength, stiffness, improved fatigue strength, workability, low density, better thermal stability and improved wear resistance, corrosion etc. [1]. Thus AMCs find their applications in aviation, automobile, structural as well as transportation industries [2]. Composites with Fe3Al reinforcement exhibits better strength in contrast with the some of the ferrite-austenite steels. Hence these become potential candidates for the applications in high temperature materials. On the other hand, the strength of these materials decreases at higher temperatures which limits their applications [3-4].

Various authors studied the Al-Fe system, and reported that as the Fe content increases the modulus of elasticity and rigidity increases. The poison's ratio of the Al-Fe system decreases nearly by 0.0023% for every 1% addition of Fe. It has been also reported that the increases in the Fe content increases the creep strength, slight improvement in the machinability and decrement the fatigue strength if the crystals size is course [5]. Wang et al [6] conducted the experimental investigation on the influence of the chemical composition of the alloy prepared using the high pressure die casting technique. The composite has prepared using the 7 wt% and 13 wt% of silicon reinforcement. From the results it is concluded that the elastic modulus of the alloy has no significant influence by the chemical composition of alloy. On the other hand, the increment in the Fe content by 0.7 to 1.6% has the slight improvement in the elastic modulus. M. Makhlouf [7] worked on the Al–Si alloys and stated that the addition of the Fe, which is insoluble, enhance the strength at elevated temperatures. Authors noted that the adding of elements Fe and Ni reduces the Cu content without reducing the high temperature strength. On the other hand, authors observed that the addition of Fe and Mn holds the crack propagation rate of the alloy. It is also reported that the adding of the Fe in an Al–Si–Mg alloy, have significant influence, improves the hot tearing resistance as it increases the high temperature strength. Maity et al [8] worked on the addition of the CuO particles in the pure aluminum for the production of the Al_2O_3 in situ particle composite. In the reaction, the metal generally reacts with aluminum and forms the intermetallic phases such as Al_2O_3 and Fe_3Al.

Sarkari Khorrami et al [9] examined the influence of the number of FSP passes on the micro-structural and mechanical properties of aluminum matrix reinforced with 4 wt% of Fe particles. The two FSP parameters used are volume fraction of the reinforcing particles and the tool rotation direction. Through this study the effectiveness of the formation of in-situ particles, grain boundaries and other strengthening mechanisms were explored.

Similar works has been carried out by many researchers about the preparation and evaluation of mechanical properties of the Al-matrix composites with Fe as a reinforcement prepared using stir casting technique [10-12], powder metallurgy route [13-16] etc. From the obtained results it is concluded that the presence of Fe in the composite enhances the mechanical properties of the material and that strengthening mechanism is related to the uniform distribution and grain refinement of the Fe particles and the generation of the intermetallic.

Stir casting method of casting the aluminium composites were widely used because of their liquid phase method, economical and ease to conduct. The other methods include the powder metallurgy, die casting, centrifugal casting, pellet method etc. which are expensive, have complex process and methodology and may require the skilled labours. By this way they were limited in application than the stir casting technique.

Various authors studied the aluminium composites produced using the stir casting method. The fabrication of aluminium composites with the iron oxide and multi wall carbon nano tubes(MWCNT) [17] reinforcements, AA6061-redmud composites [18] and AA6061-rutile composites [19] through stir casting techniques. The study of the mechanical properties of aluminium-silicon carbide composites has been carried out using the stir casting techniques [20-23] for varied weight fraction of the silicon carbide. From the results it is observed that the mechanical properties of the said composite have been increased as the increment in the reinforcement [21-23]. However, authors [22, 27] reported in their work that the addition of SiC particles up to 5 wt% in the aluminium matrix will enhance the mechanical properties of the composites.

From the comprehensive study of literature, it is identified that much work has been carried out on the aluminum matrix Fe reinforcements prepared by the powder metallurgy route. There is a scope for the study of the aluminum-iron oxide composite produced using the stir casting technique. The major objective of this work is to investigate the mechanical properties of the AA6061-Fe_2O_3particulate composites produced using the stir casting route. The composition considered for the preparation of the composite is 2wt%, 4wt%, and 6 weight fraction of the reinforcement. The tensile, bending, impact and hardness tests of the said composite will be conducted using ASTM standards. Through this work an attempt has been made to compare the AA6061-Fe_2O_3 composites with AA6061-SiC composites.

2. Materials and Preparation

2.1. Material Selection

Aluminum 6061 as matrix material is used in the present work. Its chemical composition is given in the Table 1. AA6061 is selected due to its outstanding properties such as weld ability, light weight, corrosion resistance, etc. which were necessary for the applications in the automobile and aerospace industries. The reinforcement chosen is iron oxide.

Table 1. Chemical Composition of Aluminum 6061

Element	Mg	Si	Cu	Zn	Ti	Mn	Cr	Al
Wt %	0.85	0.68	0.22	0.07	0.05	0.32	0.06	Balance

Figure1 shows the particles of the iron oxide, which is used as reinforcement, is inorganic compound with chemical name Fe_2O_3. Iron oxide may available in main three forms such as FeO, which is rare; Fe_3O_4, occurs naturally; and the Fe_2O, which is the main source of iron. The iron oxide is the ferromagnetic material, in appearance dark red in color, ready to react with the acids. Sometimes iron oxide referred as the rust, due to its similar composition and shares many properties as iron.

Figure 1. Iron oxide powder

2.2. Fabrication of composites using stir casting

The techniques used in manufacturing of the metal matrix composite are expensive, complex process and skilled labors are required. Among many methods the liquid phase routes were attract the researcher because of their ease and economical as like conventional casting processes and can produce components of complicated shapes. The most economical, widely used liquid phase route method is the stir casting method which is shown in figure 2.

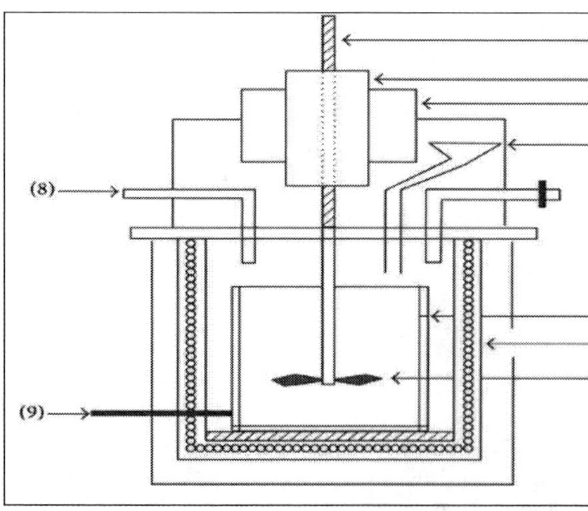

(1) Electric Motor
(2) Shaft
(3) Holder
(4) Thermocouple
(5) Particle Injection Chamber
(6) Insulation Hard Chamber
(7) Stirrer
(8) Thermocouple
(9) Thermocouple

Figure 2. Stir casting setup [12]

The 2, 4 and 6 wt% of the iron oxide particles were poured, separately, to the super-heated liquid aluminum at the temperature 720°C. The required quantity of the reinforcement is added while stirring at the speed of 550 rpm. The liquid Al6061- Fe_2O_3composite then poured to the graphite mold and left to solidify. The parts taken from the mold were now machined to prepare the specimens for different experimental testing.

3. Experimentation

To determine the mechanical properties of the AA6061- Fe_2O_3particulate composite different tests were conducted such as tensile, bending, impact and hardness test. For all the three combinations i.e. AA6061-2wt% Fe_2O_3, AA6061-4wt% $Fe_2O_3$and AA6061-6wt% Fe_2O_3, specimens were prepared to examine the material properties. All the experiments were carried out as per the ASTM standard testing procedures.

3.1. Tensile test

The tensile tests for all the three compositions have been conducted using universal testing machine (UTM). The specimens prepared are as per the ASTM E08-8 standards. The standard test specimen with dimensions is given in figure 3. The specimen diameter used is 13mm and gauge length used is 57mm, is fitted in a UTM of capacity 200kN. During testing the load applied and the displacements reading were recorded to calculate the stress and strain. From the experiments, yield strength, ultimate tensile strength etc. can be determined. The results were based on the average of three samples.

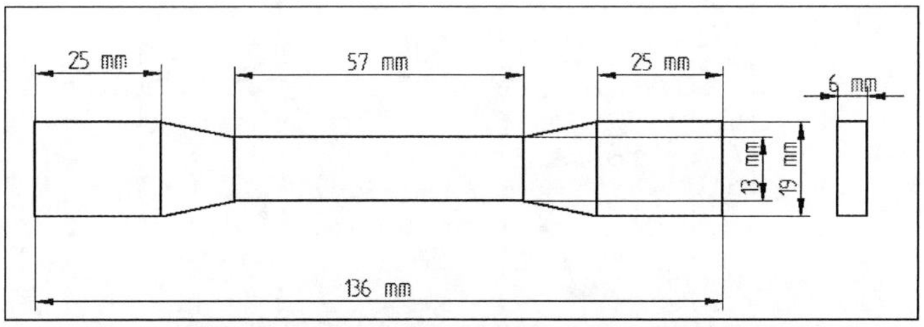

Figure 3. Tensile Test Specimen [24]

3.2. Flexural test

The bending test experiment was conducted for the all the three compositions. The specimen prepared is as per the ASTM E290 standards. The specimen (shown in figure 4) is prepared for the following dimensions to study the flexural behaviour of the composite.

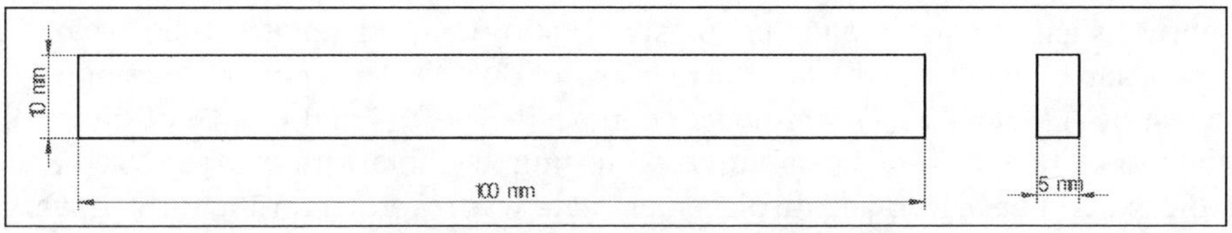

Figure 4. Flexural test specimen

3.3. Impact test

Due to the reliability and economical, charpy impact testing is widely used to test the metallic materials. The charpy impact test is conducted, for all the three compositions, to absorb energy required to initiate the crack and continue till the fracture. The specimen (shown in figure 5) is prepared as per ASTM E23 standards.

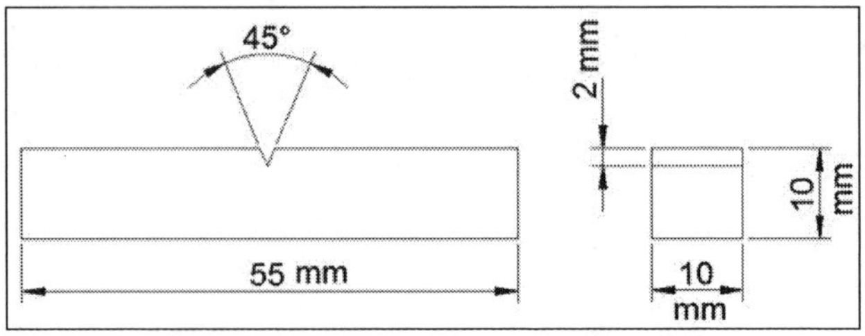

Figure 5. Impact test specimen [25]

3.4. Hardness test

The characteristic that resist to the deformation of the solid material is referred as the hardness. Hardness test can be conducted based on the depth of penetration under the application of the external uniform load. For the hardness test, ASTM E23 standards were prescribed for the specimen preparation. Prepared specimens with dimensions were shown in figure 6. The hardness tests were carried out, for all the three compositions, using the diamond indenter in the Brinell hardness testing machine with applied force of 10kgf. For each composition, three specimens were prepared and tested and average value is considered as the hardness of the composite.

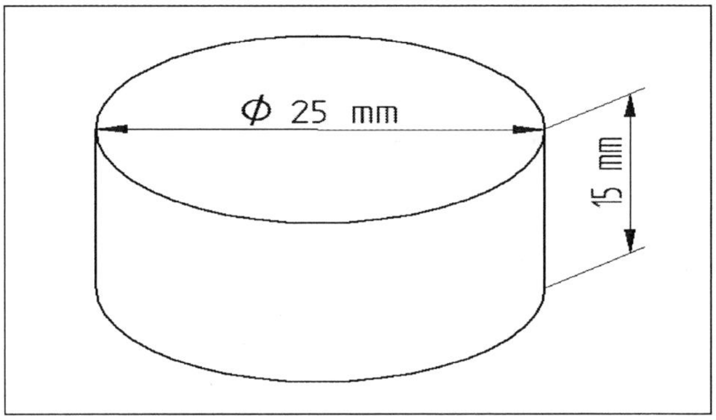

Figure 6. Hardness test specimen [26]

4. Results and Discussion

4.1. Tensile Properties

Tensile properties such as elastic modulus, poison's ratio, tensile strength, yield strength, percentage reduction in area and elongation (which is the measure of ductility) etc. can be determined by the tensile test. Tensile test of AA6061- Fe_2O_3 particulate composite for 2, 4 and 6 wt% of the reinforcement is conducted, in a room temperature, using universal testing machine with computerized data acquisition units. From the experiments, load and displacement data was recorded for the further analysis.

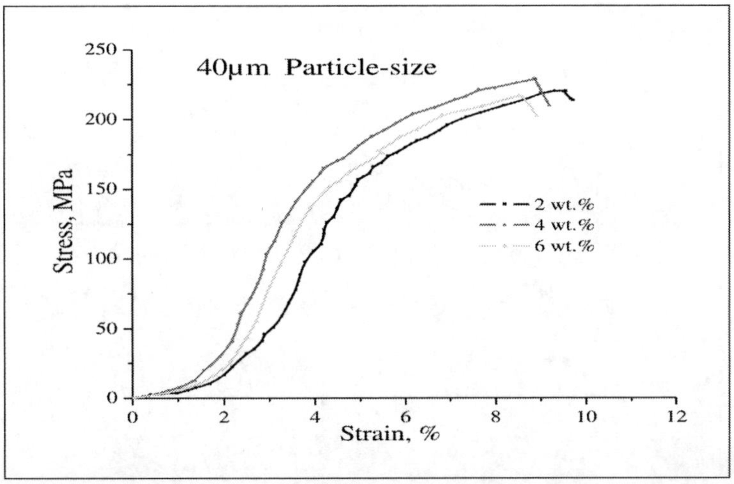

Figure 7. Tensile curves of AA6061- Fe_2O_3composite for various percentages of iron oxide

Test specimens of the composite were tested, for the fracture load, till it reaches ultimate tensile strength. From the recorded data of load and displacement, the stress and the strain values were calculated. The stress-strain graph is plotted to found the elastic modulus, yield strength and ultimate tensile strength as shown in figure 7. From the plot it can be observed that the reinforcement has significant influence on the tensile properties of the AA6061- Fe_2O_3 composite. As the composition of the reinforcement increases from 2wt% to 4wt%, the increment in the tensile properties of the composite has been observed with the decrement in the strain by 0.1%. Also further increment in the composition from 4wt% to 6wt%, the decrement in the tensile properties has been observed. From the literature, it has been also observed that the mechanical properties of the MMCs were optimum for the composition in the range 3 to 5wt% of reinforcement [12]. The determined values of the tensile properties of the AA6061-Fe_2O_3 composite have been listed in the Table.2.

Table 2. Tensile properties of Aluminum Iron oxide composites with 40μm particle.

Sl.No	Specimens	Yield strength(MPa)	Ultimate tensile strength(MPa)	Young's modulus(GPa)	Percentage Elongation
1	AA6061	110	152	60	20
2	AA6061+2% Fe_2O_3	175	215	64	10
3	AA6061+4% Fe_2O_3	186	229	70	9.8
4	AA6061+6% Fe_2O_3	178	220	66	8.9

From the table it is also observed that there is little decrement in the percentage elongation for the reinforced composite. This means that AA6061-Fe_2O_3 composite is able to maintain its ductility even after increasing the weight fraction of the reinforcement. Hence the enhancement in the tensile, yield and elongation of the composite has been observed.

4.2. Bending Properties

The three point bending specimen is used to characterize the flexural strength. Flexural strength is the stress at failure of the material under the flexural load. All the three composition of the composites have been tested as per the standard testing procedures to find the bending properties. During bending test, the applied load and displacement values have been recorded and used to calculate the values of stress and strain. The strain-strain curves were plotted using the calculated values of stress and strain.

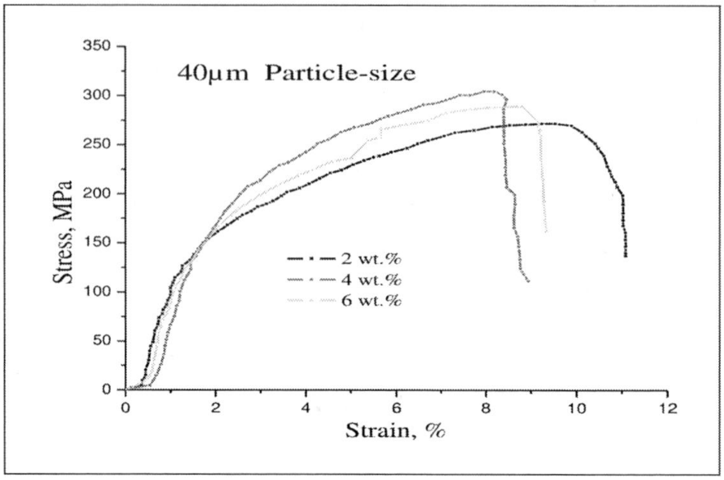

Figure 8. Bending curves of AA6061- Fe_2O_3composite for various percentages of iron oxide

Table 3. Flexural properties of AA6061- Fe_2O_3composites with 40 μm particle.

Sl.No	Specimens	Flexural strength (MPa)	Percentage Elongation
1	AA6061	152	20
2	AA6061+2% Fe_2O_3	271	10
3	AA6061+4% Fe_2O_3	305	9.8
4	AA6061+6% Fe_2O_3	289	8.9

The bending curves of AA6061-Fe_2O_3 composite for various weight percentages (2wt%, 4wt% and 6wt%) of reinforcement is shown in the figure 8. From the plot it is clear that as the increment in the composition of iron oxide the bending resistance increases. It is also observed that the bending properties of the AA6061-Fe_2O_3 composite increases with the weight fraction up to 4 wt% of the reinforcement, further increases of composition decrease the bending properties.
The calculated flexural strength and the percentage elongation are listed in the Table.3. From the outcomes it is clear that addition of iron oxide greatly influences the flexural strength of the composites. The flexural strength of the AA6061- Fe_2O_3 composite increases as the increment in the percentage reinforcement up to 4 % and the maximum flexural strength obtained is 304 MPa. There is increment of 50% in the flexural strength of the composite for 4 wt% of iron oxide in comparison with the unreinforced aluminium alloy.

4.3. Impact Properties

Using the charpy impact test, the energy required to break the sample can be determined. Charpy test is the one used to characterize the material under impact loading conditions. The Charpy impact test is conducted for all the compositions of AA6061-Fe_2O_3 composite specimens to determine its impact strength. The energy absorbed (J), in room temperature, for all the specimens is recorded.

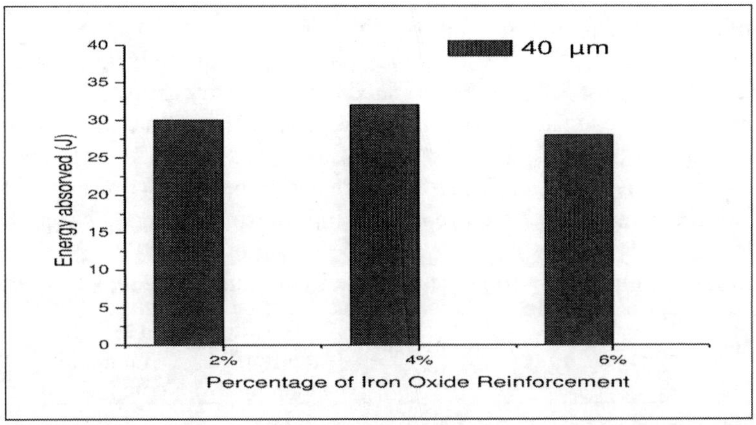

Figure 9. Variation of impact strength with percentages of iron oxide reinforcement

In view of the fact that aluminum 6061 is tougher material, its energy absorption capacity is higher. Also, since these composites were prepared using the liquid phase route methods, exhibits the higher impact toughness as compared to other methods. From the figure 9 it is clear that, as the weight percentage of iron oxide increases, impact strength increases. This increment in the impact strength is do ability of material that holds its ductility property, against the increased composition, in turn absorbs more energy to fracture. The maximum impact energy absorbed is 32 N-m for the composition 4 wt% of iron oxide.

4.4. Hardness analysis

All the three compositions of the composites were tested for its hardness using the Brinell hardness testing machine at room temperature. The specimen dimensions used are 15mm × Ø25 mm.

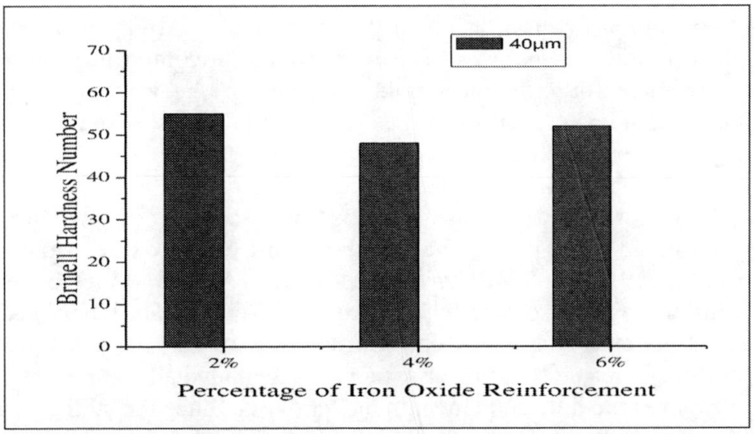

Figure 10. Variation in hardness of composites with percentage of iron oxide reinforcement

At different locations on the surface of the specimen Brinell hardness test is conducted and average value of the hardness has been calculated. Since the uniform distribution of the iron oxide particles in the aluminium 6061 matrix gives the uniform hardness values in the specimen at any location.
The average values of the hardness of the composite for all the compositions are shown in the figure 10. From the figure 10, it is observed that the increment in iron oxide influence the hardness of the AA6061-Fe_2O_3 composite. The maximum value of hardness is obtained at composition 2 wt% of reinforcement.

4.5. Comparison of AA6061- Fe_2O_3 and AA6061- SiC

The aluminium 6061 composite with silicon carbide particles, with the weight fraction of 0 to 4% [21], 0 to 5% [22] and 0 to 10% [23] of SiC, were fabricated using the stir casting process. The tensile strength of the composite has been investigated. From the results was reported that the addition of reinforcement improves the mechanical properties of the AA6061-SiC composites [21-23]. The Table.4 shows the comparison of the increment in the tensile strength of the composite with respect to the unreinforced aluminium alloy. All the composites were prepared using the stir casting technique.

Table 4. Comparison of the percentage increment in the tensile strength

Sl no	Reference	Reinforce-ment	Tensile strength MPa		% improvement
			Unreinforced aluminum	Aluminum composite	
1	Sivananthan et al. [21]	4% SiC	125	157	20.38
2	Nagendra et al.[22]	5% SiC	277	288	03.81
3	Amir Hussain et al. [23]	10% SiC	225	240	06.25
4	Present work	4% Fe_2O_3	152	229	33.62

From the Table.4 it is clear that adding of reinforcement increases the tensile strength of the composite. The tensile strength obtained by the adding the 4 wt%, 5 wt% and 10 wt% of SiC particles to the aluminium alloy improves the tensile strength by 20.38%, 3.81% and 6.25% respectively in contrast with the unreinforced aluminum alloy. The addition of 4 wt% Fe_2O_3 to the aluminium alloy improves the tensile strength of the composite by 33.62% in comparison with the unreinforced aluminum alloy. Thus the AA6061-Fe_2O_3 composite can be utilized for the automobile, aerospace, marine and structural applications where it requires high strength requirements.

5. Conclusion

The influence of the addition of the iron oxide particles in the AA6061 matrix on the mechanical properties is examined in the present work. The AA6061-Fe_2O_3 composite is processed successfully using stir casting technique for different weight fractions of the iron oxide. Using the different experimentation, the mechanical properties of the AA6061-Fe_2O_3 composite were determined and the conclusions were summarized as follows.

The addition of iron oxide enhances the tensile properties. The ultimate tensile strength of the composite was increased by 50% for 4 wt% and decrement in the strain by 51% for the 4wt% of the AA6061-Fe_2O_3 composite has been observed. From the flexural test the adding of iron oxide to aluminium increases the flexural strength from 152MPa to 305 MPa for the 4wt% of the AA6061-Fe_2O_3 composite. The addition of 4 wt% of Fe_2O_3 influences more on the impact strength of the AA6061-Fe_2O_3 composite. Since the material holds the ductility even at 4 wt% of the reinforcement, it absorbs more energy as compared to the 6 wt% of Fe_2O_3. Since the addition of iron oxide particles have a little impact on the ductility, the composite prepared possess the little decrement in the hardness. Since the ability of giving more tensile strength for the given weight fraction of the reinforcement of the AA6061-Fe_2O_3 composite is more as compared to the AA6061-SiC composite, it will be used for the automobile, aerospace, marine and structural applications where it requires high strength requirements.

References

[1] Mondolfo F 1976 *Aluminum alloys: structure and properties* 282–285, London, Butterworths.
[2] Yashpal S, Jawalkar CS, Verma AS and Suri NM 2017 Fabrication of Aluminium Metal Matrix Composites with ParticulateReinforcement: A Review *Mater. Today- Proc.* **4** 2927–2936
[3] Miracle D 2005 Metal matrix composites – From science to technological significance Compos. *Sci. Technol.* **65** 2526–2540
[4] Chawla N and Chawla KK 2006 Metal Matrix Composites *Springer* 351–384
[5] National Research Council 1996*New Materials for Next-Generation Commercial Transports* Washington, DC: The National Academies Press. https://doi.org/10.17226/5070.

[6] Wang DA and Makhlouf M 1997 High Performance Die Casting Alloys *Proc. 19thInt. Diecasting Congress* 179–189.

[7] Wang MM and Apelian D 1995 Aluminium die casting alloys: alloy composition, microstructure, and properties-performance relationships *Int. Mater. Rev.* **40** 225–252

[8] Maity PC, Chakraborty PN, Panigrahi SC 1997 Al-Al_2O_3 in situ particle composites by reaction of CuO particles in molten pure Al *Materials Letters* **30** 147-151

[9] Sarkari Khorrami M, Samadi S, Janghorban Z, Movahedi M 2015 In-situ aluminum matrix composite produced by friction stir processing using Fe particles *Mater Sci Eng A* **641** 380–390

[10] Susanta KP, Subhranshu C, Amitava BM, Debdulal D 2016 A simple stir casting technique for the preparation of in situ Fe-aluminides reinforced Al-matrix composites *Perspectives in Science* **8** 529—532

[11] Mikhaylov VI, Maslennikova TP, Ugolkov VL, Krivoshapkin PV 2016 Hydrothermal synthesis, characterization and sorption properties of Al/Fe oxide–oxyhydroxide composite powders *Advanced Powder Technology* **27** 756-764

[12] Divya Chandra, Nathi Ram Chauhan, S Rajesha, 2020*Hardness and toughness evaluation of developed Al metal matrix composite using stir casting method*, Materialstoday: Proceedings **25** 872-876, https://doi.org/10.1016/j.matpr.2019.12.026

[13] Yang X, Rujuan S, Ni S, Song M, Xiao D 2015 Fabrication, microstructure and mechanical properties of Al–Fe intermetallic particle reinforced Al-based composites *Journal of Alloys and Compounds* **618** 537–544

[14] Fathy A, El-Kady O, Moustafa MM Mohammed 2015 Effect of iron addition on microstructure, mechanical and magnetic properties of Al-matrix composite produced by powder metallurgy route *Trans. Nonferrous Met. Soc.* China **25** 46–53

[15] Zuhaib N, Kazi A, Safiulla M, Faisal M 2017 A Thorough Study: In-Situ Aluminium LM6 Metal Matrix Composites Reinforced with Iron Oxide and MWCNTs, *ICNANO 2016 Materials Today: Proceedings* **4** 11999–12006

[16] Sudeshna, Asutosh J, Patnaik SC, Sahoo SK, Omkar T 2020 A study on microstructure and mechanical properties of aluminium matrix composites with micro-sized iron fillers produced by powder metallurgy route *2nd ICPCM Materials Today: Proceedings 733.*

[17] Emin B, Fayza A, Ming JT, Ayse TB, Dhurata K 2013 Manufacturing of Aluminum Matrix Composites Reinforced with Iron Oxide (Fe_3O_4) Nanoparticles: Microstructural and Mechanical Properties *Metallurgical and Materials Transactions* B 11663

[18] Narender Panwar, Amit Chauhan, Harveer Singh Pali, Mukund Dutt Sharma 2020*Fabrication of Aluminum 6061 Red-mud Composite using Stir Casting and Micro Structure Observation*, ISFM-2018, Materials Today: Proceedings **21** 2014–2023.

[19] Subramanya RP, Arun KS, Mervin AH, Shrikantha SR 2019 Microstructure and mechanical properties of rutile-reinforced AA6061 matrix composites produced via stir casting process *Trans. Nonferrous Met. Soc.* China **29** 2229–2236

[20] Swapnil VG, Jaybhaye MD 2020 Microstructural studies on aluminium metal matrix composite (Al7075-SiC) fabricated through stir casting process *Materials Today: Proceedings*

[21] Sivananthan S, Ravi K, Samson Jerold Samuel C 2020 Effect of SiC particles reinforcement on mechanical properties of aluminium 6061 alloy processed using stir casting route *Materials Today: Proceedings* **21** 968-970

[22] Nagendra KM, Manish M, Srivastava AK, Dwivedi SP, kumar A, Chauhan S 2020 Investigation of mechanical properties of Al 6061/SiC composite prepared through stir casting technique Materials Today: Proceedings **25** 755-758

[23] Idrisi AH and Ismail AH 2019 Conventional stir casting versus ultrasonic assisted stir casting process: Mechanical and physical characteristics of AMCs *Journal of Alloys and Compounds* **805** 502-508

[24] ASTM E8/E8M−13a 2013*Standard Test Methods for Tension Testing of Metallic Materials*, ASTM International, West Conshohocken, PA, www.astm.org.

[25] ASTM E23−17 2017*Standard Test Methods for Notched Bar Impact Testing of Metallic Materials*, ASTM International, West Conshohocken, PA, www.astm.org.
[26] ASTM E10−15 2015 *Standard Test Methods for Brinell Hardness of Metallic Materials*, ASTM International, West Conshohocken, PA www.astm.org.
[27] Nagaraja KC Rajanna S, Prakash GS, and Rajeshkumar G 2020. Improvement of mechanical and thermal properties of hybrid composites through addition of halloysite nanoclay for light weight structural applications *Journal of Industrial Textiles* 1-19

IOP Conf. Series: Materials Science and Engineering **925** (2020) 012050 doi:10.1088/1757-899X/925/1/012050

A Facile Route in Controlling the Optical Absorbance of Polyvinylpyrrolidone-Capped Silver Nanoparticles *Via* Chemical Reduction Technique

Marco Laurence Budlayan[1,2], Jeanne Phyrre Lagare-Oracion[1,2], Lyka Dela Rosa[1,2], Mikee Joy Rodriguez[1,2], Rey Y. Capangpangan[1,2], Jonathan Manigo[3], Arnold Alguno[4], Eleanor Austria[5], Susan Arco[6], and Jonathan Patricio[6]

[1] Materials Science and Polymer Chemistry Laboratory, Caraga State University, Ampayon, Butuan City 8600, Philippines
[2] Center for Nanoscience and Technology for Research and Entrepreneurship, Caraga State University, Ampayon, Butuan City 8600, Philippines
[3] Physics Department, Caraga State University, 8100 Ampayon, Butuan City, Philippines
[4] Department of Physics, Mindanao State University-Iligan Institute of Technology, Tibanga, Iligan City 9200, Philippines
[5] Department of Biology, Adamson University, Manila, Metro Manila 1000, Philippines
[6] Synthetic Organic Chemistry Laboratory, Institute of Chemistry, University of the Philippines Diliman, Quezon City, Metro Manila 1101, Philippines

Abstract. This work presents the synthesis of silver nanoparticles (AgNPs) capped with polyvinylpyrrolidone (PVP) *via* a simple chemical reduction approach. A route in controlling the optical absorbance of the AgNPs by varying the concentration of silver precursors, as reflected by the shifts and broadening of the UV-vis spectra of the AgNPs is also highlighted. Dynamic Light Scattering results revealed the narrow and small hydrodynamic size for AgNPs synthesized at low precursor concentration, while multiple hydrodynamic size peaks were observed for AgNPs produced at high precursor concentration. Likewise, Fourier transform-infrared spectroscopy confirmed the successful capping of PVP on the AgNPs through the O and N atoms as potential binding sites. Transmission electron micrographs revealed the formation of spherical PVP-capped AgNPs, having no apparent aggregation observed. A closer inspection of the micrograph showed the subsequent formation of rod-like growth template of the AgNPs, that might lead to the formation of non-spherical silver nanostructures.

Keywords: silver nanoparticles, polyvinylpyrrolidone; optical absorbance; transmission electron microscopy

1. Introduction

The unique properties of silver nanoparticles (AgNPs) have been extensively studied due to their promising applications in a wide variety of technology. In fact, silver nanoparticles are now being incorporated in various products, including conductive pastes and inks, photovoltaic cells and devices, antimicrobial coatings and dressings, and molecular diagnostic devices and colorimetric sensors [1].

Unlike its bulk counterpart that has a lustrous grayish color, AgNPs may appear in diverse colors depending on the nanoparticles' size and shape [2]. For instance, AgNPs with a spherical particle size of around 5 nm has a pale yellow color while AgNPs at around 40 nm and 100 nm have a greenish and grayish color, respectively [3]. This interesting property of AgNPs is credited to the strong collective oscillation of the electron cloud on the surface of the nanoparticles in the presence of light, which is commonly termed as surface plasmon resonance (SPR) [4]. Because of their tunable SPR, AgNPs have become versatile materials for emerging researches that aim to explore the diverse and unique applications of AgNPs, particularly in the field of sensors, photovoltaics, and photonics [5].
In exploiting AgNPs as an outstanding functional material, stability and tunability are two of the most important criteria that should be considered. It has been known that AgNPs are very sensitive to the changes of its environment [6]. In fact, minimal modification on its surface chemistry *via* chemical or physical interactions might induce significant changes in its stability, leading to aggregation and deterioration in terms of physical and optical properties [7]. To enhance the stability of AgNPs, stabilizing agents are being introduced in the synthesis process that may also act as a capping agents to provide specific functionalities [6,7]. Typical capping agents consist of polar head groups that link to the nanoparticles' surface and nonpolar hydrocarbon tails that interact with the surrounding media. Polymers, dendrimers, and biological extracts are commonly utilized as capping agents depending on the nature and application of the nanoparticles [8].

Poly(vinylpyrrolidone) or PVP is a very common water-soluble polymer that is used as a binder in pharmaceutical pills. It is also found in disinfectant solutions, liquid soaps, scrubs, ointments, and other medical and commercial products [9]. Together with its availability and affordability, PVP is considered as a promising capping agent for nanoparticles due to its biocompatibility. In fact, a number of work reported the successful synthesis of AgNPs incorporating PVP as a capping and stabilizing agent using various synthesis techniques. For instance, Kittler *et al.* reported the synthesis of PVP-coated AgNPs *via* the polyol process and its bioactivity towards human stem cells [10]. Also, Xiliang, *et al.* highlighted the large-scale synthesis of AgNPs with PVP for low-temperature sintering bonding application [11]. PVP was also incorporated on AgNPs synthesized from bacterial supernatant, as demonstrated by Bhatia and colleagues [12]. Among these successful routes in synthesizing AgNPs with PVP, the chemical reduction has remained the easiest and fastest way to produce stable nanostructures [13-15]. Although the incorporation of PVP to AgNPs has been previously studied and elaborated, the influence of various synthesis parameters on the properties of PVP-capped AgNPs is still a wide area to investigate. Moreover, tuning the properties of the PVP-functionalized AgNPs by manipulating the key synthesis parameters is still an open area for further exploration. Specifically, the effect of silver precursor concentration in controlling the optical absorbance of PVP-capped functionalized AgNPs has never been fully elucidated. Hence, this work demonstrates a simple route to manipulate the optical absorbance of PVP-capped AgNPs by varying the volume of silver precursor used in the chemical reduction process.

2. Experimental details

2.1 Chemicals and Materials

Silver nitrate (Sigma Aldrich, 99.8%) was used as the silver precursor. Both sodium borohydride (Sigma Aldrich, 99.9%) and sodium citrate (Sigma Aldrich, 99.8%) were used as reducing agents. PVP was ~~also~~ used as a stabilizer and capping agents. Double deionized water was used in preparing all solutions needed in this study.

2.2 Preparation of PVP-capped AgNPs

Silver nanoparticles (AgNPs) were prepared using a simple chemical co-reduction of silver nitrate in sodium borohydride and sodium citrate solution as demonstrated by previous works [13-15] but with some modifications. In brief, a known volume of 0.02 M sodium borohydride and 0.02 M sodium citrate were mixed and stirred for 30 minutes at room temperature. A known volume of silver nitrate (0.001-0.02 M) was added dropwise to the sodium borohydride-sodium citrate mixture. After the addition of silver nitrate, the color of the solution changed from clear to yellow, indicating the formation of AgNPs. Drops of 1% PVP were then added to stabilize the AgNP solution. To tune the optical absorbance of the AgNPs, an additional volume of the silver precursor was simply supplied to the mixture until the desired color change was observed.

2.3 Characterization of AgNPs
UV-visible (UV-Vis) spectroscopy was employed to examine the optical property of synthesized AgNPs using a Perkin multimode microplate reader. Fourier transform-infrared (FT-IR) spectroscopy was used to identify the functional groups present in the samples. Furthermore, a dynamic light scattering (DLS) technique was used to determine the hydrodynamic size distribution. Lastly, transmission electron microscopy (TEM) was used to investigate the morphology and size of the PVP-capped AgNPs.

3. Results and discussion
The optical absorbance spectra of the PVP-capped AgNPs in response to the variation in the precursor concentration is presented in figure 1. The change in the appearance of the mixture from clear to yellow indicated the successful formation of AgNPs. It was observed that with lesser concentration of $AgNO_3$, a pale yellow AgNP solution was obtained, that produced a narrow and symmetric optical absorbance spectrum with a peak at 402 nm. This spectral behavior may be attributed to the SPR of small and monodisperse AgNPs [15]. Increasing the concentration of silver precursor to 0.003 M resulted in the production of intense yellow AgNP solution projecting a symmetric spectrum with a red-shifted peak at around 408 nm. It has a wider spectrum compared to the previous that might suggest the subsequent growth of AgNPs into larger sizes [16-18]. Increasing the concentration of silver precursors further resulted in the production of deep yellow to orange AgNP solution, which projected a gradual widening and asymmetry of the absorbance spectrum, suggesting the continuous growth of the AgNPs. Further increase of the precursor up to 0.02 M resulted in red-orange AgNPs solution with an-asymmetric UV-Vis spectrum with two absorbance peaks located at 405 and 555 nm, respectively. This broad and asymmetric UV-Vis profile may be associated with the overlapping SPR bands of a polydisperse population of PVP-capped AgNPs [19]. However, several results attribute the broadening of the peak at around nm 600 to the gradual formation of non-spherical and asymmetric aggregates of AgNP crystals such as nanorod structures [20-22]. It is also worthy to mention that in the absence of PVP, AgNPs synthesized using 0.02 M of precursor concentration exhibited an abrupt aggregation as manifested by the low intensity and wide absorbance profile in figure 1(b).

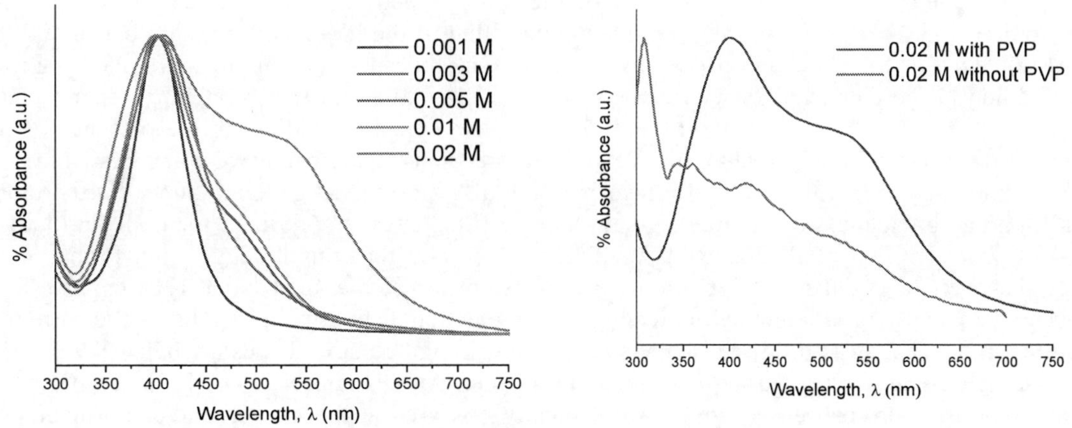

Figure 1. (a) Normalized UV-Vis spectra of the PVP-capped AgNPs synthesized using a varied concentration of silver precursor and (b) UV-vis spectra of AgNPs synthesized with 0.02 M precursor concentration with and without PVP as a capping agent.

IOP Conf. Series: Materials Science and Engineering **925** (2020) 012050 doi:10.1088/1757-899X/925/1/012050

DLS technique was employed to investigate the change in the hydrodynamic size of the PVP-capped AgNPs, as presented in figure 2. For brevity, only the DLS profile of the PVP-capped AgNPs synthesized using 0.001 M, and 0.02 M precursor was shown. The AgNP derived from low precursor concentration projected a mean hydrodynamic size of 21.10 nm with a polydispersity index (PI) of 0.264. According to literature, PI values that are less than 0.3 suggest a good dispersity of nanoparticles in an aqueous medium [23]. On the other hand, the DLS profile of AgNP produced from high precursor concentration exhibited two intensity peaks signifying two hydrodynamic sizes averaging at 11 nm and 76.2 nm, respectively. The appearance of these two peaks might indicate the population of polydisperse AgNPs as a consequence of the growth and aggregation of AgNPs. There is also a possibility that these two peaks signify the presence of rod-like structures nanoparticles diffusing the aqueous media as similar DLS profiles were observed for silver and gold nanorods, as reported in previous studies [24-26]. However, information from both UV-Vis and DLS characterizations were insufficient to confirm the definite structure obtained from the nanoparticle's nucleation and growth.

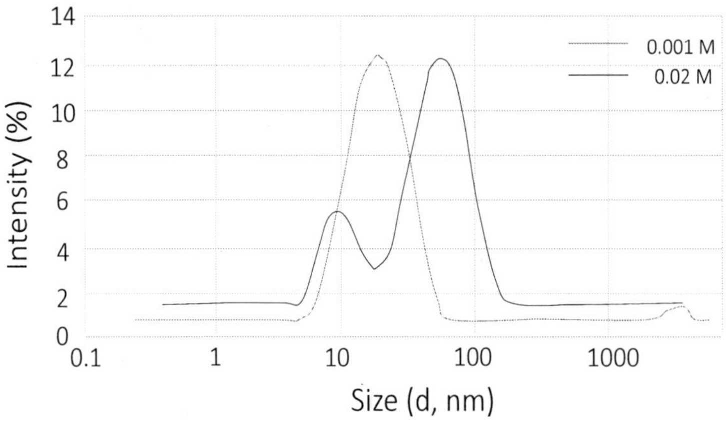

Figure 2. DLS result of the PVP-capped AgNPs synthesized using 0.001 M and 0.02 M of silver precursor.

The chemical composition of the AgNPs was investigated using FT-IR spectroscopy, as presented in figure 3. For comparison, the FT-IR spectra of the PVP-capped AgNPs at varied silver precursor concentrations (0.001 M and 0.02 M) were shown with that of the PVP. Transmission bands at 2954 cm^{-1}, 1428 cm^{-1}, 1281 cm^{-1}, and 1073 cm^{-1} appeared in the spectrum of the pure PVP sample, corresponding to the characteristic vibrations of CH_2, C-H, N-O, and C-N stretchings respectively, typical for PVP [27, 28]. A strong band at 1655 cm^{-1} also appeared, indicating the presence of the carbonyl group present in the structure of PVP [27]. Interestingly, these bands were manifested in both AgNP samples, suggesting the successful incorporation of PVP on the AgNP structures. Furthermore, the shifting and broadening of the transmission bands of the C=O at 1655 cm^{-1} for PVP (in blue) to 1680-1683 cm^{-1} for that of PVP-capped AgNPs (in black and red) might suggest the formation of linkages between the AgNPs and C=O group. It was also observed that the band at 1281 cm^{-1} for PVP disappeared in both AgNP samples indicating the interaction between the AgNP to the nitrogen-containing group. This was followed by the shifting and strengthening of bands at 1059-1079 cm^{-1} for AgNPs samples, that suggest strong coordination between AgNPs and N atoms [29]. According to previous reports, coordination of AgNPs was typically observed at the oxygen and/or nitrogen sites highly available in the chemistry of PVP [28-30]. In general, there was no significant difference between the spectra of PVP-capped AgNPs synthesized at 0.001 M and 0.02 M precursor concentration. However, there was an observed difference in the intensity of their transmission bands that might suggest a difference in the amount of functional groups coordinated on the AgNP surface.

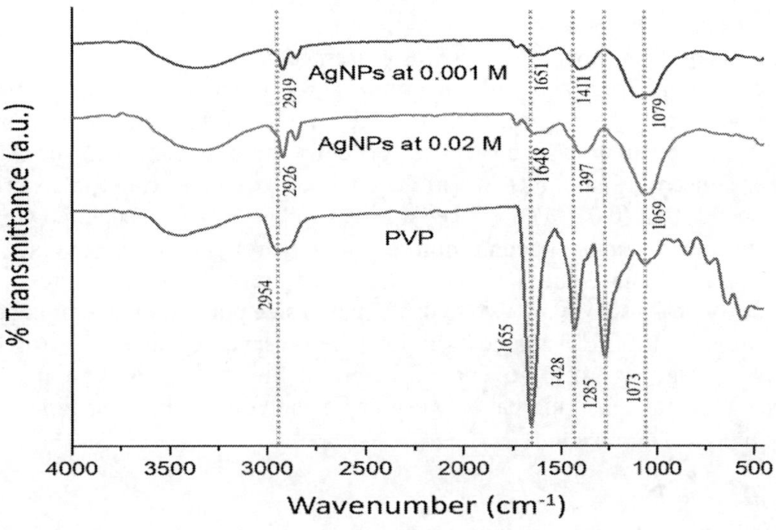

Figure 3. FT-IR spectra of the PVP and PVP-capped AgNPs produced using 0.001 M and 0.02 M precursor concentration.

A sample TEM image and size histogram of the PVP-capped AgNPs derived using low precursor concentration is shown in figure 4. The formation of spherical AgNPs was evident with a mean diameter of 14.67 ± 3.85 nm. As expected, this value was smaller with respect to the DLS result since TEM cannot detect the polymeric PVP that capped the AgNPs that was detected by the DLS technique [31]. Upon closer inspection, a rod-like growth of the nanoparticles was observed. This might suggest that the formation of rod-like or elongated nanostructures was favoured in this synthesis route and addition of more silver ions to supply the nanoparticle growth may lead to the formation of non-spherical AgNPs [32-33]. It has been reported that several nanoparticles with non-spherical structures such as nanorods were obtained in the presence of PVP [33-35]. Accordingly, the strong binding of action of PVP effectively controls the growth of nanoparticles. The growth of crystallite faces where PVP were attached becomes limited while faces with minimal linkages of PVP have grown freely leading to the anisotropic growth of the nanoparticles [36-38].

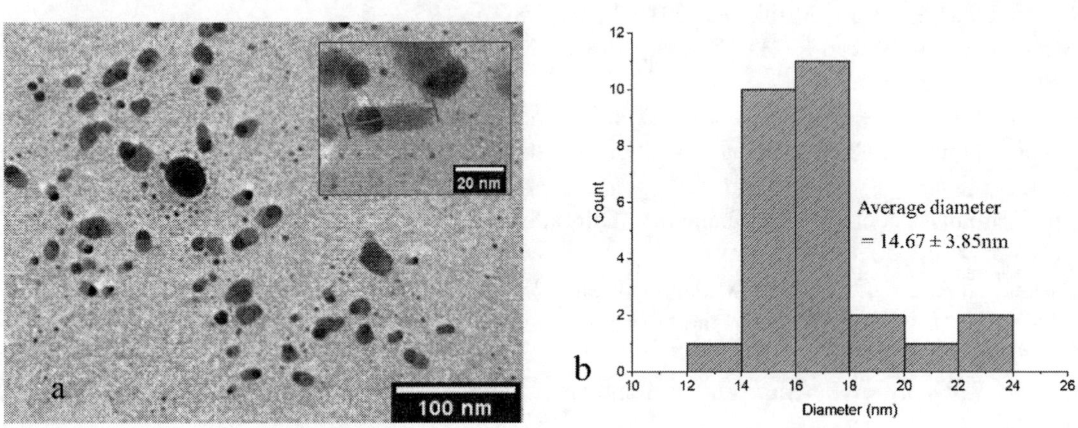

Figure 4. (a) TEM image of PVP-capped AgNP (inset shows the magnified image of the AgNPs) and (b) histogram of the AgNP diameter size.

4. Summary and conclusion

A successful synthesis of PVP-capped AgNPs was presented *via* chemical reduction of silver ions in the presence of sodium borohydride and sodium citrate as co-reducing agents. Control on the optical absorbance of PVP-capped AgNPs was demonstrated by the shifting and broadening of absorbance peaks at around 401 nm to 530 nm as we vary the concentration of silver precursor. AgNPs synthesized from low precursor concentration (0.001 M) produced a narrow and symmetric absorbance peak while high precursor concentration (0.02 M) produced a broad and asymmetric peak. DLS result of the PVP-capped AgNPs from low precursor concentration revealed hydrodynamic size average at 21.10 nm with a good dispersity (PDI = 0.264) while PVP-capped AgNPs derived from high precursor concentration exhibited two intensity peaks signifying two hydrodynamic size population averaging at 11 nm and 76.2 nm, respectively. Finally, the TEM image exhibited the presence of spherical AgNPs showing good dispersity with an average size of 14.67 nm ± 3.85 nm. Subsequent formation of rod-like AgNPs nanostructures was also noticed. This paper presented a valuable insight for tuning the properties of AgNPs for future practical applications.

Acknowledgment

The authors would like to acknowledge the support of Commission on Higher Education- DARE To of the Philippines for funding this work.

References

[1] Beyene H D, Werkneh A A, Bezabh H K and Ambaye T A 2017 *Sustainable Materials and Technologies* **13** pp 18-23
[2] Gonzalez A L, Noguez C, Beranek J and Barnard A S 2004 *J. Phys. Chem. C* **118** pp 9128-9136
[3] Agnihotri S, Mukherji S and Mukherji S 2014 *RSC. Adv.* **4** pp 3974-3983
[4] Amendola A, Bakr O and Stellaci F 2010 *Plasmonics* **5** pp 85-97
[5] Jain P, Huang X, El-Sayed I and El-sayad M 2007 *Plasmonics* **2** pp 107-118
[6] Haque Md N, Kwon S and Cho D 2017 *Korean J. Chem. Eng.* **34** pp 2072-2078
[7] de Oliveira J F A and Cardoso M B 2014 *Langmuir* **30** pp 4879-4886
[8] Sachdeva S G and Bhasin K K 2018 *AIP Conference Proceedings* 1953 (AIP Publishing)
[9] Teodorescu M and Bercea M 2015 *Polymer-Plastics Technology and Engineering* **54** pp 923- 943
[10] Kittler S, Greulich G, Koller M and Epple M 2009 *Mat.-wiss. u. Werkstofftech* **40** pp 258-264
[11] Xiliang Q, Yang C, Tiesong L, Peng H, Jun W, Ping L and Xiaolong G 2014 *Journal of Nanomaterials* **2014** p 8
[12] Bhatia D, Mittal A and Malik D K 2016 *3 Biotech* **6** pp 195-196
[13] Mirzaei A, Janghorban K, Hashemi B, Bonyani M, Leonardi S G and Neri G 2016 *J. Nanostruct. Chem.* **7** pp 37-46
[14] Munoz RV, Jimenez M J A, Lopez F D and Ribot J L L 2019 BMC *Res Notes* **12** p 6
[15] Wang H, Qiao X, Chen J, Wang X and Ding S 2015 *Materials Chemistry and Physics* **94** pp 449-453
[16] Sundarrajan M, Jeelani A, Santhaman V, Durgadevi S and Abirami S 2018 *Int. J. Cur. Res. Rev* **10** pp 25-29
[17] Ahmad N, Ang B C, Amalina M A and Bong C W 2018 *Sains Malaysiana* **47** pp 157-168
[18] Badi'ah H I, Seedeh F, Supriyanto G and Zaidan A H 2019 *IOP Conf. Series: Earth and Environmental Science* **217**
[19] Tsekov R, Georgiev P, Simeionova S and Balashev K 2017 *C. R. Acad. Bulg. Sci.* **70** pp 1237-1246
[20] Desai R, Mankad V, Gupta S K and Jha P K 2012 *Nanoscience and Nanotechnology Letters* **4** pp 30-34
[21] Hormozi-Nezhad M R, Jalali-Heravi M, Robatjazi H and Ebrahimi-Najafabadi H 2012 *Colloids and Surfaces A: Physicochem. Eng. Aspects* **393** pp 46-52

[22] Rekha C R, Nayar V U, Gopchandran K G 2018 *Journal of Science: Advance Materials and Devices* **3** pp 196-205
[23] Danaei M, Dehghankhold M, Ataei S, Hasanzadeh Davarani F, Javanmard R, Dokhani A, Khorasani S and Mozafari M R 2018 *Pharmaceutics* **10** p 17
[24] Liu H, Pierre-Pierre N and Huo Q 2012 *Gold Bull* **45** pp 187-195
[25] Kajanawarut R, Yuan B and XiaoDi S 2013 *Nucleic Acid Therapeutics* **23** pp 273-280
[26] Glidden M and Muschol M 2012 *J. Phys. Chem. C* **116** pp 8128-8137
[27] Safo I A, Dosche C and Oezaslan M 2019 *Nanoscale Adv.* **1** pp 3095-3106
[28] Kumar M, Devi P and Kumar A 2017 *Journal of Materials Science: Materials in Electronics* **28** pp 5014-5020
[29] Zhang Z, Zhan X, Xin Z, Deng M, Wen Y and Song Y 2011 *Nanotechnology* **22** p 8
[30] Wang H, Qiao X, Chen J, Wang X and Ding S 2005 *Mater. Chem. Phys* **94** pp 449-453
[31] Domingos R, Baalousha M, Ju-nam Y, Reid M M, Tufenkji N, Lead J, Leppard G and Wilkinson K 2009 *Environ. Sci. Technol.* **43** pp 7277-7284
[32] Johan M R, Aznan N A K, Yee S T, Ho I H, Ooi S W, Singho N D and Aplop F 2014 Journal of Nanomaterials 2014 p 7
[33] Jana N, Gearheart L, and Murphy C 2001 *J. Phys. Chem. B* **105** pp 4065-4067
[34] Yang X, He W, Wang S, Zhou G and Tang Y 2014 *Journal of Experimental Nanoscience* **9** pp 541-550
[35] Gu X, Nie C, Lai Y and Lin C 2006 *Mater. Chem. Phys.* **96** 217–222
[36] Silvert P, Urbina R H and Elhissen K T 1997 J. Mater. Chem. 7 pp 293-299
[37] Almeida M and Alcacer L 1983 *Crystal Growth* **62** pp 909-912
[38] Khodashenas B and Ghorbani H R 2019 *Arabian Journal of Chemistry* **12** pp 1823–1838

IOP Conf. Series: Materials Science and Engineering **925** (2020) 012051 doi:10.1088/1757-899X/925/1/012051

In-Silico Pharmacological and Molecular Docking Studies of Natural Inhibitors form *Musa Spp.* On Vaca Gene a Vacuolating Cytotoxin Autotransporter

N R Rakesh[1], H Gurumurthy[2], H K Pradeep[3], S Pradeep[4], D M Gurumurthy[5] and C L Sachin[5]

[1,2] Department of Biotechnology, GM Institute of Technology, Davangere, Karnataka India
[3] GM Institute of Pharmaceutical Sciences and Research, Davangere, Karnataka India
[4] Department of Biotechnology, BMS College of Engineering, Bengaluru, Karnataka, India
[5] GM Institute of Technology, Davangere, Karnataka India

Email: rakeshnr@gmit.ac.in

Abstract. The integration of computational with various bioinformatics software/tools/ webservers and the molecular docking process is the current key point method in reducing the time for drug discovery and drug development from a Bioactive compound. Current Insilco Study considers the Bioactive Phytoconstituents from *Musa* species for our drug discovery process by using bioinformatics software to find out the analogs by following their related physicochemical properties for the structure-based drug design. The gastric bacterium H. pylori infect the gastric mucosa, and its eradication is associated with the prevention of ulcer reoccurrence. The significant Protein target from this species was a challenging task for finding, targeting, treating. Offering hope from the Bioactive compounds considered in line with already FDA approved Drug molecules against peptic ulcer causative organism's special protein vacuolating cytotoxin autotransporter translated from *vacA* Gene. The recent challenge is to identify a drug moiety that will effectively work on the target protein. Usually, the Phytocompounds will not significantly cause a side effect. Sitosterol with docking score of -*6.417 kcal/mol* has promised to serve as vacuolating cytotoxin autotransporter inhibitor by blocking the autotransporter protein to the periplasmic space avoiding the toxin passage between periplasmic membrane which will avoid the gastric infections from H. *pylori*.

Keywords: Musa *paradisica, Helicobacter pylori*, Chemspider, SwissADME, Swissprediction Vacuolating cytotoxin autotransporter, Molecular Docking.

1. Introduction

Molecular Docking Studies have given shape for the success of Drug discovery process in Pharmaceutical research for understanding the Bioactivity between Compound and biological systems. Integration of computational and experimental strategies has been of great value in the identification and development of novel promising compounds. [7] Broadly used in modern drug design, molecular docking methods explore the ligand conformations adopted within the binding sites of macromolecular targets.

Musa is one of the two or three genera in the family Musaceae. It includes bananas and plantains. Plantains refer to India as a coarse banana. It grows 10 to 40 feet in height and has broad green leaves that grow through a hollow stem bears flowers and fruit. [24] These group of plants are commonly cultivated in all areas in India. *Musa spp.* have been reported to have many biological activities. In Ayurveda, a traditional system of medicine is citied for limited to Florida, Egypt, Southern Japan, and South Brazil for the treatment of many disorders. [26, 27] *Musa spp.* leaves are useful in treating upper respiratory disorders. Some parts of roots could help in treating hemoptysis and anthelmintic. [1] Fruits can increase renal activities, reduces the risk of kidney cancer. It contains anti oxidizer and counteracts the noxious effects of the free radicals. [23,28] The *Musa spp.* have various biological activities such as antidiabetic, Antiulcerogenic, antiatherogenic, antidiarrheic, antitumoral, antimutagenic. [10] It has also been found to be effective in the treatment of migraine, hypertension, cholesterol, and hyperoxaluria. It can be used as an antidote for snakebite, asthma, burns, excessive menstrual flow, fever, gangrene, gout, haemorrhage, inflammation, insomnia, intestinal parasites. [11, 29] A potent drug molecule must reach its target in the body in sufficient concentration, and stay there is a bioactive form long enough for the expected biologic events to occur. [3] Drug development involves assessment of absorption, distribution, metabolism and excretion (ADME) increasingly earlier in the discovery process, at a stage when considered compounds are numerous but access to the physical samples is limited. [6] In that context, computer models constitute valid alternatives to experiments. [8]

Molecular descriptors are mostly a set of chemical information from chemical structures which are considered in the field of chemoinformatics field one typical example is the Molecular Fingerprint which says about the sequence of bits present or absent of a chemical feature of the compounds. Few of the simple physicochemical descriptors like the molecular weight of a compound, molecular refractivity index of a bioactive substance, number count of specific atom types presence and polar surface area, etc. [3] Along with the chemical information of Bioactive compounds and biological targets molecular docking has become an increasingly important tool for the drug discovery process. The docking process involves two necessary steps: prediction of the ligand conformation as well as the different geometry conformations with the already know Pharmacopore features. It is the most likely process of knowing the binding site in the receptor can significantly increase the efficiency of ligand-target interactions during molecular docking.

1.1 Peptic Ulcer

A peptic ulcer is a disease caused on the lining on stomach or duodenum, at the start of the small intestine position. One in every 10 American will develop an ulcer in their life. The Common current cause of peptic ulcer is a bacterial infection other kinds of cause is drugs nonsteroidal anti-inflammatory agents (NSAIDs), like aspirin and ibuprofen for long-term use. [12] In a few cases, cancerous tumours can also cause ulcers in certain few cases in stomach and pancreas, and it is clear that spicy food or stress are not the causatives of ulcers. [12, 13] Researchers have found Helicobacter pylori one of the causatives for peptic ulcer which weaken the mucous coating of the stomach and duodenum for protection. Creating this mucous lining damage make the sensitive inner lining of the stomach to get exposer to acids and bacteria, creating irritate on the tissue lining followed by sore or ulcer. H. Pylori can secrete enzymes which assist for the existence in the stomach by neutralizing the acid. [13] Commonly, bacterial infections are treated with Antibiotics and some types of acid production suppressants are used two types of acid-suppressing drugs might be used like H2-blockers and proton pump inhibitors. [12, 25]

2. Methods

2.1 Pharmacological Prediction

The Pharmacological Physicochemical properties prediction Figure 2 of a given compound which consist of both primary and secondary properties include SMILES, Molecular formula, Molecular weight, No. of Rotatable Bonds, No. of hydrogen acceptors and donors, and TPSA. Secondary properties include Consensus Log Po/w, Leadlikeness, Lipinski, Bioavailability score which are shown in Table 1[16].

2.2 Ligand- Receptor preparation

The in-silico ligand and the receptor studies were obtained from ChemSpider, and PDB, respectively. The 3D structure of the receptor i. e. 2QV3 and five ligands Sitosterol, Fucostanol, Clionasterol, DL-Tryptophan and L-5-hydroxy-Tryptophan were prepared for docking using Molecular Operating Environment (A drug discovery software platform). Removal of hetero atoms molecules like water and non-amino acid groups and the addition of hydrogen atoms are crucial steps during ligand and receptor preparation [13, 15]

2.3. Molecular docking

The molecular docking is usually initiated with the preparation of ligand and receptor, further setting the docking grid and choosing the active site of the receptor along with Ligand conformation through energy refinement process for the best binding pose during ligand and receptor interaction as shown in Figure 2. Results are of affinity binding values are saved in Portable Document Format. Bioactive compounds and the selective receptor interactions were analyzed through Molecular Operating Environment's platform that provides integrated visualization [7, 8]

2.4. Vacuolating cytotoxin autotransporter PDB-ID: 2QV3

Vacuolating cytotoxin autotransporter is a type of protein which interacts selectively with one or more biological molecules in other organisms biological system (Target host) to pathogenesis (creation of abnormalities within the host at detectable statues). This activity is because of the expression of the *vacA* gene of *H. pylori,* which code for the structure and function of vacuolating cytotoxin autotransporter. The creation of a pore from an autotransporter pathway leading the multiple alterations with the host mucosal cell line, and finally turning the host cell line to cause of peptic ulcer disease and reoccurrence of more host cell line damage by the secreted gastric juices. [13, 18] Here, we present a 2.4-A crystal structure of the *vacA* p55 domain, which has a vital role in mediating VacA binding to host cells. The structure is predominantly a right-handed parallel beta-helix, a feature that is characteristic of autotransporter passenger domains but unique among known bacterial protein toxins. Notable features of *vacA* p55 include disruptions in beta-sheet contacts that result in five beta-helix subdomains and a C-terminal domain that contains a disulphide bond. Analysis of *vacA* protein sequences from unrelated *H. py*lori strains, including m1 and m2 forms of *vacA*, allows us to identify structural features of the *vacA* surface that may be important for interactions. [14, 18]

Table 1. Bioactive Compound from *Musa spp.* Retrieved from ChemSpider with Percentage of Similarity from FDA approved Drugs against Vacuolating cytotoxin autotransporter

Sl. No.	Bioactive Compounds	Target	FDA Approved DRUG	% Similarity
1.	Sitosterol	PDB-ID 2QV3 Vacuolating cytotoxin autotransporter	Allylestrenol	91.2
2.	Fucostanol		Menthol	92.9
3.	clionasterol		Allylestrenol	91.2
4.	DL-Tryptophan		L-Tryptophan	100
5.	L-5-hydroxy-Tryptophan		Oxitriptan	100

IOP Conf. Series: Materials Science and Engineering **925** (2020) 012051 doi:10.1088/1757-899X/925/1/012051

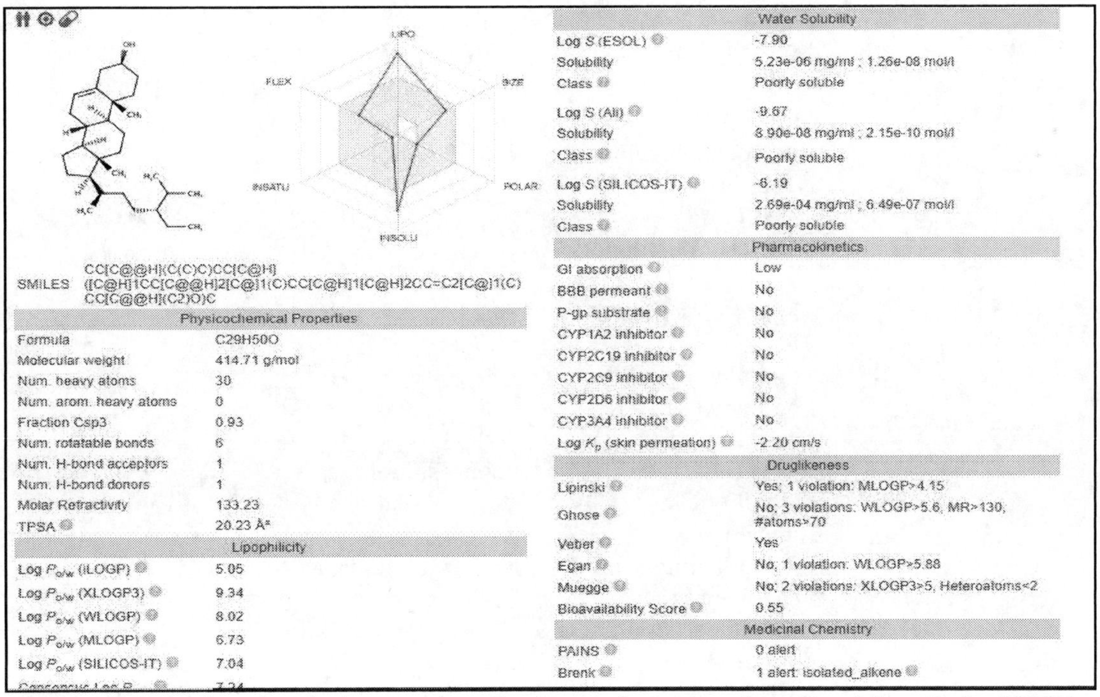

Figure 1. Physico-chemical properties of a sitosterol compound predicted from SwissADME

3. Result and Discussion

Table 2. Bioactive Ligands along with Different Molecular Descriptors and Pharmacological Properties

Ligands	Predicted Physicochemical Pharmacological properties	
	Formula:	$C_{29}H_{50}O$
	SMILES:	[H][C@@]1(CC[C@@]2([H])[C@]3([H])CC=C4C[C@@H](O)CC[C@]4(C)[C@@]3([H])CC[C@]12C)[C@H](C)CC[C@@H](CC)C(C)C
	Mol.Wt:	414.71g/mol
	No. Rotatable bond:	6
	No. of H-bond	1
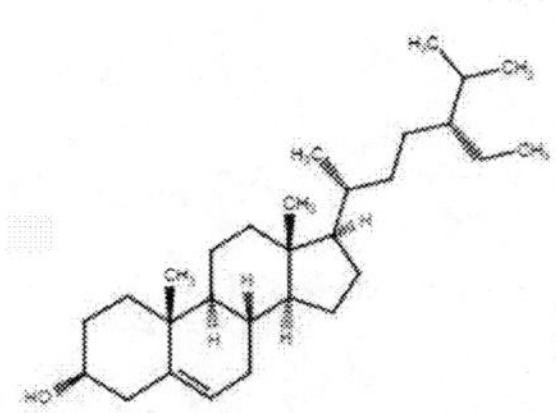	acceptors:	1
	No. of H-bond donors:	133.23
	Molar Refractivity:	20.23 A^0
	TPSA:	5.05
Sitosterol	Log Pa/w (iLOGP):	9.34
	Log P a/w(XLOGP3):	6.73
	Log Pa/w(MLOGP):	-7.90
	Log S (ESOL):	Poorly soluble
	Solubility class:	Yes; 1 voilation : MLOGP> 4.15
	Lipinski:	0.55
	Bioavailability score:	No; 2 violations; MW > 350, XLOGP3 > 3.5
	Leadlikeness:	6.30
	Synthetic accessibility:	

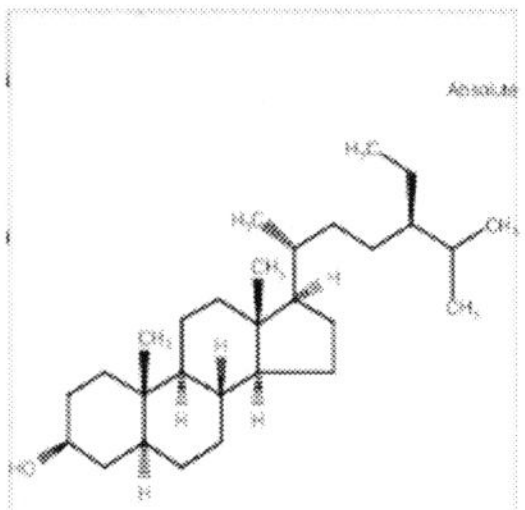

Fucostanol

Formula:	$C_{29}H_{52}O_2$
SMILES:	CC[C@@H](C(C)C)CC[C@H]([C@H]1CC[C@@
	H]2[C@]1(C)CC[C@H]1[C@H]2CC[C@@H]2[C
Mol.Wt:	@]1(C)CC[C@@H](C2)O)C
No. Rotatable bond:	416.72 g/mol
No. of H-bond	6
acceptors:	1
No. of H-bond donors:	1
	133.70
Molar Refractivity:	20.23A^0
TPSA:	5.17
Log Pa/w (iLOGP):	8.32
Log P a/w(XLOGP3):	6.88
Log Pa/w(MLOGP):	-7.27
Log S (ESOL):	Poorly soluble
Solubility class:	Yes; 1 violation: MLOGP > 4.15
Lipinski:	0.55
Bioavailability score:	No; 2 violations: MW >350, XLOGP3 > 3.5
Leadlikeness:	5.38
Synthetic accessibility:	

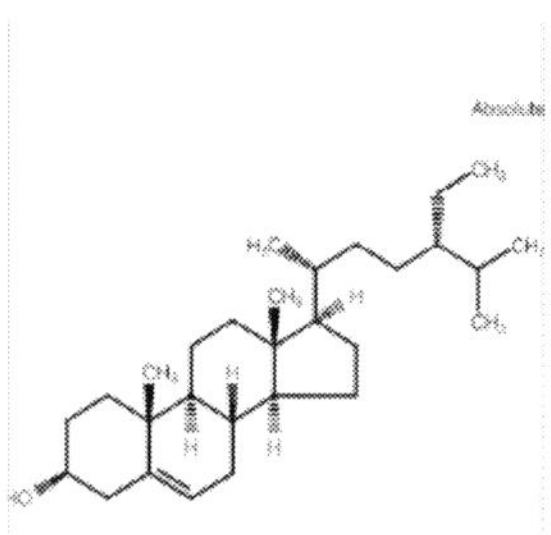

Clionasterol

Formula:	$C_{29}H_{50}O$
SMILES:	CC[C@H[(C(C)C)CC[C@H]([C@H]1CC[C@@H]
	2[C@]1(C)CC[C@H]1[C@H]2CC=C2[C@]1(C)C
Mol.Wt:	C[C@@H](C2)O)C
No. Rotatable bond:	414.71 g/mol
No. of H-bond	6
acceptors:	1
No. of H-bond donors:	1
Molar Refractivity:	133.23
TPSA:	20.23 A^0
Log Pa/w (iLOGP):	5.07
Log P a/w(XLOGP3):	9.34
Log Pa/w(MLOGP):	6.73
Log S (ESOL):	-7.90
Solubility class:	Poorly soluble
Lipinski:	Yes; 1 violation: MLOGP >4.15
Bioavailability score:	0.55
Leadlikeness:	No; 2 violations: XLOGP3>3.5, MW >350
Synthetic accessibility:	6.30

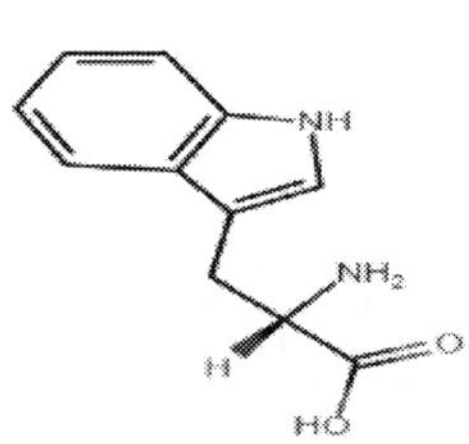

DL- Tryptophan

Formula:	$C_{11}H_{12}N_2O_2$
SMILES:	OC(=O)[C@H](Cc1c[nH]c2c1cccc2)N
Mol.Wt:	204.23 g/mol
No. Rotatable bond:	3
No. of H-bond	3
acceptors:	3
No. of H-bond donors:	57.36
Molar Refractivity:	79.11 A^0
TPSA:	0.99
Log Pa/w (iLOGP):	-1.06
Log P a/w(XLOGP3):	-1.66
Log Pa/w(MLOGP):	-0.68
Log S (ESOL):	Very soluble
Solubility class:	Yes; 0 violations

IOP Conf. Series: Materials Science and Engineering **925** (2020) 012051 doi:10.1088/1757-899X/925/1/012051

	Lipinski:	0.55
	Bioavailability score:	No; 1 violation: MW <250
	Leadlikeness:	2.09
	Synthetic accessibility:	
	Formula:	$C_{11}H_{12}N_2O_2$
	SMILES:	OC(=O)C(Cc1c[nH]c2c1cccc2)N
	Mol.Wt:	204.23 g/mol
	No. Rotatable bond:	3
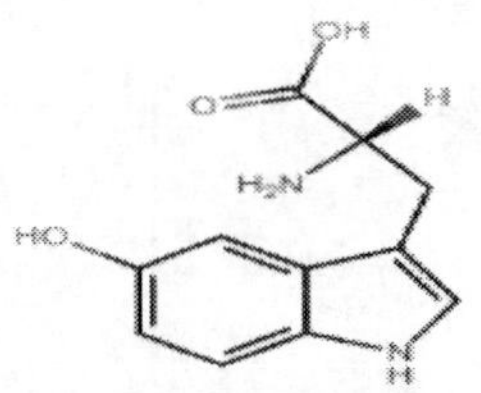	No. of H-bond	3
	acceptors:	3
	No. of H-bond donors:	57.36
	Molar Refractivity:	$79.11\ A^0$
	TPSA:	0.94
	Log Pa/w (iLOGP):	-1.06
L-5-hydroxy-Tryptophan	Log P a/w(XLOGP3):	-1.66
	Log Pa/w(MLOGP):	-0.68
	Log S (ESOL):	Very soluble
	Solubility class:	Yes; 0 violations
	Lipinski:	0.55
	Bioavailability score:	No; 1 violation: MW < 250
	Leadlikeness:	2.09
	Synthetic accessibility:	

Table 3. Docking Energy Scoring Functions and Amino Acids from receptor 2QV3_Vacuolating cytotoxin autotransporter protein

Sl. No.	Ligand	Amino Acids with Distance in Å	RMSD in Å	Rough Docking Score in kcal/mol	Refinement energy in kcal/mol	Final Docking Score in kcal/mol
1.	Sitosterol	GLY(3.9)	1.9409	-10.5921	-31.9434	-6.417
2.	Fucostanol	ASN(4.1)	1.3366	-10.1892	-15.4823	-4.637
3.	Clionasterol	SER(3.6) ASN(4.5)	1.7175	-10.9913	-29.1627	-5.965
4.	DL-Tryptophan	THR(3.0) SER(3.1)	2.2974	-9.7871	-21.9120	-4.928
5.	L-5-hydroxy-Tryptophan	ARG(2.9)	1.6200	-13.7376	-15.5269	-4.567

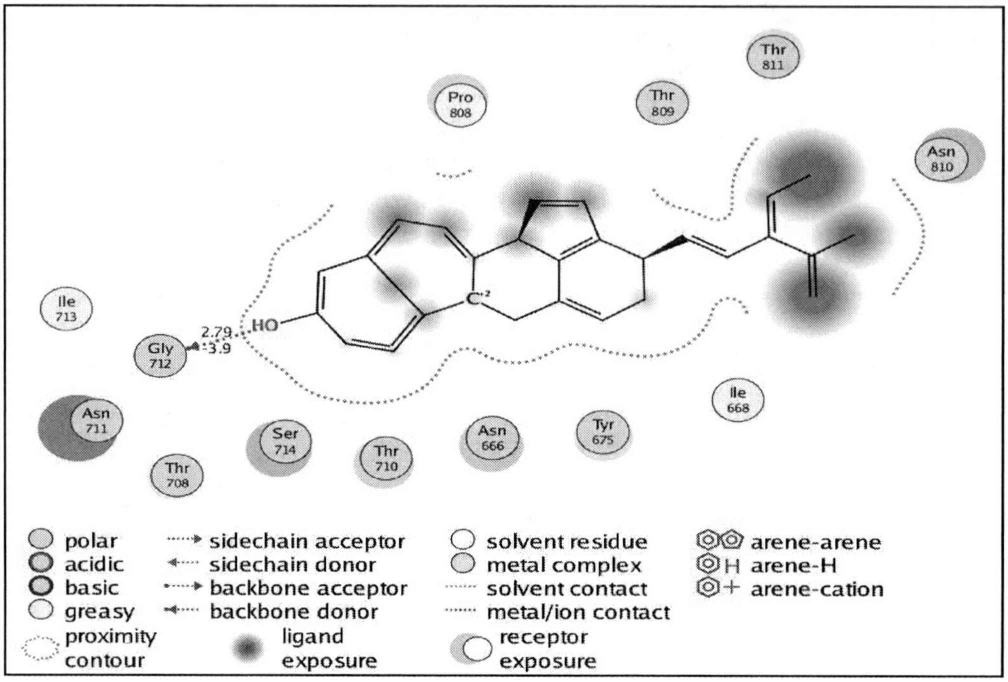

Figure 2. Molecular Interaction of Sitosterol with 2QV3_Vacuolating cytotoxin autotransporter

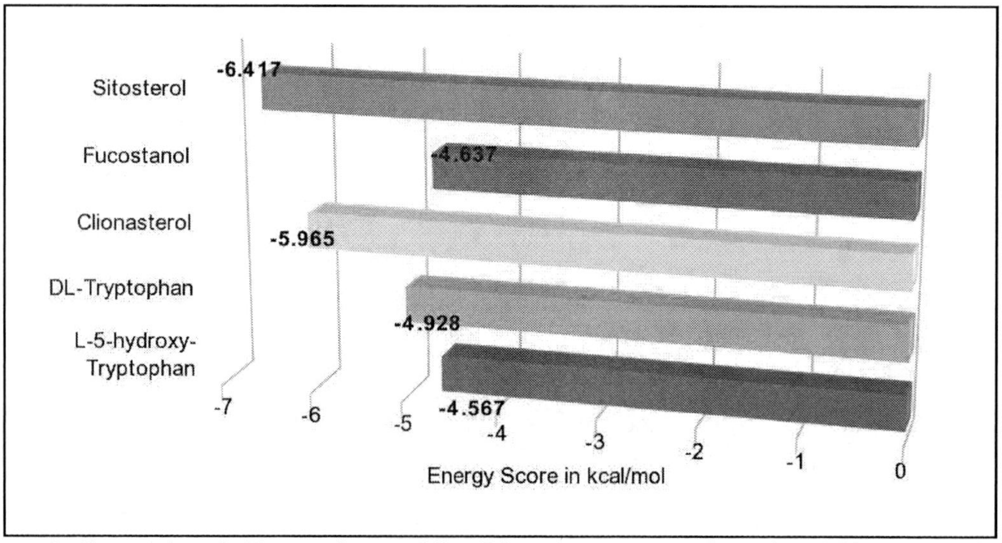

Figure 3. Comparison in Energy Score of five Bioactive compounds against 2QV3_Vacuolating cytotoxin autotransporter

3.1 Docking Score

Ligands could be more potent and likely to interact more significantly to receptor 2QV3_Vacuolating cytotoxin autotransporter based on the predicted free-energy of binding, i.e., Docking Score predicted value shown in Table 3 and comparision of energy graph in Figure 3. In this study, the Five natural Inhibitors have shown promising Binding affinity which is expressed in terms of Docking out of which Sitosterol with *-6.417 kcal/mol* has promised to serve as Vacuolating cytotoxin autotransporter inhibitor from the *Musa spp.* to prevent gastric infections with *H. py*lori. [17, 18, 30]

4. Conclusion

By the main Docking Energy Scoring when involved during in-Silico Docking Studies with Triangle matching Receptor-Ligand Movement with the Rigid receptor 2QV3_Vacuolating cytotoxin autotransporter along with five selected bioactive compounds have shown a significantly good Ligand along with the Percentage of structural similarity of FDA drugs Table 1 with an affinity towards the targeted protein of *vacA* gene product Vacuolating cytotoxin autotransporter, which could be further studied using Molecular Dynamics simulation and docking since it can even reveal the dynamic activity of target and receptors. Through this In-silico pharmacological and molecular docking studies will provide a strong recommending analysis for selected five Bioactive compounds will be able to become potent drug molecule against *H. pylori* in the clinical trial of a drug discovery process. Based on the Docking energy score predicted results, one can mostly conclude that the Bioactive compound Sitosterol from the *Musa spp.* will further promise to inhibit the activity of *vacA* gene function and its structural output vacuolating cytotoxin autotransporter.

Acknowledgements

We would like to thank all our students, department of biotechnology, GM Institute of technology for their encourage and support throughout this research. Also, we could like to thank support staff and management for providing computing facilities.

References

[1] Obolskiy D, Pischel I, Siriwatanametanon N, Heinrich M and Garcinia mangostana LA phytochemical and pharmacological review. Phytotherapy Research *An International Journal Devoted to Pharmacological and Toxicological Evaluation of Natural Product Derivatives* **23** 1047-65

[2] Lipinski CA, Lombardo F, Dominy BW and Feeney PJ 2001 Eksperimental and computational approaches to estimate solubility and permeability in drug discovery and development settings *Advance Drug Delivery Reviews* **64** 1-17.

[3] Onasanwo SA, Emikpe BO, Ajah AA and Elufioye TO 2013 Anti-ulcer and ulcer healing potentials of Musa sapientum peel *Pharmacognosy Research* **5** 3

[4] Elango et al. J. Res. Educ. Indian Med., Jan.-March, 2007; **13**, 63-69

[5] Shahin R, Mansi I, Swellmeen L, Alwidyan T, Al-Hashimi N, Al-Qarar'h Y and Shaheen O 2018 Ligand-Based Computer Aided Drug Design Reveals New Tropomycin Receptor Kinase A (TrkA) Inhibitors *J Mol Graph Model* **80** 327-352

[6] Nisha CM, Kumar A, Nair P, Gupta N, Silakari C, Tripathi T and Kumar A 2016 Molecular docking and in silico ADMET study reveals acylguanidine 7a a potential inhibitor of β-secretase *Advances in Bioinformatics* 1-6

[7] Laurie ATR and Jackson RM 2014 Q-SiteFinder: an energy-basedmethod for the prediction of protein-ligand binding siter *Bioinformatics* **21** 1908-1916

[8] Manjulika Y, Sanjukta C, Sharad KG and Watal G 2014 Alternative Therapeutics Unit, Drug Development Division, Medicinal Research Lab, Department of Chemistry, University of Allahabad

[9] Swathi D, Jyothi B, Sravanthi C 2011 Pharmacognostic studies and Pharmacological actions of Musa Paradisiaca A Review Paper *International Journal of Innovative Pharmaceutical Research* **2** 122-5

[10] Ehiowemwenguan G, Emoghene AO, Inetianbor JE 2014 Antibacterial and phytochemical analysis of Banana fruit peel *OSR Journal of Pharmacy* 4 18-25

[11] Podolski JL 1996 Recent advances in peptic ulcer disease: H. pylori infection and its treatment *Gastroenterology Nursing* **19** 128-136.

[12] Soll AH 1996 Medical treatment of peptic ulcer disease: Practice guidelines *Journal of the American Medical Association* **275** 622-628

[13] National Institutes of Health, Office of the Director. 1994 NIH consensus statement: Helicobacter pylori in peptic ulcer disease **12** Bethesda, MD.

[14] Nerkar G, Shena AK, Poorvashree JE and Hemant UC 2012 In silico screening synthesis and pharmacological evaluation of novel quinazolines as NMDA receptor inhibitors for anticonvulsant activity *Inter. J. Pharm. Pharm* Sci. **4** 449-53

[15] Zubair MS and Subehan 2010 Molecular docking of lunacridine from Lunasia amara to DNA: its inhibition and interaction study correlated with the cytotoxic activity on P388 murine leukemia cells *Indonesian Journal of Cancer Chemoprevention* **1** 108-17

[16] Agarwal PK, Singh A, Gaurav K, Goel S, Khanna HD, Goel RK 2009 *Indian J Exp Biol* **47** 32-40.

[17] Foegeding NJ, Caston RR, McClain MS, Ohi MD, Cover TL 2016 An Overview of Helicobacter pylori VacA Toxin Biology *Toxins* **8** 173

[18] Raghu PS, Antioxidant and Antiulcer activity of musa paradisica in rats *Int. J. Pharma & Ind. Res* **2**

[19] Onasanwo SA, Emikpe BO, Ajah AA, Elufioye TO 2013 Anti-ulcer and ulcer healing potentials of Musa sapientum peel, *Pharmacognosy Research*, **5**

[20] Sidhu JS and Zafar TA 2018Bioactive compounds in banana fruits and their health benefits **2**

[21] Rao T 2016 Antiulcer activity of musa paradisiaca (banana) tepal and skin extracts in ulcer induced albino mice, Malaysian journal of analytical sciences **20** 1203-1216

[22] Maharani L, Apriasari I and Eko S 2014 Bioactive Compound and Antioxidant Activity of Methanol Extract Mauli Bananas (Musa sp) Stem *International Journal of Bioscience, Biochemistry and Bioinformatics* **4** 2014

[23] Asif K, Rahman M, Annanya K 2019 Extraction and evaluation of phytochemicals from banana peels (musa sapientum) and banana plants (musa paradisiaca) *malaysian journal of halal research journal (mjhr)***2**

[24] Bohr UR, Annibale B, Franceschi F, Roccarina D ad Gasbarrini A 2007 Extragastric manifestations of Helicobacter pylori infection–other Helicobacters *Helicobacter* **24** 45-53

[25] Bishajit S, Asad U, Syed SI 2020 In Silico Analysis of Some Phytochemicals as Potential Anti-cancer Agents Targeting Cyclin Dependent Kinase-2, Human Topoisomerase IIa and Vascular Endothelial Growth Factor *Receptor-2*

[26] Laeliocattleya RA, Estiasih T, Griselda G, Muchlisyiyah J 2013 The bioactive compounds and antioxidant activity of ethanol and ethyl ecetate extracts of Candi Banana (Musa paradisiaca) *IOP Conf. Ser.: Earth Environ. Sci.* **131**

[27] Prabha P, Karpagam T, Varalakshmi B and Packiavathy AS. Indigenous anti-ulcer activity of Musa sapientum *Pharmacognosy Research* **3**

[28] Onyema CT, Ofor CE, Okudo VC and Ogbuagu AS 2016 phytochemical and antimicrobial analysis of banana pseudo stem (musa acuminata) bjpr, **10**

[29] Calister T 2018 In vivo Antiulcer Activity of Phospholipid–Based Complexes of Musa, Paradisiaca (Musaceae) Peel Extract for Improved Oral Drug Delivery *Indian Journal of Novel Drug Delivery* **11** 20-29

IOP Conf. Series: Materials Science and Engineering **925** (2020) 012052 doi:10.1088/1757-899X/925/1/012052

Synthesis, characterisation and biological investigation of some third-row elements with halo- substituted coumarin schiff base with N and O as donors

Pratiba Patil[1,4], Virupaxappa S Betageri[1], Shivashankar M Kinnal[3] and Latha M S[2]

[1] Research Centre, Department of Chemistry, GM Institute of Technology (GMIT), Davangere, Karnataka, India- 577006
[2] Department of Chemistry, K. L. E. Society's R. L. S. Institute (Autonomous), Belagavi, Karnataka, India.
[3] P. G. Department of Chemistry, Karnataka University, Dharwad- 580003, Karnataka, India.
[4]Visvesvaraya Technological University, Belagavi, Karnataka, India
Corresponding Author: betagerivs@gmit.ac.in, lathamschem97@gmail.com

Abstract. The present work involves in the synthesis, spectral characterisation and the biological analysis of the metal complexes which were derived from 3-Chloro-8-formyl-7-hydroxy-4-methylcoumarin and benzylamine. The molar conductivity values indicated that the complexes are non-electrolytes. Elemental analysis was carried out to assign the $[ML_{2}(H_2O)_2]$ stoichiometry to the metal complexes, where M= Co (II), Ni (II) and Cu(II) and L= deprotonated ligand. The IR data confirmed coordination of metal ions to the ligand. The presence of a coordinated water molecule was investigated by thermal studies and the electronic spectral results were helpful in assigning the geometry to synthesised complexes. The both Schiff base and the derived metal complexes were subjected various biological activities to check their potency.

Keywords: Halogenated Schiff Base, Fluorescence, and Antituberculosis, anthelmintic, antimicrobial.

1. Introduction

Coumarin and its derivatives have the enormous properties have attracted many researchers. As it occurs naturally in some of the plants and can be extracted using various methods [1]. When comes to the chemical synthesis, there are several routes [2-7] and one of the best method is Pechmann Condensation. Substituted phenols; can be used to relocate the various groups in the ring. 7- Hydroxycoumarin is the parent coumarin between all the complex coumarin structures. [8]. Coumarin derivatives has the ability to exhibit pharmacological property, such as antiplasmodial [9], anticancer [10], cytotoxic [11], antimicrobial [12], antioxidant [13], antitubercular [14], anti-HIV [15], anti-inflammatory and anticoagulant [16], antispermatogenic [17], anthelmintic [18]and inhibition of platelet aggregation [19] activities. The ability of coumarins to chelate [21] and form the complexes have been studied [20].

Recent survey suggested that the halo-substitution at various positions in an aromatic ring brings about more changes in their activities. These derivatives obtained can be used as potent drugs against various disorders. Such compounds are known to work against several protein kinases as inhibitors [22]. Recent work indicates that the halogen-bonded ligands probe shows increasing affinity to act like a potential cancer drugs [23]. Solubility of the compounds in water is decreased upon halogen substitution and helps to penetrate the lipid membrane. Less bulky atoms such as fluorine are capable of blocking the metabolically active positions.

Based on the above observations, in the present study it is planned to synthesise biologically active complexes of some 3rd row elements with halogen substituted coumarin Schiff Base. This part of work includes the synthesis of Nickel, Cobalt and Copper complexes within the Schiff Base obtained as a condensation product of benzylamine and 3-Chloro-8-formyl-7-hydroxy-4-methylcoumarin. The spectral characterisation of the synthesized compounds were carried out with the help of (UV-visible, IR, ESI-MS, NMR, Mass,), molar conductivity, thermal and fluorescence studies. Later various biological activities were conducted to test their potency towards some species.

2. Experimental Section

2.1 Reagents used
Chemicals utilized in the present research work are purely reagent grade and they are used without purification.

2.2 Physical measurements
The percentage composition of, Hydrogen, Carbon and Nitrogen were investigated using TRUSPEC CHN analyser. The Infrared spectra of the corresponding Schiff base and its metal complexes were recorded on HITACHI-270 IR spectrophotometer in the region 4000-400 cm^{-1}with the help of KBr disk. The metal complexes' UV spectra were reported on a VARIAN CARY 50-BIO UV-spectrophotometer in the region of 200 -1100 nm using HPLC grade DMF solvent. The proton and carbon NMR spectrum of the corresponding Schiff Base was measured using TMS as internal reference. HITACHI F-7000 Fluorescent Spectrophotometer was used to study the characteristic fluorescent property of coumarins. GCMS of the corresponding Schiff base, ESI- MS spectra of the synthesized compounds were reported using THERMO Finnegan and QP 2010S LCQ Advantage max ion trap mass spectrometer. In TGA, heating rate of 10 °C / min from a range of room temperature to 1000 °C was varied to calculate the data and the instrument used was PERKIN-ELMER DIAMOND TG / DTA. Conductivity Bridge ELICO-CM-82 T was used to record the values of molar conductivity showed cell constant 0.51 and the faraday balance were used to record the values of the magnetic moment.

2.3 Synthesis Part

2.3.1 Synthesis: 3-Chloro-7-hydroxy-4-methylcoumarin.
Mixture of 0.2 mol each dry resorcinol and ethyl-2-chloroacetoacetate was gradually added to Con. Sulfuric acid in a cool atmosphere (0-5 °C) and then kept in a refrigerator overnight. It was quenched in ice with continuous stirring until solid separated out. The solid so obtained was subjected to filtration and recrystallized using ethanol. Yield: 78% M. P. 241- 243 °C.

IOP Conf. Series: Materials Science and Engineering **925** (2020) 012052 doi:10.1088/1757-899X/925/1/012052

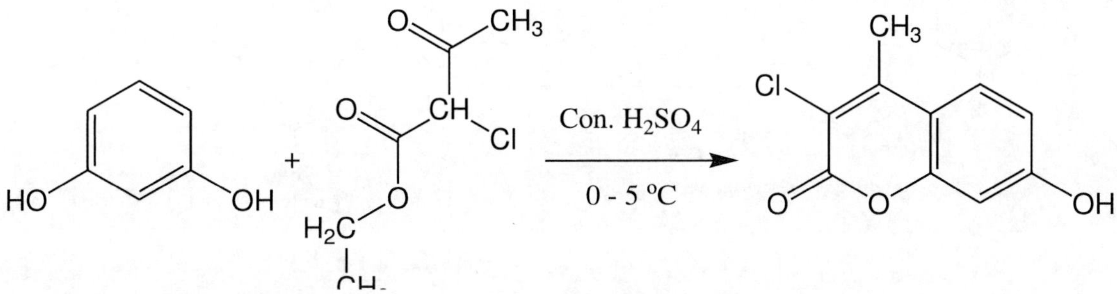

2.3.2 Synthesis: 3-Chloro-8-formyl-7-hydroxy-4-methylcoumarin.
Mixture of 0.03 M of 3-Chloro-7-hydroxy-4-methylcoumarin and 0.07 M of hexamine were subjected for heating in acetic acid for 4-5 hr followed by the addition of HCl and proceeded to heat for another 20 min. The mixture was then dissolved in ether, and removed. The desired product was the pale-yellow solid obtained after pouring in crushed ice, and it was filtered and dried. Yield: 25.0 percent; point of melting: 163-166° C.

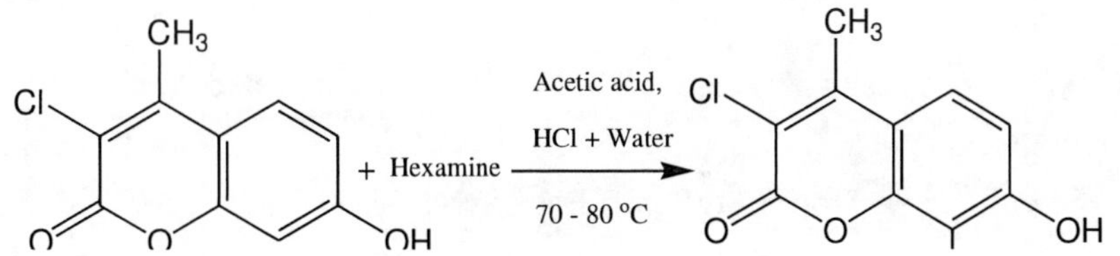

2.3.3. Schiff Base Synthesis.
The pre-hot ethanolic solution of benzylamine was combined with the 3-Chloro-8-formyl-7-hydroxy-4-methylcoumarin hot ethanolic solution, and then refluxed for 4-5 hours (few drops of acetic acid). As a result of refluxion, the precipitate which is obtained is being filtered, washed and recrystallized using acetic acid.

IOP Conf. Series: Materials Science and Engineering **925** (2020) 012052 doi:10.1088/1757-899X/925/1/012052

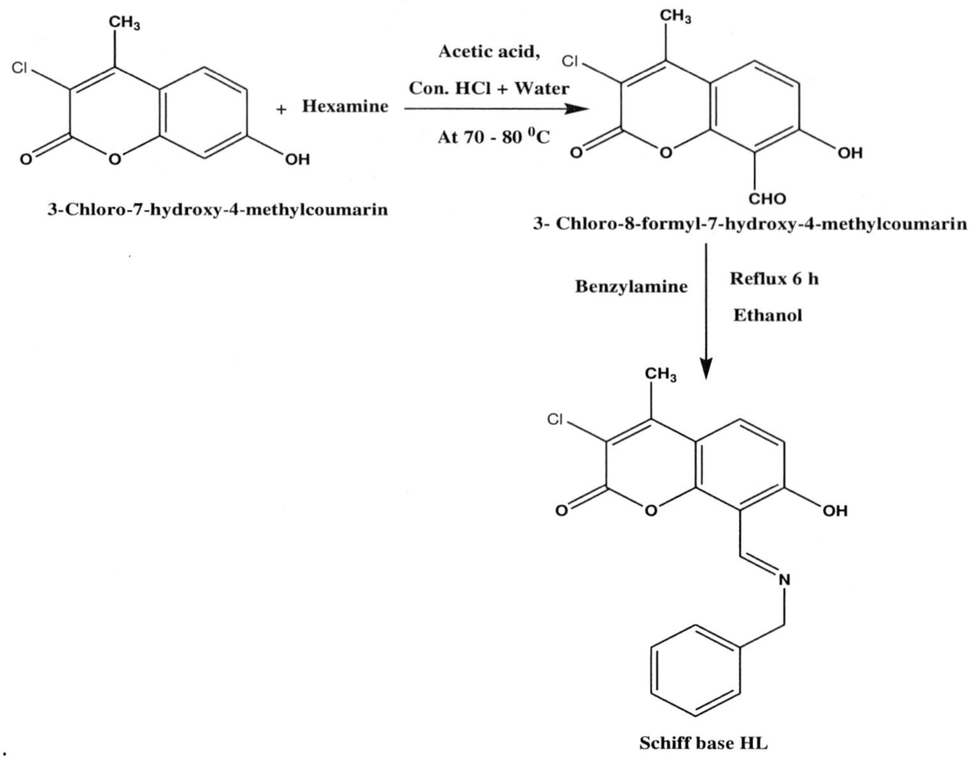

Process-1. Schiff Base Synthesis

2.3.4 NMR spectral study

^{1}H NMR

One singlet detected at range 14.90 ppm is attributed to the phenolic -OH in the (figure 1) Schiff base spectrum. A single proton resonance is -CH = N, methylene and -CH3 at 9.10, 4.85 and 2.43 ppm respectively. Furthermore, aromatic protons activate during multiplet signals between 6.05-7.63 ppm.

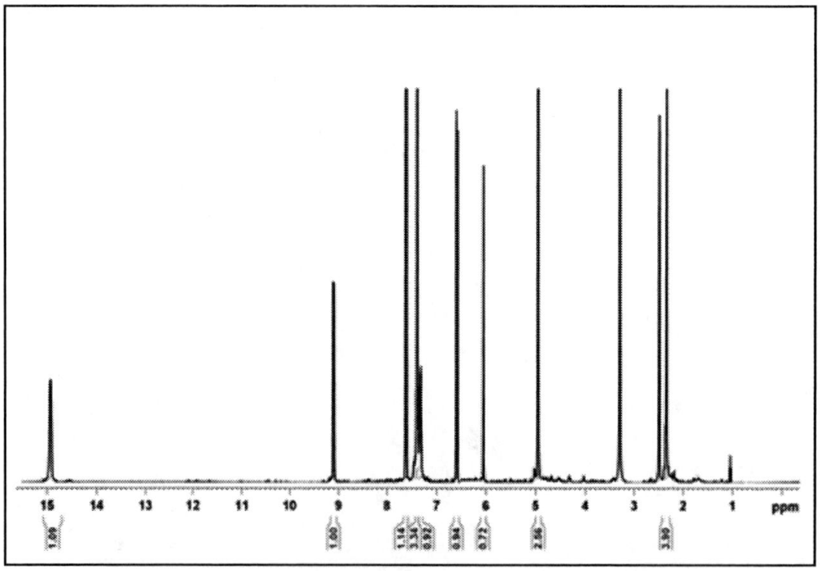

Figure 1. Schiff base-^{1}H NMR spectrum

^{13}C NMR

The **Schiff base** spectrum of signals ^{13}C NMR showed resonance at 173.22 (C=N), 160.45 (C=O), 164.011 (C-Cl), 56.84(-CH_2) and 18.33(-CH_3) ppm. Signals showed appearance in the respective range 104.039 - 159.39 ppm which were further assigned to the aromatic carbon substitutes. The ^{13}C NMR spectrum is embedded in figure 2.

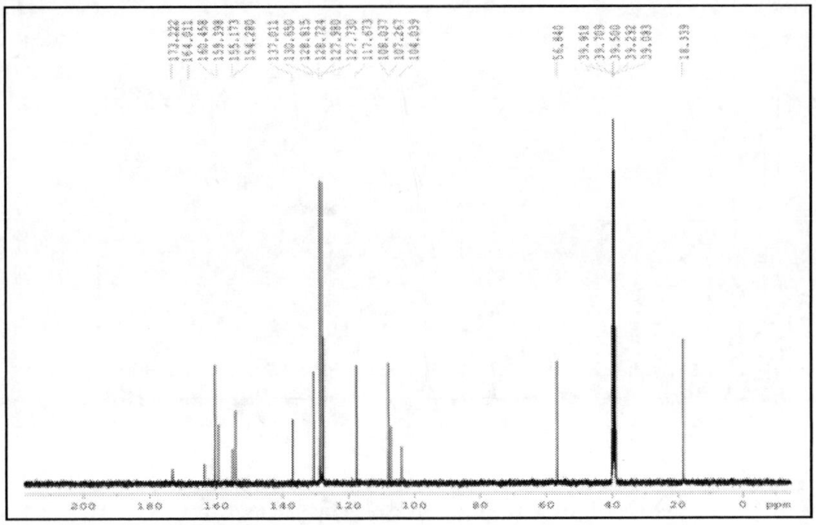

Figure 2. Schiff base-**^{13}**C NMR spectrum

2.3.4.1. Infrared Spectral Studies.

In spectrum of ligand (Fig.3) due to the presence of -C=N exhibits a very high intense band at 1626 cm^{-1} which ensures the complete formation of Schiff base. Due to the presence of *v*(O-H), another band of high intensity at 1257 cm^{-1} is assigned to the phenolic v(C-O) stretching frequencies and 1719 cm^{-1} for the lactonyl carbon. Appearance of a broad band under the region 3397 cm^{-1} observed.

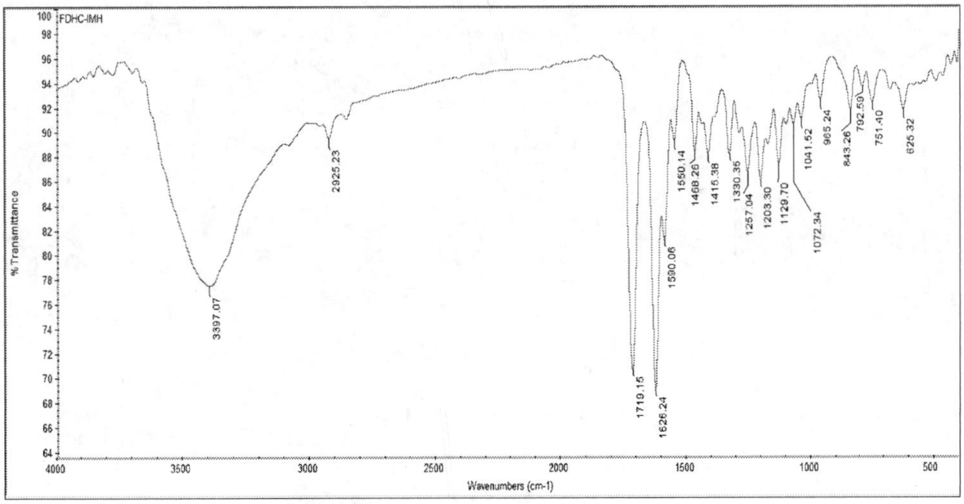

Figure 3. Schiff Base- Infrared Spectrum

2.3.4.2. GC- Mass Studies.

The Formation of the respective ligand showed confirmation by a crest at m/z 327 which clearly showed correspondence to its molecular weight. A fragmentation crest at m/z 310 was mainly by the cleavage of –OH group of coumarin (figure 4).

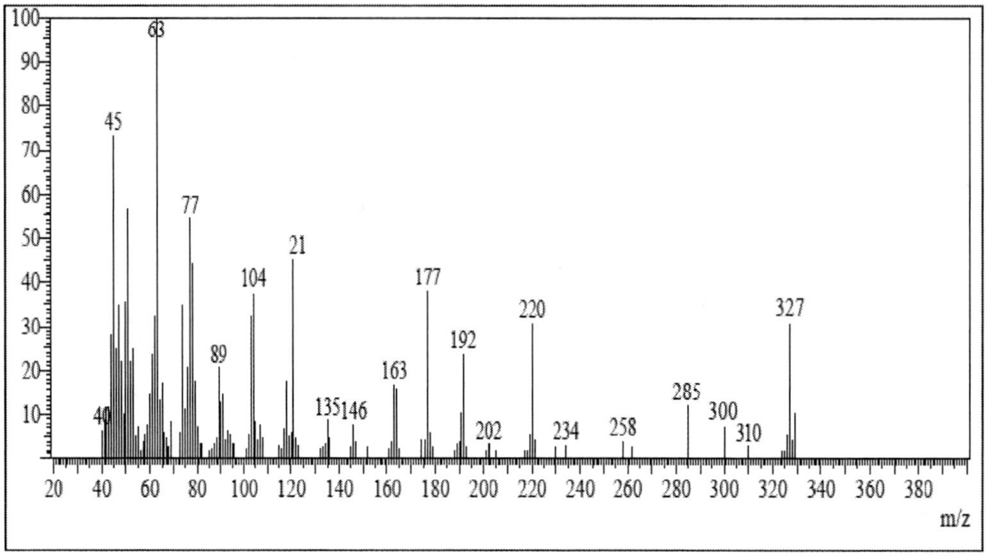

Figure 4. Schiff Base- Mass Spectrum

2.3.4.3 Synthetization of Cobalt, Nickel and Copper Complexes.
2 M Schiff base was dissolved in alcohol with combined refluxion in 1 M of the metal salts on the water tub for about 1h. Reflux is to be continued for 2 more hours after the addition of 3 h sodium acetate (1mmol). After separation the metal complex was subjected to filtration followed by washing with water; ether and alcohol. It was vacuum dried over fused $CaCl_2$ to remove any moisture.

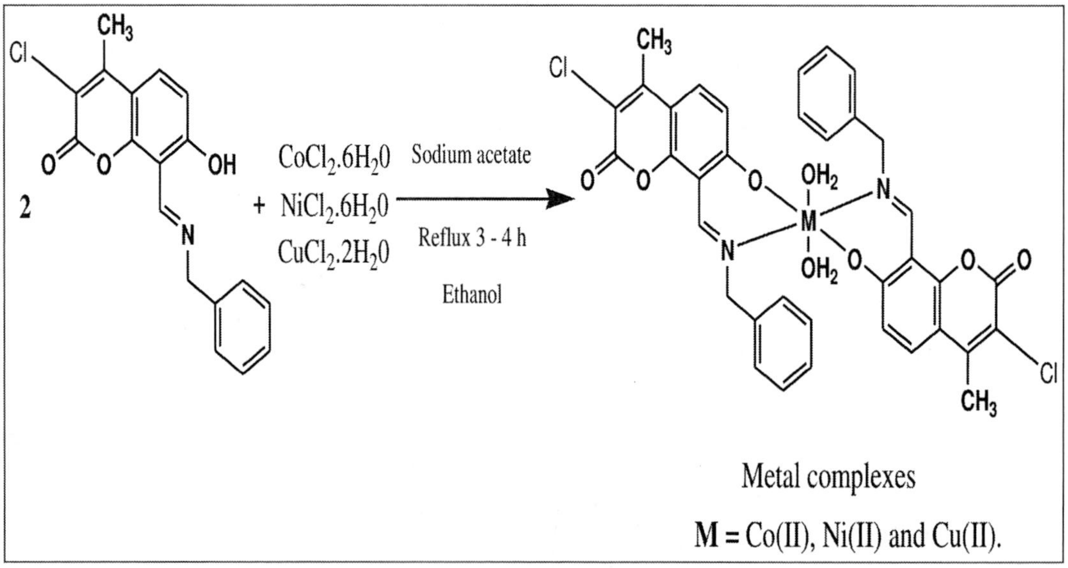

Scheme- 2. Metal Complexes- Synthesis

2.3.4.4. FTIR Studies of Metal Complexes.
The mode of binded Schiff base with the metal ion in the complexes were studied by analyzing the IR spectra of the corresponding metal complex which is further studied in comparison with the Schiff base.

The lowered values of azomethine group confirm the coordination complexes via nitrogen. The respective bands observed in Cobalt, Nickel and Copper complexes at 1598, 1592 and 1591 cm^{-1} are the shifted values of ν(C=N) from the ligand. The absence of ν(O-H) band in the metal complexes spectrum reveal that the coordination via deprotonation and formation of M-O bond. Bands which are observed at 1327 and 1719 cm^{-1} are further assigned to phenolicν(C-O) and lactonyl ν(C=O) respectively. The representative IR spectrum of Cobalt complex has been depicted in figure. 5.

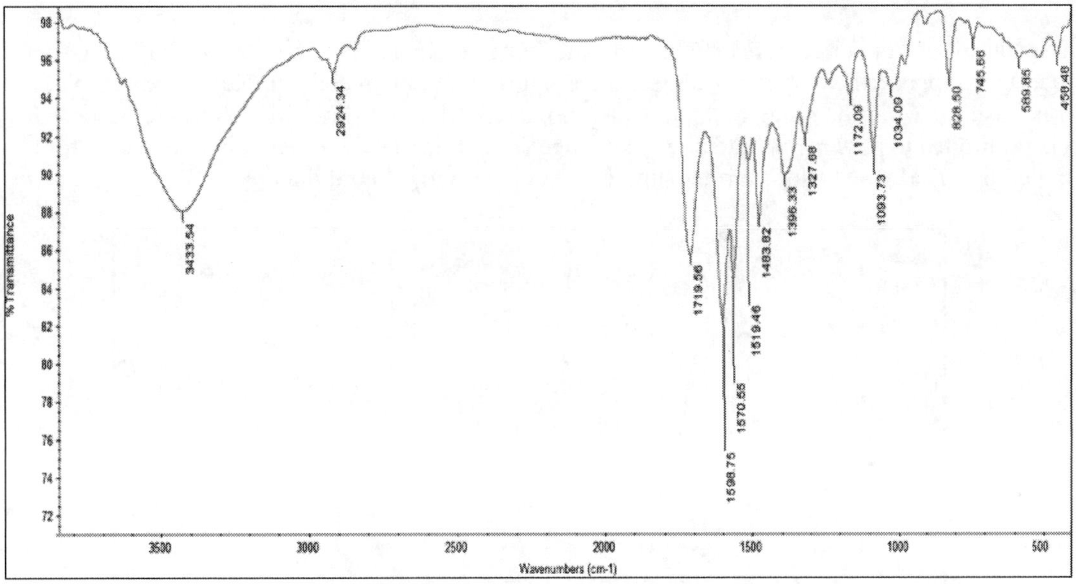

Figure 5. IR Spectrum-Cobalt Complex

The small shifts observed in the lining of ν(C=O) suggest as follows, there is no involvement of lactone carbonyl oxygen in coordination of complex. A band is observed in the range 1327 cm^{-1} is ascribed to ν(C-O), the observations helps in the formation of Metal-Oxygen bond. There is a broad band in the range 3443 cm^{-1} which supports the complete presence of coordinated water molecules. The bands metal sensitive bands are in the region 458 and 589 cm^{-1} and ascribed to ν(M-N) and ν(M-O) respectively.

Table 1. Elemental Analyses of Schiff Base and its Metal Complexes Along with Molar Conductance Data

Compound Name	Empirical Formula	Yield %	M% Obs.	M% Cal.	C% Obs.	C% Cal.	H% Obs.	H% Cal.	N% Obs.	N% Cal.	Molar Cond. (Ohm^{-1} cm^{-2} $mole^{-1}$)	μ_{eff} (BM)
Schiff Base	$C_{17}H_{12}NO_3$	80%	-	-	68.68	68.65	4.07	4.09	4.71	4.68	-	-
$Co(II)L_2.(H_2O)_2$	$Co(C_{17}H_{12}NO_3)_2.(H_2O)_2$	59%	8.57	8.54	59.40	59.43	3.81	3.79	4.07	4.04	22.10	4.90
$Ni(II)L_2.(H_2O)_2$	$Ni(C_{17}H_{12}NO_3)_2.(H_2O)_2$	63%	8.16	8.14	59..42	59.45	3.81	3.83	4.08	4.11	12.45	3.15
$Cu(II)L_2.(H_2O)_2$	$Cu(C_{17}H_{12}NO_3)_2.(H_2O)_2$	56%	9.18	9.15	59.00	59.05	4.07	4.09	4.05	4.08	27.40	1.77

2.3.4.5. Electronic Absorption Spectra and Magnetic Studies of complex.

Stereochemistry of metal ions was studied by recording the spectra along with ligand in DMF. The absorption spectra provide the valuable information to allocate the stereochemistry of metal ions, the assignment occurs in conjunction with the d-d transition peaks.

Ranging between 255 and 330 nm two bands were observed for the ligand, the bands which appear below the range 350 nm are ligand-based transitions of the azomethine group's aromatic ring and the non-bonding electrons of nitrogen [24, 25].

Table-2 contains the results of electronic spectra which are recorded in DMF. Two absorption bands of Cobalt complex occurred in the region 568 and 980 nm correspond to

$^4T_{1g}$ (F) → $^4T_{2g}$ (F) (v_1) and $^4T_{1g}$ (F) → $^4T_{1g}$ (P) (v_3) transitions respectively, which are well matching with high spinning octahedral geometry [26] and magnetic moment of 4.90 BM proves it [27].

Nickel complex exhibits following bands at 989, 674 and 365 nm attributable to 3A2g(F)3T2g(F)(1), 3A2g(F)3T1g(F)(2) and 3A2g(F)3T1g(P)(33) transitions respectively indicate the octahedral geometry [28] and magnetic momentum value of 3.15 BM is also stand by the octahedral geometry around metal ions [29, 30].

The spectrum of Copper complex (figure-6) exhibited two bands of absorption. 2Eg a 2T2 g transition, evidently it supports the distorted octahedral configuration to the metal complex [31] is given at a very low intensity band in the range 690 nm. Correspondingly the high-intensity bands are observed at 391 nm is prohibited to pass metal load due to symmetry. The magnetic moment obtained value for Copper complex is 1.77 BM, which in further supports the electronic spectral data [32, 33].

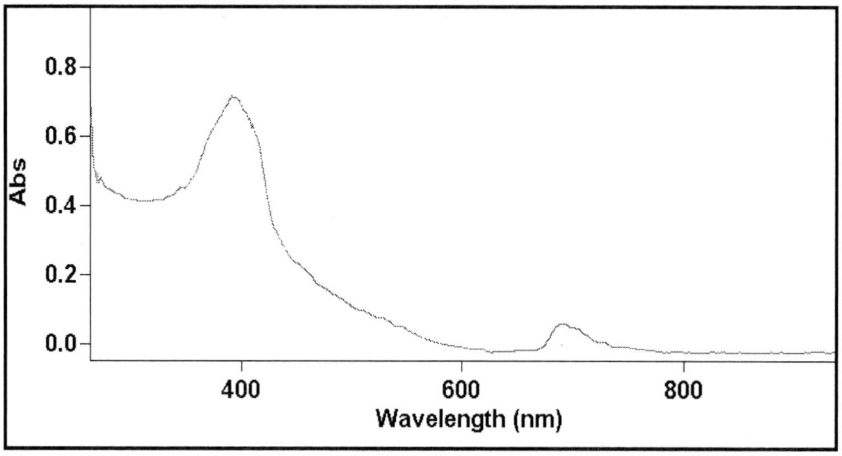

Figure 6. Electronic Absorption Spectrum-Cu(II) Complex

Table 2. Electronic Spectral Results with Magnetic Susceptibility Data of the Metal Complexes (**1-3**).

Complex.	Obs λ_{max} in cm^{-1} (nm)	Band Assignments	Geometry	μ_{eff} (BM)
1	10152 (985) 17730 (564)	$^4T_{1g}(F) \rightarrow {}^4T_{2g}(F)$ (v_1) $^4T_{1g}$ (F)→ $^4T_{1g}$ (P) (v_3)	octahedral	4.90
2	10131 (987) 14880(672) 27624(362)	$^3A_{2g}(F) \rightarrow {}^3T_{2g}(F)$ (v_1) $^3A_{2g}(F) \rightarrow {}^3T_{1g}(F)$ (v_2) $^3A_{2g}(F) \rightarrow {}^3T_{1g}(P)$ (v_3)	octahedral	3.15
3	14450 (692) 25380 (394)	$^2T_{2g} \leftarrow {}^2E_g$ L→M	distorted octahedral	1.77

2.3.4.6. Fluorescence study of Schiff base and its metal complexes.
Fluorescence being one among the spectrochemical analytical methods in which an analyte's molecules is being excited through irradiation at a particular wavelength and emits radiation from a different wavelength.

The present research is been analysed to the fluorescent nature of the base of the Schiff and its metal complexes at 27^0c temperature within the concentration of 10^{-5} M in DMSO solution. It is observed that, as a result of interaction between phenolic oxygen and metal ion, the emission band at 482 nm of Schiff base disappears upon complexation. Synthesized complexes Co (II), Ni (II) and Cu (II) shows the bands of emission at 466, 452 ,437 nm respectively. Diminishing electron density on ligand decreases the complexes' emission rate. The decrease in metal complex intensities was compared to the base of Schiff and the order is Co (II) > Ni (II) > Cu (II) is shown in figure 7.

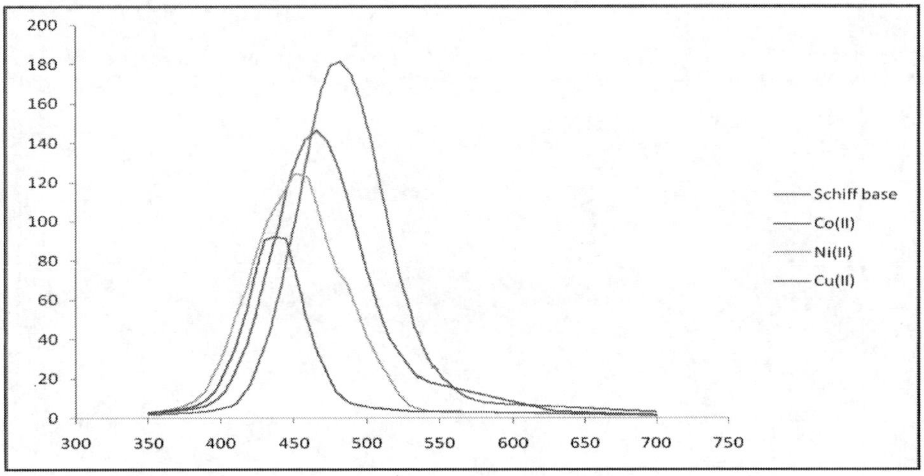

Figure 7. Emission Spectra-Schiff base and its Metal Complexes

2.3.4.7. ESI-MS Study.

The Cobalt, Nickel, and Copper complexes ESI-MS analysis were reported a M+1 crest at m / z 747, 748, 753 respectively, and it is also equal to their corresponding molecular weight. This facilitates development of $ML_2(H2O)_2$type metal complexes. The representational Cobalt complex spectra is in the figure 8.

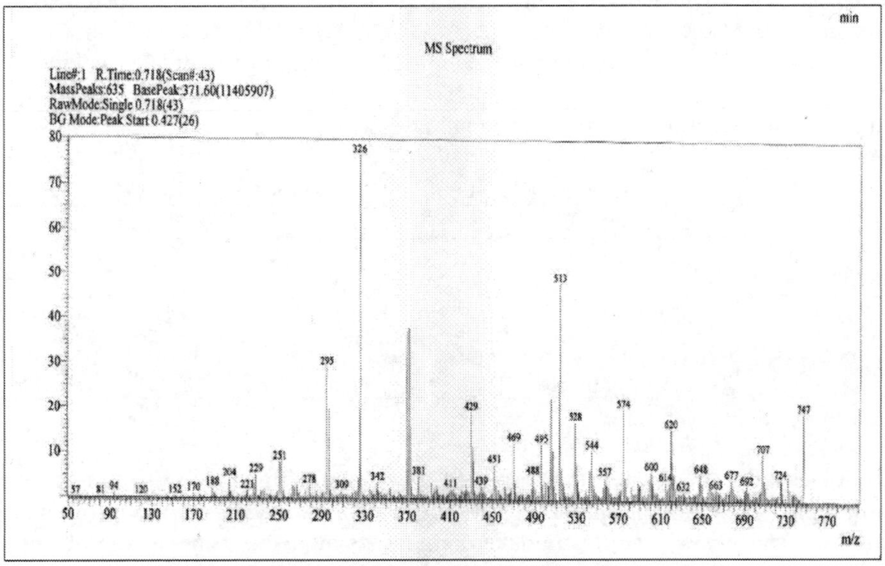

Figure 8. ESI-MS Spectra-Co (II) Complex

2.3.4.8. Thermal analysis of metal complex.

The thermal behaviour of synthesized Schiff base L complexes Co (II), Ni (II) and Cu (II) was subjected for study as a function of temperature. Decomposition was observed to occur in two stages. In other words, all the complexes have the same thermal features. The definition of the symbolic thermogram

IOP Conf. Series: Materials Science and Engineering **925** (2020) 012052 doi:10.1088/1757-899X/925/1/012052

Copper complex is discussed here. The first decomposition was observed due to the loss of the coordinated water molecules in range 180-220 °C region centred at 198 °C. Decomposition in the 350-540 °C regions based at 475 °C leading to the loss of organic moulds. The metal oxide then grew above 650 °C.

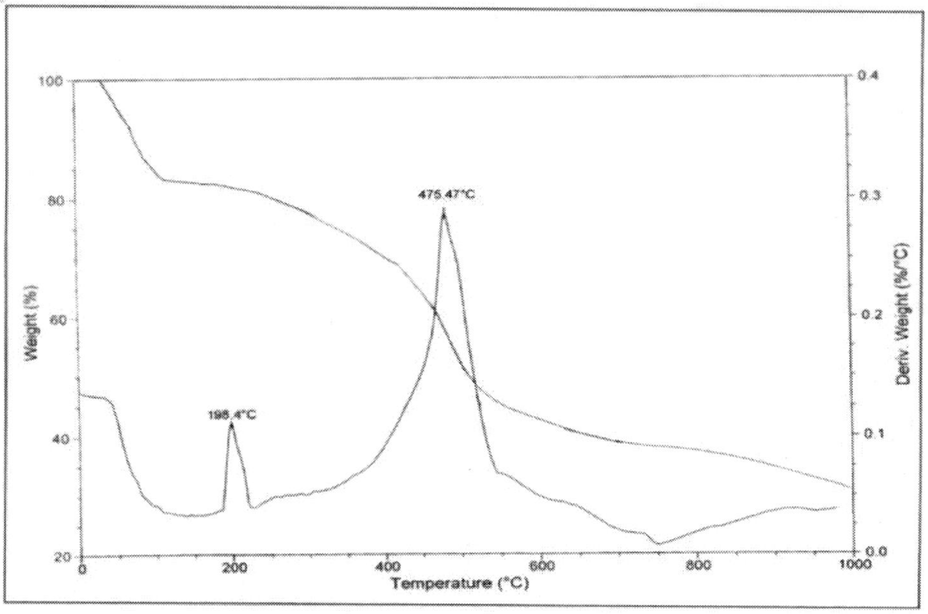

Figure 9. TG/DTA spectrum - Copper Complex

3. Biochemistry

3.1. Anti-tuberculosis Activity Study

The metal complexes which were synthesized were tested for their anti-tuberculosis study against M. Tuberculosis using Microplate Alamar Blue Assay (MABA). The results further reveal that, the metal complexes exhibited better activity than the Schiff base with different MIC values as shown in Table-3. From results we can conclude that, the Cu(II) complex exhibited the prominent activity with the MIC of 12.5 (μg/mL) when compared to other metal complexes.

Table 3. Results of anti-tuberculosis Activity

Compound	MIC Values
Schiff base	50
1	25
2	25
3	12.5
Streptomycin Std.	6.25

3.2. Antimicrobial Activity

The activity ensures that, when compared to their parent ligand, the synthesized metal complexes exhibit the greater lethal effect. The Cu (II) complex displayed the prominent antifungal and antibacterial activity against all strains among the synthesized metal complexes. The results are tabulated in below tables (4 and 5). By this it can be suspected that, presence of any metal ions may possibly show increase in the activity.

Table 4. Antibacterial Studies - Schiff Base and its Metal Complexes

Compound Name/No.	Conc. (μg/mL)	Zone of Inhibition (mm)			
		E. coli	*P. aeruginosa*	*K. pneumoniae*	*S. aureus*
Schiff base	200	6	5	6	7
1	200	9	7	9	8
2	200	9	8	8	10
3	200	10	10	11	12
Gentamicin	200	14	13	14	15

Table 5. Antifungal Studies of Schiff Base (HL) and its Metal Complexes

Compound Name/No.	Conc. (μg/mL)	Zone of Inhibition (mm)	
		P. chrysogenum	*A. niger*
Schiff base	200	5	6
1	200	8	10
2	200	9	9
3	200	10	12
Flucanozole	200	13	16

3.3. Anthelmintic Study

Table 6. Anthelmintic Studies - Schiff Base and its Metal Complexes

Compound	Concentration (μg/mL)	Paralysed Time (Min)	Lethal time (Min)
DMSO*	-	-	-
Schiff base	2	20.16±0.02	24.28±0.12
	10	9.29±0.22	14.30±0.12
1	2	12.16±0.15	19.60±0.02
	10	7.51±0.04	10.14±0.03
2	2	16.21±0.24	21.30±0.08
	10	9.10±0.01	15.60±0.01
3	2	10.30±0.03	16.41±0.05
	10	5.48±0.22	9.45±0.01
Albendazole Std.	10	3.49±0.06	7.26±0.14

Results of the ligand anthelmintic activity and the synthesized compounds are shown in Table- 6. From this result it is clear that the metal complexes show more active than its parent ligand. The Copper complex, in particular, exhibits the prominent behaviour against a standard drug Albendazole at concentration of 10μg / mL relative to the other complexes.

3.4. DNA Cleavage Activity

Schiff base's DNA cleavage activity and its metal complexes is subjected for analysis using pBR322 Plasmid DNA and using the agarose gel electrophoresis process. Schiff base's cleavage activity and its complexes is shown in figure 10. In the present investigation, ligand along with the Cobalt and Nickel complexes shows the partial cleavage; but the Nickel complex cleaves the DNA completely.

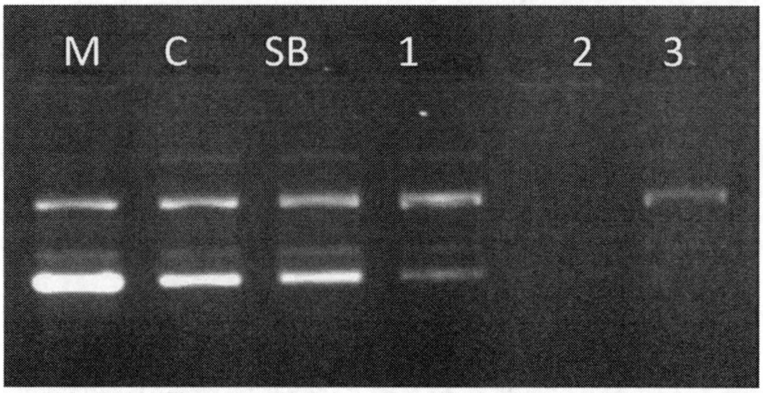

Figure 10. DNA Cleavage Activity - Schiff Base and its Metal Complexes

('M'-being the standard DNA and molecular weight marker, 'C'– being the control DNA i.e. untreated pBR322 Plasmid DNA and lanes SB and 1-3 are pBR322 treated with different Schiff base and metal complexes respectively).

4. Conclusion

The Schiff base which has been synthesised acts as a bidentate ligand that is coordinated with deprotonation to metal ions by means of oxygen atoms and azomethine nitrogen. The elemental analyzes verified that the synthesized metal complexes have form stoichiometry $[M(L)_2.(H_2O)_2]$ where [M=Co(II), Ni(II) and Cu(II)] and 'L' are doubly deprotonated ligands. The molar conductivity analysis indicates that there are poor conductivity values in all the synthesized metal complexes. Such various spectro-analytical techniques have helped to assign the following preliminary structure to the metal complexes synthesized. The newly synthesized Schiff base and its metal complexes have been subjected for screening for their *in-vitro* Cleavage studies Antitubercular, cytotoxic, antimicrobial and DNA. It has been proved to be a strong antitubercular agent among the tested compounds Cu (II) complex. Whereas antimicrobial assays illustrate Schiff Base's enhanced activity upon coordination. Schiff base's DNA cleavage activity and its metal complexes were analysed using pBR322 Plasmid DNA, an agarose gel electrophoresis method in which the Ni (II) complex showed the successful cleavage compared to any other synthesized compounds.
Considering all the above facts the following tentative structure has been assigned to the synthesised metal complexes.

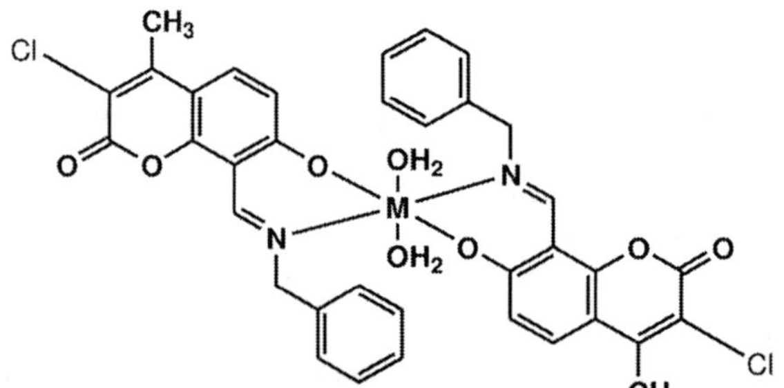

Metal complexes

M = Co(II), Ni(II) and Cu(II).

References

[1] Celeghini RMS, Vilegas JHY and Lanças FM 2001 Coumarin in Hydroalcoholic Extracts of Mikania glomerata Spreng. ("guaco") Leaves, Braz Extraction and Quantitative HPLC *Analysis of Chem. Soc.* **12** 706-709

[2] Mali RS and Tilve SG 1791 Useful Synthesis of Coumestans, *Synth. Commun.* **20** 1781-1791

[3] Saidi MR, Bigdeli K 1998 Microwave promoted and improved thermal synthesis of pyranocoumarins and furocoumarins *J. Chem. Res.* 800-801

[4] Shockravi A, Sharghi H, Valizadeh H and Heravi MM Solvent Free Synthesis of Coumarins Phosphorus, Sulfur, *Silicon Relat. Elem.* **177** 2555-2559

[5] Cairns N, Harwood LM, Astles DP, 1994 Tandem thermal Claisen–Cope rearrangements of coumarate derivatives. Total syntheses of the naturally occurring coumarins: suberosin, demethylsuberosin, ostruthin, balsamiferone and gravelliferone *J. Chem. Soc., Perkin Trans.* **1** 3101-3107

[6] Perkin WH 1875 On propionic coumarin and some of its derivatives *J. Chem. Soc.* **28** 10-15.

[7] Brufola G, Fringuelli F, Piematti O Pizzo F 1996 Simple and efficient one-pot preparation of 3-substituted coumarins in water *Heterocycles* **43** 1257-1266

[8] Gottlieb OR, Herrmann K, Murray RDH, Ohloff G, Pattenden G 1978 Progress in the chemistry of organic natural products, *Springer Verlag Wein* 200-427

[9] Marshall EM, Ryles M, Butler K, Weiss L 1994 Cancer Res. Clin. Oncol. **120** 535.

[10] Lacy A, O'Kennedy R 2004 Studies on coumarins and coumarin-related compounds to determine their therapeutic role in the treatment of cancer *Curr. Pharm. Des.* **10** 3797-3811

[11] Stanchev S, Momekov G, Jensen F and Manolov I 2008 6-Substituted 1, 4-Naphthoquinone Oxime Derivatives (III): Synthesis and Cytotoxic Evaluation. *Eur. J. Med. Chem* **43** 694

[12] Patel KS, Patel JC, Dholariya HR, Patel KD 1992 Synthesis of Cu(II), Ni(II), Co(II), and Mn(II) Complexes with Ciprofloxacin and Their Evaluation of Antimicrobial, Antioxidant and Anti-Tubercular Activity *Spectrochim. Acta Part A* **96** 468

[13] Zabradnik M 1992 The Production and Application of Fluorescent Brightening Agents *John Wiley and Sons, New York*

[14] Patel KS, Patel JC, Dholariya HR, Patel KD 1992 Synthesis of Cu(II), Ni(II), Co(II), and Mn(II) Complexes with Ciprofloxacin and Their Evaluation of Antimicrobial, Antioxidant and Anti-Tubercular Activity *Spectrochim. Acta, Part A.* **96** 468

[15] Trivedi JC, Bariwal JB, Upadhyay KD, Naliapara YT, Joshi SK, Pannecouque CC, Clercq ED, Shah AK 2007 Improved and rapid synthesis of new coumarinylchalcone derivatives and their antiviral activity, *Tetrahedron Lett.* **48** 8472–8474

[16] Khan MSY and Sharma P. Spectrometric Identification of Organic Compounds Indian J. Chem. 1993, **32** 817.

[17] Tyagi VP, Dixit BC and Joshi 2018 Synthesis, Characterization and Biological Evaluation Of Some Metal Complexes Of Transition Metal Ions With Coumarin Schiff Bases Having N And O Donor Atom *Naturwissenschaften* **67** 104–109

[18] Sandhya B, Giles D, Vinod M, Basavarajaswamy G and Rekha A 2011 Eur. J. Med. Chem. **46** 4696

[19] Cavettos G, Nano GM, Palmisano G and Tagliapietra S 2001 Tetrahedron: Asymmetry **12** 707

[20] Ketata I, Mechi L, Ben Ayed T, Dusek M, Petricek V and Ben Hassen R 2012 Open J. Inorg. Chem. **2** 33–39.

[21] Mandakmare AU, Navwade ML 1997 Orient. J. Chem. **13** 155–158.

[22] Voth AR and Ho PS 2007 Curr. Top Med. Chem. **7** 1336.

[23] Wilcken R, Liu X, ZImmermann MO, Rutherford TJ, Fersht AR, Joerger AC and Boeckler FM 2012 J. Am. Chem. Soc. **134** 6810

[24] Ramesh R and Maheswaran S. 2003 J. Inorg. Bio. Chem. **96** 457

[25] Silverstein R, Bassler C and Morrill T 1981 Spectrometric Identification of Organic Compounds *4th Ed., John Wiley and Sons Inc., New York*

[26] Georgrieva I, Trendafilova N. and Bauer G. 2006 Synthesis and characterization of tetradentate bis-Schiff base complexes of di-and tri-valent transition metals *Spectrochim. Acta Part A* **63** 403

[27] Ul-Hassan M, Chohan ZH and Supuran CT 2002 Chohan ZH, Scozzafava A, Supuran CT. Zinc complexes of benzothiazole-derived Schiff bases with antibacterial activity. *Main Group Met. Chem.* **25** 291

[28] Prakash PD and Patel MN 2004 Syn. React. Inorg. Met. Org. Chem. **34** 383.

[29] Alarcon-Payer C, Pivetta T, Chuqusillo-Lazarto D, Gonzalez Perez JM, Crisponi G Castineiras A and Niclos-Gutierrez J 2004 Inorg. Chem. Commun. **7** 1277.

[30] Rao TR and Archan P 2005 Synth. React. Inorg. Metal. Org. Chem. **35** 299
[31] Liu H, Wang H, Gao F, Niu D and Lu Z 2007 Synthesis and characterization of self-assembled coordination polymers of N-diaminomethylene-4-(3-formyl-4-hydroxy-phenylazo)-benzenesulfonamide *J. Coord. Chem.* **60**
[32] Raman N and Dhaveethu R 2007 Synthesis, spectral characterization of Schiff base transition metal complexes: DNA cleavage and antimicrobial activity studies *J. Indian J. Chem* 46A1611
[33] Dianu ML, Kriza A, Stanica N. and Musuc AM 2010 SYNTHESIS, Spectroscopic Characterization and Evaluation (Antibacterial & (GOT, G pT) Enzyme) Activity of Mixed Ligand Complexes of M(II) with Amino Acid (L-Asparagine) and Schiff Bases Derived from (Sulfamethoxazo Drug with 2,4Dimethoxybenzaldehyde) J. Serb. Chem. Soc. **75** 1515

IOP Conf. Series: Materials Science and Engineering **925** (2020) 012053 doi:10.1088/1757-899X/925/1/012053

A Study of Response Parameters during ECDM on Quartz Material

S G Dileepkumar[1], K N Bharath[2], B C Naveen[3] and Anil kumar[4]

[1,2,4] Department of Mechanical Engineering, GM Institute of Technology, Davanagere, Karnataka - 577006, India.
[3]Department of Industrial & Production Engineering, UBDTCE, Davanagere, Karnataka - 577004, India.
*Corresponding email: dileepsmk404@gmail.com

Abstract. Electrochemical discharge machining (ECDM) is a hybrid machining process that combines the possible characteristics of Electro Discharge Machining (EDM) and Electrochemical Machining (ECM) machining. In the present work, Quartz is used as work piece material and potassium hydroxide (KOH) as an electrolytic medium. Experiments were carried out by considering the three factors at three different levels. The input variables are Voltage, Electrolyte Concentration (EC) and Inter-Electrode Gap (IEG). The experimental response variable is Material Removal Rate (MRR). The process variables were analyzed and optimized using ANOVA from the Minitab 14 statistical tool. The response parameters are predicted by using Artificial Neural Network (ANN) and linear regression equation was generated to predict the experimental values. The results reveal that, voltage and electrolyte concentration have approximately equal percentage of contribution in achieving maximum MRR.

1. Introduction

To remove excess material from very hard and delicate materials, non-conventional machining methods are taking advantage of now a few days. Electric Discharge Machining (EDM) and Electro Chemical Machining (ECM) have been introduced to meet the needs of manufacturing industries to machine advanced materials with electrical conductivity properties. This can be used effectively for machining electrically non-conductive advanced engineering materials such as glass and ceramics has shown the possibility of efficiently and cost-effectively drilling micro-holes through smaller electrodes [1]. Hybrid machining means two or more machining processes are combined to maximize the capacities of the machining processes and take advantage of the efficiency of the constituent processes [2]. Hybrid machining method incorporates the features of many traditional as well as non-conventional machining processes that lead to achieving superior machining capabilities, i.e. combining the advantages of different processes. The ECDM is a mechanism which involves a combination of the electrochemical reaction and the electro discharge action. Due to the breakdown of the insulating layer of gas bubbles when the DC power is supplied between the cathode and the anode, the electrical discharge occurs between the tool and the work piece, resulting in material removal owing to melting, vaporization of the work piece material and mechanical erosion [3]. The machining is partly a chemical process and thus the nature of the electrolyte strongly influences the machining behavior. It seems that Sodium hydroxide (NaOH) electrolyte has the most fascinating properties relative to other electrolytes but molten salt electrolytes (eutectic of melting of NaOH and KOH at 170°C) will significantly enhance the smoothness of the machined surface [4]. Due to better flushing action in machining area, the low viscosity of KOH electrolyte yields high MRR. The other electrolytes, such as H_2SO_4, NaCl, and

$NaClO_3$ are less flexible due to harmful gas production and poor MRR [5].
Several researchers have studied various parametric optimization of ECDM process. Sathish N et al. discussed the influence of process variables namely electrolyte concentration, voltage and stand-off distance on MRR and tool wear rate. Experiments were done on the soda lime glass and the plan of experiment was according to the Taguchi technique. Taguchi method was utilized to optimize the response parameters. Analysis of variance was employed to determine the significance of input parameters on tool wear rate and MRR [6]. Harugade et al. reported on the effect of various electrolytes on MRR in ECDM processes was discussed. From the results it is noted that the voltage applied is important parameter affecting MRR and that the removal rate of KOH electrolyte is better compared to other solutions proposed [1]. Laio et al. made attempt to improve the quality of drilling by adding the surfactants to the electrolyte in ECDM process. Adding of surfactants to solution improves the current density and more stable bubbles are formed around the cathode electrode [7]. Liu proposed model of ECDM to analyze the influence of discharge mechanism on reinforced metal matrix composites. A set of experiments were carried out to validate the model and experimental outcome shows that the occurrence of arcing action increases with increase in the, duty cycle and current concentration of electrolyte [8]. Bhondwe et al. developed thermal model based on FEM with experimental results, in the range of accuracy. The parametric tests are also carried out for various parameters such as Concentration of the electrolytes, duty factor and partition of the electricity. The increase in MRR is found to increase with electrolyte increase concentration of soda lime glass working component material due to Electro- chemical Spark Machining (ECSM) [9]. The advanced engineering materials with high specific strength, heat-resistant capacity and high corrosion resistance have seen a rapid increase in demand in recent years. The advanced materials include glass (quartz), ceramics, silicon nitride, etc. The machining of quartz glass is very difficult by using conventional method because of its high hardness, brittle and non-conductive properties. Quartz glass is a pure form of silica with more than 99.9 % SiO_2 content. Because of this it has superior working properties such as high hardness, resistance to degradation, stability under atomic bombardment and improved optical transmission than other glass materials. The ECDM has been identified as an effective method for processing of this type of materials [10,11]. In this work an attempt has been made to develop an experimental set up for ECDM and optimize the response variable MRR.

2. Experimentation

2.1 Experimental Setup
Experimental setup of ECDM as shown in 'Figure 1', it consists of a container, cathode (tool), anode (auxiliary anode), DC power supply and tool monitoring unit. Work piece is positioned on the table and it is inflexibly secured on the work table. Work table is attached to the container. The container is made up of soda lime glass and it is having the dimension of 8″×8″×6″.

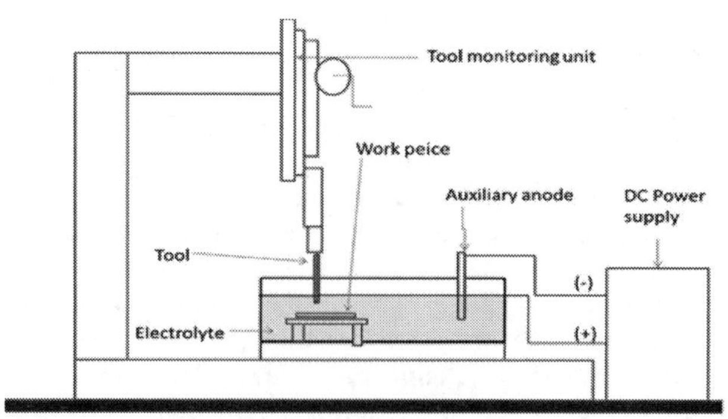

Figure 1. Schematic Diagram of Electrochemical Discharge Machining

2.1 Experimental Details

Table 1. The experimental details of ECDM

Sl. No.	Non-variable parameters	Description
1	Work piece material	Quartz (25 mm x 25 mm x 10 mm)
2	Work piece thickness	1 mm
3	Cathode tool	Stainless steel wire of 1 mm diameter
4	Auxiliary anode	Stainless steel plate (100 mm x 50 mm x 10 mm)
5	Electrolyte	Potassium hydroxide (KOH)
6	Stand-off distance	0.03 mm
7	Time	20 Min.
	Variable parameters	Description
8	Voltage	50V,60V and 70V
9	Electrolyte concentration	10%,15% and 20%
10	Inter-electrode gap	20 mm,30 mm and 40 mm

2.2 Mechanism of Metal Removal and Spark Generation

The ECDM process consists of a complex bubble are normally formed by hydrogen bubbles. Due to the failure of the non-conductive layer of gas bubbles, the electrical discharge takes place between the tool and work piece. The mechanical erosion and material removal results in the melting and vaporization of the work piece material as the DC power supply voltage is applied between the cathode and the anode. Because of the electrochemical reaction the positively charged ionic gas.

3. Results and Discussions

The experimental work on Quartz was carried out with three parameters and three stages of inputs. Experiments were performed using a well-designed series of experiments using Taguchi based DOE technique. Whereas it has been used for three stages and three L9 OA variables consisting of nine trials. The effect of process parameters such as voltage, electrolyte concentration and IEG on material removal rate (MRR) and will be evaluated in order to obtain the finest machining conditions for achieving higher MRR. W_1 = Initial Weight (mg), W_2 = Final Weight (mg).

$$MRR = \frac{(W_1 - W_2)}{t} \, mg/min \quad (1)$$

Table 2. Experimental results of ECDM

Trial no.	Voltage (volts)	Electrolyte Concentration (%)	IEG (mm)	MRR (mg/min)
1	50	10	20	0.0591
2	50	15	30	0.1178
3	50	20	40	0.1332
4	60	10	30	0.1243
5	60	15	40	0.1342
6	60	20	20	0.1857
7	70	10	40	0.1296
8	70	15	20	0.2113
9	70	20	30	0.2321

Analysis of variance is performed to oversee which machining parameter has more essential impact on the quality characteristics of ECDM process furthermore to figure out contribution of machining parameter. From the table 3 it is found that, the probability of the parameter is less than 0.05 and the percentage contribution of voltage is 46.9% which are more than IEG (45.2%) and electrolyte concentration (19.9%). Hence voltage is more significant factor and other parameters such as inter-electrode gap and electrolyte concentration have less implication.

Table 3. Analysis of variance (ANOVA)

Source	DF	Seq SS	Adj SS	Adj MS	F	P	% of Contribution
Voltage	2	0.010789	0.010789	0.005394	25.24	0.038	46.9
Electrolyte Conc.	2	0.010417	0.010417	0.005208	24.37	0.039	45.2
IEG	2	0.001364	0.001364	0.000682	3.19	0.239	5.93
Error	2	0.000427	0.000427	0.000214			1.97
Total	8	0.022997					

S= 1.615 R-Sq= 94.7 % R-Sq (Adj)= 80 %

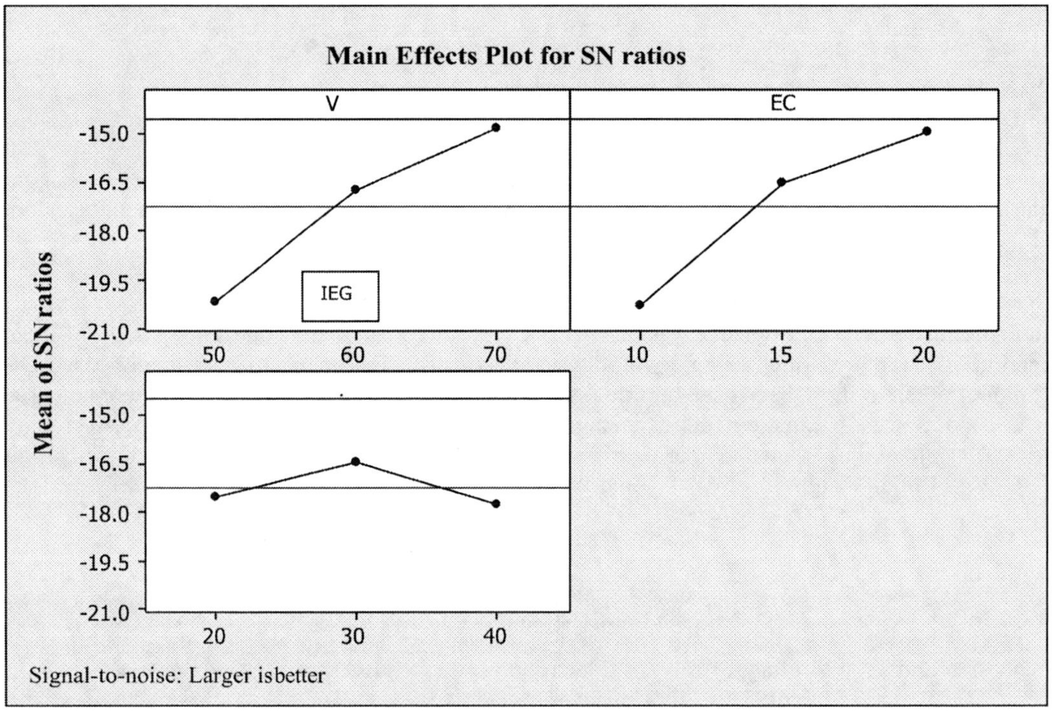

Figure 2. Main Effects Plot for S/N ratios on MRR

Main Effects Plot for S/N ratios on MRR as shown in 'Figure 2'. Material removal rate increases with increase in the voltage applied and electrolyte temperature [4]. Increase in supply voltage, rate of material removal and machined depth increases. Applied voltage increases the current intensity, hence higher material removal from higher sparks [12]. It is observed that mean value is higher for voltage compared to that of electrolyte concentration and IEG. The MRR is maximum at high frequency and high applied voltage, produces high intensity sparks. Higher level of electrolyte concentration accelerates the rate of production of gas bubbles, which in turn increases the frequency of discharge due to the accelerated forming of films around the electrode; it also facilitates chemical etching [13]. Hence it can be stated that voltage is the primary factor that impacts the MRR followed by electrolyte concentration and IEG. Also it is seen from the graph that 70V, 20% and 30 mm are the optimal combinations of Voltage, Electrolyte Concentration and IEG in achieving higher MRR.

A scientific mathematical statement has been produced by utilizing statistical tool MINITAB 14 to acquire the relationship between the material removal rate and the different input variables such as, Voltage, Electrolyte concentration and IEG. Regression analysis was performed taking into account trial results and estimations of regression coefficients are evolved. The equation for MRR of quartz is stated as,

$$MRR= - 0.205 + 0.00438 (V) + 0.00793(EC) - 0.000985(IEG) \quad (2)$$

Experimental MRR value for any given values of input variables like voltage, Concentration and Inter-Electrode Gap can be predicted by using above equation. The equation is used to calculate the error. It is found that the experimental readings and predicted readings have a very lesser amount of deviation. The comparison of experimental MRR and anticipated MRR are outlined in table 4

Table 4. Comparison of Experimental MRR values with ANN Predicted MRR

Trail no.	Experimental MRR(mg/min)	Predicted MRR (mg/min)	Error
1	0.0591	0.0736	-0.2453
2	0.1178	0.1034	0.1222
3	0.1332	0.1332	0
4	0.1243	0.1174	0.0555
5	0.1342	0.1472	-0.0968
6	0.1857	0.117	0.3699
7	0.1296	0.1174	0.0941
8	0.2113	0.1472	0.3033
9	0.2321	0.177	0.2373

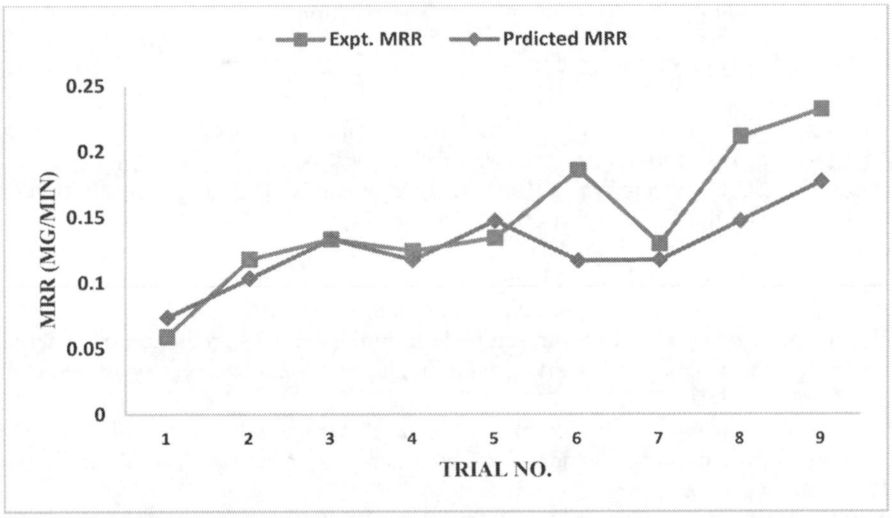

Figure 3. Plot for Experimental MRR and ANN Predicted MRR

The evaluation between of experimental MRR and predicted MRR are illustrated in Table 4. It is observed that the values obtained by using regression model are slightly diverged from the experimental values with an average error of 0.093. The correlation between the experimental MRR and the predicted MRR as shown in 'Figure 3'. It is observed that less variance occurs between the experimental and predicted values.

4. Conclusion

Experimental set up of ECDM is developed and experiments were conducted on Quartz using KOH electrolyte. The experiments were carried out using L9 orthogonal array. Finally, the prediction of the response parameters was done using regression analysis. Following conclusions were made

- Combination of 70V, 20% EC and 20mm IEG achieved a higher MRR. The shows the best combination.

- From ANOVA, voltage has the highest percentage of contribution to achieving higher MRR.
- The comparison between experimental result and ANN Predicted result are slightly deviated from the experimental result.
- Further work on ECDM is carried out to study the process variables such as tool-wear and surface roughness on quartz material by employing various electrolytes.
- Different materials can be drilled and slots can be produced by using the setup with automated feeding mechanisms

Reference

[1] Harugade ML, Kavade MV and Hargude NV 2013 An Experimental Investigation Of Effect Of Electrolyte Solution On Material Removal Rate In ECDM *International Journal of Engineering Research & Technology* **2** 1–8

[2] Singh B and Vaishya RO 2017 Analyses of Output Parameters of ECDM Using Different Abrasives-A Review *International Journal of Materials Science* **12** 307–14

[3] Vinod S, Dileepkumar SG, Bharath KN, Naveen BC 2018 Statistical Analysis of Response Parameters during Electro-Chemical Discharge Machining of Crystal Glass Statistical Analysis of Response Parameters during Electro- Chemical Discharge Machining of Crystal Glass *IJARA*

[4] Wüthrich R and Fascio V 2005 Machining of non-conducting materials using electrochemical discharge phenomenon - An overview *International Journal of Machine Tools and Manufacture* **45** 1095–10

[5] Singh M and Singh S 2019 Electrochemical discharge machining: A review on preceding and perspective research *Proceedings of the Institution of Mechanical Engineers, Part B: Journal of Engineering Manufacture* **233** 1425–49

[6] Sathisha N, Somashekhar SH, Shivakumar S and Jagannatha N 2013 Optimization of ECD process parameters using Taguchi Robust Design and Utility Concept, *International Journal of Engineering Trends in Engineering and Development* **2**

[7] Laio YS, Wu LC and Peng WY 2013 A study to improve drilling quality of electrochemical discharge machining (ECDM) process *Procedia CIRP* **6** 609–14

[8] Liu JW, Yue TM and Guo ZN 2010 An analysis of the discharge mechanism in electrochemical discharge machining of particulate reinforced metal matrix composites *International Journal of Machine Tools and Manufacture* **50** 86–96

[9] Bhondwe KL, Yadava V and Kathiresan G 2006 Finite element prediction of material removal rate due to electro-chemical spark machining *International Journal of Machine Tools and Manufacture* **46** 1699–706

[10] Liu Y, Wei Z, Wang M and Zhang J 2017 Experimental investigation of micro wire electrochemical discharge machining by using a rotating helical tool *Journal of Manufacturing processes* **29** 265–71

[11] Kumar M, Vaishya RO, Oza AD and Suri NM 2019 Experimental Investigation of Wire-Electrochemical Discharge Machining (WECDM) Performance Characteristics for Quartz Material *Silicon*

[12] Jain VK and Chak SK 2000 Electrochemical spark trepanning of alumina and quartz *Machining Science and Technology* **4** 277–90

[13] Goud M and Sharma AK 2017 On performance studies during micromachining of quartz glass using electrochemical discharge machining *Journal of Mechanical Science and Technology* **31** 1365–72

Nonlinear Radiative Heat Transfer of Magnetite(Fe_3O_4)-Water Nanofluid over Unsteady Stretching Surface with Fluid Particle Suspension

N Srikantha[1], M R Krishnamurthy[2], M M Praveena[3], K S Onkarappa[1], K Bharathi[1] and G H Pujar[4]

[1]Research Centre, Department of Mathematics, GM Institute of Technology, Davangere – 577006, Karnataka, India
[2]Department of Mathematics, JNN College of Engineering, Shimoga-577201, Karnataka, India
[3]M.S. Ramaiah Institute of Technology, Bengaluru – 560054, Karnataka, India
[4]Research Centre, Department of Physics, GM Institute of Technology, Davangere – 577006, Karnataka, India
E-mail: srikanthan@gmit.ac.in, krishnamr@jnnce.ac.in, mmpraveenamaths@gmail.com, onkarappaks@gmit.ac.in, bharathik@gmit.ac.in, gonibasappapujar@gmit.ac.in

Abstract. The exploration of the magneto hydrodynamics(MHD) flow and nonlinear radiative heat transfer of magnetite $Fe_3O_4 - H_2O$ nanoparticle with fluid particle suspension over an unsteady stretching sheet is the central theme of the present work. The similarity transformations were employed to transfer the governing partial differential equations into ordinary ones prior to solve numerically using Runge – Kutta - Fehlberg-fourth and fifth order method with shooting technique. The variations of the temperature and velocity distribution and coefficients of heat transfer for magnetic parameter, temperature ratio parameter nanoparticle volume fraction, and thermal radiation parameter are discussed graphically. Comparative analysis of reported ones and present work results was presented and found to be in good agreement.

Keywords. MHD, nanofluid, nonlinear radiative heat transfer, unsteady stretching sheet

1. Introduction

Improvement in heat transfer in liquid cooling process is achieved by introducing nanoparticles in the fluid base. The idea of nanotechnology started in a talk by Richard "Feynman's There's Plenty of Room at the Bottom". According to him nanoparticles possess unique physical and chemical properties can be used across all the science fields like physics, chemistry, biology, material science and Engineering. Nanofluids are the fluids in which particles of nanoscale were suspended in it. This technique improves fluids thermal conductivity, moreover nanoparticles are more stable in base fluids because of its large surface area which solves the many practical problems.

In recent years, heat transfer (HT) and nanofluid flow (NF) areas have become imperative hot-topics for many researchers. Sheikholeslami [1] presented the thermal behavior of the NF in cavity with tilted elliptic inner cylinder under the influence of Lorentz force. By comparing the two different models, Ahmad *et. al.* [2] addressed about effective thermal conductivity of the rotating NF induced by exponential stretching. Khan *et. al.* [3] studied the impact of inclined magnetic field on the NF over the sheet. In order to investigate the radiation affects, Poornima *et. al.* have come with some mathematical solutions for magneto hydrodynamics (MHD) free convective boundary layer NFs over non-linear stretchable sheets [4].

The problem of rotating flow of silver-copper oxide/water (Ag-CuO/H_2O, a hybrid NF) was investigated by Hayat *et. al.* [5]. The problem of micropolar NF was solved in exponentially stretchable surface by Subhani *et. al.* [6]. Numerical solutions to the problem of Cu-Al_2O_3/H_2O NF over a permeable stretchable layer under the hydro-magnetic effects were addressed by Devi *et. al.* [7].

Prasanna *et. al.* and Reddy *et. al.* [8,9] discussed the behaviour of non-linear thermal radiation on Williamson NF under different conditions. Ibrahim *et. al.* [10] examined the HT and hydromagnetic boundary layer flow of NF over a porous stretchable layer in presence of thermal velocity, and solutal slip conditions. Eegunjobi *et. al.* [11] have reported about unsteady MHD mixed convection slip flow with chemical reaction past a porous stretchable layer medium. At stagnation point, the effects of MHD flow and double-stratification of Eyring-Powell NF past a stretched cylinder were revealed by Ramzan *et al.* [12]. Prasad *et al.* [13] numerically analyzed the HT and MHD flow in a NF over an elastic sheet. Characteristics of Al2O3, TiO2 and Fe3O4 nanoparticles in water based materials are presented by Mushtaq *et al* [14], Wahed [15] and Khan *et al* [16].

A revolutionary investigation on 3-D circulating viscous flow persuaded due to stretchable surface has been reported [17] and the its extended work for unsteady case detailed by Rajeswari *et. al.*[18] and Nazar *et al.* [19]. The influence of thermophoresis and Brownian motion over a unsteady porous shrinking layer was prescribed by Kumar *et al.* [20]. Numerical investigation for an unsteady MHD laminar stream of a NF carried out over a perpendicular stretchable sheet with porous medium was [21]. Sheikholeslami *et al.* [22] obtained numerical results of HT and NF in circulating system by considering the effects of magnetic field. Turkyilmazoglu [23] detaild about the water based NFs flow of five different types. Bahirael *et al.* [24,25] have discussed transfer of heat efficiency of Mn-ZnFe_3O_4-H_2O ferrofluid in counterflow double-pipe heat-exchanger as well as in annulus by applying magnetic field. Recently, unsteady magnetohydrodynamic flow of Eyring-Powell NF over an inclined permeable stretchable layer was carried out by Bharath Kumar and Suripeddi Srinivas[26]. The thermal radiation and magnetic field influence on circular flow of MWCNTs- H_2O and SWCNTs- H_2O NF between rotatable stretchable disks was reported by Jyothi et al. [27]. Elgazery [28] have considered 4 different nanoparticles to study the unsteady 2-D NFover a vertical stretchable porous sheet under the influence of nonuniform heat sink /source and magnetic field.

In this context, the aim of this work is to analyse the nonlinear radiative HT of magnetite ($Fe_3\,o_4$)-water NF over unsteady stretchable surface with fluid particle suspension.

The similarity transformations, shooting technique and together with rkf45 method were employed to solve the controlling flow equations numerically. The effect of different significant parameters on velocity, temperature plots has been detailed through plots.

2. Mathematical Model

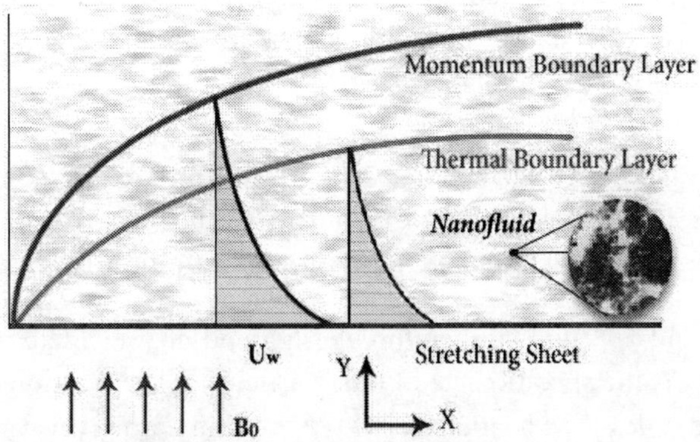

Figure 1. Typical flow diagram

Here, it is considered the incompressible dusty NFas unsteady, 2-D laminar boundary flow over stretchable sheet entrenched in permeable medium. The layer is stretched in X- direction and perpendicular to Y-axis, keeping origin point fixed. The fluid flow is restricted in the region $y > 0$. It is assumed that the velocity of stretchable sheet to be $U_w(x,t) = \frac{cx}{1-at}$ also we have considered the flow is steady at $t < 0$ and unsteady scenario starts at $t = 0$. Along with this we have assumed the following assumptions

- Nonlinear radiation and non-uniform heat sink / source has been considered.
- Volume fraction of dust and Nanoparticles and Number density of dust-particles were considered.
- The size of dust particles assumed to be uniform.
- Shape of the dust and nano particles are considered as spherical
- Water based dusty fluid with Fe_3O_4 nanoparticles is taken into account.
- The nanoparticles of uniform size and shape are also considered.
- T_f be surface temperature of sheet and monitored by convective HT,
- Finally, both NPs & fluid-phase are in thermal equilibrium state.

The equations which govern the flow of NF phase and dust phase under typical boundary layer approximations have become

$$\frac{\partial u}{\partial x} + \frac{\partial v}{\partial y} = 0, \tag{2.1}$$

$$\frac{\partial u}{\partial t} + u\frac{\partial u}{\partial x} + v\frac{\partial u}{\partial y} = \frac{\mu_{nf}}{\rho_{nf}}\frac{\partial^2 u}{\partial y^2} + \frac{KN}{\rho_{nf}}(u_p - u) - \frac{\sigma B^2{}_0}{\rho_{nf}}u \tag{2.2}$$

$$\frac{\partial u_p}{\partial x} + \frac{\partial v_p}{\partial y} = 0, \tag{2.3}$$

$$\frac{\partial u_p}{\partial t} + u_p\frac{\partial u_p}{\partial x} + v_p\frac{\partial u_p}{\partial y} = \frac{K}{m}(u - u_p) \tag{2.4}$$

$$(\rho c_p)_{nf}\left[\frac{\partial T}{\partial t} + u\frac{\partial T}{\partial x} + v\frac{\partial T}{\partial y}\right] = k_{nf}\frac{\partial^2 T}{\partial y^2} + \frac{Nc_{pf}}{\tau_T}(T_p - T) + \frac{N}{\tau_v}(u_p - u)^2 - \frac{\partial q_r}{\partial y} \tag{2.5}$$

$$\frac{\partial T_p}{\partial t} + u_p\frac{\partial T_p}{\partial x} + v_p\frac{\partial T_p}{\partial y} = \frac{c_{pf}}{c_{mf}\tau_T}(T - T_p) \tag{2.6}$$

where, (u, v) and (u_p, v_p); the x and y-components of velocity of NF, dust phases, respectively. ρ_{nf} ; effective density, μ_{nf}; effective dynamic viscosity, which are expressed as [32],

$$\rho_{nf} = (1-\phi)\rho_f + \phi\rho_s,\ \ \mu_{nf} = \frac{\mu_f}{(1-\phi)^{2.5}} \tag{2.7}$$

where ϕ; solid volume fraction, ρ_f and ρ_s; density of base fluid and NPs, respectively. μ_f; dynamic viscosity of base-fluid. Stokes drag constant is given by $K = 6\pi\mu_{nf}r$, N is number-density of particles, k' is permeability of permeable medium, m is mass concentration of dust particles, r is radius of particles.

In equations (2.5 &2.6) and T_p; fluid temperatures of NF and dust inside boundary layer. τ_T ; thermal equilibrium time. c_{pf} , c_{mf}; the specific heat of dust particles, τ_v; relaxation time of dust particle, k_{nf} ;thermal conductivity and $(\rho c_p)_{nf}$; heat capacity of NF, which are given by [32],

$$\frac{k_{nf}}{k_f} = \frac{k_s + 2k_f - 2\phi(k_f - k_s)}{k_s + 2k_f + 2\phi(k_f - k_s)},\ \ \ (\rho c_p)_{nf} = (1-\phi)(\rho c_p)_f + \phi(\rho c_p)_s \tag{2.8}$$

where, k_s and k_f; thermal conductivity of nanoparticle and base fluid, respectively, $(\rho c_p)_f$ and $(\rho c_p)_s$; heat capacity of base-fluid &NPs, respectively.

The consequent boundary conditions are expressed as,

$$\begin{gathered} u = U_w v = 0,\ \ T = T_w \text{ at } y = 0, \\ u = 0, u_p \to 0, v_p \to v, T \to T_\infty, T_p \to T_\infty \ \text{ as } y \to \infty, \end{gathered} \tag{2.9}$$

The nonlinear Rosseland diffusion approximation was used because unlike the linearized Rosseland approximation this method will give results for either small or large differences between T_f and T_∞. Radiative heat flux is simplified using Rosseland [33,34] estimation as,

$$q_r = -\frac{4\sigma^*}{3k^*}\frac{\partial T^4}{\partial y}, \tag{2.10}$$

Further, q_r for boundary layer flow over a horizontal flat plate was simplified by Pantokratoras and Fang [30] as,

$$q_r = \left(-\frac{16\sigma^* {T_\infty}^3}{3k^*}\right)\frac{\partial T}{\partial y}. \tag{2.11}$$

In view to equation (2.11), the equation (2.5) becomes

$$(\rho c_p)_{nf}\left[u\frac{\partial T}{\partial x} + v\frac{\partial T}{\partial y}\right] = \frac{\partial}{\partial y}\left[\left(k_{nf} + \frac{16\sigma^* {T_\infty}^3}{3k^*}\right)\frac{\partial T}{\partial y}\right] + \frac{Nc_{pf}}{\tau_T}(T_p - T) + \frac{N}{\tau_v}(u_p - u)^2. \tag{2.12}$$

Following similarity transformations have incorporated to convert derived equations to a set of similarity equations:

$$\left.\begin{gathered} u = \frac{cx}{1-at}f'(\eta),\ \ v = -\sqrt{\frac{c\nu_f}{(1-at)}}f(\eta),\ \ \ \eta = \sqrt{\frac{c}{\nu_f(1-at)}}y, \\ u_p = \frac{cx}{1-at}F'(\eta),\ \ v_p = -\sqrt{\frac{c\nu_f}{1-at}}F(\eta) \\ \theta(\eta) = \frac{T - T_\infty}{T_w - T_\infty},\ \ \ \theta_p(\eta) = \frac{T_p - T_\infty}{T_w - T_\infty} \end{gathered}\right\} \tag{2.13}$$

with $T = T_f(1 + (E-1)\theta)$and $E = \frac{T_w}{T_\infty}$, $E > 1$ the temperature ratio parameter.

In view of equation (2.13), the equations (2.1& 2.3) are automatically holds good and (2.2), (2.4), (2.6) and (2.12) will takes the form ,

$$f''' + (1-\phi)^{2.5}\left[(1-\phi) + \phi\frac{\rho_s}{\rho_f}\right]\left(ff'' - f'^2 - \frac{\alpha}{2}(\eta f'' + f')\right) \tag{2.14}$$

$$+(1-\phi)^{2.5}\left[l\beta_v(F'-f')-Mf'\right]=0$$

$$FF''-[F'^2]-\frac{\alpha}{2}(\eta F''+F')+\beta_v[f'-F']=0, \qquad (2.15)$$

$$\frac{k_{nf}}{k_f}[1+Nr(1+(E-1)\theta)^3\theta']'+Pr\left[(1-\phi)+\phi\frac{(\rho c_p)_s}{(\rho c_p)_f}\right]\left(f\theta'-\frac{\alpha}{2}\eta\theta'\right) \qquad (2.16)$$

$$+\frac{lPr\beta_T}{m}[\theta_p-\theta]+\frac{lPrEc\beta_v}{m}[F'-f']^2=0$$

$$F\theta'_p-\frac{\alpha}{2}\eta\theta_p'+\gamma\beta_T[\theta-\theta_p]=0 \qquad (2.17)$$

The consequent boundary conditions are now become,

$$f(0)=0, f'(0)=1, \ \theta(0)=1,$$
$$f'(\infty)=0, \quad F'(\infty)=0, \quad F(\infty)=f(\infty), \quad \theta(\infty)=0, \quad \theta_p(\infty)=0. \qquad (2.18)$$

Where $\alpha=\frac{a}{c}$, $\beta=\frac{1-at}{\tau_v c}$ and $\beta_T=\frac{1-at}{\tau_T c}$, $k_p=\frac{\nu_f(1-at)}{k'c}$, $M=\frac{\sigma B^2_0}{\rho_f}$, $Nr=\frac{16\sigma^* T^3_\infty}{3k_{nf}k^*}$ respectively are the unsteady-parameter, fluid-particle-interaction parameter for velocity and temperature, permeability parameter, magnetic parameter, radiation parameter. $l=\frac{mN}{\rho_f}$ is particles mass concentration, $\tau_v=\frac{m}{K}$ is relaxation time, $\gamma=\frac{c_{pf}}{c_{mf}}$ is specific heat ratio and $Pr=\frac{(\mu c_p)_f}{k_f}$ is Prandtl number, $Ec=\frac{U_w{}^2}{(T_f-T_\infty)c_{pf}}$ is Eckert number.

The local Nusselt number (Nu_x) is expressed as,

$$Nu_x=\frac{xq_w}{k_f(T_f-T_\infty)}\frac{\partial T}{\partial y}\bigg|_{y=0}.$$

3. Numerical Solution

Runge-Kutta-Fehlberg45 order approach usually abbreviated as RKF45 coupled with shooting technique is considered to tackle the system of nonlinear equations (2.14) to (2.17), subjected to the boundary conditions (2.18). Shooting method incorporated in Maple software used to identify appropriate initial conditions by an iterative process to obtain a precise and accurate solution to original boundaryvalue problem which is employed to crack a variety of non-linear problems successfully. The thermo-physical properties of H_2O-Fe_3O_4 nanoparticles are tabulated in Table 1. While, Table 2 collects the all comparative analysis of the obtained results with reported ones [29,31] for derived Nusselt number. From the above analysis, it is concluded that obtained results have shown good agreement with the reported ones.

Table 1. **Thermo-physical properties of H_2O -Fe_3O_4 nanoparticles.**

	$\rho(kg/m^3)$	$C_p(j/kgk)$	$k(W/mk)$
Pure H_2O	997.1	4179	0.613
Magnetite(Fe_3O_4)	5180	670	9.1

Table 2. **Comparative analysis of $-\theta'(0)$ when $\beta=\phi=Nr=E=\alpha=l=0$**

Pr	Elbashbeshy et al. [29]	Prasannakumara et al. [31]	Present work
0.72	0.76728	0.7672701	0.767290
1	0.95478	0.9547788	0.957888
2	1.47146	1.4714572	1.471472
3	1.86907	1.8690619	1.869317
5	2.50013	2.5001321	2.500134
10	3.66037	3.6603723	3.660384

4. Results and Discussion

The nature of velocity and temperature distribution for both phases on changing unsteady parameter (α) is shown in Fig. 2 and 3. It is seen that both temperature and velocity plots increases with α Fig. 4 and 5 depict the effect of fluid-particle interaction parameter for velocity (β) on velocity and temperature distribution. It is observed that velocity plot of momentum boundary layer decreases and temperature plot of thermal boundary layer augments as β increases.

Fluid particle interaction parameter for temperature β_T on velocity plots is presented in Fig. 6. It is noticed that the velocity curve of the fluid decreases whereas velocity profile of the dust phase increases as increasing the values of temperature fluid particle interaction parameter.

The Fig. 7 illustrates the temperature ratio parameter (E) effect on the both phases. As parameter E increases, an augment in both the phases of temperature was noted, which is owing to considerable increment in thermal boundary layer thickness. Further, it is seen that the behavior of Eckert number (EC) is same as parameter (E), (see Fig. 8). The nature of the temperature profile as varying in EC is attributed to stored heat energy in considered fluid due to frictional heating.

The effect of mass concentration of particle parameter (l) on both phases of velocity profile is depicted in Fig 9 and effect of l on both phases of temperature is plotted in Fig. 10. From these plots, it is examined that, the velocity and temperature profiles decreases as mass concentration of particles increases, which is true for NF as well as dust phases.

The Lorentz force produced due to magnetic field perpendicular to flow in conducting fluid acts against the flow, due to this the increasing values of magnetic parameter (M) decreases velocity profile in the boundary layer region. It is revealed in Fig. 11. But on the thermal boundary layers magnetic field influences positively which is shown in Fig. 12. Physically, it can be concluded that effect of magnetic field increases the thickness of thermal boundary layer and reduces the momentum boundary layer.

Radiation parameter (Nr) exhibits the property that temperature profiles for both phases increases as Nr increases, which is illustrated in Fig. 13. Increasing in solid volume fraction parameter (ϕ), velocity and temperature profiles of the both fluid & particle phases increases, which is graphically illustrated in Fig. 14 and 15.

Fig. 16 demonstrates the effects of Prandtl number (Pr) on θ profiles. It is observed that temperature profiles of both phases decreases as increasing in the values of Pr. Lower the Prandtl number, thicker in the thermal boundary layer structures because lesser Prandtl number implies the fluids with greater thermal conductivity.

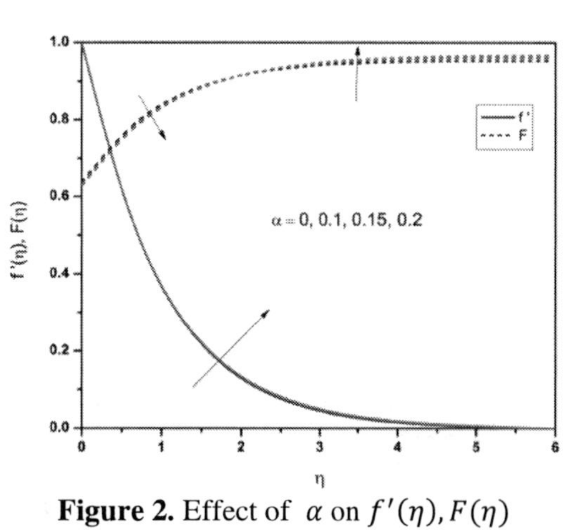

Figure 2. Effect of α on $f'(\eta), F(\eta)$

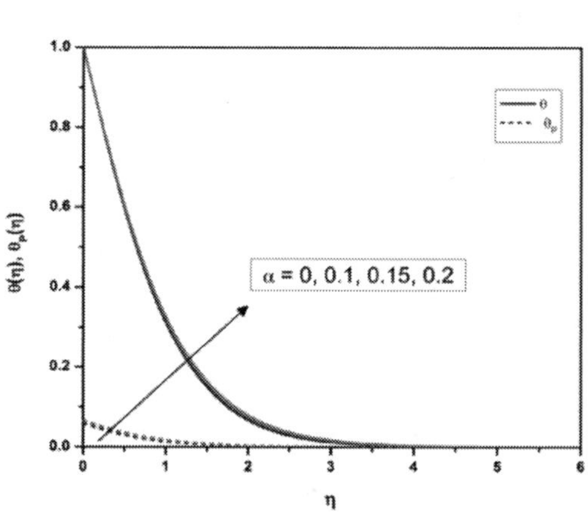

Figure 3. Effect of α on $\theta(\eta), \theta_p(\eta)$

IOP Conf. Series: Materials Science and Engineering **925** (2020) 012054 doi:10.1088/1757-899X/925/1/012054

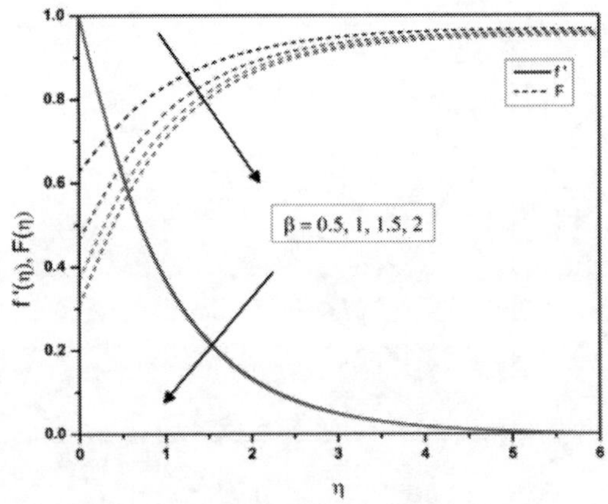

Figure 4. Effect of β on $f'(\eta), F(\eta)$

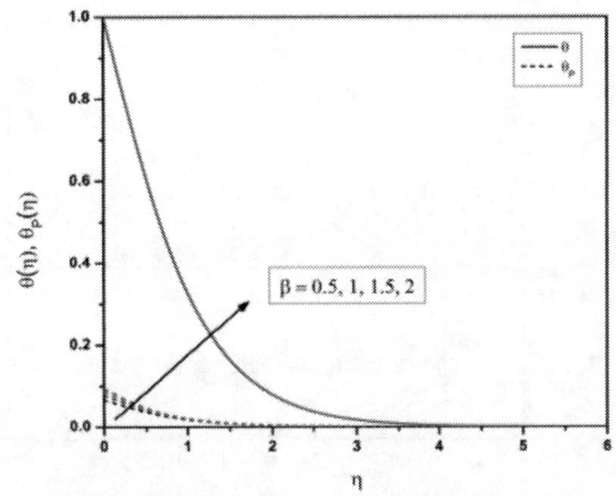

Figure 5. Effect of β on $\theta(\eta), \theta_p(\eta)$

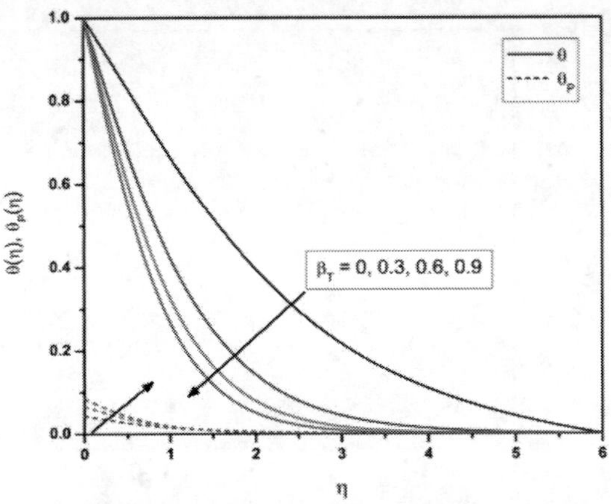

Figure 6. Effect of β_T on $f'(\eta), F(\eta)$

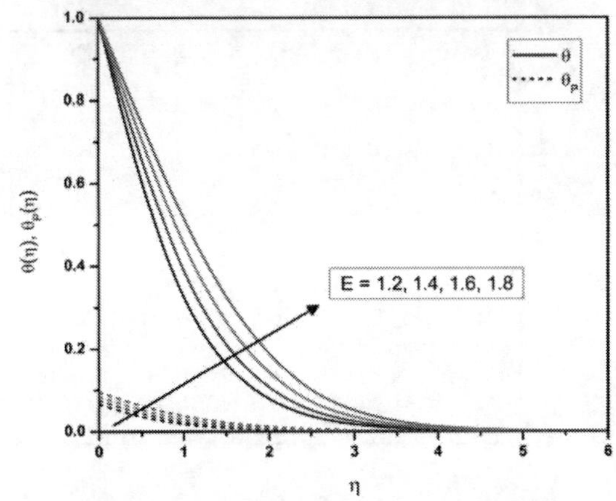

Figure 7. Effect of E on $\theta(\eta), \theta_p(\eta)$

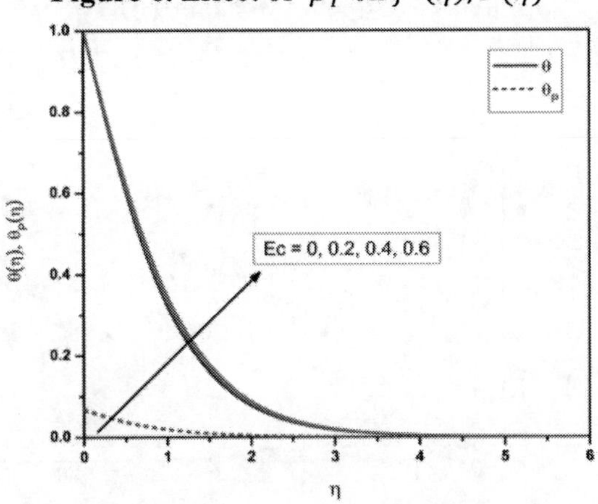

Figure 8. Effect of Ec on $\theta(\eta), \theta_p(\eta)$

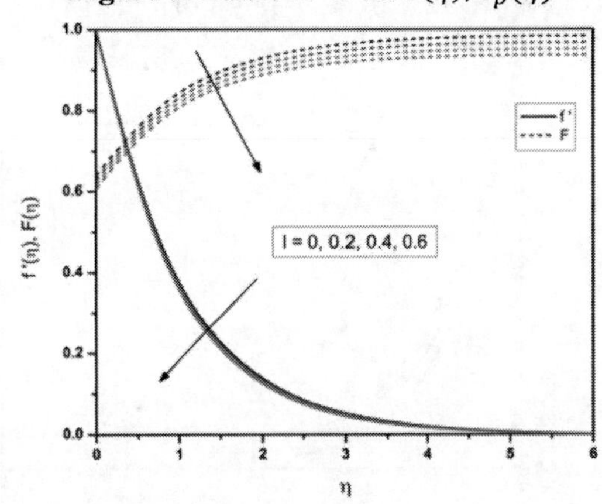

Figure 9. Effect of l on $f'(\eta), F(\eta)$

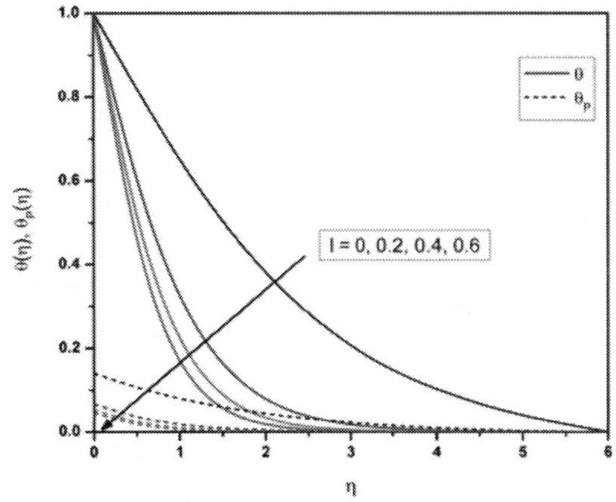

Figure 10. Effect of l on $\theta(\eta), \theta_p(\eta)$

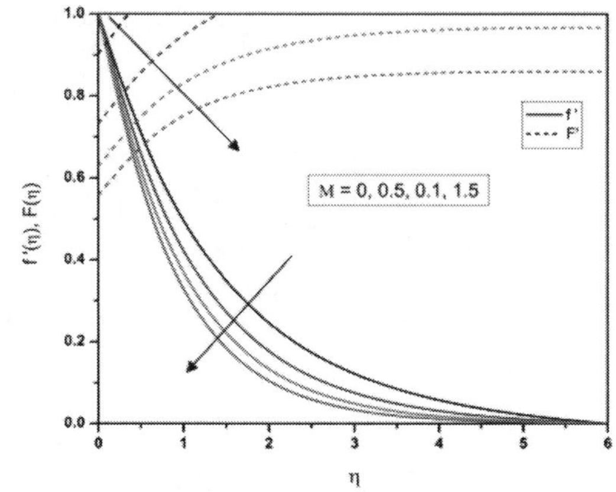

Figure 11. Effect of M on $f'(\eta), F(\eta)$

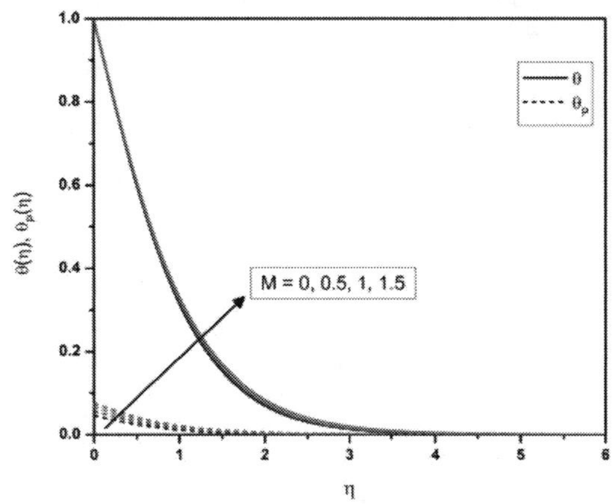

Figure 12. Effect of M on $\theta(\eta), \theta_p(\eta)$

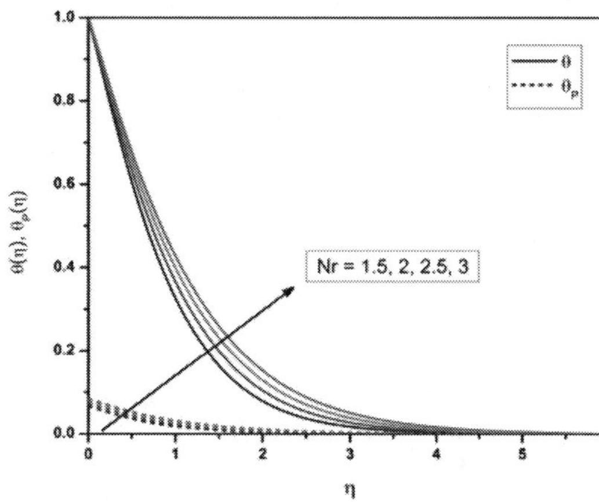

Figure 13. Effect of Nr on $\theta(\eta), \theta_p(\eta)$

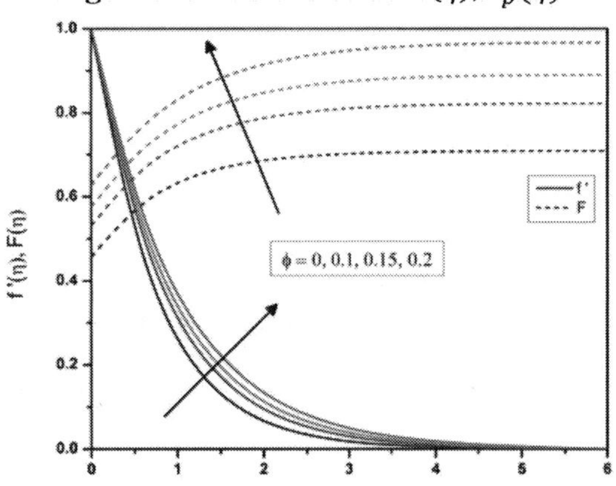

Figure 14. Effect of ϕ on $f'(\eta), F(\eta)$

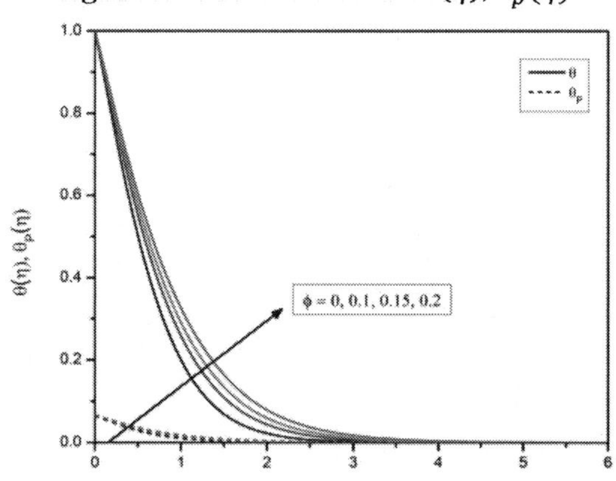

Figure 15. Effect of ϕ on $\theta(\eta), \theta_p(\eta)$

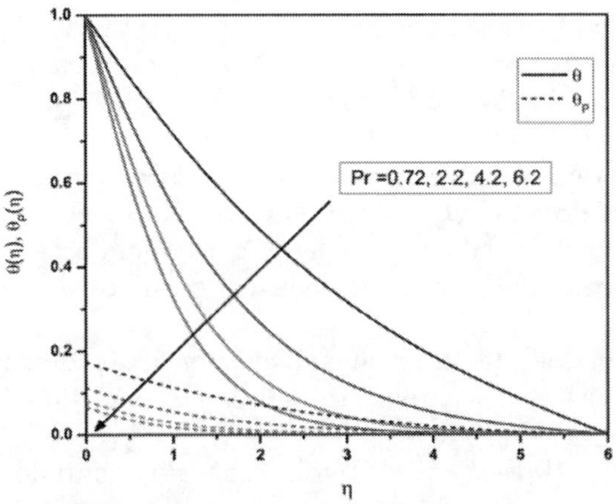

Figure 16. Effect of Pr on $\theta(\eta), \theta_p(\eta)$

5. Conclusion

The problem of an unsteady non-linear radiative & transfer of heat flow features of dusty NF over linear stretchable sheet is studied by considering volume fraction of dust and nanoparticles. Here, Fe_3O_4-water entrenched with dust particles was considered. A set of similarity transfer is presented to alter boundary layer equations into self-similar form and then numerically solved by RKF-45 approach coupled with shooting-technique. Impact of non-dimensional governing parameters on temperature and velocity plots for dust and fluid phases have been detailed through plots. It is finally summarized that:

- As volume fraction of the nanoparticles increases, an enhancement in friction factor and rate HT were observed.
- As the fluid particle interaction parameter rises, there was a increment in rate of HT and reduction in friction factor.
- Higher value of Prandtl number rapidly decreases the thermal boundary layer thickness and temperature.
- Effect of temperature ratio parameter, unsteadiness parameter and thermal radiation parameter is enhanced the temperature profile for both phases.
- An appreciable cooling performance was observed for dusty fluid conducting with Fe_3O_4 nanoparticles.
- Higher thermal conductivity has been noticed for dusty NFs, compared to NF and dusty fluid.

References

[1] Mohsen S 2017 Magnetic field influence on NF thermal radiation in a cavity with tilted elliptic inner cylinder *J. Mol. Liq* **229** 137–147.

[2] Ahmad R, Mustafa M 2016 Model and comparative study for rotating flow of NFs due to convectively heated exponentially stretching sheet *J. Mol. Liq.,* **220** 635–641

[3] Khan WA, Makinde OD, Khan ZH 2016 Non-aligned MHD stagnation point flow of variable viscosity nanofluids past a stretching sheet with radiative heat *Int. J. Heat Mass Transf.* **96** 525–534

[4] Poornima T, Bhaskar Reddy N 2013 Radiation effects on MHD free convective boundary layer flow of nanofluids over a nonlinear stretching sheet *Adv Appl Sci Res.,* **4** 190–202

[5] Hayat T and Nadeem S 2017 Heat transfer enhancement with Ag-CuO/water hybrid nanofluid. *Results in Physics*, **7** 2317–2324.

[6] Subhani M and Nadeem S 2019 Numerical analysis of micropolar hybrid nanofluid, *Applied Nanoscience* **9**, 447–459.

[7] Devi SPA and Devi SSU 2016 Numerical investigation of hydromagnetic hybrid Cu-Al2O3/water nanofluid flow over a permeable stretching sheet with suction, *Int. J. Nonlinear Sci. Numer. Simul.* **17** 249–257

[8] Prasannakumara BC, Gireesha BJ, Gorla RSR and Krishnamurthy MR 2016 Effects of chemical reaction and nonlinear thermal radiation on Williamson nanofluid slip flow over a stretching sheet embedded in a porous medium *Journal of Aerospace Engineering* **29** 4016019-1-10.

[9] Reddy SC, Naikoti K and Rashidi MM 2017 MHD flow and heat transfer characteristics of Williamson nanofluid over a stretching sheet with variable thickness and variable thermal conductivity, *Transactions of A. Razmadze Mathematical Institute,* **171** 195–211.

[10] Ibrahim W Shankar B 2013 MHD boundary layer flow and heat transfer of a nanofluid past a permeable stretching sheet with velocity, thermal and solutal slip boundary conditions, *Computers & Fluids*, **75** 1-10.

[11] Eegunjobi AS, Makinde OD, Srinivas J 2017 Unsteady MHD chemically reacting and radiating mixed convection slip flow past a stretching surface in a porous medium, *Defect and Diffusion Forum* **377** 200-210.

[12] Ramzan M, Bilal M and Chung JD 2017 Radiative flow of Powell-Eyring magneto-nanofluid over a stretching cylinder with chemical reaction and double stratification near a stagnation point *PLoS ONE*, **12** e0170790.

[13] Prasad KV, Vajravelu K, Vaidya H, Gorder RAV 2017 MHD flow and heat transfer in a nanofluid over a slender elastic sheet with variable thickness, *Results in Physics*, **7** (2017), 1462-1474.

[14] Mushtaq A, Mustafa M, Hayat T, Alsaedi, Numerical study for rotating flow of nanofluids caused by an exponentially stretching sheet. *Adv Powder Technol* **27** 2223–31.

[15] Wahed MSA 2017 Rotating ferro-nanofluid over stretching plate under the effect of hall current and joule heating *J. Magn Magn Mater* **429** 287–93

[16] Khan U, Ahmed N, Mohyud-Din ST 2017 Numerical investigation for three dimensional squeezing flow of nanofluid in a rotating channel with lower stretching wall suspended by carbon nanotubes. *Appl Therm Eng* **113** 1107–17.

[17] Wang CY 1988 Stretching a surface in a rotating fluid. *Journal of Applied Mathematics and Physics (ZAMP)* **39** 177–185

[18] Rajeswari V, Nath G, Unsteady flow over a stretching surface in a rotating fluid, *Int. J. Eng. Sci.* **30** (1992), 747–756.

[19] Nazar R, Amin N, Pop I 2004 Unsteady boundary layer flow due to a stretching surface in a rotating fluid, *Mech. Res. Commun.*, **31**121–128.

[20] Kumar Nandy S, Sidui S, Mahapatra TR 2014 Unsteady MHD boundary-layer flow and heat transfer of nanofluid over a permeable shrinking sheet in the presence of thermal radiation, Alexandria *Eng. J.* **53** 929-937

[21] Freidoonimehr N, Rashidi M M, Mahmud S, Unsteady MHD free convective flow past a permeable stretching vertical surface in a nano-fluid, *Int. J. Thermal Sciences*, **87** (2015), 136-145.

[22] Sheikholeslami M, Hatami M, Ganji D D, Nanofluid flow and heat transfer in a rotating system in the presence of a magnetic field, *J. Mol. Liq.* **190** (2014), 112–120.

[23] Turkyilmazoglu M, Nanofluid flow and heat transfer due to a rotating disk, *Comp. Fluids* **94** (2014), 139–146.

[24] Bahiraei M, Hangi M 2013 Investigating the efficacy of magnetic nanofluid as a coolant in double-pipe heat exchanger in the presence of magnetic field, *Energy Conversion and Management* **76** 1125– 1133.

[25] Bahiraei M, Hosseinalipour SM and Hangi M 2014 Numerical study and optimization of hydrothermal characteristics of Mn—Zn ferrite nanofluid within annulus in the presence of magnetic field *J. Supercond Nov. Magn.* **27** 527–534.

[26] Kumar B, Srinivas S 2020Unsteady Hydromagnetic Flow of Eyring-Powell Nanofluid over an Inclined Permeable Stretching Sheet with Joule Heating and Thermal Radiation, J. *Appl. Comput. Mech.* **6** 259-270

[27] Jyothi K, Reddy PS, Reddy MS 2018 Influence of magnetic field and thermal radiation on convective flow of SWCNTs water and MWCNTs-water nanofluid between rotating stretchable disks with convective boundary conditions, *Powder Technol.,* **331** 326–337.

[28] Elgazery NS 2019 Nanofluids flow over a permeable unsteady stretching surface with non-

uniform heat source/sink in the presence of inclined magnetic field, *Elgazery Journal of the Egyptian Mathematical Society*, **27** 1-26

[29] Elbashbeshy EM, Emam TG, Abdelgaber KM, Effects of thermal radiation and magnetic field on unsteady mixed convection flow and heat transfer over an exponentially stretching surface with suction in the presence of internal heat generation/absorption. *J. Egypt. Math. Soc.* 2012; 20:215–222.

[30] Pantokratoras A, Fang T. 2013 Sakiadis fllow with nonlinear Rosseland thermal radiation. *Phys Scr.* 87 015703

[31] Prasannakumar BC, Gireesha BJ, Krishnamurthy MR, Gorla RSR 2017 Unsteady Boundary Layer Flow and Convective Heat Transfer of a Fluid Particle Suspension with Nanoparticles over a Stretching Surface, *Journal of Modeling in Mechanics and Materials* 20170002

[32] Schlichting H and Gersten H 2000 *Boundary-layer theory*. New York: Springer, 2000

[33] Rosseland S. 1931 Astrophysik und atom-theoretische Grundlagen. *Berlin: Springer Verlag,*:41–44.

[34] Seigel S, and Howell JR 1981 Thermal radiation heat transfer *2nd ed. New York: McGraw-Hill Inc.*

IOP Conf. Series: Materials Science and Engineering **925** (2020) 012055 doi:10.1088/1757-899X/925/1/012055

Facile Green Synthesis and Characterization of Moringa Oliefera Extract-Capped Silver Nanoparticles (MO-Agnps) And Its Biological Applications

Kiran M S[1,4], Latha M S[2,4], Nagappa B Gokavi[2,4], G H Pujar[3,4], Rajith Kumar C R[1,4], Shwetha U R[1,4] and Virupaxappa S Betageri[1,4]

[1] Research Centre, Department of Chemistry, GM Institute of Technology, Davangere, Karnataka, India-577006
[2] Department of Chemistry, R L Science Institute, Belagavi, Karnataka, India- 590001
[3]Research Centre, Department of Physics, GM Institute of Technology, Davangere, Karnataka, India-577006
[4]Visvesvaraya Technological University, Belagavi, Karnataka

Corresponding Author: virupaxb@gmail.com, lathamschem97@gmail.com

Abstract. In the current study, we developed a simple, eco-friendly, quick and economical process to synthesize stable silver nanoparticles (AgNPs) using medicinal herb Moringa oliefera leaves extract. Obtained AgNPs are characterized by spectrophotometrical and microscopic techniques such as UV-Visible spectroscopy, x-ray diffraction (XRD), scanning electron microscopy (SEM) and high resolution transmission electron microscopy (HR-TEM). The study on the in vitro antidiabetic activities displayed that the particles have greater α-amylase inhibition potentials. In addition, exceptional antioxidant activity on 1, 1-Diphenyl-2-picrylhydrazyl (DPPH) is observed. Cytotoxic study against Michigan Cancer Foundation-7 (MCF-7) cell lines displayed significant anticancer activity of AgNPs and hence they can be employed as a promising, cost-reliable and non-toxic strategy for treating these diseases in future.

Keywords: Anticancer, Antidiabetic, Antioxidant, Moringa Oliefera, Phytochemicals, Silver Nanoparticles,

1. Introduction

Nanotechnology is a multidisciplinary science field that intended to produce atomic, molecular and supramolecular materials with enriched functionalities. Electronic, magnetic, catalytic and optical behaviors of nanoparticles (NPs) rely on their size, shape, physical and chemical natures [1] for which they found great importance in remarkable applications in numerous fields such as catalysis, cosmetics, photonic crystals analysis, food, agriculture, drug delivery, paints, coatings, health care and material science. Various metal and metal oxide particles in nanoscale such as Ag, Au, Pt, Cu, ZnO, NiO etc., have been synthesized and analyzed in specific domains for their interesting unusual physicochemical characteristics. The development of simple, cost-reliable and environment-friendly methods for production of NPs is most significant fields of research in nanoscience [2]. The NPs can be produced by different approaches such as chemical, physical and biological methods [3]. Electro-deposition, vapour deposition methods, laser ablation, pyrolysis, sol-gel, lithography are some examples for physical and chemical methods, many of these processes imposes toxic effects on environment and human health, which confines their enormous applications [4]. Physical and chemical approaches have advantageous in NPs uniform size distribution whereas the major drawbacks are huge energy requirement, duration, expensive and practice of toxic chemicals. In biological method, NPs are synthesized using yeast, fungi, bacteria, algae, and plants [5]. Use of plant extract or bioactive molecules as source for the NPs synthesis are most promising than bacterial and chemical approaches since there is no risk of contamination from bacteria, hazardous chemicals and optimum energy consumption provides greater applicability and simplicity. Bioactive secondary constituents like tannins, glycosides, saponins, flavonoids, alkaloids and terpenoids present in plants can act as reducing and capping agents. Moreover, plant extracts have inherent biological properties such as antihyperglycemic, antioxidant, antimutagenic, cytotoxic, antifungal, anti-inflammatory, antiviral and antibacterial potential which may apparent in the final AgNPs biological activities and may therefore be used in medical applications [6]. Silver nanoparticles (AgNPs) are specially interested among other different metal nanoparticles (MNPs) and drawing more consideration of scientists due to its various applications in bimolecular detection, diagnosis, therapeutics, catalysis and micro-electronics. Researches have also revealed their medicinal use as antifungal, antimicrobial, anti-inflammatory, antidiabetic, antioxidant agents and several studies also described the AgNPs use in cancer treatments [7].

Oxidative stress is a phenomenon induced by disturbance between production and accumulation of reactive oxygen species (ROS) including singlet oxygen ($^1O_2$), hydrogen peroxide (H_2O_2), hydroxyl radicals (•OH) and Superoxide radicals ($O2^{\bullet -}$) are commonly produced by several biochemical processes within the body. Excess free radicals overtake the natural body antioxidant defense system leading to oxidative stress. A large body of evidences displays that oxidative stress can lead to the development of age-related diseases like metabolic syndromes, dementia, osteoporosis, atherosclerosis, vascular diseases, obesity, arthritis, cancer and diabetes. The role of oxidative stress is an essential factor in understanding the complex mechanism and complications of diabetes. In this perspective, numerous researchers are focused on the characterization of the ROS sources, their activation path, scavenging and antioxidant constituents in diabetes [8].

Diabetes mellitus (DM) is an endocrine disease characterized by chronic hyperglycaemia, triggered by the combination of genetic and the environmental factors. Two major types of DM are type-1(insulin dependent) and type-2 (non-insulin dependent), type-2 DM is the most prevalent form of the disease, contributing for 90-95% of cases. Severity of diabetes is connected with permanent damage and failure of different organ systems and can cause cardiovascular disease, blindness, kidney failure and lower limb amputation [9]. Type 2 diabetes can increase the risk of bladder, breast, blood and colorectal cancers by 20% to 50%. Starches, hydrolyzed dietary carbohydrates are the major source of blood glucose. Pancreatic α -amylase breaks down the complex carbohydrates by hydrolysis to simple carbohydrates like maltose and glucose. α –glucosidase performs the end line process by which carbohydrates are absorbed by the small intestine. Identifying the inhibitors for these digestive enzymes is one of the prominent therapeutic methods to regulate the blood sugar by hindering the glucose uptake [10]. Cancer is disease characterized by unusual

IOP Conf. Series: Materials Science and Engineering **925** (2020) 012055 doi:10.1088/1757-899X/925/1/012055

cell growth with the potential to attack or spread through the body parts and hence considered as the world's most devastating diseases. There are several treatments are available such as surgery, radiation and chemotherapeutic drugs but these treatments have limitation of normal cell death and toxicity. Therefore, non-toxic, ecological and inexpensive drugs are essential to be discovered. Rapidly evolving field of different materials and devises in nanoscale incorporating has heightened the possibility for the timely identification and management of these diseases including cancer with minimum toxicity [11]. Many researches have been made to identify materials to inhibit the free radicals and carbohydrate-hydrolyzing enzymes that leads to various complications. Recently researchers demonstrated that AgNPs are very effective as antimicrobial, antioxidant, antifungal agents and antidiabetic activity. AgNPs have also proved to be the cytotoxic to different cancer cell lines. Particularly, AgNPs synthesized by green method have reported for inhibition of cancer proliferation [12].

Moringa oleifera (MO) is a member of moringaceae family, commonly known as the drumstick tree, is a wonderful plant that is used throughout the world traditionally. This serves as a stimulant to circulation and to the heart. The MO is a conventional medicinal tree that has displayed significant prospective in complementary and alternative medicine. It is historically being used in the therapy of hyperglycemia, inflammation, microbial and cancer infections. It contains high constituents of antioxidants and biologically active compounds, which performs a major role in their effectiveness. All parts of the plant were apparently used in traditional medicine. Especially, leaves are recognized for their natural remedial characteristics and are consumed universally in a variety of ways [13]. Thus, the objective of the present work was to assess the biosynthesis of AgNPs using MO leaf extract. The NPs obtained by green synthesis were characterized by analytical techniques like XRD, SEM and TEM. Further, study of its activity against a-amylase enzymes, antidiabetic properties and cytotoxic activity against human breast cancer cell line MCF-7.

2. Experimental

2.1 Materials used
Silver nitrate ($AgNO_3$) of analytical grade was purchased from Hi- Laboratories Pvt. 2,2-diphenyl-1-picrylhydrazyl (DPPH) and alcohol were procured from Sigma Aldrich. All glassware's were washed and dried in an oven, throughout the experiments double distilled water (DDW) was used.

2.2 Collection and preparation of leaves extract
MO leaves were collected from GMIT campus, healthy and free from disease leaves were picked and washed thoroughly with tap water to remove dirt and adhering substances followed by double distilled water. In 100 ml of distillate water, 20 gm of leaves were boiled at 60^0 C for 30 minutes, cooled to room temperature. The resulting mixture was filtered through Whatman grade-1 filter paper and refrigerated at 4^0C until further use

2.3 Phytochemical screening of Moringa oliefera extract
The MO extract has been subjected for qualitative analysis for various active chemical components which are accountable for their nutritional and biological application. Phytochemical analysis was carried by following standard protocol [14].

2.4 Green Synthesis of Silver Nanoparticles
The MO leaves extract of 10 milliliters was combined with 90 milliliters of 0.001 M silver nitrate solution, mixture was kept in dark for overnight to achieve complete reduction of Ag^+ to Ag^o [36]. AgNPs synthesis was performed at $25° \pm 2°C$ temperature. Change in color of the solution to brown from pale yellow indicates

IOP Conf. Series: Materials Science and Engineering **925** (2020) 012055 doi:10.1088/1757-899X/925/1/012055

completion of the reaction. The AgNPs are obtained by centrifugation of the colloidal mixture for 15 min at 8000 rpm accompanied by repeated washing with mixture of distilled water and alcohol. For further studies AgNPs powder thus obtained were deposited in airtight viols.

2.5 Characterization

Moringa oliefera extract mediated synthesized AgNPs (MO-AgNPs) were characterized by Synthesized AgNPs were characterized using Agilent Technologies-Cary 60 UV-Vis spectrophotometer. The XRD spectrum of AgNPs coated on XRD measuring grid was furnished using (Rigaku Smart Lab, Cu – Kα, π = 1.5406 Å with scanning angle 2 = 20° to 90°) X-ray diffractometer. MO-AgNPs morphology was studied using scanning electron microscopy (Jeol-JSM-IT500LA: 5 to 3,00,000, Acc. Voltage 0.3–30 kV). High resolution transmission electron microscopic (HR-TEM) examination was performed on the sample of MO-AgNPs to investigate the scale, shape and distribution of synthesized nanoparticles. Small amount of colloidal dispersion of nanoparticles was loaded on carbon standard TEM grids and enabled to dry within a vacuum dryer, scanned by TEM analyzer (Jeol-JEM 2100, LaB6/200 kV).

3. Antioxidant activity

The in vitro antioxidant capability of MO-AgNPs were determined by DPPH assay, it is a well-known and more stable free radical for evaluating the antioxidant properties of compounds [15]. For the study, in a sequence of test tubes different concentrations of MO-AgNPs and standard (20, 40, 60, 80 and 100 µg/mL) were mixed with dimethyl sulfoxide (DMSO) and volume was adjusted to 500 µL using Methanol. 5 mL of 0.1 mM DPPH dissolved in methanol was added to each tubes and shaken vigorously. Control was prepared as above excluding sample, ascorbic acid was used as standard for the analysis. The mixture was vortexed and left to stand in the dark for 30 min at room temperature to allow radical scavenging of the DPPH. Absorbance was recorded at 517 nm [16], antioxidant level was determined using the below equation.

$$\text{Free radical scavenging } (\%) = \frac{A_{Control} - A_{Sample}}{A_{Control}} \times 100$$

Where, $A_{Control}$- absorbance of control, A_{Sample} - absorbance of MOE@AgNPs.

4. Antidiabetic activity

Diabetic mellitus is a non-transmissible disease that is mostly genetic in nature but can evolve due to lifestyle changes resulting in the hyperglycemia over a sustained period of time. α-amylase inhibition is an important therapeutic target in controlling postprandial blood glucose production in diabetic patients. MO-AgNPs antidiabetic efficacy was measured in-vitro by starch hydrolysis in the presence of α-amylase enzyme.

Different concentration of MO-AgNPs (50,100, 150 & 200 µg/mL) in 1mL of sodium phosphate buffer (PBS) solution were taken in an Eppendorf tubes and mixed with α-amylase (0.5 µg/mL). The mixture of solution was incubated for 10 min at 25±2^0 C followed by the addition of 5 µg/mL starch, again incubated for 10min. The reaction was terminated by adding 1mL of dinitrosalicylic acid (DNS). Tubes were placed on water bath at 100^{0}C for 5min and cooled. Absorbance was measured at 540 nm for both sample and control (prepared as above without sample) [17]. Metformin was taken as standard, the percentage of inhibition of α-amylase activity was calculated by given formula.

$$\% \text{ of } \alpha - \text{amylase inhibition } = \frac{AC - AS}{AC} \times 100$$

where, AC and AS are the absorbance of control and sample, respectively

5. Anticancer activity

MTT (3-(4,5-dimethylthiazol-2-yl)-2,5-di-phenyltetrazolium bromide), a yellow tetrazolium assay on human breast adenocarcinoma cell line (MCF-7) was employed to establish the cytotoxic capacity of the

IOP Conf. Series: Materials Science and Engineering **925** (2020) 012055 doi:10.1088/1757-899X/925/1/012055

MO-AgNPs. The cells were cultured and maintained in Dulbecco's Modified Eagles Medium (DMEM). Then, trypsinized and aspirated to 96-well microtiter plates at density of 1×10^4 cells/well then incubated at 37 °C with 5% CO_2 atmosphere. Medium was then carefully drained from the wells and replaced by a suspension of 200 μL of selected concentrations of MO-AgNPs (20 to 100 μg/mL) and again incubated in the above atmosphere for 24Hrs. Plates were separated from the incubator and drugs comprising media was removed. Final concentration was made to 0.5 mg/mL by adding 10% MTT reagent medium (200 μL) and incubated at 37^0 for 3Hrs. Then culture media was entirely extracted without upsetting crystals formed, then 100 μL DMSO was added and thoroughly mixed to solubilize formazan. Absorbance at 570 nm was recorded using a microplate scanner. 50% inhibition concentration (IC_{50}) value was determined graphically and the effect of the MO-AgNPs on the proliferation of MCF-7 cells was expressed. Cells viability (%) was estimated using below formula [18].

$$\text{Cell viability (\%)} = \frac{OD_{Sample}}{OD_{Control}} \times 100$$

Where, $OD_{Sample} = Optical\ density\ of\ test\ sample$
$OD_{Control} = Optical\ desity\ of\ control$

6. Result and discussion

Solution obtained by the extraction of MO leaves was screened for its phytochemical constituents which are responsible for their biological importance. Qualitative analysis indicates the existence of tannins, flavonoids, glycosides and phenolic compounds. Results were presented in **Table 1**.

Table 1. Phytochemical analysis of MO leaves extract

SL No	Phytochemical	Result
1	Reducing sugars	-
2	Tannins	+
3	Saponins	-
4	Glycoside	+
5	Phenolic compounds	+
6	Flavonoids	+
7	Alkaloids	-

+ refers present and – refers absent

Confirmation of these phytochemicals in the aqueous extract was thought to be participating in reduction and capping of AgNPs. Extract (10mL) is mixed in the stoichiometric ratio of 1:9 with the freshly prepared silver nitrate solution (0.001M) for the synthesis of AgNPs. Dark brown color formation from light yellow indicates the reduction of Ag^+ to Ag^0. For a complete reduction of MO-AgNPs solution mixture was kept in dark for 24Hrs.

6.1 Characterization of silver nanoparticles synthesized with MO extract

IOP Conf. Series: Materials Science and Engineering **925** (2020) 012055 doi:10.1088/1757-899X/925/1/012055

The formation of MO-AgNPs was confirmed by UV-Visible spectroscopy. One of the most fascinating characteristics of MNPs are their optical properties which changes according to the morphology of NPs. They contain valence electrons, which produces surface plasmon resonance (SPR) absorption bands because of the collective electronic and light wave vibrations in resonance. For MO-AgNPs, sharp SPR band was observed at 442 nm as shown in **Figure 1**.

Further, AgNPs were characterized by XRD to study crystalline nature and geometric arrangements. The XRD plot (**Figure 2)** displayed four major characteristic diffraction peaks at 2θ= 32.4°, 38.0°, 64.20° and 77.40° corresponding to lattice planes (111), (200), (220) and (311) respectively confirming that the MO-AgNPs are having face-centered cubic (fcc) structure. Debye – Scherer equation was used to calculate the average crystalline size which is found to be 15.5 nm.

$$d = \frac{k\lambda}{(\beta \cdot cos\theta)}$$

Where, d- crystal size, k- Scherer constant, λ- wavelength of X-ray, β- line extension in radians; θ - Bragg angle.

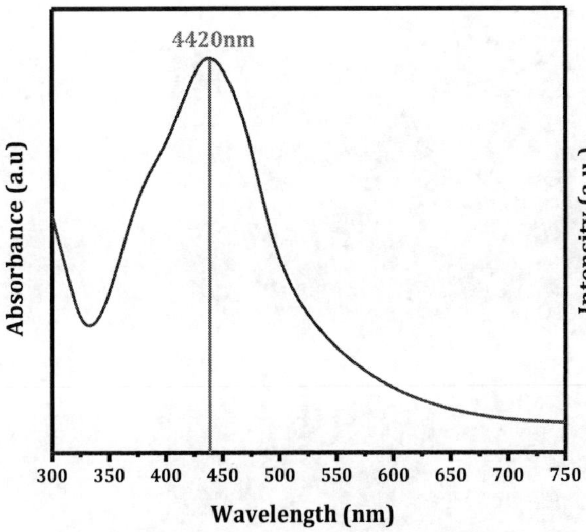

Figure 1. UV-Vis spectrum of MO-AgNPs

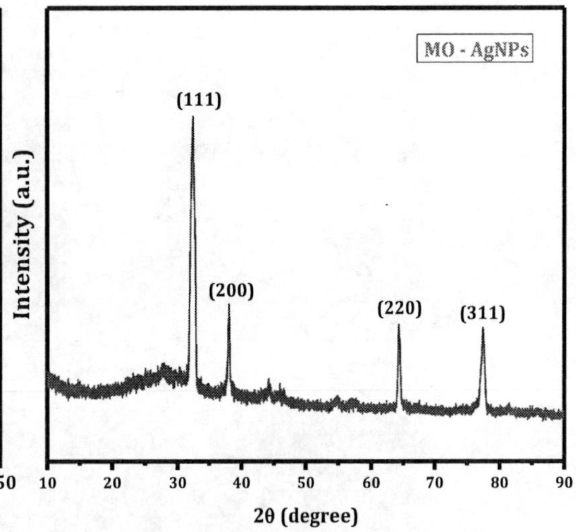

Figure 2. XRD Plot MO-AgNPs

Surface morphology, size and shape of MO-AgNPs were studied by electron microscopic techniques such as SEM (Jeol-JSM-IT500LA model) and HR-TEM. The SEM images as shown in **Figure 3** specifies that the MO-AgNPs are almost spherical in shape with slight agglomeration. However, each agglomerate was composed of small crystallites. HR-TEM results obtained by capturing images at different ranges reveals that the MO-AgNPs are in nanoscale, uniform and spherical with an average size of ≈ 21.5-40nm (**Figure 4**).

Figure 3. SEM images of MO-AgNPs

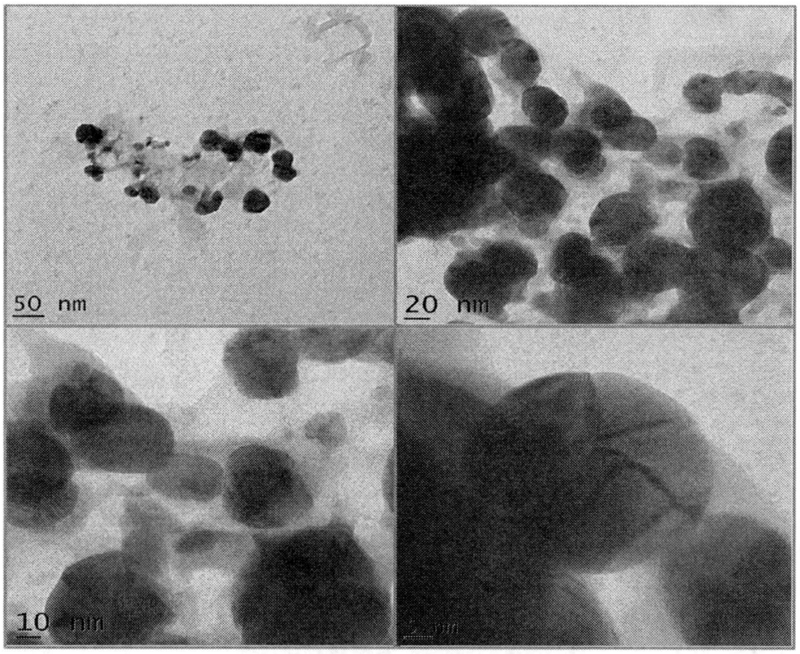

Figure 4. TEM images MO-AgNPs

6.2 Antioxidant activity

DPPH free radical scavenging assay was used to study the free radical scavenging potential of the MO-AgNPs. This method is reliant on the reduction of DPPH* to stable DPPH–H by hydrogen-contributing antioxidants. DPPH absorbance was decreased at 517 nm with rise in concentrations of NPs changing its color from violet to light yellow [19]. MO-AgNPs showed the oxidant activity of 14.77, 27.42, 36.46, 48.29 and 57.32% for the concentrations of 20, 40, 60, 80 and 100 µg/mL. Compelled with the standard ascorbic

IOP Conf. Series: Materials Science and Engineering **925** (2020) 012055 doi:10.1088/1757-899X/925/1/012055

acid which is having maximum scavenging potential of 93.88 % at 100 μg/mL. 50% inhibition concentration was found to be 62.6 μg/mL (**Figure 5**). Dose-dependent free radical scavenging behavior of MO-AgNPsis tabulated in **Table 2.**

Table 2. Antioxidant activity of MO-AgNPs

SL.No	Concentration	% of free radical scavenging activity	
		MO-AgNPs	Ascorbic acid
1	20 μg/mL	14.77	20.37
2	40 μg/mL	27.42	44.82
3	60 μg/mL	36.46	74.19
4	80 μg/mL	48.29	89.27
5	100 μg/mL	57.32	93.88

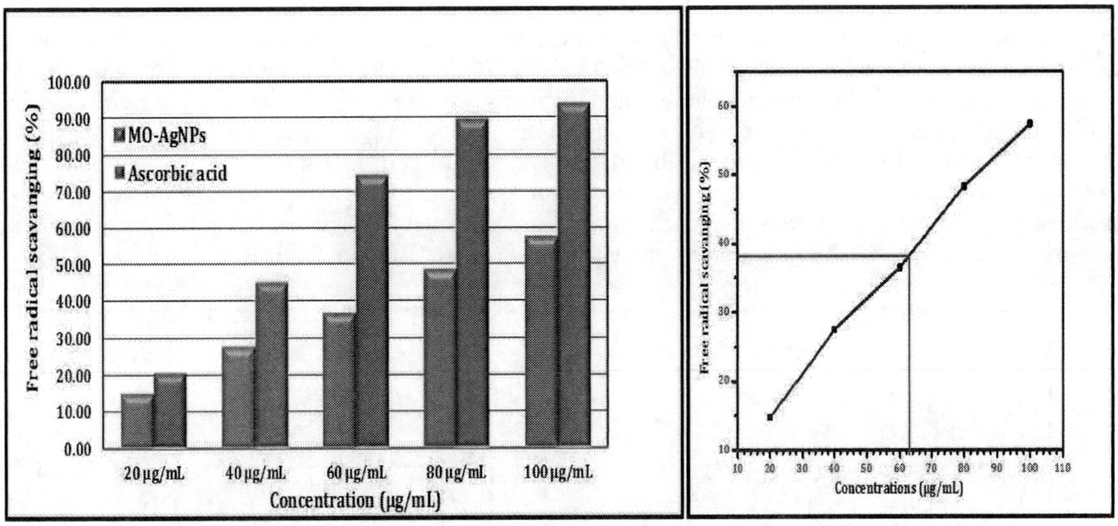

Figure 5. Free radical scavenging activity with IC_{50} value

6.3 Antidiabetic activity

The inhibition of carbohydrate digesting enzymes such as α – Amylase is a significant target for preventing a sudden rise in blood glucose levels, thereby intercepting the conversion of carbohydrates into simple sugars which is a major source for elevation in the blood sugar [20]. In the present analysis, in comparison with the standard drug metformin, the α-amylase inhibition by MO-AgNPs was observed in an increasing order of concentration as shown in **Figure 6**. It is presumed that the biomolecules present in the extract enhanced the antidiabetic potential of MO-AgNPs, hence they can be utilized in the diabetic treatments.

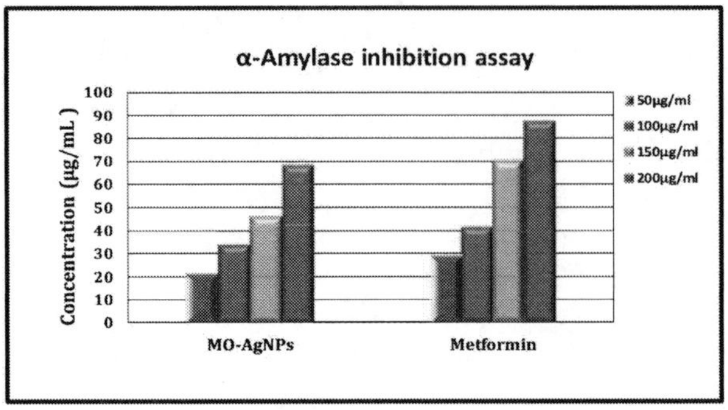

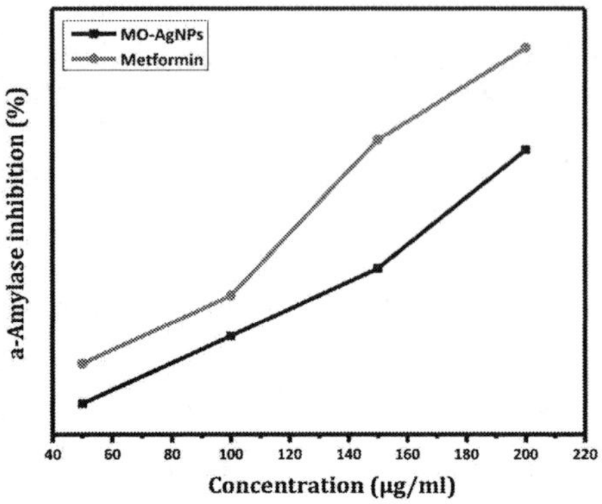

Figure 6. α – Amylase inhibition activity of MO-AgNPs

6.4 Anticancer activity

The MO-AgNPs mediated apoptosis in MCF-7 cells were examined using cytotoxicity assay of MTT dye conversion. Apoptosis is a programmed biochemical event within the cell which leads to death of cells. The intracellular production of ROS by the NPs is believed to enhance the anticancer activity [21] and these are dose dependent. Different concentrations of MO-AgNPs (20 µg/mL to 100 µg/mL) were treated with MCF-7 cell line to assess their cytotoxic potential and cisplatin (15 µg/mL) was used as the reference drug for study. Minimum and maximum inhibitions were observed at 20 µg/mL and 100 µg/mL respectively in comparison with the standard which exhibited 77.54% inhibition of cell growth at 15 µg/mL. IC_{50} value calculated from dose-response curve is found to be 68.03 µg/mL. The findings are shown in **Figure 7** and the viabilities of the cells at various concentrations are depicted in **Figure 8**. Comparison of anticancer activity of AgNPs synthesized by various plants extracts are tabulated in **Table 3**.

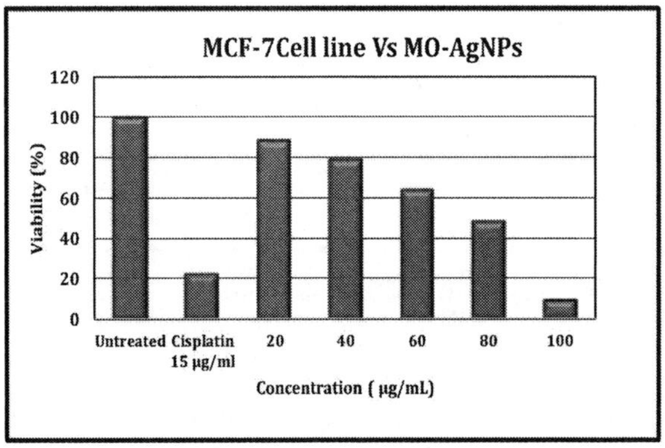

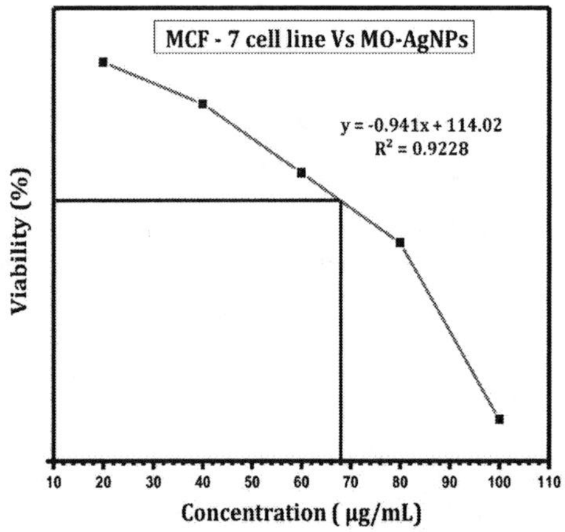

Figure 7. Anticancer activity of MO-AgNPs

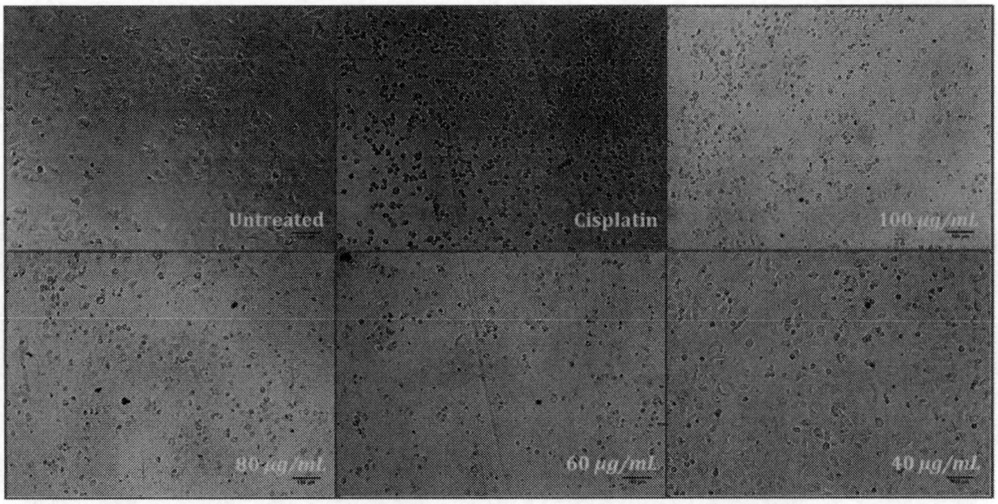

Figure 8. MCF-7 cell viabilities treated with MO-AgNPs

Table 3. Comparison table of Anticancer activity of AgNPs synthesized using other plant extracts.

SL. No	Plant name	AgNPs sizes	IC_{50} Value	References
1	*Syzygium aromaticum*	5 – 50 nm	70 µg/mL	[22]
2	*Adenium obesum*	10 – 30 nm	217 µg/ mL	[23]
3	*Artocarpus integer*	5.76 -19.1 nm	90 µg/mL	[24]
4	*Multiple fruits*	25 nm	500 µg/mL	[25]
5	*Moringa oleifera*	21.5- 41 nm	68.03 µg/mL	Present work

7. Conclusion

We successfully synthesized the AgNPs using aqueous extract of *Moringa oleifera* leaves which is well known for their therapeutic applications. Applied green synthesis method offered simple, rapid, and ecological advantages for MO-AgNPs synthesis. In order to assess their purity, size and shape, AgNPs were then characterized by standard analytical techniques such as XRD, SEM, and HR-TEM. The XRD analysis demonstrates that the AgNPs are in face-centered cubic (fcc) lattice. SEM and TEM examinations revealed that the AgNPs are spherical and size ranging from 21.5 nm to 41.0nm. MO-AgNPs displayed strong α-amylase inhibition activity, DPPH radical scavenging ability with IC_{50} value of 62.6 µg/mL and concentration-dependent cytotoxicity towards MCF-7 cell line exhibiting the half maximal inhibitory concentration of 68.03 µg/mL. MO-AgNPs are therefore considered to be a good candidate for biomedical and environmental bioremediation.

References

[1] Behravan M, Panahi AH, Naghizadeh A, Ziaee M, Mahdavi R and Mirzapour A 2019 *Int. J. Biol. Macromol.* **124** 148-54
[2] Kumar C R, Betageri V S, Nagaraju G, Pujar G H, Onkarappa H S and Latha M S 2020 *J. Inorg. Organomet. Polym. Mater.* 1-8
[3] Singh A and Dutta P K 2020 *Int. J. Biol. Macromol.*

[4] Dakshayani SS, Marulasiddeshwara MB, Kumar S, Golla R, Devaraja SRHK and Hosamani R 2019 *Int. J. Biol. Macromol.* **131** 787-97
[5] Hemmati S, Rashtiani A, Zangeneh MM, Mohammadi P, Zangeneh A and Veisi H 2019 *Polyhedron.* **158** 8-14.
[6] Orlowski P, Zmigrodzka M, Tomaszewska E, Ranoszek-Soliwoda K, Czupryn M, Antos-Bielska M, Szemraj J, Celichowski G, Grobelny J and Krzyzowska M 2018 *Int. J. Nanomed.* **13** 991
[7] Raj S, Mali SC and Trivedi R 2018 Biochem. Biophys. Res. Commun. **503** 2814-19
[8] Liu Z, Zhou T, Ziegler A C, Dimitrion P and Zuo L 2017 *Oxid. Med. Cell. Longev.*
[9] Baldea LAN, Martineau LC, Benhaddou-Andaloussi A, Arnason J T, Lévy É and Haddad P S 2010 *J. Ethnopharmacol.* **132** 473-82
[10] American Diabetes Association 2016 *Diabetes Care 39 Supplement* 1 S4-S5
[11] Saratale RG, Shin HS, Kumar G, Benelli G, Ghodake GS, Jiang YY, Kim DS and Saratale GD 2018 *Environ. Sci. Pollut. Res.* **25** 10250-263
[12] Saratale GD, Saratale RG, Benelli G, Kumar G, Pugazhendhi A, Kim D S and Shin H S 2017 *J. Clust. Sci.* 28(3) 1709-27
[13] Aisida S O, Madubuonu N, Alnasir M H, Ahmad I, Botha S, Maaza M and Ezema F I 2020 *Appl. Nanosci.***10**(1) 305-15
[14] Harborne JB 1973 *Phytochemical methods* Chapman Hall 49-188
[15] Jacob SJP, Mohammed H, Murali K and Kamarudeen M 2012 *Colloids Surf. B.* **98** 7-11
[16] Zangeneh MM, Bovandi S, Gharehyakheh S, Zangeneh A and Irani P 2019 *Appl. Organomet. Chem.* **33** 4961
[17] Popli D, Anil V, Subramanyam AB, Rao SN, Rai RV and Govindappa M 2018 *Artif. Cells. Nanomed. Biotechnol.* **46**(sup1) 676-83
[18] Kiran MS, Betageri VS, Kumar CR, Vinay SP and Latha MS 2020 *J. Inorg. Organomet. Polym. Mater.*1-10
[19] Saratale RG, Shin HS, Kumar G, Benelli G, Kim DS and Saratale GD 2018 *Artif. Cells. Nanomed. Biotechnol.* **46** 211-22
[20] Du L, Suo S, Wang G, Jia H, Liu KJ, Zhao B and Liu Y 2013 *Chemistry–A European Journal* **19** 1281-87
[21] Mortazavi-Derazkola S, Ebrahimzadeh MA, Amiri O, Goli HR, Rafiei A, Kardan M and Salavati-Niasari M 2020 *J. Alloys Compd.* **820** 153186
[22] Venugopal K, Rather HA, Rajagopal K, Shanthi MP, Sheriff K, Illiyas M, Rather R A, Manikandan E, Uvarajan S, Bhaskar M and Maaza M 2017 *J. Photoch. Photobio. B.* **167** 282-89
[23] Farah MA, Ali MA, Chen SM, Li Y, Al-Hemaid FM, Abou-Tarboush FM, Al-Anazi KM and Lee J 2016 *Colloids Surf. B.* **141** 158-69
[24] Majeed S, Bakhtiar NFB, Danish M, Ibrahim MM and Hashim R 2019 *Sustain. Chem Pharm.* **12** 100138
[25] Naganathan K and Thirunavukkarasu S 2017 *In IOP Conf. Ser.: Mater. Sci. Eng.* **191** 012009

Experimental Analysis of Luffa Composite Material

Kumar Vijayendra Gopal[1], Fredrick Gnanaraj F[2], Vijaya Kumar K R[3] and Andrew Nallayan W[4]

[1,2,3,4] Department of Mechanical Engineering, Dr. M.G.R. Educational and Research Institute, Chennai, India.
[1]kumarvijayendragopal@gmail.com

Abstract. In recent times maximum research is based on the composite materials. In the future, many engineering developments are based on the composite material. This research is entirely based on natural composite material. In this work, Luffa material is used for making composite materials in different proportions and its mechanical properties calculated. Luffa is a natural material with less cost and easily available material. Thus the main objective of this work is determining the best composition by comparing the properties of different compositions.

Keywords. Luffa, composite, natural composite, mechanical properties.

1. Introduction

The reason for the more composite material requirement in recent time is that because of engineering development is based on the lightweight materials with less cost. Composite is defined as combining two materials or more than two materials. Recently various types of composite materials are available, which are based on the raw materials used [1]. The different categories of composite materials used now are synthetic composite, natural composite and hybrid composite. Composite materials are fabricated by various methods, but they are used to replace the ferrous materials. This research is based on natural composite. Natural fibers available are ramie, coir, luffa, hair, etc. [2].

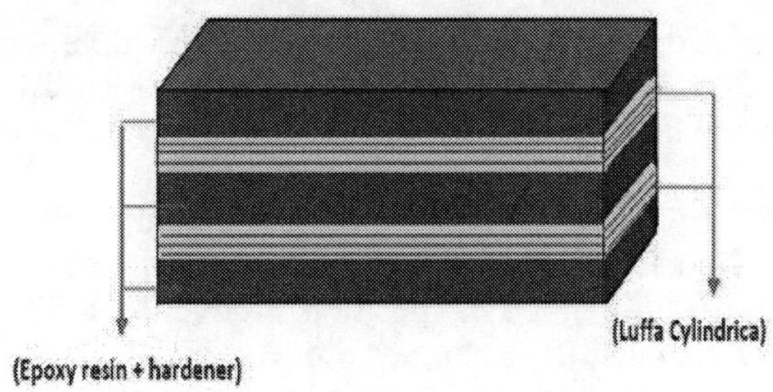

Figure 1. Composite Material

The chemical like epoxy bonds natural composite materials. The natural composite materials have excellent strength with less cost and easily available [4]. The main advantage of natural composite material is eco-friendly with good strength. The figure shows the composite material with two layers.

2. Materials

Luffa fiber is the primary raw material used to make the luffa composite material. The bonding is made with hardener and epoxy resin. Among the different types of epoxy and hardener available, hardener HY 951 and epoxy resin LY 556 are the best bonding chemicals suitable for the luffa composite. Luffa fiber is obtained from the plant. The luffa fiber fabrication process is shown in figure 2.

Figure 2. Luffa fiber

3. Methodology

The cylindrical shape luffa is available in the market. The cylindrical luffa is cut at its centre rib and all seeds are removed. Then the luffa is flattened, as shown in figure 2. The scissors or sharp knife is used to cut the luffa with correct finishing and the luffa fiber is chapped into different sizes for different composites.

The plastic tray is used as a mould for the composite. PVA is applied on the plasitic tray and thoroughly dried. PVA chemical is used for insulation between composite and mould. The hand lay-up is suitable for fabricating natural composite and hence in this work, this method is used to fabricate the luffa composite.

After the drying of PVA, the LY 556 and HY 951 mixing is applied on the tray over the PVA layer. The LY 556 and HY 951 are mixed in 10:1 ratio, which is obtained from previous research work. Above the chemical mixing, the luffa is placed in a vertical position. Above the luffa fiber, the chemical mixing is applied again. For a certain period, the luffa is rammed properly and the composite is dried for two or three days. The composite is cut according to the ASTM standard and the samples are prepared according to the specification. The samples are shown in figure 3.

Figure 3. Luffa Composite

Specimen preparation

Specimen 1: The specimen is made by the chapped luffa.
Specimen 2: This specimen is made by flatten luffa.
Specimen 3: This specimen is made by cylindrical luffa.

4. Results and Discussion

4.1. Tensile strength

The main mechanical property of the luffa composite is stress and strain value as per the load. The tensile test determines the stress value of the luffa composite and a universal testing machine is used for the tensile test. The result of the three different composite values is shown in figure 4.

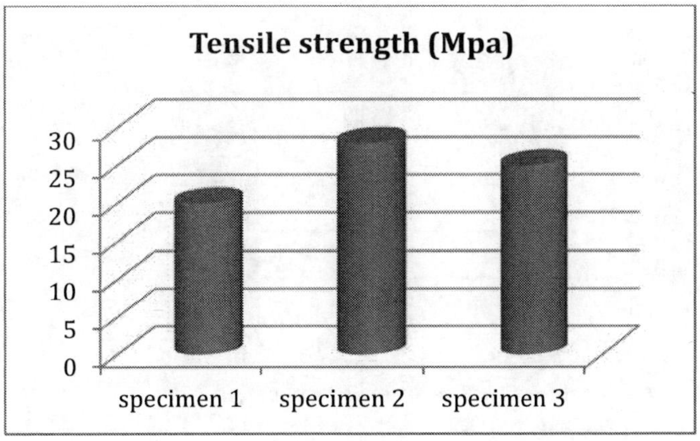

Figure 4. Tensile test results

4.2. Bending strength

The flexural testing determines the bending strength of the luffa composite. The luffa composite is cut in the shape of a dog bone for testing [1]. The dog bone shape luffa composite material is placed in the flexural testing machine. The point load acts on the centre of composite material to find the bending strength of the luffa composite material. The values are compared in figure 5.

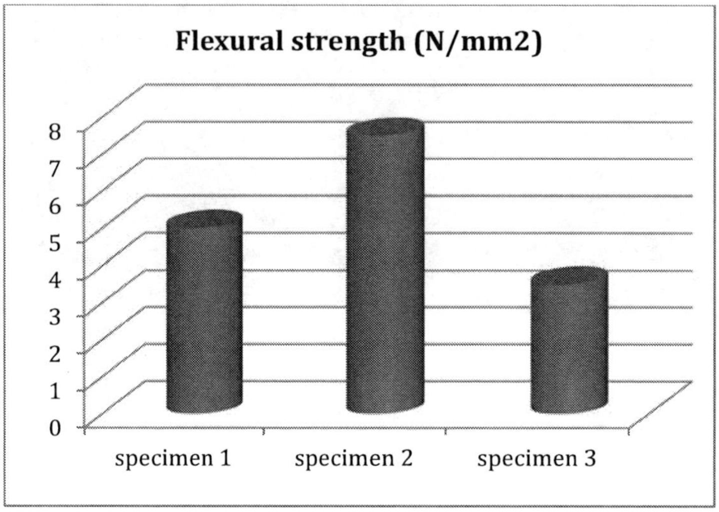

Figure 5. Flexural test results

4.3. Rockwell test

This Rockwell testing finds the hardness of the luffa composite material. Luffa composite is cut in the square shape and placed in the Rockwell machine. The machine has ball on the tip, which gives the load on various places. The material hardness of all the three materials are compared in figure 6.

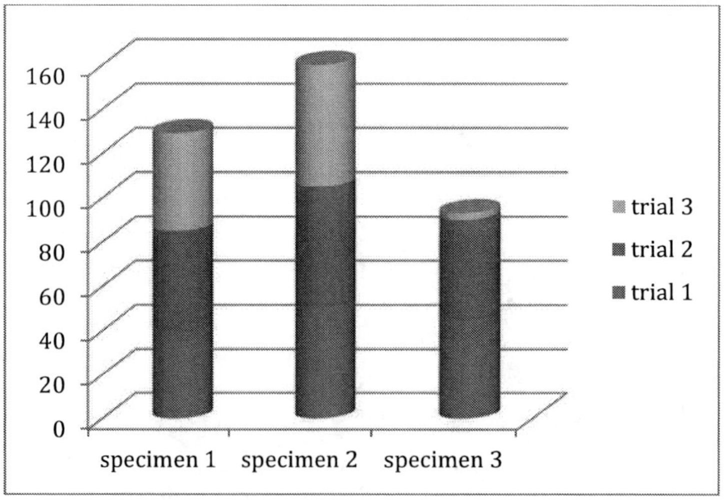

Figure 6. Hardness test results

IOP Conf. Series: Materials Science and Engineering **925** (2020) 012056 doi:10.1088/1757-899X/925/1/012056

4.4. *Water absorption test*

This test is performed by the kitchen scale machine. Luffa composite is cut into a small square, weight is measured and dumped in water for one day. After one day, the weight is determined to find the water absorption level of the luffa composite material. The water absorption results are shown in figure 7.

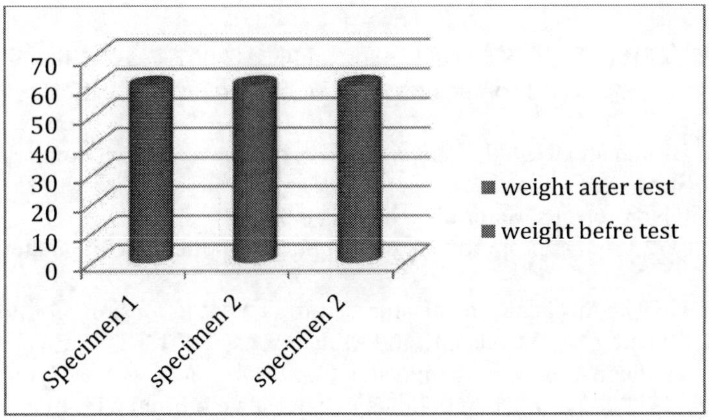

Figure 7. Water absorption test results

4.5. *Impact test*

This test helps to find the fracture value of the luffa composite as per load using the Charpy test. The luffa composite is cut in the small square shape and takes v groove in the centre. The results of all three composites are shown in figure 8.

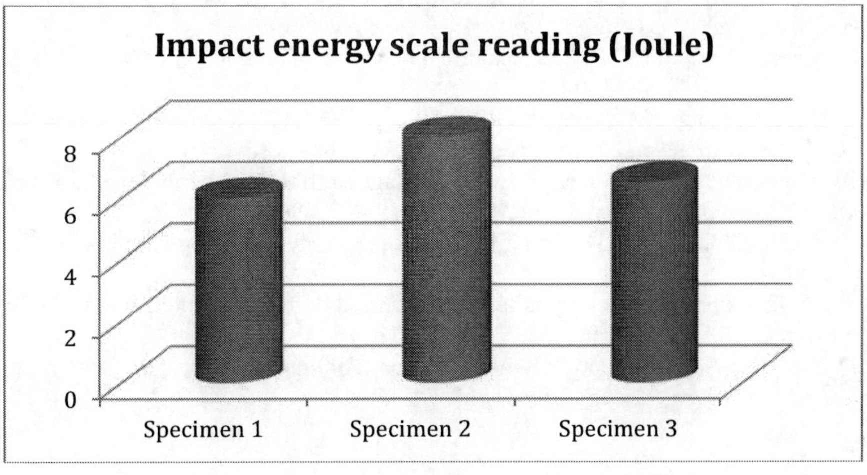

Figure 8. Impact results

5. Conclusion

The luffa composite is prepared in three different forms of specimens using the hand lay-up method. The specimens are taken out appropriately from the mould and cut for the testing. All three specimens have

IOP Conf. Series: Materials Science and Engineering **925** (2020) 012056 doi:10.1088/1757-899X/925/1/012056

excellent mechanical properties. Here, the Rockwell hardness test, impact Charpy test, water absorption test, flexural test and bending test are performed to determine its mechanical strength. Comparing the results, it proves that specimen 2 consists of good mechanical strength compared with the other two specimens. This specimen 2 has 7.5-joule impact strength, 7 N/mm2 flexural strength and 25 Mpa tensile strength. These values are high compared to the other two specimens.

Reference

[1] Khanam PN, Reddy MM, Raghu K, John K and Naidu SV 2007 Tensile, flexural and compressive properties of sisal/silk hybrid composites *Journal of Reinforced Plastics and Composites* **26** 1065-1070

[2] Agarwal BD and Broutman LJ 1990 Analysis and performance of fiber composites, Second edition, John wiley & Sons, Inc 2-16.

[3] Berghezan A 1966 Non-ferrous Materials, Nucleus **8** 5–11

[4] Van Suchtelen J **1972** Product properties: a new application of composite materials *Philips Res. Rep* **27** 28-37

[5] Outwater Jr JO 1956 The mechanics of plastics reinforcement in tension *Modern Plastics* **33** 156

[6] Rout J, Misra M, Tripathy SS, Nayak, SK and Mohanty AK 2001 The influence of fiber treatment on the performance of coir-polyester composites *Composites Science and Technology* **61** 1303-1310

[7] Li Q and Matuana LM 2003 Surface of cellulosic materials modified with functionalized polyethylene coupling agents *Journal of Applied Polymer Science* **88** 278-286

[8] Mohanty AK, Misra M and Drzal LT 2002 Sustainable bio-composites from renewable resources: opportunities and challenges in the green materials world *Journal of Polymers and the Environment* **10** 19-26.

[9] Wambua P, Ivens J and Verpoest I 2003 Natural fibers: can they replace glass in fiber reinforced plastics *Composites science and technology* **63** 1259-1264

[10] Gamstedt E and Almgren K 2007 Natural fiber composites–with special emphasis on effects of the interface between cellulosic fibers and polymers, *In Proceedings of the 28th International Symposium on Materials Science.*

[11] Mukherjee PS and Satyanarayana KG 1986 Structure and properties of some vegetable fibers *Journal of materials science* **21** 51-56

[12] Jain S, Kumar R and Jindal UC 1992 Mechanical behavior of bamboo and bamboo composite *Journal of Materials Science* **27** 4598-4604

[13] Nishino T, Hirao K, Kotera M, Nakamae K and Inagaki H 2003 Kenaf reinforced biodegradable composite *Composites Science and Technology* **63** 1281-1286

[14] Vazquez A, Dominguez VA and Kenny JM 1999 Bagasse fiber polypropylene based composites. *Journal of Thermoplastic Composite Materials* **12** 477-497

[15] Holbery J and Houston D 2006 Natural-fiber-reinforced polymer composite in automotive applications *Jom* **58** 80-86

[16] Clemons C 2002 Wood-plastic composites in the United States: The interfacing of two industries. *Forest Products Journal* **52** 10

[17] Plackett D 2002 The Natural Fiber-Polymer Composite Industry in Europe- Technology and Markets *Progress in Woodfibre-Plastic Composites*

IOP Conf. Series: Materials Science and Engineering **925** (2020) 012057 doi:10.1088/1757-899X/925/1/012057

Performance features enhancements and trade off study of NOx emission of diesel engine by using ternary blends of karanja - cotton seed biodiesel -diesel.

K Rajesh[1] and D B Ganesh[2]

[1] Department of Mechanical Engineering, STJ Institute of Technology, Ranebennur, Karnataka, India.
[2] Department of Mechanical Engineering, GM Institute of Technology, Davanagere, Karnataka, India.

Abstract: In this work, Karanja and cotton seed biodiesel were blended with the mineral diesel fuel. The effect of addition of the two biodiesels at different proportion were investigated for the performance and oxides of nitrogen emission of the diesel engine. All the blends exabits the higher break thermal efficiency than that of the diesel. Trade- off analysis was done with respect to efficiency, fuel consumption and oxides of nitrogen to reflect the emission pattern.

Key words: Biodiesel, Cotton, Karanja, Trade-off analysis.

1. Introduction

Compression ignition engines enjoy prominence in internal combustion engines owing to comparatively better fuel efficiency, lesser hydrocarbon and carbon monoxide tail end discharge compared to petroleum fuelled engines. However, nitrogen oxides and particular matter tail end discharge were high [1-3]. Fast depletion of the oil reserves combined with increase in the pricing, environmental issues and new technologies in automobile sector led to the look for the different fuel combination along with the diesel [4]. Biodiesel is a more appealing comparison to the other sources of oil, because it forms the lower pollution and smoke in line with environmental concerns. The tail-end pollution generated from the compression ignition engine may be greatly decreased by utilizing biodiesel [4,5]. Complete replacement of the diesel in the diesel engine is task which involves many technical modification and lot of economic implications. However partial replacement of the diesel by using the one or two biodiesels along with diesel needs no modifications of the engine. Karanja biodiesel-diesel blend and cotton seed biodiesel-diesel blend exabits the better thermal properties, performance and emission pattern for diesel engine which are in line with the diesel operations [6-10]. In this work, Karanja and cotton seed biodiesel were blended with the mineral diesel fuel. The effect of addition of the two biodiesels at different proportion was investigated for the performance and NOx emission of the diesel engine.

2. Related works

In this section different ternary blend investigations carried out by the researchers has been reviewed. Pankaj et al. [11] conducted experiments in compression ignition engine fuelled with 10% and 20% of Karanja and Roselle biodiesel with diesel. The studies indicated that chemical energy conversion efficiency (BTE) reduced by 1.82% and 4.36% for 20% blends of karanja (KB20) and Roselle (LA20) biodiesel with diesel respectively. Specific fuel consumption with respect to break power reduced by 6.48% for KB20 and 8.58% for LB20. Tail end exhaust temperature reduces by 4.43OC and 6.11OC for KB20 and LA20 blend respectively indicating the lower peak cylinder temperature. Lower peak cylinder temperature reduces the NOx emission by 3.83% and 6.01% for KB20 and LA20 blend respectively when compared to diesel fuel. Ignition delay reduced by 13.49% for LA20 and 13.4% for KB20 when compared to diesel at full load. They attributed this trend to higher cetane number of the biodiesel which in turn reduces ignition delay.

Prem et al. [12] conducted the experiments by using waste cooking biodiesel, Pongamia and diesel (WCB: PB:D). Prepared three blends containing 10% ,20% and 30% of the biodiesel each with the diesel fuel. The 10% of the ternary blends have showed almost similar pour and cloud point as of diesel fuel. The BTE of 10% ternary blend is 12% which is very near to diesel fuel (12.63%). Concluded the research by stating that the ternary blend are suitable alternatives for the CI engine.

Prabhu et al. [13] operated the diesel engine by using ternary blend consist of diesel, jatropha biodiesel and pentanol blends. Prepared two fuel combinations first combination consist of 70% diesel, 20% jatropha biodiesel, 10% pentanol and second consist of 60% diesel, 20% jatropha biodiesel, 20% pentanol. Reported that the little reduction in the output power and torque for the pentanol addition. Finally concluded that the blend with 20% biodiesels were shown the reduction in the tail end emissions.

Saeid et al. [14] evaluated the viscosity and density of the binary and ternary biodiesel diesel blends. Investigation concluded that the viscosities and densities of blends decreased nonlinearly and linearly with temperature which will affect the behaviours of the blends at different operating conditions.

Saumitra et al. [15] conducted the experiments on the diesel engine by using ternary blends consist of waste Cooking Oil, diesel and ethanol. Investigation reported that the 10 % raise in the BTE as compared to that of diesel. As compared to the diesel fuel decrease of 13%,30%,43% and 45% in NOx, CO, HC and smoke respectively.

3. Preparations of the fuel combinations

Three fuel combinations were prepared by volume percentage viz. 5K10C85D (5% Karanja biodiesel +10% cottonseed biodiesel + 85% Diesel), 5K15C80D (5% Karanja biodiesel +15% cottonseed biodiesel + 80% Diesel) and 5K20C75D (5% Karanja biodiesel +20% cottonseed biodiesel + 75% Diesel). Elementary biodiesel properties and their blends with the diesel were presented in\ table 1.

Table 1. Biodiesel and blends characters

Properties	Unit	Diese l	Karanja biodiesel	Cotton Seed biodies el	5K10C85 D	5K15C80 D	5K20C75 D
Density	Kg/m^3	850	840	887	853	855	856
Kinematic viscosity	cSt	2.46	3.11	4.34	2.63	2.71	2.78
Calorific value	KJ/kg	4400 0	37700	39234	43194	42948	42704

IOP Conf. Series: Materials Science and Engineering **925** (2020) 012057 doi:10.1088/1757-899X/925/1/012057

4. Methods

The experiments were carried out on the single cylinder, water cooled engine and specification of the setup were tabulated in the table 1. The experimental setup is as shown in Figure 1. For all experiment Eddy Current dynamometer reading, time taken for the 10 ml of fuel consumption, exhaust gas and ambient temperature were recorded. On the exhaust side all the emissions were recorded by tail end gas analyzer. Measurements time interval of fifteen minutes was given for the subsequent test to ensure that no previous fuel residue was left in the fuel injection system. With all the variations of fuel all the tests were replicated four times.

Figure 1. Experimental setup

Table 2. Instruments and device Specification

Make and type	Kirloskar, single cylinder, 4 stroke diesel engine
Rated power output	3.37 Kw 1500 RPM
Bore and stroke	85 X 110 mm
Compression ratio	16.5:1
Starting	By hand cracking
Injection system	Mechanical Injection
Intake system	Natural aspiration
Dynamometer	
Make	Powermag Control Systems(P) Ltd
Type	FTAC
Duty	S-1
Torque	2.4 kg-m
RPM	1500 rpm
Excitation max	80 V
Insulation	Class -F
Exhaust Gas Analyzers	
Make	AVL
Type	AVL Digas 444
Power supply	110V-220V 15 W
CO Measuring Range	0-0.15% Volume
CO2 Measuring Range	0-20 %
HC Measuring Range	0-30000 ppm Volume
NO Measuring Range	0-5000ppm Volume

5. Outcome and discussion

5. 1 Engine performance:

5. 1. 1 Brake specific fuel consumption

The SFC (specific fuel consumption) of an engine is the rate of fuel burnt to produce a unit of power. As load raises the decrease in BSFC trend was noticed as seen in figure 2,3 and 4. All ternary blend BSFC were found to be less than diesel and pure biodiesel. It is detected that 5K15C80D blend gives the minimum BSFC of 0.16 kg/kW-hr which is less than the mineral diesel at full load.

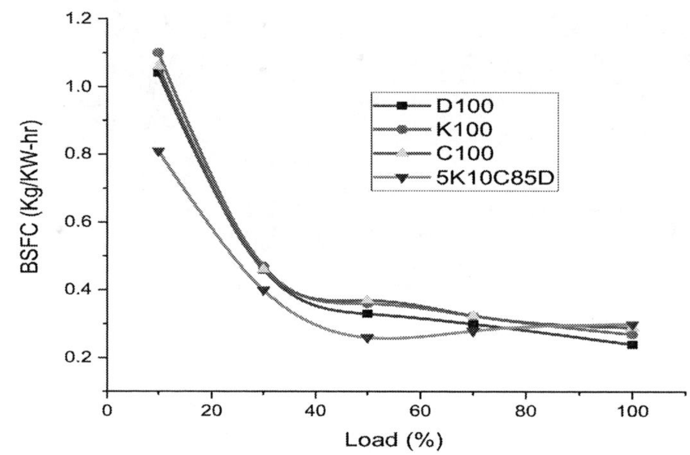

Figure 2. BSFC for 5K10C85D at different load

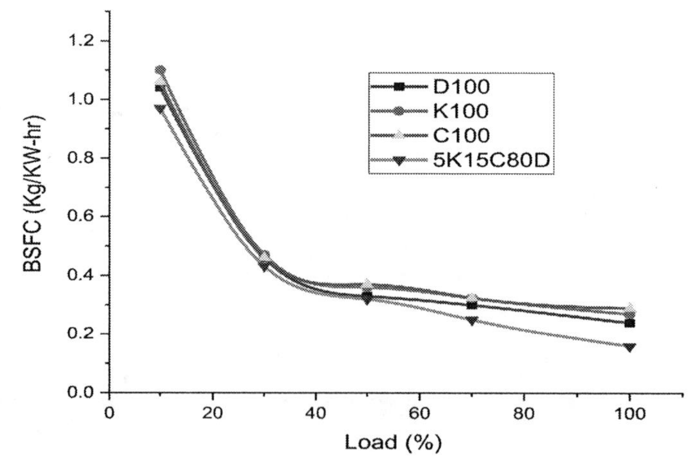

Figure 3. BSFC for 5K15C80D at different load

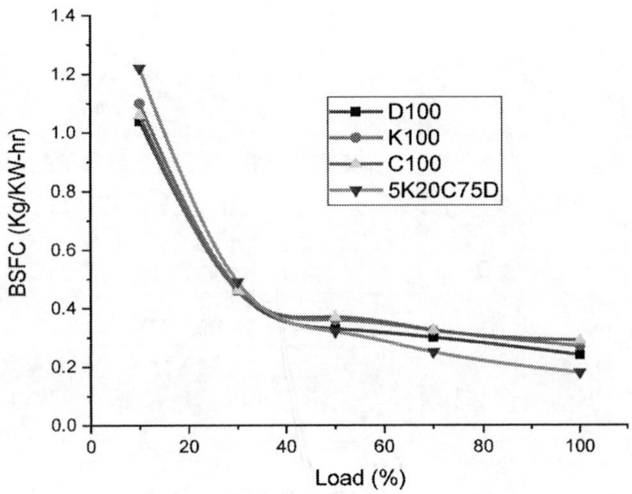

Figure 4. BSFC for 5K20C75D at different load

5. 1. 2 Conversion efficiency (BTE)

For all fuel combinations used in the test, BTE increases with load as seen from the Figure 5, 6 and 7. All the fuel blends exabits higher BTE then the mineral diesel. It is observed that the blend 5H20C75D generates the peak BTE of 37% at peak load which is greater than the BTE of the mineral diesel.

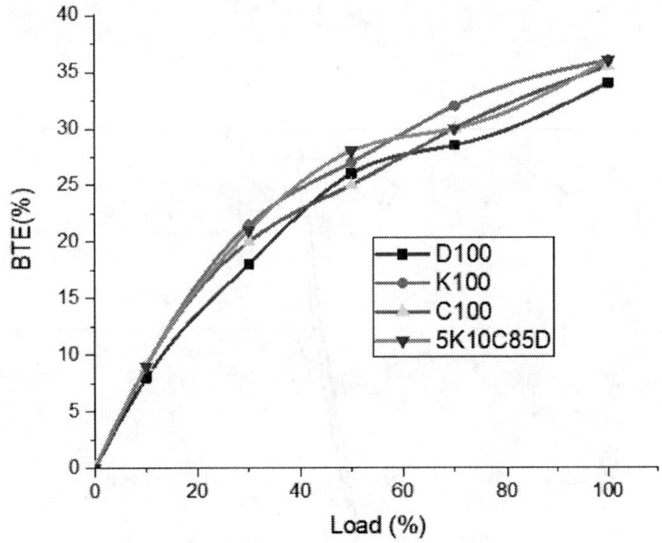

Figure 5. BTE for 5K10C85D at different load

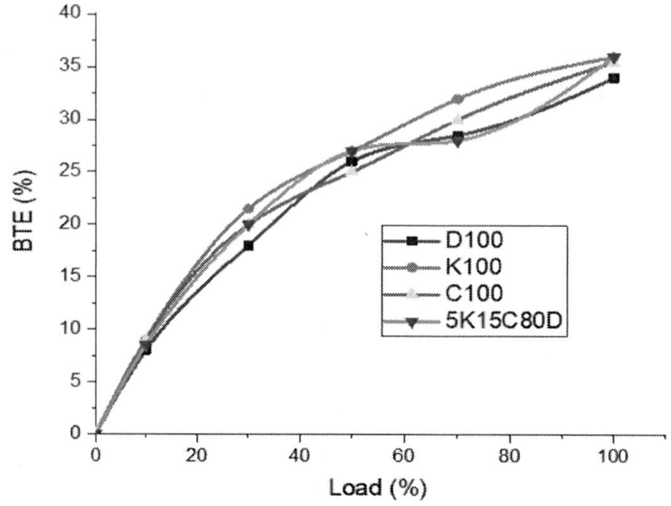

Figure 6. BTE for 5K15C80D at different load

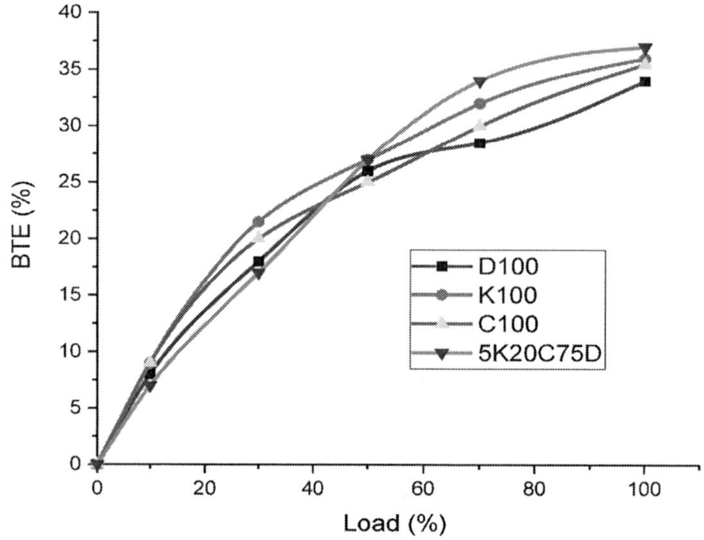

Figure 7. BTE for 5K20C75D at different load

5.2 Trade-off analysis
Trade off analysis involving NOx emission, BSFC and BTE has been carried at 50% and full load which has been presented in the figure 8 and 9. NOx emission for biodiesel operation found to be low as compared to the diesel operations. Same decrease in the trend was noticed by the Hao [16], attributed to the fact that the higher density imparted the negative effect on the spray pattern which results in the lower chamber temperature intern reduces the NOx at medium loads. The study mainly deals with determine the best possible fuel combination at 50% and full load which produces the less NOx with

optimum BSFC and BTE. The fuel blend 5K20C75D was the better fuel combination which produces less NOx compared to the other two blends with optimum BTE and BSFC for both 50% and full load.

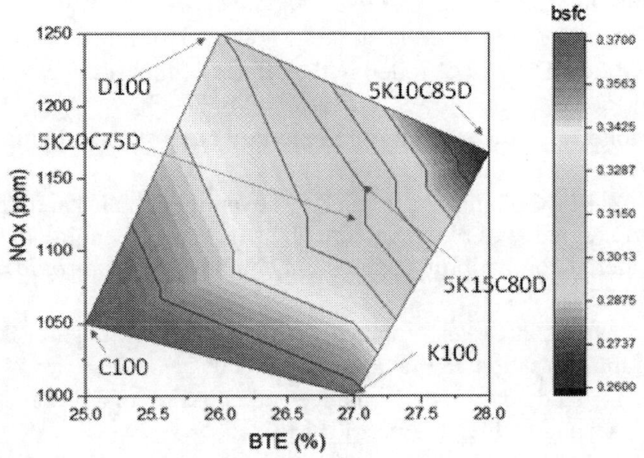

Figure 8. NOx-BTE -BSFC trade-off at 50% load

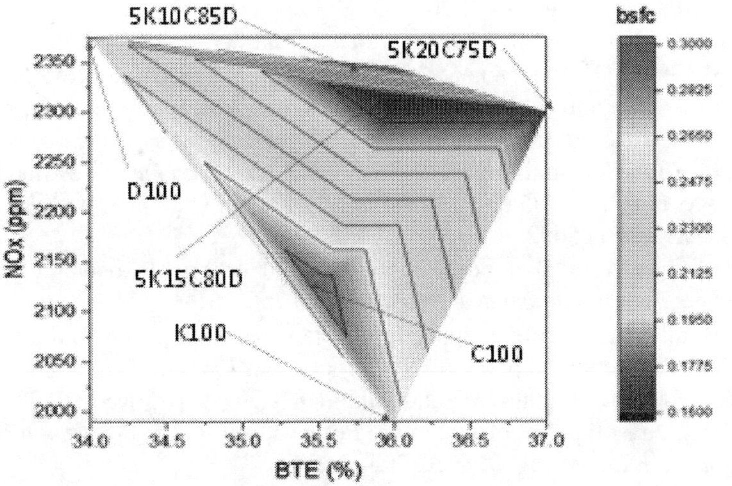

Figure 9. NOx-BTE -BSFC trade-off at full load

6. Conclusion

The motivation behind the current test study was to examines the performances and emissions of Karanja and cotton seed biodiesel combinations with mineral diesel. The subsequent derivations are evolved from the test study results:

- The BTE of all the fuel combinations are improved with respect to mineral diesel. BSFC is lesser for all testing fuel combinations then that of the diesel,100% Karanja and cotton seed biodiesels at all load conditions
- In trade-off investigation it has been detected that at 50 % and full load condition 5K20C75D is the optimal fuel combination for the less NOx emission with the optimum BTE and BSFC.

Reference

[1] John BH 2011 *Internal Combustion Engine Fundamentals.* McGraw Hill Education Pvt Ltd
[2] Ganesan V 2012 *Internal combustion engines*. McGraw Hill Education (India) Pvt Ltd
[3] Patterson MA and Rolf DR 1998 Modeling the effects of fuel spray characteristics on diesel engine combustion and emission *SAE transactions* 27-43
[4] Janaun, J and Naoko E 2010 Perspectives on biodiesel as a sustainable fuel *Renewable and Sustainable Energy Reviews* **14** 1312-20
[5] Shahir SA, Masjuki HH, Kalam MA, Imran A, Rizwanul Fattah IM and Sanjid A 2014 Feasibility of diesel–biodiesel–ethanol/bioethanol blend as existing CI engine fuel: An assessment of properties, material compatibility, safety and combustion *Renewable and Sustainable Energy Reviews* **32** 379-395
[6] Dhar, A and Avinash KA 2014 Performance, emissions and combustion characteristics of Karanja biodiesel in a transportation engine *Fuel* **119** 70-80.
[7] Chauhan BS 2013 A study on the performance and emission of a diesel engine fuelled with Karanja biodiesel and its blends *Energy* ***56*** 1-7
[8] Shrivastava P, Verma TN and Pugazhendhi A 2019 An experimental evaluation of engine performance and emisssion characteristics of CI engine operated with Roselle and Karanja biodiesel *Fuel* **254** 115652
[9] Nabi N, Mustafizur R and Akhter S 2009 Biodiesel from cotton seed oil and its effect on engine performance and exhaust emissions *Applied thermal engineering* **29** 2265-70
[10] Karabektas M, Ergen G and Hosoz M 2008 The effects of preheated cottonseed oil methyl ester on the performance and exhaust emissions of a diesel engine *Applied Thermal Engineering* **28** 2136-43
[11] Shrivastava P, Verma TN and Pugazhendhi A 2019 An experimental evaluation of engine performance and emisssion characteristics of CI engine operated with Roselle and Karanja biodiesel *Fuel* **254** 115652.
[12] Prem K, Pal Sharma M and Dwivedi G 2016 Impact of ternary blends of biodiesel on diesel engine performance *Egyptian Journal of Petroleum* **25** 255-261
[13] Appavu P, Madhavan VR, Venu H and Jayaraman J 2019 Experimental investigation of an unmodified diesel engine operated with ternary fuel *Biofuels* 1-7
[14] Baroutian S, Aroua MK Abdul AAR and Sulaiman NMN 2010 Viscosities and densities of binary and ternary blends of palm oil+ palm biodiesel+diesel fuel at different temperatures *Journal of Chemical & Engineering Data* **55** 504-507
[15] Mishra S, Chauhan A, and Bhushan K 2020 Role of binary and ternary blends of WCO biodiesel on emission reduction in diesel engine *Fuel* **262** 116604
[16] Chen H, Xie B, Ma J and ChenY. NOx emission of biodiesel compared to diesel: Higher or lower *Applied Thermal Engineering* **137** 584-593
[17] Magesh P and Ramesh GP 2017 Fuzzy logic control implementation of ultra-sparse matrix converter for renewable energy applications *In Proc. Int. Conf. Inf. Commun. Embedded Syst. (ICICES)*, 1-3.

CFD Simulation of Producer Gas Fuelled SI Engine

Amit Kumar[1], Rajakumar D G[1], Mownesh G K[1] and Basavarajappa[1]

[1]Department of Mechanical Engineering, GM Institute of Technology Davangere, India.

Abstract. Syngas generated from the thermochemical conversion of biomass in a gasifier is increasingly being used for fuelling internal combustion engines, especially for distributed power generation. Producer gas as a fuel has thermo physical properties significantly different from those of conventional fuels. As of date, there are no dedicated engines for alternative fuels in general and producer gas in particular. A review of the available literature indicates experimental experience with producer gas but very little information on modelling and simulation studies are available. While experimental investigations provide actual performance parameters, the information is essentially spatial or temporal average only. Apart from that, fluid dynamic and combustion progress parameters cannot be acquired inexpensively. The current work primarily addresses the combustion progression parameters in a typical spark ignited engine under naturally aspirated and turbocharged after cooled configuration at varying mixture quality using the ANSYS FLUENT computational fluid dynamics software. The CHEMKIN software is used for determining the laminar flame speed for producer gas as a function of the mixture quality. Spatially averaged pressure traces from the simulation are compared with literature pressure traces at the desired condition and various simulation parameters are tuned till a match between the literature and simulation pressure is obtained. Once the simulation is validated, the progress of combustion parameters is derived from the simulation in post processing.

Keywords: Producer gas, Laminar flame speed, CHEMKIN, IC engine, CFD simulation.

1. Introduction

In the current scenario, all the energy conversation devices have been fuelled with hydrocarbon based fuels formed over millions of years. Country like India which completely depends on the import of the hydrocarbon based fuels. According to the estimates oil peak have been already been surpassed, in feature oil importing countries like India are expected to worst hit due to the major dependence on the import of the hydrocarbon based fuel. Increase in the hydrocarbon based fuels is in one of the major reason for atmospheric pollution and the drastic changes in the weather pattern are being attributed to the large amounts of CO_2 released by the combustion of hydrocarbon fuels [1]. Currently to address the above mention twin challenges various studies have been carried out by research teams across the globe. Renewable energy sources will play an important role in address the issues related to the fossil fuels. Renewable energy has the potential to provide zero emission of pollutants, also it can address the meet the small scale and rural needs like decentralized power system in a reliable and environmentally sustainable way. Is as been estimated that by 2040 renewable sources will contribute 47.7% of the total global energy requirement [2] [3].

Biomass is one of the alternative fuel to replace the conventional hydrocarbon based fuels and also addressed environmental concerns the fact that with bio derived fuels retention time of CO_2 in the atmosphere is much lower than the time required for CO_2 to start contributing to global warming. Biomass can be broadly based on the moisture and lignin content, those with less lignin (less than 10%) and high moisture beyond 70% can be used for bacterial conversion processes – the processes of biological conversion root when largely yields in the gas consisting of CH_4 and CO_2. Biomass with high lignin (more than 10%) with reasonable less moisture (less than 30%) can be used for thermo-chemical conversion processes which yields gas known as synthetic gas or commonly known as syngas. Syngas basically consists of H_2 and CO and some trace of methane as combustible species [4] [5].

Producer gas (PG), a bio-derivative alternative gaseous fuel generated from thermo-chemical conversion of biomass. Producer gas typical composition of 20% CO and H_2, 2% CH_4 and balance incombustibles has thermo-physical properties significantly different from standard gaseous fuels as highlighted in table 1. The comparison of the PG with CNG is more vital because CNG is considered as one of the standard gas fuels with regards to commercial availability of thermal conversion devices like IC engine, industrial burners etc., which can be easily be modified for the other gas fuel operations [6]. The difference in the thermo-physical properties would potentially require study of certain fuel specific features. The air fuel ratio of the PG gas is about 1.30 compared to that of Natural gas (CNG) is 17.0 on mass bases which differ in terms of magnitude when compared to CNG. In the same line PG calorific value also varies in the ratio of 1:10 when compared to that of CNG gas. Laminar flame speed for PG is 20% higher compared to that of the CNG. Even though the calorific value of the PG is very less when compared to CNG the mixture caloric value is less by 23%. The adiabatic flame temperature (AFT) are also lower for PG when compared to CNG and Hydrogen [7].

Table 1. Thermo-physical properties of some fuels

Fuel Property	H_2.	CNG	Bio-gas	PG
Air to Fuel (kg/kg)	34.4	17.0	10.5	1.30
Calorific Value (MJ/kg) at ϕ =1	121.0	50.0	23.6	5.0
Laminar flame speed (m/s) at ϕ =1	2.70	0.40	0.26	0.50
Mixture LCV (MJ/kg) at ϕ =1	3.41	2.78	2.30	2.17
Adiabatic Flame Temperature (K) at ϕ =1	2360	2214	2102	1873

Literature survey in the field of producer gas can classified into generation and its application. Generation of producer gas has been reported since World War II, European nations which as exploited the gasification technology during the World War II because of the petroleum oil crisis. National Swedish Testing Institute of Agriculture Machinery, Sweden has reported work on the design and development of closed top gasification systems [8]. In the recant time Indian institute of Science, Bangalore has worked on open top downdraft gasification technology, which is able to generate the engine quality producer gas with low contaminates and also gasifier is able to handle any solid biomass with moisture content less than 20% [9]. In the contest of producer gas application, Dasappa et al have reported major factors influencing the power output of the diesel frame engine converted to operate on producer gas, also presented some actual field studies [10]. Sridhar et al as reported experimental study of SI engine operated for high possible compression ratio to establish the engine performance, he has also reported the performance analysis of small and medium scale engine derived from diesel frame converted for producer gas operation by replacing spark plug in the place of fuel injector [9]. Shivapuji et al reported the experimental and quasi dimensional analysis of multi-cylinder SI engine operated on producer gas operated in both natural aspirated and turbocharge mode to establish the peak load operation [11]. Biomass to electricity efficiency has been reported to be at 18% for naturally aspirated mode which increases to 22% under turbocharged mode of operation. Emissions under both the modes of operation are reported to be well within the international permissible limits [12] [13]. Yarasu et al has reported on the turbulent combustion of producer gas in closed vessels and engine cylinders. The main objective the work has been multi-dimensional simulation of turbulent combustion in the bowl-in-piston engine operating on producer gas fuel and to observe the fame and flow field interaction [14]. Ulugbek Azimov et al has reported on a CFD simulation of syngas fuelled turbocharged duel fuel system. The modelled results were validated by comparing predictions against corresponding experimental data for a supercharged dual-fuel engine [15].

A review of the available literature on experimental and simulation studies indicates that which significant amount of experimental work has been done covering a broad range of aspects dealing with producer gas fuelled operation, the modelling and simulation has largely been restricted to zero and quasi dimensional studies. No significant work is reported on full 2D computational fluid dynamics simulation. The motivation for the current work stems from this lack of availability of information in the 2D CFD simulation area which is sought to be addressed over the full cycle of a four stroke cycle on fuelling with producer gas. the scope of the current work spans obtaining the geometric specifications of the engine of interest, modelling the 2D geometry, deriving the 2D mesh using appropriate scheme, arriving at the initial and boundary conditions, establishing of laminar flame speed correlations as a function of engine operating parameters; primarily the equivalence ratio and finally the full simulation of the engine for peak load conditions. The simulation results are validated using literature data from reported on the in-cylinder investigation of a multi-cylinder engine fuelled with producer gas by Shivapuji et al. Post the validation, spatial analysis of various parameters of interest related progress of combustion are extracted from the simulation in post processing exercise and reported.

2. Multi-dimensional model formulation

Conservation equation formulation

The modelling approach in based on the unsteady Reynolds Average Navier-Stokes (RANS) to solve the mass and momentum conservation equations. Mass and momentum equation are solved together for the proper calculation of pressure gradient in the momentum equation. Also energy conservation equation is solving to account of the energy conservation in the system. The compressible mass and momentum equation are given as follows,

$$\frac{\partial \rho}{\partial t} + \frac{\partial \rho u_i}{\partial x_i} = S \tag{1}$$

$$\frac{\partial \rho u_i}{\partial t} + \frac{\partial \rho u_i u_j}{\partial x_j} = \frac{\partial P}{\partial x_i} + \frac{\partial \sigma_{ij}}{\partial x_j} + S_i \tag{2}$$

In the above equation, ρ is fluid density term, u is velocity term, P is pressure term, σ_{ij} is viscous tress tensor, and S is the source term. Redlich-Kwong equation is used to couple density pressure and temperature terms for solving compressible mass and momentum equation [16].

$$\frac{\partial \rho e}{\partial t} + \frac{\partial u_i \rho e}{\partial x_j} = -\text{P}\frac{\partial u_j}{\partial x_j} + \sigma_{ij}\frac{\partial u_i}{\partial x_j} + \frac{\partial}{\partial x_j}(K\frac{\partial T}{\partial x_j}) + \frac{\partial}{\partial x_j}(\rho D \textstyle\sum_m h_m \frac{\partial Y_m}{\partial x_j}) + S \tag{3}$$

Where, ρ is density, P is the pressure, D is the mass diffusion, Y_m is the mass fraction of species m, coefficient, e is the specific internal energy, K is the conductivity, hm is the species enthalpy, σ_{ij} is the stress tensor, S is the source term and T is temperature, Transport equation describing conservation of chemical species consisting of convection, diffusion and chemical reaction source for each component of species is solved [16], Where c is the chemical concentration of the species, D is the diffusion co-efficient and S is source term.

$$\frac{\partial c}{\partial t} + \frac{\partial u_i c}{\partial x_j} = \frac{\partial}{\partial x_i}\left(D\frac{\partial c}{\partial x_i}\right) + S \tag{4}$$

Combustion formulation

The flame may be freely propagating as it when flame is initiated in a tube containing combustible gas mixture. Laminar flame speed is function of pressure and temperature for given equivalence ratio, the

IOP Conf. Series: Materials Science and Engineering **925** (2020) 012058 doi:10.1088/1757-899X/925/1/012058

flame approaches the unburned mixture at laminar speed, S_L. The unburned gas enters the flame in a direction normal to the flame sheet. Since the heats the products, the burnt density is less than the unburned density. Continuity thus requires that the burned gas velocity is higher than the unburned gas. The flame consists of two regions they are the preheat zone and reaction zone In preheat zone where little heat is released, the temperature of unburned gas is raised mainly due to the conduction from the reaction zone and the reaction zone where the bulk of the chemical energy is released in the form of heat results in the zone. The region between the temperature where exothermic chemical reaction begins and the hot boundary at the downstream equilibrium burned gas temperature is called the reaction zone. The thickness of preheat zone can be calculated for one dimensional flames from conservation equations of mass and energy [16]. The thickness of the preheat zone δ_L is

$$\delta_L=\left[\frac{4.6k}{c_p\rho_u S_L}\right] \tag{5}$$

The laminar flame speed at pressure and temperature typical of unburned mixture in engines are usually measured in spherical closed vessels by propagating a laminar flame radially outside from vessel centre [16]. The laminar speed is then given by

$$S_L=\left[\frac{dm_b}{A_f\rho_u}\right] \tag{6}$$

Where the mass rate is determined from the rate of pressure rise in the vessel and the A_f is the flame area.

Metghalchi and Keck experimentally determined the laminar flame speed for various fuels over a range of temperature and pressure typical conditions associated with the internal combustion engines. The laminar flame speed for different fuel-air mixture have fitted by simple power law [17].

$$S_L=S_{L,ref}\left(\frac{T_u}{T_{u,ref}}\right)^{\alpha}*\left(\frac{P}{P_{ref}}\right)^{\beta}*(1\text{-}2.1*Ydil) \tag{7}$$

Where $T_{u,ref}$= 298K and P_{ref} = 1atm are the reference temperature and pressure, $S_{L,ref}$, α and β are the constants for given fuel, equivalence ratio, and burned gas diluent fraction the term Y_{dil} is the mass fraction of diluent present in the air-fuel mixture [12][13]. For Methane, propane, isooctane and methanol these constants can be represented as follows,

$$S_{L,ref}= B_M+B_2(Fr-FM)^2 \tag{8}$$

Where B_M, B_2 and FM are constants depending upon the fuel type. The temperature and pressure depends on the value of α and β which are the function of equivalence ratio (Fr), expressed as follows

$$\alpha=2.18-0.8(F_r-1) \tag{9}$$

$$\beta=-0.16+0.22(F_r-1) \tag{10}$$

The reference laminar flame speed is calculated using the equation 8, for different equivalence ratio α and β values are calculated using equation 9 and 10. By knowing $S_{L,ref}$, initial pressure and temperature Laminar flame speed (S_L) can he calculated from equation (8). Analysis by Damkohler also known as the Damkohler paradigm, puts the turbulent combustion within a spark ignited (SI) engine in the wrinkled laminar flame (flamelet) regime. In this regime, the cylinder content is divided into burned and unburned regions separated by an infinitesimally thin but highly wrinkled reacting flame sheet. This thin sheet is basically the laminar flame sheet with flame thickness (δ_L) less than the regime for the Kolmogorov micro-scale. In the flamelet regime, a reaction progress variable 'c' having a value between 0 (unburned mixture) and 1 (burned gas) describes the progress of combustion and is described as in equation 11 where 'n' is the number of products, Yi is the mass fraction and $Y_{i,eq}$ is the equilibrium mass fraction of the ith specie [18].

$$c = \frac{\sum_{i=1}^{n} Y_i}{\sum_{i=1}^{n} Y_{i,eq}} \tag{11}$$

In ANYSY FLUENT, the flame front propagation is modelled by solving the transport equation where S_{c_t} and S_c represents the turbulent Schmidt number and the reaction progress source term respectively as presented in equation 12.

$$\frac{\delta(\rho c)}{\delta t} + \nabla.(\rho \bar{v} c) = \nabla.\left(\frac{\mu_t}{S_{c_t}} \nabla c\right) + \rho S_c \tag{12}$$

The mean reaction rate in the above equation 'ρS_c' is modelled along the Zimont model and is given by equation 13 where ρ_u and U_t stand for the density of the unburned mixture and the turbulent flame speed respectively.

$$\rho S_c = \rho_u U_t |\nabla c| \tag{13}$$

The turbulent flame speed U_t is modelled considering the highly wrinkled and thickened front as in equation 14 below where, $u'$the root mean square velocity, $U_l$the laminar flame speed, $\alpha$the thermal diffusivity and l_t the turbulent scale [18].

$$U_t = A(u')^{3/4} U_l^{1/2} \alpha^{-1/4} l_t^{1/4} \tag{14}$$

3. Simulation methodology

Initially laminar flame speed was calculated using the CHEMKIN simulation software for different initial pressure and temperature and by least square fits the constants were determined for the equivalence ratio when are necessary inputs for the engine combustion simulation in ANSYS-Fluent. The determined laminar flamed speed values were compared against the available literature data. Once the laminar flame speed constants values are in expectable error band width i.e is within 2 to 3% the values α and β were used in the engine combustion simulation.

Main prerequisite for CFD simulation is the geometry modelling and discretization of the fluid domain. In the current work engine geometry were modelled from the literature report by Shivapuji et al [19]. 2D model was created consisting of intake, exhaust and piston geometry was modelled and meshed in ICEM-CFD software. 2D simulation was chosen because it is of simple geometry, and take less time take for the simulation when compared with the time taken for 3D simulation, but it fails to capture the turbulence parameters. Once the mesh is created it is imported to ANSYS-FLUENT wherein the dynamic mesh was created i.e. moving boundary was created for intake, exhaust and piston parts and respective profile were assigned. After the creation of dynamic mesh the mesh is previewed to verify the mesh proper moment according to the profile assigned to different regions. The simulation initiated by assigning initial values for the variable, namely initial pressure, temperature. The fuel properties such as laminar flame speed un-burnt fuel mass fraction, lower calorific value of fuel and spark timing are also assigned.

Simulation was carried out initially to validate the literature data and then parametric of load variation was carried out for natural aspirated and turbocharged mode conditions. In all the cases the laminar flame speed coefficients were same, laminar flame speed coefficients were defined using User defined functions (UDF) in ANSYS-FLUENT interface. Unburnt fuel fractions were calculated based on the operating equivalence ratio and defined in the ANSYS-FLUENT. Initial pressure, temperature were defined based on the simulation load conditions.

4. Results and Discussion

Laminar flame speed coefficients calculation

CHEMKIN results for estimation of laminar flame speed for different initial pressure and temperature and by least square fits the constants were determined for the equivalence ratio of the range 0.8 to 2.01.4,

Fig. 1 show the plot of laminar flame speed of producer gas determine using CHEMKIN for different equivalence ration (Φ) for atmospheric conditions at 1 bar and 300 K, engine like conditions at 25.4 bar and 716 K. S_L_Tinaut are the values from the literature [20] and S_L_Chemkin are the values obtained from the CHEMKIN software. Laminar flame speed is high at the rich conditions.

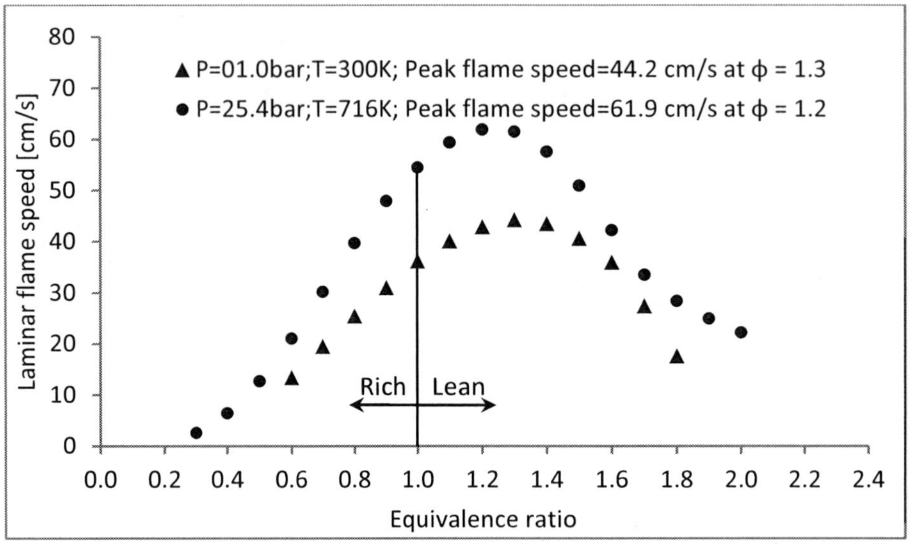

Figure 1. Laminar flame speed vs. equivalence ratio (Φ).

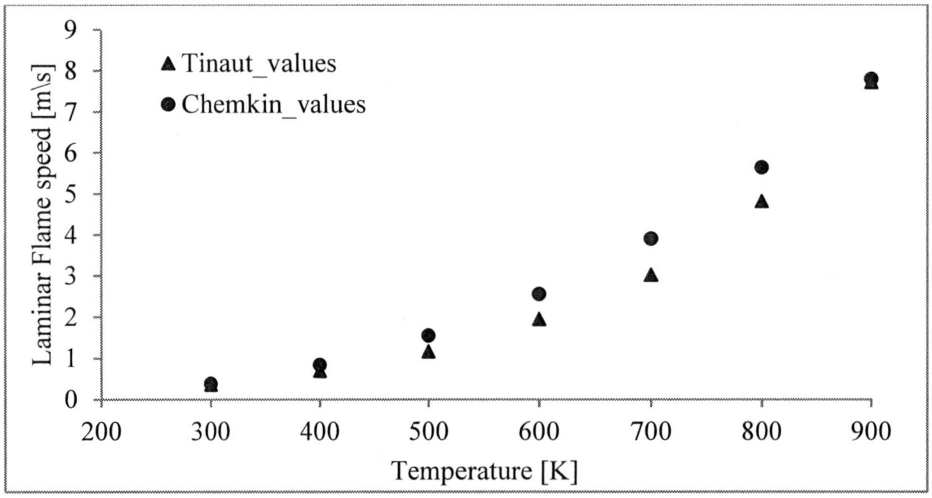

Figure 2. Laminar flame speed for different inlet temperature at inlet pressure 1 bar.

The variation of laminar flame speed for varies inlet temperature by maintaining the inlet pressure constant were calculated. It was observe the laminar flame speed at low pressure and high temperature the laminar flame speed is high. But the laminar flame speed goes on decreasing as one increases the inlet pressure and temperature. It can be observed that the laminar flame speed at 10bar and 900K inlet condition the laminar flame speed is about 2.2 m/sec. Fig.2and 3 show the plot of laminar flame speed for different inlet temperature at inlet pressure conditions 1 and 10 bar respectively and the obtained values are compared with the literature (Tinauts) values.

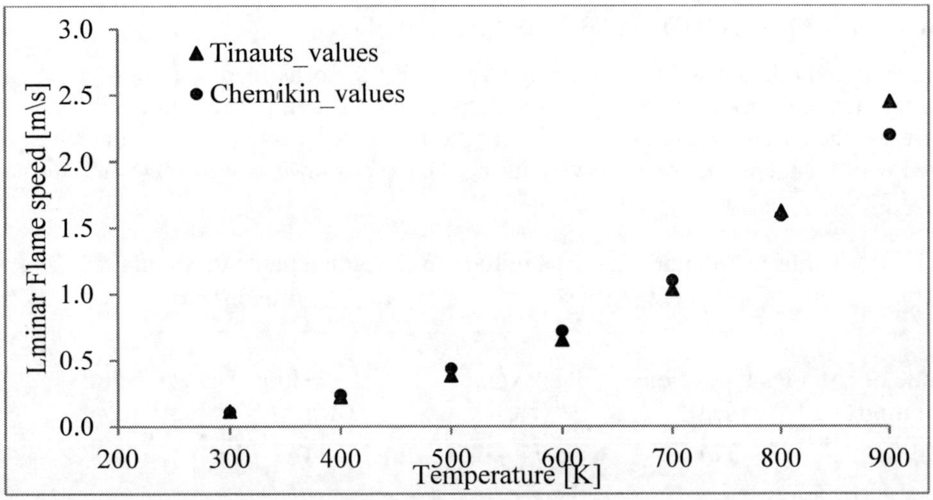

Figure 3. Laminar flame speed for different inlet temperature at inlet pressure 10bar.

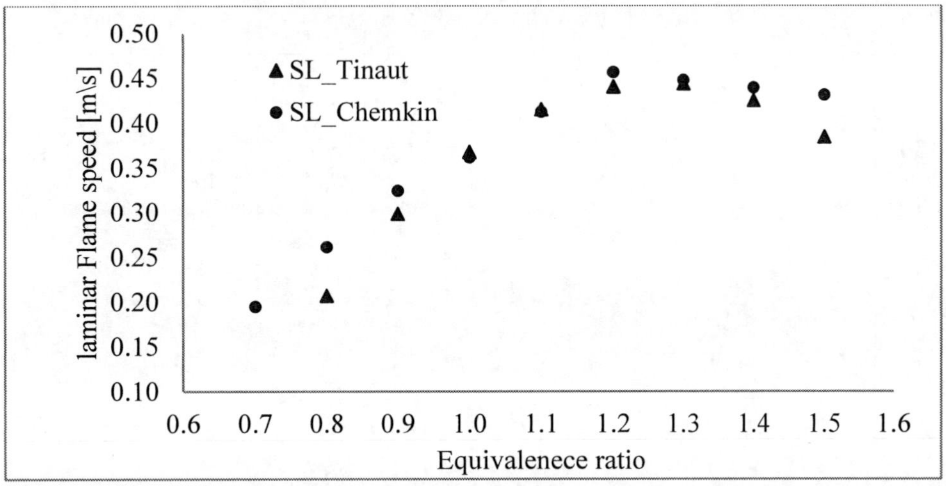

Figure 4. Comparison between the CHEMKIN and literature values for different equivalence ratio.

Fig.4 gives the comparison of laminar flame speed at different equivalence ratio for the values obtained from CHEMKIN and literature (Tinauts) values. In Table - 2 shows the constants values obtained from least square fits methods the first row represent the values obtained from the CHEMKIN, and second row represents the from the literature.

Table 2. The details of the constants determined by using CHEMKIN and literature

Fuel	FM	BM	B2	α	β	S_L,ref (m/sec)
Producer Gas (CHEMKIN)	1.365	48.00e-02	-150.00e-2	2.18	-0.56	0.43
Producer Gas (F.V.Tinaut)	1.186	50.06e-2	-82.70e-2	2.0	-0.4	0.47

IOP Conf. Series: Materials Science and Engineering **925** (2020) 012058 doi:10.1088/1757-899X/925/1/012058

Validation of 2D CFD simulation result with literature data

2D simulation in-cylinder pressure – crank angle results for the peak load operation of the engine under naturally aspirated and turbocharged operation respectively. Literature reported was the experimental results represent the ensemble average of 250 consecutive cycles, near complete match of the traces can be observed with a small deviation observed during the expansion stroke for 2D simulation.

Table 3. Comparison of magnitude for Experimental and Simulation

	Literature		**Simulation**		Difference
Mode of operation	Peak position (deg)	Peak value (bar)	Peak position (deg)	Peak value s(bar)	Peak value (%)
TA	374	74.4	374	76.1	2.25
NA	375	34.4	374	34.0	1.17

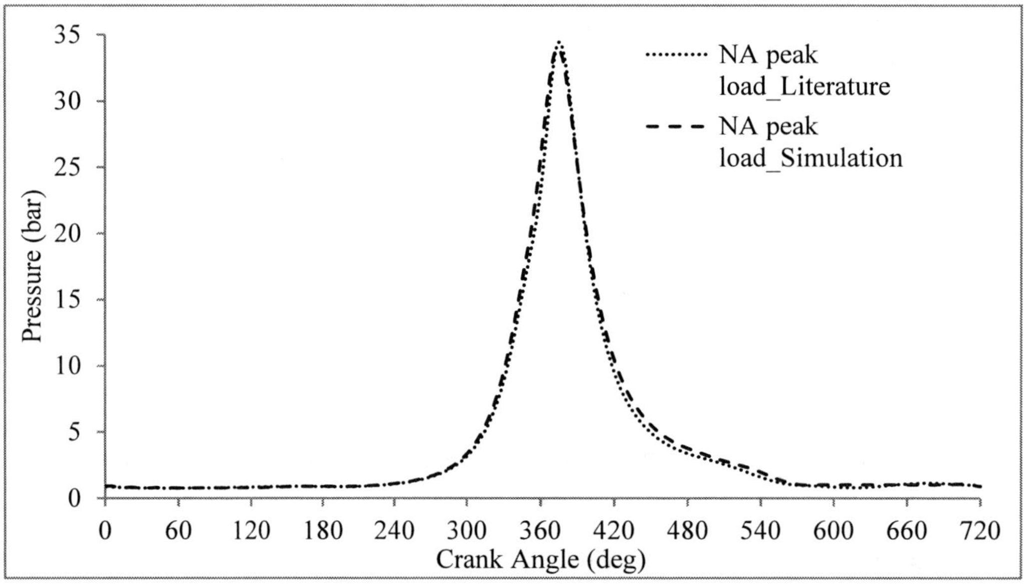

Figure 5. Plot of Pressure-Crank angle for Literature and simulation values for NA mode.

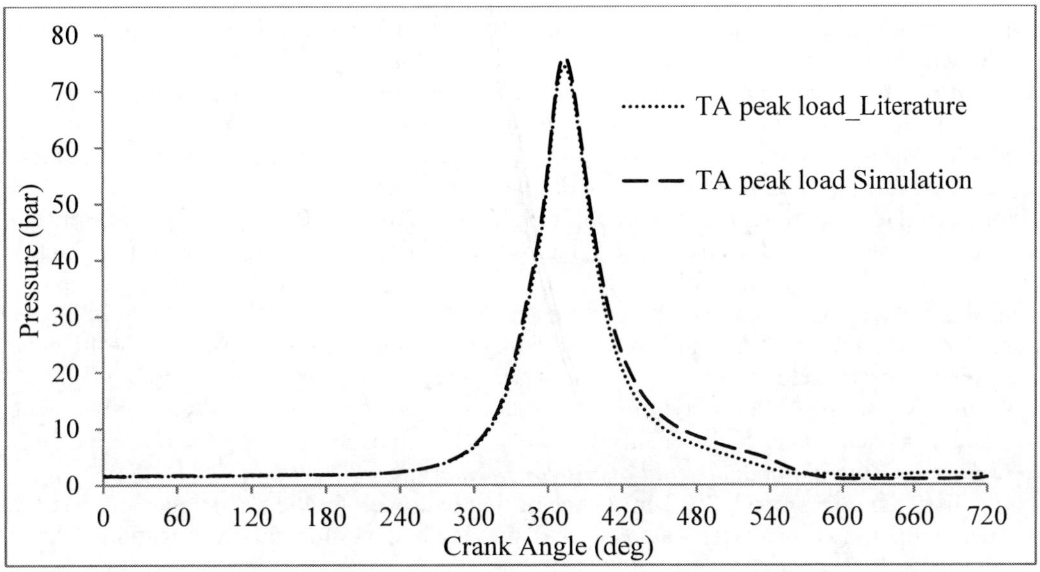

Figure 6. Plot of Pressure-Crank angle for Literature and simulation values for TA mode.

The outcome of the experimental and 2D simulation results are consolidated in Table – 3, where the position and magnitude of peak pressure are compared for the literature and simulation results. It can be observed that the deviation in the peak pressure at is at around 2.25% for TA and 1.17% NA mode of operation respectively. The position of peak pressure differs by 1 degree which is well within the cyclic variability limits of engine operation.

5. Conclusion

In the current work, Laminar flame speed calculations using CHEMKIN package at various equivalence ratio, pressure and temperature are used to establish the expression for laminar flame speed as a function of equivalence ratio, pressure ratio and temperature ratio. Using the established expression for laminar flame speed, 2D simulations were developed for computational fluid dynamics model using the initial conditions and boundary conditions from Literature. Comparing literature and simulation cylinder average pressure indicate a very close match validating the computational fluid dynamics simulation.
The simulations have been carried out considering the in-cylinder mixture as an ideal gas and in place of the chemical kinetics established expression for laminar flame speed was used as user defined functions. Overall 2D simulation were carried and achieved good match with the literature data within expectable error range of 2.5%.

Reference

[1] Panwar NL, Kaushik SC and Surendra K 2011 Role of renewable energy sources in environmental protection: A review *Renewable and Sustainable Energy Reviews* **15** 1513-1524
[2] Copa R, José T, Silveira C, Mancebo J, Brito RP, Valter S and João C 2020 Techno-Economic Assessment of the Use of Syngas Generated from Biomass to Feed an Internal Combustion Engine *Energies* **13** 10.3390
[3] Richard B and Klaus D 2017 Syngas Use in Internal Combustion Engines - A Review *Advances in Research* **10** 1-8
[4] Daniele P, Michela C, Zvonimir P, Milan V, Villetta L, Caputo M, and Domenico C 2018 CFD Modelling of a Spark Ignition *Internal Combustion Engine Fuelled with Syngas for a mCHP System*
[5] Mukunda HS 2011 Text book on Understanding Clean Energy and Fuels from Biomass
[6] Gururaja R, Sridhar HV, Sridhar S, Paul D, Rajan NK and Hanasoge M 2005 Development of

producer gas engines. Proceedings of The Institution of Mechanical Engineers *journal of Automobile Engineering - Proc Inst Mech Eng D-J Auto* **219** 423-438
[7] Anand MS 2009 Experimental studies on multi-cylinder natural gas engine fueled with producer gas, Selected for presentation,*19th European Biomass Conference and Exhibition, Berlin, Germany.*
[8] Anon 1986 Wood gas as engine fuel, A report of mechanical wood products branch od's FAO paper *72, Food and agriculture organization of united National*
[9] Sridhar GR 2003 Experiments and Modelling Studies of Producer Gas based Spark-Ignited Reciprocating Engines, Department of Aerospace Engineering, Indian Institute of Science,
[10] Dasappa S 2003 The estimation of power from a diesel engine converted for gas operation – a simple analysis, ASTRA, Indian Institute of Science
[11] Anand MS 2012 Experiments and zero D modeling studies using specific Wiebe coefficients for producer gas as fuel in spark-ignited engines *Journal of Mechanical Engineering Science*
[12] Shivapuji AM Dasappa S 2014 In-cylinder investigations and analysis of a SI gas engine fuelled with H2 and CO rich syngas fuel: Sensitivity analysis of combustion descriptors for engine diagnostics and control *International Journal of Hydrogen Energy*
[13] Shivapuji AM and Dasappa S 2015 Influence of fuel hydrogen fraction on syngas fueled SI engine: Fuel thermo-physical property analysis and in-cylinder experimental investigations *International Journal of Hydrogen Energy* **40** 10308-10328
[14] Ravindra B 2009 Premixed turbulent combustion of producer gas in closed vessel and engine cylinder, Combustion, Gasification and Propulsion Laboratory, Department of Aerospace Engineering, Indian Institute of Science.
[15] UlugbekAzimov (2011), Multidimensional CFD simulation of syngas combustion in a micro-pilot-ignited dual-fuel engine using a constructed chemical kinetics mechanism, Department of Mechanical Engineering, Okayama University, 3-1-1 Tsushima-naka, Kita-ku, Okayama 700-8530, Japan.
[16] Ansys Fluent Theory Guide 2011
[17] Metghalchi M 1977 Laminar burning velocity of isooctane-air, methane-air, and methanol-air mixtures at high temperature and pressure, Massachusetts Institute of Technology, Dept. of Mechanical Engineering,
[18] Chemkin 2012.Theory Manual CK-THE-10113-1209-UG-1, Reaction design.com
[19] Anand MS 2015 doctoral thesis on In-cylinder experimental and modeling studies on producer gas fuelled operation of spark ignited gas engines, Indian Institute of Science Bangalore.
[20] Francisco VT 2006 Method for predicting the performance of an internal combustion engine fuelled by producer gas and other low heating value gases, Department of Energy and Fluid Mechanics Engineering, University of Valladolid, Paseo del Cauce s/n, E-47011 Valladolid, Spain.

Performance and Emission Characteristics of VCR Diesel Engine with Pre Heated Lemon Grass Biodiesel as Fuel

Kolla Kotaiah[1], Periyasamy P[2] and Prabhahar M[3]

[1,2] Department of Mechanical Engineering, St. Peter's Institute of Higher Education and Research, Chennai, India.
[3] Department of Mechanical Engineering, Aarupadai veedu Institute of technology, Chennai, India.
[1]kollakotaiah1@gmail.com

Abstract. Under this experimental analysis the lemon grass oil was considered, it's converted into biodiesel as lemon grass biodiesel methyl ester (LGB) through transesterification process. The Kirloskar TV1 was used as the reference fuel in the fuel injection system 4-stroke diesel engine. The lemon grass biodiesel methyl ester (LGB B20) blend B20 is selected because it nearest to the diesel value. The lemon grass biodiesel is preheated at 60°C temperature that is (LGB B20 preheating @ 60°C). The results of the test showed that LGB B20 kinematic viscosity and density gradually decreased with fuel temperature rise. The performance characteristics of Preheating biodiesel is increased the BTE and slightly reduce the BSFC at noted in the full load condition. And the emission characteristics at full load condition of CO, HC and smoke density as decreased, it was observed slight increase of NOx emission.

Keywords: LGB, Preheating, VCR engine, performance and emission

1. Introduction

The fossil-fuel vehicle population increases annually, leading to high emission rates. Fossil fuels are a threat in the long run because of their inconsistencies in demand and supply and their fluctuating fuel prices and their pollution problems. Similar physical, molecular and thermal features are contained in the edible and inedible oil near the diesel engine. However, the diesel engine is not used specifically because of its high viscosity in edible and inalienable oils. The preheating biodiesel data was referred Abdulfatah Abdu Yusuf et.al. [1] A preheated gasoline is processed at various temperatures (65, 75 & 80 ° C) in order to achieve better performance and pollution characteristics. Fish Oil Ethyl Ester (FOEE) tested in diesel without altering the engine setup. There are numerous attempts to examine the characteristics of a preheated biodiesel from various vegetable oils that are divided into edible and non-edible oils. V. Rambabu et.al [2] & Said M. A. Ibrahim et.al [3] reported that the preheated biodiesel which leads to higher brake thermal efficiency and slightly decreased in BSFC of performance declared, Co & HC pollution characteristics have reduced and improved significantly in the pollution of NOx. The Cottonseed Methyl Ester (CSME) & jatropha oil as used for this experiment. The inbuilt biodiesel oxygen lowers the adiabatic flame temperature, resulting in a linear reduction in NOx emission. M. Saravanakumar et.al. [5] & Krishnamoorthi S et.al [9] reported that potassium hydroxide (KOH) catalyst and cooking oil biodiesel were examined for emission characteristics. Biodiesel viscosity obviously exceeds that of diesel, especially at low temperatures. The biodiesel relative gravity is around 6.1 per cent more than that of diesel. Dinesh P et.al. & Prabhahar M et.al [4, 10] investigates the efficiency & pollution characteristics at B40 at 60, 75, 90 and 110o C preheated pongamia ester & Jatropha Biodiesel and B40 mixtures, Jatropha Biodiesel and its blends with Ethanol were used. The preheating of high-temperature biodiesel mixture thus increases viscosity & other characteristics dramatically and enhances performance and pollution. Prabhahar M et.al [6] reported the dual biodiesel (pine & Palm). [7, 8] Vegetable oil for the elimination of toxic emissions, as tested by direct injection diesel engine the oils consist primarily of triglycerides (90 to 98 per cent) and minor quantities of mono- and di-glycerides. Triglycerides contain three groups of fatty acids and one of glycerol. Methyl ester Vegetable Oils for better results for thermal efficiency and diesel fuel consumption have been recorded [9, 11] for potassium hydroxide (KOH), sodium hydroxide, and dimethoxy-methane (C3 H8O2) catalytic ester. Edible oil, for comparative efficiency and emission studies, like vegetable oil, Eucalyptus oil, pumpkin seed oil biodiesel pre-heated gasoline and gasoline free. [12, 14]. Amir Khalid et.al [13] was studied the comparison of preheats fuel characteristics of biodiesel and straight vegetable oil, Applied Mechanics and Materials. Mixing vegetable oil with diesel decreases the viscosity and increases stability. This enhanced property lead to better mixture formation and penetration of the spray. A variety of researchers tested the vegetable oils with diesel in different amounts.

Prakash et.al [14] studied the efficiency and pollution profiles of biodiesel engine comportments with specific exhaust gas pollution recirculation levels (10%, 20% and 30%), describe a decrease in BTE, BSFC and NOx emissions, but marginally increase in other emissions. [16, 17] were examined Combustion, efficiency & pollution characteristics of a biodiesel & diesel fuel combination without fuel modifications. They revealed that decline in BTE, increment in nitrogen oxides (NOx) emission and decrease in particulate issue (PM), carbon monoxide (CO) and hydrocarbons (HC) discharges when contrasted with diesel. The problems associated with low volatility and high viscosities are compensated by subjecting the oil to the transesterification cycle, and the high viscosity can be minimized. Methyl and ethyl esters in vegetable oil (called biodiesel) have identical physical and chemical properties to those of ethanol. The different investigations describe the similarities of the single cylinder diesel engines fuelled with Jatropha biodiesel traditional diesel fuel for four stroke direct injection. Pre heating the vegetable oil reduces the viscosity and improves combustion characteristics.

From the literature it was observed that very less work carried out with lemon grass biodiesel methyl ester, in these experimental studies the LGB B20 with unheated/preheating condition. The preheating fuel intake as 60°C temperature of LGB B20 is used in this experimental work. The goal was to define the LGB B20 pre-heating temperature that provides better efficiency & polluting characteristics.

2. Materials and methods

Right now, Lemongrass oil methyl ester delivered from lemon grass oil. Lemongrass (Cymbopogon flexuous) is a local sweet-smelling tall sedge (family: Poaceae) which develops in numerous pieces of tropical and sub-tropical South East Asia and Africa. It develops to around 2 meters and has red base stems. The pollutions present in lemon and lemon grass oil will be evacuated including 10 percent of hexane. The lemongrass oil is mixing it for 15 min at 85ºC and leaves an undesirable natural issue to make due with 30 minutes. The properties of test fuel as shown in Table 1.

Table 1. Properties of test fuels

Characteristics	ASTM Std	Diesel	LGB
Kinematic Viscosity @ 40°C (cSt)	D445	3	4.34
Calorific value (MJ/kg)	D240	40	38.79
Fire point (°C)	D93	63	58
Cetane Number	D976	45	64
Flash point (°C)	D93	55	50
Specific gravity	D1298	0.8	0.87
Density (kg/m^3)	D4052	840	830

3. Experimental test setup

Fig 1 shows the test engine setup. A single chamber, quad strokes and the pressure ratio with the vortex current dynamometer is used in the engine configuration. A Kirloskar motor converts to a diesel engine with a variable pressure proportion. The principal components of the frame include a fuel infusion siphon, gas exhaust analyser, dynamometer & turbocharger. Burning temperature, cylinders and fuel flow and load calculations are a critical tool. The rotary meter is calculated to use water to cool and calorimeter water to detect the flow of water. The engine was executed with a brake power, torque, showed power, grinding power, explicit fuel brake use, explicit use of fuel, warm brake skills, warm effectiveness & high productivity. The engine was executed.

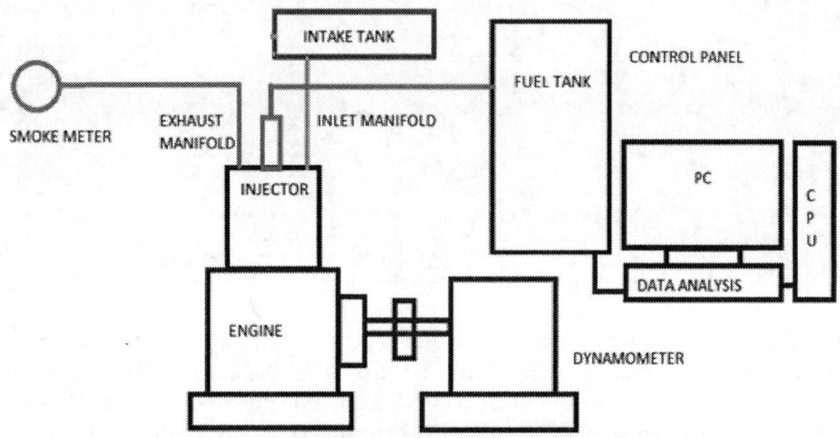

Figure 1. Block Diagram of VCR Diesel Engine

4. Results and discussion

4.1. Performance Characteristics

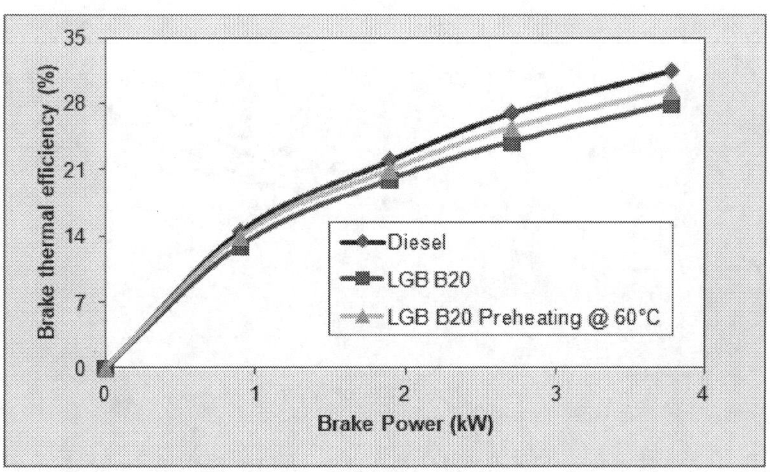

Figure 2. BTE for diesel, LGB B20 and LGB B20 preheating at 60°C

Fig 2 gives the Brake variable fuel consumption varies with the pre-heating of BGB B20, LGB and LGB B20 at 60 ° C. In maximum conditions of load, BTE is 30%, 25% & 29% respectively for preheating diesel, LGB B20 & LGB B20. The preheating of LGB B20 shows higher BTE than without preheating of biodiesel. Heating the LGB B20 at 60°C will stronger affect how it burns in the cylinder than how much energy it adds to the framework by recouping waste heat. Preheating of LGB B20 at 60°C rely on pressure to heat and light the fuel including heated fuel would make it disintegrate sooner and predetermination would be conceivable. The BTE was increased due to preheating temperature and lower heating value.

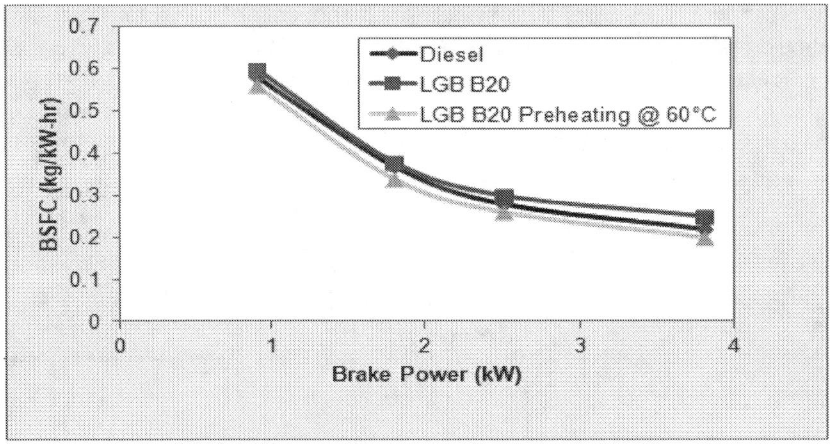

Figure 3. BSFC for diesel, LGB B20 and LGB B20 preheating at 60°C

Figure 3 indicates the discrepancy in the usage of diesel fuel, LGB B20 & LGB B20 preheating at 60 ° C. The BSFCs for diesel preheating, LGB B20 & LGB B20 respective were measured at full load condition as 0.22, 0.25 & 0.20 Kg / Kilowatt-hr. BSFC has been seen to decline as the engine load rose. The results show that the use of unheated LGB B20 by the BSFC is higher over varying rate. The preheating LGB B20 was decreased by 20% respectively as compared to the BSFC of unheated LGB B20. This is primarily due to decreased kinematic viscosity and improved atomization of fuel.

4.2. Emission Characteristics

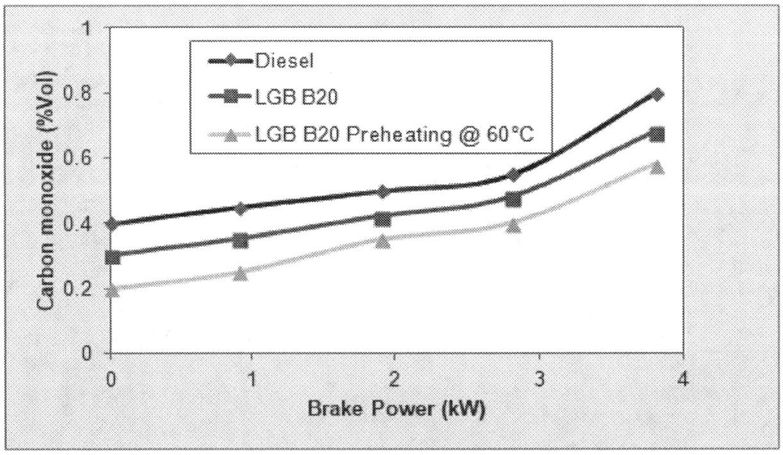

Figure 4. CO Emission for diesel, LGB B20 and LGB B20 preheating at 60°C

Figure 4 demonstrates CO emission difference at 60 ° C respectively in the pre-warming processes LGB B20 & LGB B20. At maximum load the CO emission for petrol, LGB B20 and LGB B20 preheating respectively was measured as 0.8 and 0.68 and 0.58 (percent volume). It's clearly shows the preheating LGB B20 as decreased the CO emission as compared to the unheated LGB B20. At varying loads, CO emissions using unheated / heated LGB B20 were found to be lower than diesel. The presence of oxygen molecules in LGB B20 decreases the efficiency of preheated LGB B20 in combustion. LGB B20 emissions were reduced by heating due to improved fuel atomization, distributing, vaporization and mixing levels of injected fuel, resulting in better air- fuel mixing.

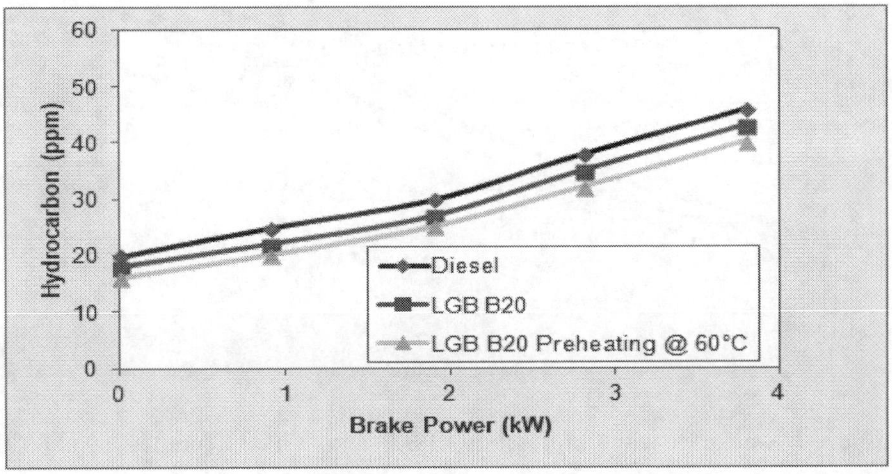

Figure 5. HC Emission for diesel, LGB B20 and LGB B20 preheating at 60°C

Figure 5 shows the max power conditions for petrol, LGB B 20 and LGB B20 preheating the hydrocarbon emissions were measured as 46, 43 & 40 ppm. It was observed that the preheating LGB B20 as decreased the HC emission as compared to the unheated LGB B20. Such reductions state that additional combustion of the fuel occurs at high load and thus the HC level drops significantly. It is mostly due to higher biodiesel oxygen content. Decrease in HC emissions was due to increased fuel atomisation and full preheated combustion of LGB B20.

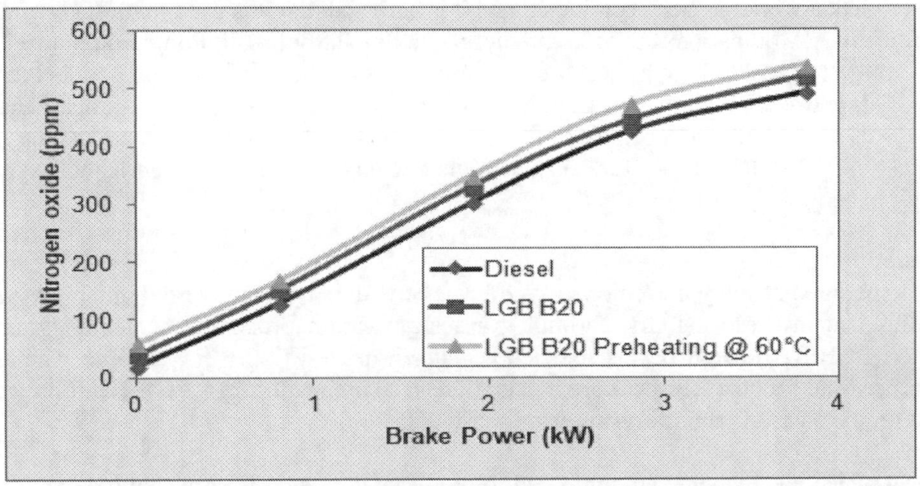

Figure 6. NOx Emission for diesel, LGB B20 and LGB B20 preheating at 60°C

Figure 6 shows the NOx emission as plays major role in the atmospheric condition as increase the risk of respiratory condition of poisonous emission. Figure 6 indicates the NOx pollution difference of the 60 ° C preheating LGB B20 & LGB B20. NOx emissions for diesel, LGB B20 & LGB B20 pre-heating at full load were found as 494, 520 & 540 ppm. This is because NOx formation is based on the temperature of the cylinder gas. It is observed that the preheating LGB B20 as increased 17% compare to the unheated LGB B20. At varying loads, NOx emissions using unheated / heated LGB B20 were found to be higher than diesel.

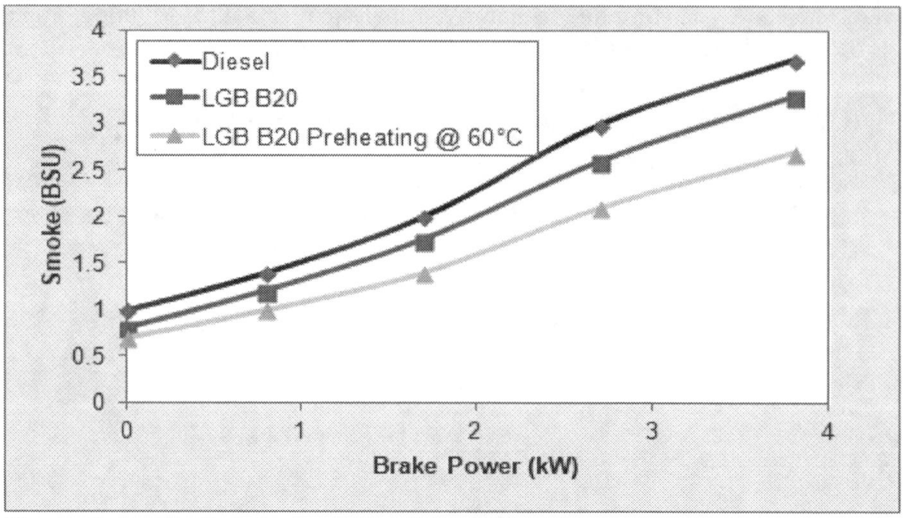

Figure 7. Smoke Density for diesel, LGB B20 and LGB B20 preheating at 60°C

Figure 7 illustrates a difference of the density of smoke preheating gasoline, LGB B20 & LGB B20 at 60 ° C. At full load condition the Smoke Density was observed as 3.7, 3.3 and 2.7 BSU for diesel, LGB B20 and LGB B20 preheating respectively He noticed that because of the presence of water in biodiesel, the smoke density is reduced. The Preheating LGB B20 shows lower smoke density as compared to the diesel and unheated biodiesel LGB B20. It shows the smoke density as reduced due to low fuel viscosity at preheating time.

The state of the art in diesel engine efficiency and emission characteristics fuelled with lemongrass biodiesel methyl ester (LGB) and its blends as alternative diesel engine fuels. LGB B20 preheating @ 60 ° C experimental work has been conducted and it is believed that it is used in diesel engine with little to no modifications. Compare to the other biodiesel such as Jatropha oil, Pongamia pinnata, Algae oil, Pine oil methyl ester, Fish oil methyl ester etc. Lemongrass biodiesel preheating and its blends contain the higher BTE and BSFC is lowered in diesel fuel. The comparative analyses indicate that biodiesel and its mixtures contain reduced exhaust emissions such as CO, HC and particulate matter / smoke and higher levels of NOX than diesel fuel due to higher levels of cetane and enriched oxygen content contributing to better combustion.

5. Conclusion

Under this experimental study the lemon grass oil was considered, it's converted into biodiesel as lemon grass biodiesel methyl ester (LGB) through transesterification process. The lemon grass biodiesel methyl ester (LGB B20) blend B20 is selected because it nearest to the diesel value. The lemon grass biodiesel is preheated at 60°C temperature that is (LGB B20 preheating @ 60°C). The following findings were taken on the basis of laboratory studies:

- Diesel which has higher BTE compared to biodiesel, however BTE was increased 29.5% due to preheating temperature and 28% for unheated biodiesel.
- BSFC is lowered in diesel compared to biodiesel, however the BSFC was decrease by preheating Biodiesel LGB B20 preheating @ 60°C
- The emission characteristics at full load condition of CO, HC and smoke density as decreased, it was observed slight increase of NOx emission.

References

[1] Abdulfatah AY and Karthickeyan V 2020 Data set on performance and relative exhaust emission (REE) characteristics of fish oil ethyl ester with fuel preheating in diesel engine *Data in Brief* **29** 105397

[2] Rambabu V, Subramanyam T and Prasad VJJ 2017 Evaluation of Performance and Emissions of A VCR DI Diesel Engine Fuelled with Preheated CsME, *Global Journal of Researches in Engineering, Mechanical and Mechanics Engineering* **12**

[3] Said MA, Ibrahim KA, Abed L Experimental Study on the Effect of Preheated Egyptian Jatropha Oil and Biodiesel on the Performance and Emissions of a Diesel Engine *International Journal of Mechanical & Mechatronics Engineering* **20**

[4] Mohanan P and Dinesha P Experimental investigations on the performance and emission characteristics of diesel engine using preheated pongamia methyl ester as fuel, *International Journal of Advances in Engineering & Technology.*

[5] Saravanakumar M and Prabhahar M 2018 Emission characteristics of biodiesel derived from used cooking oil blended with diesel in the presence of potassium hydroxide (KOH) catalyst as alternative fuel for diesel engines *Rasayan Journal of Chemistry* **11** 372-377

[6] Prakash S, Prabhakar S and Saravana Kumar M 2015 Experimental study of using vegetable oil as renewable fuel in diesel engine *Journal of Chemical and Pharmaceutical Sciences* 252-255

[7] Prabhahar M, Sangeetha K 2018 Experimental Investigations on Dual Bio-Fuel (Pine Biodiesel and Palm Biodiesel) Blended with Diesel on a Single Cylinder Diesel Engine *International Journal of Mechanical and Production Engineering Research and Development* **8** 87-92

[8] Govind N, Kumar Sureddy K 2018 Preheating of Biodiesel for the Improvement of the Performance Characteristics of Di Engine, *American Journal of Mechanical Engineering* **6**

[9] Prabhahar M, Sassykova L and Sendilvelan S 2015 Studies on pongamia oil methyl ester fueled direct injection diesel engine to reduce harmful emissions *Indian Journal of Environmental Protection* **38** 269-277

[10] Prabhahar M, Sendilvelan S and Francis Xavier J. 2017 Effect of Dimethoxy-Methane (C3H8O2) on Combustion Characteristics of a Direct Injection Diesel Engine with Variable Compression Ratio Fuelled with Biodiesel Blends with Diesel (C10.8H18.7). *Oriental Journal of Chemistry* **33** 2937-2941

[11] Prabhahar M, Krishnamoorthi S and Saravana Kumar M 2018 Yield characteristic of biodiesel derived from used vegetable oil methyl ester (UVOME) blended with diesel, in the presence of sodium hydroxide (NAOH) and potassium hydroxide (KOH) catalyst, as alternative fuel for diesel engines *International Journal of Mechanical and Production Engineering Research and Development* **8** 9-16

[12] Prabhahar M, Sendilvelan S and Prakash S 2017 Investigation of pine oil methyl ester blends with diesel on a compression ignition engine to control oxides of nitrogen and soot particles *Rasayan Journal of Chemistry* **10** 1075-1079

[13] Amir K and Norrizal M 2014 The Comparison of Preheat Fuel Characteristics of Biodiesel and Straight Vegetable Oil *Applied Mechanics and Materials* 465-466

[14] Prakash S, Prabhahar M and Saravana Kumar M 2020 Experimental analysis of diesel engine behaviours using biodiesel with different exhaust gas recirculation rates *International Journal of Ambient Energy*

[15] Muneeswaran R, Thansekhar MR and Viswanathan K 2018 Comparative studies on the performance and emissions of a direct injection diesel engine fueled with neem oil and pumpkin seed oil biodiesel with and without fuel preheater *Environmental Science and Pollution Research* https://doi.org/10.1007/s11356-017-0838-9

[16] Prakash S, Sendilvelan S and Prabhahar M 2019 Experimental studies on the performance and emission characteristics of an automobile engine fueled with fish oil methyl ester to reduce environmental pollution *Energy Procedia* **160** 412-419

[17] Saravana Kumar M, Prabhahar M, Prakash S 1980 Effect on performance and exhaust emission using lemongrass biodiesel as fuel on vcr direct injection diesel engine *International Journal of Mechanical and Production Engineering Research and Development* **9** 951-964

[18] Shahada S, Abu A, Mohammed SH, and Shamsudheen S 2019 IoT Based Garbage Clearance Alert System with Gps Location Using Arduino *International Journal of MC Square Scientific Research* **11** 1-8

[19] Badawi WA 2019 Underground Pipeline Water Leakage Monitoring Based On Iot, *International Journal of MC Square Scientific Research* **11** 01-08

[20] Tamilselvan P, Nallusamy N, Rajkumar S 2017 A comprehensive review on performance, combustion and emission characteristics of biodiesel fuelled diesel engines *Renewable and Sustainable Energy Reviews* **79** 1134–1159

Hybrid Renewable Energy Based CFSI for and Motor Application using ANFIS Based MPPT and IFOC Controller

G P Ramesh[1] and Shaik Mohammed Irshad[2]

[1] Department of Electronics and Communication Engineering, St. Peter's Institute of Higher Education and Research, Chennai, India.
[2] Department of ECE, SPIHER, Avadi, Chennai, India.
[1]rameshgp@yahoo.com, [2] irshadphd123@gmail.com

Abstract. In many industrial applications, the varying speed of the motor drives has become essential for the fast dynamic load conditions. Therefore, the detailed analysis of Field Oriented Control (FOC) with the sensor-less technique of BLDC is proposed where the reference speed is obtained from the MPPT based ANFIS controllers for control the speed. The FOC not only control the magnitude but also provides the phasor control. With Back-EMF estimation process, the motor is applied to obtain the torque ripple minimization and correct speed control using simpler IFOC. It is used to estimate speed in a sensorless and to map the continuous velocity variation. This approximate velocity is used to activate and to predict hall switching output less BLDC motor signals with low tension MATLAB / Simulink tool is using for results verification.

Keywords: ANFIS Controller, BLDC Motor, Back EMF Method, Indirect Field Oriented Controller (FOC)

1. Introduction

In recent days, electric power utilization is exploiting by motor load applications. The huge transformation of human styles and the development of the use of electric motors have increased the interest in electrical vitality. The CFSI converter is designed to extract the power from PV source and increase the voltage gain. CFSI's implementation guides PV to load and WECS. In order to control speed, and is proposed using IFOC method EMF estimate back. The proposed CFSI feeds both the BLDC motor and the RL load, using the new IFOC to control the speed of the BLDC motor. The demand for electrical energy is based on renewable sources. Solar, wind and fuel cells are the most commonly used renewable resources. Solar and wind energy is the most important source of energy for power generation [1]. Hybrid renewable energy uses continuous energy flow and meets sufficient energy demand. Efficient energy consumption is a serious issue in the use of a sustainable source. The best possible use of MPPT techniques is based on the efficiency of the power flow. Various control strategies exist and the algorithm-based MPPT is proposed by researchers [2-4]. In [4-6] the MPPT based on ANFIS makes use of the Properties and advantages of Artificial Neural Network (ANN) and Fuzzy Logic Control (FLC). The FLC properties open the way to follow the rules produces a non-linearity property. There are different types of algorithms used in MPPT techniques. The INC algorithm is very accurate than the P&O calculation. Incremental conduct can be realized at a high cost because it includes

progressively complex methods; it is worth it when more energy needs to be exchanged from the source. Hybrid renewable energy sources have been effectively exchanged through MPPT [7-9].
The principle pact is to plan the inverter design on the basis of a consistent quality, productive and cost-effective system. The dispute is to configure a high gain. In this way, the CFSI is proposed and a high output gain ratio can be created inverter. In the earlier period, implementation of the system is increasingly difficult and exorbitant. However, it does have the disadvantage of leakage of current flows. Power quality is reduced as a result of the infusion of leakage currents afterwards, downsides of less output gain inverter [10]. Efficient energy from a hybrid renewable source is obtained from the CFSI control technology used by the BLDC engine. The efficient energy from the hybrid renewable source is obtained by the control techniques of CFSI which is given to the BLDC motor [11-12].
The BLDC motors are ordinarily utilized in numerous applications, for example, car, PC, mechanical, aviation and so on. In BLDC there is no appearance of brushes henceforth maintenance of BLDC is less. This has the Due to low inertia; BLDC motor has the benefit of fast response to model reference waveform of trapezoidal voltage, and a quasi-rectangular waveform of voltage. The BLDC engine has many favorable circumstances over other motor drives and is suitable for higher output responsibilities. Various control methods are proposed in that back EMF and IFOC controller behaves well under dynamic load conditions. [13] Operate on electromagnetic, and mechanical commutation is not required. The electronic commutation requires. Using Hall Impact location sensors is achieved for the continuous monitoring of rotor location [14-15]. The BLDC motors are mostly favored for high-speed applications and work is undertaking the production of BLDC power. This is operated by CFSI system is implemented to improve the output of motors. IFOC approach is used and BLDC motor speed is measured instead of being collected from a sensor, the back EMF measurement method is used [16-17].

2. Proposed system

In this proposed system consists of hybrid renewable resources struggling to produce efficient electricity, and the ANFIS-based MPPT control system for CFSI and IFOC motor control controllers appears to be completed. The successful converter power is transferred via the CFSI to the BLDC motor. The new IFOC gives the inverter switches the pulses of the gate for engine speed control. The proposed BLDC engine load scheme is shown in figure. 1.

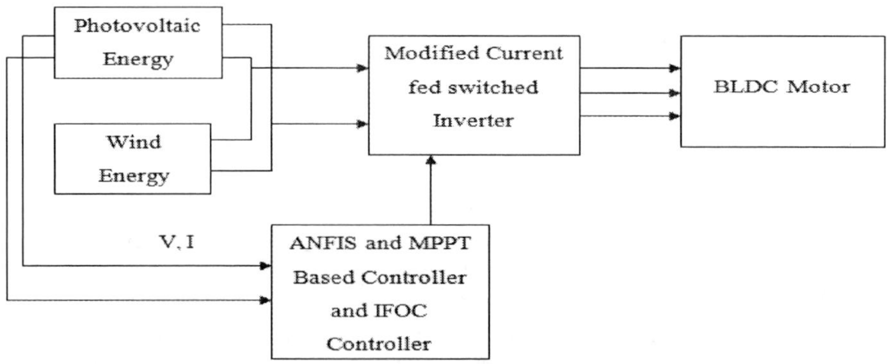

Figure 1. Architecture of the proposed system for BLDC Motor

3. Modelling

3.1 PV Energy Generation

There's just a harmful quantity of fossil fuels and minerals for electricity generation recently. As a result, the use of sustainable energy source assets is rapidly expanding step by step. Owing to sun-powered energy is most widely used as renewable sources. The Photovoltaic (PV) system utilizes solar energy. The photovoltaic device converts directly into electricity at daylight. PV cells, which form a key for PV generation. A PV cell is a semiconductor diode which includes diode. The photoelectric effect is the theory used in system. The theory includes the cycle of electrons forced out each time the solar radiation is absorbed by the PV cell. PV device has two types of layout. They are 1) single diode model 2) two-diode model. The proposed single diode model has less complexity and more rapid results the proposed have faster outputs. Figure 2. Shows the single PV model. Simulation of a single model can be achieved by using the equations shown in the following: A single diode configuration is equated by

$$I_0 = I_{pc} - I_{rs}\left(e^{\frac{V_{pv} + I_{pv}R_{series}}{a}} - 1 \right) - \frac{V_{pv} + IR_{series}}{R_{shunt}} \tag{1}$$

The following equation shows that PV model's current (I) and voltage (V) (I-V) characteristics:
Where
Io - output current,
Ipc -photocurrent,
R_{series}- series resistance, and
R_{shunt} -shunt resistance
I_{rs}- saturation current which is reversed,
V_{pv} - voltage of the PV cell,
I_{pv} -current of the PV cell,

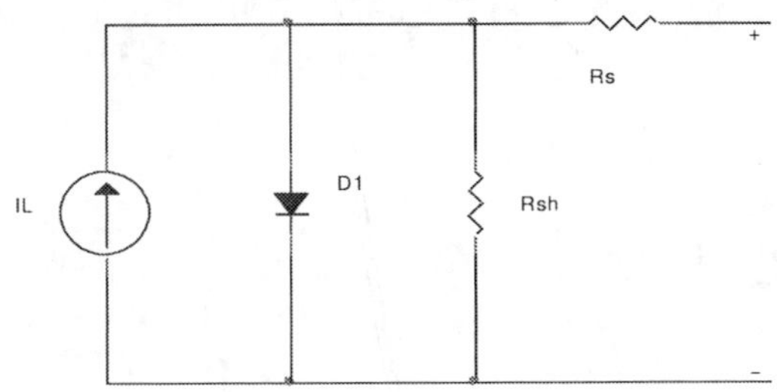

Figure 2. Mathematical Modeling of the Proposed PV Model

3.2 Incremental & Conductance (Inc) Maximum Power Point Tracking (MPPT) Method

The MPPT is a tracking device that tracks the maximum power because renewable sources cannot achieve the use of efficient power, especially in PV panels. The MPPT is then run in a PV panel to improve performance. The P&O approach has the advantages of simplicity but can't precisely control the full power point it faces the issue of wandering around the maximum power point without correctly pointing it and, however, it will almost certainly not respond to minor and regular changes in irradiance

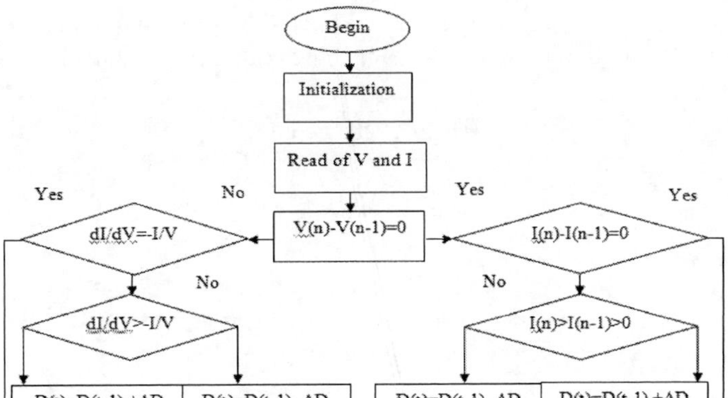

Figure 3. Flowchart of the INC MPPT Algorithm

3.3 Wind Energy Generation

The Wind energy is a vitality commodity for eco-companions even as it needs the less space for installations. The use of wind turbines makes use of wind power. Different designs for wind turbines are suggested for the successful task of generating wind power. The power network has a combination of the transfer of power DC and AC. In comparison to variable speed wind turbines, the fixed turbine is less efficient, just as it has the maximum speed run. The WECS wind energy transmission shown at figure.4. Design of the construction, four fundamental parameters is considered. They are engines, windscreens, gearboxes and power tools. The method of extracting wind energy involves the wind turbine that absorbs wind energy, and the wind energy that has been captured using a generator, is converted to wind power. Turbine modeling is performed by partnership, and is provided by,

$$P_m = \frac{1}{2}\rho S_{sa} C_{power}(\lambda, \beta)\upsilon^3 \tag{2}$$

Where Pm is the mechanical power, where Ssa is the air density, where ssa is the rotor tip, where Power is the strength coefficient, where wind speed is . The rotor blade TSR, β the rotor blade pitch angle. Configuration of PMSG turbine used for producing wind. The synchronous permanent magnet generator has advantages such as no requirement for excitation winding, no electrical indulgence, increasingly stable & safe, no slip ring supply, simple construction, reduced maintenance, most resistant limit and highly efficient

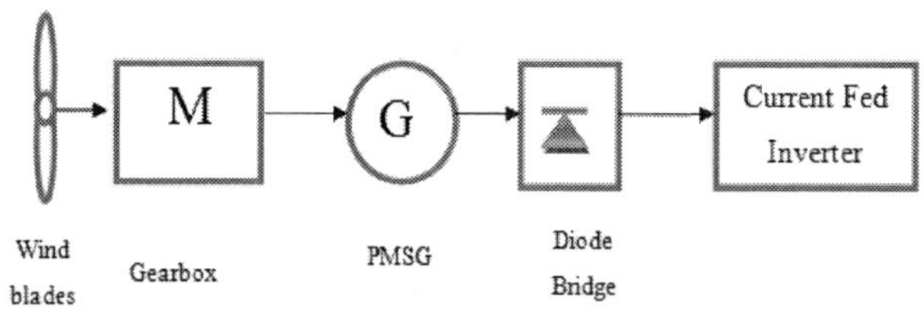

Figure. 4. Wind Energy Conversion System (WECS)

3.4 Proposed CFSI

The CFSI shown in figure 5. The power grid has a mix of power transmission DC and AC, primary diodes D1 and D2, four switch inverters S1-S4 and C3 and C4 split condensers. The current source inverter is worked in four modes. The primary Sa switch is switched ON in mode 1. The L1 input inducer charges the power supply. Hence the current in L1 is slowly rising. . The primary switch is switched to ON in mode 3. Input inducer L2 charges the energy of the signal. Then the current in L2 slowly rises. In mode 4 both the main Qa and Qb switches are switched ON. Input inductors L1 and L2 charge during this mode. Even the current in the inductors is rising. The equation gives the raise up ratio service cycle.

$$\text{Duty cycle} = 1 - \frac{\text{Vin}}{\text{Vout}} \tag{3}$$

$$V_{in} = V_{C1} + V_{C2} \tag{4}$$

$$V_{L1} = V_{C1} - V_{Qa} \tag{5}$$

During ON time of Qa and Qb, the VQa and VQb are given by,

$$V_{Qa} = 0, (0 < t < \text{sampling time}) \tag{6}$$

$$V_{Qb} = 0, (0 < t < \text{sampling time}) \tag{7}$$

IOP Conf. Series: Materials Science and Engineering **925** (2020) 012060 doi:10.1088/1757-899X/925/1/012060

During OFF time of Qa and Qb, the VQa and VQb are given by

$$V_{Qa} = \frac{V_{in}}{2}, (0 < t < \text{sampling time}) \tag{8}$$

$$V_{Qb} = \frac{V_{in}}{2}, (0 < t < \text{sampling time}) \tag{9}$$

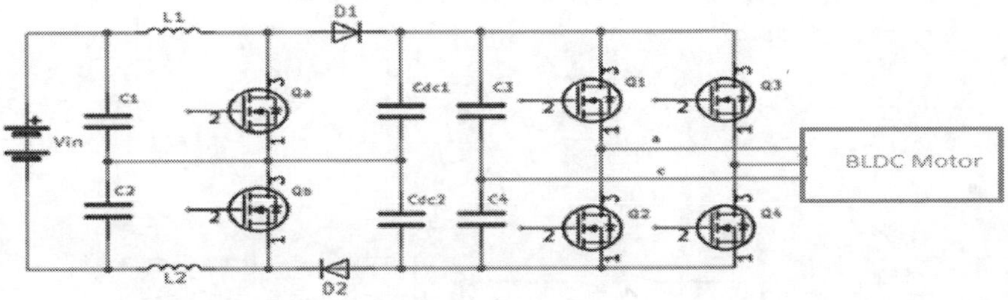

Figure 5. Circuit diagram of CFSI for BLDC motor

3.5 Adaptive neural fuzzy inference program and its algorithm

There are numerous control systems that analysts suggest. The PID controller has the primary impediment of inadequate monitoring of the reference value when the complex conditions are substantially extended. The neural network and the fuzzy inference method are drawn in because it has the advantage of following the reference value when large aggravations and changes in input occur, but they face complexity disadvantage. And they can't have user-friendly experience. ANFIS has introduced a hybrid technology that combines the advantages of both fuzzy and neural networking techniques. ANFIS integrates the process of obtaining input values and processing them as per fuzzy process rules, and the simulated output from the fuzzy rules is again sophisticated by specifying rules for the accuracy of the system via the neural network. Layer 1 shall include the input date for processing purposes. Inputs are defined by membership functions. Layer 2 performs AND operates on the basis of fuzzy rules with input data. The layer 3 performs the same AND operation based on the rules for accuracy enhancement described by the neural network. The output parameter is modified through layer 4. Layer 5 provides the desired performance by processing the input data with the rules provided. Figure 6. displays the general block diagram for the current ANFIS scheme and the ANFIS model.

The assignment of the learning algorithm for the ANFIS architecture is to adjust all the parameters that can be modified. It uses the hybrid algorithm which combines the method of least squares with the method of gradient descent. This combination addresses the slowness problems in training the values. The system of the least squares acts as forward pass, and the system of gradient descent acts as backward pass. The parameters modifiable in the first layer are considered parameters of the principle. In the fourth layer, the parameters which can be modified are called consequent parameters. The forward pass is used to combine the resulting parameters with the fixed parameters of the premise. The reverse transfer invokes the process of optimizing the parameters of the premises according to the fuzzy set of values input. Through ANFIS architecture, the back propagation algorithm is often used to address the error in the output side by optimizing only the parameters for the assumption. The hybrid algorithm has greater efficacy than the other algorithm. Table.1. Represents regulation laws for the ANFIS.

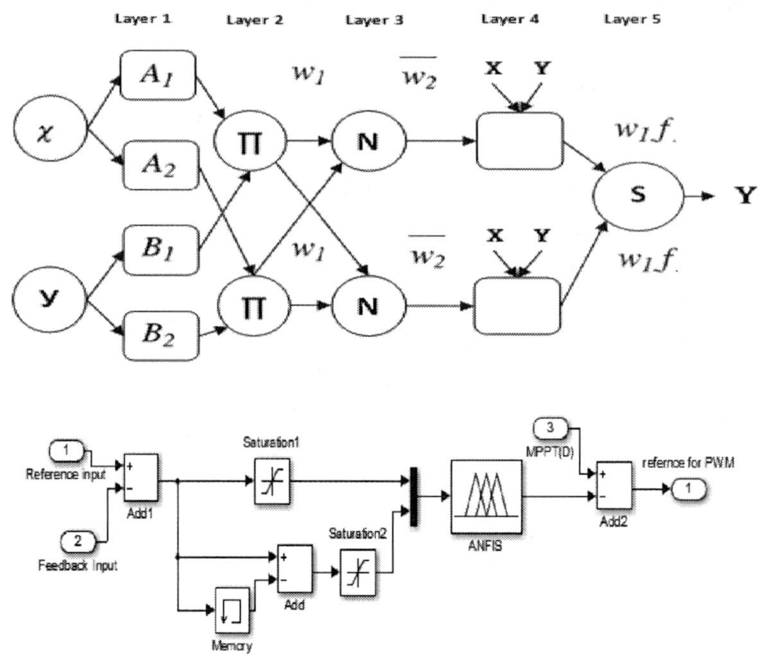

Figure 6. Architecture of ANFIS and MPPT based ANFIS

Table 1. ANFSI rules

Δe \ e	NB	NM	NS	ZE	PS	PM	PB
NB	NZ	NZ	NS	NM	PS	PM	PB
NM	NZ	NZ	NZ	NS	PM	PM	PB
NS	NZ	NZ	NZ	NZ	PM	PM	PB
ZE	NB	NS	NZ	NZ	PS	NZ	NZ
PS	NB	NM	NS	NZ	NZ	NZ	NZ
PM	NB	NM	NS	PS	NZ	NZ	NZ
PB	NM	NM	NS	PM	NZ	NZ	NZ

3.6 Back EMF and IFOC controller

It is important to calculate three phase voltage and current flowing from the inverter to the current flowing from the inverter to the BLDC motor which is used for measuring the BLDC motor back EMF. Calculation of line voltages and line currents from determined values. Three phase EMF equation can be estimated from mathematical modeling of the BLDC motor back and the absolute value of the back EMF is obtained by equation (3), which helps to determine the motor's rotor speed.

$$\begin{bmatrix} V_a \\ V_b \\ V_c \end{bmatrix} = \begin{bmatrix} R & 0 & 0 \\ 0 & R & 0 \\ 0 & 0 & R \end{bmatrix} \begin{bmatrix} i_a \\ i_b \\ i_c \end{bmatrix} + \begin{bmatrix} L-M & 0 & 0 \\ 0 & L-M & 0 \\ 0 & 0 & L-M \end{bmatrix} \frac{d}{dt} \begin{bmatrix} i_a \\ i_b \\ i_c \end{bmatrix} + \begin{bmatrix} e_a \\ e_b \\ e_c \end{bmatrix}$$

$$\omega_r = \frac{E_m}{2 * P * \phi}$$

IOP Conf. Series: Materials Science and Engineering **925** (2020) 012060 doi:10.1088/1757-899X/925/1/012060

$$I_{ref} = \frac{T_{ref}}{2 * P * \phi}$$

Where, V_a, V_b, V_c phase voltage, I_a, I_b, I_c Phase currents, e_a, e_b, e_c back EMF of BLDC motor

The early research into program is to measure the speed of the motor engine and to be re-calculated using the EMF equation process. After assessing the approximate controller, the reference current is determined on the basis of the it is related to the equation (5). Estimates are created using an essential rotor angled by the operator. To produce the gate pulses must be measured, which is combined with the reference current. The Inc& Con calculation is used to extract of the obligation duration is determined from the current reference value's MPPT algorithm balances. Hysteresis Controller. The present controller function properly only when we are correctly characterizing the two category viz. Higher band hysteresis and lower relay band hysteresis. Present error is fed as the controller input and when the error is below or when the pulse exceeds the band limit, low gate pulse is produced to shut down switches. Unless the band limit is selected, it will be low.

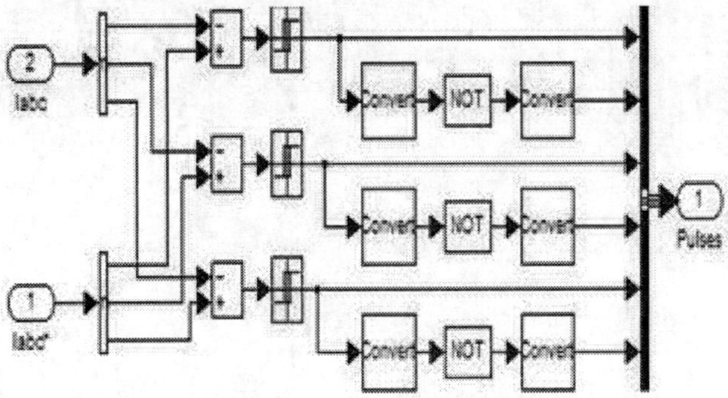

Figure. 7. Proposed Hysteresis Controller

4. Simulink Results

Figure. 8. Shows the PV and WECS powered CFSI powered BLDC engine drive simulink implementation. Hybrid renewable energy source is built which contains solar parameters like speed and torque PV and wind-generated energy is fed to CFSI. Figure. 9 represents the waveform of DC connection voltage, and the BLDC motor stator current. Speed is measured using rear EMF calculation, sensing phase voltage and phase currents. EMF is calculated with mathematical simulation machine back. Rotor speed is determined on the basis of the relationship between back EMF and predicted engine speed

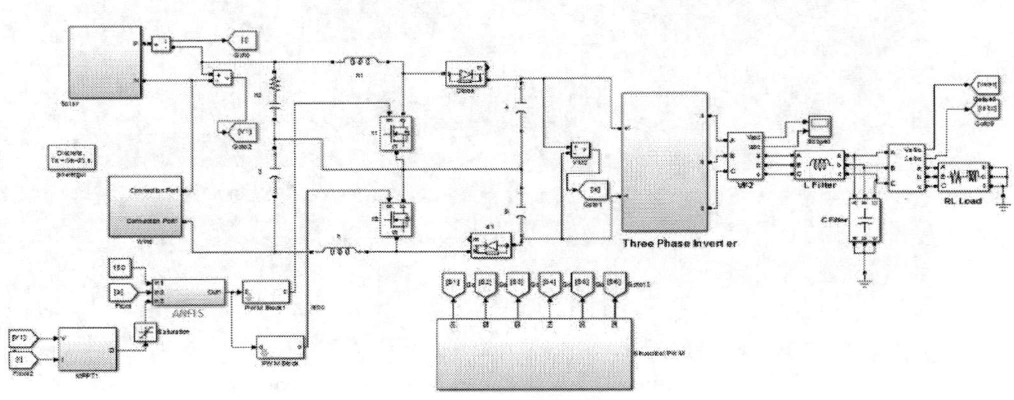

Figure 8. Overall Matlab/Simulink diagram for BLDC Motor

IOP Conf. Series: Materials Science and Engineering **925** (2020) 012060 doi:10.1088/1757-899X/925/1/012060

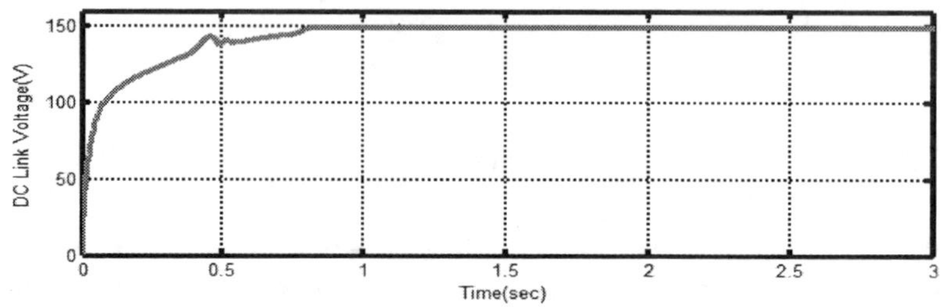

Figure 9. DC bus voltage

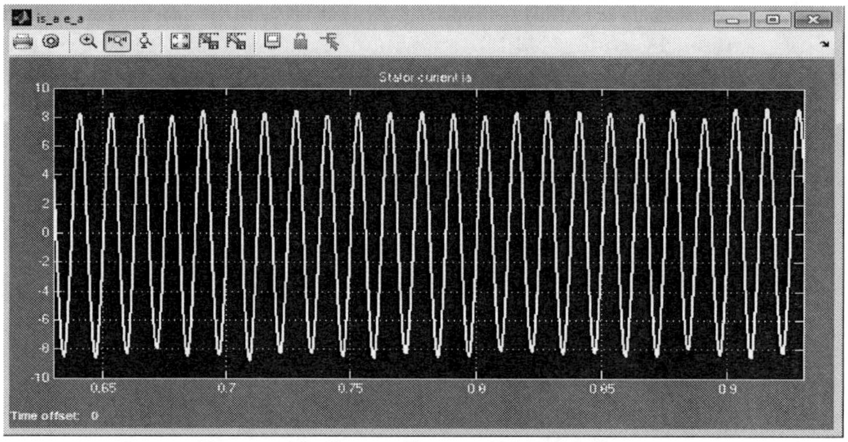

Figure 10. Stator Current BLDC Motor

The IFOC has removed harmonics of current and THD value to the lowest compared to the control techniques used previously. For a reference speed of 600 rpm, the response of systemis shown in Figure. 11 at Figure. 13. Plot of actual speed, electro-magnetic torque and approximate speed of the BLDC motor is shown.

The findings show that projected velocity is equal to the real velocity at all Times also at different load commands and different reference speeds.

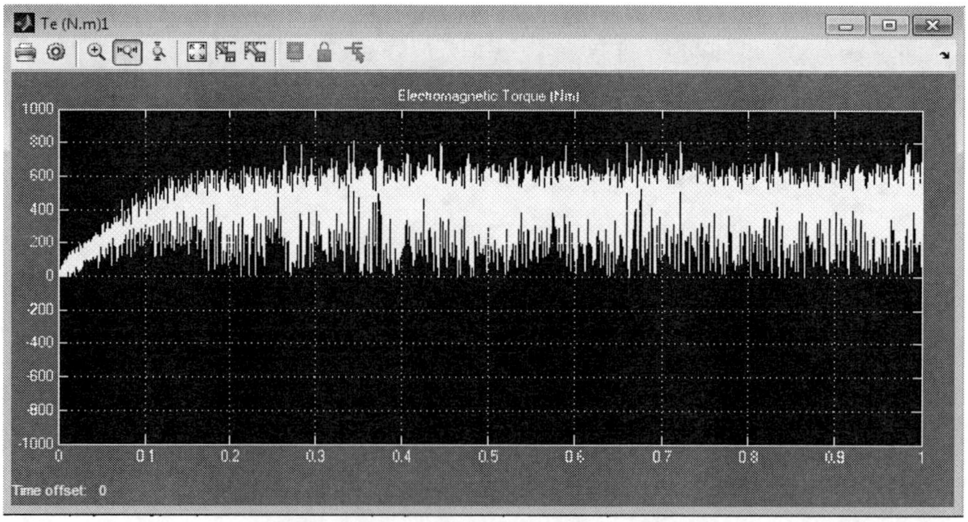

Figure 11. Estimated Speed of the Motor

IOP Conf. Series: Materials Science and Engineering **925** (2020) 012060 doi:10.1088/1757-899X/925/1/012060

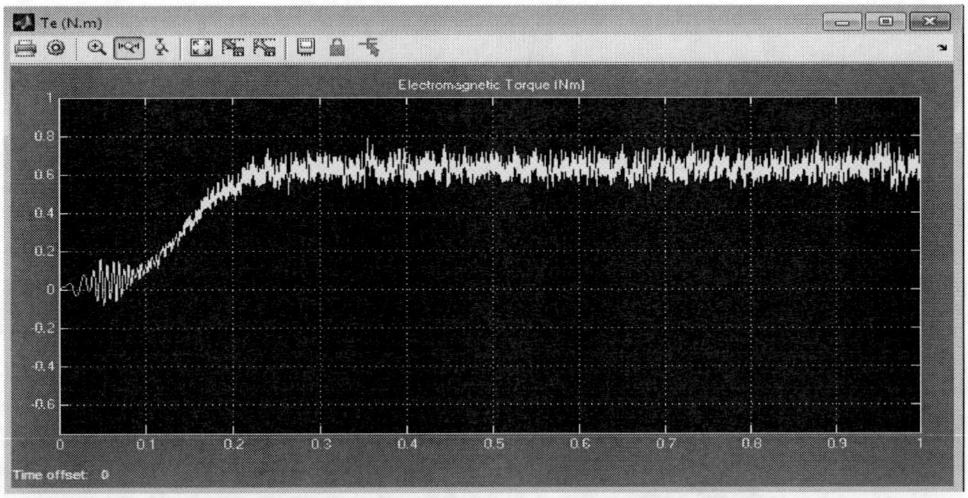

Figure 12. Electro-Magnetic Torque of the Motor

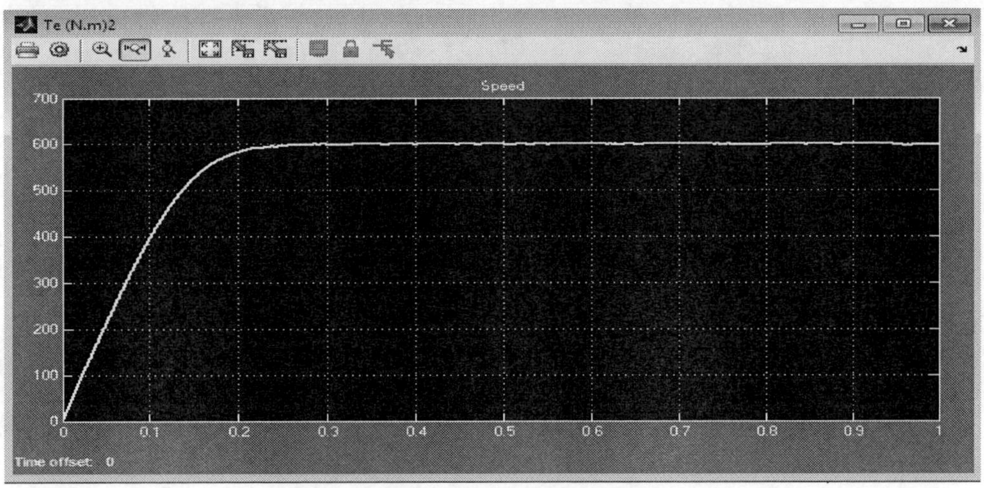

Figure 13. Speed Response of the Motor

5. Conclusion

In this paper, improvement of the efficiency of a sensor with a lower BLDC engine fed from hybrid renewable energy sources such as PV and a current charged inverter using IFOC control technology paper. Using FFT stator current study, improved sensor performance less BLDC motor with IFOC control technique is tested to reduce THD of the drive. The DC contact voltage, stator current harmonics are speed and torque are achieved using the IFOC control method MATLAB / simulink verifies the outputs of PV and wind driven CFSI.

References

[1] Abdelkhalek O, Benachaiba C, Gasbaoui B and Nasri A 2010 Using of ANFIS and FIS methods to improve the UPQC performance *International Journal of Engineering Science and Technology* **2** 6889-6901

[2] Sankar K 2017 Design of Intelligent Controller for a Hydraulic Solar PV System *International Journal of MC Square Scientific Research* **9** 66-71

[3] Sundaram A 2017 Sensor less Control of BLDC Motor using Fuzzy logic controller for Solar power Generation *International Journal of MC Square Scientific Research* **9** 70-79

[4] Abu-Rub H, Iqbal A, SkMoin A, Peng FZ, Li Y and GeBaoming 2013 Quasi-Z-source inverter-based photovoltaic generation system with maximum power tracking control using ANFIS, *IEEE Transactions on Sustainable Energy* **4** 11-20

[5] Gregory Joseph K, Lee JJ, and Lehn PW 2012 Modelling and control of photovoltaic panels utilising the incremental conductance method for maximum power point tracking *IET Renewable Power Generation* **6** 259-266

[6] Singh R, Bajpai SK and Sandhu HS 2015 Comparative study of PWM control and PI control of Induction motor *Bulletin of Electrical Engineering and Informatics* **4** 53-58

[7] Singh B, Sharma U and Kumar S 2018 Standalone photovoltaic water pumping system using induction motor drive with reduced sensors *IEEE transactions on industry applications*, **54**(4), pp.3645-3655,2018.

[8] DXu, B Wang, G Zhang, G Wang and Y Yu A review of sensorless control methods for AC motor drives *CES Transactions on electrical machines and systems* **2** 104-115

[9] Singh S and Tiwari AN 2017 Various techniques of sensorless speed control of PMSM: A review In 2017 *Second International Conference on Electrical, Computer and Communication Technologies (ICECCT)* 1-6

[10] Tir Z, Malik OP and Eltamaly AM 2017 Fuzzy logic based speed control of indirect field oriented controlled Double Star Induction Motors connected in parallel to a single six-phase inverter supply *Electric Power Systems Research* **134** 126-133

[11] Ahmed AA, Koh BK and Lee YI 2017 A comparison of finite control set and continuous control set model predictive control schemes for speed control of induction motors *IEEE Transactions on Industrial Informatics* **14** 1334-1346

[12] Irshad SM, and Ramesh GP 2020 ANFIS-Based MPPT Control in Current-Fed Inverter for AC Load Applications *Advances in Intelligent Systems and Computing* **1125**

[13] Cho Y, Bak Y, and Lee KB 2017 Torque-ripple reduction and fast torque response strategy for predictive torque control of induction motors *IEEE Transactions on Power Electronics* **33** 2458-2470

IOP Conf. Series: Materials Science and Engineering **925** (2020) 012061 doi:10.1088/1757-899X/925/1/012061

Adaptive Noise Cancellation Using Kalman Filter for Non-Stationary Signals

Murugendrappa N[1], A G Ananth[2] and Mohanesh K M[3]

[1] Department of Electronics and Communication, GM Institute of Technology, Davangere, Karnataka, India.
[2] Department of Electronics and Communication, NMAM Institute of Technology, Karkal, Udupi Karnataka, India.
[3] Department of Electronics, Sahyadri Science Collage, Shimoga, Karnataka, India.

Abstract. The present paper states with the Adaptive Noise Cancellation (ANC) of speech signal corrupted with additive Gaussian white noise. A new method is proposed based on adaptive Kalman filtering. The probabilistic approach of kalman filters over a packet-delaying network given the postpone distribution and decide the minimum required buffer length and algorithm has been used for estimation of unknown state variables within the system. Any work has been achieved for structures with fractional-order dynamics. The new techniques based totally at the kalman filter proposed within the beyond, function in steps: first the noise variance and the parameters of the signal version is estimated and secondly the speech signal expected. The strategies provided inside the paper gives an exclusive method consequently it does not require estimation of the noise variance. The noise variance estimation is always will become part of the kalman advantage calculation. The results of the application of Kalman filter on a non-stationary acoustic signal indicated that SNR of the ~ 1.17 dB and MSE`0.032 can be achievable using Kalman filters and Kalman filter can be efficiently used for noise cancellation in place of other adoptive filters.

Key Word: Adaptive filters, Adaptive Noise Cancellation (ANC), Gaussian noise, Kalman filter, Signal to Noise Ratio (SNR),

1. Introduction

The overall performance of any speech sign processing gadget can be degraded within the presence of noise (either additive or convolution). Numerous techniques for extracting the clatter from a speech waveform were studied. Most of these techniques are based upon the concept of adaptive filtering that seeks benefits from quasi-periodic speech signals which acts as reference for adaptive filter. The designed clear out has been applied on non -stationary speech alerts to have a look at its noise suppression capability measured because the distinction among enter signal snr and filter output sign SNR and the Mean Square Error (MSE). The efficiency of every filter out for suppression of noise in SNR(dB) is decided and in comparison to assess the relative performance of numerous adoptive filters consisting of kalman filter out which has precise applications

The techniques even enhanced the output of linear prediction analysis. It makes the noise separation technique have a huge frequency variety and aren't without problems separated from noise the use of filtering strategies. Application of Kalman filter is the reduction of noise in communication channel. When it is non-stationary, recursive filter allows estimation of the useful signal in noise from time series [1],

A noise itself is a facts bearing sign that conveys facts concerning the sources of the noise and the surroundings wherein it propagates. Cosmic radiation presents data on formation and form of the universe and historical past speech conversations in a crowded venue can represent interference [2]. If the speech signal is considered as an output of a system adaptive kalman filter is used to estimate the speech sign. [3-4]. SNR optimization benefit procedure. So, this new approach appears very appealing in comparison to the earlier strategies [5]. For linear but time variation structures the kalman clear out is based on a kingdom area method of a non-stop or discrete time device. The system ought to be best considered in discrete time. The kalman filter will gives the estimate of the system given fast of outputs. It will additionally minimize the output errors of the clear out [6]. Many approaches the use of kalman filtering had the typically perform in steps: first, the noise and riding method variances and parameters of speech model are predicted; then, the speech sign is anticipated with the aid of kalman filtering. In fact, these strategies fluctuate best by the selection of the set of rules used to estimate model parameters and the selection of the models followed for the speech sign and the additive noise [7]. The signal it will be presence of noise degraded within (either additive convolution).it creatively due to the acoustic mismatch between the speech functions used to teach and take a look at this gadget and the capability of the acoustic fashions to describe the corrupted speech. [8-9]. the factors had been organized in a matrix technique so that the favored signal from the noisy can be separated. The evaluation will outcomes in designed kalman filters showing that it can correctly dispose of the noises [10]. In the present paper the kalman clear out-which counts as one of the notable filters- has been surveyed whose factors is being calculated to design an efficient filter [11-13]. Initially a pattern signal is randomly decided on which may be similar to an auto regressive sign. Then random Gaussian noise is implemented on auto regressive signal; and therefore the noisy sign is suppressed. A Mat lab software simulation has been carried out and the results are presented.

2. Kalman Filter

The kalman filter has played a critical function in systems concept and has observed extensive programs in lots of fields consisting of signal processing. The probabilistic technique of kalman filters over a packet-delaying community given the postpone distribution and decide the minimal required buffer length. The algorithm has been used for estimation of unknown country variables inside the device and work has been executed for structures with fractional-order dynamics

The work is primarily based on monitoring best a limited wide variety of strongest interference's assumptions of synchronous interferes operation with overlapping however extraordinary schooling alerts. Kalman filtering is used for interfering users channel estimation and following calculation of interference correlation matrix. Such in-time correlation matrix estimate exploited in MMSE primarily based developed algorithms may be utilized in subsequent technology

$$x(n+1) = \Phi(n+1,n).\mathrm{X(n)} + \mathrm{V_1(n)} \tag{1}$$

$$K(n) = \mathrm{K(n,n-1)} - \Phi(n+1,n).G(n).c(n).K(n,n-1)$$

$$K(n+1,n) = \Phi(n+1,n).K(n).\Phi^{\mathrm{H}}(n+1,n) + Q_1(n) \tag{2}$$

IOP Conf. Series: Materials Science and Engineering **925** (2020) 012061 doi:10.1088/1757-899X/925/1/012061

The Kalman filter gives the solution to the following problem. Given the state space model in equation (3) or figure 1 where **Q1** (n) , **Q2** (n) , **C**(n) , $\Phi(n+1,n)$ and **y**(n) are known quantities, find the best estimate $\hat{X}(\frac{n}{Y^n})$ to the state vector **x**(n) , in the expected least errors squares sense.

$$Y(n) = \mathrm{C(n).X(n)} + \mathrm{V_2(n)} \tag{3}$$

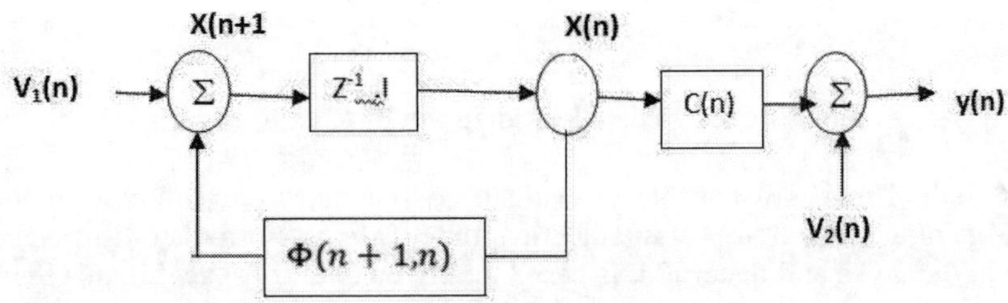

Figure 1. Space state model used for the system in the Kalman filter.

$$E[V_1(n).V_1^H(n)] = Q_1(n), E[V_2(n), E[V_2(n).V_2^H(n)] = Q_2(n) \tag{4}$$

$$G(n) = \Phi(n+1,n).K(n,n-1).C^H(n)[(n).K(n,n-1).C^H(n) + Q_2(n)]^{-1} \tag{5}$$

$$\alpha(n) = y(n) - C(n).\hat{X}(\frac{n}{Y_{n-1}}) \tag{6}$$

$$\hat{x}\left(\frac{n+1}{y_n}\right) = \Phi(n+1,n).\hat{x}\left(\frac{n}{y_{n-1}}\right) + G(n).\alpha(n) \tag{7}$$

$$\hat{x}\left(\frac{n}{y_n}\right) = \Phi(n,n+1).\hat{x}\left(\frac{n+1}{y_n}\right)$$

$$K(n) = \mathrm{K(n,n-1)} - \Phi(n+1,n).G(n).c(n).K(n,n-1)$$

$$K(n+1,n) = \Phi(n+1,n).K(n).\Phi^{\mathrm{H}}(n+1,n) + Q_1(n) \tag{8}$$

There has been much interest in fast convergence algorithms, but is fast convergence really needed in ANC. Fast convergence means that the time the algorithm takes from its initialization to the point it reaches an optimal value is short. These algorithms are of great interest in telecommunications system where the goal is to reduce the size of the training sequence and the Corresponding overhead.

3. Kalman Filter Automatic Noise Cancellation (KFANC)

Acoustic noise cancellation ANC is best suited to remove ambient noise. The conventional algorithms based on ANC had superior performance at low frequency bands and with increase in frequency and bandwidth the performance disorients. Generally, the noises that affect the system have wider frequency range and only a small amount of energy is not present near low frequency region have relatively high frequency components. When ANC is cascaded with different methods. The frequency dependent noise cancellation will discard any effect on speech signal. However, the ANC systems fall back in performance when noise frequency level goes high. The noise in real world has higher broadband and high frequency component.

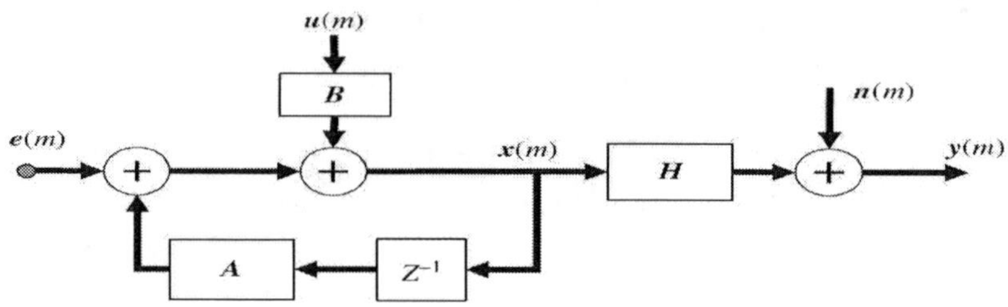

Figure 2. The Block diagram of Kalman filter

The use of this method makes it feasible to evaluate a sign transmitted through manner of a distorted channel, even as gaining a few noises at the identical time (2-five). Formula of this clear out, in order that it could dispose of the noise and distortion, is based totally on country space. In this regard, x(m) vector is taken to expose the favored vector and y(m) vector indicates the output noisy vector. Equations 1 and pair of display the vectors of the country area and output filter.

$$x(m) = A\, x(m-1) + B\, u(m) + e(m) \tag{9}$$

$$y(m) = H\, x(m) + n(m) \tag{10}$$

U (m) is the P-dimensional control input;

E (m) is the P-dimensional equation,

Figure.2 is the block diagram Shows Kalman filter Vectors is showed in this figure Y (m) and X (m).

4. Kalman Filter Algorithm

Step 1: Read the input signal (speech signal) using wave read i.e. input signal

Step 2: Parameters of input signal are Fs = 8000, number of bits encoded is 16bits (39500 samples = 39500/8000 = x seconds)

Step 3: Generate noise signal of length of the input signal using Gaussian white noise i.e. noise

Step4: Calculate input SNR in DB

Step 5: Add input signal with noise signal to generate a estimated signal i.e. input_signal+noise = Estimated signal

Step 6:

(i) Calculate AR parameters for both input signal and noise signal of 16 co-efficient using Burg's algorithm

(ii) Give this AR parameter to unidirectional non-stationary Kalman filter, output of this function is output signal

Step7: Now calculate output SNR in DB

Step8: Calculate mean square error (MSE)

5. Result and Methods

The input non stationary acoustic input signal chosen for the present analysis is shown in figure 3.

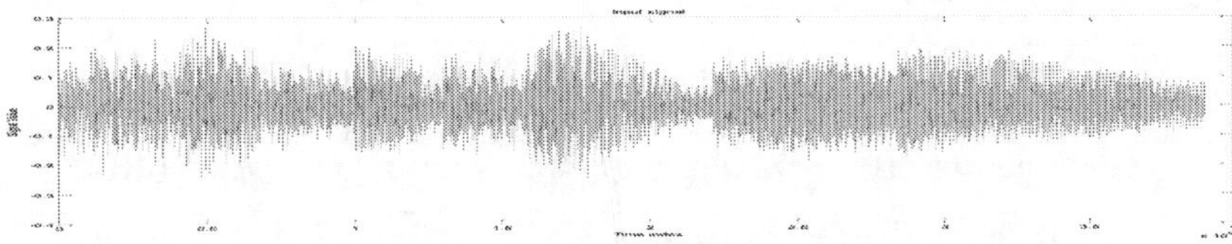

Figure 3. Input non stationary acoustic signal

It can be seen from the figure 3 that the signal is also associated with minimum amount of noise.
Further the acoustic input signal is combined with the white Gaussian noise and the estimated input signal is shown in figure 4.

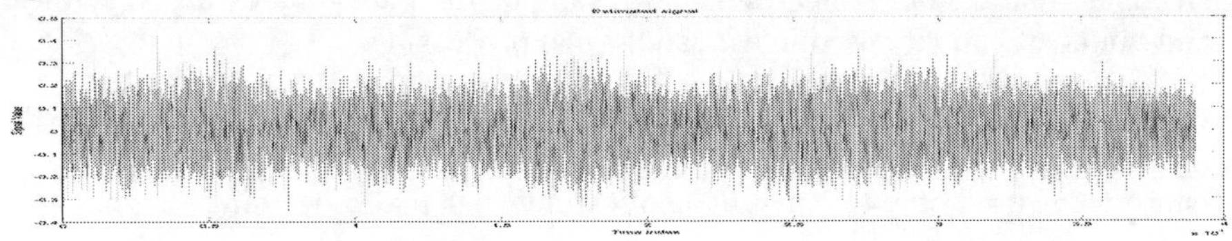

Figure 4. Estimated signal (input signal + Generate noise = ES) signal

It can be seen from the figure 4 that the signal is almost merged within noise The estimated input signal generated with by adding noise to the signal (ES=Signal + Noise). The SNR for the estimated input signal is determined is ~87dB. The estimated input signal is passed through Kalman filter for the cancellation of noise.
The output signal derived after the Kalman filter using the algorithms mentioned above is shown in figure 5.

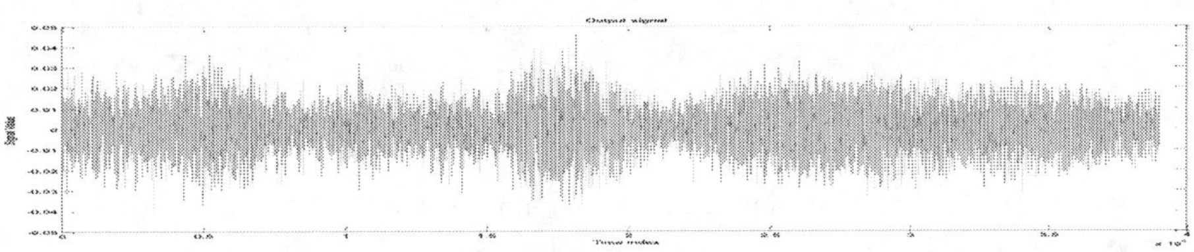

Figure 5. Output signal for Kalman filter

It is evident from the figure 5 that the noise cancellation in the Kalman filter output Signal is very substantial. The SNR Determined form the Kalman filter signal output ~ 69 dB. It is indicated that the noise reduction by the Kalman filter is very significant and SNR ~ 1.16 dB

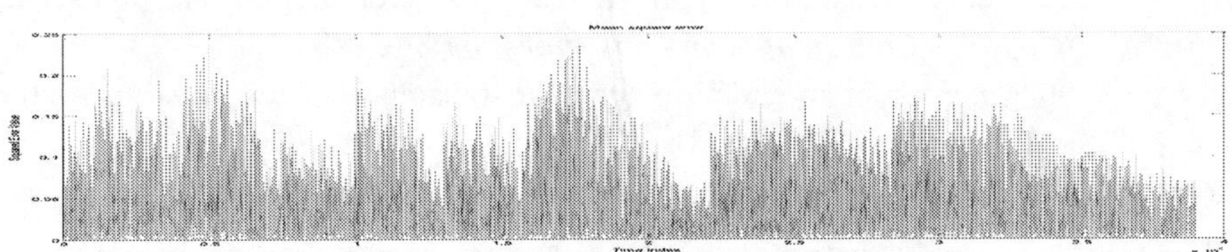

Figure 6. MSE of output Kalman filter signal

The mean square error distribution derived for the acoustic signal using Kalman filter is shown in figure 6. The MSE for the acoustic signal is found to be ~ 0.032 suggesting that the filtering process in very efficient.

The noise suppression for real time signal using the Reference of few paper of Kalman filters in found to be SNR ~0.2 dB and MSE~0.0030 for the non-stationary speech signal. The new method of the adoptive Kalman filter techniques are effective in noise suppression can be used only for specific applications in signal processing is show in table 1.

Table 1. Results of Non-Stationary Kalman filter

signal	SNR before(dB)	SNR after(dB)	Noise Suppression (dB)	MSE
Real time signal	-87.4679	86.2998	1.1681	0.0032

The result of the Kalman filtering of the non-stationary acoustic signal is summarized in table 1. It is clearly seen from the table that the noise cancellation ~1.17 dB using the Kalman filter for the non-stationary acoustic signal input. The results indicate that the Kalman filter techniques are very efficient in noise cancellation and can be effectively used in acoustic signal processing.

The noise suppression for real time signal using the Kalman filters in found to be SNR ~0.2 dB and MSE~0.0030 for the non-stationary speech signal. The adoptive Kalman filter techniques are effective in noise suppression can be used only for specific applications in signal processing.

6. Conclusions

From the results presented above the following conclusions can be drawn

1. The Kalman filter is found to be very efficient can effectively replace the adoptive filters for noise cancellation in non- stationary signals.

2. The noise cancellation is found to be very significant~ 1.17 dB and MSE ~ 0.032 are achievable for the acoustic non stationary signals.

Reference

[1] Kalman RE 1960 A new approach to liner filtering and prediction problems *Transaction of the ASME-Journal of Basic Engineering* **82** 35-45
[2] Saeed VV 2008 Advanced Digital Signal Processing and Noise Reduction *Fourth Edition, 2008, John Wiley & Sons, Ltd* 35-36.
[3] Morikawa H and Fujisaki H 1988 Noise Reduction of Speech Signal by Adaptive Kalman Filtering Special *M. Najim* **22** 1988, 53-68
[4] Mehra RK 1970 On the Identification of Variances and Adaptive Kalman Filtering *IEEE Trans. on Automatic Control* 175-184
[5] Stuart JF 1993 Fast Adaption Algorithms in Active Noise Control *Second Conference on Recent Advances in Active Noise Control of Sound and Vibration* 802 -810
[6] Bingham JAC 1990 Multicarrier modulation for data transmission: an idea whose time has come *Communications Magazine, IEEE* **28** 5–14
[7] Tekale PB and Kulkarni SR 2007 Modified Kalman Based NLMS Algorithm for Noise Cancellation *IJST* 0974-0107
[8] Lakshmikanth S, Natraj KR, Rekha KR 2014 Noise Cancellation in Speech Signal Processing, *International Journal of Advanced Research in Computer and Communication Engineering* **3** 1 – 12
[9] Josephine Sathya MA and Victor SP 2915 Noise Reduction Techniques and Algorithms for Speech Signal Processing *IJRSET* **2**

[10] Soleyman S 2016 Noise Removing of Audio Speech Signals by Means of Kalman Filter *International Journal of Advanced Biotechnology and Research (IJBR)* **7** 98-103

[11] Murugendrappa N and A G Ananth 2016 Interference Cancellation Using Adaptive Filter For Fractional Demine Methods *International Journal of Advances in Engineering & Technology* **9** 545-550

[12] Murugendrappa N and Ananth AG 2016 Noise Cancellation in DSSS by Using Adaptive LMS Filter in Fractional Demine Methods International *Journal of Innovative Research in Science Engineering and Technology* **5** 2319-8753

[13] Murugendrappa N and G Ananth 2017 Efficient Noise Suppression in Real Time Speech Signals Using Adaptive RLS Algorithms *International Journal of signal processing & image processing* **6** 2318-8679

IOP Conf. Series: Materials Science and Engineering **925** (2020) 012062 doi:10.1088/1757-899X/925/1/012062

Comparative Analysis to Identify Efficient Technique for Interfacing BCI System

Keerthi Kumar M[1], Parameshachari BD[2], Prabu S[3] and Silvia liberata Ullo[4]

[1,2] GSSS Institute of Engineering and Technology for Women Mysuru, India.
[3] Mahendra Institute of Technology, Tamilnadu, India.
[4] Engineering Dept., University of Sannio, Italy.
[1] keerthim.k123@gmail.com, [2] parameshbkit@gmail.com, [3] vsprabu4u@gmail.com, [4]silvullo@unisannio.it

Abstract. Brain Computer Interfacing (BCI) allows user to control certain devices or objects using human brain. With the availability of various sensors, it is possible to obtain the data from the human brain which controls the physical activities in the brain itself. BCI users can access a system capable of communicating efficiently in a special manner that works based on the brain signals. BCI gives a better platform for physically challenged people to do certain significant functions in their daily routine. The use of brain imaging technologies in BCI helps to enhance the signal quality during the communication among the machines and the human beings. The paper describes some of the research works carried on BCI and its advancements to perform a better communication between the computers and the human brain, by using emerging technologies. This paper also covers the applications, challenges, and future directions of BCI.

1. Introduction

The technological advancements lead to the development of different kinds of interactive systems that use different parts of the body to acquire the data. Example, Nam et al. [1] presented a paper on how a wheelchair is controlled by the movements of tongue. Liu et al. [2] proposed the human-computer interaction system that makes use of eye-tracking technique. BCI lets users to communicate with computer only through their brain signals. BCI studies mainly focused on people suffering from motor impairments i.e. paralyzed persons. BCI is an emerging field of research been studied since 1970s in various zones like biomedicine, neuroscience, computer science, automation and control engineering, but it is quiet relatively immature. The applications of BCI range from medicine to entertainment. The Human brain comprises a complex network of around hundred billions of neuron cells. Electrical signals generate in the brain when the person moves, feel, remember, or think about something. Sensors are used to detect and record these electrical signals which can be used later to control a device. BCI make use of brain signals captured by specialized equipment to control the devices and it doesn't involve any muscular activity, hence it forms a separate interaction mode between the humans and computer [26].

BCI captures signals directly from human brain. The three possible methods to acquire the brain signals are: (i) non-invasive technique, (ii) partially invasive technique and (iii) invasive technique. The non-invasive type allows the system to capture signals without any implant. Here the electrodes are placed on surface of the scalp. The partially invasive method captures information by means of implants that are placed below the skull. The risk factor is low to health compared to invasive method. Invasive method of capturing introduces the implants directly into subject's encephalic mass i.e. into the grey matter. This approach provides very good quality of signal recording. This method not only causes major risks to human health but also inconvenience for the user while using it. Non-invasive BCIs are more suitable and easy to use, and the advancements technology provides good quality of signal capturing. BCI forms a non-muscular way to express a user's intentions and it is particularly helpful for people with severe disabilities like amyotrophic lateral sclerosis,

2. Related Previous Work

In the previous research work, researchers in the field have scripted several survey papers to complete their work. In the field of computer science, automation, control engineering neuroscience, and biomedicine research work were carried out during the mid of 1970s [3] as per survey done by Ferreira et al.

Keum-Shik Hong et al. [4], this paper provides significant published articles on BCI in the year from 2009 to 2016 which is shown in Fig. 1. The figure describes the increase in number of articles published in this field. Arafat's [5] survey provides the complete history of Brain Computer Interface work carried out from 1929. The author concluded that this is the most challenging field of research. This BCI approach offers best result for the treatment of lunatic and differently abled. From the survey done by Dhabale et al. in [6] on BCI for controlling robots. The interface controls the robot using different techniques that include pattern classifier and from alpha waves of brain etc.

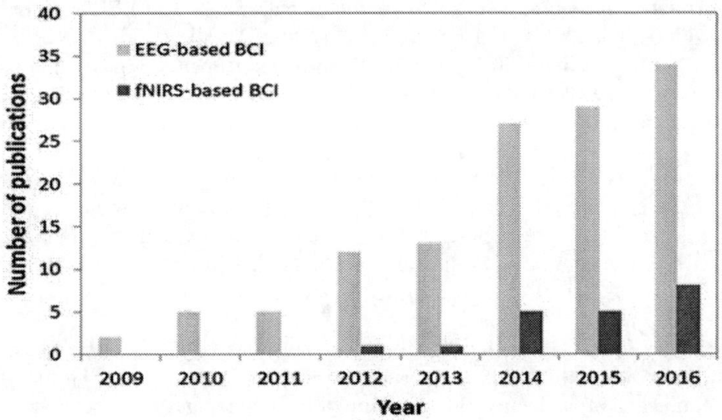

Figure 1. Published BCI articles from 2009 - 2016 [4].

3. BCI System

BCI system has four different stages, such as Signal acquisition, Signal Pre-Processing, Feature extraction & Classification.

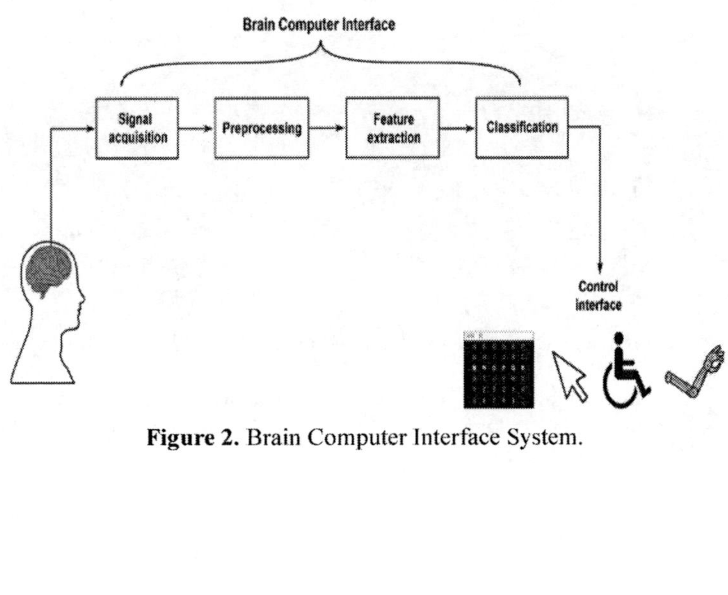

Figure 2. Brain Computer Interface System.

3.1. Signal Acquisition and Signal Pre-Processing

Data collection and filtering are carried out at the primary stage in BCI. Current Approaches concentrate on EEG signals [7], However, there are many other techniques which can record neural activity. Every technique has its own merits and demerits while capturing various parts of brain signals. And here are some common techniques for obtaining brain information.

3.1.1. Functional Magnetic Resonance Imaging (fMRI). This method captures activity of brain via flow of blood and blood oxygenation that increases particular region of mental activity. The method needs the use of large- scale equipment scanner that works on very high magnetic field. This technique involves analyzing the change in blood stream (hemodynamic response) of a subject in a magnetic resonance imaging machine (MRI). The response produced by the active neurons processes is referred as Blood Oxygen Level Dependency-BOLD [8]. But slow system response time becomes the main disadvantage of this method. The delay is related to BOLD response time of brain that can usually be delayed from 3 to 6 seconds. [9]. The drawbacks of this method include the following reasons: expensive, difficulty in understanding the working principle and fMRI can only provide the level of information about the blood flow in the brain.

3.1.2. Near-Infrared Spectroscopy (NIRS). This strategy additionally measures mind action through blood oxygenation and stream, however it depends on distinguishing variety of optical behaviors in cerebrum pictures. Near- infrared light passed through subject's forehead and the reflected lights are collected by light detectors. The output of detector is correlated to precise level of oxygen concentrations. The LED emits near infrared light that penetrate into the cerebrum tissue and it is collected by the detectors. The limitation of NIRS method is dependency of light passing via brain; i.e. hair like things can significantly interrupt the signals and gives flawed readings. Example of NIRS method:"On-line BCI spelling machine [10]. Limitation of NIRS method: The technique is unable to distinguish the attenuation level between the signals due to the absorbency of signals that overlap in the NIR range.

3.1.3. Magnetoencephalography (MEG). The method offers more sensors, hence provides more spatial information. To record the signal, user is placed in chair inside the magnetic shielded room equipped with a range of superconducting quantum interference devices (SQUIDs). The disadvantage of this method is the dependency of magnetic shielded room and the enormous brainwave sensor [11]. Limitation: This technique has difficulty in recording the activities of paroxysmal movement confined to the medial temporal lobe.

3.1.4. Electrocorticography (ECoG). Electrocorticography is an invasive type of approach that needs surgery to insert sensors straightforwardly onto the brain surface to acquire the signals. High spatial resolution, less vulnerability, high amplitude and broad bandwidth are some of the advantages of these techniques [12]. Limitations: Partial sampling period, recording is subjected to the stimulus of narcotic, analgesics/ anaesthetics, and the surgery.

3.1.5. EEG. The possible way to capture the activities of the brain neurons is by fixing the electrodes on the surface of the scalp. There is a standard technique that defines the position of electrodes, example: 10/20 positioning [13] standard. While placing the electrodes, a conductive gel is used as a medium that helps to capture the data with more accuracy.

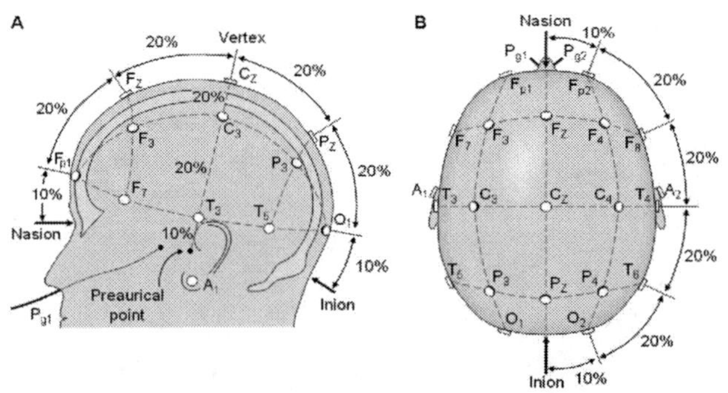

Figure 3. The 10/20-International positions and labeling [14]

EEG based BCI is a most popular method of capturing brain signals is only because of the following factors: standardized electrode placement, plenty of techniques available to acquire brain signals, and the comparatively low cost than other methods.The signal acquired typically much smaller compared to the surrounding signals, and those signals require special filtering methods to identify and extract the required information. The signal generated from the brain neurons is called as event related potential (ERP). Hence the recognition of this evoked response is the major objective of a BCI system. Identification of required signal is very difficult since the occurrence of artifacts in the EEG signal that created from facial muscle movement, eye-blinks or movements. The data related to ERP is normally analyzed in frequency domain; hence the information is transformed to frequency domain from time domain using Fourier transforms. The transformed signals then represent the spectral power of the signal.

3.2. Feature Extraction

A feature represents a unique property. Several linear and non-linear approaches for feature extraction have been described in the literature. Linear methods include Principal Component Analysis (PCA), Fast Fourier Transform (FFT), Independent Component Analysis (ICA), Wavelet transform (WT), Wavelet Packet Decomposition (WPD), Eigenvector, Autoregressive (AR). Non-linear methods consist of largest Lyapunov exponent (LLE), correlation dimension (CD), Fractal Dimension (FD), Higher Order Spectra (HOS), recurrence plots and phase space plots.

According to the Heisenberg uncertainty principle, it is hard to compute the signal in both frequency and time domains concurrently, because increasing accuracy in time domain will leads to decrease in accuracy in frequency domain, and vice-versa. A combination of features in both the domains may result in better results, when compared with features in each domain, separately [15]. Bidiagonalization[23] method can also be used for feature extraction of motor imagery EEG, where bidiagonalization is one of unitary (orthogonal) matrix decompositions.

3.3. Signal Classifications

Below table provides the information about commonly used classifiers in BCI [16].

Table 1. Some commonly used classifiers in BCI

Classifiers	Categories	Field of application
Linear Classifiers	Linear Discriminant Analysis (LDA) Support Vector Machine (SVM)[24].	A P300-based BMI system developed for remote writing using human brain-actuated robot arm. [17]. A spelling application for paralyzed patients (suffering from focal epilepsy) [18].
Neural Networks	Back Propagation BP network Multilayer Perceptron Learning Vector Quantization (LVQ) Neural Network	A new EEG recognition algorithm for integrating discrete wavelet transform (DWT) with BP neural network[19]
Nonlinear Bayesian classifiers	Bayes quadratic Hidden Markov Model	Virtual Reality BCI Application
Nearest Neighbor classifiers	Nearest Neighbors Mahalanobis distance	An application to extract features that the user can control and translate into device commands [20]

4. Types of BCI

This section provides the brief note on BCI types.

4.1. Active BCI

It is a commonly used prototype where the subject actively generates the brain signals that can be used for control applications. One such application is illustrated in [21], the active BCI is directly adopted to control the movement of paddles to left or right in a pinball game. The prototype was developed to distinguish two different types of motor-imagery, like right or left hand movement. Motor imagery is the most commonly used paradigm in the research. The short-lasting attenuation of brain oscillations and localized amplitude decrease of rhythmic activity is termed as event-related desynchronization (ERD).It is often found in the alpha and beta bands of individuals during imagery of movement and observation. It is considered to reflect the action-perception coupling and cortical motor activity.

When the subject visualizes the movement of right hand, the amplitude level of the mu-rhythm in the sensory-motor area of the human brain increases in the right hemisphere and decreases in the left hemisphere. For the left hand motor imagery, the reverse phenomenon occurs. The motor imagery thus results in spatially different patterns for brain activity according to which the hand is used. The use of signal processing and classifiers allow BCI to notice the spatially different patterns which decides the movement direction of the subject. The direction information then used to control left or right paddles, which is similar to conventional interfaces like left or right keyboard arrows. Here the subject initiates activity of brain; therefore, the information embedded in the captured signal is then used to control the paddle movements.

4.2. Reactive BCI

Information regarding the user's intention is embedded in the stimulation response signal. The user can activate their brain signals by simply taking care of certain stimuli or looking at them. The steady state visual evoked potential (SSVEP) signals are natural responses to visual stimulation at specific frequencies. The SSVEP signal is used to control some application like "Mind Balance" game. The 17 and 20 Hz flickering checkerboards are placed on the right and left side of a game icon called avatar. The user maintains balance of the avatar by focusing on either the right or left checkerboard. The brain activity in the occipital region contains the correlated frequency information in response to visual stimulation. The P300 (P3) wave is an event related potential (ERP) component stimulated in the process of decision making. As an example of SSVEP, P300 paradigm is widely used in the development of reactive BCI.

4.3. Passive BCI

The function of the passive BCI system is not to give the user control and no effort on the part of the user is required; instead, it automatically monitors the mental state of the user. It quantifies the attention level or distinguishes emotional conditions to facilitate communication between the system and the user. Example: for attention index quantification, frequency band ratios such as sensory- motor rhythm (12–15 Hz), beta (13–30 Hz) and theta(4–8 Hz) are used; for relaxation level, alpha(8–13 Hz) power is used.

5. Applications of BCI

The BCI technology in different forms have proved by its performance in some extensive applications such as medical, communication, security, marketing, entertainment etc. This technology was initially introduced to support physically challenged people and now expanded enormously in all the fields.

5.1. BCI for Prosthetic Applications

5.1.1. Vision. Invasive technologies were utilized initially for treating blindness. Scientist William Do belle successfully implemented a prototype of a functioning brain interface implanted in a man blinded visual cortex in the year 1978 that enclosed nearly 69 electrodes and produced a light sensation without light affecting the eye. In a 2008 surgery, American scientists used important hardware to insert electro-biological parts and sensors into a volunteer's brain. The person blind from birth voluntarily agreed for this research to fix a camera to electro biological organs to replace the eye. This research worked after a tedious surgery of nearly 48 hours and 7 days after medication. The surgery was successful and the person began to identify, observe everything; even though the camera used was colour her vision was black and white.

5.1.2. Movement. The BCI is used for people with paralysis so that they can activate their paralyzed body parts.It is usually accomplished by giving them dedicated devices to move their paralyzed muscles. The device provides various kinds of interfaces of computers with knee joints, legs, hands, robot arm, and other parts of the body for control. Till 20th century, the wooden hands/legs were used as a substitute for any injury. Later the advancement in technology helps to develop the metallic arm which can be controlled by user with their brain signals. The invention of robotic arm gave birth to the development of "Mind Controlled Prosthetic Robotic Arm". The prosthetic arm is coupled to sensory cortex and motor cortex using wires; as a result, when user touches object it feels and responds to the stimulus. Signals from motor cortex helps to control the pressure and motion of the robotic arm. Sensory cortex helps to provide the sensation when the arm touching something. Fig. 4 depict the schematic for controlling prosthetic arm using brain signals.

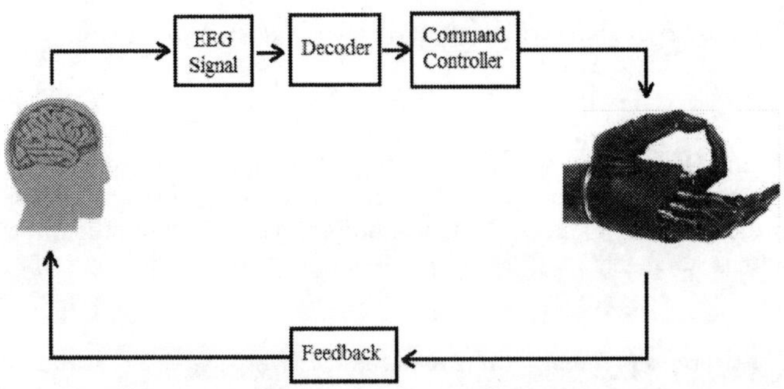

Figure 4. The mind controlled prosthetic robotic arm.

5.2. Entertainment Applications

The connection between the brain and the computer has developed into a cutting- edge technology as neuro-gaming. Some of the neuro-games include, "Mind Balance video game", "NeuroMage", "NeuroRacer" and "Throw trucks with your mind". Brain Computer Interface offers enormous applications with signal acquisition techniques, feature extraction techniques and different classifiers. Some applications in this area are listed in Table 2:

Table 2. Research works carried out in the field of BCI [22]

Sl. No	Topic	Description
1	Driving a wheel chair using BCI	Brain Computer Interface is used to control the movement of an autonomous wheelchair
2	The life of Rats	Controlling Rat's brain using Brain Machine Interface (BMI), Electroencephalography (EEG).
3	Human thought to control robots	BMI is used to control human thoughts and later used to control ASIMO- Honda's Humanoid Robot.
4	EPOC- A neuro headset for gaming.	It detects non-conscious thoughts, expressions and conscious emotions based on electrical signals around the brain. It opens up a plethora of new applications The application can be controlled with our expressions, thoughts, and emotions.
5	BCI based Brain to brain Communication.	A experiment had a person using BCI headset to send the thoughts, translated as a series of binary values, over the internet to other person whose PC receives the binary digits and sends them to the second user's through flashing an LED light. The ciphered information is then extracted and deciphered to know whether a zero or a one was transmitted.
6	Gesture recognition.	Here the person allowed wearing "data gloves" with illuminated LEDs. The information sent through the flashing LEDs are tracked by two pairs of webcams interfaced to a computer working to produce an all-round binocular view. This allows the PC to monitor the person's shoulder or hand movements. This input can then be fed to a program, a game, or simulator, or to control a character, an avatar, in a 3D virtual environment.

In brain-to-brain communication system the data extracted in the form of images from the activities of brain neuron can be transferred from one system to another. Partial image encryption [25] algorithm can be used to secure these data.

6. Challenges Faced by BCI Technology

Each and every success is followed by several challenges. Similarly, researchers faced many challenges, in overcoming uncertainties and creating something extraordinary and new. However, brain- computer interfaces are not new to us, but yet lot of research has to be done.

6.1. Human Brain Activity

For engineers, it is very difficult to understand the functionality of the human brain.

The user must first identify the key regions which should be emphasized when working on prosthesis or other BCI devices, to find a solution for challenging job.

6.2. Usefulness
In all cases, person wants the BCI system to be accessible. Developing a system to be user-friendly with such a complexity is challenging and tedious task. The EEG process involves applying a cream to the sensors to measure the accurate signal amplitude values. The recent EEG technology works without the injection of gel into the sensors and also gives 70% accurate results.

6.3. Hardware Problem
The challenging task of hardware implementation is that, the hardware system must be capable of operating outside the labs and in sensitive areas. These BCI systems must be intelligent enough to work outside the appropriate environment especially noisy locations. They must be capable to distinguish between random inputs and actual inputs in such positions.

7. Conclusion
The work gives the information about the most remarkable techniques and examples in the field of brain computer interfacing system. The information here highlights the most common uses. According to the review, a great deal of improvement is needed in order to create a system which can be worked in real conditions. EEG based BCI is a most popular and best suitable method of capturing brain signals is only because of the following factors: standardized electrode placement, plenty of techniques available to acquire brain signals, and the comparatively low cost than other methods.

References
[1] Nam Y, Zhao Q, Cichocki A and Choi S 2012 Tongue-Rudder: aglossokinetic-potential-based tongue-machine interface *IEEE Transactions on Biomedical Engineering* **59** 290–299
[2] Liu SS, Rawicz A, Rezaei S, Ma T, Zhang C, Lin K and Wu E 2012 An eye-gaze tracking and human computer interface system for people with ALS and other locked-in diseases *Journal of Medical and Biological Engineering* **32** 37–42
[3] Ferreira ALS, de Miranda LC, de Miranda EEC and Sakamoto SG 2013 A survey of interactive systems based on brain-computer interfaces *SBC Journal on Interactive Systems* **4** 3-13
[4] Keum-Shik H and Jawad Khan M 2017 Hybrid Brain–Computer interface Techniques for improved Classification Accuracy and increased Number of Commands: *A Review published* **24**
[5] Arafat I 2013 Brain-Computer Interface: Past, Present & Future, *International Islamic University Chittagong (IIUC).*
[6] Dhabale M. I. S. S. R. U. T. U. J. A Literature Survey on BCI Conrolled Robot.
[7] Bekcer H. 1932 Uber das Elektrenkephalogramm des Menschen. IV, Mitteilung. Arch. f. Psychiat 278 1932
[8] Logothetis NK 2002 The neural basis of the blood-oxygen-level-dependent functional magnetic resonance imaging signal *Philosophical transactions of the Royal Society of London Series B, Biological sciences* **357** 1003–37
[9] Weiskopf N, Mathiak K, Bock SW, Scharnowski F, Veit R, Grodd W, Goebel R and Birbaumer N Principlesof a brain-computer interface (BCI) based on real-time functional magnetic resonance imaging (fMRI) *IEEE transactions on bio-medical engineering* **51** 966–70
[10] Sitaram R, Zhang H, Guan C, Thulasidas M, Hoshi Y, Ishikawa A, Shimizu K and N. Birbaumer, Temporal classification of multichannel near-infrared spectroscopy signals of motor imagery for developing a brain-computer interface *NeuroImage* **34** 1416–27
[11] Mellinger J, Schalk G, Braun C, Preissl H, Rosenstiel W, Birbaumer N and K¨ubler A 2007 An MEG-based brain-computer interface (BCI) *NeuroImage* **36** 581–93
[12] Leuthardt EC, Miller KJ, Schalk G, Rao RPN, and Ojemann JG 2018 Electrocorticography-based brain computer interface–the Seattleexperience *IEEE transactions on neural systems and rehabilitationengineering*
[13] Jurcak V, Tsuzuki D, and Dan I 2007 10/20, 10/10, and 10/5 SystemsRevisited: Their Validity As Relative Head-Surface-Based Positioning Systems *NeuroImage* **34** 1600–11
[14] John, Where Do The Electrodes Go?[Online]. Available: http://www.diytdcs.com/tag/1020-positioning/
[15] Mehdi H, Mehrani M, Mehranzade A and Forootan F 1996 Proposing an Effective Feature Extraction Model for EEG Signals to Enhance Quality of Hand's MotionDetection *Journal of Academi.*
[16] Bishop C 1995 Neural Networks for Pattern Recognition Oxford Univ

[17] Wolpaw JR, Birbaumer N, McFarland DJ, Pfurtscheller G, and Vaughan TM 2002 Brain–computer interfaces for communication and control Clinical *Neurophysiology* **113** 767–791
[18] Support vector machine (http://en.wikipedia.org/wiki/support vector machine).
[19] Zhang L, Cichocki A, Amari S 2004 Self-adaptive blind source separation based on activation functionsadaptation *IEEE Transactions on Neural Networks* **15** 233– 244
[20] Li MA, Rui W, Dong-Mei H, and Jin-Fu Y 2009 Feature Extraction and Classification of Mental EEG forMotor Imagery *Proceedings of the 2009 Fifth International Conference on Natural Computation* **139**
[21] Tangermann M, Krauledat M, Grzeska K, Sagebaum M, Blankertz B, Vidaurre C, Müller KR 1648 Playing inball with Non-Invasive BCI *In Advances in Neural Information Processing Systems* **21** 1641–1648
[22] http://www.airlab.elet.polimi.it., and www.Sciencedaily.com.
[23] Parameshachari BD 2020 A Block Bi-Diagonalization-Based Pre-Coding for Indoor Multiple-Input-Multiple-Output-Visible Light Communication System *Energies* **13** 1-16
[24] Parameshachari BD 2016 Demand side management of small scale loads in a smart grid using glow-worm swarm optimization technique *Microprocessors and Microsystems* **71** 102886
[25] Parameshachari BD 2019 Controlled Partial Image Encryption Based on LSIC and Chaotic map, 4th International Conference on Multimedia and Image Processing (ICMIP 2019), *University of Malaya, Kuala Lumpur, Malaysia* 60-65
[26] Nithya V, and Ramesh GP 2020 Wireless EAR EEG Signal Analysis with Stationary Wavelet Transform for Co Channel Interference in Schizophrenia Diagnosis *In Recent Trends and Advances in Artificial Intelligence and Internet of Things* 253-265

IOP Conf. Series: Materials Science and Engineering **925** (2020) 012063 doi:10.1088/1757-899X/925/1/012063

Design of MUX based Flash ADC for Reduction in Number of Comparators

Tejaswini J Kutre[1], Sujata N Patil[2] and Sheela Kiran Kore[3]

[1,2,3] Department of Electronics & Communication Engineering, KLE Dr. M S Sheshgiri College of Engineering & Technology, Belgaum, Karnataka, India
[1]kutretejaswini@gmail.com, [2]sujata.patil@klescet.ac.in, [3]sheelakoreys@gmail.com

Abstract. This paper presents the design of multiplexer-based Flash ADC with a reduced number of comparators to achieve less area and less power consumption with increased resolution. As the number of bits increases, flash ADC needs a huge number of comparators which increases the area of the chip, and also the power consumption will increase. The conventional N-Bit Flash ADC requires a 2^N number of resistors and 2^N-1 number of preamplifiers and comparators. Here the number of comparators is reduced by using multiplexers by providing reference voltage through multiplexers. And also, the encoder is designed using multiplexers. The 6-bit Flash ADC is designed utilizing multiplexers and a reduced number of comparators. The Simulation is done by using the proteus 8.9 design suite with 1v supply voltage.

Keywords: Comparators count, Flash ADC, Multiplexer, Power consumption.

1. Introduction

In all the digital systems data conversion is the main process that is done with the help of Analog to digital converter. There are many types of ADC's available for this conversion but among those Flash ADCs are more advantageous because of their high speed. Flash ADC has a high speed due to the simultaneous conversion. The addition of the comparator delays and the logic delays (logic delays are negligible) gives the conversion speed of flash ADC. Flash ADC is provided with input voltage which needs to be converted and the unique reference voltage. The conventional flash ADC includes the resistor ladder network, preamplifiers, comparators, and priority encoder. The specified figure 1 exhibit the structure of conventional Flash ADC, For N-bit transformation, 2^N number of resistors and 2^N-1 number of comparators and preamplifiers are needed [1]. Here reference voltage is provided by using a resistor ladder network for each comparator [2-3]. Each comparator is provided with a reference voltage of 1 LSB (least significant bit). The comparator yields logic "1" if the signal applied on the input line is more prominent than the reference voltage. Conversely, the comparator yields logic "0", if the signal applied on the input line is smaller than the reference voltage and then all the comparators outputs are provided to the priority encoder for final conversion into digital output [4].
In conventional Flash ADC increase in resolution increases the number of comparators. we need to reduce the comparators without limiting the resolution of the flash ADC. To achieve less area and less power consumption.

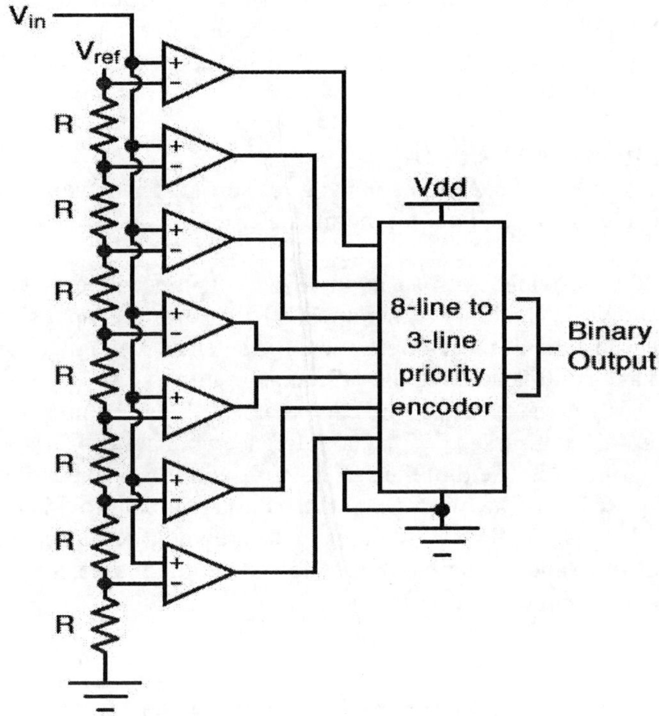

Figure 1. Conventional Flash ADC.

The above figure 1 shows a 3bit conventional flash ADC with 7 comparators, 7 registers, and 1 (8 to 3) priority encoder. Here the input is given to one of the input lines of all the comparators and reference voltage to all the comparators is provided with register ladder network. The outputs from all the comparators are given to the priority encoder for the final conversion and the output of the priority encoder is the final digital output of the conversion.

The reference voltage is divided into steps and the step size is given by vref / 2^N. The resolution of Flash ADC is defined by the step size. The conventional Flash ADC has the disadvantage of limited resolution, larger area, and high power consumption so it can be improved by using multiplexers in the design of Flash ADC. The proposed design of Flash ADC has many benefits compared to conventional Flash ADC. In the proposed work Flash ADC needs only ($2^{(N-2)} + 1$) number of comparators, which is the need for high-resolution ADCs [5-6].

2. Literature Review

The flash ADC with low power and low resolution with different sampling rates is used in many versatile applications because of its effortlessness and quick activity. The flash ADCs also suffer from huge input-referred noise and comparator offsets. To remove these problems many strategies have been presented. Preamplifiers count can be reduced by using interpolation techniques [1-2]. For N-bit Flash ADC needs a 2^N number of registers and 2^{N-1} number of preamplifiers and comparators so the fabrication area on-chip will increase as the number of bits increases which also increases the power consumption due to a large number of components. So low power and low voltage comparators are designed like double tail CMOS comparator to reduce the power consumption [3-4]. The high speed and high-resolution comparators are designed by using BiCMOS and CMOS VLSI technologies. We will have the option to accomplish a higher goal and speed by keeping up low power, less information capacitance, and low intricacy. With 12-bit resolution and 10 MHz of comparison rates, the BiCOMS and CMOS comparators are designed [5]. comparator operation can also be improved by modifying the threshold of the comparator circuit. In this TIQ comparator can also be used to replace the resistor network but it suffers from input frequency to overcome this improved CMOS inverter comparator is proposed to reduce area and power requirement. A flash ADC with a bit reference encoder can be designed for high-speed applications [6-18]. The Flash ADC can also be designed using domino CMOS logic for reducing power consumption and increasing the speed [7].

3. The architecture of Proposed Flash ADC

The proposed 6-bit MUX based flash ADC contains a resistor ladder network, a (2x1) multiplexer, 8 (4x1) multiplexers, 17 comparators, and a 6-bit thermometer to the binary encoder. The below figure 2 shows the optimized design of flash ADC using multiplexers.

The proposed Flash ADC is provided with an input signal which needs to be converted into a digital signal and the reference voltage. The first comparator (C1) compares the input signal with the reference voltage level 32/64 vref. The yield of this comparator C1 is taken as the digital outputs most significant bit (MSB) and provided as a control signal to the preceding multiplexers (M1-M9). Similarly, the same input signal is simultaneously given to the second comparator (C2) and compared with either 16/64 vref or 48/64 vref by using a 2x1 multiplexer [7-9]. The reference voltage for C2 is selected depending on the first comparator (C1) yield. The yield of the second comparator is again taken as the digital outputs second most significant bit (MSB). The two MSB's of the digital output provided from comparators (C1 & C2) acts as the control signal to all 4x1 multiplexers to provide a reference voltage to all the other comparators. The output from all the comparators is fed to the thermometer to the binary encoder to convert the input signal into digital output [10].

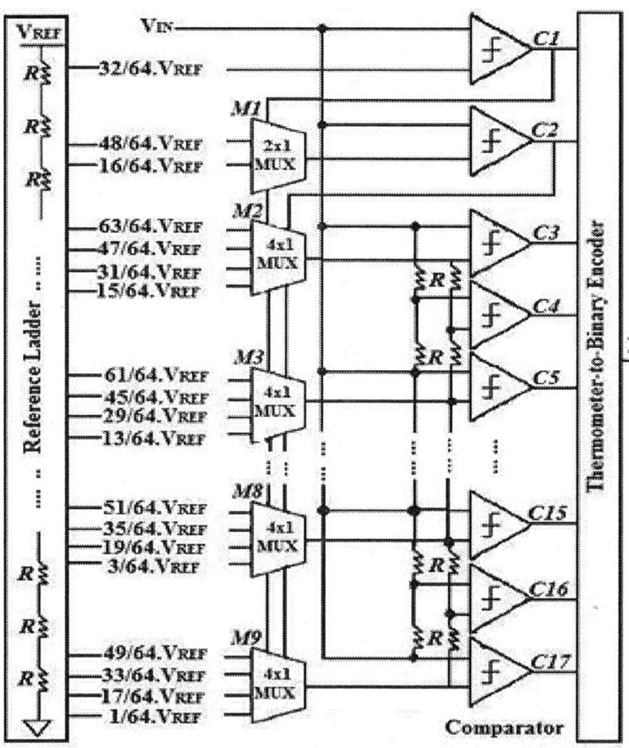

Figure 2. The architecture of 6-bit MUX based Flash ADC.

In this work, the new low power, low offset CMOS comparator is introduced. With this comparator, the elimination of preamplifiers and track and hold circuit is achieved. Because comparators are the sampling devices here no need to have a track and hold circuit at the input side. The number of comparators is also reduced; therefore, less fabrication area and reduced power consumption are achieved.

Table 1. The total preamplifiers and comparators required.

Resolution (Bit) N	Flash ADC		Proposed Flash ADC	
	Pre-amp $2^N - 1$	Comparators $2^N - 1$	Pre-amp No use	Comparator $2^{N-2} + 1$
4	15	15	-	5
6	63	63	-	17
8	255	255	-	65
10	1023	1023	-	257
12	4095	4095	-	1025

Where N denotes the Resolution.

Table 1 shows the comparators and preamplifiers count in the conventional and MUX based flash ADC with different resolutions. From the above table, we can observe that for 6-bit conventional flash ADC 63 comparators and preamplifiers are required whereas in proposed flash ADC only 17 comparators are required and preamplifiers are eliminated.
Thus, we can observe that number of comparators has been subsequently reduced and due to a reduction in comparators count area and power consumption will be reduced. Also, the preamplifiers are eliminated. so, this can be used in the various applications where the power and area are of major interest [11].

4. Applications of MUX Based Flash ADC

These digital systems find applications in various high-speed devices that range from audio communications to medical applications. Numerous remote versatile applications require fast information converters with the high sign to commotion ratio, wide transmission capacity, and variable resolution with streamlined power and decreased expense. so, we need to redesign the ADC to meet prerequisites as indicated by developing advances. The digital signal p: where N s the number of resolution effective way of signal processing in various applications the input and output signals will be in the analog form so conversion from analog to digital is required at interfaces. so, this flash ADC will act as the interface between the analog signal coming from the real world and the functional block of DSP. This flash ADC can be used in video-imaging systems and personal communication systems. The high speed and high-resolution flash ADCs can be used in wireless communication and image processing. It also finds applications in Radar detection, Wideband radio receivers, Electronic test equipment, optical communication links, Satellite communication, Sampling oscilloscope, High-density disk drivers, and universal DSP receivers, etc. [12-16].

5. Components in MUX Based Flash ADC

5.1 Comparator

The comparator is a very important component in Flash ADCs, here it acts as a quantizer in the ADCs. It is of 1 bit having two levels either a '1' (implies VDD = +2.5) or '0' (implies VSS = - 2.5). Whenever input applied to the comparator is more prominent than the reference voltage at that point yield is 1 and whenever the contribution of input voltage of the comparator is not as much as the reference voltage at that point yield is 0. Here we have utilized two-stage CMOS OPAMP as a comparator to accomplish faster operation and low power utilization. The comparator comprises current mirrors, current sinks, active load, and constant current source. Transistors W/L proportions are chosen so that comparators give the exact and ideal outcomes. Parasitic impacts that influence the performance of the comparator are limited in this structure. This helps to get a low power and high-speed comparator [13-14].

IOP Conf. Series: Materials Science and Engineering **925** (2020) 012063 doi:10.1088/1757-899X/925/1/012063

The comparator sampling speed is obtained by its regeneration time constant which is contrarily proportional to the square of gate length for a CMOS technology [15]. Due to very low matching requirements, the comparator can be optimized for high speed, low power, and low area requirements. The generation of bubble errors is eliminated [16].

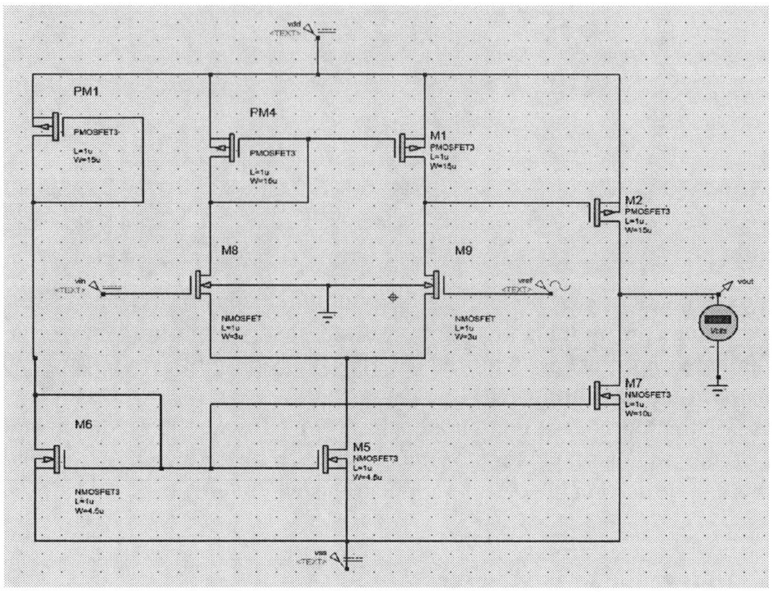

Figure 3. Schematic of comparator.

Figure 3 shows a schematic of the CMOS comparator made using a two-stage operational amplifier.

5.2 Thermometer to Binary Encoder

It is used to create binary data that can be processed in real-time. There are many ways to design the encoder but most of the designs have the disadvantages like less speed, high complexity, the occurrence of bubble error problem, high power consumption, and unbalanced signal paths so to avoid these all problems multiplexer based thermometer to the binary encoder is designed with lower bubble error suppression, less delay, lower critical path, and less area. Due to these all benefits this encoder can be used in many low power applications. Here we have designed an 8-to-3 MUX based thermometer to a binary encoder that has 7 OR gates and 4 (2X1) multiplexers as shown in figure 4. The outputs of all the comparators which is the thermometer code is fed to the OR gates and the OR gate outputs are provided to the multiplexer inputs. The mux gives the output according to the control signal provided to it .one output bit is taken directly from the OR gate and the other two bits are taken from the two Multiplexers makes the final digital output. This works like a priority encoder by selecting higher-order active inputs from comparators output and converts them into binary and ignores all other active inputs. Here multiplexers are used to correct the bubble errors that occurred due to mismatches in the comparator's inputs. Table 2 denotes the 8 to 3 encoder truth table.

IOP Conf. Series: Materials Science and Engineering **925** (2020) 012063 doi:10.1088/1757-899X/925/1/012063

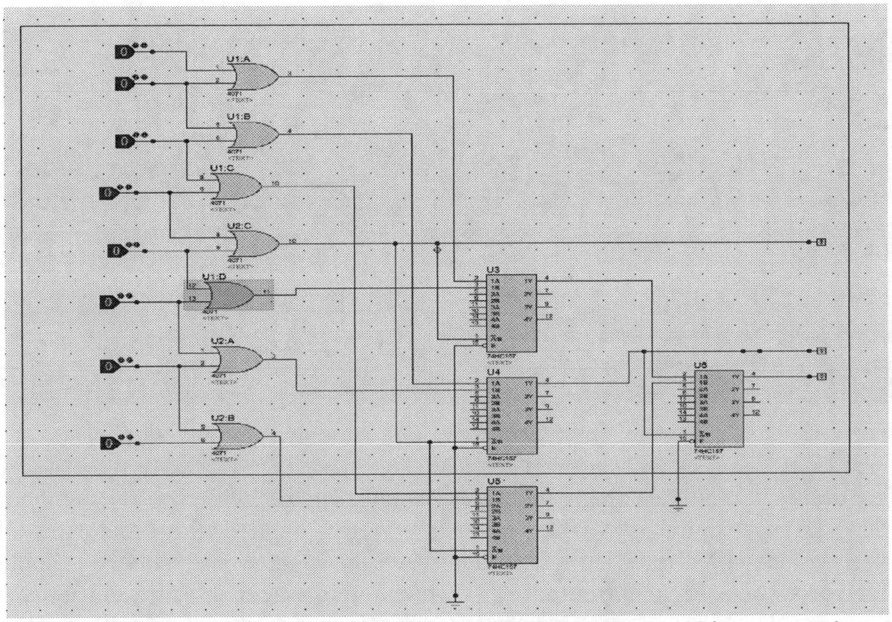

Figure 4. Block diagram of a 3-bit thermometer to a binary encoder.

Table 2. The 8 to 3 Encoder truth table.

D7	D6	D5	D4	D3	D2	D1	D0	B2	B1	B0
0	0	0	0	0	0	0	1	0	0	0
0	0	0	0	0	0	1	X	0	0	1
0	0	0	0	0	1	X	X	0	1	0
0	0	0	0	1	X	X	X	0	1	1
0	0	0	1	X	X	X	X	1	0	0
0	0	1	X	X	X	X	X	1	0	1
0	1	X	X	X	X	X	X	1	1	0
1	X	X	X	X	X	X	X	1	1	1

Where X denotes don’t care.

Comparator simulations

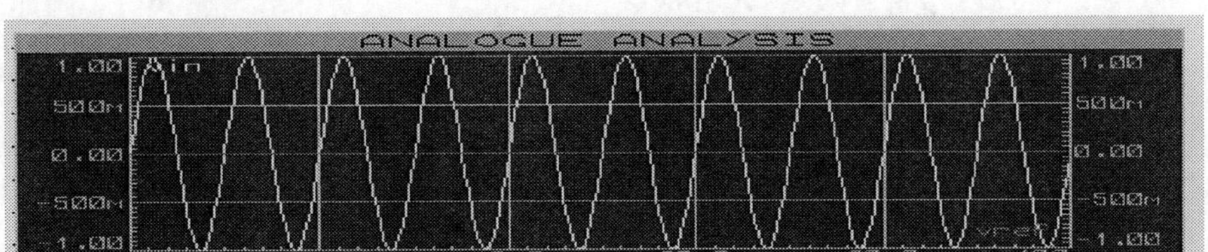

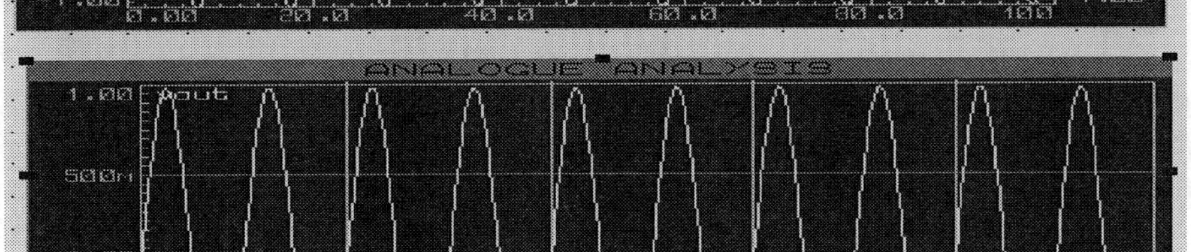

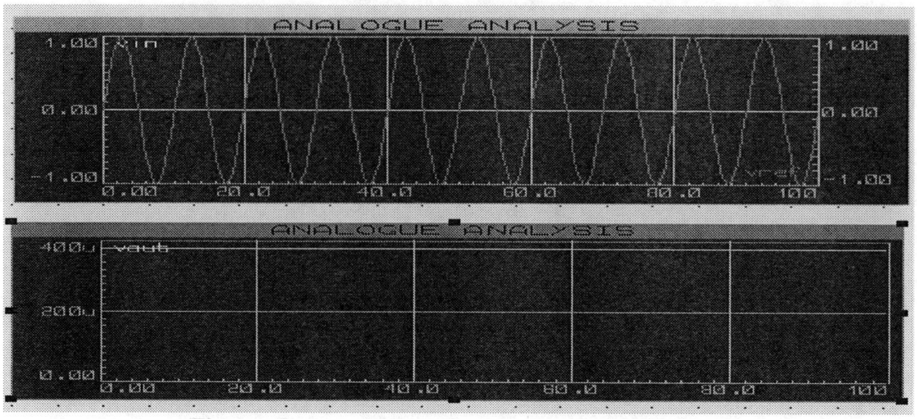

Figure 5. waveforms of comparator simulations

The comparator design is simulated on the proteus 8.9 simulation tool with a power supply of +2.5v and -2.5v. The sinusoidal input signal of 1v with a frequency 5kHz is applied with a reference voltage of 0v, we get the output as 1v which refers to +2.5 v and when the input signal with 0v is applied with 1v sinusoidal reference voltage we get the output as 0v which refers to -2.5v as shown in above waveforms. When the input signal is prominent compared to the reference signal then the positive half cycle of the input signal become visible at the output which represents +2.5v VDD supply and when the input signal is smaller than the reference signal the output we get as 0v because here we have used reference signal as 0v which represents the -2.5v VSS supply.

Table 3. Comparison of flash ADC's

Specification (unit)	[7]	[17]	[11]	This work
Publication title	CICC 2011	2012	TENCON 2013	-
Technology	90nm	180nm	180nm	90nm
Resolution (Bit)	6	4	6	6
Supply voltage(v)	1.2	2.5	1	1
Power (mw)	12	bandwidth 2.511E6 Hz	0.4	1.19

6. Conclusion

In this paper, MUX based Flash ADC is designed by using Proteus 8.9 simulation tool. The number of comparators is reduced and the preamplifiers are eliminated. So, for the design of N-bit flash ADC, $2^{N-2}+1$ number of comparators are required instead of $2^N - 1$. And in the 6-bit Flash ADC, the comparators are reduced by 73%. Hence, the proposed MUX based Flash ADC can have less area and less power consumption.

7. Future Scope

The proposed flash ADC has reduced the number of comparators from $2^N - 1$ to $2^{N-2} + 1$, further it can be reduced to increase the resolution. In the proposed design comparators are reduced by using multiplexers and designing low power, the high-speed comparator in the future can be reduced by using any other methods. In the present design, the encoder is designed by using multiplexers which convert the thermometer code into binary output, in future ADC with different encoders can be designed which can be used in high-frequency applications.

IOP Conf. Series: Materials Science and Engineering **925** (2020) 012063 doi:10.1088/1757-899X/925/1/012063

References

[1] Lee JI and Song JI 2013 TENCON IEEE Region Conference 1-4
[2] Aditya A, Rakesh Chowdary G, Meenakshi J, Praveen TB, and Vamsi Krishna MS 2013 Signal Processing Image Processing & Pattern Recognition ICSIPR 107-111
[3] Babayan- SM and Lotfi R 2014 IEEE Transactions on Very Large-Scale Integration (VLSI) Systems **22** 343-352
[4] Razavi B and Wooley BA 1992 IEEE Journal of Solid-State Circuits **27** 1916-1926
[5] Sail E and Vesterbacka M 2004 TENCON IEEE Region 10 Conference **4** 250-253
[6] Lee WT, Huang PH, Liao YZ Lee JI and Song JI 2013 TENCON IEEE Region Conference 1-4
[7] Aditya A, Rakesh Chowdary G, Meenakshi J, Praveen TB, and Vamsi Krishna MS 2013 Signal Processing Image Processing & Pattern Recognition ICSIPR 107-111
[8] Babayan- SM and Lotfi R 2014 IEEE Transactions on Very Large-Scale Integration (VLSI) Systems **22** 343-352
[9] Razavi B and Wooley BA 1992 IEEE Journal of Solid-State Circuits **27** 1916-1926
[10] Sail E and Vesterbacka M 2004 TENCON IEEE Region 10 Conference **4** 250-253
[11] Lee WT, Huang PH, Liao YZ and Hwang YS 2007 Electron Devices and Solid-State Circuits EDSSC 2007 IEEE Conference on, Tainan, 341-344
[12] Mikael G, Jacob WJ and Nianxiong NT 2000 CMOS Data Converters for Communications, Kluwer Academic Publishers
[13] Sail E and Vesterbacka M 2004 TENCON IEEE Region 10 Conference **500** 250-253
[14] Yang DXD and Bidermann WR 2001 Multiplexed multichannel bit serial analog-to- digital converter Google Patents
[15] Jamal SM, Fu D, Chang NJ, Hurst PJ, and Lewis SH 2002 IEEE Journal of Solid-State Circuits 37 1618-1627
[16] Boni S, Kumar MN International Journal of Electronics & Communication Technology 8
[17] komali V, Balaraju G 2017 International Journal of Advance Engineering and Solid-State Circuits **37** 1618-1627
[18] Kale S and Gamad RS 2010 International Journal of Electronic Engineering Research **2** 29-34
[19] Daly DC and Chandrakasan AP 2009 IEEE journal of solid-state circuits **44** 3030-3038
[20] Sunghyun P 2007 In IEEE Journal of Solid-State Circuits **42** 1865-1872
[21] Aida V 2013 IEEE Journal of Solid-State Circuits, 48 3038- 3048
[22] SD Panchal, Gajre SS, Ghanwat VP 2012 International Journal of Advanced Research in Computer and Communication Engineering **1**
[23] Senthil Sivakumar M, Sowmya Priya M 2019 International Journal of Recent Technology and Engineering (IJRTE) **7**
[24] Hwang YS 2007 Electron Devices and Solid-State Circuits EDSSC 2007 IEEE Conference on, Tainan, 341-344
[25] Mikael G, Jacob WJ and Nianxiong NT 2000 CMOS Data Converters for Communications, Kluwer Academic Publishers
[26] Sail E and Vesterbacka M 2004 TENCON IEEE Region 10 Conference **500** 250-253
[27] Yang DXD and Bidermann WR 2001 Multiplexed multichannel bit serial analog-to- digital converter Google Patents
[28] Jamal SM, Fu D, Chang NJ, Hurst PJ, and Lewis SH 2002 IEEE Journal of Solid-State Circuits **37** 1618-1627
[29] Boni S, Kumar MN International Journal of Electronics & Communication Technology 8
[30] komali V, Balaraju G 2017 International Journal of Advance Engineering and Solid-State Circuits **37** 1618-1627
[31] Kale S and Gamad RS 2010 International Journal of Electronic Engineering Research 29-34
[32] Daly DC and Chandrakasan AP 2009 IEEE journal of solid-state circuits 44 3030-3038
[33] Sunghyun P 2007 In IEEE Journal of Solid-State Circuits **42** 1865-1872
[34] Aida V 2013 IEEE Journal of Solid-State Circuits **48** 3038- 3048
[35] Panchal SD, Gajre SS, Ghanwat VP 2012 International Journal of Advanced Research in Computer and Communication Engineering **1**

[36] Senthil Sivakumar M, Sowmya Priya M 2019 International Journal of Recent Technology and Engineering (IJRTE) **7**

Real Time Design and Implementation of FPGA Based Timing and Control System for Wireless Radar Communication

P Kamaraj[1] **and R Jagadeesh Kannan**[2]

[1] National Atmospheric Research Laboratory (NARL), Tirupati, India
[2] School of Computer Science & Engineering, Vellore Institute of Technology Chennai Campus, India.
[2]jagadeeshkannan.r@vit.ac.in

Abstract. This paper introduces real time application of FPGA based Timing Control Signal Generation card for Active Phased Array Radar Transmit Receive modules wireless radar communication. Active phased array radars usually have Transmit-Receive (TR) modules located near to the antenna elements and all the TR modules are to be synchronized for proper operation of the radar system. The TCSG is the heart of radar system which is designed to provide interface for digital controls, analog parameter monitoring, and timing and control signals generation required for 1 kW TR Module operation. The main objective of TCSG is to control and monitor the field located TR modules remotely from the control and instrumentation room, usually far away from the antenna field. Each TCSG card is provided with unique IP address and port number to distinguish each card though they perform same kind of operations. After remote connection establishment, the TCSG card responds to the commands given for controlling or monitoring the TR module parameters. A PC based user interface allows sending controls and monitoring the health of the TR modules. The functionality of the TCSG card is verified with a test bench setup and the results shows that the TCSG card is working properly for all the user parameters. The proposed framework comprises of detailed design, architecture, visualization, implementation and integration on a 1 kW high power TR module for the radar communication. Simulation of various control pulses for radar real time operations are carried out using Vivado and updated to the real time application using Artix-7 FPGA.

Keywords: Artix-7 FPGA, TR-Module, Radar, Vivado, Analog to digital Converter (ADC), Ethernet PHY

1. Introduction

Active Phased Array Radars [1] have become the modern radar systems since their architecture overcome the major problems with the passive array. The active array radars have dedicated transmit-receive (TR) modules [2] for each antenna which are distributed in the antenna array field. The complexity of such radar lies in achieving the synchronous operation of the TR modules as they have to generate all the required pulses for the proper operation. National Atmospheric Research Laboratory (NARL) has indigenously developed and realized Active Array MST radar system with the latest technologies, such as state-of-the-art high power solid state Transmit-Receive modules (TRM)with

IOP Conf. Series: Materials Science and Engineering **925** (2020) 012064 doi:10.1088/1757-899X/925/1/012064

Atrix Field-Programmable Gate Array (FPGA) [3] based distributed Timing and Control Signal Generator card (TCSG), distributed control signal system to each TRM through optical fiber. Atrix-7 based TCSG system card located in the TR modules, receives the master reference pulse (IPP) from the master TCSG located in the control and instrumentation room and generates all the pulses such as the phase shift, digital attenuator, transmit (Tx), blanking and gating for power amplifier required for the normal operation of the TR module. Atrix-7 FPGA has both analog and digital parameter monitoring facility for monitoring the health of the TR modules. The communication between the radar control system and Artix-7 based FPGA is through the optical Ethernet interface. The FPGA generates real time pulses according to the experimental specifications file (ESF) provided by the user via radar controller system. Implementation of Artix-7 FPGA in the TR module control card to generate the control pulses and sample outputs are discussed in this paper.

2. Methodology and design

TCSG Card is an Artix7 XC7A35T with 33,280 logic cells FPGA based board, which provides interface for digital controls, analog parameter monitoring and timing generations required for the normal operation of the TR module. Artix-7 TCSG Card hardware is provided with about 27 digital I/O lines (for controlling and monitoring), four Analog parameters monitoring and Optical Ethernet interface for remotely controlling or monitoring a particular TR Module. All the parameters are controlled and monitored using Ethernet based protocol from radar controller PC. Each Artix-7 TCSG card is provided with a unique IP and Port No. to distinguish each card though they perform same kind of operations. After remote connection establishment the Artix-7 TCSG card will respond to the commands given for controlling or monitoring the RF parameters.

Artix-7 TCSG card is provided with Clock (16 MHz), IPP +5V (DC) as inputs. TCSG card will generate four pulses for TR module operation and: Transmit Receive control, RF on/off, Cover Pulse, Blanking Control with respect to IPP reference. It will also provide controls for phase shifter (6bits) [4], attenuator (5bits) and +60V ON/OFF (one bit) of RF Circuit remotely using Optical Ethernet. TCSG monitors Power supplies; +60V, +28V and +5V (3bits) of the RF Circuit. This card generates interlock condition by monitoring five TTL parameters: VSWR, Excess RF IN, Excess Duty, two Junction Temperatures and monitors Forward Power and Reverse Power of the TR Module.

TCSG has got other features like In System Programmability, Watchdog functionalities. The former one is implemented using Ethernet Connectivity to erase and reprogram the Serial Flash, and other is implemented using a Microcontroller which will operate independently. Three 8-bit ADC channels of Microcontroller are used for analog parameter monitoring. The design of Artix-7 based FPGA is shown in the Figure.1, which includes the following components. The Xilinx Artix-7 XC7A35T FPGA in a 324 pin LFBGA package with 33,208 logic cells and about 210 user I/O lines are used to make the TCSG card. It has 256MB of DDR3 SDRAM, 16 data interfaces with a clock speed of 166.667 MHz Apart from the DDR3, it has 128MB of SPI serial Flash memory, which store the FPGA configuration and MicroBlaze code shadowing. The communication between the Radar controller and the Artix-7 FPGA is made via EthernetPHY. ATMEGA16L 8-bit microcontroller with three ADC channels with reference voltage 2.5V is also embedded in the TCSG card.

IOP Conf. Series: Materials Science and Engineering **925** (2020) 012064 doi:10.1088/1757-899X/925/1/012064

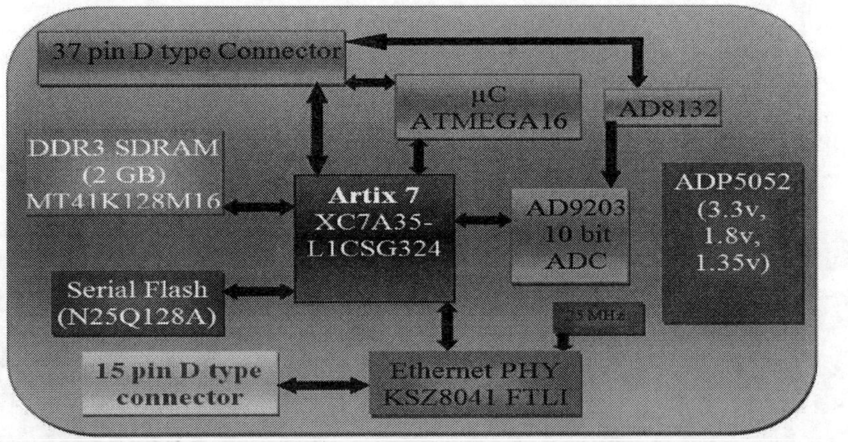

Figure 1. Complete block diagram of Artix-7 FPGA control card

3. Hardware implementation

3.1. EthernetPHY
The communication between the Radar Controller and TCSG is implemented using 10/100 Ethernet PHY [5] with media-independent interface (MII) support whose Ethernet media access control (MAC) is implemented in FPGA. Ethernet PHY uses an external 25MHz clock generated from FPGA for it operation.The Ethernet PHY interface is shown in the figure. 2

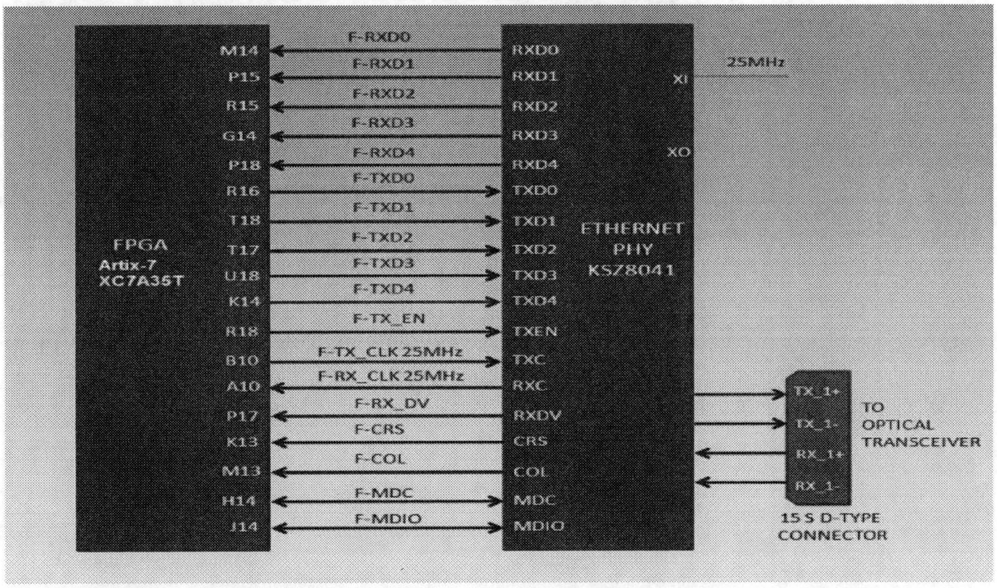

Figure 2. Ethernet PHY (KSZ8041FTLI) interface circuit on the TCSG card

The error–free transmission between FPGA and media access control (MACs) [6] is achieved via EthernetPHY KSZ8041FTLI. The EthernetPHY acts as a bridge between the MAC and FPGA by controlling the Ethernet transmits and receive data rate.
Ethernet PHY uses an external 25MHz clock generated from FPGA for its operation. EthernetPHY can be used for Optical as well as Copper Ethernet based interfaces. EthernetPHY provides high speed transceivers for higher band width support.

3.2. Analog to Digital Converter (ADC)

The Coupled forward power received from the TR modules is given as an input to the single ended AD 8132 and the amplified differential output is received from the output pin. This differential output is given to the AD 9203 multistage 10 bit ADC [7] as an input. This analog input is converted to a 10 bit (D0-D9) digital output and which is fed to the Artix-7 FPGA as an input signal for forward power level measurement. ADC interface with other circuit on the TCSG card is shown in figure. 3. This ADC is working on 20MHz clock frequency, with 40Msps sampling rate.

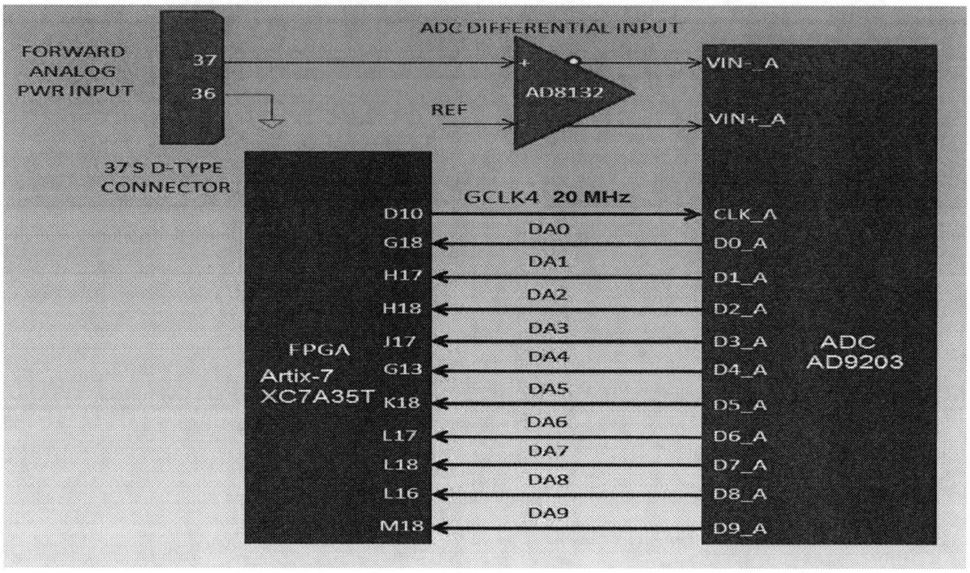

Figure 3.ADC interface with other circuit on the TCSG card

3.3. Micro-controller

ATMEGA 16L 8-bit microcontrollers [8] is used to read out the 60V DC voltage level and junction temperatures of the power amplifier MRF157 (pair) at the TR module as a monitoring circuit. In order to perform this, the 60V from the TR module is taped and it is step down to 3.3V by a potential divider circuit. This voltage and a reference voltage of 2.8V are given to the single supply quad comparator IC LM339m via pin no. 7 and 6 respectively. After comparing the two input voltages levels the non-inverting comparator output via pin no.1 given to the ATMEGA 16L in the microcontroller via pin no. 37. This voltage level is compared with the pre-defined voltage levels in the Artix-7 FPGA memory and it displays the corresponding voltage at radar controller via EthernetPHY. Figure.4 shows the Microcontroller interface with Artix-7 FPGA.

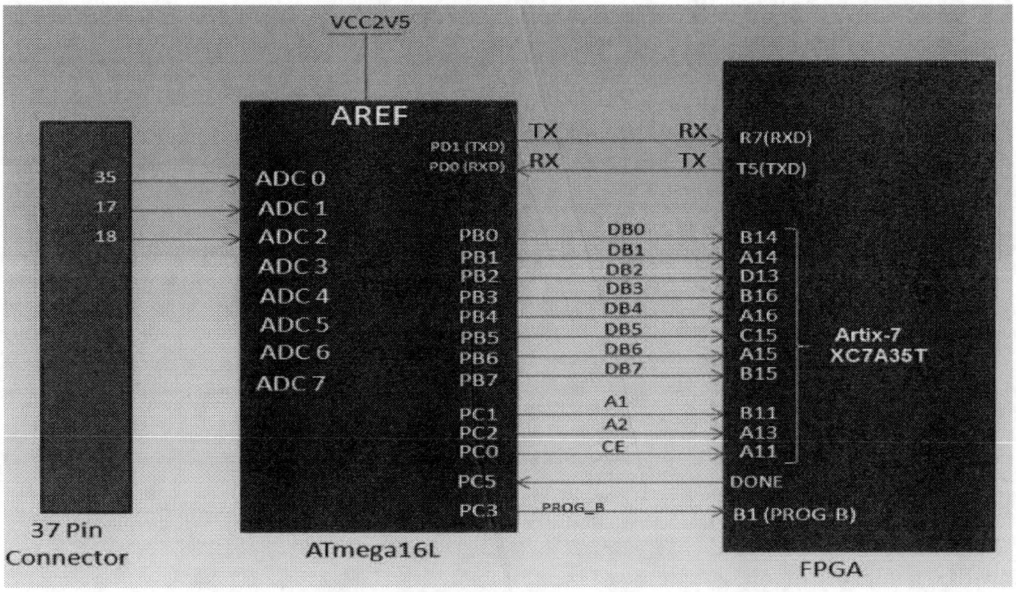

Figure 4. Microcontroller interface with Artix-7 FPGA

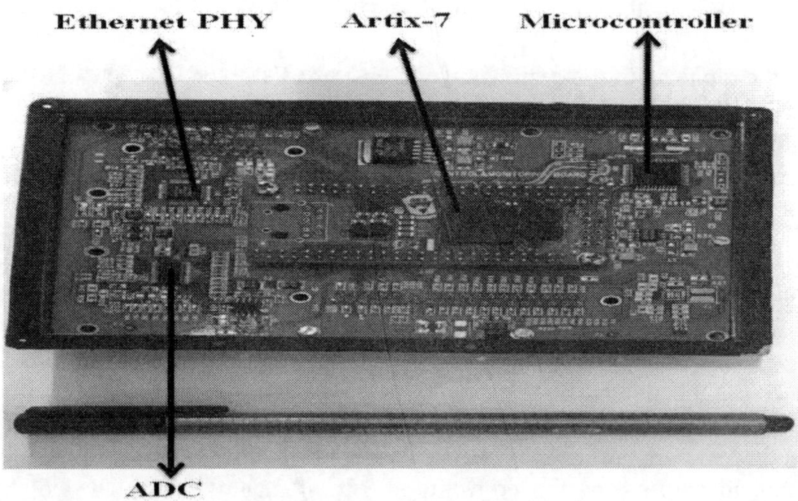

Figure 5. Hardware implementation of Artix-7 FPGA based control card

Figure.5 shows that hardware implementation of Artix-7 FPGA based control card which includes micro controller, Ethernet PHY, ADC and other circuitry. The junction temperatures of the MRF 157 power amplifier of both the IC's at TR module is continuously measured by positive temperature coefficient thermistor. This output is fed into the Low Power Programmable Temperature Controller (LPPTC) IC TMP01FS through pin no.5. The corresponding voltage level from the LPPTC is given to the microcontroller via pin no.35 and 36 respectively. These voltage levels are compared with the predefined values in the microcontroller and it will display the corresponding temperature in the radar controller PC. Figure.6 shows the Temperature scale and real time measurement.

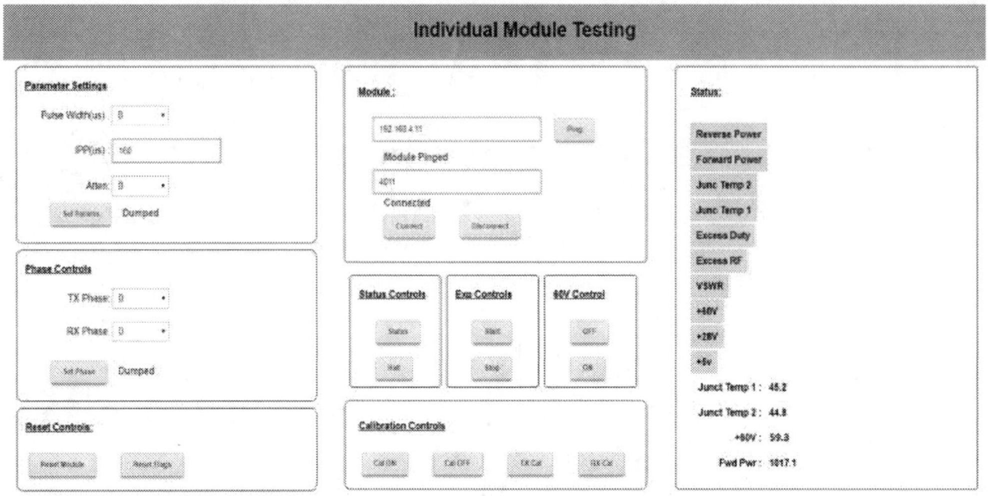

Figure 6.GUI screenshot of individual TR module testing

4. Microblaze microcontroller system software

All the peripherals (e.g. Ethernet PHY, UART), memory (e.g. HYPERRAM) and interfaces are implemented in MicroBlaze soft-core [9] CPU using Xilinx Platform Studio (XPS). There are two ways to setup FPGA architecture using an embedded MicroBlaze processor.

1. MicroBlaze be the top level of the design and add custom VHDL code as a MicroBlaze peripheral.
2. Custom VHDL [10] logic is the FPGA top level design and instantiate the MicroBlaze as a component.

The FPGA in the TCSG card is intended to be a logical array, which needs a CPU for some control or communication functions hence the second method is followed for FPGA architecture. The embedded processor system is then instantiated as a component in the top level VHDL design which is written in ISE Project Navigator. To communicate between Micro Blaze processor and top level VHDL design, a GPIO block is created in MicroBlaze as a memory address map and connected those signals in VHDL top level wrapper to do write/read from VHDL design. The proposed architecture is synthesized using Vivado 2017.4and is implemented on Artix-7 XC7A35T-2CSG324C customized FPGA board. The necessary software for this design is written using VHDL Behavioralmodeling for top level wrapperand C language is used for programming MicroBlaze processor, includes special C commands belongs to MicroBlaze's cores in addition to the conventional C language. Finally the combined bit file is downloaded to FPGA and initial simulations are carried out in Vivado.

5. Simulation and Results

Micro The output from the Artix-7 FPGA should be validated before implementing it in the real time operation. This has been achieved by carrying out simulations using the Vivado [11]. The simulated output from Vivado is shown in the figure. 7.

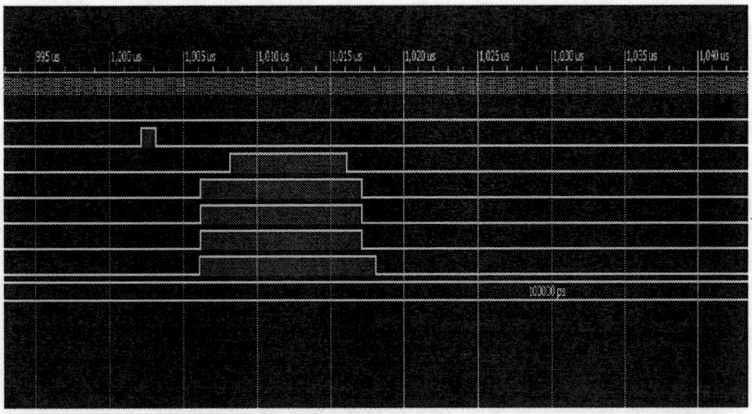

Figure 7. Simulated output from Vivado.

Artix-7 FPGA real time outputs of IPP, Tx pulse and Gate bias pulse for MRF157 power amplifier, RF ON/OFF, blanking pulse [12][13], T/R control and phase shifter are shown in figure.8 and figure.9.

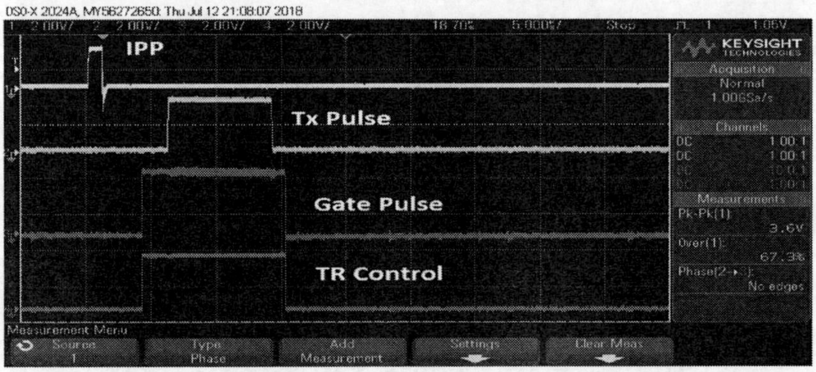

Figure 8. Real time output from Artix-7 FPGA

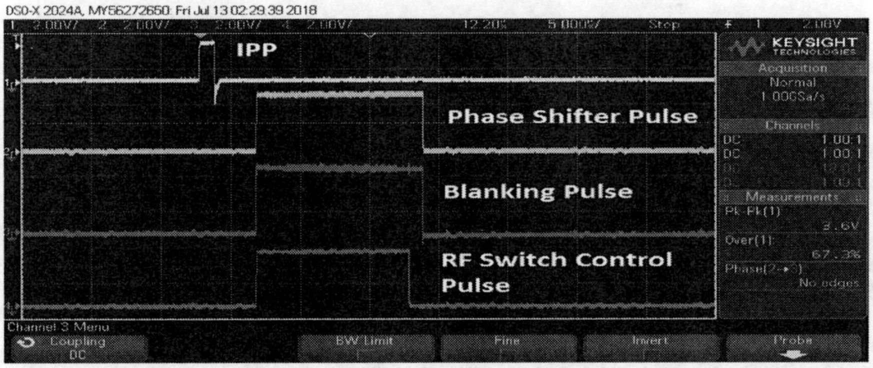

Figure 9. Real time output from Artix-7 FPGA

The messages are sent in JSON format and the message telemetry is storage in storage endpoint server for further analysis. The IoT hub can be connected to less than 100 devices (in simulation) and handles MQTT, HTTP protocols. The refresh time to update device state is also set in the configuration file.

IOP Conf. Series: Materials Science and Engineering **925** (2020) 012064 doi:10.1088/1757-899X/925/1/012064

6. Implementation details

The design and configuration of Artix-7 FGPA is tested and satisfactory performance has been noted. Thus, the Artix-7 FPGA is embedded in the TR module for the real time operation [14]. Individual 1kW TR module is shown in the figure. 10 and the hardware is shown in figure. 11.

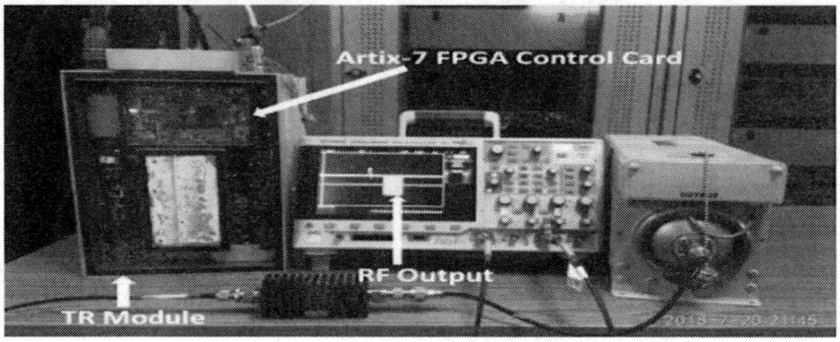

Figure 10. Individual 1kW TR Module

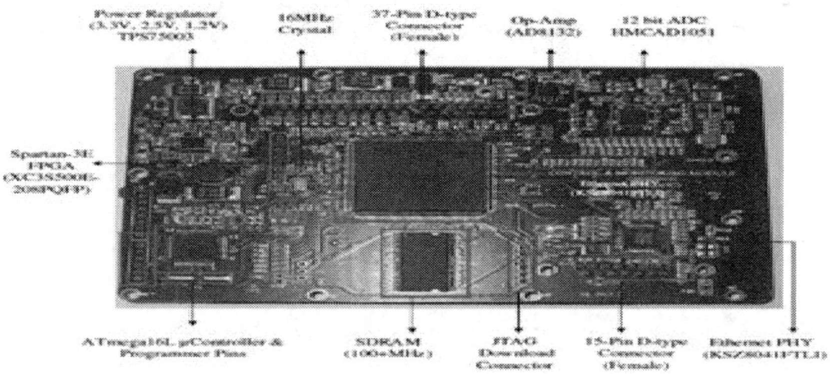

Figure 11. Hard ware picture

Functionality of the card is thoroughly tested with test bench setup equivalent to real time system. The results are found to be satisfactory as shown in figure.12.

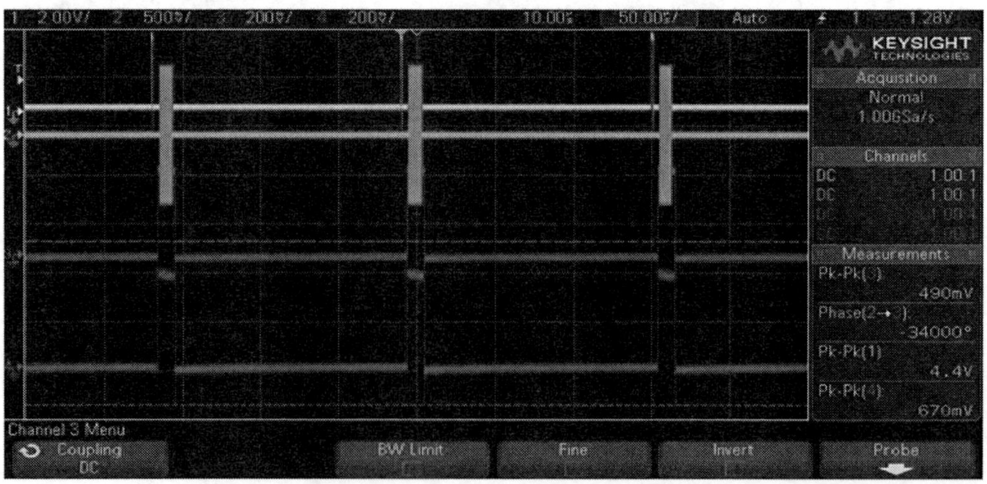

Figure 12. Existing TCSG card Spartan 3E based

IOP Conf. Series: Materials Science and Engineering **925** (2020) 012064 doi:10.1088/1757-899X/925/1/012064

7. Conclusion

In this paper the full potential of Artix-7 FPGA in generation of control pulses is presented. Atrix-7 FPGA successfully embedded in the TR module for the real time operation. Ethernet PHY is used for the communications between the radar controller and the field located TR module. Fast and robust radar operation can achieve through the Artix-7 FPGA based control card.

References

[1] Srinivasulu P 2012 1280-MHz active array radar wind profiler for lower atmosphere: System description and data validation *J. Atmos. Ocean Technol* **29** 1455–70

[2] Qureshi AA, Afzal MU, Tauqeer T and Tarar MA 2011 Signal analysis, design methodolgy, and modular development of a TR module for phased array radars *in Emerging Technologies (ICET),* 1–6

[3] Przybus B and Xilinx 2010 Redefines power, performance, and design productivity with three new 28 nm fpga families: Virtex-7, kintex-7, and artix-7 devices *Xilinx White Pap*

[4] Kenichi M, Hieda M, Tarui Y, Hatamoto M, Kanaya K, Kasahara M and Takagi T 2002 A 6-18 GHz 5-Bit Phase Shifter MMIC Using Series/Parallel LC Circuit *Microwave Conference* 1-4

[5] Zhang L and Zhang B 2011 The Novel Frame Boundary Detection and Fast Frame Synchronous Structure for 10 Gb/s Ethernet Phy FEC Sub-Layer VLSI Implementation *Wireless Communications, Networking and Mobile Computing (WiCOM)* 1–4

[6] Luo X, Zheng K, Pan Y and Wu Z 2004 A Sensor Media Access Control Protocol Based on TDMA *Embedded Software and Systems ICESS*

[7] Kwuang-Han C and Hsieh CC 2019 A Calibration-Free 13-Bit 10-MS/s Full-Analog SAR ADC with Continuous-Time Feedforward Cascaded Op-Amps *Solid-State Circuits IEEE Journal of,* **54** 2691-2702

[8] Rosiek S and Batlles FJ 2008 A microcontroller-based data-acquisition system for meteorological station monitoring *Energy Convers Manag* **49** 3746–54

[9] Sauriol B and Landry RJ 2007 FPGA-based architecture for high throughput, flexible and compact real-time GNSS software defined receiver *Proceedings of the 2007 National Technical Meeting of the Institute of Navigation* 708–17

[10] Goossens K 1995 Reasoning about VHDL using operational and observational semantics,"In: P E Camurati, H Eveking (eds.) *CHARME* **987**

[11] Parasuraman S and Ramesh GP Design of MML Reduction in Super High Frequency Antenna Using HFSS for Wireless Communication System *Indian Journal of Public Health Research & Development* **8** 288

[12] Annemieke T 2013 Dual axis multi-beam radars Radar, *International Conference* 412-16

[13] Rajesh D Ch Panel 2016 Based Routing Scheme for Mobile Wireless Sensor Network," *International Journal of MC Square Scientific Research* **8** 183-198

[14] Snehalatha N and Amudha S 2016 Remote display access using remote frame buffer and io streaming *International Journal of MC Square Scientific Research* **8** 23-40

IOP Conf. Series: Materials Science and Engineering **925** (2020) 012065 doi:10.1088/1757-899X/925/1/012065

Area Efficient Reconfigurable Buffer for NoC Router

N C Sendhilkumar[1] and R Murugesan[2]

[1] Department of Electronics and communication Engineering, Sri Indu college of engineering and technology, Hyderabad, India.
[2] Department of Electrical & Electronics Engineering, Annamacharya Institute of Technology & Sciences, Tirupati, India.
[1] sendhilkumarnc@gmail.com, [2]rmurugesandr@gmail.com

Abstract. In the current research, FPGA-based architecture of the Noc device reconfigurable router is suggested. Proposed router specification entry is done with the Verilog Hardware Description Language. The latest research has a five-channel router and a crossbar switch (east, west, north, south and local). Each channel has buffers and multiplexers. FIFO buffer stores data, and multiplexer monitors the input and output of data. The channel contains FIFO architecture and multiplexers. The crossbar switch is then planned and five other channels. Reducing the number of LUTs in the proposed work decreased the router area. A report is given for the number of LUTs used in a router. Results obtained indicate the area efficiency of the proposed router over existing method.

Keywords: Buffer, FIFO, LUTs, NOC.

1. Introduction

The Conventional Chip System (SOC) consists of bus, cable delay, noise, problems with power dissipation and signals. A network on a chip, which provides a high-performance networking infrastructure, is called an on-chip communication interconnection architecture. Networks-on-Chip has all the requirements required for the production of future SoCs. It is managed by a network of communicating routers and point-to-point networks. Colliding between packaging is considered an unavoidable issue by a conventional computer network and computer networks' ultimate aim is to reduce area probability. Management of buffers is not a recent issue, but still needs consideration. This increases the output to immediately use the unused buffers. In the experiments performed by the buffers, NoCs were the biggest consumers of leakage power, with crossbars, arbiters, buffers and connections, dissipating about 64% of the overall power budget. This considered the buffers as candidates to maximize leakage capacity, since there were still 85% of idle buffers, even at high loads. The consumption of buffers is also high with regard to dynamic capacity, and it is increasingly increasing as the packet flow increases.

Our basic contribution is designed to provide the router with some reconfiguration logic that makes adjustments to the use of buffers on each input channel in compliance with communication needs. The idea is that each input channel should provide buffer units to and from nearby channels for a given bandwidth. If the whole buffer is not needed for a channel, it can loan buffer word slots to nearby

IOP Conf. Series: Materials Science and Engineering **925** (2020) 012065 doi:10.1088/1757-899X/925/1/012065

channels. The findings show that the buffers used by the same router are unreliable and the improvements made with the new strategy can be achieved.

2. Related work

The buffer share of the latest router architectures has recently been analyzed and compared. The chip system has been suggested by a number of recent works, [1-3] the failing to regularly use a buffer is abused through the network and its power gate is extended to buffers that are rarely in use. This design choice has a significant performance advantage instead of shutting down the buffers, because the virtual channel buffer still is able to accept network packages [4-6]. Installation options are not yet available. The use of buffers in a chip network varies widely across the network. Power gates allow the turn off of buffers that are rarely used. A new pipelined router design [7-8], They based on reducing latency in the router. Primarily certain router components have been defined by taking critical paths that bind the router frequency. Limitations placed by a part called, arbiter, on router output. Thus, the builder has made many smaller arbiters. [9] Solution has been provided for the critical communication problem between multiple IP cores. The initial router architecture on chip interconnect network NoC was integrated in the Module. Device has a parametric router design on chip interconnect network. [10-11]. Rising electronic systems requirements are one of the main factors for the advancement of integrated circuit technology. Multiprocessing is the solution for meeting potential application requirements. Efficient on-board communication is important for multiple processing over heterogeneous operational units [11]. Network-on-Chip (NoC) is the overall on-chip connectivity principle that offers high performance, the essential prerequisite for addressing modern systems' complexity. All links in NoC can be used for the simultaneous transmission of data that ensures high parallelism and makes the replacement of traditional communication architectures such as joint buses attractive.

Comparison of packet delay and XY routing algorithm results with OE routing algorithm. The problem of architectural hindered contact and clock. In the architecture built, in which the buffer size can be reconfigured. This approach will reduce the excessive latency, power dissipation. The architecture is also an productive field. [12-13]. It defined the integration of different components into the NoC architecture as well as the modeling of reconfigurable components, such as IP cores and fixed IPs. This work provides a fresh design method for customizing the routers in a network-on chip for reconfigurable buffer systems over [13], More coordination mechanisms are increased as the complexity of the NoC architecture increases. To reduce the buffer demand and increase the device output, a wormhole flow control has been proposed. But a packet can take several intermediate switches simultaneously. On the other hand [14-19]. When a packet occupies a channel buffer in traditional NoC architectures, only blocking the original message does not allow the actual channels to be used on other channels. This raises the issue of deadlock and livestock.

3. Proposed system

A new add-on module will decide which FIFOs will accept the received packet that are available on the network. Finally, communicating with both output ports is responsible for passing the received packet onto its downstream routers according to the route method employed.

Here, a FIFO can accept packets from the directly connected upstream router as FIFOs for the Base routers, but from other ports. Transfer logic: Apply a routing algorithm to select the correct output port for the packet path. The Flexible router's operations are the same as the base router, but in the case of a dispute it will be special. In terms of contention, the Flexible Router shall not wait until one or more free slots are available to the required full FIFO, such as the Base Router, Nevertheless, the FFC searches for a free slot in any suitable non-full FIFO throughout the router by asking for FIFOs that are not complete in other input ports and returns the application to the upstream router when it finds a free slot. The kit is then passed to the chosen FIFO. The Robust Router will then function in the same way as the Base Router does in the same way as the Base Router does. Every port of entry only has one physical channel linking all channels of virtual input port.

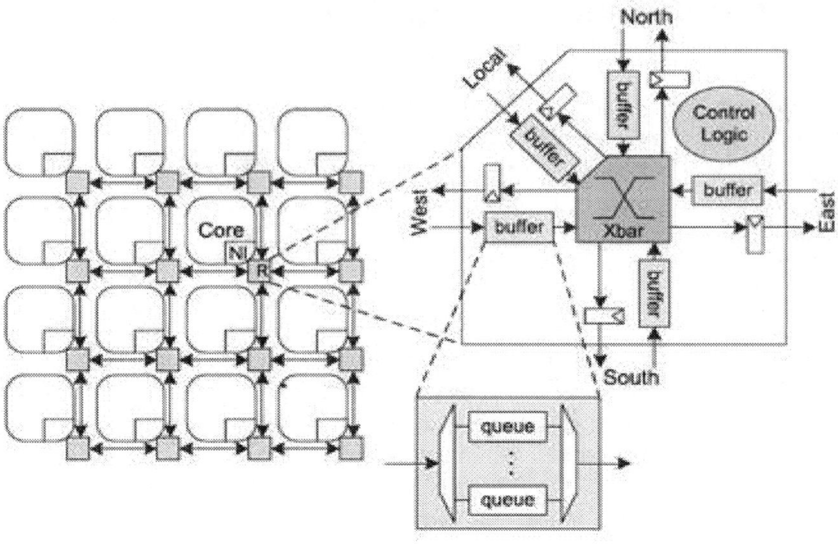

Figure 1. Generic router

These have no completely specific, virtual channels and the buffer length can be altered in real time for each virtual channel. In addition, flits on the next input port that have the same output are shared with the VC buffer. The buffer area between the current input port and the next port is split and the split area can also be actively requested based on real demand for traffic. The buffer input is divided in a clockwise direction in order to keep it simple. Eastern port, for example, only the south port buffer tool can be used. Further buffers are used by the port to transfer further data, utilizing the buffer resources entirely. application-oriented NoC, the traffic at the router's ports is special.

Some regulations are established to allow other adjacent channels to use buffers from one channel. If a channel fills its FIFO, it can give its neighbors more buffer terms. The channel first asks the right Neighbour for buffer terms and attempts to borrow from the left neighbor FIFO if it is still in need of more buffers. In order to monitor its stored flows, certain signals from each channel must be sent to the nearby channels. Every channel must therefore know how many buffer terms its own channel and its adjacent channels use and also how much its own buffer is used by the next channels. This number is informed by a control block.

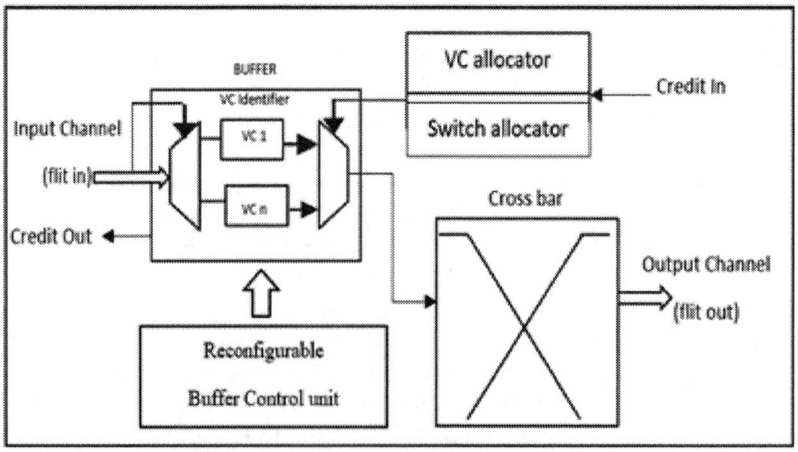

Figure 2. Proposed reconfigurable buffer

IOP Conf. Series: Materials Science and Engineering **925** (2020) 012065 doi:10.1088/1757-899X/925/1/012065

The entire router therefore needs to support dynamic configuration to satisfy the generality requirements. Architecture of the router includes mechanism for allocating the buffer inter port. At the one side, as per the data needs of intelligent allocation of buffer resources, the channel can be automatically configured by various ports. In another hand. VC in same ports takes the flexible linked list as a means of realizing changeable buffer length and storing the flits based in calculated output.

The first two bits are used in an 8-bit packet to show the address of the next NOC router. In this job, the block diagram as shown is displayed. Four new directions have added fewer storages to the support buffer that will boost performance. 1. It's the first step that has been taken. The block was built in theory with the previous iterations of homogeneous routers, which had all resources automatically allocated, and which could not be modified during design times, which presented problems of a more sluggish rejection of data when used in different channels. Four other directions (North East, North West, South East and South West) are presented with the planned architecture which will contribute to multi-directional steps. Newly added details from your neighbour shares fifo to save your data. In contrast with the previous one, dates from one direction can now be route to seven directions, where the data from one direction can be routed only in three directions. The newly developed architecture offers more routes for data routing so that the critical path time of a network can decrease and the IC speed can be helpful. We now follow the rectilinear path of stonier tree to cut the data path to the router with this newly developed architecture

The input port includes an actual channel, which is connected to all virtual input port channels and does not have a fully separate virtual channel and the buffer length can be modified in real time for any virtual channel. Only flits with the same output are shared with the VC buffer at adjacent input port. Based on the actual traffic request, a dedicated buffer zone can be requested between the current port and the nearest port of entry. The input buffer is shared in the rectangular direction to keep it simple. For instance, eastern port can use a buffer resource only for south port. More buffers will complete a port requiring more data, so that the buffer functions shown in figure 2 can be used entirely.

Table 1. Comparison with routers

Parameters	Generic router	Virtual channel router	Proposed router
Buffer/Buffer less	Buffered	Buffered	Buffered
Types of buffer	Simple buffer	Virtual channel as buffer	Flexible buffer
Probability of data loss	High	Medium	Low or negligible
Hand shaking	No	No	No

The two adjacent ports set up common buffer space. The buffer state will change in time when a buffer space is borrowed or loaned. The buffer is connected in a clockwise direction on the entire router. One system uses two multiplexers to automatically monitor the input tampon and to share the tampon with the nearby terminal only. When a port receives large quantities of data well outside its independent buffer, it will ask the data to be processed from the specific port. Based on fixed exclusive buffer sources, the inter-port buffer allocation process prevents changing internal programming mechanisms and ensures communication requirements. We can see that the buffer value is reconfigurable. Uses an improved control logic buffer.

IOP Conf. Series: Materials Science and Engineering **925** (2020) 012065 doi:10.1088/1757-899X/925/1/012065

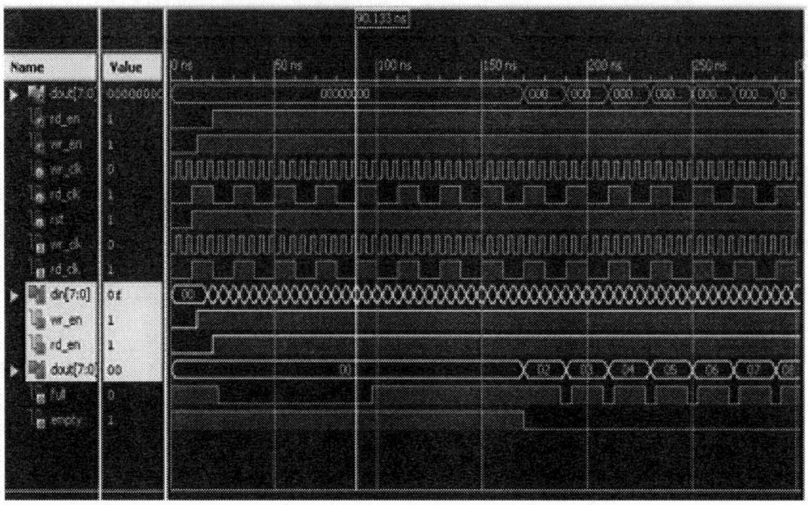

Figure 3. Simulation results of FIFO

4. Results and Discussion

The benefit of reconfigurable buffered is realized in the proposed design. The packet traversal is performed by a buffer network at medium load condition the packet traversal is through reconfigurable buffer.

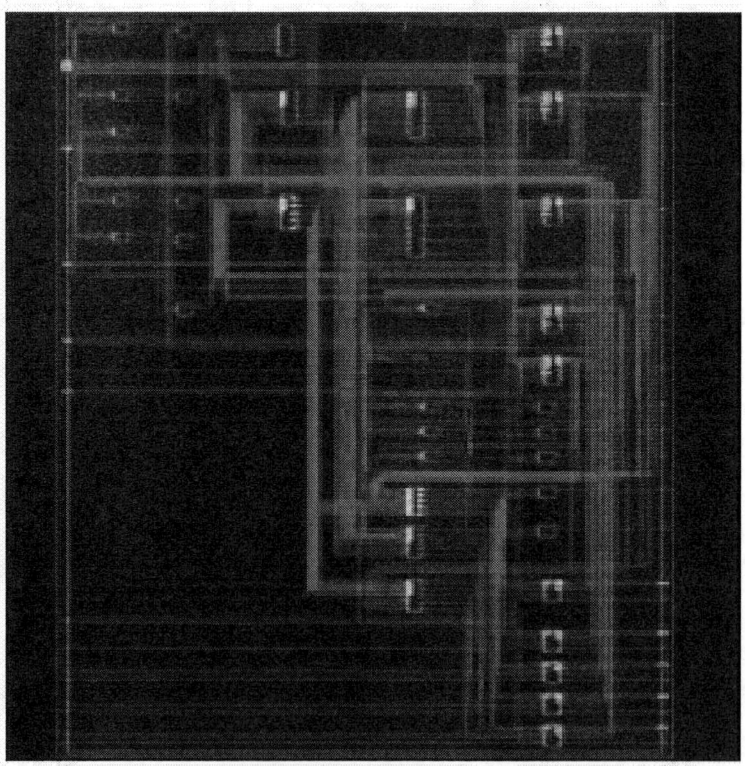

Figure 4. RTL view of proposed router

It offers better performance than generic NoC router. This article introduced a resource efficient Noc that would support rearrangement for mode changes. A NoC expands a current NoC and offers guaranteed-service inter processor communication.

Figure 5. Simulation result of proposed method

Introduction of the planned design is measured in terms of the worst-case reconfiguration time, cost of hardware and overall service frequency. Compared to generic router, proposed router area shown in table 2. The proposed router received percent less delay than the generic router and when compared to the virtual channel router in Table1. Reducing pipeline stages in each router leads to reduced delay for packets, which in effect reduces average latency shown in figure 3a. RTL view of proposed method is shown in figure 3b and simulation results of proposed router shown in figure 4.

Table 2. LUT's comparison of proposed with existing method

Parameters	Existing method	Proposed method
Slices	338	306
LUT's	1267	1124
FF's	924	908

The processing nodes can be separated by location into three different groups with traditional NoC mesh corner, edge and central nodes. The comparison between proposed with existing method shown in fig 5. Three, four and five I / O ports need routers based on these categories. Thanks to the distributed and independent controlling logic, routers can be designed to comply with this requirement with the required hardware without any additional effort.

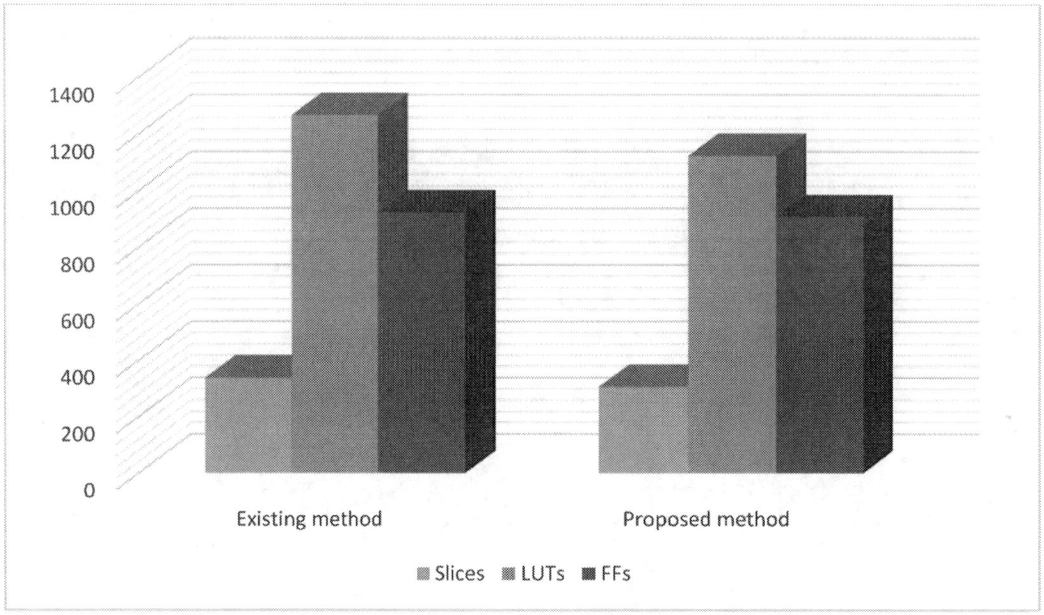

Figure 6. Comparison between existing and proposed method

Therefore, for any number of I / O ports the Architecture is completely configured. By simply doubling the input portion of the unit and increasing a crossbar on the output side the number of I / O ports can be increased to any amount. FIFO input buffer and cross-point input buffer to increase performance and reduce HOL. The architecture of the cross-point input buffer will optimize output. For each input output pair, there is one buffer. However, we can streamline buffer design in the cross-point buffer architecture for an application such NoC features. The IP core must be mapped in the basic application design flow of NoC. Router design using reconfigurable buffer is implemented at the RTL level using Verilog HDL in this paper. Design is synthesized using Xilinx and modeled using ISIM 14.7 for router-proposed region evaluation. In conventional static router is used on chip throughout the network. Nevertheless, it does not acknowledge the applications' need for traffic. This will result in lower buffer usage and will drain valuable buffer space on the router.

5. Conclusion

This paper proposes efficient router architecture to boost network efficiency including mechanism for internal-port allocation. The new router architecture will increase buffer efficiency without impacting network output by modifying the Buffer input unit of Noc router. Future work an application for real-time analysis can be used with different techniques of routing available and comparable to existing routers. In addition, different technologies may be implemented for efficiency enhancements such as multicast and network surveillance. Recently a domain language was implemented in [12] for a particular distributed architecture. In order that we increase the degree of abstraction from application layers, we expect a similar approach for further creation of NoC systems includes improving the latency process by which reconfigurable buffers are associated with input ports.

References

[1] Chen Y, Gurumani ST, Liang Y, Li G, Guo D, Rupnow K and Chen D 2015 FCUDA-NoC: A scalable and efficient network-on-chip implementation for the CUDA-to-FPGA flow *IEEE Transactions on Very Large Scale Integration (VLSI)* Systems **24** 2220-2233

[2] Tan W, Gu H, Yang Y, Fadhel M, and Zhang B 2016 Network condition-aware communication mechanism for circuit-switched optical networks-on-chips *Journal of Optical Communications and Networking* **8** 813-821

[3] Sorensen RB, Pezzarossa L and Sparso J 2016 An area-efficient TDM NoC supporting reconfiguration for mode changes *10th IEEE/ACM International Symposium on Networks-on-Chip (NOCS)* 1-4
[4] Kasapaki E, Schoeberl M, Sørensen RB, Müller C, Goossens K and Sparsø J 2015 Argo: A real-time network-on-chip architecture with an efficient GALS implementation *IEEE Transactions on Very Large Scale Integration (VLSI) Systems* **24** 479-492
[5] Liu J, Harkin J, Maguire LP, McDaid LJ, Wade JJ and Martin G 2016 Scalable networks-on-chip interconnected architecture for astrocyte-neuron networks *IEEE Transactions on Circuits and Systems I: Regular Papers* **63** 2290-2303
[6] Yu Z, Wang X and Shen K 2016 Conditional forwarding: simple flow control to increase adaptivity for fully adaptive routing algorithms *The Journal of Supercomputing* **72** 639-653
[7] Zoni D, Flich J and Fornaciari W 2015 Cutbuf: Buffer management and router design for traffic mixing in vnet-based nocs *IEEE Transactions on Parallel and Distributed Systems* **27** 1603-1616
[8] Jindal N, Gupta S, Ravipati DP, Panda PR and Sarangi SR 2019 Enhancing Network-on-Chip Performance by Reusing Trace Buffers *IEEE Transactions on Computer-Aided Design of Integrated Circuits and Systems* **39** 922-35
[9] Kamali M, Petre L, Sere K and Daneshtalab M 2015 Towards correct and reusable Network-on-Chip architectures *In Modeling and Simulation of Computer Networks and Systems* 357-392
[10] Hassan H, Shalaby A and Kim H 2015 DPSB: Dual port shared buffer mechanism for efficient buffer utilization in Network on Chip routers *Int. SoC Design Conf. (ISOCC)* 135-136
[11] Becker DU, Jiang N, Michelogiannakis G and Dally WJ 2012 Adaptive backpressure: Efficient buffer management for on-chip networks *IEEE 30th Int. Conf. on Computer Design (ICCD)* 419-426
[12] Seitanidis I, Psarras A, Dimitrakopoulos G and Nicopoulos C 2014 Elastistore: An elastic buffer architecture for network-on-chip routers *Design, Automation & Test in Europe Conference & Exhibition (DATE)* 1-6
[13] Manivasagam M A 2017 An Efficient Self-Reconfiguration and Route Selection for Wireless Sensor Networks *International Journal of MC Square Scientific Research* **9** 192-199
[14] Clerk Maxwell MJ 1892 A Treatise on Electricity and Magnetism, 3rd ed. *Oxford: Clarendon* **2** 68-73
[15] Young M 1989 The Technical Writer's Handbook *Mill Valley, CA: University Science*
[16] Benini L and Micheli GD 2002 Networks on chips: A new SoC paradigm *computer* **35** 70-78
[17] Sharma N and Gadag S 2013 An efficient way to increase performance by using low power reconfigurable routers *IOSR Journal of Electronics and Communication Engineering (IOSR JECE)* **8** 39-44
[18] Véstias MP and Neto HC 2006 Area and performance optimization of a generic network-on-chip architecture *In Proc. of the 19th annual symp. on Integrated circuits and systems design* 68-73
[19] Nithya V and Ramesh GP 2019 EAR-EEG signal transmission in WSN to evaluate co-channel interference of OLSR routing through residual of Fractional spline wavelet for auditory hallucination diagnosis *IJAST* **28** 271 - 279

Probabilistic Analysis of Dynamic Power and Area in Network on Chip

G P Ramesh[1]

[1]Department of Electronics and Communication Engineering, St. Peter's Institute of Higher Education and Research, Chennai, India.
[1]rameshgp@yahoo.com

Abstract. The power used by digital integrated circuits has become a key restriction for the design and development of VLSI as the device complexity and transistor density increase. Clock-gate synthesis methods are applied to circuits to prune changes in registers by modifying the next-state register functions in order to reduce complex power dissipations. Therefore, a sequential inspection of the circuits is needed for testing this form of synthesis. A new design to increase the efficiency of network-on-chip buffers and minimize the overall area and power consumption of the router is proposed. The non-uniform use of buffers in the network is leveraged, and control is used in unusual buffers. A part of the buffer is turned on instead of shutting down the buffer completely. For some recent studies, power-gating inactive components have been used as a promising approach to minimize static and dynamic power consumption, although its disadvantages, like wake-up delay and power overhead, should also be properly addressed. Throughout this work, a probabilistic experiment was carried out to estimate the power-gating performance and a control unit was proposed to handle the confirmation of the sleep signal to buffer entries based on the results of the study. The analysis results show an average increase in static power savings. For the research community to tackle the Contest of improved power performance and efficiency, including control gate, operating insulation, storage slicing, power gate and intensity scaling, respectively. The analysis is also examined other systems and strategies for optimization built to reduce power further without noticeable loss of performance.

Keywords: Buffer, Density, Efficiency, Network on chip, Optimization, System control methods

1. Introduction

The Traditional chip system (SoCs) is based on bus, cable delay, noise, power dissipation problems, and signal synchronization problems. With an on-chip communication interconnect architecture is called Network on Chip, which offers a high-performance networking infrastructure [1]. Networks-on-Chip has all the requirements needed to be potential SoCs. It is constructed in a coordinated way by a collection of interconnecting routers and point-to-point channels. Traditional computer network considers collision between packets to be an unavoidable problem, and the ultimate aim of computer networks is to reduce the likelihood of collision, but the possibility of dropped packets is very small in on-chip networks. This is related to the fact that the router's communication links within a Network on chip are shorter than those in computer networks, thereby allowing tight router synchronization. Some of an on-chip network's characteristics are low power consumption, small area (in routing nodes [5]), cheap cables and low interconnect latency. Though traditional computer network characteristics are long wires, high communication latency, and a lot of complex routing nodes.

2. Related works

All of them are based on the routing algorithm. For example, using a dynamic routing algorithm, we can minimize power, but it adds complexity to the design. Several works in the chip network have focused on the architecture of the topology [1, 2], Pointed out that eliminating some linkages between the core nodes could effectively prevent the flow of data from the core nodes, thus reducing traffic congestion[3], found that removing some linkages between the large Internes nodes could also reduce congestion[4], considered both the transmission capability of the node and the bandwidth constraints, and found that allocation of resources based on algorithm [5],

Several recent works have proposed the chip network, in [1,2,3] the non-uniformity of the usage of the buffer is leveraged through the network and the power gating is extended to rarely used buffers. Instead of turning off the buffers entirely, the buffer part is switched on [4], This design option has a major performance advantage since the virtual channel buffer is still capable of receiving network packets [5-12]. The usage of buffers in a Network on chip differs considerably across the network and buffers that are rarely used can be turned off through a power gate [13-17]. Power-gating consists of switching off the NMOS footer linked in series with logic to reduce leakage current flow. Design and implementation of a configurable virtual channel in the router intended as a dedicated Network on chip support module in an FPGA is presented. The router is built using crossbar switching and is used to interconnect routers in the communication protocol [17-18]. Chip Network interconnects IP to reduce the power consumption and die size for consumer electronics, mobile, automotive and other applications on chip (SoC) devices. In addition, only leakage savings and network on chip efficiency are recorded in several recent works, while aspects such as the area and the leakage of the power gate are addressed in our proposed system.

3. Proposed system

Primarily, focus on techniques that are designed to reduce the Area and Dissipation of power by modifying the virtual channels in the input ports of router from [3] connecting the network to the chip as well as by routers and network interfaces, and their subscription is expected to rise as per current technologies. In particular, designed a Set of data correction programs running at flit rate but on an end-to-end base system that could allow us to minimize the changing activity and the switched coupling activity together. Communication complexity is the virtual channel Classification area throughout the flits, which adds a lot of bits to the flit contents. data between the nodes must also be shared on a virtual channel basis. The nodes should also contain information per virtual channel condition. The transmission node holds state information such free buffer count, virtual channel allocation, and packet priority. The receiving node also maintains virtual channel designation, input and output pointers for each path buffer and channel status. Suggested the computational level and are evaluated by simulation on synthesis and actual traffic situations. The inquiry considers many aspects and metrics of the architecture of the system in the information sharing subsystem, and also the gate area, power dissipation and energy usage. Outcomes have shown that using suggested error correction schemes in the communications system, energy and power could be saved with little or no significant deterioration in performance but with an area operational costs in the Network

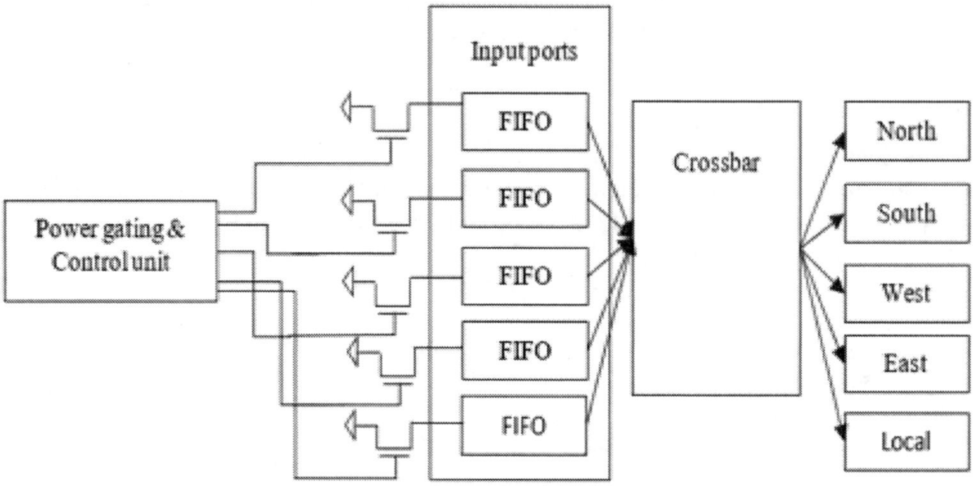

Figure 1. Block diagram of router using power gating

FIFO Flexibility Controller: A new add-on module is responsible for determining some appropriate FIFO available on the network to carry a received packet. Finally, it is accountable for communicating with both the output ports to pass the obtained packet to their downstream routers according to the route method used.

FIFO buffer: A FIFO here can accept packets from the explicitly attached upstream router as the Base routers FIFOs, but from other input ports. Forwarding logic: Apply the routing algorithm to decide the path of the packets in order to select the correct output port. The activity of the Versatile router is the same as the operation of the Base router, but it will be unique in the event of a conflict. As far as contention is concerned, the Versatile router does not delay until the demanded full FIFO has one or more free slots like the Base router does, yet the FFC does look for a free slot in any appropriate not full FIFO throughout the router by Asking for FIFOs that are not complete in other input ports, and when it finds a free slot, it returns the application to the upstream router. The package is then moved to the chosen FIFO. Following that, the Robust Router should work exactly the same as Base Router.

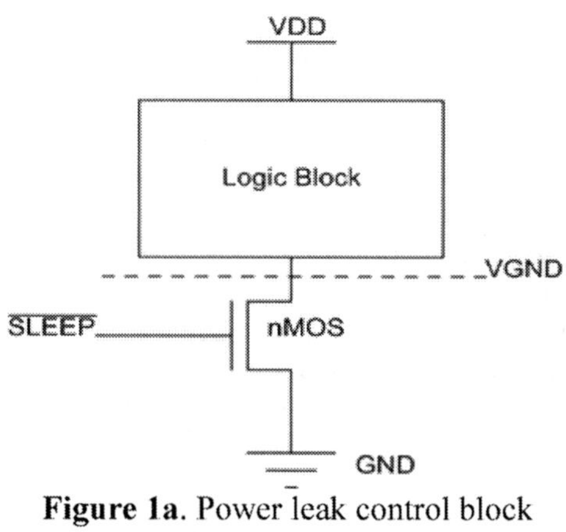

Figure 1a. Power leak control block

A proper sizing of the sleep transistor is required for effective use of power gating, as it affects the output of all the connected gates. A big one means an significant overhead in area and an ineligible capacity (i.e., capacity and delayed) for ON$OFF transitions while an unacceptably small transistor slows the circuit down in active mode because of its high resistance. Another problem is that the cumulative current injected into the sleep transistor is calculated via the sleep transistor drain-source route and consequently leads to deterioration of the sleep delay during active operation shown in fig 1.

IOP Conf. Series: Materials Science and Engineering **925** (2020) 012066 doi:10.1088/1757-899X/925/1/012066

This finer technique is referred to as the insertion of the clustered sleep transistor, a very efficient solution that connects many cell subsets, i.e. clusters, to dedicated transistors in the entire organized sleep network. Cylindered energy gating is not new in the low-power design area and a large number of clustering solutions have been suggested. The solution for groups of gates with mutually exclusive discharge patterns allows for optimal sleep transistor dimensions. Instead, the authors demonstrate a timely clustering technique that can manage the time and the region simultaneously

For efficient use of power electricity, the sleep transistor needs to be properly designed. In addition, even though the circuit in active mode can be slowed by a small sleep transistor due to its high resistance, a larger one requires a wide area and a substantial energy cost for driving. Designers typically establish an IR-drop threshold (for example 10% of VDD) as a limit to be met by the sleep resistance size transistor. Eq can be used for the measurement of the average sleep channel resistance.

$$R_{st}\ D\ \frac{V_{DD\_drop}}{I_{on}}$$

When the VDD remainder of the sleep transistor is allowed to decrease the voltage, expressed as a percentage of the voltage supply, while the ion is the highest discharge current

Power-gated cells insert in the active mode into the sleep transistor. Estimating the active current is not an easy task. An incorrect estimation of Ion actually indicates a sleep transistor size that is sub-optimal and can increase the area and power or, worse yet, timing violations during active mode. The sleep transistor is working in a stable region with knowledge of Rst and its size must be measured correctly.

The size and physical efficiency of the sleep transistor is strongly affected by carrier mobility, threshold voltage, and the gate duration. Consequently, the type of MOS device is needed for the proper sleep transistor size nMOS. While both devices can be used without distinction as power switches (i.e. retaining the same reduction in leakage), nMOS is typically better suited. transistors are more suitable. PMOS devices are generally less leaky than nMOS devices, but have a lower carrier range, reducing the ongoing current. Consequently, pMOS sleep transistors require more silicone area than nMOS, to ensure a certain current efficiency.

Power gating is indeed a widely recognized method for raising the loss potential of unused electronic circuitry products. In order to efficiently introduce energy gating to a digital loop, the inactive duration of both the circuit must be large enough yet to account for both the energy of the wake-up below. While Network on chips are designed to allow strong inter-node connections to prevent congestion performance, actual-world implementations generally experience steady traffic, contributing under-utilized on-chip properties. Our review of mesh topology benchmarks reveals that on average, 92.7% of routers are inactive during runtime applications. Entirely, the power gating is essential to Network on chips due to the inclusion of its leak capacity to overall processor capacity but, secondly, this appears to be encouraging due to the abundance of indolence in the modules.

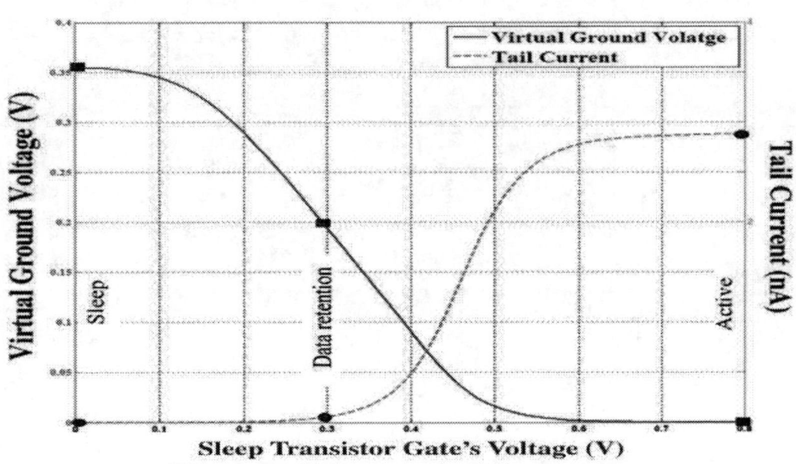

Figure 1b. Tail current in different modes

Various sleep transistor modes for one cell are shown in Figure 1b. The sleep saving leakage and the data retention modes are in spite of the circuit leakage. The penalty for power gating (energy and time delay) is increased by the the circuit leakage. shown by the use of the can gate in the GND. Fig 1c indicates the wake up time and energy exhaustion

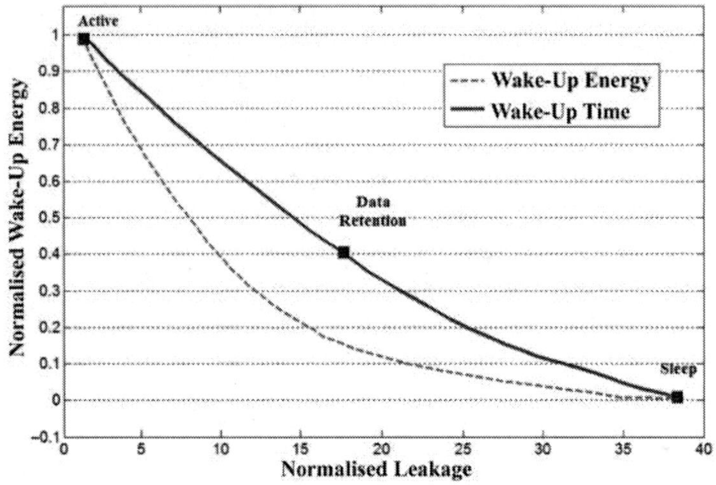

Figure 1c. Wake up energy and wake up time

The parallelizing view of the network breaks down the data forwarding phase in a few short steps. Standard flow control techniques, such as wormhole routing, first allocate a buffer and keep it from the time the data flit moves away the current node to the time the flow control signal returns to notify the current node that the buffer can be released. The buffer can also be reused after this turn-around time. And the credit turns around the time of the buffer is at least the amount of the delay in the spread of the data flit in forward direction and the flow control back.

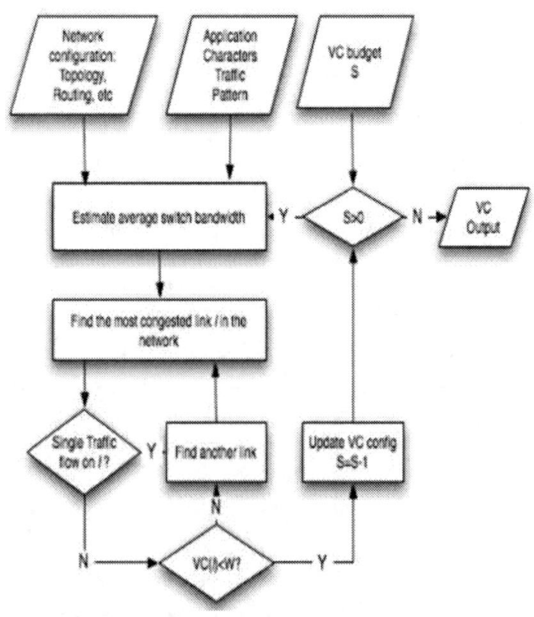

Figure 2. Virtual channel planning

The approach presented in the paper addresses overhead with respect to the optimization of the buffer. Complex distribution of these buffers to various*s virtual channels must be considered. It would also improve the efficiency of the buffer usage, as not all virtual channels need the same amount of buffering. Getting dynamic virtual channels provides a basic structure for the implementation of various schemes. Dynamic buffer allocation to virtual channels is more complicated and hence needs more overhead in terms of control logic complexity. Such complex schemes aim to optimize buffer performance. Neighboring routers, however, should also be informed about a power-closed router, then they may send packets that are tossed out of the network. Therefore, handshake alerts are needed to warn the neighbor if the router is switched down. Throughout this scenario, that competitor marks relevant outlet ports as being worked off and makes this inaccessible while assigning a change. Likewise, neighboring routers want handshake signals that alert the router when sending a message. The router on the right is power gated but alerts its neighbors when using PG signals. As a result, the former neighbors tag the corresponding actual output safety buffer as either power-off. Nearby routers are encouraging this router to be operated by a wake-up message

4. Results and discussion

The accuracy of both the gate is largely limited by the precision of all its based timing mechanisms and the energy-based modules. The gate itself is specifically modeling the actions of the Network on chip power gate.

Table 1. Comparison with existing methods

Parameters	Existing method	Proposed method
Area of consumption of virtual channel	38.94%	27.4%
Power consumption of virtual channels	55.73%	43.6%
Network topology	Mesh	Mesh
Routing algorithm	XY	XY

Almost every idle time cycle produced somewhere in Garnet is still controlled and otherwise reported as a power cycle even if certain conditions have been met and the static and dynamic power of each Network on chip element is taken into account shown in the table 1. given that the dependent control modules correctly model the energy consumption of the modules. One disadvantage in a gate was that the specificity of the power gate is at the per router stage. Although older Network on chip power-gating works explored the potential for power-gating of individual elements within a router given the difficulty of handshaking and command, the most famous works favor power-gating to specificity per router.

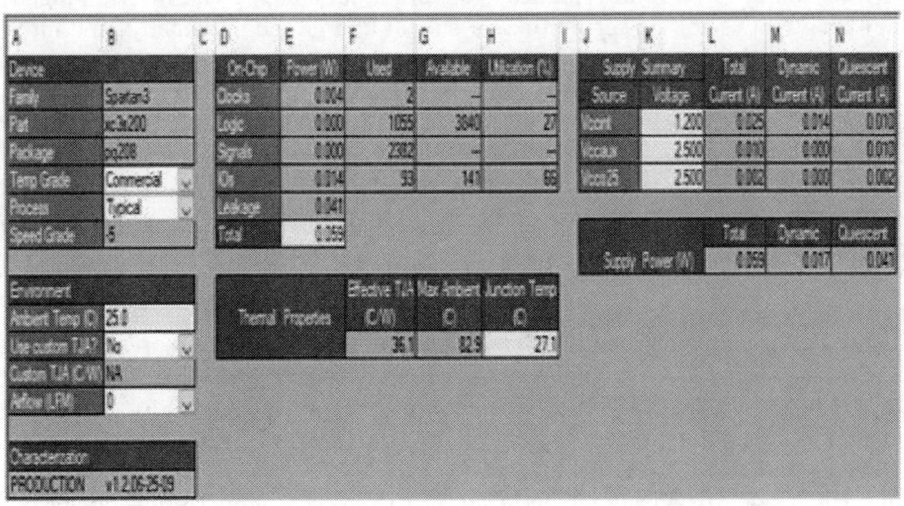

Figure 3. Power analysis of proposed method

The area and power consumption of virtual channels in proposed method is 27.4% and 43.6% when compared to existing method [3] 38.94% and 55.73% respectively.

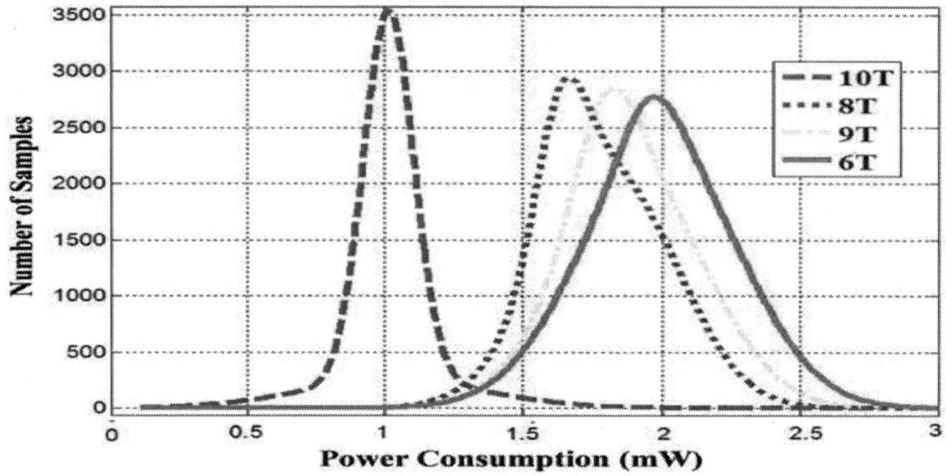

Figure 4. Power consumption of proposed method

The best possible optimization rates for various cell topologies, 6 T, 8 T, 9 T and 10 T are taken. The reduction in power of design, stability and loss of time is thoroughly investigated. Where the average area and power of the router can be possibly improved by around 11% without having a major effect on the overall Performance which is compared to [3], shown in table 1 and figure 3, future work can be undertaken to help to reduce the latency using power-gating method.

5. Conclusion

The Virtual channel flow control is one of the most common flow control methods used in interconnecting networks due to its various advantages powerful buffer usage, maximum throughput, low latency and network capacity regardless of the number of stages. Throughout this paper, we addressed the importance of the allocation of buffer for virtual channel flow control. Various different buffer allocation strategies that are valid at node and network level have been addressed. Higher buffer utilization one should suggest using schemes such as flit-reservation flow control, this system completely avoids idling buffer utilization due to credit turnaround period. Power-gating is a reasonable way of minimizing NoC static power but may result in a large output penalty. Within this paper we discuss the essential problem of simulating NoC power management systems correctly and efficiently. In order to achieve accuracy and efficiency of the simulation we define main requirements which should be fulfilled by a No C power-gating simulator. Power-gating as a symbolic low-power technique can be added to Network on chip to reduce this increase in leak capacity. Nevertheless, the cancellation issue significantly limits the power gate to be used effectively due to its negative effect on efficiency. In this paper, we suggest a creative limited power-gating solution to avoid the issue of power-gated separation in Network on chip.

References

[1] F Chang W, Yubai L, and Song C 2008 Design and simulation of a Torus topology for network on chip *Journal of Systems Engineering and Electronics* **19** 694-701

[2] Kearney DA and Veldman G 2003 Evaluation of network topologies for a run time re-routable network on a programmable chip *In Proc. IEEE Int. Conf. on Field-Programmable Technology (FPT)* 178-185

[3] Mondal HK, Gade SH, Kishore R and Deb S 2015 Power-and performance-aware fine-grained reconfigurable router architecture for Network on chip *6thInt. Green and Sustainable Computing Conf. (IGSC)* 1-6

[4] Zhang GQ, Wang D and Li GJ 2007 Enhancing the transmission efficiency by edge deletion in scale-free networks *Physical Review E* **76** 017101
[5] Huang W and Chen S 2011 Epidemic metapopulation model with traffic routing in scale-free networks *Journal of Statistical Mechanics: Theory and Experiment* **12** P12004
[6] Casu M R, Yadav MKY and Zamboni M 2013 Power-gating technique for network-on-chip buffers *Electronics letters* **49** 1438-1440
[7] Chen L, Zhu D, Pedram M and Pinkston TM 2015 Power punch: Towards non-blocking power-gating of Network on chip routers *21st International Symposium on High Performance Computer Architecture (HPCA)* 378-389
[8] Kadeed T, Tobuschat S, Kostrzewa A and Ernst R 2019 Safe and efficient power management of hard real-time networks-on-chip *Integration* **65** 1-17
[9] Farrokhbakht H, Kamali HM and Jerger NE 2019 Muffin: Minimally-Buffered Zero-Delay Power-Gating Technique in On-Chip Routers *IEEE/ACM International Symposium on Low Power Electronics and Design (ISLPED)* 1-6
[10] Ahmed AB, Fujiki D, Matsutani H, Koibuchi M and Amano H 2018 AxNetwork on chip: Low-power approximate network-on-chips using critical-path isolation *12th IEEE/ACM International Symposium on Networks-on-Chip (NETWORK ON CHIPS)* 1-8
[11] Narayan S, Vats S and Mani S 2019 Method for power gating for wide dynamic voltage range operation *U.S. Patent 10, 396,778*
[12] Zhu D, Li Y, and Chen L 2019 On Trade-off Between Static and Dynamic Power Consumption in Network on chip Power Gating *Int. Symp. on Low Power Electronics and Design (ISLPED)* 1-6
[13] Gupta Y, Bhargava L, Sharma A and Gaur MS 2020 Hybrid buffers based coarse-grained power gated network on chip router microarchitecture *Int. Journal of Electronics* **107** 272-287
[14] Parasuraman S and Ramesh GP 2017 Design of MML Reduction in Super High Frequency Antenna Using HFSS for Wireless Communication System *Indian Journal of Public Health Research & Development* **8** 288
[15] Zheng H, and Louri A 2017 Ez-pass: An energy & performance-efficient power-gating router architecture for scalable Network on chips *IEEE Computer Architecture Letters* **17** 88-91
[16] Mondal HK, Gade SH, Kishore R, Kaushik S and Deb S 2016 Power efficient router architecture for wireless Network-on-Chip *17th International Symposium on Quality Electronic Design (ISQED)* 227-233
[17] Manivasagam MA 2017 An efficient self-reconfiguration and route selection for wireless sensor networks *Int. Journal of MC Square Scientific Research* **9** 192-9
[18] Benini L and Micheli GD 2002 Networks on chips: A new SoC paradigm *Computer* **35** 70-78
[19] Sharma N and Gadag S 2013 An efficient way to increase performance by using low power reconfigurable routers *IOSR Journal of Electronics and Communication Engineering (IOSR JECE)* **8** 39-44
[20] Véstias MP, and Neto HC 2006 Area and performance optimization of a generic network-on-chip architecture *In Proc. of the 19th annual symp. on Integrated circuits and systems design* 68-73
[21] Hemanth Kumar G and Ramesh GP 2019 Reducing Power Feasting and Extend Network Life Time of Iot Devices Through Localization *IJAST* **28** 297-305

IOP Conf. Series: Materials Science and Engineering **925** (2020) 012067 doi:10.1088/1757-899X/925/1/012067

Design of Optimized Radix-4 FFT Processor with Multiplier Sharing Method

A. Manimaran[1] and S. Parasuraman[2]

[1,2] ECE Department, Karpaga Vinayaga College of Engineering and Technology, Chennai, India.
[1]lims0181@gmail.com, [2]parasuengineers@gmail.com

Abstract. To implement the parallel pipelined FFT architecture in OFDM applications. The OFDM has developed in wide band Digital communication. The radix4 algorithm can achieve to reduce the half computational steps compare than the radix2 algorithm and R4MDC is reduces the slices and power consumption than the single path delay feedback. It can be converted into real and imaginary values. To reduce the twiddle factor complexity by using the bit parallel multiplier. It can be exploits the constant twiddle factor when the twiddle factor multiplicatier relocate the twiddle factor multiplications in timing process. The changes can be appeared in multiplications so have to perform the TF at clock signals in two ways data path.

Keywords OFDM, FFT, MDC, BPM, SDF.

1. Introduction

In General, the FFT can make simple stand most effective optimizations when the number of samples to be transformed is an exact power of two, for which it can eliminate many unnecessary operations. The results of the FFT are the same with DFT, the only difference is that the algorithm usually runs much faster and less resources [1]. In algorithm terms, the DFT takes O(N^2)arithmetic operations to be computed, whereas the FFT takes O($N\log_2 N$) arithmetic operations. There are two basic DFT algorithm decimation in time and decimation in frequency [2-4].
This algorithm decomposes even and odd-indexed frequency samples

$$.X(2k) = \sum_{n=0}^{N-1}[x(n)\, W_N^{2nk}]$$

$$= \sum_{n=0}^{N/2-1}[x(n)\, W_N^{2nk}] + \sum_{n=0}^{N/2-1}[x(n+n/2)\, W_N^{2k(n+\frac{N}{2})}]$$

$$= \sum_{n=0}^{N/2-1}[x(n)\, W_N^{2nk}] + \sum_{n=0}^{N/2-1}[x(n+n/2)\, W_N^{2kn}]$$

$$\sum_{n=0}^{N/2-1}[(x(n) + x\left(n+\frac{N}{2}\right))\, W_{N/2}^{kn}]$$

$$= \text{DFT}_{N/2}\,[x(n) + x\left(n + \tfrac{N}{2}\right)]$$

$$X(2k+1) = \sum_{n=0}^{N-1}[x(n)\,W_N^{(2k+1)n}]$$

$$\sum_{n=0}^{N/2-1}[x(n) + W_N^{N/2}x\left(n + \frac{N}{2}\right)W_N^{(k+1)n}]$$

$$\sum_{n=0}^{N/2-1}[(x(n) - x\left(n + \frac{N}{2}\right))\,W_N^n W_{N/2}^{kn}]$$

$$= \text{DFT}_{N/2}\,[(x(n) - x\left(n + \tfrac{N}{2}\right))W_N^n]$$

FFT is a fastest method for finding DFT. There are many kind of algorithms and architecture available for FFT. This section will review commonly using algorithm and architectures. Decomposing the frequency samples into 2 bins (one each for odd and even samples) gives the Radix-2 Algorithm. Increasing the number of bins to 4 gives the Radix-4 Algorithm.

$$X(4k_1 + l) = \sum_{k=0}^{N}[x(n) + x\left(n + \tfrac{N}{4}\right)W_4^l + x\left(n + \tfrac{N}{2}\right)W_4^{2l} + \left(n + \tfrac{3N}{4}\right)W_4^{3l}]W_N^{nl}W_{N/4}^{nk_1}$$

$$l = 0,1,2,3;\ k_1 = 0,1,2,3 \cdots \frac{N}{4} - 1$$

While radix-2 and radix-4 FFTs are the most widely used FFT algorithms, it is also possible to design FFTs with even higher radix butterfly. The reason they are not often used is because the control and data flow of their butterfly are more complicated and the additional efficiency gained diminishes rapidly for radices greater than four. The radix-8 algorithm is described here for completeness. The other algorithms try to achieve the benefit of higher-radix FFTs (lower multiplications required) even though they use structures which have lesser complexity similar to radix-2 or radix-4 divided algorithms [5] .

Table 1.1 Twiddle factor multiplier required for Radix-4FFT

Stage I	Stage II
W_{16}^0	1
W_{16}^1	0.9238-j0.3826
W_{16}^2	0.707-j0.707
W_{16}^3	0.3826-j0.9238
W_{16}^4	-j
W_{16}^5	-0.3826-j0.707
W_{16}^6	-0.707-j0.707
W_{16}^7	-0.9238-j0.3826

Radix-4 FFT processor reduces half of the computational steps than Radix-2 FFT. Hence, Radix-4 FFT is preferred in this research work for alleviating the architectural performances of frequency modulation technique of OFDM system. The FFT architecture basically divided into the Two types. There is sequential processor, pipeline processor. Here using pipeline processor in R4MDC Multipath Delay Commutator (MDC) architecture reduces the hardware slices and power consumption of the FFT processors than Single Path Delay Feedback (SDF [3]). As shown in architecture of R4MDC, input data is shifted according to their appropriate delay positions. Commutator unit is used to re-arrange the one form of signals into another form of signal. On the other hand, Butterfly unit is used to add and subtract

IOP Conf. Series: Materials Science and Engineering **925** (2020) 012067 doi:10.1088/1757-899X/925/1/012067

the signed bit values. Appropriate delays are used to determine proper frequency response in appropriate delay positions. If there is more number of samples are included in FFT processes, signal strength should be increased while performing data communication. Due to merging the real and imaginary values of complex input values, it is possible to reduce the half of the accumulation and subtract or units while calculating FFT processes. [1] However, it requires more delay for exhibiting frequency transformation process.

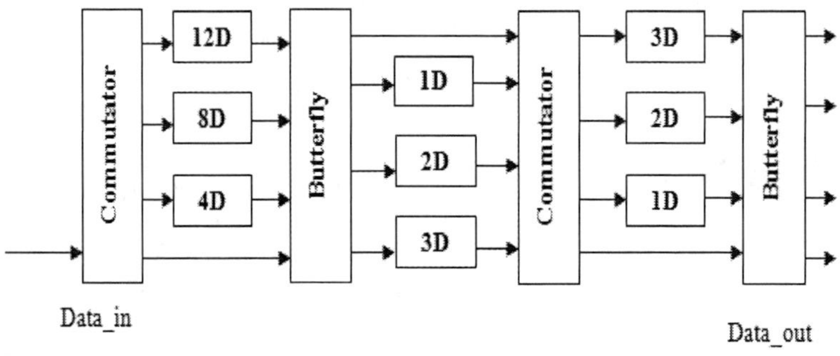

Figure.1. Architecture of MDC

Radix-2 Multi-path Delay Commutator is the most classical approach for pipeline implementation of radix-2 FFT algorithm. As shown in Fig.3.7, the input sequence has been broken into two parallel data stream flowing forward, with correct distance between the data elements entering the butterfly scheduled by proper delays. Both butterfly and multipliers are in 50% utilization. In certain applications like Multiple-input Multiple-output (MIMO) Antenna systems, the first commutate at the input data stream can be can be replaced with 2 such commutates to keep the pipeline components 100% utilized. The output data rate for this structure is twice the clock frequency and that is what makes it appealing for high-speed applications [10-11].
Multiplication of twiddle factor value 0.707 requires 4 numbers of shifters and three number of adder to perform multiplication.

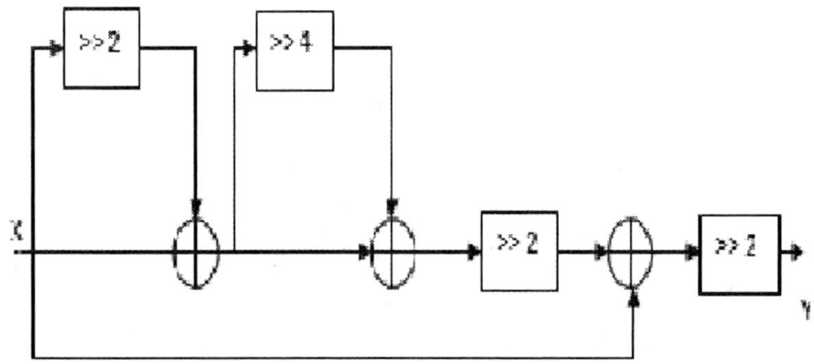

Figure 2. Bit parallel multiplication for twiddle factor 0.707

Similarly, twiddle factor multiplication of 0.9238 and 0.3826 with constant input x is illustrated in fig. 8 and fig. 9 respectively. Multiplication of twiddle factor value 0.9238 requires 5 numbers of shifters and single adder to perform multiplication. Similarly, 0.3826 require two shifters and single adder to perform multiplication. In this research work, we can realize these BPM based complex multiplier to further reduce the hardware components. For the twiddle factor multiplication of 0.707, Equation (11) can be further simplified based on taking common factor for reducing the number. Also equation (11) can be write combinational circuit design is described. This technique allows reducing power consumption, propagation delay and area of digital circuits while maintaining low complexity of logic design.

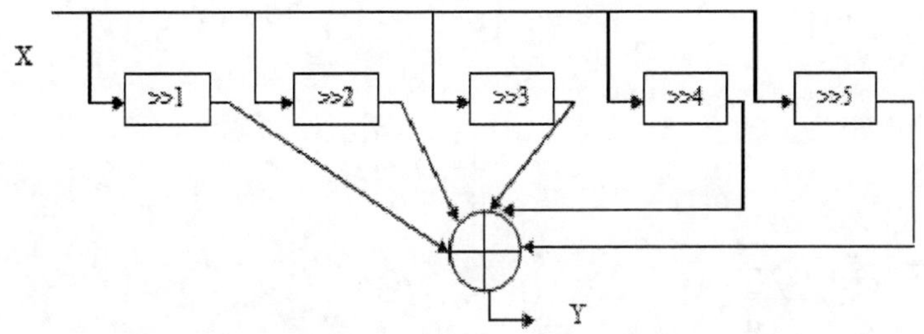

Figure 3. Bit parallel multiplication for twiddle factor 0.9238

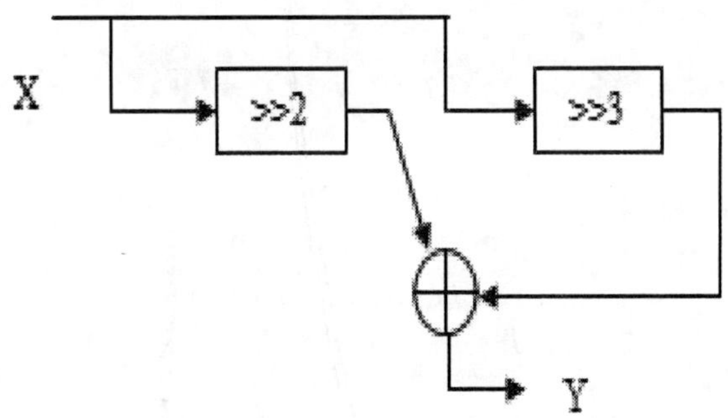

Figure 4. Bit parallel multiplication for twiddle factor 0.3826

2. Architecture analysis and Design

As we mentioned in this section, the data flow structure of Radix-4 FFT has been illustrated in figure 2. Each stage of Radix-4 FFT performs two stage of Radix-2 FFT. For Radix-4, $N = r^v = 4^v$ where v is number of stage and r is the radix of FFT [6].

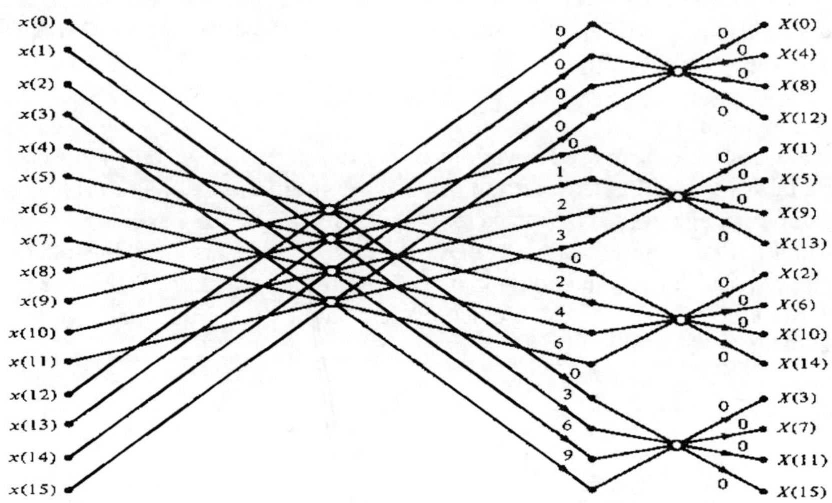

Figure 5. Data Flow structure of 16 point Radix-4 FFT

$$Y[K]=\sum_{n=0}^{\frac{N}{4}-1} y(n)W_N^{nk}+\sum_{n=N/4}^{\frac{N}{2}-1} y(n)W_N^{nk}+\sum_{n=N/2}^{\frac{3N}{4}-1} y(n)W_N^{nk}+\sum_{n=3N/4}^{N-1} y(n)W_N^{nk}$$

$$Y[K]=\sum_{n=0}^{\frac{N}{4}-1} y(n)W_N^{nk}+\sum_{n=0}^{\frac{N}{4}-1} y\left(n+N/4\right)W_N^{\left(n+\frac{N}{4}\right)k}+\sum_{n=0}^{\frac{N}{4}-1} y\left(n+N/2\right)W_N^{\left(n+\frac{N}{2}\right)k}+$$
$$\sum_{n=0}^{\frac{N}{4}-1} y\left(n+3N/4\right)W_N^{\left(n+\frac{3N}{4}\right)k}$$

$$Y[K]=\sum_{n=0}^{\frac{N}{4}-1} y(n)W_N^{nk}+W_N^{Nk/4}\sum_{n=0}^{\frac{N}{4}-1} y\left(n+N/4\right)W_N^{nk}+W_N^{Nk/2}\sum_{n=0}^{\frac{N}{4}-1} y\left(n+N/2\right)W_N^{nk}+$$
$$W_N^{3Nk/4}\sum_{n=0}^{\frac{N}{4}-1} y\left(n+3N/4\right)W_N^{nk}$$

$$Y[K]=\sum_{n=0}^{\frac{N}{4}-1} y(n)W_N^{nk}+W_N^{Nk/4}\sum_{n=0}^{\frac{N}{4}-1} y\left(n+N/4\right)W_N^{nk}+W_N^{Nk/2}\sum_{n=0}^{\frac{N}{4}-1} y\left(n+N/2\right)W_N^{nk}+$$

where the twiddle factor values are given as,

$$W_N^{Nk/4}=\left[\cos\left(\frac{2\pi}{4}\right)-j\sin\left(\frac{2\pi}{4}\right)\right]^k=(-j)^k \quad (1)$$

$$W_N^{Nk/2}=\left[\cos\left(\frac{2\pi}{2}\right)-j\sin\left(\frac{2\pi}{2}\right)\right]^k=(-1)^k$$

$$W_N^{3Nk/4}=\left[\cos\left(\frac{2\pi*3}{4}\right)-j\sin\left(\frac{2\pi*3}{4}\right)\right]^k=(j)^k$$

Radix-4 FFT processor reduces half of the computational steps than Radix-2 FFT. Hence, Radix-4 FFT is preferred in this research work for alleviating the architectural performances of frequency modulation technique of OFDM system [7-10]].

3. Simulation and Results

This OFDM application has been used in various categories of digital system in this design technologies in present state of developed implementation of power consumption is major problem in the design to overcome the disadvantage of the twiddle factor complexity, increase speed, increase speed. Simulation of Proposed Modified BPM based Radix-4 MDC FFT architecture has been illustrated in figure 5.1. As shown in Figure 6, to push the sequential inputs starting from 0+0i, 1+1i, 2+2i, and 3+ 3i ...63+ 63i in first 16 successful clock cycles. After 16 clock cycles, outputs are sequentially getting from designed Radix-4 MDC FFT.

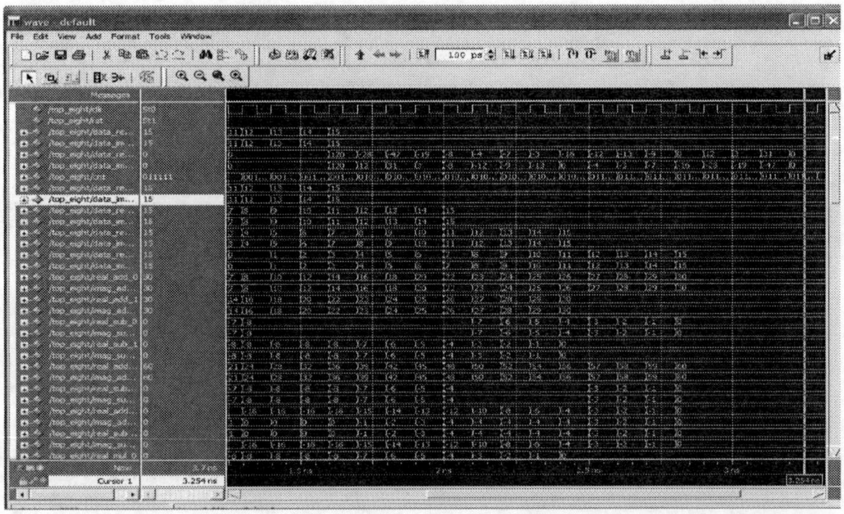

Figure 6. shows the simulation result of the Radix4 16 point FFT

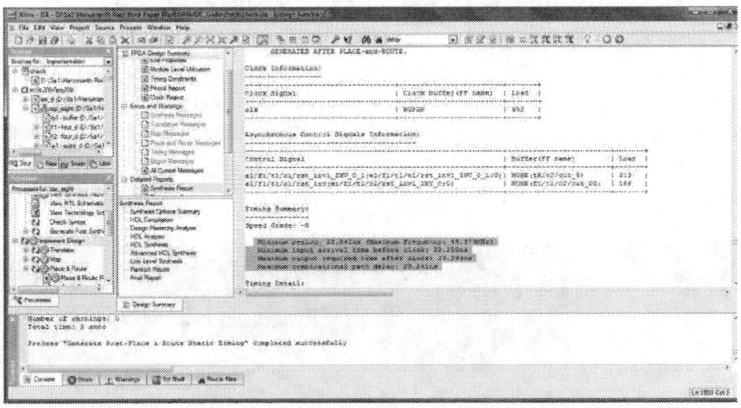

Figure 7. synthesis report in power summary

The figure 1.6 shows the synthesis report in Xilinx and gives the minimum period 22.04ns compare then the existing and the frequency in before and after clock is 23.24 computational delay is 23.241.

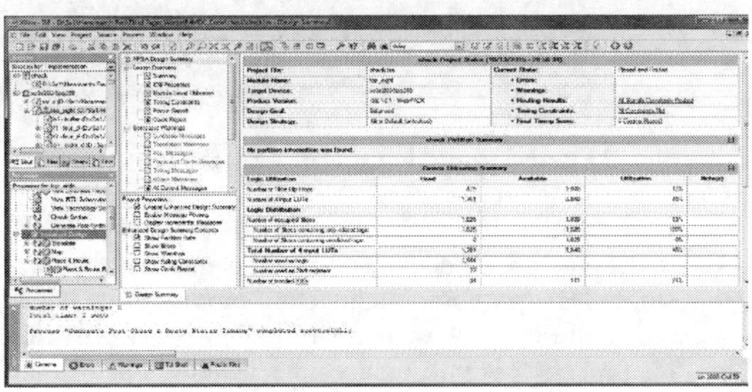

Figure 8. Area analyzer summary

Using the bit parallel multiplier it has the four no of adder and three no of shifters.it shows the area analysis report in Xilinx.and gives reduced slices and LUT compare existing in slices is 1299 and proposed is 1025 have been reduced and LUT is 2193 into 1761 have been reduced. The XPower Analyser tool is used to measure the amount of power consumed during the operations. Here it used to calculate both the static and dynamic power. The power level can be increased or modified based on the switching activities of CMOS in hardware. So it's necessary to choose the good SDK tool to implement this. The screen shots of power simulation is shown in Fig 5.3.

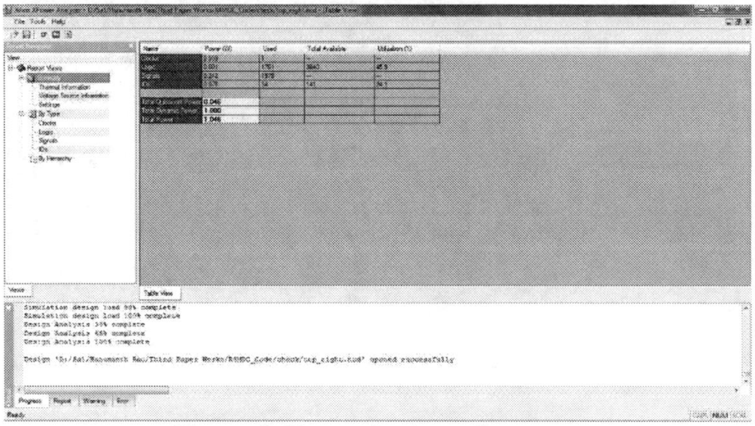

Figure 9. Power Analyzer Summary

Fig.9 shows the output analysis of power consumption to compare than existing SDF architecture in reduced eight percentages. An RTL description is usually converted to a gate-level description of the circuit by a logic synthesis tool [11]. The synthesis results are then used by placement and routing tools to create physical layout. Logic tools may use a design's RTL description to verify its correctness. The RTL view of top level module is shown in Figure 10.

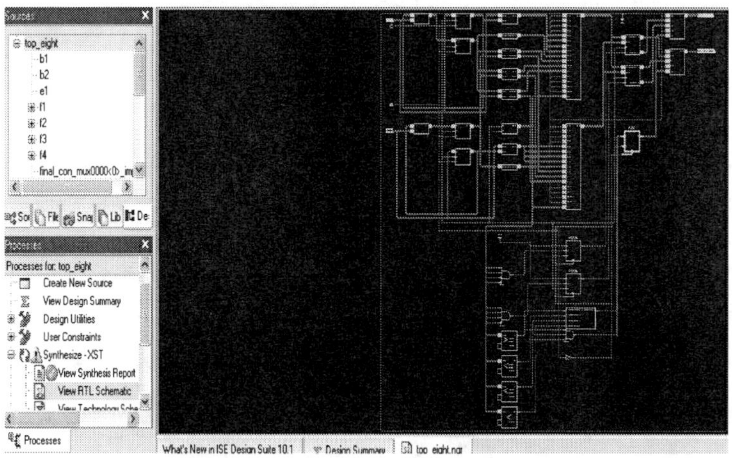

Figure 10. RTL View of FFT Design

4. Conclusion

By using the FFT processor, the chip area reduced and speed is increased and frequency is achieved by reducing the computation stages. The proposed architecture provides low complexity, high speed and also reduces the chip area. In future this low complexity FFT processor will be used to implement in Radix-64 FFT. By using this processor is will reduces the Twiddle Factor multiplication complexity, speed and frequency is increased and reduction in power consumption are achieved. So this will be used in various research applications.

IOP Conf. Series: Materials Science and Engineering **925** (2020) 012067 doi:10.1088/1757-899X/925/1/012067

References

[1] Badar S and Dandekar DR 2015 High speed FFT processor design using radix− 4 pipelined architecture *Int. Conf. on Industrial Instrumentation and Control (ICIC)* 1050-1055

[2] Li W, Ma Y and Wanhammar L 2000 Word length estimation for memory efficient pipeline FFT/IFFT processors. Univ.

[3] Malathi K and Rabi BJ 2014 Design of optimized M-R2MDC FFT with wave pipelined technique for OFDM *2nd Int. Conf. on Current Trends In Engineering and Technology-ICCTET 2014* 426-429

[4] Kim EJ, Lee JH and Sunwoo MH 2014 Novel shared multiplier scheduling scheme for area-efficient FFT/IFFT processors *IEEE Transactions on Very Large Scale Integration (VLSI) Systems* **23** 1689-99

[5] Swartzlander EE and Yu S 2010 A new pipelined implementation of the Fast Fourier Transform. IEEE Trans. Circ. Syst.

[6] Lee H and Shin M 2007 A high-speed low-complexity two-parallel radix-2 4 FFT/IFFT processor for UWB applications *IEEE Asian Solid-State Circuits Conf.* 284-287

[7] Lin BC, Wang YH, Huang JD and Jou JY 2010 Expandable MDC-based FFT architecture and its generator for high-performance applications *23rd IEEE International SOC Conf.* 188-192

[8] Lin YW, Liu HY and Lee CY 2005 A 1-gs/s fft/ifft processor for uwb applications *IEEE Journal of solid-state circuits* **40** 1726-35

[9] Hasan M, Arslan T and Thompson JS 2003 A novel coefficient ordering based low power pipelined radix-4 FFT processor for wireless LAN applications *IEEE Transactions on Consumer Electronics* **49** 128-34

[10] Yogeeswaran S and Ramesh GP 2017 Design of Reconfiguration Based Manchester coding Technique for Li-Fi System *Indian Journals of Public Health Reasearch & Developments* **8** 630-4

[11] Yogeeswaran S and Ramesh GP 2017 Carrier Less Amplitude and Phase (CAP) odulation Technique for OFDM system *International Journal of MC Square Scientific Research* **9** 145-53

IOP Conf. Series: Materials Science and Engineering **925** (2020) 012068 doi:10.1088/1757-899X/925/1/012068

An Evaluation of Effective Intrusion DoS Detection and Prevention System Based on SVM Classifier for WSN

Prabakaran K[1], Kumaratharan N[2], Suresh G[3] and Epsiba P[4]

[1] Abdul Hakeem College of Engineering and Technology, Tamil Nadu, India
[2] Department of ECE, Sri Venkateshwara College of Engineering, India
[3] Department of ECE, Sri Indu College of Engineering and Technology, India
[4] Department of ECE, Pallavi Engineering College, India
[1]prabuswamy@yahoo.com, [2]kumaratharan@rediffmail.com, 3geosuresh@gmail.com.
[4]epsisuresh@gmail.com

Abstract. The trending technology of Wireless Sensor Networks (WSNs) are broadly used in this recent earth for a greatest rate of purpose for the better worth of person living however owing to their energy constrains as well as utilization and problems as mentioned above, these WSN has their restriction in their intention as well as their practicality. The wireless sensor networks face main threat challenges such as Denial of Service (DoS) attack, flooding attack, replay attack. In this approach, we proposed an Intrusion Detection (ID) technique using Support Vector Machine (SVM). In the ID algorithm, we employ feature vector to constitute nodes. Such components consist of routing message number, varies destinations in sending and receiving routing packets, and similar source node. The similar source node consists of similar characteristics; the node of the similar category has the same features. Therefore, irregular nodes will be notable. The measurement result shows an introduced scheme attain higher rate of accuracy. Here, Ant colony Optimization (ACO) technique to choose best route between entire adjacent feasible routes from source to receiver node for data communication. The proposed approach enhances the network lifespan as well as discovers intrusion nodes in WSN.

Keywords: Ant colony Optimization, Denial of Service Attack, Intrusion Detection, Support Vector Machine, Wireless Sensor Networks.

1. Introduction

Generally, WSNs consisting of quantitatively great amount of dispersed sensor nodes interconnected together to examine and determine the status of real atmospheric conditions. The evaluated variables are communicated to various distributed end users through Wireless Network (WN). WSNs are miniaturized, power consumption is considerably good and have been extensively established for real time applications as compared to the existing networks. Even then, it faces lot of storage constraints, message delivery distance, processing complexity and energy harvesting. Conversely, embedded system plays a major role in the past decade. These are very small in size, cheap, consumes less power and they had many sensors which works on various environment [3]. A typical WSN consists of various sensor components. These components combined to a simple small device and this is implemented into various application scenarios [1] [2]. These advance specifications make them to be flexible when compared to

the ordinary traditional networks. The applications of WSNs are abundant: they are extremely scalable and a high density range, due to their cheap rate and their smaller size [15], [8]. Normally, all the sensor network protocols are designed with their own organizing capacity. Hence these sensor network has an effective maintenance and they have forbearance against communication failure and the topology could change based on the nodes that could break down, or be mobile or there can be depletion of energy. Moreover, sensor nodes could be implemented in the rough environment to maintain the given task without any man power.
Moreover, supplementary applications are inadequate due to some disadvantages of WSN. For instance, the restricted giving out capability and minimum rate of data in the nodes of the sensor couldn't warranty the maximum performance in some situations particularly for instantaneous applications [16]. When the communication range is very small, then there could be maximum energy consumption and network incompetence, since the requirement of multi-hop communications is often used for the transportation of data [17]. The high drive restriction also direct to bad inspection: the development of data processing, as the energy gets drained rapidly and makes the network ineffective for the particular job. Low amount of energy consumed maintains multi-hop communications work for a maximum duration.
WSN has become most favoured networking solution, though; it's tremendously incomplete resources face serious threats. In addition, the nodes of sensors are very sensitive to some attacks which could be caused due to its lack of tamper-resistance. Several threads which occur in WSN are DoS attack, differential power attack, node balancing attack, Scheme-dependent attack as well as imaging attack [7], [11]. DoS (Denial of Service) attack is typically anticipated attack of opponent for wiping out the sensor network. It limits or eliminates the anticipated performance of the sensor network than expected. DoS attack intrudes the efficiency of the target networks by consuming the resources, destructing infrastructure configuration, and the network components.

1.1 Architecture of WSN
As a reliable technology, the WSN has been employed widely in other technologies such as Internet of things. Numerous associated uses have been performed. The function of the WSN is gaining reliability and fashionable in network. WSN is considered as most significant technology in the 21st century, WSN is considered as a vital component in linking the rationality entropy as well as the prevailing world. Though, the exposed working principle of WSN and entropy media as well as the deprived organization of surroundings possess many threats to the security as well as it is extremely limiting the claim of WSNs.

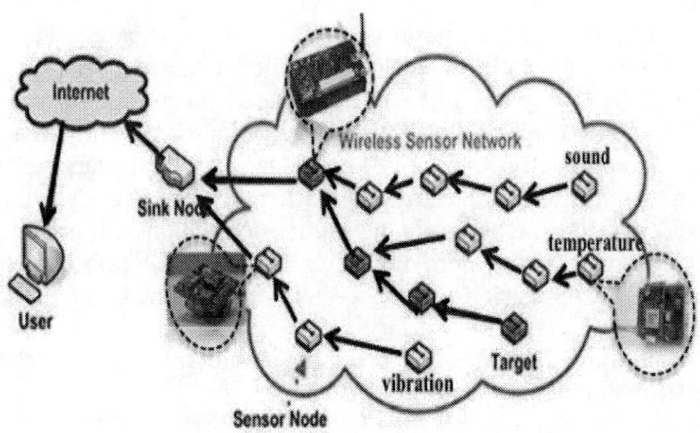

Figure 1. Architecture of WSNs

The trending technology of Wireless Sensor Networks (WSNs) are broadly used in this recent earth for a greatest rate of purpose for the better worth of person living however owing to their energy constrains as well as utilization and problems as mentioned above, these WSN has their restriction in their intention as well as their practicality. In spite of its accuracy and energy constrain analysis from real exploitation can suffer restrictions, since the data acquired by the sensors from a wide area is very tough to obtain. On the other hand, analytical/theoretical models obtained from the mathematical calculation offers huge exploration capacity but it can simplify through idealized hypotheses which could cause trouble during the practical implementation

2. Related work

WSN works in several environments which could consist of effective and limited processing sensor nodes that is low in cost, extremely small and consumes less power. This wireless sensor networks differs from the traditional network due to its wide range application specifications. The applications of WSNs are enormous: they are extremely scalable and have a high density range, extremely cheap in cost and their smaller size. In [13] proposed a technique on the basic principles of least cut set and less cluster head was carried out in a distributed intrusion detection module, with the help of signature based detection technique. Detection of black hole by centralized IDS procedure and the cluster-based wireless sensor networks that works by selective forwarding has been scheduled [6]. In [15] proposed partially distributed IDS that work on minimum power and less memory occupation. Detection carried by anomaly is also called as behavioral based detection in certain cases. Irregularities can be separated into three classes called point, appropriate as well as finally collective unusual person [12]. The WSN networks can be used as an alert system for the flood detection and it was suggested at Lancaster University [14] by a research. They made a protocol and did its functionality by a platform composed of powerful GridStix sensor [5] which has certain modules like GPRS to monitor the flood arrival and estimation in UK. This system was implemented at river ribble in the Yorkshire Dales which covered about 1 km. The system was deployed with 13 modes and the implementation is carried out as a real-time. The attacks in DoS based approaches affects the most important threads in OSI layer [4], [9] as described. More often these impairments are caused due to the system weakness or protocol defencelessness.

3. Support vector machine

In this context, SVMs [5] [1] are one of the supervised machine learning approach which employed for analysis. For given finite set of training samples, the SVM marks them for categories and thus it builds the training model which later assigns new test samples into the relevant category. It is binary linear and non-probabilistic classifier. In SVM model the training/test samples are represented as dot/points in the space and they are mapped in such a way that the clear gap among the categories appears which separates the samples. New test samples/examples are later mapped on the Using the kernel trick SVMs are also be able to perform non-linear categorization in addition to linear categorization. SVMs mechanically chart the remarks on towering multidimensional quality distances. In usual SVM defends to build anxious aircraft in any gap as well as this can be engaged for any jobs for example failure, forecast. SVM hyper plane is the only responsible for good separation of training data with the largest distance to its nearby data. This approach is generally known as functional margin. The state of the rule is if margin is larger than the generalization error of classifier will be lower. The characteristics of SVM are established to be exceedingly valuable for higher dimensional planes. It is highly preferable in less data samples provided as compared to the more dimensions. SVMs are flexible, memory efficient and which improves efficiency in critical decision making even more than one kennel method.

The important intend of rating the nodes is to precisely predict the objective class from the data in every container. The classifier engine procedure admits two treads:

1. Classifier construction: This approach is to construct a taking stage. The word that developed from tuition sure information examples is built through the categorization methods. Individuality instance which is collected of the training cluster is observed as a cluster. These tuples can as well as remarked to as data details.

2. Utilisation of Classifier – The training classifier rendered applying the data set will categorize test data preset objects.

• Significance Analysis: Information might contain several distinct properties. Correlation analysis will recognize two specified properties are connected.

• Data alteration as well as diminishing: The subsequent procedure assists in data communication:

Normalization: This communication concerns scaling each value which will build them fall inside a precise series. This approach is mostly utilized in the discovering stage while the neural networks otherwise these procedures developing evolution are utilized. Simplification: It is a main idea of communication. At this point we can employ the hierarchy idea. If the preparation nodes of n points is given and the method of it is (−x→1 , y1) .... (−x→n , yn) and here yi is 1, this indicates the category to that the point →-xi is present. Each →-xi is a real vector of P-dimension. Our interest is to discover the "extreme- border hyper plane" which classifies the GoS/points since the group of points →−xi here yi =1. It is understandable that H1 usually cannot distinguish the classes. While, H2 divide them with a small margin, on the other hand H3 depart them with the highest margin. The hyper plane is defined as the cluster nodes →-x fulfilling. The SVM consists the sample on the margin. The offset of the hyper plane from the origin with normal vector →−w is determined by the parameter →−w .Hard-margin two parallel hyper planes which separate two classes of data can be selected, when the training data sets are linearly separable components. So, the distance between them is as possible as large. The "margin" is nothing but the area bounded by two hyper-planes. And the maximum margin hyper plane lies between these planes. Where, the hyper planes are in and . →−w is the region distance of the two hyper planes, by minimizing w we can maximize the distance between the planes. By adding the constraint: for each either or , if yi = -1, hence the data points can be prevented from falling into the margin

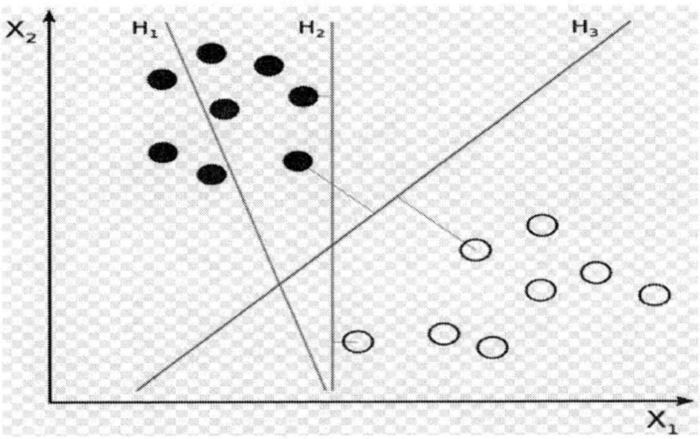

Figure 2. Categorization based on linear SVM

According to the constraints/conditions each and every data point should be the actual side of the region of two hyper planes. The rewritten equation is as observes:

$$y_i(-\vec{w}*-\vec{x}-b)\geq 1, forall 1 \leq I \leq n \quad (1)$$

To get the optimization problem we can put all these together: "Minimize →*w* subject to →−*w* * →−*x* - b ≥ 1 for I = 1, 2, 3,. N". Completely the maximum margin hyper plane is firmed by over right arrow(x_i) which lies nearer to the region. Here, the →−*x* *is* known support vectors.

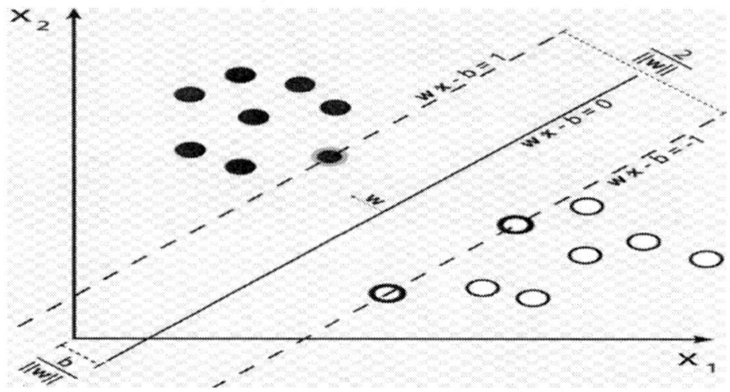

Figure 3. Categorization based on Hard SVM.

Soft-Margin is the loss function with the equation $max(0,1-yi(-\vec{w}*-\vec{x}-b))$. Especially, this was introduced to enhance Support Vector Machine where the data are non-linearly separable. If the constraint in (1) is satisfied, then this function value is zero. The function with minimization is as below:

$$high(0,1-y_i\ (-\vec{w}*-\vec{x}-b))+\lambda-\vec{w} \qquad (2)$$

Here the parameter λ finds the trade-off between increment the values of margin-size and making sure that the $\rightarrow -x$ is fall/separated with the actual side of it. Therefore, the soft margin Support Vector Machine may behave equally same as hard margin Support Vector Machine for enough small values of λ in the case when linearly classifiable test data are available (Yang, 2016).

4. Proposed intrusion detection

In the ID algorithm, we employ feature vector to characterize nodes. Such components consist of routing message number, varies destinations in sending and receiving routing packets, and similar source node. The similar source node consists of similar characteristics; the node of the similar case has the similar features. Therefore, irregular nodes will be notable. The feature weights and SVM parameter chromosomes are formed based on optimal feature subset. The best optimal chromosome is fixed by analyzing the chromosome with the greatest categorization precision of the chromosome; the best attributes as well as the characteristic weights are decrypted. The information is partitioned addicted to k sub-sets with the equal size as initial, next, ... k sub- parts, as well as the residual k-1 sub-parts as training data sets and for categorization the sub portions are combined k prediction results, where k=10 is chosen to evaluate the performance and to authenticate the robustness of the proposed SVM classifier [10].

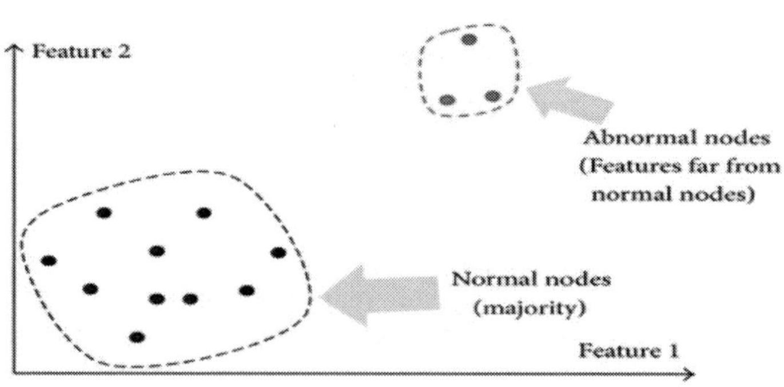

Figure 4. The representation of SVM ID technique

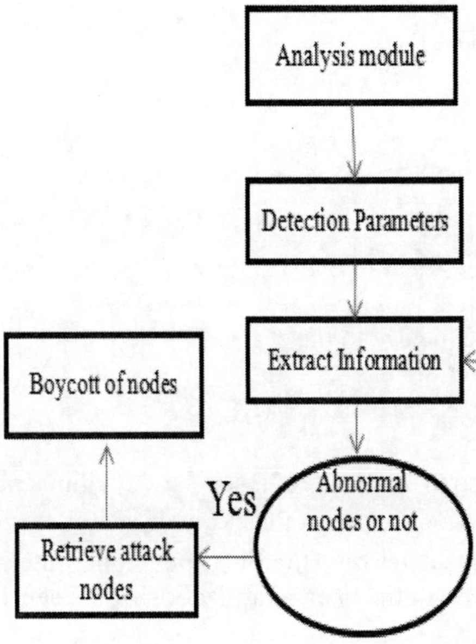

Figure 5. Architecture of Proposed System

We employed two important parameters, namely, limit value and score. The score represents total values of most adjacent nodes. Limit value represents the limit or threshold employed for classifying abnormal nodes from normal nodes. The RREQ (Route Request Packet), entails the target identity of all specified nodes of the network is spread. Thus, the SVM detection method categorizes from where irregular nodes transmit RREQ messages extra repeatedly than normal nodes. By connecting the incidence to transmit RREQ messages of every node, we can categorize as well as notice the irregular nodes. The characteristic collection marks the node with the incidence of RREQ messages. For the intention of categorization, it is essential to utilize a *best quantity of characteristics* for machine learning simulating. Afterwards a vast quantity of character elevate the calculation necessitates; it turns additional disputing to outline precise result borders inside a big multidimensional gap. This represents that a best separation of characters' desires to be certain for the intention of machine learning. The achievement of a categorization method mostly based on the characters particular as well as the information they offer for their function in the replica. Throughout the *categorization* period, the rules of the incidence of RREQ messages arriving through the nodes are separated as whichever irregular otherwise regular nodes on the foundation of the character particular as well as employ assurance evaluates to specify which a flooding attack is happened.

The boundary value honestly impresses the discovery achieves. The apposite boundary value ought to create the error rate of recognition algorithm as small. For the unnamed allocation representation of the network, best possible boundary value is not received however during examination as well as information; we can get the choice of the appropriate boundary value.

Proposed Energy Efficient Routing

In this strategy, we concern Ant Colony Optimization (ACO) procedure that detects the shortest route among source to receiver. Here, an ant tries to determine the smallest distance to face from nest to food source. Before route seeking we have occupied several initialization factors as traces. The n is the surrounding phenomenon which illustrates the lesser quantity of pheromones in the communication route. The conversion possibility is utilize for an ant a to development from sender node x to receiver node r that denotes the routing information. The possibility decision rule.

$$P_{xr}^{a}\left|\frac{[\lambda_{xr}]^{\gamma}[\beta_{xr}]^{\rho}}{\sum[\lambda_{xr}]^{\gamma}[\beta_{xr}]^{\rho}} \quad if\ r \in L^{a}\right|$$

Here, λ_{xr} the quantity of pheromone as well as β_{xr} is the heuristic function. The γ and ρ are the constant factor utilize to control the manipulate of the λ_{xr} and β_{xr}. In ACO technique ants restrain the memory L^{a} that has the information about nodes which previously inflicted or not. The heuristic function β_{xr} is the mutual of Euclidean distance (d_{xr} among x as well as r receiver node. After closing of ant tours the quantity of pheromones updated consequently to the

$$\lambda_{xr}(\mathrm{t+N}) = (1-\theta)\cdot\lambda_{xr}(\mathrm{t}) + \Delta\lambda_{xr}$$

Here, θ denotes the local pheromone destroy factor, t represents the previous pheromone update time, pheromone updates at period of time t+N. $\Delta\lambda_{xr}$ denotes the extra pheromone amount at the t + N point. This iteration procedure is repetitive till ants determine the finest potential route from source to receiver. The lifespan of the network computed through the energy utilization of the sensor node at best route as well a whole energy utilization of the ant r.

$$Energy = \frac{Ec^{2}}{\sum_{x=1}^{k}\sum_{r=1}^{k} e_{xr}^{\gamma} W_{r}^{\rho} P_{xr}^{a}}$$

Here k is the amount of the network as well as e_{xr}^{γ} is the energy utilization among source to receiver. W_{r}^{ρ} is the available energy receiver node. If ant a has travel edge (x,r) next the value P_{xr}^{a} is 1 else 0.

5. Evaluation

So as to measure the function; we engaged the GAINZ nodes as well as terminal device ready on wired network card to obtain transmit blacklist as well as traffic. The terminal device is applied to rule manage method as well as check the irregular nodes, also reply to attacks. The Tiny operating system and AVR Studio integrated expansion surrounding as illustrated in figure 4. The serial transaction helper is applied to exchange the manage data message

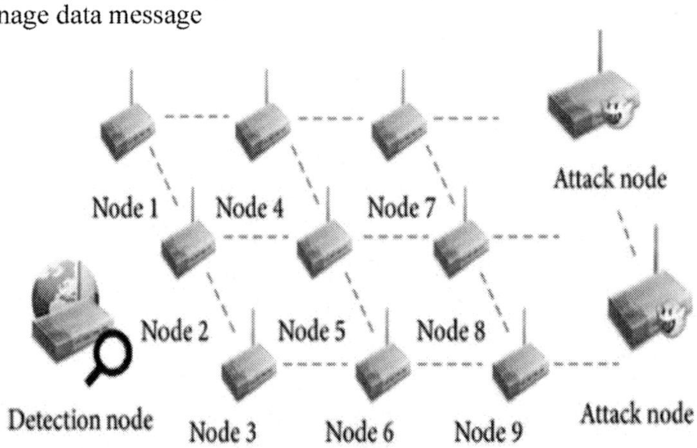

Figure 6. Representation of Test Scenario

The flooding attack occurrence, amount of flooding attack node as well as regular node as attack attributes are arbitrarily particular. There is over flow in one node as well as flooding attack frequencies are observed. The first case situation is accomplished under varies flooding attack frequencies on the equivalent amount of flooding attack nodes as well as regular nodes as well as irregular nodes. We selected three regular nodes and one flooding attack node.

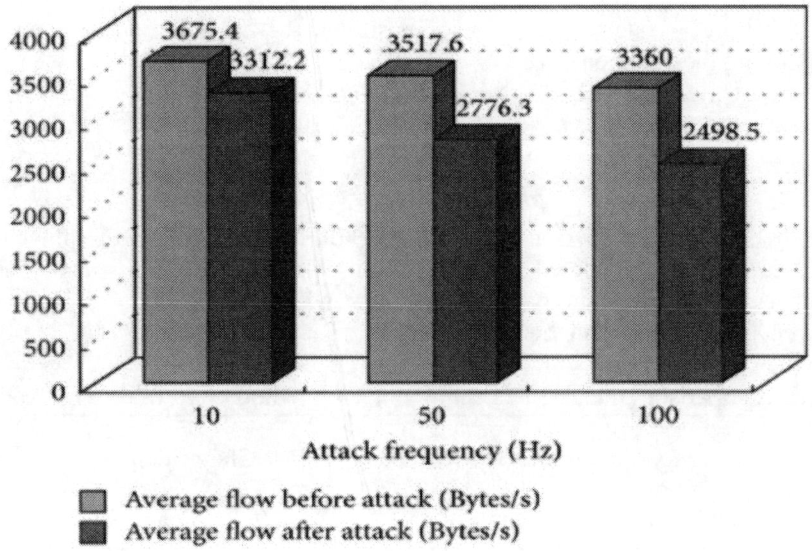

Figure 7. Flooding attack frequency varies

Now, when there are 60 RREQ packets every second, there is change flow of 35 byte/s and when flooding occurs with 150 RREQ packets every second, the flow changes to 34 Bytes/s. We also observed that the number of normal nodes rises when the flow declines. The reason is that many nodes participate in progressing RREQ packets. Hence, with large number of nodes the bigger the effect in the network. We also evaluated and compared the measurable parameters to authenticate the function of the introduced method with Naïve Bayes. The amount of True Positive(T_P), False Positives(F_P), False Negatives(F_N), as well as True Negatives(T_N) is offered via uncertainty matrix that contains two rowing as well as two columns. Such allows extra complete examination than insufficient fraction of precision. For every rowing in the uncertainty matrix means an investigational class, every column means an accepted class as well as every cell corresponds the amount of samplings in the association of those two classes. The uncertainty matrix is as given below:

Table 1. uncertainty matrix

	Yes	No
Yes	True Positive (TP)	False Negative (FN)
No	False Positive (FP)	True Negative (TN)

The true positives as well as true negatives categorizations are measured as precise categorizations which dwell alongside the mainly significant slanting line in the uncertainty matrix. The model errors signify the residual areas. We can obtain function attributes from the uncertainty matrix. At the end, the characteristic weights as well as measurable attributes of SVM are all together optimized based on feature subset. The experimental results exhibit that the proposed scheme speeds up the convergence, enhances the TP rate, reduces the error rate, as well as shrinks the categorization period. While comparing other schemes, SVM depend ID algorithm increase the detection rate and the FP as well as

FN rates are lowered. We can decide the precision in the %age of whole amount of guess is positive by employing equation 3:

$$Accuracy = ((TP + TN) / (TP + TN + FP + FN)) \times 100\% \quad (3)$$

Where, True Positive (T_P) rate is the %age of positiveness which is appropriately recognized, as determined employing the equation:

$$Recall = (T_P / (T_P + F_N)) \times 100\% \quad (4)$$

The Precision is the %age of the expected positiveness which is precise, as obtained by employing the equation:

$$Precision = (F_P / (F_P + T_N)) \times 100\% \quad (5)$$

Detection time: In order to recognize an attack in packet during detection, we employed detection time, and a short detection time is considered as better detection time. The assessment of all the classifiers applying the process of 10 folds annoyed establishment with attain attribute as well as lacking score characteristic is specified in the table below.

Table 2. Comparison of classifiers employing the method of 10 folds cross validation

Evaluation Criteria	Classifiers (Without Score Feature)			Classifiers (With Score Feature)		
	Naïve Bayes	Decision Tree	PC	Naïve Bayes	Decision Tree	PC
Accuracy (%)	92.53	95.52	97.01	74.62	79.10	82.08
Sensitivity (%)	87.87	96.96	96.96	60.60	81.81	81.81
Specificity (%)	97.05	94.11	97.05	74.66	82.35	82.35
MSE	0.0139	0.0108	0.009	0.0126	0.0174	0.0093

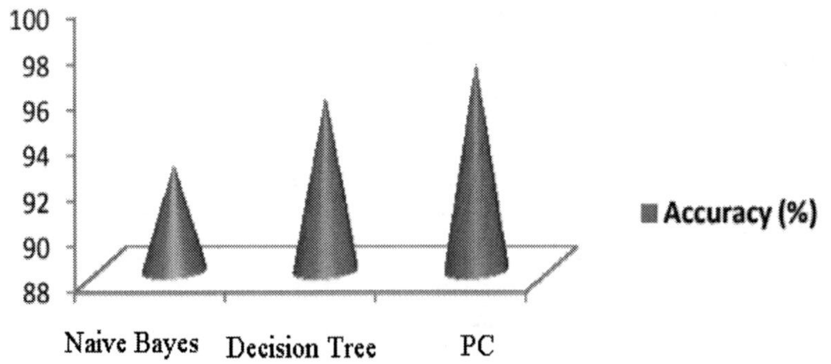

Figure 8. Comparison of Categorization accuracy

Following ten-fold cross confirmation is discover the greatest function in every classifier is 92.53% for Naïve Bayes, 95.52% for decision tree as well as 97.01% for Proposed Classifier (PC). It is discover that amongst classifiers, Naïve Bayes as well as Decision tree, the introduced classifier executes better with the categorization precision of 97% when tested with 10 folds cross confirmation. Furthermore, the model as well enhances categorization quality in terms of both sensitivity as well as particularity employing a better feature vector.

IOP Conf. Series: Materials Science and Engineering **925** (2020) 012068 doi:10.1088/1757-899X/925/1/012068

6. Conclusion

The WSN endure from several security disputes. In this paper, we introduced an ID technique established SVM. In the ID technique, we employ feature SVM vector to represent nodes. These components consist of routing message number, varies destinations in sending and receiving routing packets, and similar source node. The similar source node consists of similar characteristics; the node of the identical variety has the similar characteristics. Here, Ant colony Optimization (ACO) technique to choose best route between entire adjacent feasible routes from source to receiver node for data communication. The proposed approach enhances the network lifespan as well as discovers intrusion nodes in WSN. Therefore, irregular nodes will be notable. Hence, the proposed scheme achieved higher rate of accuracy from the evaluation results

References

[1] Lim B and Uddin MS 2005 Statistical-based SYN-flooding detection using programmable network processor 3rd *Int. Conf. on Information Technology and Applications (ICITA'05)* **2** 465-470

[2] Schuba CL, Krsul IV, Kuhn MG, Spafford EH, Sundaram A and Zamboni D 1997 Analysis of a Denial of Service Attack on TCP *IEEE Symp. on Security and Privacy* 208-223

[3] Karlof C and Wagner D 2003 Secure routing in wireless sensor networks: Attacks and countermeasures *Ad hoc networks* **1** 293-315

[4] Cho EJ, Hong CS, Lee S and Jeon S 2013 A partially distributed intrusion detection system for wireless sensor networks *Sensors* **13** 15863-15879

[5] Stavrou E and Pitsillides A 2010 A survey on secure multipath routing protocols in WSNs *Computer Networks* **54** 2215-2238

[6] Butun I, Morgera SD and Sankar R 2013 A survey of intrusion detection systems in wireless sensor networks *IEEE communications surveys & tutorials* **16** 266-282

[7] Akyildiz IF and Stuntebeck EP 2006 Wireless underground sensor networks: Research challenges *Ad Hoc Networks* **4** 669-686

[8] Duraipandian M and Palanisamy C 2014 An intelligent agent based defense architecture for ddos attacks *Int. Conf. on Electronics and Communication Systems (ICECS)* 1-7

[9] Yi P, Zhu T, Zhang Q, Wu Y and Li J 2012 Green firewall: An energy-efficient intrusion prevention mechanism in wireless sensor network *IEEE global communications conference (GLOBECOM)* 3037-3042

[10] Yang Q, Fu H and Zhu T 2016 An optimization method for parameters of SVM in network intrusion detection system *Int. Conf. on Distributed Computing in Sensor Systems (DCOSS)* 136-142

[11] Mitchell R and Chen R 2014 A survey of intrusion detection in wireless network applications *Computer Communications* **42** 1-23

[12] Malhotra S, Bali V and Paliwal KK 2017 Genetic programming and K-nearest neighbour classifier based intrusion detection model 7th *Int. Conf. on Cloud Computing, Data Science & Engineering-Confluence* 42-46

[13] Chandola V, Banerjee A and Kumar V, 2009 Anomaly detection: A survey *ACM computing surveys (CSUR)* **41** 1-58

[14] Geng X and Whinston AB 2000 Defeating distributed denial of service attacks *It Professional* **2** 36-42

[15] Yu Y, Li K, Zhou W and Li P 2012 Trust mechanisms in wireless sensor networks: Attack analysis and countermeasures *Journal of Network and computer Applications* 35 867-880

[16] Hemanth Kumar and Ramesh GP 2020 Energy efficiency and Data packet security for wireless sensor networks using African Buffalo Optimization *IJCA* **13** 944-954

[17] Shahada SAA, Hreiji SM and Shamsudheen S 2019 IOT Based Garbage Clearance Alert System with Gps Location Using Arduino *International Journal of MC Square Scientific Research* **11** 1-8

Congestion aware Adaptive Reverse Routing Strategy for Improving QoS in WSN

Elangovan G[1] **and Kumanan T**[2]

[1,2] Department of Computer Science and Engineering, Meenakshi Academy of Higher Education and Research Chennai, India.

[1]elangovangphd@gmail.com

Abstract. General purpose routing techniques for a WSN cannot be capable to give adequate functioning for various communication. This creates low adaptively provided with a general purpose routing technique ensuing in several connections reaching extremely congested. Ant Colony Optimization-Based Fault-Aware Routing (ACO-FAR) is used to balance the load in WSN. In this scheme, the node can calculate the present routes as well as deviation packets via a least congested path [17]. However, this scheme lacking the efficient awareness of bandwidth distribution, guaranteed quality of service in the WSN. To overcome these problems in this paper, Congestion aware Adaptive Reverse Routing strategy (CARS) for Improving Quality of Service (QoS) in WSN is proposed. It is a highly adaptive congestion free routing algorithms that also allocate traffic more equally and reduce network congestion. Reverse routing is used for diminishes the energy utilization and avoid the multiple retransmission in the network. Average delay as well as throughput is the two parameters that are utilized to calculate the performance of routing strategy. Simulation results demonstrate that throughput 36.76% enhanced dramatically whereas preserving 40.31%lesser latency performance.

Keywords: Congestion-aware Dynamic Routing, Energy Efficiency, Load balancing, Reverse Routing, Simulation Analysis, Wireless Sensor Networks.

1. Introduction

WSN is probable to be utilized in better functioning multi systems by future. Numerous elements concern the routine reached by an application on a WSN program. For applications which necessitate thorough transaction between cores, the major element that impresses the entire function of a WSN is constituted by its mechanism [14]. Conventional path protocols have been intended lacking mention to the features of the system traffic. The major motivation was which the transaction traffic cannot be correctly characterized as a result the routing mechanism are planned to offer deadlock freedom under kind of traffic as well as offer better performance.

Application Specific Routing Algorithms for WSN (APSRA) efforts transaction data to enhance the updating when ensuring deadlock-free path for designing WSN system. Although, it cannot receive into report the transaction attributes similar to the transaction bandwidth necessities of dissimilar transmitting charge pairs represented on dissimilar network nodes. therefore, chosen of the tracks to be evade to limit the routing task as well as to assurance deadlock discharge, is expressed out in a unsighted

fashion. It is corresponding to accepting that entire the transactions have the identical bandwidth necessities. This unknowingness might guide to a dreadful dispersion from the travel load more the system. It is mainly accurate whereas the choice of the bandwidth necessities of dissimilar transaction is great [11]. Ant Colony Optimization-Based Fault-Aware Routing (ACO-FAR) is used to balance the load in WSN. In this scheme, the node can calculate the present routes as well as deviation packets via a least congested path. However, this scheme lacking the efficient awareness of bandwidth distribution, guaranteed quality of service in the WSN.
Our paper is aim to recover WSN performance such as network delay as well as throughput. A Congestion Aware Adaptive Reverse Routing strategy for WSN approach is additional valuable to applications that can endure not working release of packets. This congestion control scheme which contains bandwidth aware route chosen is used to balance the traffic load distribution. The section is formed in to two phases. The initial phase introduces the strategy utilized to diminish the total of bandwidth which should be spread between the residual routing. The next phase covers with the trouble of assuring as well as retrieving while collected bandwidth on several connections greater the capacity. Thus this scheme is used to minimize the probable serious congestion situations around these regions so reduce the network delay. Section 2 introduces the related work. Section 3 depicts the Congestion Aware Adaptive Reverse Routing strategy for WSN approach. Section 4 shows the simulator for measuring the performance of CARS. Section 5, concludes the paper.

2. Related work

Local congestion evading is typically happening at a fraction of nodes in an area. Congestion is drastically minimizing the performance as well as principally affects several exacting transaction pairs. This scheme is assist to resolve limited congestion with dealing dissimilar area size, depend on Divide-Conquer method as well as routing pressure. This scheme evades congestion in each local area with maintaining routing force of each local area lowest [1]. A process variation delay and congestion aware routing (PDCR) method is enclosed no synchronous WSN design. PDCR is adaptative, least expenditure, as well as ascendible. This method outperforms dissimilar adaptiative routing algorithms in the latency as well as dissemination throughput for different traffic rules [2]. WSNs can attain superior operation than bus schemes for chip multiprocessor schemes. An efficient adaptative routing can assist diminish route congestion via load balancing. The remodel path congestion data is utilized to depict hidden spatial congestion data as well as recover the efficiency of pathway election. Route congestion awake election approach that concurrently conceives switches congestion as well as channel congestion. Also argument forecast method which utilizes the charge of modify in the buffer level to forecast probable switch debate [3]. Selection strategy (SS)is a necessary element of an adaptative path method which determines the routine of the WSN. A SS is utilized for choosing the greatest yield channel since the accessible channels depend on the network condition. Destination intensity as well as congestion aware (DICA) which utilizes both local as well as area congestion details from neighbouring also two hops absent neighbours on the pathway to receiver according to the switch as well as channel details. This scheme disseminates traffic extra evenly throughout the network. This scheme enhances the throughput as well as minimizes the latency with overhead [4]. A low-distance route based multicast technique for WSNs that utilize partitioning, better receiver prescribing as well as the even odd twist pattern adaptive routing method for both the multicast as well as unicast information. In addition, the method raises non-congested routes information to avoid making extremely congested fields. This is achieved by considering the congestion condition of the input ports. This scheme provides less delays as well as minimum highest energy utilization for dissimilar message insertion ranks [5]. A self-reconfigurable channel information buffering method as well as circuit design for next-generation WSNs. It provides better for energy capability as well as throughput. Throughout congestion, the buffering method recognizes adaptative flow assure with reconfiguring the channel buffers. The congestion is relieved, information communication summarizes from the primary buffer phase, thus recovering throughput. In addition, it attains energy utilization via co deign strategy [6]. Incorporating congestion-awake multicast routing with network coding, the WSN is capable to powerfully deal weighty multicast injections. A broadcast heavy Hammer cache consistency scheme attains decrease delay [7]. Dynamic traffic guideline to enhance the WSN functions as well as chip patterns. It can be employed for logical attribute incorporation in an open-loop way with interjecting traffic based on its run-time represented features. It can furthermore be functional in a closed-loop mode with including traffic entirely adaptive to the traffic as well as network nations [8]. Single-cycle multi hop asynchronous repeated traversal (SMART) by a proficient node-bypass method, also enhancing the WSN clock frequency takes down the amount of hops which can be got around in a single cycle. It cooperating on-chip wireless connections, a look-ahead bypass-appeal method as well as congestion-free transmit routing method. It is probable to allow less delay as well as energy well-organized data transfers in the WSN [9]. An arbitration method which contributes to diminish congestion delay in nodes as well as the node delay. This method is well-matched with avoid and baseline pipeline in nodes. [10]. An adaptive routing mechanism depend on limited traffic load entropy is utilized for attain better operation. The introduced node circuits are depend on the compacted data process by MOUSETRAP similar to evolution indicating scheme that programmable wait factors for the corresponded latency are utilized so as to accept inactive as well as active latency versions. Repeated manage circuits to enduring passing mistakes are moreover intended as well as assessed. This scheme is used to evade the readjusting phase overhead [12]. This scheme utilized for minimizing the switch delay. This scheme discovers the more taking functions beside the significant route of the switch, as well as introduces the techniques to minimize the overhead [13]. The throughput is enhanced with data-aided valuation method by means of transmit choice, channel then transmit obligation [15]. Frequency Band Suppression and Throughput Enhancement approach is assists for minimizing idleness as well as rises the bandwidth additional proficiently [16].

3. Bandwidth-aware routing algorithm

In this paper, we propose a Congestion aware Adaptive Routing strategy (CARS) for improving QoS in WSN. The reverse routing of this strategy is to launch a path with evade the highest amount of retransmission messages that attempts to diminish the packet losses in the network. Here, the network G(n,l) contains n number of nodes as well as l represents the network channels. The transmission diagram, CG =G (T, C), here T is the position of jobs as well as C represents the set of transactions. Every transaction $c_{i,j}$= (ti ,tj) ∊ C. For a transaction c ∊C, the operation B(c) precedes the bandwidth necessity that represents the least bandwidth which must be apportioned with the system to gather the presentation constraints for transaction c. As we are distributing among adaptative routing, the necessary bandwidth for transaction c is divide more several paths.

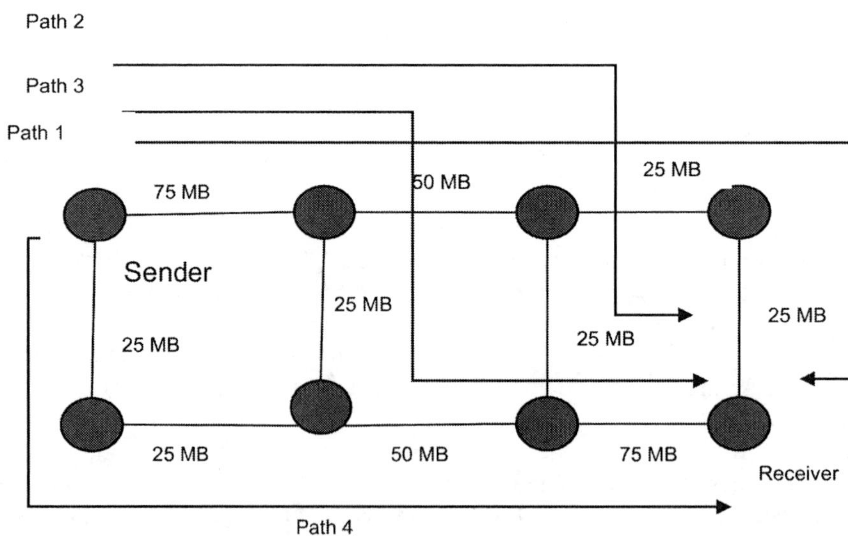

Figure 1. Topography

Figure. 1 illustrates topography. Assume transaction sender to receiver needs a bandwidth of 100 MB/sec as well as the routing permits the minimal paths from sender node to receiver node (4 paths).

3.1. Load Evaluation

The balance load situation is identical to water flowing in a Pipe. Because the highest flowing of the complete WSN is insufficient to the connection ability as well as the node before the connection will expand into the traffic jam. Properly, efficient channel bandwidth l L since a transaction c calculated as follows.

$$EB(c,l) = \frac{|CT(c,l)|}{|\mu(c)|} \tag{1}$$

Here represents the specify smallest routes included with the routing operation to transaction c, as well as CT(c,l) is the communicate throughout connection set which represents the set of routes of c that include the connection l. Then compute the combined bandwidth CB(l) that is given below.

$$CB(l) = \sum_{c \in C} EB(c,l) \tag{2}$$

Utilizing these functions, the bandwidth-aware routing algorithm satisfies the tracing constraint

$$\forall l \in L \Rightarrow CB(l) \le Cap(l) \tag{3}$$

Such as the transmission load of channel l, should not exceed its capacity (l). The link load analysis is executed to recognize connections in that combined bandwidth exceeds the link capacity. In this situation we utilize the bandwidth reallocation. It attempts to allot alternative paths in this a scheme that load is disseminated approximately equally between links. Thus we get deadlock-free routing function.

3.2. Load Bandwidth Reallocation

This scheme creates a set of routes with offering more adaptivity to transactions features with better transaction bandwidth. Though, it is probable that the combined bandwidth on system connections exceeds the ability of these connections during utilizing Path2dis, otherwise we using the Paths2en.The algorithm of BRT is illustrated below. Here, the input attributes are the set of connections, threshold, set of path as well as set of transactions. The output denotes the routing paths. For every connection l as well as for every transaction c that has as a minimum one path utilizing l, as well as two list known as Path2dis, Paths2en for more than one path. Path2dis consist of entire paths for c that should be disconnected as they use network links whose load exceeds the threshold. Paths2en consist of those paths which can be utilized by other transactions as they utilize connections whose load is lower the threshold. Then, the listpaths2dis is read as well as routing paths going to it are disconnected from P. Certainly, disconnected a path induces rearrangement of the bandwidth allocated on it to the other paths going to paths2en. Therefore, the path disconnection stops while there is as a minimum one path in paths2en that comprises a connection whose load exceeds the threshold.

Algorithm*:*

```
Band_Realloc (L:set of trans, C:set of conn, P:set of path, thr: Threshold)
While (CB(l)>thr) do
For l ϵ L
For c ϵ C:PT(c,l) {
If |P(c)>1|
For P ϵ P(c)
If l ϵP:CB(l)>thr
Path2dis
Else
```

3.3. Energy Efficient Reverse Routing

Usually, traditional strategies to launch a routing with fewer route request messages that attempts to reply utilizing a several route replies to increase energy utilization throughout the transmit of control packets. But, this strategy to evade drop of route replies as well as enhance the function of routing.

In this strategy, we believe the present energy stage of a node as well as speed of energy utilization at particular invariable time (IT) period. This speed of energy utilization calculation is given below.

$$S_{EU}(\mathrm{t}) = \frac{A_{RE}(\mathrm{t}-1) - A_{RE}(\mathrm{t})}{IT}$$

Where A_{RE} (t) is the approximate remaining energy calculated at time t as follow:

$$RE(\mathrm{t}) = \max\left\{ E_{present}(\mathrm{t}) - \sum_{k=1}^{N\ pkts} E_{Tx}(\mathrm{k}), 0 \right\}$$

Here, $E_{present}(t)$ represents the present energy as well as E_{Tx} denotes the energy that will be utilized to communicate the N packets residual packets in the buffer.

Next we measure the required remaining lifespan (t) computation is given below.

$$R_{Lifespan}(\mathrm{t}) = \frac{A_{RE}(\mathrm{t})}{S_{EU}(\mathrm{t})}$$

Therefore, reverse routing evades the highest amount of retransmissions and diminishes the energy utilization of the network.

4. Performance evaluation

The performance of the CARS is analyzed by using the Network simulator (NS2). The nodes are distributed in the simulation environment. The parameters used for the simulation of the CARS are tabulated in Table 1. The simulation of the proposed scheme has 30 nodes are deployed in the simulation area 500×500. The nodes are communicated with each other by using the Transmission Control Protocol (TCP). The traffic is handled using the Variable Bit Rate (VBR). The radio waves are propagated by using the propagation model two-ray ground. All the nodes receive the signal from all direction by using the Omni directional antenna. The performance of the proposed scheme is evaluated by the parameters packet loss ratio, average delay, throughput and residual energy.

Table 1. Simulation parameters of CARS and ACO-FAR

Parameter	Value
Number of node	16
Routing scheme	CARS and ACO-FAR
Traffic model	VBR
Simulation Area	500x500
Channel	Wireless Channel
Transmission range	250m
Communication Protocol	TCP

This analysis of Figures 2–5, the x-axis notes the traffic load that is changed from 0 to 0.6. The y-axis acts the tempered value for every execution condition. Figure 2 equates the function of the residual energy of wireless nodes. The figure.2 demonstrates that the 20.40 % increases the energy of CARS method when compare to the ACO-FAR. To exploit a lifespan, the residual energy is a significant operation parameter. The proposed scheme efficiently chooses the next node connection with regarding the residual energy information. As a result, we manage greatly residual energy less than heavily traffic load intensities; it assures a longer node lifespan

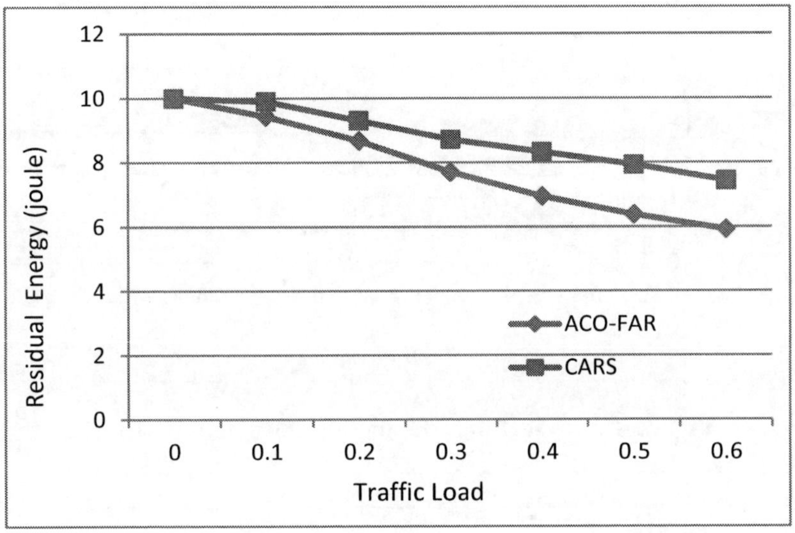

Figure 2. Residual Energy of CARS and ACO-FAR

The execution evaluation of throughput is shown in figure 2. The figure 2 demonstrate the CARS method 36.76% increases when compare to the ACO-FAR method. Generally, the throughput represents the range of productive information release via transaction channel. The throughput is typically evaluated in bps. In this scheme, the throughput is specified by the fraction of data obtained at the receiver nodes to the entirety rendered data. The above process are repetitive till the load on every connection does not greater the threshold. This process exhausts if the path removal step needs to eliminate a path that is unique for a definite transaction

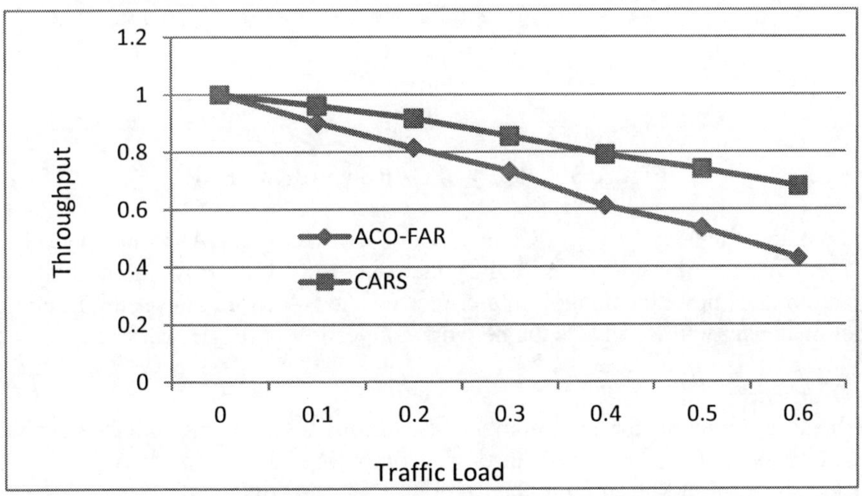

Figure 3. Throughput of CARS and ACO-FAR

Packet drop represents the fault of one or extra communicated packets to get at their receivers. As the proposed traffic load raises, the nodes flow out of the energy otherwise capability for data communications as well as data packets are possible to be losses. Thus, the packet drop possibility rises linearly by the load of traffic. 31.65% reduces packet drop ratio when compare to the ACO-FAR method.

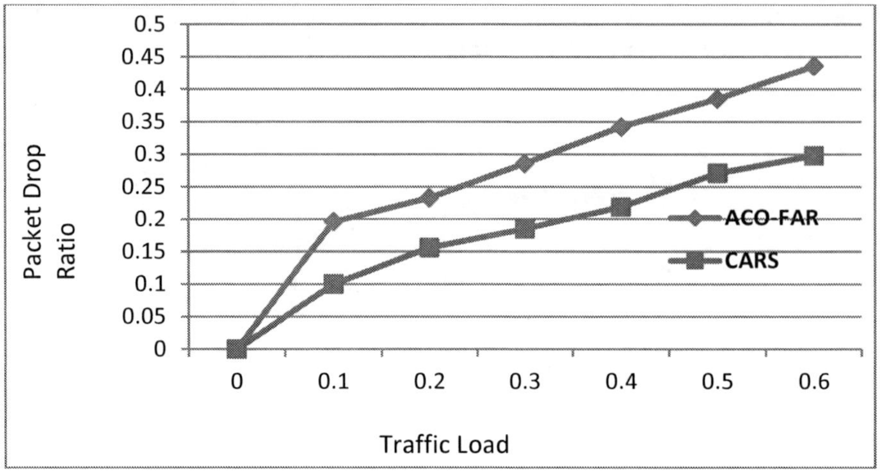

Figure 4. Packet Dropped Ratio of CARS and ACO-FAR

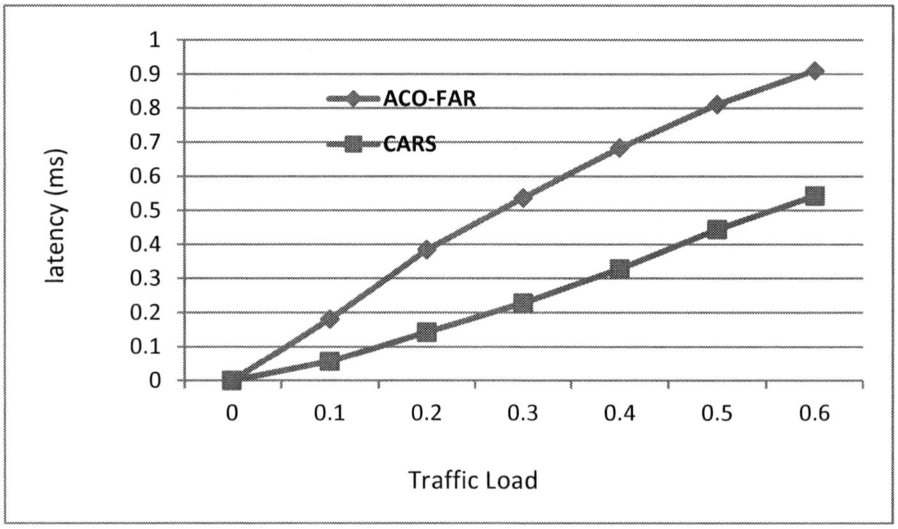

Figure 5. Latency of CARS and ACO-FAR

Figure 5 demonstrates the latency of CARS and ACO-FAR. Here, in CARS the network of traffic load raises the data communication latency is slightly increase. But, ACO-FAR method latency is 40.31 % increased owing to does not allocate the route properly. The CARS method using multiple paths based on the connection capacity thus reduces the network congestion in the network.

4.1. Routing Overhead

Routing Overhead is depicting the sum number of control packets communicated for yielding a data route per data packet. It is received through computing the ratio among the sum numbers of control packets transmit to the sum number of control packets obtained.

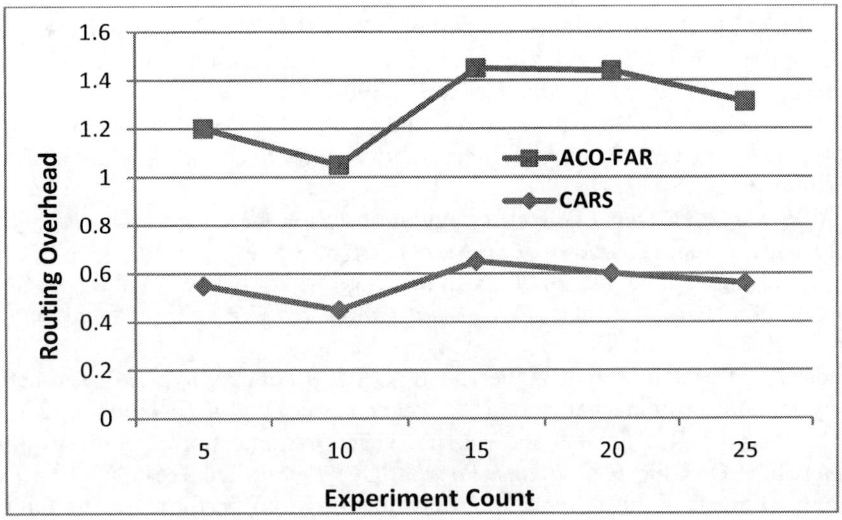

Figure 6. Routing Overhead of CARS and ACO-FAR

Figure 6 illustrates the proposed strategy CARS has lowest routing overhead than the existing strategy ACO-FAR. Since, the CARS choose the relay based on reverse routing. So, avoid the congestion in the network. But, ACO-FAR increase the routing overhead because of it does not select the congestion aware routing.

5. Conclusion

Congestion aware Adaptive Reverse Routing strategy for WSN designs to enhance the QoS performance in terms of latency and throughput. The section is formed in to two phases. The first phase presents the strategy utilized to diminish the amount of bandwidth that must be redistributed among the residual routing paths. The second phase deals with the problem of checking as well as recovering when combined bandwidth on some connection greater the capacity. Thus this scheme is used to eliminate potential heavy congestion situations around these regions so reduce the network latency. To confirm the proposed algorithm, via simulation finished to show it's achieve ability as well as authority. Compared to a baseline approach, the proposed congestion management method provides better scalability performance of WSN platforms. Simulation results demonstrate that throughput 36.76% enhanced dramatically whereas preserving 40.31% lesser latency performance.

References

[1] Tang M, Lin X and Palesi M 2015 Local congestion avoidance in Network-on-Chip *IEEE Transactions on Parallel and Distributed Systems* **27** 2062-2073

[2] Ezz-Eldin R, El-Moursy MA and Hamed HF 2015 Process variation delay and congestion aware routing algorithm for asynchronous NoC design *IEEE Transactions on Very Large Scale Integration (VLSI) Systems* **24** 909-919

[3] Chang EJ, Hsin HK, Lin SY and Wu AY 2013 Path-congestion-aware adaptive routing with a contention prediction scheme for network-on-chip systems *IEEE Transactions on computer-aided design of Integrated circuits and systems* **33** 113-126

[4] Mehranzadeh A, Khademzadeh A, Bagherzadeh N and Reshadi M 2019 DICA: destination intensity and congestion-aware output selection strategy for network-on-chip systems *IET Computers & Digital Techniques* **13** 335-347

[5] Daneshtalab M, Ebrahimi M, Mohammadi S and Afzali-Kusha A 2009 Low-distance path-based multicast routing algorithm for network-on-chips *IET computers & digital techniques* **3** 430-442

IOP Conf. Series: Materials Science and Engineering **925** (2020) 012069 doi:10.1088/1757-899X/925/1/012069

[6] Bondade R and Ma D 2010 Self-reconfigurable channel data buffering scheme and circuit design for adaptive flow control in power-efficient network-on-chips *IEEE Transactions on Circuits and Systems I: Regular Papers* **57** 2890-2903

[7] Duraisamy K, Xue Y, Bogdan P and Pande PP 2016 Multicast-aware high-performance wireless network-on-chip architectures *IEEE Transactions on Very Large Scale Integration (VLSI) Systems* **25** 1126-1139

[8] Lu Z and Yao Y 2016 Dynamic traffic regulation in noc-based systems. *IEEE Transactions on Very Large Scale Integration (VLSI) Systems* **25** 556-569

[9] Duraisamy K and Pande PP 2017 Enabling high-performance SMART WSN architectures using on-chip wireless links *IEEE Transactions on Very Large Scale Integration (VLSI) Systems* **25** 3495-3508

[10] Noghondar AF, Reshadi M and Bagherzadeh N 2017 Reducing bypass-based network-on-chip l atency using priority mechanism", *IET Computers & Digital Techniques* 12 1-8

[11] Palesi M, Holsmark R, Kumar S and Catania V 2009 Application Specific Routing Algorithms for Networks On Chip *IEEE Trans. Parallel Distrib. Syst.* **20** 316-330

[12] Imai M and Yoneda T 2011 Improving dependability and performance of fully asynchronous on-chip networks *17th IEEE Int. Symp. on Asynchronous Circuits and Systems* 65-76

[13] Roca A, Flich J, Silla F and Duato J 2011 A low-latency modular switch for CMP systems *Microprocessors and Microsystems* **35** 742-754

[14] Yoneda T, Imai M, Onizawa N, Matsumoto A and Hanyu T 2012 Multi-chip WSNs for automotive applications *IEEE 18th Pacific Rim Int. Symp. on Dependable Computing* 105-110

[15] Santhakumar R 2017 Resource Allocation in Wireless Networks by Channel Estimation and Relay Assignment Using Data-Aided Techniques *Int. Journal of MC Square Scientific Research* **9** 40-47

[16] Hemanth Kumar and Ramesh GP 2020 Energy efficiency and Data packet security for wireless sensor networks using African Buffalo Optimization *IJCA* **13** 944-954

[17] Hsin HK, Chang EJ, Lin CA and Wu AYA 2014 Ant colony optimization-based fault-aware routing in mesh-based network-on-chip systems *IEEE Transactions on Computer-Aided Design of Integrated Circuits and Systems* **33** 1693-1705

Congestion and SINR Evaluation for Improving Traffic Capacity in Ad Hoc Wireless Networks

Nandakumar S D[1] and Thirunadana Sikamani K[2]

[1,2] Department of Computer Science and Engineering, St. Peter's Institute of Higher Education and Research, Chennai, India

[1]nandakumarsdphd@gmail.com

Abstract. More than the several years, multi-hop wireless ad hoc network shave obtained significant attention. This network is usual to have pervasive pertinences for the intention of transactions as well as disseminated calculation. Generally, the ad hoc network uses the hybrid routing protocols to handle the traffic flow in the network. However, the hybrid protocols did not consider the congestion related problems. Therefore, we propose Congestion and Signal to Interference Noise Ratio (SINR) Evaluation for Improving Traffic Capacity in Ad Hoc Wireless Networks (CETC) aiming to maximize the throughput and improve the energy efficiency. In this scheme, the congestion measurement is used to evaluate demanding time for determining a route with an expected lower level congestion. In addition, the SINR measurement is used to determine the node link performance. Thus, it improves the energy-efficiency and network throughput. The Simulation results show that the proposed CETC improves the network performance and reduce the delay as well as energy consumption.

Keywords: Ad Hoc wireless Networks, Congestion Measurement, Energy Efficiency, Network Simulation, Signal to Interference Noise Ratio.

1. Introduction

Wireless ad-hoc networks (WANETs) comprise a numerous of mobile wireless nodes which can travel in an arbitrary manner with the ability to connection otherwise depart the network. It can origin congestion which ensues in enhancing communication latency as well as loss of packets. This problematic is extra simple in greater networks by additional traffic as well as extraordinary motion which applies active topography [13]. The straight pathway routing cannot equilibrium the routing capacity on the connections based their capability therefore some connections involvement extra traffic than they can adapt that congests the system. Therefore, there is an essential to stability the traffic load on entirely connections based on their capabilities [14]. Congestion Control is additional significant dispute when leading for QoS attainment precisely with extremely movable mobile stations. The congestion control effects introduce an raised traffic method in Transmission Control Protocol (TCP) utilizing the token bucket traffic shaping method through packet transmitting at the middle nodes [9].AD-hoc Transmission Control Protocol, the receiver notices the feasible current condition as well as communicate this details to the sender as response [10]. However wireless networks necessity effective energy controlling, because communication energy is a valuable reserve. A wireless system is associated to an energy channel, energy is yet significant as it straight involves the amount of interfering formed in the network, as well as accordingly affects the entire throughput that the network. Hence, power control is a significant element in a wireless network [12]. To diminish the energy utilization in a wireless network, several methods for example power management, congestion minimization and energy efficient routing techniques have been studied. Energy-efficient techniques discover the path which diminishes the entire energy utilization. Routing as well as congestion control at the network layer to fulfil quality-of-service (QoS) necessities otherwise diminish the energy depletion.

2. Related work

The Delay-based load-balancing routing (DLBR) algorithm is utilized for evaluating the congestion level. This scheme select the node has less congestion and offer high capacity [1]. A bandwidth aware routing scheme (BARS), which can evade congestion based on supervising remaining bandwidth capability. The sum of present as well as utilized bandwidth together with remaining accumulation necessity be functioned out earlier communicating messages. This scheme operates the response method to close the traffic for correcting the data rate by the present bandwidth as well as queue in the route. However, this scheme cannot control the transmission energy in Ad Hoc Wireless Networks [2]. History based scheme applies modern effective communication factors for the present communication effort where Priority based controls every node's transmitting time period. The Priority based approach endures from a largest communication latency owing to priority, however attains a greater packet received rate. Instead, the History-based approach accomplishes a least queuing delay at the expenditure of fewer effective communication efforts per part time [3]. An iterative double auction regard to energetically divest the traffic of their mobile employers. Every mobile node (MN) can apply several access points (Aps) as well as every AP can simultaneously assist traffic from several MN. This scheme provides effectual function whereas MN can exploit their discharging gains and APs diminish their offloading costs [4]. Multicast tree construction is used to diminish the multicast traffic. The NP-hard as well as effective heuristic technique for minimizing the multicast traffic [5]. Flash Lin Q (FLQ) Device-to-Device (DtD) transaction Scheme can recover operation of wireless medium through creating a DtD link. Communication power control system for FLQ that operates every DtD link's communication power regarding its forces to the FLQ's medium access method [6]. A delay-guaranteed geographic routing which can endure several end-to-end delay levels. This scheme alertness of the existence of a routing hole as well as it evades the hole [7]. A hybrid multipath energy as well as QoS-aware enhanced routing scheme which is established to handle with the disputes accessible with imperfect energy resourcefulness, nodes mobility, as well as traffic congestion through data communication. It employs a node rank rate based on multi criteria node rank metric (MCNR). It combines several metrics associated to QoS as well as energy into a complete metric to intensely diminish the difficulty of several reserved concerns as well as evade the control overhead. The MCNR parameter is applied with a link quality valuation operation for several route calculations [8]. Framework and implementation agent for controlling the congestion in the network. This technique, the nodes are categorized into four classes depend on its traffic. It measures acquire in extent of the numerous traffic modules as well as the channel difference also approximates the congestion parameter to discover the least congestion level in the network. This parameter is utilized for electing the least congested path [11]. Agent based congestion control method that the details about congestion is gathered as well as disseminated by mobile agents (MA). This method is utilized for choosing the least congested path [12]. Modified Russian Peasant Multiplier based on Divide as well as conquer method is utilized for multiplication procedure. Minimizing the chip size, enhancing the quickness as well as minimizing the energy depletion are key critical issues in System strategy [15]. The Resource Aware and Link Quality (RALQ) routing strategy [16] formed the route by link quality and energy utilization for data transmission and receiving. It improves the routing efficiency.

Modified Russian Peasant Multiplier based on Divide as well as conquer method is utilized for multiplication procedure. Minimizing the chip size, enhancing the quickness as well as minimizing the energy depletion are key critical issues in System strategy [17]. An opportunistic routing by responsiveness of energy to accept by dynamic environment. While the sender transmits the information to a multicast group, the sender transmits the information via greater energy between vicinity thus enhance the life span [18].

3. Congestion and SINR Evaluation for Improving Traffic Capacity in Ad Hoc Networks

A Generally, the ad hoc network uses the hybrid routing protocols to handle the traffic flow in the network. However, the hybrid protocols did not consider the congestion related problems. Therefore, we propose Congestion and SINR Evaluation for Improving Traffic Capacity in Ad Hoc Networks (CETC) aiming to maximize the throughput and improve the energy efficiency. The congestion evaluation of a node connected with a given receiver offers evaluate the best probable demanding time period of a packet incoming at that node till it reaches the receiver. Every node is accountable to inform its congestion evaluation as well as communicate this information to its neighbors. Also every node compute the SINR value is used to improve the energy efficiency.

3.1. Congestion Measurement

We consider a network of D nodes labelled by $\Omega = \{1......D\}$. While a node obtains extra data than its presented capability, congestion occurs. If the node's queue is complete, the extra packets have to be lost. Here, N(m) represented as a set of neighbours and the node n denoted as the neighbour node of m. The routing table is updated using a "virtual routing table" at the end of every computation cycle: an interval of T_c units of time. More specifically, node m at regular intervals calculates its congestion and subsequently updates its neighbours by control packets at intervals of $T_s < T_c$ sec. Finally, the actual routing table is updated using the entries in the virtual routing table after every seconds.

The congestion calculates from node m to receiver R at time t is given below.

$$V_m{}^R(\mathrm{t}) = \mathrm{L}_m{}^R(\mathrm{t}) + D_m{}^R(\mathrm{t}) \tag{1}$$

Where,

$\mathrm{L}_m{}^R(\mathrm{t})$ → Local demanding time

$D_m{}^R(\mathrm{t})$ → Demanding time since its next hop to the receiver

In order that calculate $\mathrm{L}_m{}^R(\mathrm{t})$ as well as $D_m{}^R(\mathrm{t})$, node m relies on the subsequent measures

P_{mn} represented as Probability that a packet transmitted via node m is obtained by node n.

$S_m(t)$ represented as neighbour nodes obtained the packet from node m at time t.

$\overline{Q}_m{}^R(\mathrm{t})$ Denoted as average number of packets at node m intended for R on the final calculation cycle, $\overline{Q}_m{}^R(\mathrm{t})$ that is informed as

$$\overline{Q}_m{}^R(\mathrm{t}) = \frac{T_s}{T_c} \sum_{m=0}^{\frac{T_s}{T_c}-1} Q_m{}^R(\mathrm{T(t)} - l) \tag{2}$$

Note that $\overline{Q}_m{}^R(\mathrm{t})$ is a constant on a single computation cycle, that is,

$\overline{Q}_m^{\,R}(\mathrm{t}) = \overline{Q}_m^{\,R}(\mathrm{t'})$ for $\forall$ t, t' s.t. T(t) = T(t')

$N_k^{(\mathrm{m,R})}(\mathrm{t})$ is mentioned as a Neighbours of node m whose sensed congestion evaluates are smaller than that of node k, that is

$$N_k^{(\mathrm{m,R})}(\mathrm{t}) = \left\{ n \in N(\mathrm{m}) : \overline{\mathrm{V}}_n^{(\mathrm{m,R})}(\mathrm{t}) < \overline{\mathrm{V}}_k^{(\mathrm{m,R})}(\mathrm{t}) \right\} \tag{3}$$

$N^{(\mathrm{m,R})}(\mathrm{t})$: Neighbours of node m that are utilized by node m to route packets intended for node R, i.e.,

$$N^{(\mathrm{m,R})}(\mathrm{t}) = \begin{cases} N_m^{(\mathrm{m,R})}(\mathrm{t}) & \mathit{if}\ N_m^{(\mathrm{m,R})}(\mathrm{t}) \neq 0 \\ \arg\min\limits_{n \in N(\mathrm{m}) - \{\mathrm{m}\}} \overline{\mathrm{V}}_n^{(\mathrm{m,R})}(\mathrm{t}) & f\ N_m^{(\mathrm{m,R})}(\mathrm{t}) = 0 \end{cases} \tag{4}$$

$P_{succ-k}^{(\mathrm{m,R})}(\mathrm{t})$: possibility that node k is chose as the next hop for a packet intended for node R and communicated by node m such as

$$P_{succ-k}^{(\mathrm{m,R})}(\mathrm{t}) = \mathrm{p_{mk}} \prod_{n \in N_k^{(\mathrm{m,R})}(\mathrm{t})} (1 - \mathrm{p}_{mn}) \tag{5}$$

Note that, between the neighbours of the packet node k is chose as then next hop, if and only if none of the other higher priority nodes $N_k^{(\mathrm{m,R})}(\mathrm{t})$ received the packet.

$P_{succ}^{(\mathrm{m,R}))}(\mathrm{t})$: Possibility that a packet transmitted by node and intended for node moves on the way to the receiver,

$$P_{succ}^{(\mathrm{m,R}))}(\mathrm{t}) = \sum_{k \in N^{(\mathrm{m,R}))}(\mathrm{t})} P_{succ-k}^{(\mathrm{m,R}))}(\mathrm{t}) \tag{6}$$

With these parameters and assuming a FIFO discipline, we proceed with the relay selection rule. In particular, for each packet, the next hop $K_{CM}^{(m,R)}(\mathrm{t})$ is selected as

$$K^{(\mathrm{m,R})}(\mathrm{t}) = \left\{ \arg\min_{k \in S_m(\mathrm{t}) \cap N^{(\mathrm{m,R})}(\mathrm{t})} \overline{\mathrm{V}}_n^{(\mathrm{m,R})}(\mathrm{T(t)}) \right. \tag{7}$$

Then we offer the calculation details of L_m^R as well as D_m^R .The local demanding time, L_m^R , relies on the fact that, while a packet gets at a node m, its awaiting time is equivalent to the time expended in demanding the packets that have arrived earlier plus its own transmission time. The required communication time at node m for the packet can be estimated by $\frac{1}{P_{succ}^{(\mathrm{m,R})}(\mathrm{t})}$, the local demanding time for node m to receiver R at time t is

$$\mathrm{L}_m^R(\mathrm{t}) = \frac{1}{P_{succ}^{(\mathrm{m,R})}(\mathrm{t})} + \sum\nolimits_{\mathrm{R'} \in \Omega} \frac{\overline{Q}_m^{\,R}(\mathrm{T(t)})}{P_{succ}^{(\mathrm{m,R})}(\mathrm{t})} \tag{8}$$

This scheme calculates the expected congestion $D_m^R(t)$ for every node $k \in \Omega$ using the latest congestion measures $\overline{V}_n^{(m,R)}(t)$ received from nodes $k \in \Omega$ with lower congestion measure. More specifically, the accepted congestion $D_m^R(t)$ can be given as

$$D_m^R(t) = \frac{1}{P_{succ}^{(m,R)}(t)} + \sum_{k \in N^{(m,R)}(t)} P_{succ-k}^{(m,R)}(t)\,\overline{V}_n^{(m,R)}(t) \tag{9}$$

3.2. Data Transmission

When a sender wants transmit the data to receiver through the relay nodes it first broadcasts a Route Request (RREQ), this message comprises the broadcast id, receiver id, receiver sequence number, sender id, sender sequence number, accepted congestion and count of hop. When a node obtains, this RREQ then it sends the RREP message back to the source node. This RREP message contain receiver id, receiver sequence number, SINR value, with accepted congestion.

This SINR has been used as a link quality as follows.

$$SINR = \beta \times SINR + (1-\beta) \times SINR_{new} \tag{10}$$

Where $\beta \rightarrow 0.75$

$SINR_{new} \rightarrow$ Most recent $SINR$ value

The lower acceptance congestion value and less SINR value based select the relay node for determining the best route path selection in the network. Finally, the data send through the better relay node. Therefore, it increases the energy efficiency and enhances the network throughput.

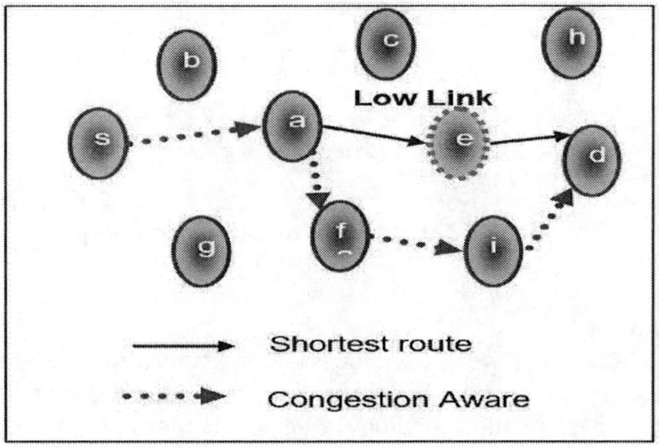

Figure 1. SINR and Congestion based Route Selection

In this figure, the route from sender to receiver is s-a-e-d. But the node e link quality is low. Therefore, the proposed scheme finds the new route s-a-f-i-d. The a-e-d path is replaced by a-f-i-d, thus the packets are received successfully. This established path is enhancing the network performance. The figure 1 indicates the working flow of the proposed scheme. The link quality is measured based on the SINR value. The highest link quality node is select as the next hop node. Finally, the source sends the data to receiver through the highest link quality nodes. Figure2 illustrates the CETC Approach.

IOP Conf. Series: Materials Science and Engineering **925** (2020) 012070 doi:10.1088/1757-899X/925/1/012070

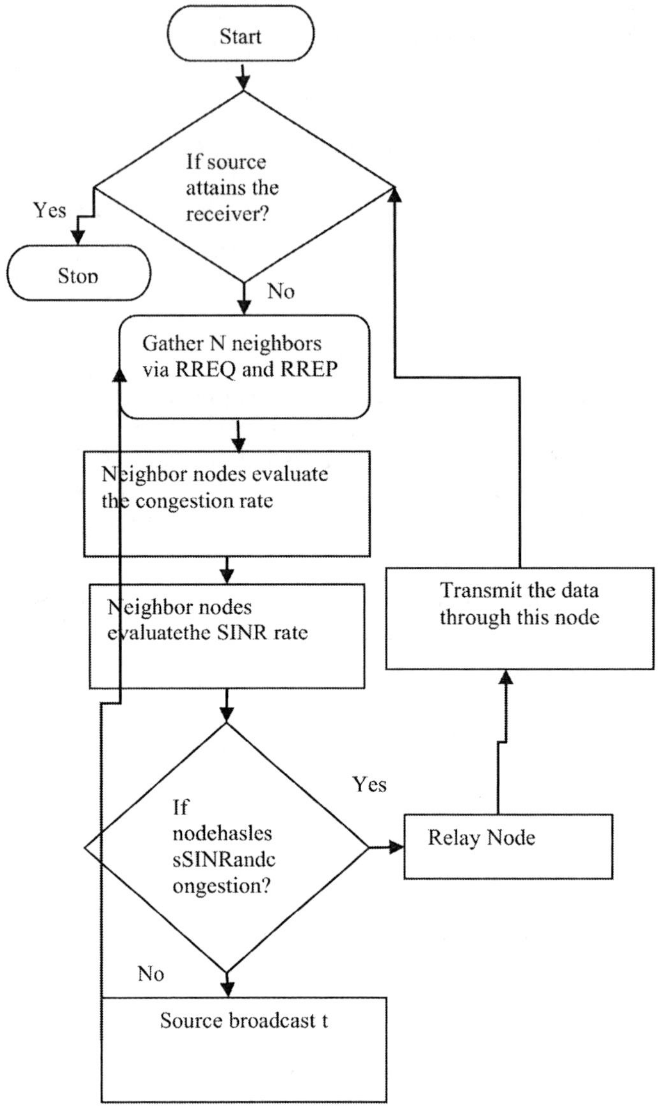

Figure 2. Flow chart of CETC Approach

4. Performance evaluation

To execute simulation examination, we utilized the NS-2.35 that incorporates the 802.11 MAC protocol. This protocol is functioning on the 2.4 GHz physical layer and each sensor node transaction range is pre-set to 250 meter. In this simulation topology, every node plays an owner of complete network as well as every wireless node function is data forwarding as well as obtaining the data. The packet traffic is yielded with variable bit rate (VBR). This simulation examination, we measured the following metrics namely Packet Obtained Rate, Latency, Loss Packet Rate, Average Remaining Energy, Throughput.

4.1. Packet Obtained rate

It is defined as the ratio among the amount of data packets perfectly obtained by the receiver and the amount of data packets yielded by entire wireless nodes. This metric mutually constitutes the less congestion and energy efficiency of the data gathering procedure.

IOP Conf. Series: Materials Science and Engineering **925** (2020) 012070 doi:10.1088/1757-899X/925/1/012070

$$POR=\frac{\sum_{0}^{n}\text{Packets Obtained}}{\text{Time}} \quad (11)$$

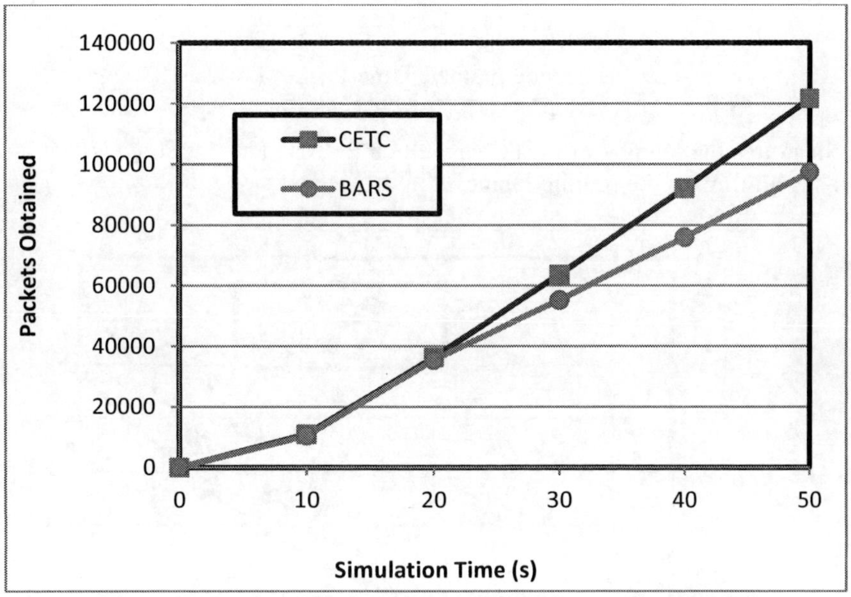

Figure 3. Packet Obtained Rate of CETC as well as BARS

From this figure 3, the proposed method CETC is greater compare to the existing method BARS. Because, the CETC pick the route by the energy efficient path. Thus, enhances the routing efficiency.

4.2. Packet Drop Rate

PDR is denoted as the distinction among the forward packets and obtained packets in the communication WANET per particular time. PDR is measured by Equation 12.

$$PDR=\frac{\sum_{0}^{n}\text{Forward Packets - Obtained Packets}}{\text{Time}} \quad (12)$$

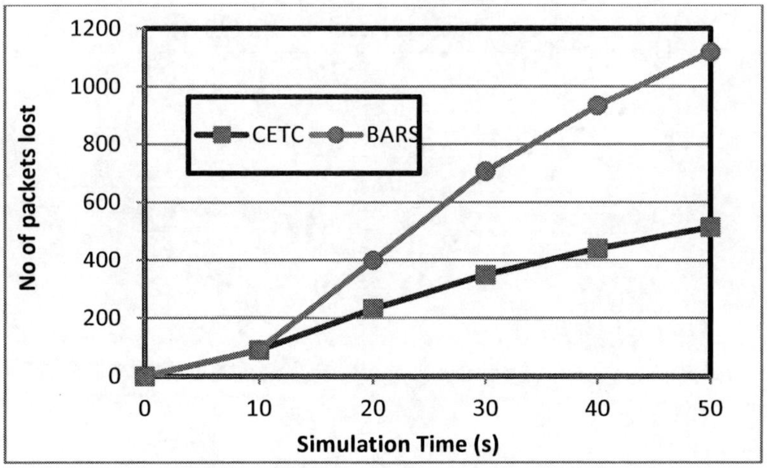

Figure 4. Packet Drop Rate of CETC as well as BARS

From figure, 4 illustrate the PDR of CETC as well as BARS. The CETC method selects the relay node by link consistency factor. Thus, reduces the network PDR in the system.

IOP Conf. Series: Materials Science and Engineering **925** (2020) 012070 doi:10.1088/1757-899X/925/1/012070

4.3. Latency

It *is* defined as the time period from while the data packet communication is initiated at the source to while the data packet is perfectly obtained by the receiver. We evaluated both its usual value and its allotment.

$$\text{latency}=\frac{\sum_{0}^{n}(\text{Packet Obtained Time-Packet Forward Time})}{n} \quad (13)$$

The figure, 5 illustrates the latency of CETC as well as BARS. The data traverse via link consistency factor nodes. Thus minimizes the routing latency.

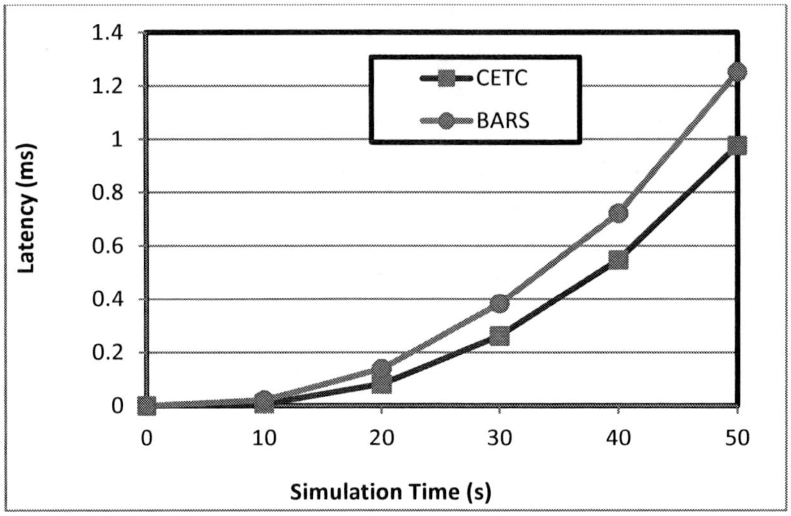

Figure 5. Latency of CETC as well as BARS

4.4. Throughput

It is the vital parameter for measuring the operations of network. In this protocol, the throughput is specified as the amount of data packets effectively obtained at the receiver.

$$\text{Throughput}=\frac{\sum_{0}^{n}\text{Packets Received(n) * Packet size}}{Time} \quad (14)$$

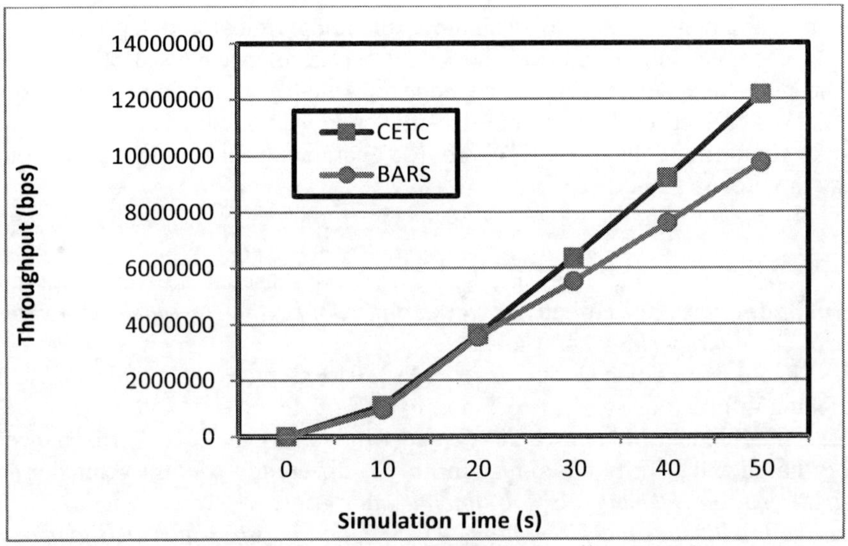

Figure 6. Throughput of CETC as well as BARS

The figure 6 proves the CETC scheme has higher throughput. Because of, BARS scheme concentrate the congestion based on the bandwidth. But, CETC focuses the less congestion and energy efficiency path thus CETC has provide better performance in WANET.

4.5. Average energy

It is defined as the whole energy exhausted by every sensor node divided by the amount of data packets perfectly distributed to the receiver. This metric evaluates the energy efficiency of the WANET.

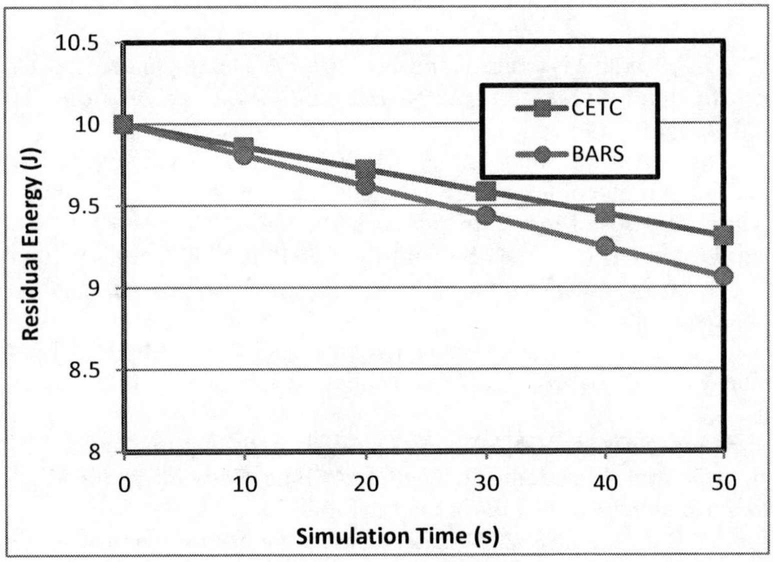

Figure 7. Residual Energy of CETC as well as BARS

Figure 7 explains the residual energy of CETC as well as BARS. The CETC has highest throughput since the energy is an important parameter during compute the congestion factor. Thus CETC raises the residual energy in the network.

5. Conclusion

In this paper, we introduced Congestion and SINR Evaluation for Improving Traffic Capacity in Ad Hoc Networks aiming to maximize the throughput and improve the energy efficiency. In this scheme, the congestion measurement is used to evaluate demanding time for determining a route with an expected lower level congestion. In addition, the Signal to Interference Noise Ratio measurement is used to determine the node link performance. The node ink quality is measures based on the SINR to guarantee optimal path selection. Thus, it improves the energy-efficiency and network throughput. The simulation results demonstrate that the CETC provide guarantee packet delivery rate and increases the Residual energy in wireless ad hoc network.

References

[1] Habib I, Badruddin N and Drieberg M 2015 Delay-based load-balancing routing (DLBR) algorithm for wireless ad-hoc networks *Int. Conf. on Computer, Communications, and Control Technology (I4CT)* 450-454

[2] Akhtar N, Khan MA, Ullah A and Javed MY 2019 Congestion avoidance for smart devices by caching information in MANETS and IoT *IEEE Access* **7** 71459-71471

[3] Kurniawan IF, Rahman MA, Asyhari AT and Bhuiyan MZA 2018 Performance evaluation of history-based and priority-based mac for traffic-differentiated intra-vehicular network *IEEE 16th Int. Conf. on Dependable, Autonomic and Secure Computing, 16th Int. Conf. on Pervasive Intelligence and Computing, 4th Int. Conf. on Big Data Intelligence and Computing and Cyber Science and Technology Congress (DASC/PiCom/DataCom/CyberSciTech)* 689-694

[4] Iosifidis G, Gao L, Huang J and Tassiulas L 2013 An iterative double auction for mobile data offloading *11th Int. Symp. and Workshops on Modeling and Optimization in Mobile, Ad Hoc and Wireless Networks (WiOpt)* 154-161

[5] Chuang CC, Yu YJ, Pang AC, Tseng HW and Lin HP 2015 Efficient multicast delivery for data redundancy minimization over wireless data centers *IEEE Transactions on Emerging Topics in Computing* **4** 225-241

[6] Bae SJ, Kim DH, Choi BG and Chung MY 2013 Transmission power control for FlashLinQ device-to-device communication system *IEEE 77th Vehicular Technology Conference (VTC Spring)* 1-5

[7] Le Nguyen P, Ji Y, Trung NT and Hung NT 2017 A Delay-Guaranteed Geographic Routing Protocol with Hole Avoidance in WSNs *IEEE 14th Int. Conf. on Mobile Ad Hoc and Sensor Systems (MASS)* 135-143

[8] Jabbar WA and Saad WK and Ismail M 2018 MEQSA-OLSRv2: A multicriteria-based hybrid multipath protocol for energy-efficient and QoS-aware data routing in MANET-WSN convergence scenarios of IoT *IEEE Access* **6** 76546-76572

[9] Rath M, Rout UP, Pujari N, Nanda SK and Panda SP 2017 Congestion control mechanism for real time traffic in mobile adhoc networks *In Computer communication, networking and internet security* 149-156

[10] Sreenivas BC, Prakash GB and Ramakrishnan KV 2013 M-ADTCP: An Approach for Congestion Control in MANET *In Advances in Computing and Information Technology* 531-540

[11] Bhadauria SS and Sharma VK 2011 Framework and Implimentation of an Agent Based Congestion Control Technique for Mobile Ad-hoc Network *In Int. Conf. on Advances in Computing, Communication and Control* 318-327

[12] Sharma VK and Bhadauria SS 2011 Agent based congestion control routing for mobile ad-hoc network *In Trends in network and communications* 324-333

[13] Nishiyama H, Ngo T, Ansari N and Kato N 2012 On minimizing the impact of mobility on topology control in mobile ad hoc networks *IEEE Transactions on Wireless Communications* **11** 1158-1166

[14] Moussaoui A, Semchedine F and Boukerram A 2014 A link-state QoS routing protocol based on link stability for Mobile Ad hoc Networks. *Journal of Network and Computer Applications* 39 117-125

[15] Sendhilkumar NC 2017 Design and implementation of power efficient modified Russian peasant multiplier using ripple carry adder *Int. Jour. of MC Square Scientific Research* **9** 154-165

[16] Gungor VC, Sastry C, Song Z and Integlia R 2007 Resource-aware and link quality based routing metric for wireless sensor and actor networks *IEEE Int. Conf. on Communications* 3364-3369
[17] Keerthi Anand VD 2015 Power Efficient Multicast Opportunistic Routing Protocol (Pemor) To Optimize Lifetime of Manet *Int. Journal of MC Square Scientific Research* **7** 109-127
[18] Hemanth Kumar and Ramesh GP 2020 Energy efficiency and Data packet security for wireless sensor networks using African Buffalo Optimization *IJCA* **13** 944-954

Credential Rate based Black hole attack Detection in MANET

Rameshkumar S G[1] **and Mohan G**[2]

[1,2] Department of Electrical Engineering, Faculty of Engineering and Technology, Annamalai University, Chidambaram, India.

[1]rameshkumarsg17@gmail.com

Abstract. Mobile Ad-hoc Networks (MANETs) are growingly being distributed in protection-essential applications. Since their integral resource-restrained features, they are prostrate to several security attacks as well as a black hole attack is a category of which critically involves data gathering. To defeat that dispute, Credential based Black hole attack Detection in MANET (CRBD) is anticipated for MANETs. In this scheme, the vicinity account (VA) checks node cooperativeness among neighbourhood nodes as well as obtained data-packet account (ODA) checks the data packet forward rate. The black hole attacker has least both ODA as well as VA. In addition, the black hole attacker has highest outstanding energy compared to the neighbour nodes. These parameters isolate the black hole attacker nodes in the network. The simulation examination demonstrates that the least packet loss as well as network delay.

Keywords: Credential Rate, Black hole attack Detection, vicinity account, QoS account, obtained data-packet account, energy account, Mobile Ad-hoc Networks.

1. Introduction

W MANETs contains group of mobile nodes that vigorously swap data between themselves lacking the confidence on centralized management otherwise permanent base station. Self-preparing feature alters MANETs to be simply demonstrated in an extensive selection of dissimilar positions, for example liberate, tragedy performances, and battle ground transactions. Though, node motion and self-preparing features of MANETs reason modify of topography in an unpredictable method. Generally, every node with imperfect communication range has to get help of its vicinity nodes for data communications. Therefore, the execution of MANETs mainly based on the consistent path between nodes [1]. Though, the entire routing approaches are intended with a presumption that all nodes are completely reliable and prepared to collaborate with all other. As a result, they are susceptible to path interruption attackers that are not mutual or violate the path principles. Usually, dynamic black hole attack, inactive black hole attack as well as gray-hole attack is simply initiate the MANET [2]. In dynamic black hole, the attackers forever maintain that they have the shortest route to the receiver still it do not have any appropriate path information. Also, this types of attack losses the sum of data packets soundlessly. In inactive black hole, the attacker assists communicating path messages but remove the entire communicating data packets [3]. In gray-hole attacks, instantly losing entire communicating data packets, gray-hole attackers might selectively transmit data packets that can exploit their own concerns. However, both black hole attackers as well as gray-hole attack will not insert fake data into system, therefore they go to inactive path interference attackers. The stay of this work is prepared as follows. In Section 2, the related work is surveyed. The Credential Rate based Black hole attack Detection in MANET (CRBD) is anticipated for MANETs is depicted in Section 3. In Section 4, present performance analyses. At last, finish in Section 5.

2. Related work

The node duplication attack recognition scheme [4] uses Danger Theory for finding duplicate nodes in the network. It recognizes the affected region via discovering the unnatural behavior of the nodes. It contains two major security phases such as attack recognition as well as security control. These phases execute a multi-level recognition that is dependable to recognize, alleviate as well as defend the duplication nodes in the network. Energy awake trust scheme [5] is provide sufficient security as maintain overhead by utilizing a game theoretic technique. This technique is functional to the trust origin method to diminish the overhead. Although, this scheme enhances the overhead of routing in the network. Evolutionary self-cooperative trust (ESCT) approach which emulates person cognitive procedure as well as swears on trust-level data is used to avoid different routing interruption attacks. In ESCT, movable nodes will swap trust details and examine obtained trust details by their own cognitive opinion. Finally, every node energetically acquires its cognition to eliminate malevolent units. The major appealing attribute of ESCT is that they cannot cooperation the organism even if the interior attackers identify how the safety method forms [6]. Generally, Vampire attacks are not specific to any type of protocol, but rather rely on the properties of many popular classes of routing protocols. The Vampire attack that hard to notice as well as simple to take out malevolent insider forwarding only tractable messages. This method is assists to alleviate these cases of attacks, considering a validation idea scheme that obviously restrictions injure induced with Vampires through the packet transmitting period [7]. The Spatial-Temporal provenance Assurance with Mutual Proofs (STAMP) is planned for movable users giving position proofs for every other in disseminated scenery. STAMP assures the honesty and non inter changeability of the position proofs as well as defends users secrecy. Semi-trusted guarantee ability is utilized to allocate cryptographic keys and protector users against complicity with a light-weight entropy-based trust assessment method [8]. The jamming and the black hole attack as well as Denial of service attacks, are coverage of multi-tier HWNs. Every tier contains normal nodes, jammers as well as black holes. Assuming that productive examination function that involves a least obtained signal quality threshold to be achieved, get precise terms and concert bounces on the communication range possibility of random multi-tier with combined black hole attacks as well as jamming that based on the ability of nodes to identify also evade relationship with malevolent nodes [9]. Cooperative bait detection scheme (CBDS) technique assists for discovering malevolent node by applying a reverse tracing strategy. In this strategy, neighboring node's tackle plays a bait receiver that assists to obtain a path respond since bait malevolent nodes. To discover the location of the malevolent node employs reverse retracing plan as well as detected malevolent node are stored in a list identified as black hole list. Then, noticed to entire nodes to break forwarding with that node rest in the list. Thus avoid black hole nodes in the network [10]. Throughput is enhanced with data-aided valuation method by means of transmit choice, channel then transmit obligation [11]. Authentication enables a node to ensure the identity of peer node. Authentication is essentially assured that participants in communication are authenticated and not impersonators. Authenticity is ensured because only the legitimate sender can produce a message that will decrypt properly with the shared key [12]

3. Credential Rate based Black hole attack Detection in MANET

A black hole malevolent mobile node disseminates short route replies thus it may be chosen as majority appropriate node to route traffic. However, it loss entire obtained packets. To overcome this problem, this paper Credential Rate based Black hole attack Detection in MANET. Every node observes the packet communicating activities for every it's one hop neighbors via discovering process. Here, every node has same resources like, transmitting power, queue size, energy as well as interface. In this scheme, every node disseminates hello messages in its transmission range nodes periodically. It consists of transmitter ID as well as credential rate. This credential rate is computed based on the node Energy account (EA), Obtained Data-Packet Account (OPA), QoS account as well as Vicinity Account (VA). Figure 1describes the structure of CRBD Scheme.

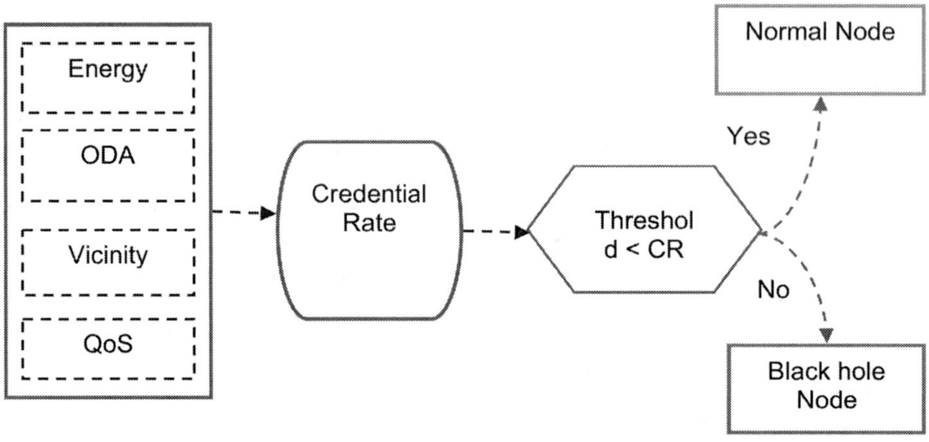

Figure1. Structure of CRBD Scheme

3.1. Energy Account

The energy account ($EA_{p,q}$) represents the outstanding energy in a mobile node. This computation is given below.

$$EA_{p,q} = Initial\ Energy_{p,q} - Current\ Energy_{p,q} \quad (1)$$

If the node energy is above the average outstanding energy that node is a black hole attack node.

3.2. Obtained Data-Packet Account (ODA)

It denotes the rate of data packets obtained from the transmitter node in the network. The obtained data-packet account is evaluating amount of exactly communicated packets to the amount of packets to be obtained. This computation is given below.

$$ODA_{p,q} = \frac{ODR_{p,q}}{CDR_{p,q}} \quad (2)$$

$CDR_{p,q}$ Indicates the sum amount of data packets communicated by node p as well as $ODR_{p,q}$ indicates the sum amount of data packets obtained by node q. If the $ODA_{p,q}$ is greater than 0.5 that node is normal node otherwise that node is a black hole**.** This computation is given below.

3.3. Vicinity Account

It denotes the rate of cooperative as well as uncooperative rate among two nodes. That identifies node characters in the network. This computation is given below.

$$VA_{p,q} = \begin{cases} PC_{p,q} = \dfrac{Coop_{p,q}}{Coop_{p,q} + UnCoop_{p,q} + 2} \\ PU_{p,q} \dfrac{UnCoop_{p,q}}{Coop_{p,q} + UnCoop_{p,q} + 2} \end{cases} \quad (3)$$

where $PC_{p,q}$ denotes the possibility that mobile node p believes mobile node q will perform cooperatively during hello packet dissemination action, as well as $PU_{p,q}$ represents the possibility that mobile node p believes mobile node q will perform not cooperatively during hello packet dissemination action. $Coop_{p,q}$ as well as $UnCoop_{p,q}$ are the cooperative and uncooperative rates that mobile node p controls for mobile node q. If $PC_{p,q}$ or $PU_{p,q}$ is larger than the threshold α, that mobile node represents the normal or black hole attack node in the network.

3.4. QoS Account

Quality of Service (QoS) is utilized for evaluating the function of the network. In this strategy, the node Qos is computed based on expression (4).

$$Qos_{p,q} = \frac{ODR_{p,q} - DDR_{p,q}}{ODR_{p,q} + DDR_{p,q}} \tag{4}$$

where

$ODR_{p,q}$ - indicates the sum amount of data packets obtained by node q

$DDR_{p,q}$ - indicates the sum amount of data packets dropped by node q

Next, the Credential Rate (CR) is computed through the expression (5) given below

$$CR_{p,q} = \frac{= ODA_{p,q} + VA_{p,q} + EA_{p,q} + Qos_{p,q}}{4} \tag{5}$$

If the CR is greater than the Threshold that mobile node is a normal node otherwise that node is deceitful node.

Node: CR	
1	0.3
2	0.8
3	0.56
4	0.2
5	0.75
6	0.1

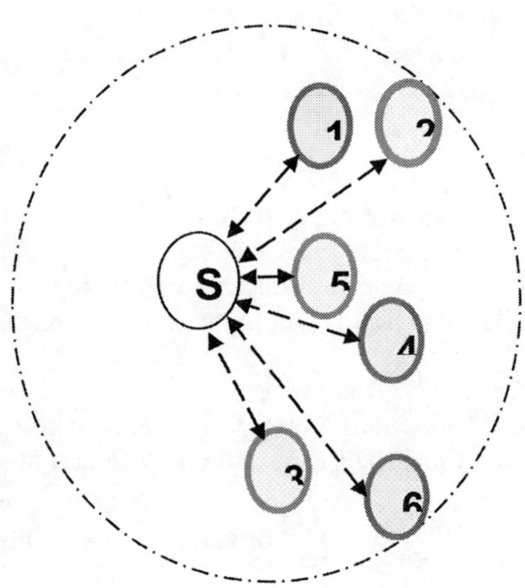

Figure 2. Example of Evaluating CR value in a Node

Every mobile node its credential rate (CR) is stored in a table. This rate is denoted by 0 to 1. Where, 0.5 above trust rate represents the normal node and below 0.5 trust rate represents the malevolent node in the network. The CRBD approach assists getting the CR and it is exemplified in the figure 2. Here, the source S which is gets the CR value about its necessitating transmitting data. The mobile node with the highest CR value is selected as the credential hop and data is communicated to that special node.

4. Performance evaluation

The complete estimation of CRBD is executed using network simulation (NS-2.35). The primary energy of mobile nodes occupied as 1 joule as well as the CR threshold is 0.5. This topography contains 50mobile nodes distributed arbitrarily region of 700x700m^2. Here, we using the traffic is constant bit rate as well as the node packet size is 512 bytes.

*4.1. **Packet Obtained rate Packet Obtained Rate (POR) Analysis***

POR is denoting as the number of packets received at receiver per particular time. POR is assessed by Equation 6.

$$\text{POR}=\frac{\sum_0^n \text{Packets Obtained}}{\text{Time}} \tag{6}$$

Where, n = Node count

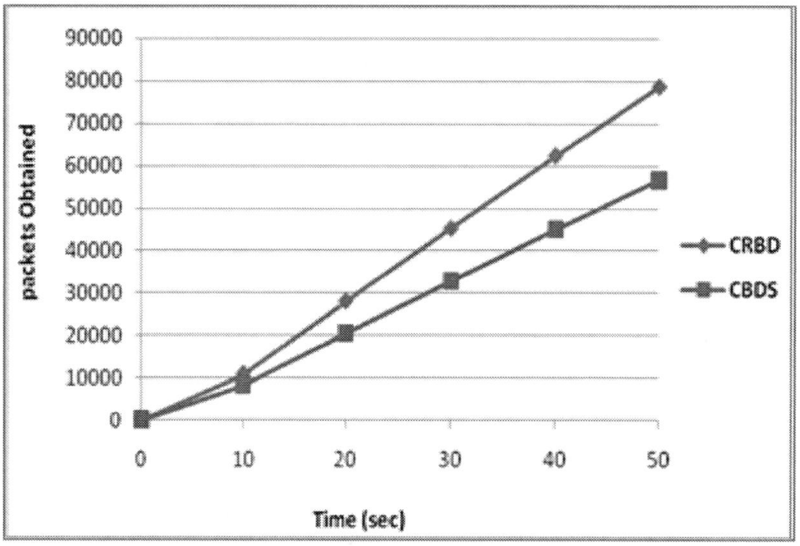

Figure 3. Packet Obtained Rate of CRBD and CBDS

The POR of CRBD and CBDS are diagrammed in Figure 3. It illustrates that the proposed scheme CRBD has 27.86% better POR when compared to the existing strategy CBDS.

4.2. Packet Drop Rate (PDR) Analysis

PDR is denoted as the distinction among the forward packets and obtained packets in the communication MANET per particular time. PDR is measured by Equation 7.

$$\text{PDR}=\frac{\sum_0^n \text{Forward Packets - Obtained Packets}}{\text{Time}} \tag{7}$$

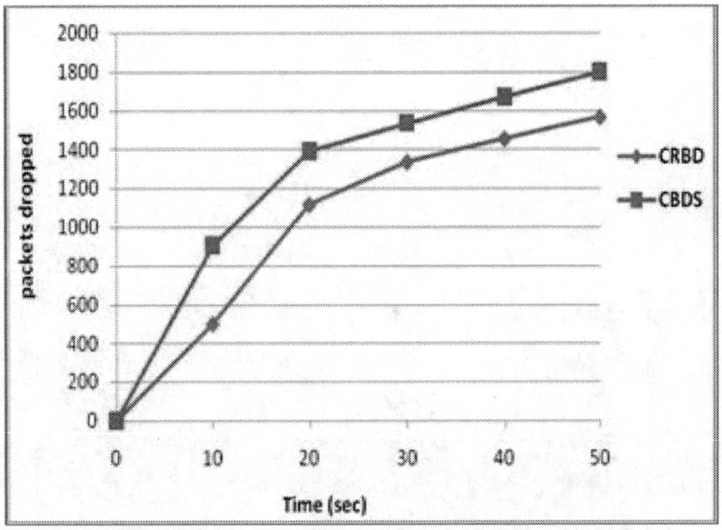

Figure 4. Packet Dropped Rate of CRBD and CBDS

Figure 4 shows the PDR values obtained from the simulation analysis of CRBD and CBDS. It indicates that PDR of CRBD is higher by 12.89% when compared with CBDS.

4.3. Outstanding Energy(OE) Analysis

Amount of energy outstanding in a node at the present occurrence of period is called as OE. A determine of outstanding energy commits at which energy is enthusiastic by the MANET functions.

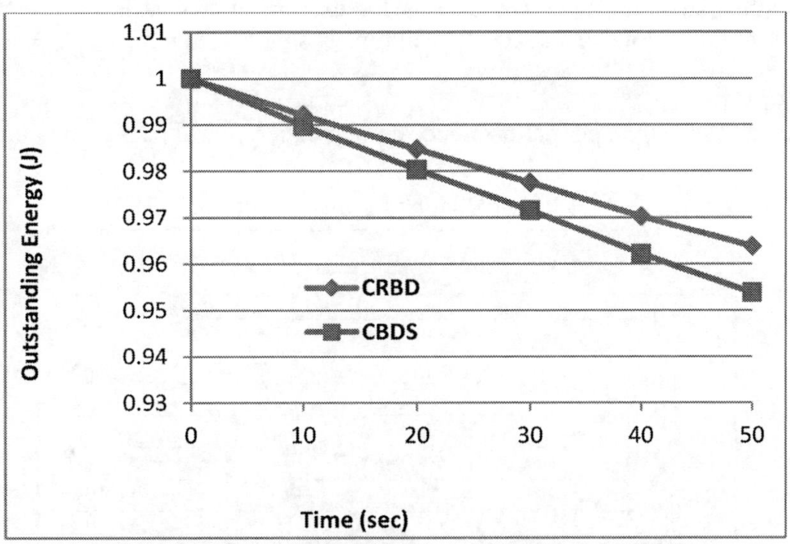

Figure 5. Outstanding Energy of CRBD and CBDS

Figure 5 indicates OE of the MANET is enhanced for proposed scheme CRBD when compared with CBDS. Around 0.09 joule of energy is saved per node by using the CRBD protocol for routing.

4.4. Throughput Analysis

Throughput denotes to an entire number of packets successfully delivered across network per unit time. Throughput is obtained using Equation 8.

$$\text{Throughput}=\frac{\sum_{0}^{n}\text{Packets Received(n) * Packet size}}{Time} \quad (8)$$

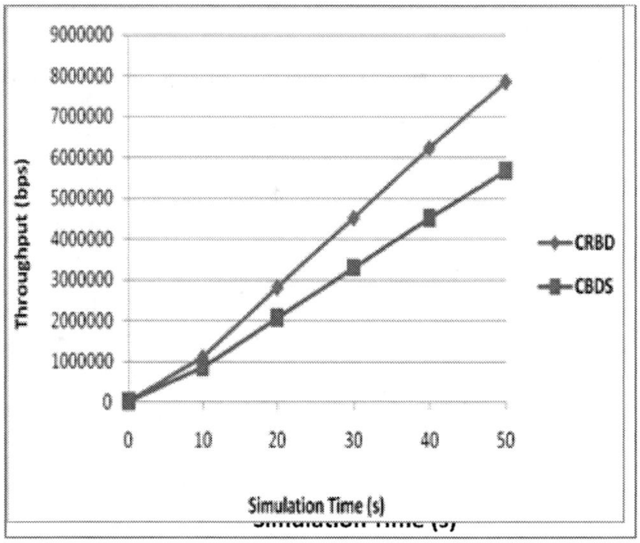

Figure 6. Throughput of CRBD and CBDS

Figure 6 indicate the outcomes for throughput execution of existing strategy CBDS and CBDS versus CBR traffic. It is mentioned that the CRBD approach has 27.87% highest throughput while equated to the existing strategy CBDS. CRBD executes the node ODA, VA as well as EA as a result detect malevolent node then communicate the data. But, the existing strategy CBDS offers independence to malevolent nodes to loss extra data packets thus diminishes the throughput.

4.5. Delay Analysis
Delay is represented as the time period difference among data sent and packets obtained. It is measured by Equation 9.

$$. \text{ Delay}=\frac{\sum_{0}^{n}\text{(Packet Obtained Time-Packet Forward Time)}}{n} \quad (9)$$

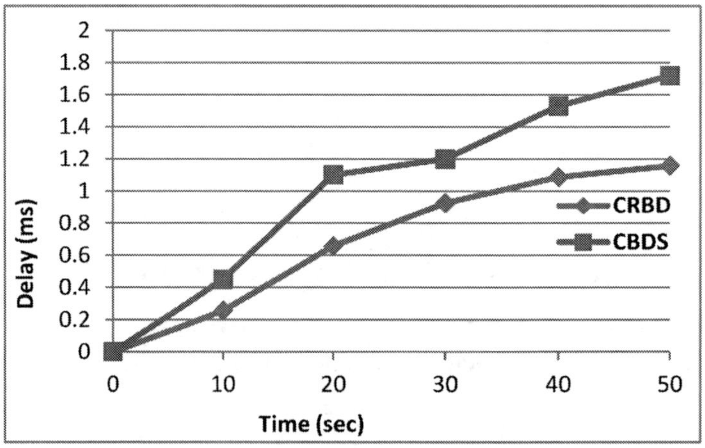

Figure 7. Delay of CRBD and CBAD

Figure 7 demonstrate the execution estimation for average delay for CBAD as well as existing strategy CBDS. It is mentioned that the CBAD approach has 32.64% less delay while equated to the existing strategy CBDS. In existing approach increases the malevolent nodes raise in the system, the delay further more rises owing to enhance the amount of system disconnections. The CBAD obtains better execution owing to it select the path is consistency.

5. Conclusion

In this paper we deliberate a prototype for black hole detection utilizing Credential Rate in MANET. In this strategy, the mobile node credential rate is computed based on the node Energy account, Obtained Data-Packet Account as well as Vicinity Account. The black hole node has lowest vicinity account rate as well as node cooperativeness rate also highest energy account rate. These parameters confirm the black hole nodes in the network. Then the source sends the notification message and discover the reliable route via credential rate based route selection in the network. The simulation examination depicts that the CRBD strategy offers highest throughput and outstanding energy when compared to the CBDS strategy

References

[1] El Houssaini MA, Aaroud A, El Hore A and Ben-Othman J 2016 Detection of Jamming Attacks in Mobile Ad Hoc Networks Using Statistical Process Control *Procedia Computer Science* **83** 26-33

[2] Sarma KJ, Sharma R and Das R 2014 A survey of black hole attack detection in manet *Int. Conf. on Issues and Challenges in Intelligent Computing Techniques (ICICT)* 202-205

[3] Shahabi, M G and Bakhtiarian M 2016 A modified algorithm to improve security and performance of AODV protocol against black hole attack *Wireless Networks* **22** 1505-1511

[4] Shaukat HR, Hashim F and Sali A 2014. Danger theory based node replication attacks detection in mobile wireless sensor network *IEEE Symp. on Computer Applications and Industrial Electronics (ISCAIE)* 18-23

[5] Duan J, Gao D, Yang C, Foh CH and Chen HH 2014 An energy-aware trust derivation scheme with game theoretic approach in wireless sensor networks for IoT applications *IEEE Internet of Things Journal* **1** 58-69

[6] Cai RJ, Li XJ and Chong PHJ 2018 An Evolutionary Self-Cooperative Trust Scheme Against Routing Disruptions in MANETs *IEEE Transactions on Mobile Computing* **18** 42-55

[7] Vasserman EY and Hopper N 2011 Vampire attacks: Draining life from wireless ad hoc sensor networks *IEEE transactions on mobile computing* **12** 318-332

[8] Reddy PLK, Reddy BRP and Krishna SR 2016 Spatial-Temporal Provenance Assurance with Mutual Proofs for Location Based Services *IJCSEST* 1-5

[9] Hemanth Kumar and Ramesh GP 2020 Energy efficiency and Data packet security for wireless sensor networks using African Buffalo Optimization *IJCA* **13** 944-954

[10] Dumne PR and Manjaramkar A 2016 Cooperative bait detection scheme to prevent collaborative blackhole or grayhole attacks by malicious nodes in MANETs *5th Int. Conf. on Reliability, Infocom Technologies and Optimization (Trends and Future Directions) (ICRITO)* 486-490

[11] Santhakumar R 2017 Resource allocation in wireless networks by channel estimation and relay assignment using data-aided techniques *Int. Journal of MC Square Scientific Research* **9** 40-47

[12] Lingeshwari S 2014 Provisioning of efficient authentication technique for implementing in large scale networks (PEAT) *Int. Journal of MC Square Scientific Research* **6** 34-42

IOP Conf. Series: Materials Science and Engineering **925** (2020) 012072 doi:10.1088/1757-899X/925/1/012072

Detect and Prevent Active and Passive Attack in Autonomous Mobile Network

Sivasankar C[1] and Kumanan T[2]

[1,2] Department of Computer Science and Engineering, Meenakshi Academy of Higher Education and Research Chennai, India.

[1]Sivashankar.phd12@gmail.com

Abstract. In Autonomous Mobile Networks (AMNs), nodes regularly collaborate as well as transmit all packets to facilitate out of range transaction. Though, in unfriendly surroundings, several nodes might reject for preserve their own senders or for intentionally disrupting regular communications. Ina addition, the node mobility to handles the active and passive attacks. To overcome these problems, in this article Detect and Prevent Active and Passive Attack in Autonomous Mobile Network (DPAP) is introduced. Here, faithful forwarder node is chosen based on the vicinity information as well as connectivity of node. The active attacker node is detected by three hop verification method. In addition, the Binomial Pascal triangle function (BPTF) is applied for concealing the node uniqueness as a result; it recognizes the passive attack in the AMN. The DPAP approach is measured by applying a network simulator-2 for recognizing the active and passive attacker in the AMN. Simulation output illustrates the DPAP offer better throughput and detection ratio in the AMN.

Keywords: Autonomous mobile Networks, Three hop verification, Fibonacci Pascal triangle, Active and Passive attacker detection.

1. Introduction

Autonomous Mobile Networks (AMN) [1] are endlessly, without infrastructure, self configuringcreated of a accumulationof mobile nodeslacking centralized direction. In this networks, every node can act a part as a node collaborating with other nodes [2]. As the main objective of routingto demonstrate aright as well as effective route by the disseminated nodes themselves also can communicate the data quick correctly. Besides the inactive attack in that the interlopers undertake to listen on the transaction, severaltypes of active attacks incline toharm the AMN by altering the data that is moreusual as well asdestructive in AMNs.

A distinctive event is that the individual node malevolent otherwise the ordinate malevolent will compromise the AMN through directly attack otherwise indirect attack next leading the routing.In the direct attack, the malevolentcan fiddle or loss the packets to construct the networkparalytic, for example wormhole as well as the black hole attack [3].

In indirect attack, the malevoents advisedly update their own logarithms lacking determining the data communicating to disconcert the rectifying scheme. The capacity to detect this malevolent otherwise

the nodes assailed by these breaking nodes is further important to a various setof management jobsfor example executing network answer ability [4], distinguishing malevolent as well asm isconductingnodes, and applying faith policies in disseminated systems [5].
Thus, security is the significant subjects for data transaction over AMN. In the previous methods, several security- raised assesses have been introduced to enhance the security of data transaction over AMN. Traditional methods on security raised data communications admits the intentions of cryptanalytics techniques as well as system structures also security- raised routing protocols [6]. Their general goals are frequently to overcome several menaces over the Internet, admitting listening, faking, session pirating, etc. Resistive to selective drop attack (RSDA) strategy to supply efficient security versus selective drop attack. It achieves dependability in routing withinvaliding the connection with the most prominent weight as well asapplying the ellipticcurve digital signature technique for checking the node authentication. RSDA protocol rejects selective dropattacks through precluding the nodes from acquiring overloaded. However, This strategy cannot prevent the data from the eavesdropper attacker [7]. To overcome this problem, In this paper we propose, Enhanced Audit-based malevolent Node revealing and secret Data Communication in Autonomous Mobile Network. Here, the Audit-based malevolent revealing (AMR) is introduced for discovering and separating malevolent nodes in AMNs. In AMR contains reputation phase, route discovery phase, audit phase. It constructs route comprising of greater faithful nodes. In addition, the Bivariate pascal triangle function (BPTF) is utilized for hide the node originality in the network.

2. Related work

A trust based security protocol that reaches authentication of data packets. In this scheme, every node keeping trust counter for bserving and seperating the malevolent node in the network [8]. However, this scheme cannot provide confidential data transmission in the network. Security of Cryptographically Generated Addresses using Adjustable Key [9] mechanism using AES algorithm.In this scheme, the every sub key is used to encrypt one AES block, using Mix-Column transformation in the original AES-128 to diminish the encryption time. In addition, the S-box and Inverse S-box are substituted by mere S-box utilized for encryption and decryption. However, this scheme does not provide node authentication in the network. Quorum time slot adaptative concentrating protocol [10] is introduced for attaining delay reduction and energy efficiency. the Quorum time slot is a packet communication period based on the features of data transmission that as a result extends the duty cycle of the nodes and diminishes the latency. Dij-Huff Method (DHM) is the compounding Dijkstra's and Huffman coding. Dijkstra's algorithm is utilized for discovering the nodes with best distance path and maximal energy and. Huffman coding is utilized for receiving security code. To offer security at every hop a individual security is being supplied known as Binary Hop Count [11]. Key distribution scheme adapt secure property by giving pairwise keys among nodes. Secure Connectivity scheme inquire the results of key distribution possibility on route length, queue size, and energy looseness while minimal cost routing is employed [12]. A key agreement scheme based on public discussion was described that attain a convinced secret key range while an opponent has more faveourable situation. A secret key agreement protocol introduced the reliable relay regarded in the discourse to attain both reward distillation and secrecy elaboration [13]. Numerous key management approaches inserted to initiate symmetric keys [14]. This approach applying arbitary seed dissemination with transient master key which offer pairwise keys. This approach approach discover the main disadvantage is providing less security. Masquerading Attack detection system (MADS) using the received signal-strength (RSS). In this method, the Signal strength variation is used todetect the attack accurately. The RSS is a lightweight cryptographic certification that as well does not necessitate extra charge to the present wireless technology. Although, the RSS is not quite authentic since it changes as well as fluctuates over time owing to several elements, for example multipath passing, deflection, etc [15]. Modified Russian Peasant Multiplier based on Divide as well as conquer method is utilized for multiplication procedure. Minimizing the chip size, enhancing the quickness as well as minimizing the energy depletion are key critical issues in System strategy [16]. An opportunistic routing by responsiveness of energy to accept by dynamic environment. While the sender transmits the information to a multicast group, the sender transmits the information via greater energy between vicinity thus enhance the life span [17].

IOP Conf. Series: Materials Science and Engineering **925** (2020) 012072 doi:10.1088/1757-899X/925/1/012072

3. Detect and Prevent Active and Passive Attack in Autonomous Mobile Network

This method faithful forwarder node is chosen based on the vicinity information as well as connectivity of node. It initiates throughout the reply of route packet communication explicitly a packet utilized to inform the sender which the route is detected. Here, the active attack is detect by three hop and passive attacks

3.1 Active Attack Detection

The figure 1 illustrates the three hop verification of nodes in the chosen route for communication. The receiver is the initiating point. It is executed among adjacent nodes.

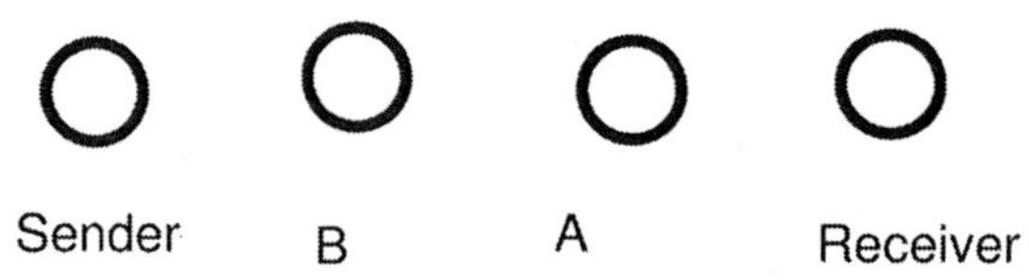

Figure 1. Three hop verification

Primary it is employed among the receiver as well as mobile node A, next among node A and node B, then among node B and the sender. Adjacent nodes which should transmit frankly seek for a substitute minimum distance among them with an utmost length of three hops. The substitute least route is explored from vicinity lists. This scheme, the active attack detected by two phases of examination that is executed among adjacent nodes in the chosen route. Assume the transfer node in the chosen route is A as well as the receiver of A is **B;** B is the next node to A in the route. The two phases of examination are executes among node A and B. Winning the initial phase represents the mobile node is faithful then no necessitate to go by the next phase. The initial phase of examination is to construct an initial hop vicinity table for A as well as an initial hop vicinity table for B. Next, to verify if exist a common node among the two tables. If not present, it goes via the next phase of verification. In the next phase, A verifies whether exist is a three hops route to node B also B performance the identical. This process is done via the node A utilizing the one hop vicinity list it is established in the initial phase. When, B constructs its two hop vicinity list. B constructs it via calling for each node in its earlier one hop vicinity list to transmit their individual one hop Vicinity lists

3.2. Passive Attack Detection

Here, the Sender does not display actual identity of in-between, receiver and itself. Data communication routing nodes are reproducing of its BPTF values. The malicious node supervises the transaction although does not recognize Sender, in-between and receiver of the route. The BPTF task is determined as

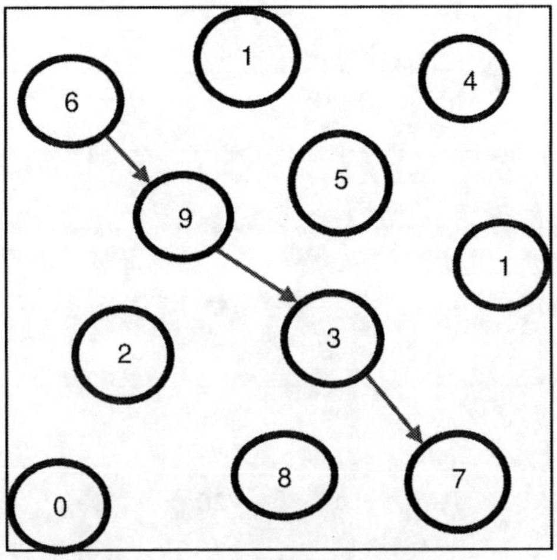

Figure 2. Fake Route creation

$$\text{BPTF} = (Q - R)^k \quad (1)$$

Where, Q denotes the Source R denotes the destination as well as k represents the hop count of in between nodes.

The figure 2 demonstrates the fake route creation in AMN. Here, the node 6 is a Sender, 7 is a receiver also in between nodes are 9, 3. These nodes are makes a fake route through the BPTF.

$$(6-7)^3 = 6^3 - 3(6)^2(7) + 3(6)(7)^2 + (7)^3$$
$$= 216 - 126 + 882 + 343$$

From figure 2, the unique identity of the sender is 6 but, BPTF makes the fake identity is 216 also the receiver unique identity is 7 though fake identity is 343. Equally, the in between nodes 9, 3 maintains the fake identities are 126 and 216. This fake path creation creates additional competent as well as the passive attacker dos not identify the unique identity of the nodes. So, passive attacks are diminished throughout data communication.

4. Performance evaluation

To execute simulation examination, we utilized the NS-2.35 that incorporates the 802.11 MAC protocol. This protocol is functioning on the 2.4 GHz physical layer and each sensor node transaction range is preset to 180 metre. In this simulation topology, the BS node plays Owner of complete network as well as every sensor node function is sensing, data forwarding as well as obtaining the data. The packet traffic is yielded with constant bit rate. This simulation examination, we measured the following metrics namely Packet Obtained Rate, Latency, Packet **Dropped Rate,** Throughput **and detection ratio**.

4.1. Packet Obtained rate (POR)

It is defined as the ratio among the amount of data packets perfectly obtained by the receiver and the amount of data packets yielded by entire mobile nodes.

IOP Conf. Series: Materials Science and Engineering **925** (2020) 012072 doi:10.1088/1757-899X/925/1/012072

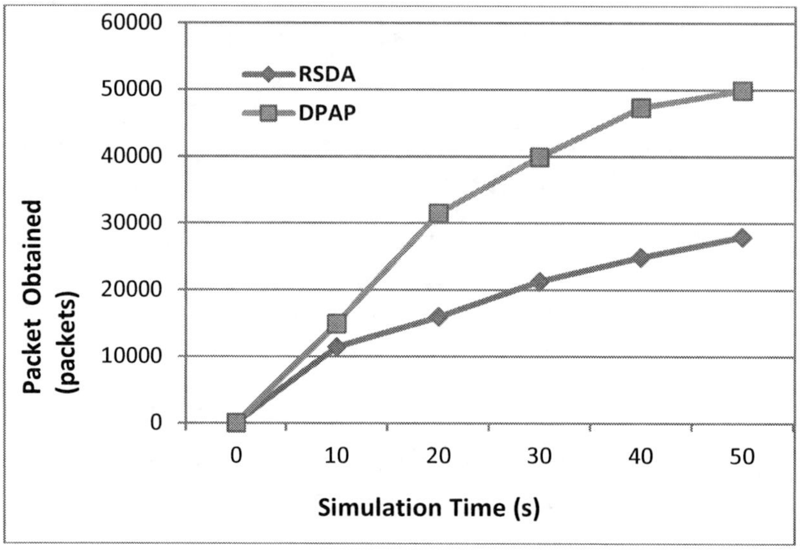

Figure 3. Packet Obtained Rate of RSDA and DPAP

$$\text{POR}=\frac{\sum_{0}^{n}\text{Packets Obtained}}{\text{Time}} \tag{3}$$

From this figure 3, the proposed method DPAP is greater compare to the existing method RSDA. Since, the DPAP pick the route by the consistency path. Thus, enhances the routing efficiency.

4.2. Packet Drop Rate (PDR) Analysis

It is denoted as the distinction among the forward packets and obtained packets in the communication MANET per particular time. PDR is measured by Equation 4.

$$\text{PDR}=\frac{\sum_{0}^{n}\text{Forward Packets - Obtained Packets}}{\text{Time}} \tag{4}$$

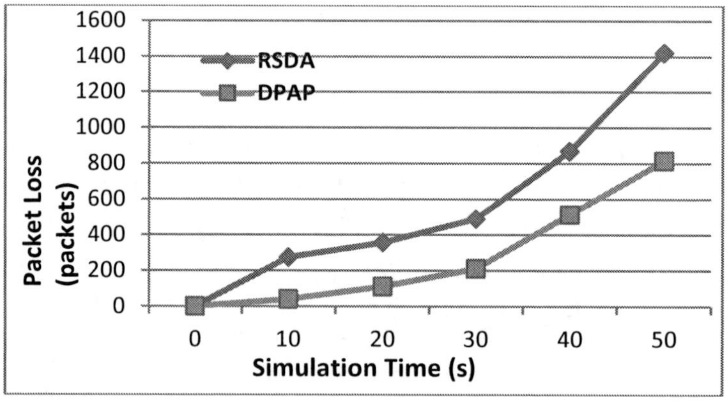

Figure 4. Packet Drop Rate of RSDA and DPAP

Figure, 4 illustrates the PDR of RSDA as well as DPAP. It indicates the successful transaction between nodes in the AMN. From this figure, DPAP execute lesser PDR of the RSDA scheme. DPAP scheme observes the active and passive attacker nodes from the path; thus it dilutes the PDR in the AMN.

IOP Conf. Series: Materials Science and Engineering **925** (2020) 012072 doi:10.1088/1757-899X/925/1/012072

4.3. Latency

It is defined as the time period from while the data packet communication is initiated at the source to while the data packet is perfectly obtained by the sink.

$$\text{Latency}=\frac{\sum_{0}^{n}(\text{Packet Obtained Time-Packet Forward Time})}{n} \quad (5)$$

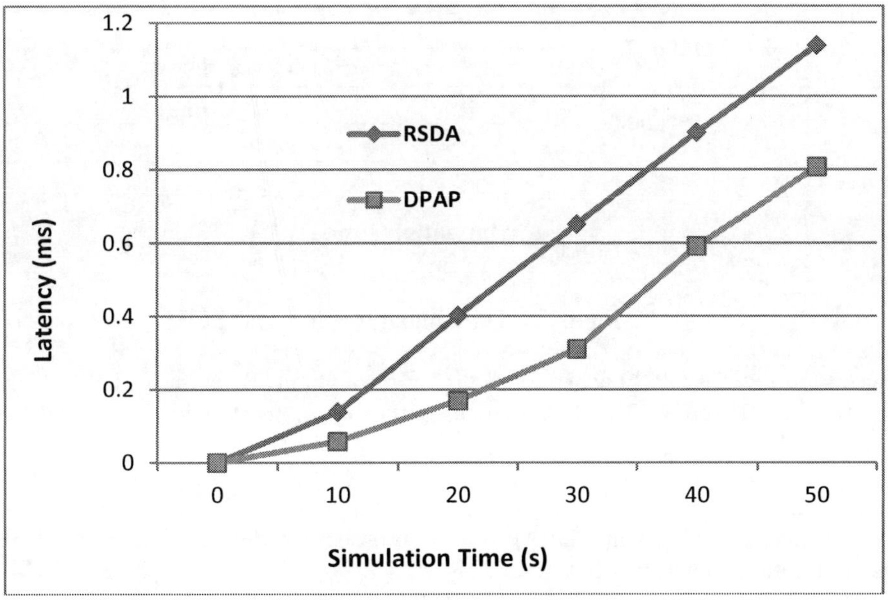

Figure 5. Latency of RSDA and DPAP

The figure, 5 illustrates the latency of RSDA as well as DPAP. The data traverse via link consistency factor nodes. Thus minimizes the routing latency.

4.4. Throughput

It is the vital parameter for measuring the operations of network. In this protocol, the throughput is specified as the amount of data packets effectively obtained at the receiver.

$$\text{Throughput}=\frac{\sum_{0}^{n}\text{Packets Received(n)} * \text{Packet size}}{Time} \quad (6)$$

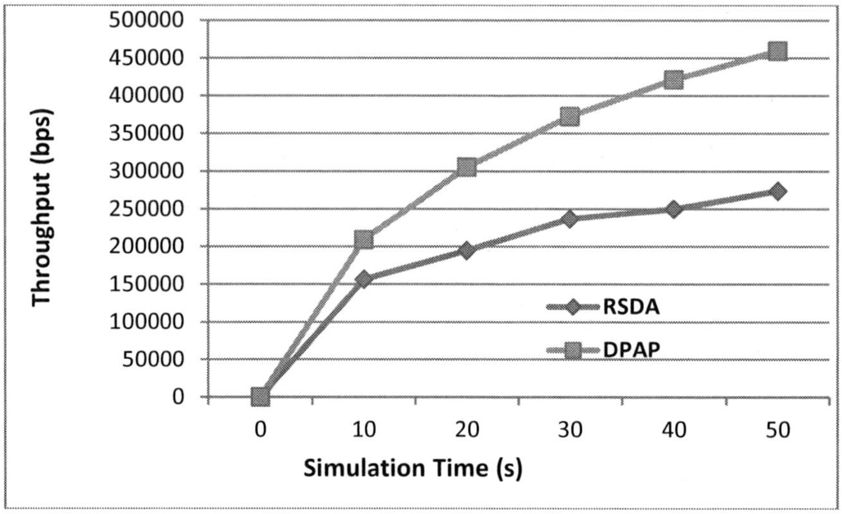

Figure 6. Throughput of IEET and LCEE

The figure 6 proves the DPAP scheme has higher throughput. Because of, RSDA scheme concentrate only the energy efficiency. But, DPAP focuses the link consistency thus DPAP has provide better performance in AMNs.

4.5. Detection Ratio

Here, we observe the RSDA and DPAP routing protocol for detection ratio and false detection ratio. The detection ratio is determined as given below:

$$Detection\,Ratio = \frac{Detected\ nodes}{Whole\ nodes} \tag{7}$$

$$False\ Detection\,Ratio = \frac{Mistook\ attac\text{ker}\ nodes}{Whole\ nodes} \tag{8}$$

Detection ratio represents the count of attacker nodes discovered through usual nodes. False detection ratio represents the number of usual nodes mistook as the attacker nodes through usual nodes.

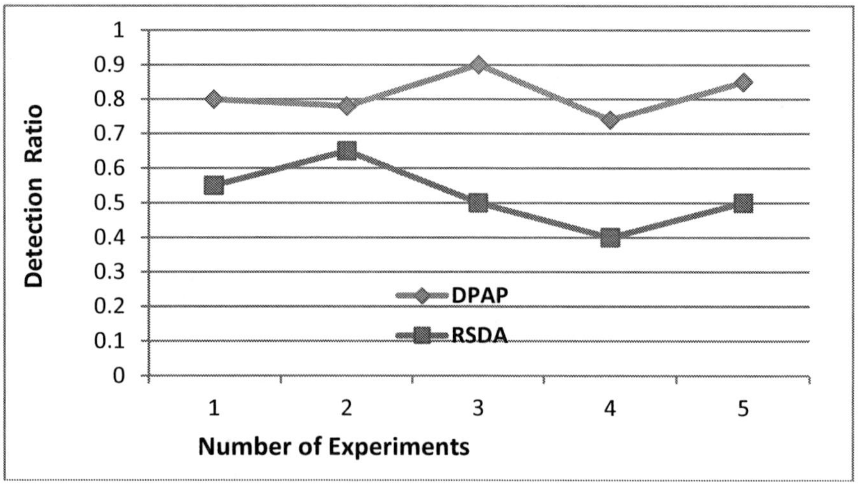

Figure 7. Detection Ratio of RSDA and DPAP

The detection ratio of RSDA and DPAP protocols are projected in Figure 7. It illustrates that the detection rate of DPAP is better than RSDA protocol due to the FPTF as well as three hop verification methods are recognize the active and passive attacks in the AMN.

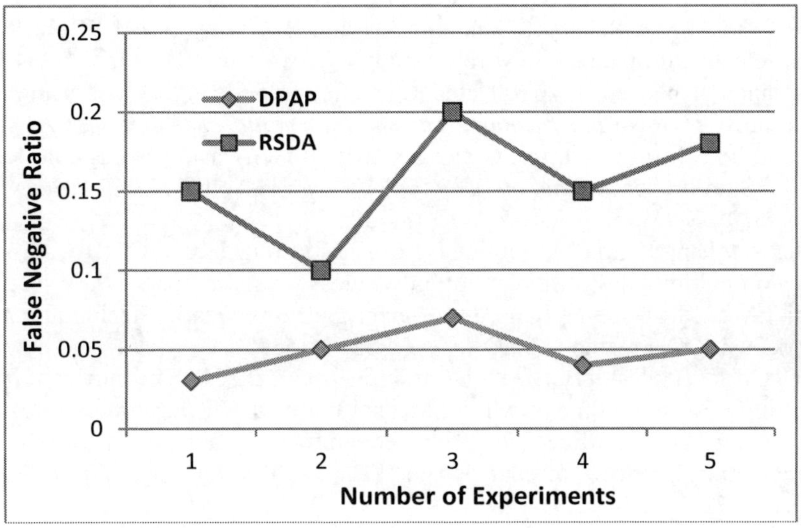

Figure 8. False Detection Ratio of RSDA and DPAP

Also the false positive ratio is diagrammed for the RSDA and DPAP schemes in Figure.8. It illustrates the DPAP has lesser false negative ratio than the RSDA scheme.

4.6. False Detection Ratio

Here, we observe the RSDA and DPAP routing protocol detection ratio. The detection ratio is determined as given below:

$$Detection\,Ratio = \frac{Detected\ nodes}{Whole\ nodes} \tag{9}$$

5. Conclusion

AMN is an important domain in current years owing to rapid dissemination of wireless devices thus the safety is a vital effect. The three hop verification is employed for identifying active attack. The Fibonacci Pascal Triangle Function is applied for recognizing passive attack in the network. Fibonacci Pascal Triangle Function method makes the fake route to defend the data from the passive attacker. Consequently, the sender transmits data through the normal nodes and receiver gets the secret routing information throughout data communication. Simulation outcomes illustrate both the highest packet obtained rate and detection ratio and lesser delay in the AMN.

References

[1] Yu W and Liu KR 2007 Game theoretic analysis of cooperation stimulation and security in autonomous mobile ad hoc networks *IEEE Transactions on Mobile Computing* **6** 507-521

[2] Shen WL, Chen CS, Lin KCJ and Hua KA 2012 Autonomous mobile mesh networks. *IEEE Transactions on Mobile Computing* **13** 364-376

[3] Jain S and Khuteta A 2015 Detecting and overcoming blackhole attack in mobile Adhoc Network *Int. Conf. on Green Computing and Internet of Things (ICGCIoT)* 225-229

[4] Thamilarasu G and Ma Z 2015 Autonomous mobile agent based intrusion detection framework in wireless body area networks *16th Int. symp. on a world of wireless, mobile and multimedia networks (WoWMoM)* 1-3

[5] Chen CL, Chen SC, Chang CR and Lin CF 2014 Scalable and autonomous mobile device-centric cloud for secured D2D sharing *Int. Workshop on Information Security Applications* 177-189
[6] Caiti A, Calabro V, Dini G, Lo Duca A and Munafo A 2012 Secure cooperation of autonomous mobile sensors using an underwater acoustic network. *Sensors* **12** 1967-1989
[7] Poongodi T, Khan MS, Patan R, Gandomi AH and Balusamy B 2019 Robust defense scheme against selective drop attack in wireless ad hoc networks *IEEE access* **7** 18409-18419
[8] Rajaram A and Palaniswami S 2010 Detecting malicious node in MANET using trust based cross-layer security protocol *Int. J Comput Science Information Technologies* **2** 130-137
[9] Patil KT and Patil ME 2014 Improve the Security of CGA using Adjustable Key Block Cipher based AES, to Prevent Attack on AES in IPV6 over MANET *IEEE Global Conf. on Wireless Computing & Networking (GCWCN)* 148-152
[10] Liu Y, Ota K, Zhang K, Ma M, Xiong N, Liu A and Long J 2018 QTSAC: An energy-efficient MAC protocol for delay minimization in wireless sensor networks *IEEE Access* **6** 8273-8291
[11] Alghamdi TA 2018 Secure and energy efficient path optimization technique in wireless sensor networks using DH method *IEEE Access* **6** 53576-53582
[12] Yildiz HU, Ciftler BS, Tavli B, Bicakci K and Incebacak D 2016 The impact of incomplete secure connectivity on the lifetime of wireless sensor networks *IEEE Systems Journal* **12** 1042-1046
[13] Wang N, Zhang N and Gulliver TA 2013 Cooperative key agreement for wireless networking: Key rates and practical protocol design *IEEE Transactions on Information Forensics and Security* **9** 272-284
[14] Hemanth Kumar and Ramesh GP 2020 Energy efficiency and Data packet security for wireless sensor networks using African Buffalo Optimization *IJCA* **13** 944-954
[15] Abbas S, Faisal M, Rahman HU, Khan MZ and Merabti M 2018 Masquerading attacks detection in mobile ad hoc networks *IEEE Access* **6** 55013-55025
[16] Keerthi Anand VD 2015 Power Efficient Multicast Opportunistic Routing Protocol (Pemor) To Optimize Lifetime of Manet. *Int. Journal of MC Square Scientific Research* **7** 109-127
[17] Sendhilkumar NC 2017 Design and implementation of power efficient modified Russian peasant multiplier using ripple carry adder *Int. Jour. of MC Square Scientific Research* **9** 154-165

Link Consistency and Energy Efficiency for Improving Throughput in Wireless Sensor Network

Vijayabaskar K[1] and Kumanan T[2]

[1,2] Department of Computer Science and Engineering, Meenakshi Academy of Higher Education and Research Chennai, India.
[1]vijayabaskar.phd@gmail.com

Abstract. Wireless Sensor Networks (WSNs) necessitate lifelong links as well as energy efficiency. Also the raising commercialization of WSN, supervising applications visits for increased quality of service (QoS). The formation of the energy proficient as well as consistent link is a vital to maintain the enhancement of the WSN performance. In this object, we introduce a Link Consistency and Energy Efficiency (LCEE) for Improving Throughput in Wireless Sensor Network. In this scheme, the link consistency factor (LCF) is used to measure the node link consistency. This LCF is computed by node consistency, Link quality, node remaining energy as well as Link healthiness. The LCEE method primarily enhances the network lifespan durability by minimizing sensor node energy utilization with almost logical throughput as well as the packet obtained rate while the improvement of the QoS parameters in the WSN.

Keywords: Wireless Sensor Network, Link Consistency Factor, Remaining Energy, Node Consistency.

1. Introduction

WSN is a sort of systems made up of thousands of small nodes. The nodes in WSN have restricted processing, storing and conveying capacity. The objective of most WSN is to distinguish functions and it gives careful consideration to the QoS of settlement information packets.

WSN are most attractive developing advancements. WSN can be utilized to perform observing, observation, detecting and estimating tasks. WSN comprises of sensor nodes to detect ecological or physical information. These sensors are little, with constrained preparing and processing assets. These sensors can detect, measure, and accumulate data from the surrounding and transmit that information to the sink. WSN is a developing field in the more extensive zone of remote networking; with appliance varies from observation to social insurance [1].

Link consistency may be pretended with the severe packet conflict, link loss out, as well as node out of communication range. Whilst mostly accessible routing protocols, generally concentrate on energy proficient routing as well as long life. These protocols offer small achievement diluted packet conflict as well as packet forward achievement rates. In common, routing protocols should reflect on the trade-offs among performance enhancement as well as energy proficiency that in the majority cases are inflexible to attain together at a time. The major QoS deprivation is final hop bottle neck in WSN, which

outcomes in short performance and important vitality exhaustion of the nodes; bringing about dishonesty of generally lifetime of the network and inertness [2].
Improvement of Energy-Efficient Technique (IEET) is used to evaluation of network lifetime that attains an energy proficient system. This method is used for energy protection based on the sleep as well as awake scheduling. This protocol minimizes the inactive listening time, thus minimizes the network latency [3].

2. Related Work

Link quality evaluation is a fundamental WSN application since it radically impacts the achievement of transmissions. Prediction of Link Quality turn to machine learning strategies to anticipate the transient advancement of link quality, with the end goal to switch the information transmission on a superior quality link. The Link-Quality Indicator (LQI) is measures the link quality. The projected learning and expectation display introduces an incredible adaptability: it is a common model that can be effectively adjusted to various link quality measurements [4]. Enhancing QoS of WSN technique calculate throughput of the Round Trip Path (RTP) and Round Trip Time (RTT) and contrast the measured throughput and RTT and porch esteems dependent on an examination result will locate a defective sensor inside the system [5].
The purpose of the plan will be on achieve low lifelessness, and to satisfy different QoS in various conditions of a network. The projected calculation abuses numerous nodes' ascribes like separation to the Sink, vitality, history of bare deferral of information to neighbors and likelihood of link accessibility to decide the finest node for conveying distinctive movement provisions. The commitment of the projected calculation is its ability to recognize the information identified with each kind of movement and give different needs to different information classifications so every sort of activity arrangement guarantees to meet the prerequisites of the application provisions [6]. Link Adaptation-based Optimization approach, the life utilization over a link is estimated. The advancement component is subject of particular QoS necessity in term of aggregate end-to end bit error rate and delay time [7].
Link quality for WSN is anything but difficult to be exasperates by the surrounding. Link quality not just influences the unwavering quality of information transmission in the system, yet in addition decides the execution of related Upper-layer arrange convention. Hence, link quality assessment turns into thinking about inquiry. The connection quality assessment techniques incorporate Link Quality Index (LQI) and information package delivery rate [8]. The effectiveness and unwavering quality of remote system mostly relies on the conveyance intensity of its links. So it is essential to gauge which interface is more dependable and can convey information successfully over an extensive stretch of time. We think about the nature of a links dependent on the accessible estimators fundamentally the Signal to Noise Ratio (SNR), the Received Signal Strength Indicator (RSSI), the Packet Reception Rate (PRR) of the CC2420 radio in an absolutely programming based condition [9].
QoS Analysis intends calculation and recreates the execution frameworks examination of QoS parameters for coordinated robotization of open utility administration, for example, gas, water and power. It is utilized for information combination and information investigation, robbery and spillage recognition [10]. A Link Quality Reliable Routing Protocol (LQRRP) which gives connects unwavering quality and vitality effectiveness for correspondence among nodes is intended. This LQRRP chooses paths on the dependent on the connection quality measurements incorporates packet delivery ratio, quantity of re-transmission count and load. The LQRRP gives connect strength to directing procedure amid the transmission [11].
QoS establish an open investigate concern inside authorised researchers in light of the fact that the fast improvement of microelectronics multiplies innovative appliances for WSNs. These appliances frequently need an association of end-to-end delay and lofty dependability for different contact designs. As the utilization of question driven WSNs winds up adaptable, the requirements to build up QoS bolster turns out to be seriously communicated, especially for medium and lofty traffic loads. In active WSNs, the QoS weakens upon expanded movement stack because of blockage and the mutual medium. Improving QoS in Query-Driven WSN scheme intends a cross layered handover (CLD) calculation with the end goal to enhance disintegrated execution and the QoS by methods for multi-channel excess [12].

Improving Quality of Service approach gives inside and out recreation and investigation of sink portability techniques to beat the final hop bottle neck concern [13].
The QoS measurements in a WSN of different sensor types rely upon the execution of the network convention layers, propelling a far reaching cross-layer configuration way to deal with advance QoS. Advances in vitality reaping strategies empower increments in lifetime of the WSN by drawing out activity of the remote nodes. While the essential target of vitality gathering is to draw out lifetime of the network, it might cause bring down estimations of different QoS measurements amid that life span. The cross-layer constraints that decide QoS esteems in the set up regarding answers for stochastic unique programming circumstances got from multivariate point-process models of transient data streams [14].
An opportunistic routing by responsiveness of energy to accept by dynamic environment. While the sender transmits the information to a multicast group, the sender transmits the information via greater energy between vicinity thus enhance the life span [15]. Modified Russian Peasant Multiplier based on Divide as well as conquer method is utilized for multiplication procedure. Minimizing the chip size, enhancing the quickness as well as minimizing the energy depletion are key critical issues in System strategy [16].
Link channel optimization that needs a channel interface for has the most excellent connection condition, and along these lines, better nature of the signal limitation specified by the least Bit Error Rate (BER). The intended enhancement incorporates control channel task reasonable and least BER requires on characterizing the quantity of transfer of bits. In this scheme, we build up an exploratory portrayal and approve our plan of scientific model with various advances utilized and over an extensive variety of system situation by conveying a real-world ecological checking application with multi-point organize estimation signals that ensure the QOS as vitality utilization, dormancy, packet delivery rate, and throughput [17]. An Efficient Learning Technique to pertains learning strategies to foresee link quality advancement in a WSN and exploit remote connections with the most ideal quality to enhance the packet delivery rate.
End-to-End Data Delivery Reliability prototype intend a system level unwavering quality model, in particular, end-to-end information conveyance dependability (E2E-DDR), for assessing as well as enhancing the dependability execution of WSNs. In this method, an outline is exhibited for catching the planning capacity among the packet gathering proportion, foundation clamour, and received signal strength (RSS). This scheme is used to evaluate the system level dependability as well as advance the WSN arrangement arguments [18].

3. Problem Statement

The majority of the reason for link failure due to degrades the link quality. The link quality degrades owing to link stability, available bandwidth, link life time.

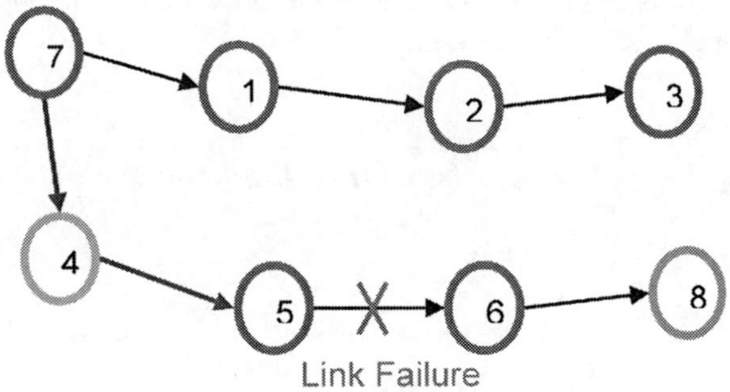

Figure 1. Link Failure in the route

The figure 1 demonstrates the link failure in the route. In this figure, the path from sender node 7 to receiver node 8 and intermediate nodes are 4, 5 and 6. Here, the link failure occurs between 5 and 6. Thus, the source cannot reach the data to destination. In WSN, the node breakdown happen owing to deficit of resources in several events vital reason is less link quality.

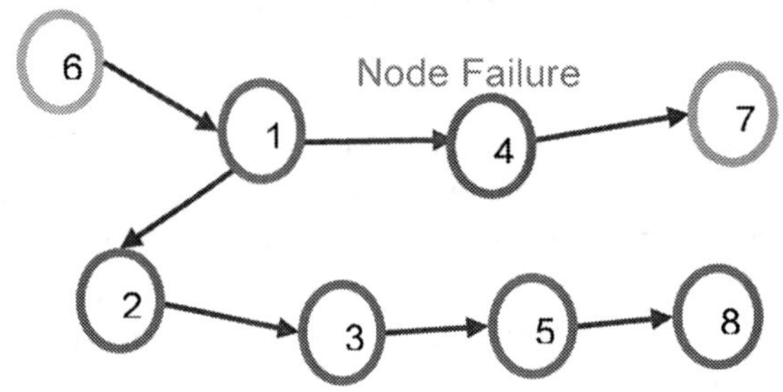

Figure 2. Node Failure in the route

The figure 2 exemplifies the node failure in the route. Here, the node 6 is a sender and 7 is a receiver. The sender node forwards the data to receiver node. But, in this route the 4th node failed due to deplete the energy. As a result, the sender data cannot reach the receiver.

Thus, energy efficiency and highest link quality is a significant factor in the network. To determine this problem, we propose Link Consistency and Energy Efficiency for Improving Packet Delivery in WSN.

4. Proposed Method

In WSN, beginning to present time the major dispute is feasible route for data communication lacking link consistency. In this paper, we concentrate on increasing a link consistency in WSN which fulfills the consistency necessities. Quality of Wireless link is a significant as well as established marker of consistency at every sensor node. The link quality is utilized for WSN coverage as well as node position presume two probable stages of the wireless link such as connected or disconnected.
In WSN, every node updates its node Identity, location, Time to live with link consistency factor (LCF) periodically. This information is stored in the table and the BS maintains all node information and LCF value. The sender checks each node LCF then makes a route path with highest LCF. During route invention procedure the sender forward a join request comprising the receiver ID's to every the nodes in a WSN. After that, a join-reply is produced by the receiver, added by the nodes in the path and are sent out back to the sender along with their position.

4.1 Link Consistency Factor

The Link Consistency Factor is calculated by four individual components regarding the link consistency of the node specified in the formula.

$$LCF = \begin{vmatrix} NC & LQ \\ LH & NRE \end{vmatrix} \tag{1}$$

Where

NC→ Node Consistency

LQ→ Link Quality

NRE→Node Remaining Energy

LH→ Link Healthiness

4.2 Node Consistency (NC)

Node Consistency denotes the node character. The character is determined on the feedback of adjacent nodes. If the node is reliable, unselfishness, then an encouraging feedback is received otherwise get discouraging feedback. The node consistency determines whether the node is a consistency or inconsistency node. It is measured by the equation 2.

$$NC = \frac{\lambda}{\lambda + \gamma} \tag{2}$$

where

λ = encouraging feedback of adjacent node

γ = discouraging feedback of adjacent node

The node Consistency value rests between 0 and 1. If the node Consistency is higher than or identical to 0.5, next the node is believed to be the consistency node Otherwise it is an inconsistency.

4.3 Link Quality

The link quality (LQ) among two nodes is inversely proportional to the Bit Error Rate (BER). Here α is a constant time. The BER is measured by wireless channel interference and communication range. The link Quality computation is given below.

$$LQ = \alpha - \frac{1}{BER} \tag{3}$$

Residual Energy

RE is the remaining energy (RE) in a node at a present time and is a vital factor for choosing a node as a relay node. The remaining energy is computed by the equation 4.

$$\text{Remaining Energy=Preliminary Energy-present Energy} \tag{4}$$

4.4. Link Healthiness

The Link Healthiness (LH) inversely proportional to the rates at that route request (R_{REQ}) is transmitting via the nodes and route reply (R_{REP}) is yield as an acceptance for the representing R_{REQ}. It is evaluated by the Equation 5.

$$LH = \frac{1}{(Transmit_{R_{REQ}} \times Accept_{R_{REP}})} \tag{5}$$

4.5 Link Consistency based Validation

The link Consistency Factor validation is executed to evaluate the next relay election as component of the route finding itself and the majority link consistency nodes are validated and separated for the current hop to choose the next relay.

In order to confirm the expected values, we evaluate the next value both on the previous LCF value, current LCF value and then predict the expected LCF values. This value is denoted as a next value as indicated in figure the expected value is modified the table.

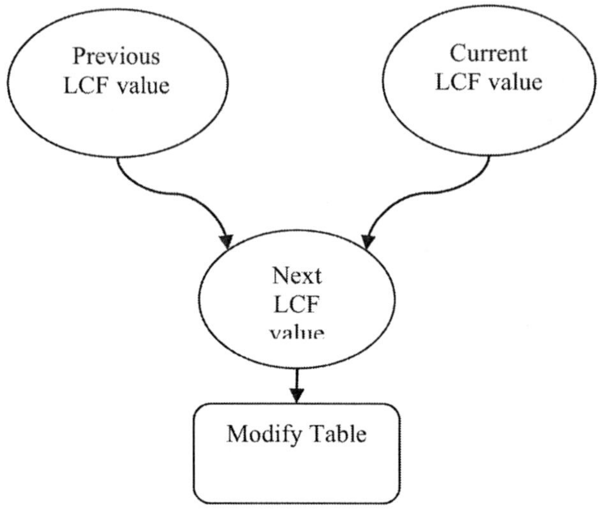

Figure 3. LCF Validation of expected value

This value is utilized for discover the next transmitting nodes from sensor to BS. Suppose, transmitting nodes LCF value is less than the threshold value during sensor node take another route to transmit the data.

5. Performance Evaluation

To execute simulation examination, we utilized the NS-2.35 that incorporates the 802.15.4 MAC protocol. This protocol is functioning on the 2.4 GHz physical layer and each sensor node transaction range is preset to 150 metre. The simulation of the proposed scheme has 30 nodes are deployed in the simulation area 500×500. In this simulation topology, the BS node plays an Owner of complete network as well as every sensor node function is sensing, data forwarding as well as obtaining the data. The packet traffic is yielded with constant bit rate (CBR). This simulation examination, we measured the following metrics namely Packet Obtained Rate, Latency, Loss Packet Rate, Average Remaining Energy, Throughput.

5.1. Packet Obtained Rate (POR)

It is defined as the ratio among the amount of data packets perfectly obtained by the BS and the amount of data packets yielded by the sensor nodes. This metric mutually constitutes the link reliability of the data gathering procedure.

$$\text{POR}=\frac{\sum_0^n \text{Packets Obtained}}{\text{Time}} \tag{6}$$

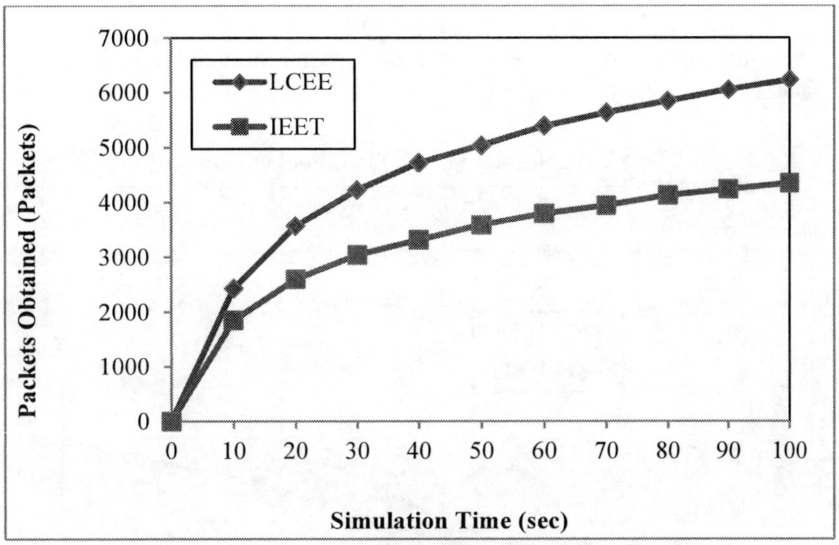

Figure 4. Packet Obtained Rate of IEET and LCEE

The POR of IEET and LCEE are diagrammed in Figure 4.6. It illustrates the POR of LCEE is better than the IEET for simulation time 100 seconds. There is an exciting enhance in obtaining the packet in LCEE. The LCEE method selects the relay node by link consistency factor. Thus, enhances the network POR in the system.

5.2. Packet Drop Rate (PDR)

PDR is denoted as the distinction among the forward packets and obtained packets in the communication MANET per particular time. PDR is measured by Equation 7.

$$\text{PDR}=\frac{\sum_{0}^{n}\text{Forward Packets - Obtained Packets}}{\text{Time}} \quad (7)$$

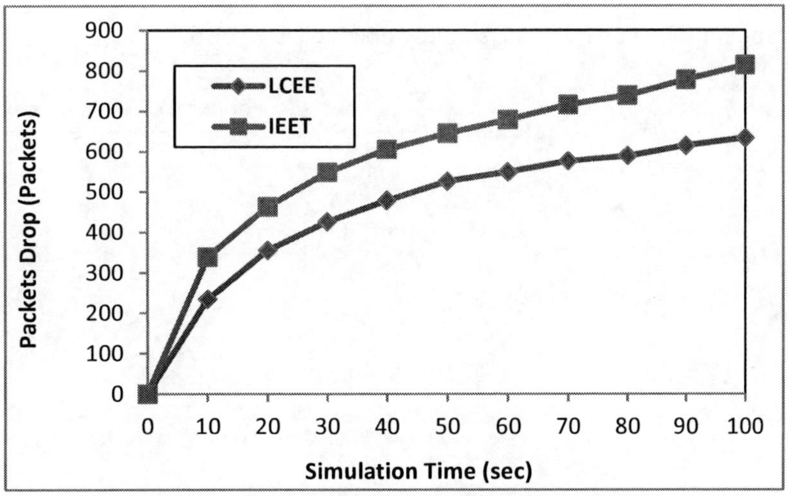

Figure 5. Packet Drop Rate of IEET and LCEE

From figure, 5 illustrate the PDR of LCEE as well as IEET. The LCEE method selects the relay node by link quality, link healthiness, and node consistency. The LCEE has smaller PDR in account of the improved function of the WSN.

5.3. Latency

It is defined as the time period from while the data packet communication is initiated at the source to while the data packet is perfectly obtained by the sink.

$$\text{Latency}=\frac{\sum_0^n(\text{Packet Obtained Time-Packet Forward Time})}{n} \quad (8)$$

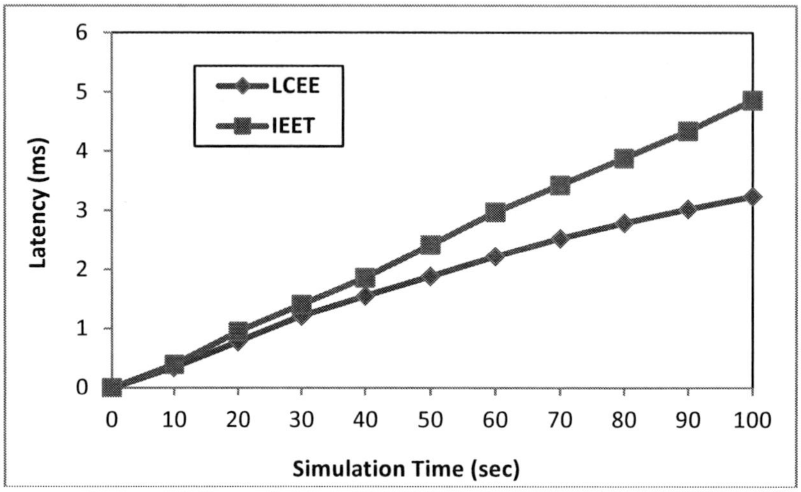

Figure 6. Latency of IEET and LCEE

The figure, 6 illustrates the latency of IEET as well as LCEE. It illustrates the latency of LCEE is lesser compared to the IEET. In LCEE, each sensor checked via node consistency factor thus chooses the relay node is a consistent. So, transmit the data to the destination with in a time period. Therefore, time delay is diminished when compared to the PPMA method. The diminution of latency is enhancing the WSN performance.

5.4. Throughput

It is the vital parameter for measuring the operations of network. In this protocol, the throughput is specified as the amount of data packets effectively obtained at the BS.

$$\text{Throughput}=\frac{\sum_0^n\text{Packets Received(n) * Packet size}}{\textit{Time}} \quad (9)$$

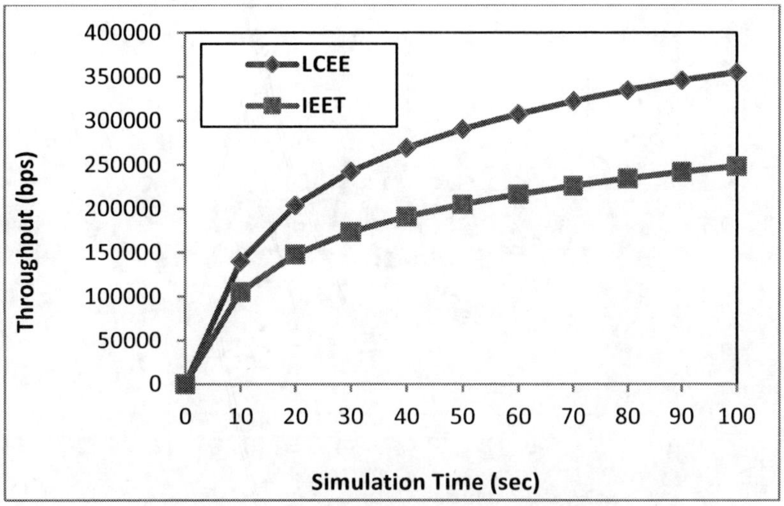

Figure 7. Throughput of IEET and LCEE

The figure 7 proves the LCEE scheme has higher throughput. Because of, IEET scheme concentrate only the energy efficiency. But, LCEE focuses on the link quality, node consistency, link healthiness and node remaining energy thus LCEE has provide better performance in WSN.

5.5. Residual Energy

It is defined as the whole energy exhausted by every sensor node divided by the amount of data packets perfectly distributed to the BS. This metric evaluates the energy efficiency of the WSN.

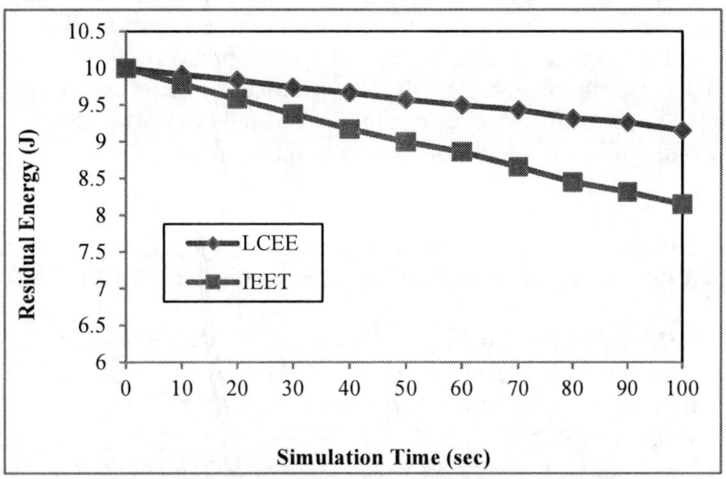

Figure 8. Residual Energy of IEET and LCEE

Figure 8 explains the residual energy of IEET as well as LCEE. The LCEE has highest residual energy since the energy is an important parameter during compute the link consistency factor. Thus LCEE raises the residual energy in the network.

5.6. Consistency Ratio

The consistency rate graph illustrates the average consistency values received for several nodes at the simulation time 100 second. This consistency ratio assists for selecting the best forwarder node.

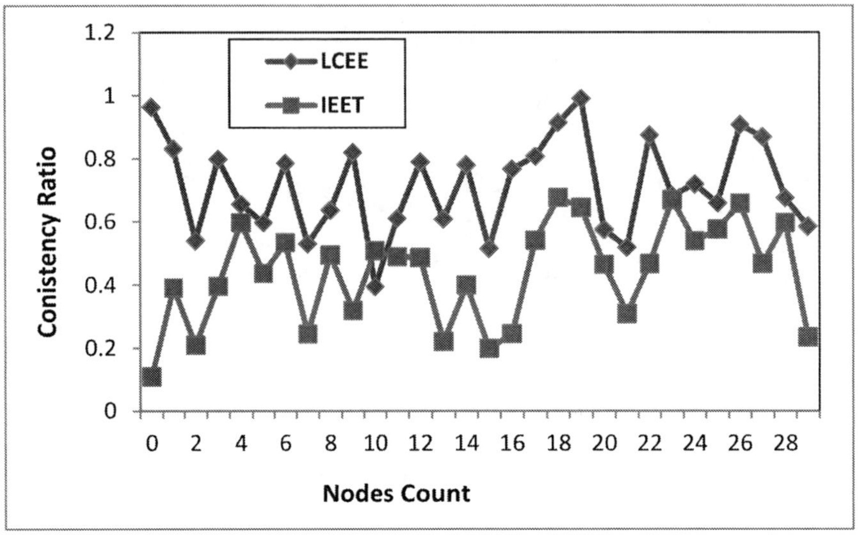

Figure 9. Consistency Ratio of IEET and LCEE

Figure 9 illustrates the Consistency Ratio of IEET and LCEE. It describes the LCEE scheme nodes are have highest consistency ratio. As a result, this node diminishes the packet losses in the network. But IEET scheme nodes have less consistency ratio.

6. Conclusion

In this protocol, we introduced a Link Consistency and Energy Efficiency for Improving Throughput in WSN. Here, the link consistency factor is used to measure the node link consistency. This LCF is computed by node consistency, Link quality, node remaining energy as well as Link healthiness. The introduced protocol fulfils the link consistency as well as additional energy saving than IEET based on our simulation results, correspondingly. Usually, this protocol provides the consistency of link as well as appreciably enhanced network performance. Simulation analysis proves that the LCEE enhanced the throughput as well as minimized the network latency in the network.

References

[1] Yildiz H U, Ciftler B S, Tavli B, Bicakci K and Incebacak D 2016 The impact of incomplete secure connectivity on the lifetime of wireless sensor networks *IEEE Systems Journal* **12** 1042-1046

[2] Petrov V, Lema MA, Gapeyenko M, Antonakoglou K, Moltchanov D, Sardis F, Samuylov A, Andreev S, Koucheryavy Y and Dohler M 2018 Achieving end-to-end reliability of mission-critical traffic in softwarized 5G networks *IEEE Journal on Selected Areas in Communications* **36** 485-501

[3] Quintero V L, Estevez C, Orchard M E and Pérez A 2018 Improvements of energy-efficient techniques in WSNs: a MAC-protocol approach *IEEE Communications Surveys & Tutorials* **21** 1188-1208

[4] Marinca D, Minet P and Hassine NB 2014 An efficient learning technique to predict link quality in WSN *IEEE 25th Annual Int. Symp. on Personal, Indoor, and Mobile Radio Communication (PIMRC)* 1564-1569

[5] Phatak T and Sawarkar SD 2016 Enhancing QoS of Wireless Sensor Network by detection of faulty sensor node *Int. Conf. on Computing, Analytics and Security Trends (CAST)* 100-105

[6] Bamatraf A, Abd Latiff MSB and Coulibaly Y 2017 Priority based QoS protocol for heterogeneous traffic routing in WSN *6th ICT Int. Student Project Conference (ICT-ISPC)* 1-5

[7] Chehri A and Mouftah HT 2012 Link adaptation-based optimization for wireless sensor networks routing protocol *26th Biennial Symposium on Communications (QBSC)* 142-145

[8] Mi X, Zhao H and Zhu J 2011 Research on EWMA based link quality evaluation algorithm for WSN *Proc. of 2011 Cross Strait Quad-Regional Radio Science and Wireless Technology Conference* **1** 757-759

[9] Deb M, Roy S, Saha B, Das P and Das M 2017 Designing a new link quality estimator for sensor nodes by combining available estimators *7th Int. Advance Computing Conf. (IACC)* 179-183

[10] Nayaka RJ and Biradar RC 2015 QoS analysis of WSN based cluster tree data fusion for integrated public utility management *IEEE Int. Advance Computing Conf. (IACC)* 579-584

[11] Priya D 2017 An energy efficient link stability based routing scheme for wireless sensor networks *Int. Conf. on Communication and Signal Processing (ICCSP)* 1828-1832

[12] Horvat G, Vlaović J and Žagar D 2015 Improving QoS in query-driven WSN using a cross-layered handover algorithm *4th Mediterranean Conf. on Embedded Computing* 236-239

[13] Ddin WU and Khan BM 2014 Improved quality of service in wireless sensor Network with mobile sink *17th IEE Int. Multi Topic Conf.* 36-40

[14] Hortos WS 2012 Effects of energy harvesting on quality-of-service in transient wireless sensor networks *IEEE Military Communications Conference* 1-9

[15] Keerthi Anand VD 2015 Power Efficient Multicast Opportunistic Routing Protocol (Pemor) To Optimize Lifetime Of Manet. *International Journal of MC Square Scientific Research* **7** 109-127

[16] Hemanth Kumar G, Ramesh GP 2019 Reducing power feasting and extend network life time of IoT devices through localization *IJAST* **28** 297-305

[17] Suárez C, Gaona P, Montenegro C and Parra J 2018 IOT quality of service based in link channel optimization in Wireless Sensor Networks *IEEE Int. Conf. on Smart Internet of Things (SmartIoT)* 172-177

[18] Sun W, Yuan X, Wang J, Li Q, Chen L and Mu D 2017 End-to-end data delivery reliability model for estimating and optimizing the link quality of industrial WSNs *IEEE Transactions on Automation Science and Engineering* **15** 1127-1137

Forecast Function Based Congestion Control in MANET Routing

Suresh G[1], Senthil Kumar[2], V Kavitha[3] and S Lekashri[4]

[1] Department of CSE, Kings Engineering College, Chennai, India.
[2,3,4] Department of ECE, Kings Engineering College, Chennai, India.
[1]suresh@kingsedu.ac.in, [2]senthilkumar@kingsedu.ac.in,
[3]Chennaikavitha@kingsedu.ac.in, [4]lekashri@kingsedu.ac.in

Abstract. In a MANET, there is not necessitating arrange any structure to create nodes to transmit with every other. The recurrent updates of the topography as well as the mutual scenery of the wireless channel create important disputes. Hence, several congestion control methods have been introduced. MANETs have a numerous of approaches to encourage the routing function through inactive as well as active routing schemes. Though, there is not an enduring resolution which subsists to resolve the fundamental congestion trouble in the MANETs. This paper proposes a Forecast function based Congestion Control in MANET Routing (FFCC). In this scheme, the congestion rate is applied for measuring the highest congested node and isolates that node. Next we choose the less congested node thus the data reach the destination correctly. In addition, the Forecasts Congestion Rate is used to predict the congested node perfectly based on the five transactions. Thus transmits the data via less congested route efficiently. Network simulator is applied for examine the simulation experiments. The results demonstrate the minimized network delay as enhances the network throughput

Keywords: MANET, Congestion Rate Forecasts Congestion Rate, traffic, healthcare, throughput.

1. Introduction

The MANETs are typically distinct as a self-governing system of nodes associated by wireless connections as well as transmitting in a multi-hop manner. The essential jobs that an ad hoc network must execute is congestion control as well as its major goal is to boundary the delay also buffer run over induced through the network congestion also it offers better function [1].

The production of novel wireless forces on a concentrated foundation has produced a growing require for strength, capability as well as throughput. Additionally, amount of restraints to function accurately are compulsory. Besides to bandwidth necessities, these restraints can be articulated include packet drop rate, latency, jitter or energy utilization [2].

For example, healthcare system requires continuous monitoring with both periodic updates as well as emergency messages flowing through the network. The major problem observed in such types of situations is the congestion occurring in the network causing unnecessary delays and hazardous effects. The causes for the occurrence of traffic congestion and methods adaptable to mitigate the traffic [3].

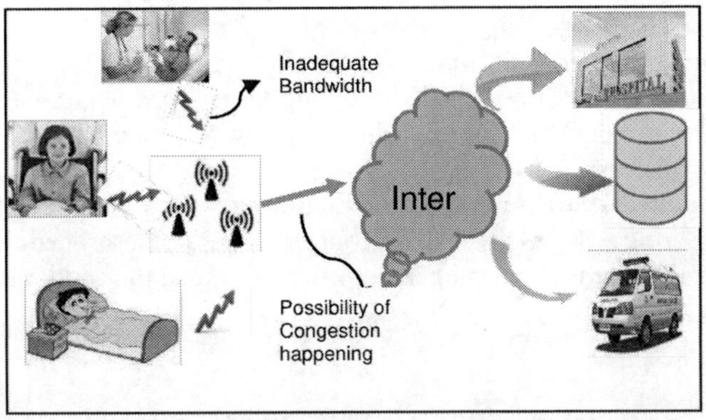

Figure 1. illustration of Congestion Occur in Healthcare System

Congestion may appear to be a very simple issue among the networks because the delayed packets ultimately try to deliver within the allocated time period. However, healthcare applications require timely responses from the site under test to avoid the entire breakdown of the network. Figure.1 indicates the illustration of Congestion occur in Healthcare system [4].

In MANET, there are numerous sources of Congestion for example communications Competitors, packet confrontation as well as overflow of buffers. Between the troubles induced through congestion, such as delay details, packets drops which occasionally restrain details as well as extreme bandwidth [5]. These are disgraces the function of MANET. This difficulty prompts necessitate to team congestion control approach to enhance the network function.

2. Related Works

Several authors have introduced approaches depend on reactive routing that are congestion adaptative as well as compact with the congestion throughout the network. The combined procedure minimizes the energy utilization also diminishes the amount of packets being missed, hence getting better the function [6]. Inadequate bandwidth as well as altering topography affects the trouble of Congestion. Observing Congestion is hard in MANET, because there might be some causes following reducing of packets. In this network, unsuccessful release should be owing to the path break also the nodes movement. Mobile Clustering enhances the received packet rate as well as diminishing the network delay [7].
Congestion Control is additional significant dispute when leading for QoS attainment precisely with extremely movable mobile stations. The congestion control effects introduce a raised traffic method in Transmission Control Protocol (TCP) utilizing the token bucket traffic shaping method through packet transmitting at the middle nodes [8]. AD-hoc Transmission Control Protocol, the receiver notices the feasible current condition as well as communicates this detail to the sender as response [9].
Several operation factors are used detect as well as defeat the non-congestion drops [10]. The cooperative attempt of mobile nodes, lacking any preset structure. Transaction between the mobile nodes is a single hop otherwise multi-hop manner. Because of their active topography as well as imperfect bandwidth, preserving guarantee facility is a tricky job. It believes channel free period as well as obtainable bandwidths when choosing route [11].
Framework and implementation agent is applied for controlling (FIAC) the congestion in the network. This technique, the nodes are categorized into four classes depend on its traffic. It measures acquire in extent of the numerous traffic modules as well as the channel difference also approximates the congestion parameter to discover the least congestion level in the network. This parameter is utilized for electing the least congested path [12]. Agent based congestion control method that the details about congestion is gathered as well as disseminated by mobile agents. This method is utilized for choosing the least congested path [13].

IOP Conf. Series: Materials Science and Engineering **925** (2020) 012074 doi:10.1088/1757-899X/925/1/012074

3. Forecast function based Congestion Control in MANET Routing

The aim of Forecast function based Congestion Control in MANET Routing applied for isolate the more congested nodes and picked the less congested route in the network. Here, isolates the congested node based on the congestion rate (CR) and forecast the after few communication achievement rates (CAR) to enhance the network function.

Figure 2 illustrates the network structure in which numerous nodes are there. The nodes are arbitrarily distributed and moving arbitrarily in X×Y size. Routing is an active procedure assembly the essential parameters which will facilitate in the arrangement of routes since the sender to the receiver.

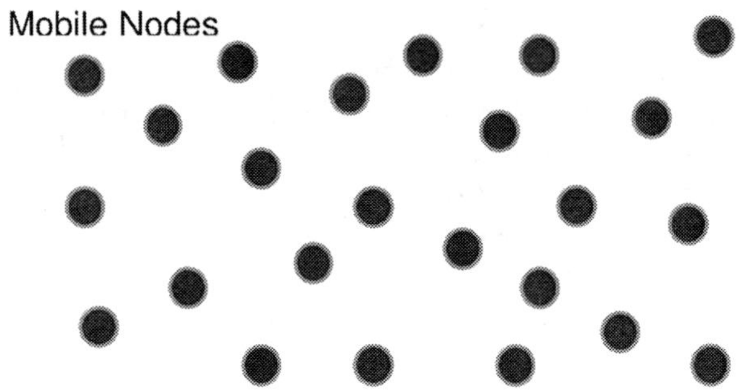

Figure 2. Network Structure

3.1. Congestion Rate Evaluation

The auto-functioning method is used to categorize the nodes as a CR. A CR is evaluating for every node throughout the routing procedure to mutually measure the present level of congestion also forecast the congestion level in the further some communications. A CR is considered via the incoming as well as outgoing packets for every node vigorously. This CR assists find out whether the mobile node is overcrowding or not. The more congested node has the maximum CR throughout the assessment procedure. These nodes are hence inaccessible from the routes in that way. The other usable mobile nodes are threaded into the data communication paths through that data is sent. The CR of node i is calculated for applying the expression 1 below.

$$CR_i = \frac{incoming\ pkts_i}{Outgoing\ pkts_i} \tag{1}$$

Here, pkts represents the packets. The present incoming as well as outgoing packets are a middling of the previous few communications. Thus it offers additional perfect incoming as well as outgoing packets. The node with the lowest CFR value is chosen as the next hop and data is transferred to that particular node. The CR_{Th} rate is set via conceiving the CR rates of entire mobile nodes here in the MANET. The computation of CR_{Th} is specified in the expression 2.

$$CR_{Th} = \frac{\sum_{i=0}^{n} CR_n}{Sum\, of\ mobile\, nodes} \tag{2}$$

Where, n represents the total number of mobile nodes.

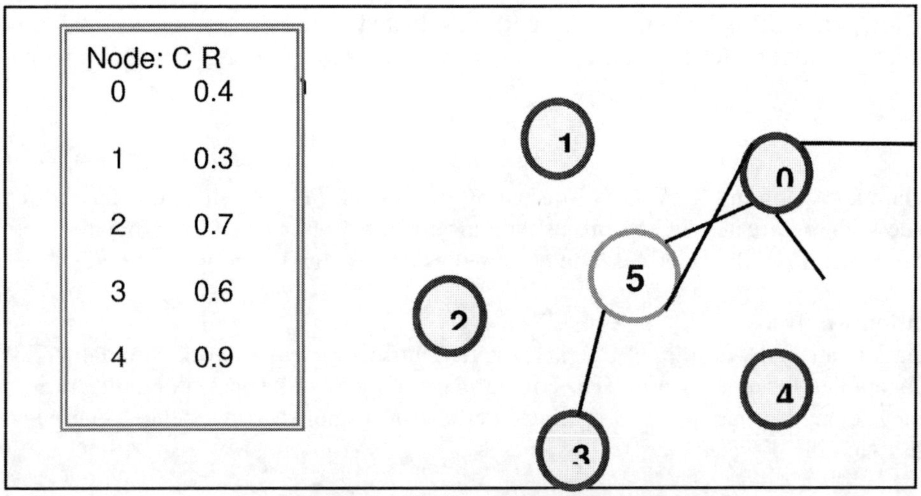

Figure 3. CR Evaluation in MANET

Figure 3 illustrates the CR evaluation in MANET. Here, the node 5 is a source node and its neighbour nodes are 0, 1,2,3,4. These nodes CR rates are 0.4, 0.3, 0.7, 0.6, and 0.9. The neighbouring nodes estimate the CFR using their own current incoming packets and outgoing packets. The nodes 4 and 2 CR rates are greater than the CR_{th}. Thus, these nodes are not selecting the relay node. Here, the 1 is a lowest CR than the CR_{th} value therefore that node is elected as a relay node. The series of procedures accepting position in the algorithm therefore is given in below.

3.2. Algorithm

For {n = 0 to h} {
  Launch $CR_h = 0$
}
While {Sender! =Destination}
  Evaluate present incoming packet
  Evaluate present outgoing packet

$$CR = \frac{Incoming\ pkts}{Outgoing\ pkts}$$

  Separate the node with CR > CR_{th}
  Discover adjacent node with CR < CR_{th}
  Transmit the data
  Eliminate high CR nodes from path
  Attend step four till data achieves receiver
} stop process

The Congestion Rate Evaluation not only discovers the present CR value however in addition forecasts the following CR as Forecasts Congestion Rate (FCR) as well as it disseminates the most horrible case CR node to every other mobile node. In this approach, the node has both the highest CR as well as the highest FCR are separate from the paths. The standard deviation process is applied to calculate the CR. Standard value of the previous five outgoing communications ($0 < h \leq 5$) is taken as an outgoing packets. The standard deviation of the previous five CR from the outgoing packet is evaluated as the σ_{CR}. Here, h indicates the neighbour nodes. The expression applied to compute σ_{CR} is below.

$$\sigma_{CR} = \sqrt{\frac{1}{h}\sum_{h=0}^{h}(CR_h - Outgoing\ packets)} \quad (3)$$

Thus the forecasting CR is known as the FCR is evaluated as provide in the expression (4), with the alternative preservative if there is a rising packet rate subtractive in casing of a declining packet rate.

$$FCR_i = FCR_i \pm \sigma_{FCR_i} \tag{4}$$

Utilizing the expression in (4) FCRi is forecast by the sender end as well as the sender not only takes out the node with the greatest FCR from its path arrangement but as well disseminates the node with a extremely elevated FCR thus alertness the other nodes concerning the congested node.

4. Simulation Analysis

The simulation analysis of the FFCC is executed utilizing the network simulator-2.35 which is expansively applied for investigation. The comparisons of factors like packet obtained ratio, packet drop ratio, latency, delay, throughput and routing overhead to establish competence. The packet traffic is yielded with variable bit rate.

4.1. Packet Obtained Ratio

POR is denoting as the ratio of packets received at receiver per particular time. POR is assessed by Equation 5. Where, n = Node count

$$\text{POR}=\frac{\sum_0^n \text{Packets Obtained}}{\text{Time}} \tag{5}$$

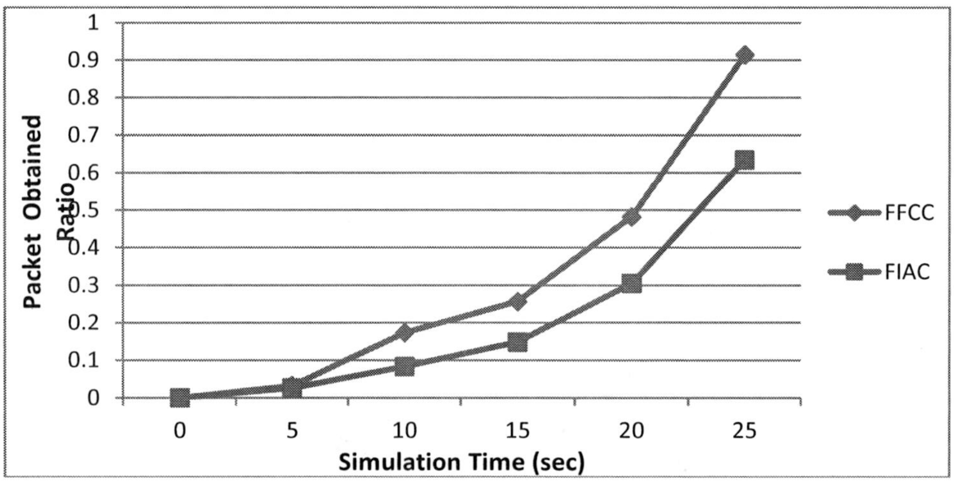

Figure 4. Packet Obtained Ratio of FIAC and FFCC

The POR of FIAC and FFCC are diagrammed in Figure 4. It illustrates that the proposed scheme FFCC has better POR when compared to the existing strategy FIAC.

4.2. Packet Drop Ratio

PDR is denoted as the ratio of distinction among the forward packets and obtained packets in the communication MANET per particular time. PDR is measured by Equation 6.

$$\text{PDR}=\frac{\sum_0^n \text{Forward Packets - Obtained Packets}}{\text{Time}} \tag{6}$$

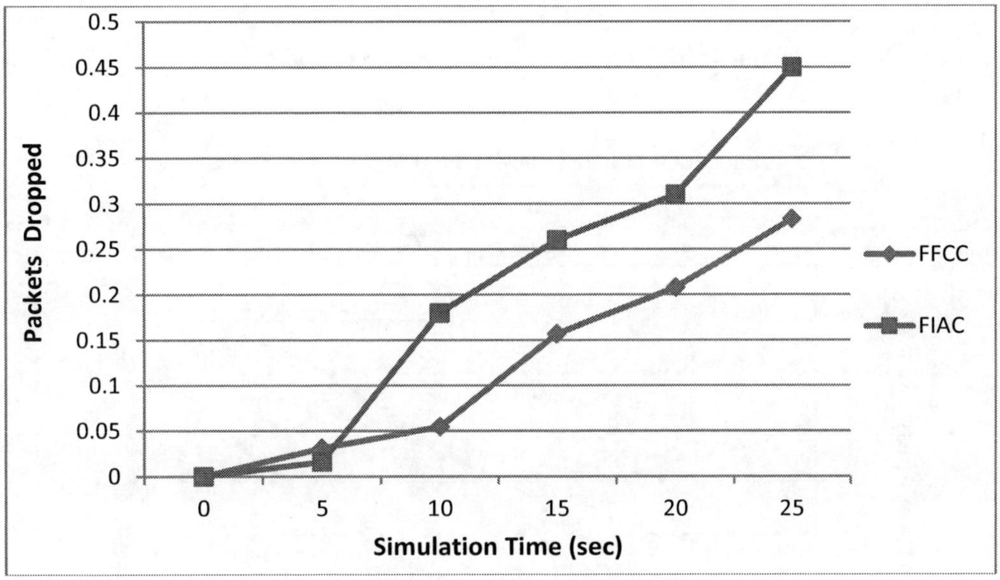

Figure 5. Packet Dropped Ratio of FIAC and FFCC

Figure 5 shows the PDR values obtained from the simulation analysis of FIAC and FFCC. It indicates that PDR of FIAC is higher when compared with FFCC.

4.3. Throughput

Throughput denotes to an entire number of packets successfully delivered across network per unit time. Throughput is obtained using Equation 7

$$\text{Throughput}=\frac{\sum_{0}^{n}\text{Packets Received(n)} * \text{Packet size}}{Time} \quad (7)$$

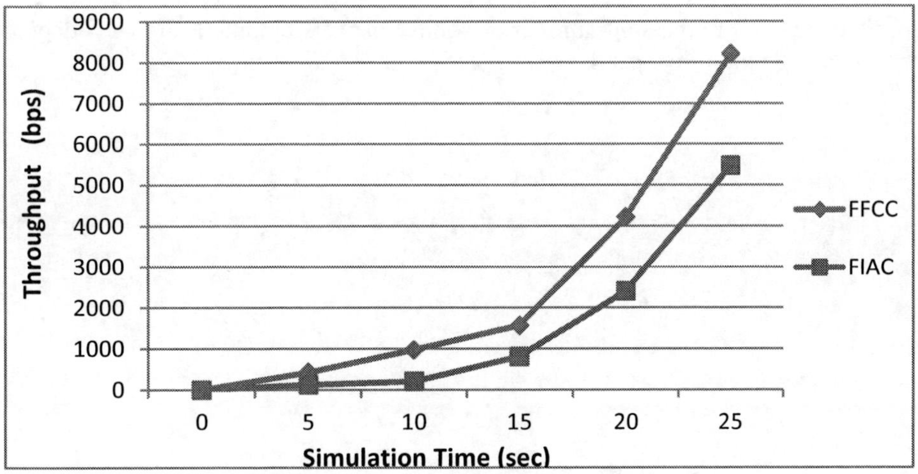

Figure 6. Throughput of FIAC and FFCC

Figure 6 indicate the outcomes for throughput execution of existing strategy FIAC and FFCC versus VBR traffic. It is mentioned that the FFCC approach has highest throughput while equated to the existing strategy FIAC.

4.4. Delay Analysis

Delay is represented as the time period difference among data sent and packets obtained. It is measured by Equation 8.

$$\text{Delay}=\frac{\sum_{0}^{n}(\text{Packet Obtained Time-Packet Forward Time})}{n} \quad (8)$$

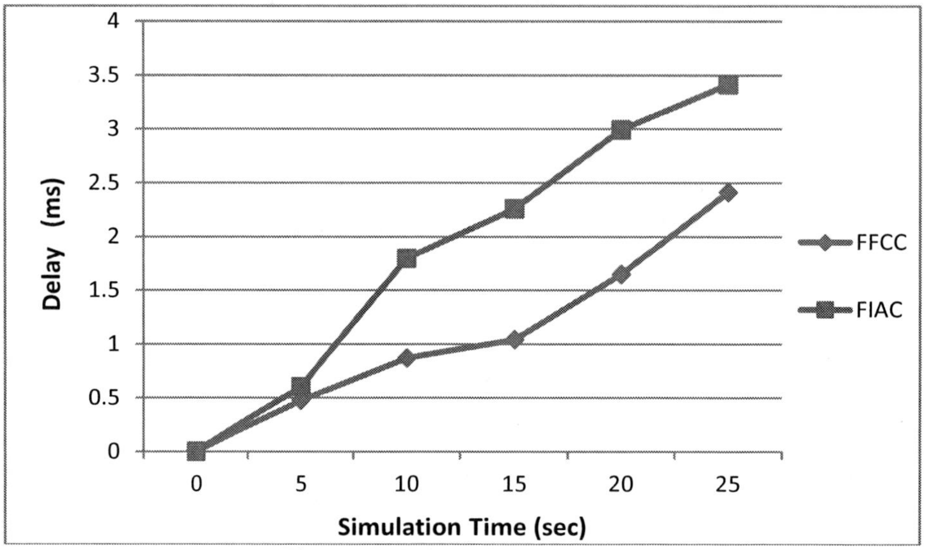

Figure 7. Delay of FIAC and FFCC

Figure 7 demonstrate the execution estimation for average delay for FFCC as well as existing strategy FIAC. It is mentioned that the FFCC approach has less delay while equated to the existing strategy FIAC. The FFCC obtains better execution owing to it select the path is less congestion.

4.5. Overhead Analysis

Routing Overhead is depicting the sum number of control packets communicated for yielding a data route per data packet. It is received through computing the ratio among the sum numbers of control packets transmit to the sum number of control packets obtained. Figure 8 depict the routing overhead based on node mobility speed.

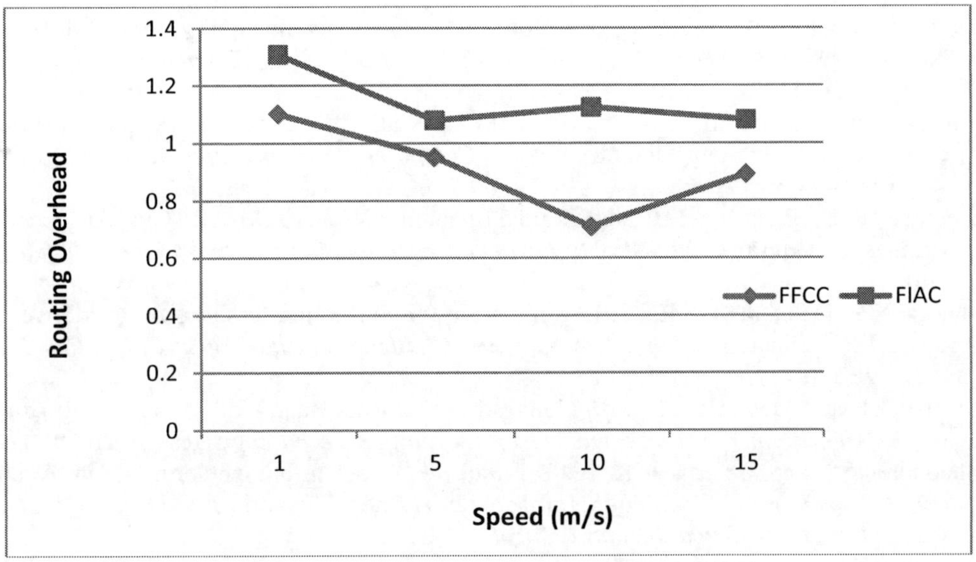

Figure 8. Routing Overhead of FIAC and FFCC

Figure 8 illustrates the proposed strategy FFCC has lowest routing overhead than the existing strategy FIAC. Since, the FFCC choose the relay is a lesser congestion rate. So, avoid the congestion in the network. But, FIAC increase the routing overhead because of it does not select the congestion aware routing in the MANET.

5. Conclusion

In this paper, we proposed Forecast function based Congestion Control in MANET Routing directing to minimize the network congestion and minimize the network delay. In this scheme, the congestion rate is applied for measuring the highest congested node and isolates that node. Next we choose the less congested node thus the data reach the destination correctly. In addition, the Forecasts Congestion Rate is used to predict the congested node perfectly based on the five transaction. Thus the source transmits the data via less congested route efficiently. The simulation results illustrate that the FFCC offer better packet obtained rate as well as minimizes the network delay.

References

[1] Gupta H and Pandey P 2013 Survey of routing base congestion control techniques under MANET *IEEE Int. Conf. on Emerging Trends in Computing, Communication and Nanotechnology (ICECCN)* 241-244.

[2] Sharma N, Gupta A, Rajput SS and Yadav VK 2016 Congestion control techniques in MANET: a survey *2nd Int. Conf. on Computational Intelligence & Communication Technology (CICT)* 280-282

[3] Keerthi Anand VD 2015 Power Efficient Multicast Opportunistic Routing Protocol (Pemor) To Optimize Lifetime of Manet *Int. Journal of MC Square Scientific Research* **7** 109-127

[4] Rao EJ, Ramanjaneyulu T and Kumar KJ 2018 Advanced Multiplier Design and Implementation using Hancarlson Adder *Int. Conf. on Intelligent and Innovative Computing Applications (ICONIC)* 1-5

[5] Bhatia B 2015 Performance analysis of AODV based congestion control protocols in MANET *Int. Conf. on Futuristic Trends on Computational Analysis and Knowledge Management (ABLAZE)* 453-458

[6] Janhavi V and Yuvaraju BN 2014 A unified method to solve the energy depletion and congestion problem using mobile relay in manets *7th Joint Int. Information Technology and Artificial Intelligence Conference* 330-335

[7] Yadav S and Firdaus T 2019 Congestion Aware Routing in AODV based Mobile Ad-Hoc Network (MANET) *Int. Conf. on Computing, Power and Communication Technologies (GUCON)* 452-457
[8] Rath M, Rout UP, Pujari N, Nanda SK and Panda SP 2017 Congestion control mechanism for real time traffic in mobile adhoc networks *Computer communication, networking and internet security* 149-156
[9] Sreenivas BC, Prakash G B and Ramakrishnan KV 2013 M-ADTCP: An Approach for Congestion Control in MANET *In Advances in Computing and Information Technology* 531-540
[10] Sharma N and Chakrawarti RK 2014 Simulation for congestion-less losses control over MANET using TCP scheme *Int. Conf. on Issues and Challenges in Intelligent Computing Techniques (ICICT)* 410-415
[11] Dsouza MB and Manjaiah DH 2019 Congestion Free And Bandwidth Aware Multipath Protocol f or MANET *1st Int. Conf. on Advances in Information Technology (ICAIT)* 267-270
[12] Bhadauria SS and Sharma VK 2011 Framework and Implimentation of an Agent Based Congestion Control Technique for Mobile Ad-hoc Network *Int. Conf. on Advances in Computing, Communication and Control* 318-327
[13] Sharma VK and Bhadauria SS 2011 Agent based congestion control routing for mobile ad-hoc network *In Trends in network and communications* 324-333

Elliptic Curve Digital Signature technique based Abnormal Node Detection in Wireless AD HOC Networks

P Deepalakshmi[1] and T Kumanan[2]

[1,2] Department of Computer Science and Engineering, Meenakshi Academy of Higher Education and Research Chennai, India
[1]deepa.mca85@gmail.com

Abstract. This Security, as well as routing Performance, is two significant operations of Wireless ad-hoc networks (WANETs). Even though, several efforts to enhance its secure routing scheme of WANETs continue susceptible to attacks. Contrasting the majority of conventional solutions which avoid the precise problems, this scheme inclines to distinguish the misconduct as well as recognize the abnormal nodes in WANETs. In this paper, the Elliptic Curve Digital Signature technique (ECDST) is used to detect the abnormal node in the WANET. This technique function recognizes the abnormal node as well as isolate that node them from the route. During route discovery, the ECC method utilizing the Weierstrass Elliptic functions for detecting the abnormal node. The abnormal nodes are discovered by signature verification technique on data communication. It attains consistency in routing with disqualifying the link by the utmost weight as well as validates the nodes. This scheme assures node availability, integrity, as well as improves network performance. The observational analyses illustrate that the introduced scheme can notice the abnormal as well as a compromised node in the network.

Keywords: Wireless Ad-Hoc Networks, Reliability, The Weierstrass Elliptic Function, Elliptical Curve Digital Signature Technique, Security, Abnormal Node Detection.

1. Introduction

WANET is a self-structured multi-hop system constituted through several mobile wireless nodes with peer-to-peer association. WANET cannot have present communications as well as wireless nodes that travel robustly lacking edge restrictions [1]. Benefits are speedy to establish, allowance fault acceptance, association as well as movement. Though, the WANET's disputes are active topography, unlock medium, embarrassed bandwidth, changeable capacity connections, energy-restrained function as well as imperfect physical security [2]. The absent of central administration as well as undependable wireless connection, susceptibility, safety coercion conducts to enhance the security. It routinely persists from attacks by self-interested otherwise abnormal nodes. For instance, irregularity actions attack, on-off attack, black hole attack, badness mouthing attack, grey hole attack and so on [3].

To overcome the above difficulty, we pertain an Elliptic Curve Digital Signature technique (ECDST) is used to detect the abnormal node in the WANET. This technique function recognizes the abnormal node as well as isolate that node them from the route. During route discovery, the ECC method utilizing the Weierstrass Elliptic functions for detecting the abnormal node. The abnormal nodes are discovered by

signature verification technique on data communication. It attains consistency in routing with disqualifying the link by the utmost weight as well as validates the nodes.
The stay of this work is prepared as follows. In Section 2, the related work is surveyed. The ECDST is depicted in Section 3. In Section 4, present performance analyses. At last, finish in Section 5.

2. Related works

First, Energy awake trust scheme [4] is provide sufficient security as maintain overhead by utilizing a game theoretic technique. This technique is functional to the trust origin method to diminish the overhead. Although this scheme enhances the overhead of routing in the network. Key distribution scheme adapt secure property by giving pair wise keys among nodes. Secure Connectivity scheme inquire the results of key distribution possibility on route length, queue size, and energy looseness while minimal cost routing is employed [5]. Authentication enables a node to ensure the identity of peer node. Authentication is essentially assured that participants in communication are authenticated and not impersonators. Authenticity is ensured because only the legitimate sender can produce a message that will decrypt properly with the shared key [6].
A trust based security protocol that reaches authentication of data packets. In this scheme, every node keeping trust counter for observing and separating the malevolent node in the network [7]. However, this scheme cannot provide confidential data transmission in the network. A key agreement scheme based on public discussion was described that attain a convinced secret key range while an opponent has more favourable situation. A secret key agreement protocol introduced the reliable relay regarded in the discourse to attain both reward distillation and secrecy elaboration [8].
Throughput is enhanced with data-aided valuation method by means of transmit choice, channel then transmit obligation [9]. Masquerading Attack detection system (MADS) using the received signal-strength (RSS). In this method, the Signal strength variation is used to detect the attack accurately. The RSS is a lightweight cryptographic certification that as well does not necessitate extra charge to the present wireless technology. Although, the RSS is not quite authentic since it changes as well as fluctuates over time owing to several elements, for example multipath passing, deflection, etc. [10].
Numerous key management approaches inserted to initiate symmetric keys [11]. This approach applying arbitrary seed dissemination with transient master key which offer pair wise keys. This approach discovers the main disadvantage is providing less security.

3. Proposed systems

In this approach, the ECDST is applied for recognizing abnormal nodes in WANET. It contains normal nodes, as well as abnormal nodes. Figure 1 illustrates the network Structure of approach Scheme. This approach comprises three parts such as generation of key, generation of signature, also verification of signature.

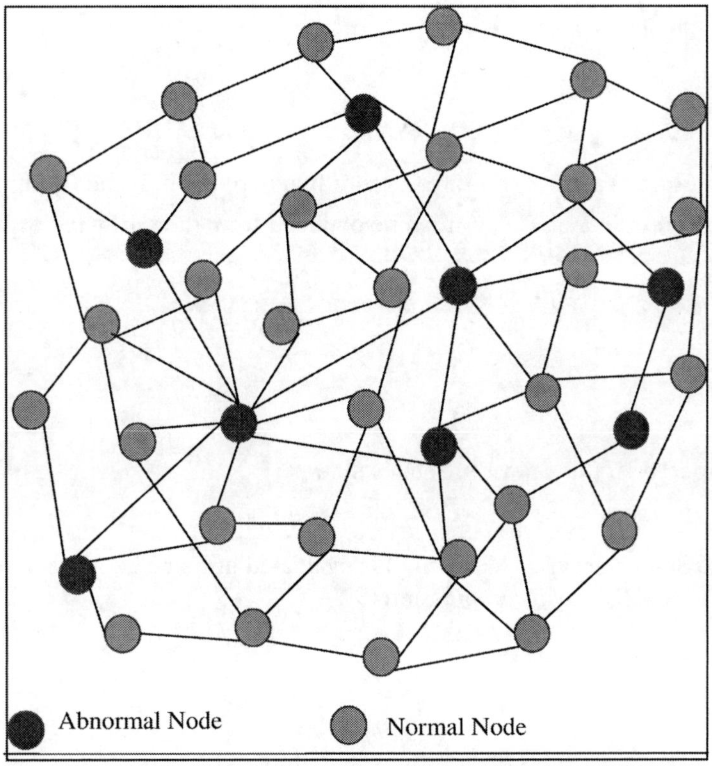

Figure 1. Network Structure of ECDST approach

3.1. Secure Route Discovery

The ECC method is utilizing the Weierstrass Elliptic purpose. Conceive the co-ordinate points of the sender as well as the Next relay node R be E as well as F correspondingly; there is another point G that makes a straight line as exemplified in the figure 2.

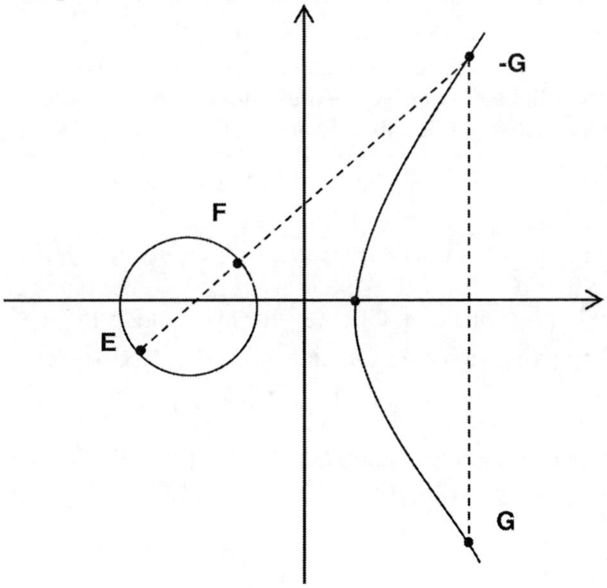

Figure 2. Elliptic Curve

The Weierstrass Elliptic procedure is definite in (3),

$$y^2 = x^3 + lx + m \quad (3)$$

$$E + F = G \quad \text{Where } E \neq F \text{ and } \forall E, F \in G \quad (4)$$

Where (x_E, y_E) and (x_F, y_F) and (x_G, y_G) are the coordinates of the E, F and G points forming an elliptic curve. Therefore, the co-ordinates (x_G, y_G) can be obtained from the following expressions,

$$x_G = \beta^2 - x_E - x_F \quad (5)$$

$$y_G = \beta(x_E - x_G) - y_E \quad (6)$$

where, $\eta = \dfrac{y_F - y_E}{x_F - x_E}$

The independent property of this procedure states that,

$$E + F = F + G \quad (7)$$

For the moment, the Source generated key (S_{key}) is evaluated utilizing expression (8) and the next relay node responds to with the R_{key} by the expression (9).

$$S_{key} = E + F \quad (8)$$

$$R_{key} = F + G \quad (9)$$

During route discovery, the source S equates its S_{key} with the R_{key} to finish that the next relay node is an abnormal node. When S_{key} is not equal as R_{key} next the node is available as an abnormal and it broadcast the whole network.

3.2. Generation of key Part

The ECDS factors are depicted as: T = (Fp, EC, C, H, m, k, R), where Fp is the finite field, EC represents the curve of elliptic, C denotes the point of Creator, m is the order of C a great prime number as well as k is coefficient factor. This approach applies a hash function H.

In addition, data communication, the source checks the abnormal based on the ECDS confirmation method. If the node success the signature verification next it communicates the data to the destination otherwise transmit the abnormal node details to whole network as well as eliminate that node from the WANET.

3.3. Generation of signature Part

The source picked out an arbitrary number A_s among 1 to m − 1 that represents a private key PR_k as well as it Calculates its public key PU_k = C. A_s. For every node j, the source creates an arbitrary number γ among $1 < \gamma < m - 1$. It calculates public key for every node PU_j = C. γ_j. The source builds a signature for every node applying signature function $s_j(\gamma_j, ID, A_s)$ is given below.

$$S_j = \gamma_j^{-1}[\gamma_j.k(\text{ID}) + \text{A}_s] \quad (10)$$

Every node is payload with its precise signature S_j public credential of source A_s, as well as public credential of every node γ_j is prepared public to the WANET.

3.4. Signature verification part

This part verifies through the node to make sure which the signature sj concerns to it as well as has been accurately yielded. The node applies its node identity, public credential, as well as the public credential of the source to confirm the accuracy of the node j signature.

$$S_j = \gamma_j^{-1}[\gamma_j.k(\mathrm{ID}) + \mathrm{A_s}] \tag{11}$$

$\gamma_j = S_j^{-1}[\gamma_j.k(\mathrm{ID}) + \mathrm{A_s}]$ Reproduce all part with C

$$C \cdot \gamma_j = C \cdot S_j^{-1}[\gamma_j \cdot k(\mathrm{ID}) + \mathrm{A_s}]$$

$$S_j^{-1} C \cdot \gamma_j \cdot k(\mathrm{ID}) + S_j^{-1} C \cdot A_s$$

$C \cdot \gamma_j = S_j^{-1} \cdot k(\mathrm{ID})$. Public certificate of node j+ sj−1 and public credential of the source and Node j confirms its signature value Sj as certain at (1) by calculating the RHS significance of expression (11) also calculating it with LHS. If the expression (11) is fulfilled next the signature of node j is confirmed. This process is attempted through the other nodes in a same way.

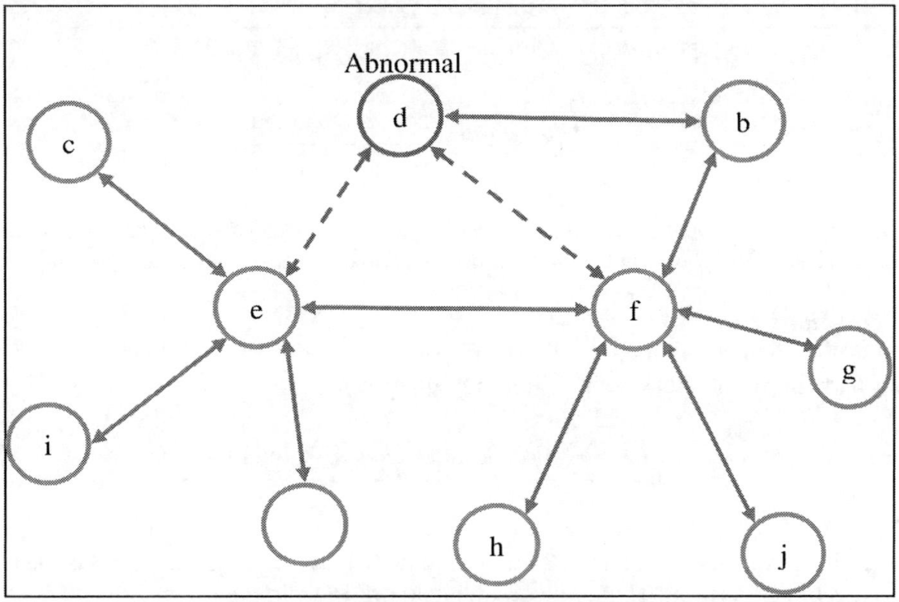

Figure 3. ECDS Confirmation method

The Figure 3 illustrates the ECDS confirmation between nodes. Here, the signatures of source as well as entire nodes are verified, except the signature of node d. Then, the source disseminates the notification message to WANET. This double checking process accurately detects the abnormal node in the WANET.

4. Performance Evaluation

To execute simulation examination, we utilized the NS-2.35 that incorporates the 802.11 MAC protocol. The packet traffic is yielded with constant bit rate. This simulation examination, we measured the following metrics namely Packet Obtained Rate, Latency, Loss Packet Rate, Average Remaining Energy, Throughput as well as detection ratio.

4.1. Packet Obtained Rate(POR)

It is defined as the ratio among the amount of data packets perfectly obtained by the receiver and the amount of data packets yielded by entire mobile nodes.

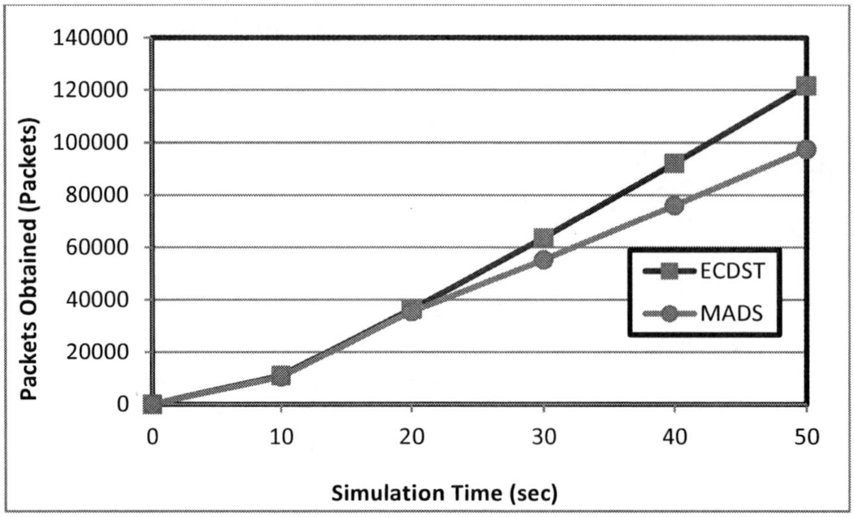

Figure 4. Packet Obtained Rate of ECDST and MADS

$$POR=\frac{\sum_0^n \text{Packets Obtained}}{\text{Time}} \qquad (12)$$

From this figure 4, the proposed method ECDST is greater compare to the existing method MADS. Since, the ECDST pick the route by the normal nodes. Thus, enhances the routing efficiency.

4.2. Packet Drop Rate(PDR)

It is denoted as the distinction among the forward packets and obtained packets in the communication WANET per particular time. PDR is measured by Equation 13.

$$PDR=\frac{\sum_0^n \text{Forward Packets - Obtained Packets}}{\text{Time}} \qquad (13)$$

From figure, 5 illustrate the PDR of ECDST and MADS. The ECDST method selects the relay node by signature verification algorithm. Thus, reduces the network PDR in the system.

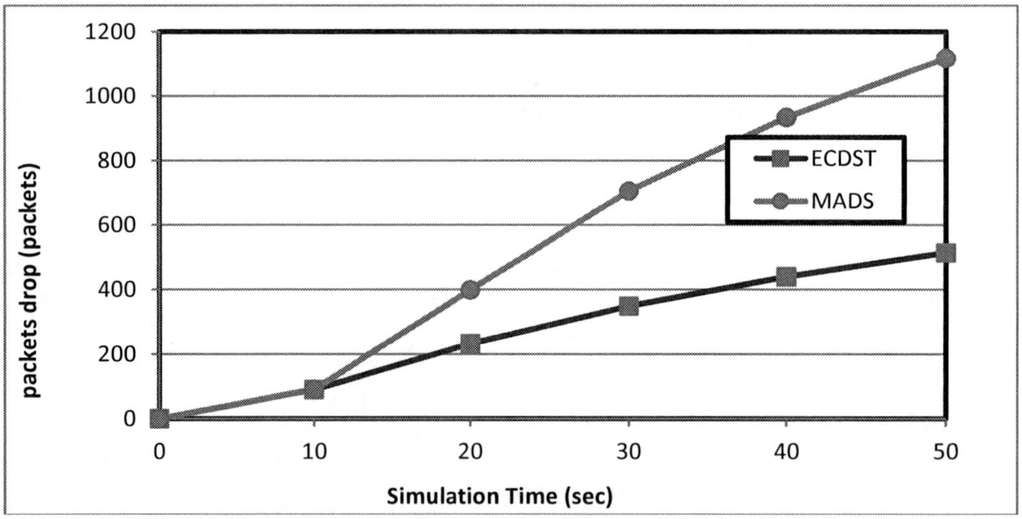

Figure 5. Packet Drop Rate of ECDST and MADS

4.3. Packet Drop Rate(PDR)

It is defined as the time period from while the data packet communication is initiated at the source to while the data packet is perfectly obtained by the destination. We evaluated both its usual value and its allotment.

$$\text{Latency}=\frac{\sum_{0}^{n}(\text{Packet Obtained Time-Packet Forward Time})}{n} \tag{14}$$

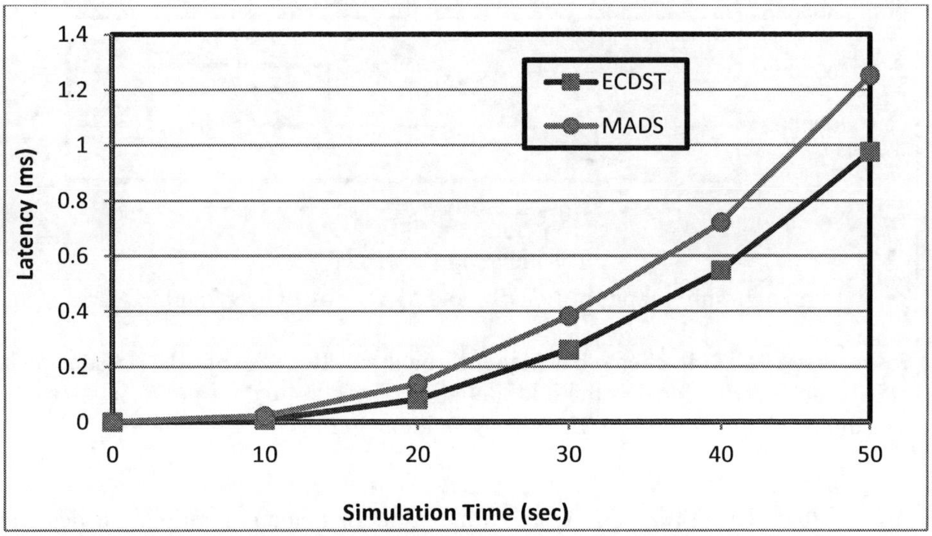

Figure 6. Latency of ECDST and MADS

The figure, 6 illustrates the latency of ECDST and MADS. The data traverse via link normal behavior nodes. Thus minimizes the routing latency.

4.4. Packet Drop Rate(PDR)

It is the vital parameter for measuring the operations of network. In this protocol, the throughput is specified as the amount of data packets effectively obtained at the receiver.

$$\text{Throughput}=\frac{\sum_{0}^{n}\text{Packets Received(n) * Packet size}}{\mathit{Time}} \tag{15}$$

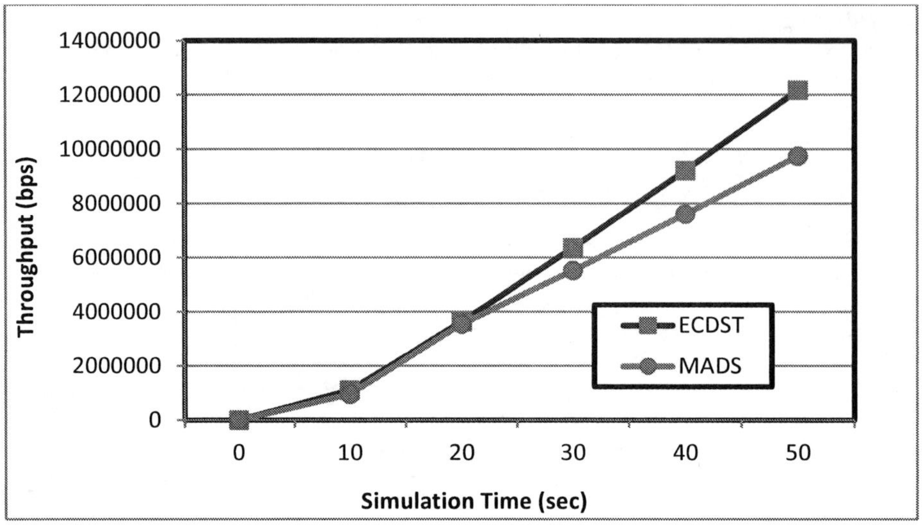

Figure 7. Throughput of ECDST and MADS of ECDST and MADS

The figure 7 proves the ECDST scheme has higher throughput. Because of, MADS scheme concentrate mainly the RSS. But, ECDST focuses the node reliability by elliptical curve Weierstrass function and elliptical curve digital signature thus ECDST has provide better performance in WANET.

4.5. Detection Ratio

Here, we observe the ECDST and MADS routing protocol detection ratio. The detection ratio is determined as given below WANET.

$$Detection\,Ratio = \frac{Detected\ nodes}{Whole\ nodes} \tag{16}$$

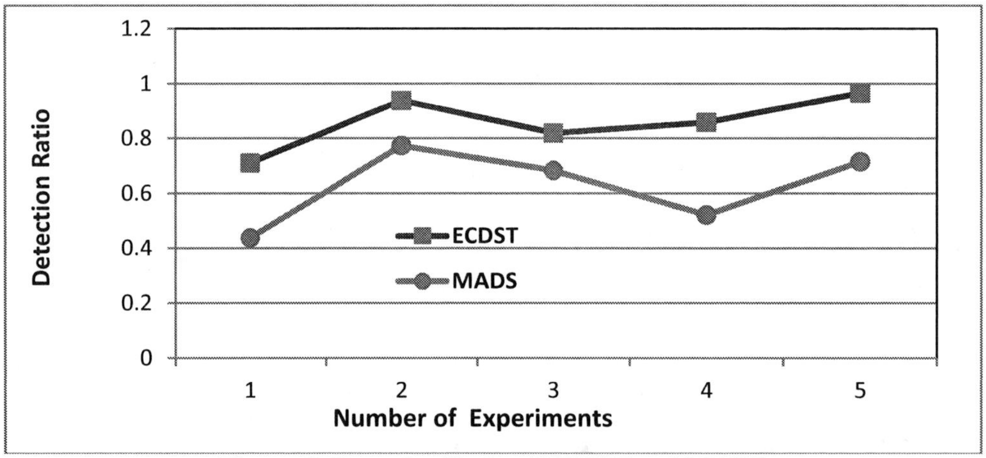

Figure 8. Detection Ratio of ECDST and MADS

The detection ratio of ECDST and MADS protocols are projected in Figure 8. During route discovery, the Weierstrass Elliptic function is used to detect the abnormal nodes in ECDST. It illustrates that the detection rate of ECDST is better than MADS protocol due to the signature verification method distinguish the abnormal nodes in the WANET.

5. Conclusion

In this paper, we proposed Elliptic Curve Digital Signature technique based Abnormal Node Detection in WANET. This approach observes the abnormal nodes as well as it offers reliable data communication in the WANET. During route discovery, the ECC method is utilizing the Weierstrass Elliptic function for detecting the abnormal node. The abnormal nodes are discovered by signature verification technique on data communication. Thus, it renders better consistency as well as expands the network function while equating to the existing MADS approach. Our simulation outcomes illustrate that this approach offers assured packet obtained rate, enhances both the throughput, as well as detection ratio also dilutes the latency in the WANET.

References

[1] Umamaheswari S and Radhamani G 2015 Enhanced ANTSEC framework with cluster based cooperative caching in mobile ad hoc networks *Journal of Communications and Networks* **17** 40-46

[2] Canêdo D R C and Romariz A R S R 2019 Intrusion Detection System in Ad Hoc Networks with Artificial Neural Networks and Algorithm K-Means *IEEE Latin America Transactions* **17** 1109-1115

[3] Poongodi T, Khan MS, Patan R, Gandomi A H and Balusamy B 2019 Robust Defense Scheme Against Selective Drop Attack in Wireless Ad Hoc Networks *IEEE access* **7** 18409-18419

[4] Duan J, Gao D, Yang C, Foh CH and Chen HH 2014 An energy-aware trust derivation scheme with game theoretic approach in wireless sensor networks for IoT applications *IEEE Internet of Things Journal* **1** 58-69

[5] Yildiz HU, Ciftler BS, Tavli B, Bicakci K and Incebacak D 2016 The impact of incomplete secure connectivity on the lifetime of wireless sensor networks *IEEE Systems Journal* **12** 1042-1046

[6] Lingeshwari S and Natchadalingam 2014 Provisioning of Efficient Authentication Technique for Implementing in Large Scale Network *Int. Journal of MC Square Scientific Rssearch* **6** 34-42

[7] Patil KT and Patil ME 2014 Improve the Security of CGA using Adjustable Key Block Cipher based AES, to Prevent Attack on AES in IPV6 over MANET *IEEE Global Conf. on Wireless Computing & Networking* 148-152

[8] Wang N, Zhang N and Gulliver TA 2013 Cooperative key agreement for wireless networking: Key rates and practical protocol design *IEEE Transactions on Information Forensics and Security* **9** 272-284

[9] Santhakumar R 2017 Resource Allocation in Wireless Networks by Channel Estimation and Relay Assignment Using Data-Aided Techniques," *International Journal of MC Square Scientific Research* **9** 40-47

[10] Abbas S, Faisal M, Rahman HU, Khan MZ and Merabti M 2018 Masquerading Attacks Detection in Mobile Ad Hoc Networks *IEEE Access* **6** 55013-55025

[11] Gandino F, Montrucchio B and Rebaudengo M 2013 Key management for static wireless sensor networks with node adding *IEEE Transactions on Industrial Informatics* **10** 1133-1143

AUTHOR INDEX

Institute of Physics
Dirac House, Temple Back
Bristol BS1 6BE UK

ISSN: 1757-8981
ISBN 978-1-7138-4764-9

9 781713 847649

CURRAN ASSOCIATES INC.
proceedings.com